W9-DDK-126

Information Arts

LEONARDO

Roger F. Malina, series editor

Information Arts

Intersections of Art, Science, and Technology

Stephen Wilson

The MIT Press
Cambridge, Massachusetts
London, England

© 2002 Stephen Wilson

All rights reserved. No part of this book may be reproduced in any form by any electronic or mechanical means (including photocopying, recording, or information storage and retrieval) without permission in writing from the publisher.

This book was set in Bell Gothic and Garamond by Achorn Graphic Services, Inc., and printed and bound in the United States of America.

Library of Congress Cataloging-in-Publication Data

Wilson, Stephen, 1944–
  Information arts: intersections of art, science, and technology / Stephen Wilson.
    p. cm.—(Leonardo)
  ISBN 0-262-23209-X (hc: acid-free paper)
  1. Art and science. 2. Art and technology. I. Title. II. Leonardo (Series) (Cambridge, Mass.).

N72.S3 W55   2002
701′.05—dc21

10 9 8 7 6 5 4 3 2   00-038027

Contents

Contents

Selected Artists

Artists include:

James Acord
Kristi Allik
Mark Amerika
Suzanne Anker
Marcel.li Antunez Roca
Margo K. Apostolos
Roy Ascott
Franko B
Nicolas Anatol Baginsky
Harlyn Baker
Bill Barminski
Gregory Barsamian
Louis Bec
Konrad Becker
Tony Belaver
Guy Ben-Ary
Maurice Benayoun
Bob Bingham
Trevor Blackwell
Marc Böhlen
Christian A. Bohn
Karl Bohringer
Tom Bonauro
Johnny Bradley

Benjamin Britton
Shawn Brixey
Peter Broadwell
Sheldon Brown
Leif Brush
C5
Patrice Caire
Jim Campbell
Bruce Cannon
Oron Catts
Shu Lea Cheang
Mel Chin
Richard Clar
Mary Anne Clark
Harold Cohen
Brent Collins
Tim Collins
Pierre Comte
Luc Courchesne
Donna Cox
Jordan Crandall
Trevor Darrell
Char Davies
Joe Davis

Walter De Maria
Paul DeMarinis
Louis-Philippe Demers
Andrea Di Castro
Stewart Dickson
Elizabeth Diller
Mark Dion
Diana Domingues
Abbe Don
Kitsou Dubois
John Dunn
Hubert Duprat
Christopher Ebener
Stephan Eichhorn
Arthur Elsenaar
Brian Evans
Ken Feingold
Gregory Fischer
Monika Fleischmann
Bill Fontana
Felice Frankel
Masaki Fujihata
Rebecca Fuson
Ulrike Gabriel
Kit Galloway
Paul Garrin
George Gessert
Bruce Gilchrist
JoAnn Gillerman
Ken Goldberg
Guillermo Gómez-Peña
George Gonzalez
Gaile Gordon
Reiko Goto
Sharon Grace
Group Spirale
Kazuhiko Hachiya

Newton Harrison
Helen Harrison
Emily Hartzell
Steven Hartzog
Grahame Harwood
Agnes Hegedüs
Michael Heivly
Nigel Helyer
Lynn Hershman
Paul Hertz
Jerry Hesketh
Perry Hoberman
Eric Hobijn
Carsten Höller
Bart Hopkins
Lisa Hutton
Tjark Ihmels
Toshio Iwai
Natalie Jeremijenko
Jodi.org
Ludwig John
Eduardo Kac
Ned Kahn
Paras Kaul
Yves Klein
Vitaly Komar
Milton Komisar
Richard Kriesche
Myron Krueger
Ted Krueger
Gregory Kuhn
Mierle Landerman Ukeles
Jaron Lanier
Eve Andrée Laramée
Brenda Laurel
Ray Lauzzanna
George Legrady

Webster Lewin
Carl Loeffler
Rafael Lozano-Hemmer
Dirk Luesebrink
James Luna
Chico MacMurtie
John Maeda
Judy Malloy
Steve Mann
Wojciech Matusik
Delle Maxwell
Alex Melamid
Laurent Mignonneau
Seiko Mikami
MIT Media Lab Aesthetics and
  Computation Group
Bonnie Mitchell
Christian Möller
Gordon Monahan
Knut Mork
Iain Mott
Robert Mulder
Antonio Muntadas
Fakir Musafar
Ken Musgrave
Rob Myers
Michael Naimark
Ikuo Nakamura
Eric Nyberg
Orlan
Karen O'Rourke
Ed Osborn
Randall Packer
Eric Paulos
Kate Pendry
Simon Penny
Jean-Marc Philippe

Clifford Pickover
Ian Pollock
Sherrie Rabinowitz
John Randolph
Sonya Rapoport
Alan Rath
Thomas Ray
Mark Reaney
Catherine Richards
Peter Richards
Ken Rinaldo
Don Ritter
Keith Roberson
Sara Roberts
Alexis Rockman
Bryan Rogers
Kathleen Rogers
David Rokeby
David Rosenboom
Teri Rueb
Eric Samakh
Joachim Sauter
Remko Scha
Stefan Schemat
Julia Scher
Thecla Schiphorst
Barry Schwartz
Ricardo Scofidio
Jill Scott
Bill Seaman
Seemen
Carlo Sequin
Paul Sermon
Jeffrey Shaw
Roberto Sifuentes
Todd Siler
Janet Silk

Karl Sims
Joel Slayton
David Small
Alexa Smith
Nina Sobell
Christa Sommerer
Alan Sonfist
Martin Spanjaard
Richard Stanford
Ed Stastny
Stelarc
Stahl Stenslie
Wolfgang Strauss
Rachel Strickland
Susan Stryker
Athena Tacha
Ed Tannenbaum
Project Taos
Nell Tenhaaf
Rob Terry
Mark Thompson
Mark Thorpe
Bruce Tombs
John Tonkin
Naoko Tosa

Michael Touma
Trimpin
Normal Tuck
Tom Van Sant
Paul Vanouse
Roman Verostko
Bill Vorn
Tamas Waliczky
Marius Watz
Peter Weibel
Barry Brian Werger
Norman White
Tom White
Neil Wiernik
Gail Wight
Stephen Wilson
Uli Winters
Andrea Wollensak
John Woodfill
Arthur Woods
Kirk Woolford
Kenji Yanobe
Pamela Z
Ionat Zurr

Selected Technologies

Research areas include:

Bioengineering

Microbiology and genetics

Smell, taste, and touch sensors

Biosensors

Bioidentification and Biosurveillance

Neuroscience

Bionics

Parapsychology and bioelectricity

Biological warfare

Plant and animal behavior

Ecology

Technological stimulation of the body

Surgery

Tissue culture

Brain monitoring

Heartbeat and breath monitoring

Body modification

Perception and cognition

Body imaging

Death processes

Nonlinear systems, complexity, and chaos

Particle physics

Nanotechnology

Materials science

Rapid prototyping

Global positioning system (GPS)

Geology and seismic activity

Fluid dynamics

Meteorology

Solar energy

Fire and magnetics

Theoretical math

Fractals

Algorithms

Artificial Life

Genetic Algorithms

Robot vision and motion

Robot autonomy

Subsumption architecture

Humanoid robots

Living architecture

Wireless communication and mobile computing

Location sensitive communication

Electromagnetic spectrum

Desktop video

Computer Telephone Integration (CTI)

Virtual communities and telecollaboration

Telepresence

Remote museums

Telemedicine

Synthetic image and telecommunication integration

Visualizing net activity

Autonomous flying vehicles

Parapsychological communication

Gesture and touch recognition

Speech recognition and synthesis

Face and facial expression recognition

Object tracking and identification

Video surveillance

Music recognition

Haptics

3-D sound

Motion simulation

Force feedback

Immersive virtual reality

Automatic video interpretation

Synthetic scene generation

Ambient sound and sound localization

Speaker identification and tracking

Earcons and sonification

Autonomous software agents

Social and emotional computing

Artificial intelligence

Information visualization and foraging

Augmented reality systems

Tangible bits

Ubiquitous/pervasive/invisible computing

Wearable computing

Smart spaces, houses, highways

Series Foreword

Editorial Board: Roger F. Malina, Denise Penrose, and Pam Grant Ryan

The cultural convergence of art, science, and technology provides ample opportunity for artists to challenge the very notion of how art is produced and to call into question its subject matter and its function in society. The mission of the Leonardo book series, published by the MIT Press, is to publish texts by artists, scientists, researchers, and scholars that present innovative discourse on the convergence of art, science, and technology.

Envisioned as a catalyst for enterprise, research, and creative and scholarly experimentation, the book series enables diverse intellectual communities to explore common grounds of expertise. The Leonardo book series provides a context for the discussion of contemporary practice, ideas, and frameworks in this rapidly evolving arena where art and science connect.

To find more information about Leonardo/ISAST and to order our publications, go to Leonardo Online at ⟨http://mitpress.mit.edu/e-journals/Leonardo/isast/leobooks.html⟩ or send e-mail to ⟨leonardobooks.mitpress.mit.edu⟩.

Joel Slayton
Chairman, Leonardo Book Series

Book Series Advisory Committee: Annick Bureaud, Pamela Grant Ryan, Craig Harris, Roger Malina, Margaret Morse, Michael Punt, Douglas Sery, Allen Strange.

Leonardo/International Society for the Arts, Sciences, and Technology (ISAST)

Leonardo, the International Society for the Arts, Sciences, and Technology, and the affiliated French organization Association Leonardo have two very simple goals:

1. to document and make known the work of artists, researchers, and scholars interested in the ways that the contemporary arts interact with science and technology, and
2. to create a forum and meeting places where artists, scientists, and engineers can meet, exchange ideas, and, where appropriate, collaborate.

When the journal *Leonardo* was started some 35 years ago, these creative disciplines existed in segregated institutional and social networks, a situation dramatized at that time by the "Two Cultures" debates initiated by C. P. Snow. Today we live in a different time of cross-disciplinary ferment, collaboration, and intellectual confrontation enabled by new hybrid organizations, new funding sponsors, and the shared tools of computers and the Internet. Above all, new generations of artist-researchers and researcher-artists are now at work individually and in collaborative teams bridging the art, science, and technology disciplines. Perhaps in our lifetime we will see the emergence of "new Leonardos," creative individuals or teams who will not only develop a meaningful art for our times but also drive new agendas in science and stimulate technological innovation that addresses today's human needs.

For more information on the activities of the Leonardo organizations and networks, please visit our Web site at ⟨http://mitpress.mit.edu/Leonardo⟩.

Roger F. Malina
Chairman, Leonardo/ISAST

ISAST Board of Directors: Barbara Lee Williams, Martin Anderson, Mark Resch, Mark Beam, Sonya Rapoport, Stephen Wilson, Lynn Hershman Leeson, Joel Slayton, Penelope Finnie, Curtis Karnow, Mina Bissell, Rich Gold, Beverly Reiser, Piero Scaruffi.

Foreword

Stephen Wilson's *Information Arts: Intersections of Art, Science, and Technology* is the first comprehensive international survey of artists working at the frontiers of scientific inquiry and emerging technologies. The scope of *Information Arts* is encyclopedic: it is both a critical text and practical resource guide. The expansive taxonomy of art and research is accentuated by theoretical perspectives, analysis, and art commentaries that address a diverse range of ideological positions. *Information Arts* also provides resources on organizations, publications, conferences, museums, educational programs, and research centers.

Steve Wilson provides a mirror that captures the essence and agendas represented by contemporary efforts of artists to integrate scientific research into their work and enterprise. Offering a critical context of their agendas, he examines research that crosses the intellectual terrain of biology, physics, materials science, astronomy, cognitive science, engineering, medicine, architecture, and social and information science. This research indicates that although art and science share many characteristics, a special role for the arts exists in the evolution and deployment of technology—the implication being that by operating outside the conventions of traditional practice, unique and significant research enterprises can and will unfold. *Information Arts* helps us understand on a deeper level that experimental research is culturally necessary and serves to transform how to simulate, interact with, and experience the world. *Information Arts* is about the unfolding of this conceptual frontier, a frontier in which art informs research and research informs art.

For more information about the book, author, and Leonardo book series, as well as updated links to artists, theorists, and researchers working at the intersection of art, science, and technology, please visit <http://mitpress.mit.edu/Leonardo/isast/leobooks/swilson/>.

Joel Slayton

Preface

In my last semester at Antioch College, all students were expected to complete an integrative final thesis. Since it was the late 1960s, most students at this experimental college focused on the political and cultural structures undergoing upheaval in that era. I approached the foment differently.

Radio, television, amplified music, and cinema were everywhere. They figured prominently as the arbiters of change even in the lives of those of us focused on the arts and humanities. Yet it struck me as strange that almost no one outside of engineering understood how devices such as the radio worked. How did it magically manage to send sounds thousands of miles through the ether? This acquiescence to ignorance seemed a critical gap in our literacy and ultimately our capacity to act in a technological world.

For my final thesis, I proposed to teach myself how radio worked even though I lacked any significant technical background. Ultimately, I did learn how radio worked. I also learned some things that may be more important: that the mystification of science and technology was unjustified; that scientific principles were understandable, just like ideas in other fields; and that technological imagination and scientific inquiry were themselves a kind of poetry—a revolutionary weaving of ideas and a bold sculpture of matter to create new possibilities.

Over the years these insights have guided my teaching and my work as an artist. They are also the foundation for this book. There is a major categorical flaw in the way we commonly think about scientific and technological research as being outside the major cultural flow, as something only for specialists. We must learn to appreciate and produce science and technology just as we do literature, music, and the arts. They are part of the cultural core of our era and must become part of general discourse in a profound way.

Many artists have begun to engage the world of technological and scientific research—not just use its gizmos—but rather to comment on its agendas and extend its possibilities. Their work can be seen as part of this essential rapprochement and as a clue to what art may look like in the twenty-first century. I wrote this book because no resource surveying this remarkable body of art and its relationship to research exists. *Information Arts* includes the following:

- It surveys artistic work related to biology (microbiology, genetics, animal and plant behavior, ecology, the body, and medicine); the physical sciences (particle physics, atomic energy, geology, physics, chemistry, astronomy, space science, and Global Positioning System (GPS) technology); mathematics and algorithms (algorists, fractals, genetic art, and artificial life); kinetics (conceptual electronics, sound installation, and robotics); telecommunications (telephone, radio, telepresence, and Web art); and digital systems (interactive media, virtual reality (VR), alternative sensors, artificial intelligence, 3-D sound, speech, scientific visualization, and information systems). Using summaries from the artists' writings, it introduces their rationales and explanations of their work.
- It considers artists approaching research from a variety of ideological stances and reviews theoretical writing related to artistic work in these areas.
- Exploring the idea of techno-scientific research as cultural acts, it also reviews the research projects, agendas, and future plans of scientists and technologists working at the frontiers of inquiry.
- It also lists resources (organizations, publications, conferences, museums, research centers, and art-science collaborations); books useful for further study; and Web sites for artists, theorists, and research centers.

The author wishes to thank the many who have helped to make this book possible: the artists and researchers who have created these extraordinary works and graciously allowed us to use images of their work. The technologists who created the Web, which allows us all to access each other's work. Students in my courses in the Conceptual/ Information Arts Program at San Francisco State University, whose enthusiasm and honesty have helped hone my ideas. Student research assistants Joseph Schecter, Max Kelly, Lisa Husby, and Torrey Nommesen for helping with image and permission research. The reviewers and editors of MIT Press's Leonardo series who recognized the value of the book and offered suggestions for its improvement. Production editor Deborah Cantor-Adams and production and graphics coordinator Sharon Deacon Warne at MIT Press who helped give the book its present polished form. Doug Sery, my editor at MIT Press, for his support and willingness to pursue these ideas. Catherine Witzling, my wife, for editing the first two chapters and her frankness in questioning the topics of this book. My daughter Sophia, for rescuing me from the obsession with the book via demands to play. Sally and Julius Wilson for teaching me to be curious about everything.

1

Introduction, Methodology, Definitions, and Theoretical Overview

1.1

Art and Science as Cultural Acts

What do art and science have to do with each other? *Information Arts* takes an unorthodox look at this question, focusing on the revolutionary work of artists and theorists who challenge the separations initiated in the Renaissance. It points toward a possible future in which the arts can reassume their historical role of keeping watch on the cultural frontier and in which the sciences and arts inform each other.

Research has become a center of cultural innovation: its results are radically influencing life and thought. Our culture needs to participate in defining research agendas, conducting inquiries, and analyzing their meanings. Artists should be hungry to know what researchers are doing and thinking, and scientists and technologists should be zealous to know of artistic experimentation. The future will be enriched if this expansion of zones of interest becomes a part of the definition of art and science.

Scientific and technological research should be viewed more broadly than in the past: not only as specialized technical inquiry, but as cultural creativity and commentary, much like art. It can be appreciated for its imaginative reach as well as its disciplinary or utilitarian purposes. Like art, it can be profitably analyzed for its subtexts, its association to more general cultural forces, and its implications as well as its surface rationales.

Art that explores technological and scientific frontiers is an act of relevance not only to a high-brow niche in a segregated corner of our culture. Like research, it asks questions about the possibilities and implications of technological innovation. It often explores different inquiry pathways, conceptual frameworks, and cultural associations than those investigated by scientists and engineers. (I have adopted the convention of referring to scientific inquiry and technological innovation as techno-scientific research, even though their activities can be quite different.)

Anthropologists claim that we increasingly live in an "information society" in which the creation, movement, and analysis of ideas is the center of cultural and economic life. In our culture, scientific and technological information is a critical core of that information. This book is called *Information Arts* because the art of such a culture must address that information if it is going to be vital.

Here, then, are the questions *Information Arts* is attempting to answer:

What kinds of relationships are possible among art, scientific inquiry, and technological innovation? How might art and research mutually inform each other?

How are artists investigating techno-scientific research? How have they chosen to relate to the world of research? How does research further their artistic agendas?

How do art historians and cultural theorists understand the interactions between culture and research?

How do researchers conceptualize? What agendas motivate their work? What future developments are likely to call for cultural commentary and artistic attention?

A Quiz

We are at an interesting place in history, in which it is sometimes difficult to distinguish between techno-scientific research and art—a sign that broader integrated views of art and research are developing. The section below offers the reader a "quiz" to illustrate this point. It briefly describes research activities mentioned in this book. The reader is invited to determine which activities have been carried out by persons describing themselves as artists and which by those describing themselves as researchers. (For the sake of the quiz, all are identified as "researchers." Answers are provided at the end of this chapter.)

Research	Art	
—	—	Researcher J.T. developed a method of using genetic engineering to encode messages in bacteria.
—	—	Researcher S. developed an arrangement so that persons far away could control his body through electrical stimulation.
—	—	Researcher E.K. created a system in which several geographically dispersed participants shared the body of a robot that they mutually controlled.
—	—	Researchers C.E. and U.W. bred a line of mice with a special proclivity for eating computer cables.
—	—	Researcher P.D. developed a method for modulating sound onto the flow of dripping water.
—	—	Researcher J.M. developed a computer display that could visualize the underlying intellectual structure of a group of articles and books.
—	—	Researcher R.B. developed colonies of small robots with a repertoire of simple behaviors that can evolve complex intelligence skills through learning and communication.
—	—	Researcher H.S. developed a "fertility bra" that used pheromone receptors to flash indicators when the woman wearing it was in a fertile period.
—	—	Researchers created a video composite representation of participants in a video conference in which nonactive participants faded with the level of their activity.
—	—	Researchers at M.R. developed a device that is sensitive to hugs and can react to things it hears on the television.
—	—	Researcher R.G. invented a toilet with biosensors that provides instant urine-based analysis of biological characteristics, such as drug presence or emotional arousal levels.

Which is which? The confusion is a significant cultural event.

Revisiting the Relationship of Art and Techno-Scientific Research

Historical Separations

The arts and the sciences are two great engines of culture: sources of creativity, places of aspiration, and markers of aggregate identity. Before the Renaissance, they were united. Science was called natural philosophy. Philosophers were as likely to speculate about art and science as about religion and truth. Similarly, in tribal societies the philosopher, shaman, and artist were likely to be the same person. Visual and performance arts were integrated into the fabric of rituals and daily life. The artist who sang stories or carved ritual objects was likely to be the person who was especially observant and wise about the ways of the heavens, the weather, animals, plants, the earth, and life and death.

In the West, the Renaissance initiated an era of specialization. Science became codified as a segregated set of processes and worldviews. While its accomplishments in providing new understanding of old mysteries increased confidence in its claims, art moved in its own direction, largely ignoring the agendas of science. During the Industrial Revolution, science inspired technology and technology inspired science. Research and invention spread into every corner of life, but mainstream art seemed oblivious. Increasingly, it became less likely that an educated person would be well versed in both areas of culture. In the 1960s commentator C. P. Snow developed his influential "Two Cultures" theory[1] that concluded that those in the humanities and arts and those in the sciences had developed sufficiently different languages and worldviews that they did not understand each other. Note that this book will concentrate on the arts, but much of the analysis holds more generally for the humanities.[2]

The Urgency for Reexamination

Can art and science/technology remain segregated in the twenty-first century? *Information Arts* seeks to revisit the relationship of art to scientific and technological research, exploring the pioneering work of artists with emerging research and the prospects for future mutual influences. Several cultural forces combine to make a reexamination of the disconnection critical.

Influence on Life Technological and scientific research are spreading their influence into every corner of life, from medicine, communication, and government to domestic life, education, and entertainment. Commercial innovators scan for research in hopes of creating new industries and fortunes. In earlier eras, the influence of research seemed more limited; there were long periods of continuity in everyday life. How can the arts keep watch on the cultural frontier if they ignore such omnipresent features of life?

Influence on Thought Science and technology are changing basic notions about the nature of the universe and the nature of humanity. New communications technologies challenge ancient ideas about time, distance, and space. New probes peer into the biological heart of life and identity and the origins of the stars. All fields that ask philosophical questions, such as art, must take heed.

Critical Studies and Cultural Theory These disciplines challenge traditional ways of studying culture and question the wisdom of trying to understand the arts, humanities, and sciences in isolation from each other and of segregating "high" and "low" culture. Critical theory deconstructs long-standing sacred cows, such as science's privileged claims to truth and objectivity, as well as art's claims to a special elevated sensitivity. Artists and scientists are seen as creatures of culture, and their work is understood within larger psycho-political-economic-cultural frameworks. Critical theory takes on concepts such as truth, progress, reality, nature, science, gender, identity, and the body. The compelling energy of this analysis is one important indicator of the wisdom of tearing down the walls between disciplines such as art and science.

Artistic Activity The increasing level of artistic activity using computers, the Internet, and other areas of scientific interest suggests the impossibility of understanding the future of the arts without devoting attention to science and technology. Twenty years ago, when I first started my artistic experiments with computers, it was hard to find similarly involved artists or relevant critical perspectives. Now there is an explosion of interest. Some artists want to assimilate the computer to traditional artistic media, for example, by treating it as a fancy paintbrush or camera. Many others, however, recognize the computer as the tip of a techno-cultural iceberg. They understand that the most interesting work is likely to derive from a deeper comprehension of the underlying scientific and technological principles that have guided the computer's development, and from participation in the research flow that points to the technological future.

Organization of the Book

Information Arts aims to be a resource in the reexamination of the relationship between research and art. It proposes to accomplish this in several ways.

Presentation of Artists   Artists have begun to engage the concepts, tools, and contexts of scientific and technological research, and their work is provocative and intriguing. No unified compendium of this work exists, yet this is the best source of information about new kinds of relationships between art and research in the future. I have conducted extensive research to identify artists working with scientific and technological research and have included both established and emerging artists. Where possible, I have incorporated excerpts of the artists' own statements, descriptions, and images. I have also offered commentary by others when useful.

Overview of Theory   Cultural theorists, art historians, and artists have begun to write about many issues in techno-culture that are germane to the discussion of the relationship between art and science/technology. For each of the major sections of the book, I have presented brief overviews of theoretical writing on the topic and indications of controversies where they exist.

Overview of Research Agendas   This book explores the possibility of viewing art and research as a unified cultural enterprise and of understanding researchers' worldviews—their goals, category systems, and visions of the future. For each section of the book, I present overviews of what practitioners in those fields see as the most important research agendas. Indeed, it is a basic premise of this book that art practice and theory in areas of science and technology can best proceed only with profound investigation of these agendas.

Methodology   Creation of this book raised a wide variety of methodological questions: How does one locate exemplary artists and researchers working at the frontiers of inquiry? How does one assess the quality of works? How have my own biases affected the choices and analysis? These questions are considered in Appendix A: Methodology.

Sections of the Book   I have organized the book using categories of research to differentiate sections. Sections cover major branches of scientific inquiry, such as biology, physical sciences, and mathematics, and areas of technological foment, such as computers, alternative interfaces, telecommunications, and robotics. Within each section, chapters focus on particular research arenas.

The Deficiency of Categorization

Artists resist categorization. Artworks are typically multilayered, addressing many themes simultaneously. Many artists purposely try to confound preexisting categories. The technology used may not be the most important element.

Why was this book organized in accordance with scientific disciplines and technological categories? How were artists and artworks placed in particular categories? As an author I confronted the challenge of developing an organizational system for considering art and artworks. Since *Information Arts* investigates the role of scientific and technological research, I adapted practical, low-inference categories focused on scientific disciplines and areas of technology. Thus, if an artwork used biological materials or sought to comment on biological issues such as genetic engineering, I placed it in the biology section of the book. The artist may or may not consider the link with biological research as important as many other issues addressed in the artwork besides biology. As an aid to preserving the way the artists framed their work, I have included artists' own descriptions and rationales wherever possible. Also, the overviews of theory relevant to the areas of research provide additional interpretive perspectives. The book attempts to cross-reference works that explore multiple research areas simultaneously.

How Does Research Function in Various Artists' Works?

The artists in the following chapters integrate techno-scientific research in a variety of ways. For some it is a central focus of their art; for others it is an incidental feature. Even for those for whom the connection is central, a variety of theoretical orientations shape their work. Here is a brief overview of the variety of approaches, starting with those in which the research is central. Note that any given artwork might mix several of these approaches.

Exploration of New Possibilities   The artist's work itself functions as research into the new capabilities opened up by a line of inquiry. For example, in investigating artificial intelligence and speech recognition technology, artist Naoka Tosa created Neuro Baby, a computer-generated character that attempted to read the emotional tone of a visitor's speech and react appropriately (see chapter 7.6).

Exploration of the Cultural Implications of a Line of Research   The artists use the new capabilities to create work that explores the narratives and conceptual frameworks that underlie the research. For example, artists David Rokeby and Paul Garrin created an installation called *White Devil,* which used motion detection technology to create a video projection of a guard dog that snapped at visitors wherever they moved. In part, the installation commented on the implications of surveillance technology by using the technology itself (see chapter 7.4).

Use of the New Unique Capabilities to Explore Themes Not Directly Related to the Research   The technologies provide a new way to address any number of issues not directly related to the technology. For example, my *Father Why* installation used motion detection to explore a variety of emotions related to my father's dying. Visitors' movements into the places of sadness, anger, nostalgia, and resignation activated sound events related to each emotion. The longer they stayed there, the deeper the exploration of that emotion. The event was mostly about these conflicting emotions; the movement detection provided a visceral way to ask visitors to confront them: How long would they stay with a particular emotion before they would need to flee by moving their body? (See chapter 7.4.)

Incidental Use of the Technology   Research provides a wealth of new images and materials. Some artists find the new images intriguing or beautiful but are not especially interested in the underlying inquires that led to those outcomes or in their cultural implications. The power of the work presented in the following chapters suggests that all levels of involvement with the technology are valuable.

What Areas of Technological Art Are Included? Which Are Not?

When I started this project I hoped to create a comprehensive compendium of science-and-technology–inspired art. I defined art broadly to include media and the performing arts in addition to visual arts. However, I quickly realized that this comprehensive approach was impractical. The difficulties I encountered raised interesting questions in thinking about techno-scientific art.

What Is Technology? What Is High-tech Art?

Where should one draw the line? Every creation system beyond the basic apparatus of the body is a technology. At various points in history, charcoal, paints, sculpting tools and techniques, ceramics, and printmaking apparatus were state-of-the-art technologies. More recently, photography, cinema, electric machines, lights, radio, recording technology, and video were considered high technology. Now, however, when people talk about high-tech arts, they are not talking about these technologies.

Technological art is a moving target. The artistic gesture to move into an area of emerging technology that is radical in one era can end up being unnoteworthy a few years later. It takes an act of artistic vision and bravery to decide to work with techniques,

What Areas of Technological Art Are Included? Which Are Not?

9

tools, and concepts from a still raw area of technology not yet accepted as a valid area for the arts. It is a challenge to work with a medium before anyone defines it as a medium. Yet several years later, when the technology has matured and a body of artistic work and commentary has appeared, the choice does not have the same meaning. At the early stages of an emerging technology, the power of artistic work derives in part from the cultural act of claiming it for creative production and cultural commentary. In this regard, the early history of computer graphics and animation in some ways mimics the early history of photography and cinema.

Information Arts generally focuses on art that addresses research activity emerging in the last seven years. I did not extensively consider video art, kinetic and light sculpture, sound art, electronic music, laser art, and holography. Although there continues to be experimental work in these fields, they are not currently considered emerging technologies, and they have well-developed aesthetic and analytic traditions of their own. *Information Arts* does not consider the popular media of science fiction, literature, cinema, and television, which offer interesting arenas of mutual influence between science and art but call for an analysis outside the scope of this book.

Because of the accelerated pace of technological innovation, even newer technologies are rapidly passing into the stage of institutionalization. Fields such as computer graphics, computer animation, 3-D modeling, digital video, interactive multimedia, and Web art, which were revolutionary a few years ago, have become part of the mainstream. Enormous amounts of work are being produced, the variety of aesthetic rationales has multiplied, and the technologies have been integrated into commercial software and media production. Artistic experimentation is quickly being assimilated. For example, computer graphic visual effects that represented innovative artistic exploration a few years ago are now part of the standard Photoshop filters available to the millions who own the software. Computer animations in 3-D and effects that were known only by a few media experimenters are now becoming standard features of movies and commercials. Interactive computer events that were of interest only to experimental artists fifteen years ago are now part of fields such as computer-assisted education and games. In one of the most remarkably speedy transformations, Web art experiments are devoured by the steamrolling commercial and media expansion of the World Wide Web almost as soon as they are invented.

This book will not consider computer graphics, computer animation, and digital video except at their more experimental fringes. Also, although it does consider artistic work with interactive computer media and Web art, a comprehensive analysis of these rapidly expanding and commercializing genres is beyond the scope of this book.

The Assimilation of Art into Research and Commercial Production

The pattern of sequential technological invention, artistic experimentation, and commercial assimilation is a fascinating part of the story of how the worlds of art and research relate to each other, and is only partially analyzed in this book. Some of the artists described in the following chapters eagerly pursue product development for their artistic ideas, and some are supported as part of corporate research labs whose ultimate goal is economic exploitation. Others resist these connections and passionately defend their independence.

In part, this book is an examination of these questions: Where do researchers and artists get their ideas? How do they explore their ideas? How are techno-scientific research and art research different? What happens to the explorations over time? Does mainstream assimilation somehow destroy the validity of the work as art?

Definitions and Theoretical Reflections

Art, science, and technology are culturally laden terms. Indeed, debates over the boundaries of the terms *art* and *science* regularly engage philosophers and historians of art and science. What is art? What is science? What is technology? What are the similarities and differences among the three? What does it mean to call someone a high-tech artist? What is art that is influenced by science? What is science that is influenced by art? This chapter examines these questions, offers a brief clarification of my usage, and identifies shifting criteria that make a definitive answer elusive.

In recent years, critical theory has been a provocative source of thought about the interplay of art, media, science, and technology. Each of the major sections of this book presents pertinent examples of this analysis. However, in its rush to deconstruct scientific research and technological innovation as the manifestation of metanarratives, critical theory leaves little room for the appearance of genuine innovation or the creation of new possibilities. While it has become predominant in the arts, it is not so well accepted in the worlds of science and technology. This chapter analyzes the special problems that this disjunction poses for techno-scientifically influenced artists and examines various stances that artists can take in working with research.

Science and technology are sometimes conflated together; even scholars of the fields acknowledge some lack of clarity. Similarly, artists working with emerging technologies and those inspired by scientific inquiries are often lumped together. This section explores these confusions.

What Is Science?

Science textbooks and philosophers and commentators on science propose a number of defining elements. This set of core ideas includes the following: an attempt to understand how and why phenomena occur; focus on the "natural" world; a belief in empirical information; a value placed upon objectivity, which is sought through detailed specifications of the operations that guide observation; the codification into laws or principles (wherever possible precisely expressed in the language of mathematics); and the continuous testing and refinement of hypotheses.

The underlying assumptions of the scientific approach are that the natural, observed world is real, nature is essentially orderly, and objectivity can be achieved through self-discipline and the reliance on techniques such as the calibration of instruments, repeatability, and multi-observer verification.[3]

This core encompasses variations in emphasis. For example, empiricists emphasize the role of observations, while rationalists focus on the logical processes of theory construction and derivation. Some stress induction built from observation; others focus on deduction drawn from theory.

Critical theorists see science as a modernist delusion. They see the self-constitution of scientist/observer as a continuation of cultural texts focused on domination and exploitation. They challenge the possibility of objectivity, noting the pervasive influences of gender, social position, national identity, and history. They focus on issues such as the social forces and metanarratives that shape the questions and paradigms used in inquiry; the role of socially constructed frameworks at all stages; and the interaction of the observer and the observed phenomenon. Radical constructivists doubt our ability to discover truths applicable across all times and cultures.

Many analysts have contributed to the critique of science. For example, in *The Structure of Scientific Revolutions,* Thomas Kuhn notes the way dominant paradigms shape the questions that get acceptance and support. In *Against Method,* Paul Feyerabend critiques assumptions of scientific rationality, noting that nature gives different answers when approached differently. In *Simians, Cyborgs, and Women: The Reinvention of Nature,* Donna Haraway analyzes the metaphoric language of science, its authoritative voice, and its unacknowledged patriarchal underlife. Having ethnographically studied life in laboratories, Latour in *Science in Action* proposes an actor-network theory of science in which organizations, persons, animals, and inanimate materials combine to shape scientific theorization. In *Picturing Science, Producing Art,* Peter Galison and Caroline Jones investigate the way representation deeply influences the conceptualization and processes of research.

In the humanities, this kind of critique predominates. Scientists and technological innovators, however, believe in the ability to discover universal truths and assert that reform can overcome those places where scientific process falls short of its aspirations to universality and objectivity. As evidence of science's validity, they point to the accomplishments of the scientific worldview in building robust, cross-substantiating theoretical structures, and in predicting and controlling the material and organic world.

Any attempt to cross the disciplinary borders between art and science will confront this disjunction—today's incarnation of C. P. Snow's "Two Cultures" theory. Some of the artists in the following chapters have created works that join the critique, creating installations that highlight aspects of science that fail the classical hygienic view. Others implicitly accept the power of the canon, building on the formulations of prior research and using processes of experimentation and theoretical elaboration.

What Is Technology?

High-tech artists do not necessarily engage science. An examination of the relationship between technology and science is useful for understanding the range of artistic work related to research. Technology is seen as "knowing how," while science is seen as "knowing why." Engineers and technologists are seen as primarily interested in making things or refining processes, not in understanding principles. Many histories of technology are essentially histories of invention—the objects, tools, and machines that people made and the processes that made them.[4] Melvin Kranzberg and Carroll Pursell believe that this definition is too broad. In *Technology in Western Cultures,* they define it more narrowly, as

man's effort to cope with his physical environment—both that provided by nature and that created by man's own technological deeds, such as cities—and his attempt to subdue or control that environment by means of his imagination and ingenuity in the use of available resources.[5]

The relationship of science to technology is quite complex; it became a focus for philosophers of science and technology. Contemporary definitions of technology sometimes call it applied science—the application of scientific principles to solving problems. However, since technology predates science, it should be seen broadly, as human attempts to shape the physical world: "[technology] for much of its history had little relation to science, for men could and did make machines and devices without understanding why they worked or why they turned out like they did."[6]

Developers of technology used many techniques in refining their methods, including learning from other practitioners, observing all aspects of their environment, and experimenting based on instinct, and trial and error. The goal was rarely the development of scientific principles. Certainly, the experiments of many artists in finding appropriate innovations to accomplish their artistic goals could fit this description.

With the Industrial Revolution and the refinement of science in the eighteenth century, technology began to draw more on scientific understanding to help solve its problems. In the twentieth century, scientific research became a major source of new technologies, and most manufacturers included scientists in their industrial research labs.

Historically, technological research is considered somehow less "pure," and less lofty than science.[7] The origins of these attitudes lie deep in the history of Western culture. Among the Egyptians and the Greeks, fabrication was done by slaves or low artisans, and concern with the material world was considered less important than focus on more essential qualities:

Making, even in the form of art, was often mistrusted as inimical to virtue or the pursuit of the highest good because it focused attention on material reality . . . [it] was not considered important as a contribution to the understanding either of the ends of human life or of the first principles of being.[8]

The distrust of "making" continued into the Christian Middle Ages. Just before the Renaissance, however, philosophers started to reexamine these notions. For example, in *City of God,* St. Augustine noted that technological accomplishments were the exercise of "an acuteness of intelligence of so high an order that it reveals how richly endowed our human nature is," as well as a sign of divine benevolence.[9]

With the Enlightenment came a positive attitude toward technological prowess. For example, Francis Bacon proposed that science should serve technological innovation, and suggested that the understanding of nature often becomes clear only when trying to manipulate it technologically:

Bacon proposes a reconstruction of science to produce "a line and race of invention that may in some degree subdue and overcome the necessities and miseries of humanity." . . . Mind must utilize art and hand until nature "is forced out of her natural state and squeezed and molded" because "the nature of things betrays itself more readily under the vexations of art than in its natural freedom.[10]

Currently, science and technology work together and inform each other. Technology developers often must work in areas where scientific understanding is not sufficient. Attempts to develop real-world devices and solutions result in new scientific questions and understanding. For example, the development of new instruments—such as a more powerful collider—may give rise to new categories of questions in physics; the development of new medications may result in information about physiology and organic chemistry.

As researchers attempt to create technologies that simulate human psychic functioning, they create possibilities that then call out for scientific study. For example, computers, once created, become part of the natural world. Cognitive scientists and artificial-intelligence researchers create new insights about the nature of mind and society; user-interface researchers study the methods by which humans and machines can interact. Scientists are confronted by new questions about the nature of mind and the relationship of material reality to human thought. Technology and science goad each other into a parade of new disciplines.

Philosophers of science and technology continue to grapple with the nature of this relationship. Edwin Layton proposes an interactive model in which science and technology are seen as "mirror images" of each other, using common methods and drawing on common intellectual heritages; technology does not only exploit the "golden eggs" created by science.[11]

This interactive model of technology probably comes close to describing what is meant when something is called high technology, or high-tech art. High-tech artists, like their counterparts in technology development settings, are engaged with the world of science. They draw on theoretical formulation and research results from scientific inquiry. They use systematic methods of experimentation borrowed from science to advance their agendas. The results can inform further work by technologists and scientists.

Cyril Stanley Smith, a historian of science and technology, reflected on the relationship of technology and science and the role of artists in the process in his book *From Art to Science: Seventy-Two Objects Illustrating the Nature of Discovery*. In it, he observes that in the areas of chemistry, physics, and materials sciences, artists and artisans discover and use "subtle properties of matter" before they are even noticed by research scientists.[12]

This is the type of interaction that engages many of the artists in this book. Stanley wrote his book in 1978, before the digital technologies of communication, simulation, representation, and information had accelerated to their current levels.

One way to differentiate between science and technology is by intention. Technology developers usually focus on specific utilitarian goals, while scientists search for something more abstract: knowledge. So what is the best way to describe the research undertaken by the artists described in the following chapters? Many focus on the interface between science and technology. A few concentrate more specifically on more classical "scientific" inquiries. Some act like technologists, seeking utilitarian applications of scientific knowledge and processes to further artistic goals. Others engage the scientific world in more open-ended inquiries analogous to those of scientists.

Throughout the book, the artists' work will return to questions about science and technology. What is the relationship of thinking and doing? What does it mean to view the analysis of mind and society as science? How pure can science be? What can we really know of the physical world, since it is seen through the lens of our conceptual frameworks?

What Is Art?

The art presented in this book is best understood within the context of the radical shift in the boundaries of "art" over the last century. Previously, art was produced in historically validated media, presented in a limited set of contexts for a circumscribed set of purposes, such as the search for beauty, religious glorification, or the representation of persons and places. Within a view that stresses conventional media and contexts, it is easy to wonder how the activities described in this book can be called art. However, this century has generated an orgy of experimentation and testing of boundaries. New technological forms, such as photography and cinema, have already raised questions about art. Artists have added new media, new contexts, and new purposes. The art world has assimilated much of this experimentation, of which a partial list follows:

- Extension beyond "realistic" representation (e.g., abstract painters)
- Incorporation of found objects (e.g., Picasso's collage and Duchamp's urinal)
- Movement into non-art settings and intervention in everyday contexts (e.g., Schwitters's Merzbau apartment and Russian AgitProp)
- Presentation of live art (e.g., Dada and Futurist performance)
- Use of industrial materials, products, and processes (e.g., Bauhaus, photography, kinetic art, electronic music, and Warhol's Brillo boxes)
- Conceptual art (ideas as art, with deemphasis of sensual form)
- Earth art (work with natural settings with resident materials)
- Interactive art (dissolution of the border between the audience and artists, for example, living theater and interactive installations)
- Performance and happenings (e.g., Allan Kaprow)

- Public art (work with site-specific materials, social processes and institutions, and community collaborators)
- Exploration of technological innovations (e.g., video, copiers, lasers, and holography)

This experimentation has left the philosophy of art in turmoil. It has become difficult to achieve consensus on definitions of art, the nature of the aesthetic experience, the relative place of communication and expression, or criteria of evaluation. However, there is some agreement on these features: art is intentionally made or assembled by humans, and usually consists of intellectual, symbolic, and sensual components. For example, the Getty Museum Program in Art Education offers this definition:

Art-making may be described as the process of responding to observations, ideas, feelings, and other experiences by creating works of art through the skillful, thoughtful, and imaginative application of tools and techniques to various media. The artistic objects that result are the products of encounters between artists and their intentions, their concepts and attitudes, their cultural and social circumstances, and the materials or media in which they choose to work.[13]

Many of the artists described in this book use unorthodox materials, tools, and ideas inspired by the worlds of science and technology. Some are present in non-art contexts, such as laboratories, trade shows, the Internet, and the street. Some intend to intervene in everyday life or the worlds of science and technology. For many, the artistic rationale guiding their work is alien to the art world. *Information Arts* investigates these artists' work as a continuation of the expanding inclusiveness of the definition of art. Some of the work could even be viewed as the attempt to revisit unresolved issues from movements, such as conceptual art, and art and life interventions. Since the book explores the boundaries between art and techno-scientific inquiry, understanding the limits of art is significant. For example, on what basis can the work of researchers and techno-scientific artists be differentiated, or is such a distinction even important? The work and analysis of the artists described in this book contributes to this ongoing debate.

Although the institutional theory has many adherents, it is not universally accepted, as philosophers struggle to identify theories that can work in the face of the last century's experimentation. Some radical revisionists, claiming that the concept of art has been corrupted by the easy inclusion of the experiments, seek to reestablish some core definitions more closely related to historical forms. Others continue to search for a phenomenological basis for the aesthetic experience. Critical theory wants to explode the concept of art and questions its continued usefulness. The artists who work at the frontiers of science and technology throw fuel on the fire as they seek to move the definition of art

Although the art world has assimilated much of the historical experimentation, the gestures do raise perplexing conceptual problems: (1) Much of the public has not yet accepted these extensions as valid art. How can this long-lasting resistance be explained? (2) Some critical theorists deconstruct the actions of the avant-garde not as radical breaks, but as part of the cultural and economic structure of the art world. They point out the function of these gestures in generating novelty and note the ease with which the mainstream art world can assimilate them. They suggest that high art is not so different from popular media and question the viability of the category, "art."

to include their activities. *Information Arts* can be seen as an investigation of these moving boundaries and the cultural significance of including techno-scientific research in a definition of art.

Similarities and Differences between Science and Art

How are science and art similar? How are they different? This analysis is useful for understanding the prospects for future relationships.

Differences between Art and Science	
Art	Science
Seeks aesthetic response	Seeks knowledge and understanding
Emotion and intuition	Reason
Idiosyncratic	Normative
Visual or sonic communication	Narrative text communication
Evocative	Explanatory
Values break with tradition	Values systematic building on tradition and adherence to standards

Similarities between Art and Science

Both value the careful observation of their environments to gather information through the senses.
Both value creativity.
Both propose to introduce change, innovation, or improvement over what exists.
Both use abstract models to understand the world.
Both aspire to create works that have universal relevance.

In "Principles of Research," Albert Einstein stated that the artist and the scientist each substitute a self-created world for the experiential one, with the goal of transcendence.[14] In "The Contribution of the Artist to Scientific Visualization," Vibeke Sorensen describes artists as "organizers of large amounts of data"; "people who find unusual relationships between events and images"; and "creative interdisciplinarians." She continues:

Artists are . . . people who create something completely original and new, something beyond the known boundaries of the information base. By using or inventing new tools, they show new uses and applications that synergize and synthesize fields. Artists push the limits of technologies, bringing them to previously unattained goals. Artists as well as scientists work with abstract symbols, representations for various realities and working tools. Even the language used by the two groups is similar. Scientists working with mathematics frequently describe a particularly good explanation or solution as "elegant" . . . The intellectual bridge of abstraction and aesthetic consideration is fundamental to both groups.[15]

A less benign critical analysis asserts that science and art both make questionable truth claims and attempt to create privileged positions, but in reality participate in the system of symbols and narratives that shape the culture.

In a paper entitled "Theoreticians, Artists, and Artisans," Feyerabend observes that scientists play a large role in creating the phenomena they study, suggests that science could benefit from art's awareness of absurdity and paradox, and notes the dilemma surfaced by Plato. The only way we can know pure being is by making inferences about it through imperfect senses that observe base matter. Plato considered artisans and artists as lowlifes who worked far from the core of a universe accessible only by contemplation. Feyerabend traces this distrust of observation to the present day, in which theoreticians are accorded higher status than empiricists.[16]

Feyerabend notes that scientists must create massive theoretical structures to link observation and the underlying "reality." Although scientists pride themselves on objectivity, they are similar to artists in their construction of artificialities.[17] He further asserts that difficulties arise from the extraordinary faith that science places in theoretical structures and the manipulations derived from them. He questions the wisdom of distrusting the world of real things and actions.[18] Feyerabend concludes that science is in many ways very similar to art, in which researchers build research structures and operations to represent their thoughts:

In a way, individual scientists, scientific movements, tribes, nations function like artists or artisans trying to shape a world from a largely unknown material, Being. . . . Or researchers are artists, who, working on a largely unknown material, Being, build a variety of manifest worlds that they often, but mistakenly, identify with Being itself.[19]

Finally, we return to the question of what makes the works described in this book art? How is the artist who explores unusual topological structures different from the mathematician focusing on similar topics? How is the algorist who develops a new algorithm for graphic output or a new way to get interesting output out of a plotter different from a technology developer? How is the artist exploring the limits of genetic inheritance by breeding mice to eat computer cables different from a biologist? In parallel fashion, one could ask why couldn't much of the work of scientists, researchers, inventors, and hackers be considered art? The reader is asked to think about the art and research described in the book with these questions in mind.

Critical Theory and Problematic Issues in the Integration of Art and Techno-Scientific Research

This book explores the ways in which art and techno-scientific research are currently being integrated, and how they might be integrated in the future. Currently, critical theory and cultural studies have become a dominant discourse in analysis of the function of the arts in today's technology-dominated world. Many sections in the following parts of this book feature theorists working from these perspectives. Important themes in critical theory include: the rejection of the modernist idea of one dominant cultural stream; the impact of mediated images and representation on ideology and behavior; the emphasis on deconstructing the language systems and metanarratives that shape culture; critiques of the narratives of progress; and challenges to science's claim to universal truth and art's claim to an elevated, privileged, avant-garde vision. In the discussion that follows, I examine issues in the application of these analyses, including their limitations, and consider other models for the integration of art and techno-scientific research.[20]

The impact of technology on contemporary life and culture is a vital issue in our age. Critical theory and cultural studies attempt to link the arts, literature, media studies, politics, sociology, anthropology, philosophy, and technology in an interdisciplinary search for relevant concepts and frameworks with which to understand the current world. While art practice and theory are being radically reshaped by this activity, the techno-scientific world in general has not deeply engaged the concepts from cultural studies.

This section attempts to elucidate some reasons for this, and the implications of this difference in worldviews. The introductory chapters to each section consider theoretical writing more specifically focused on topics such as biology, telecommunications, and digital culture.

Critical analyses have gained attention in the world as tools for understanding the work of artists who work with emerging technologies. The technologies explored by artists are the very ones that some analysts see as key to structuring postmodern, postindustrial society. They are essential components in creating the mediated vortex of free-floating significations and the implosion of meaning. They are also crucial in the creation of new cultural niches in which issues such as information flow, control, the body, and war become prominent. Many of these artists have feet in both the art world and popular culture.

Disjunctions between Scientific Worldviews and Critical Theory

Many who work in science and technology still maintain faith in progress, the universal claims of their operations, and the independent status of the phenomena they work with. They can point to an impressive record of ideas tested by methods of verification that approach objectivity, and to new knowledge, understanding, investigative tools, and new technologies that have transformed life in almost every corner of the earth. In the fields of theoretical and applied sciences, there is an optimism very different from the skepticism that marks deconstructive thought. Scientists believe they can refine theory and make universally valid discoveries, and technologists believe they can create technologies that better human life and transform culture in positive ways.

The role of computers and information technologies is one area in which the views of cultural critics and scientists diverge. Many critical theorists emphasize the insidious nature of pervasive, smoothly functioning information technologies that control and promote superficial thought and life. For example, Constance Penley and Andrew Ross note in *Technoculture* that technology is so much a part of the basic structure of society that innovations are immediately co-opted by the mainstream; thus, they dismiss the liberatory fantasies of the new technologies. In "Eclipse of the Spectacle," Jonathan Crary notes the self-delusion of those who believe in positive revolutionary effects:

The charade of technological "revolution" is founded on the myth of the rationality and inevitability of a computer-centered world. From all sides a postindustrial society is depicted that renders invisible the very unworkability and disorder of present "industrial" systems of distribution and circulation. [21]

Crary further calls into question the very notion of a smoothly running, technologically ruled world. He critiques Jean Baudrillard's writings, saying that they exclude "any sense of breakdown, of faulty circuits, of systemic malfunction . . . of disease, and of the colossal dilapidation of everything that claims infallibility and sleekness."[22]

Along these lines, the movie *Blade Runner* is often cited as an example of a cyberpunk dystopia in which technology has helped to erode order and a sense of history:

[T]he city of *Blade Runner* is not the ultra modern, but the postmodern city. It is not an orderly layout of skyscrapers and ultra-comfortable, hypermechanized interiors. Rather, it creates an aesthetic of decay, exposing the dark side of technology, the process of disintegration, postindustrialization, and quick wearing out.[23]

Optimists, however, see information technologies as democratizing access to information, humanizing labor, deepening thought, building community, and empowering people throughout the world. Stewart Brand propounds some of these beliefs in his account of MIT's Media Lab:

We have already seen the arrival of personal computers make multitudes broader in their skills and interests, less passive, less traditionally role-bound. That's renaissance. We've seen people use VCRs to stop being jerked around by the vagaries of network scheduling, build libraries of well-loved films, and make their own videos. We've seen satellite dishes by the quasi-legal million employed to break the urban monopoly on full-range entertainment. . . . Each made audiences into something else—less "a group of spectators, listeners, or readers" and more a society of selectors, changers, makers.[24]

John Sculley, the former CEO of Apple Computer, described a related vision of the technologically enabled future in the book *Interactive Multimedia*:

Imagine a classroom with a window on all the world's knowledge. Imagine a teacher with the capability to bring to life any image, any sound, any event. Imagine a student with the power to visit any place on earth at any time in history. Imagine a screen that can display in vivid color the inner workings of a cell, the births, and deaths of stars. . . . And then imagine that you have access to all of this and more by exerting little more effort than simply asking that it appear. . . . They are the tools of a near tomorrow and, like the printing press, they will empower individuals, unlock worlds of knowledge, and forge a new community of ideas.[25]

Those who work in any number of emerging technologies would describe the probable implications of their work in similar terms. Conferences, trade shows, and journals burn with intellectual foment, excitement, and eagerness to invent the future. The scientific research agendas described in the following chapters are full of optimism and methodological self-assurance.

Do these scientists and technologists live in the same world as the culture analysts? The discordance between the worldviews of culture theoreticians and those who work with new technologies may be essential for understanding the contemporary era in a unified cross-disciplinary way. One conceptualization is that each group lacks information. For example, a critical theorist might note that technologists delude themselves about the amount of autonomy they have in their research, the underlying metanarratives shaping their behavior, and the ultimate cultural ramifications of technology. It is also possible that the interpretative tone of culture theorists stems from their experience of being acted upon by new technologies, while the optimism of scientists and technologists reflects their engagement in the processes of imagining, inventing, developing, and enabling the new technologies.

Artists who work with emerging technologies face a dilemma: they stand with feet in both worlds. On one side they are invited to help create the new technologies and elaborate new cultural possibilities; on the other, they are asked to stand back and use their knowledge of the technology to critically comment on its underrepresented implications. This bifurcation causes critical discord in regard to the work of these artists because of the different stances they can assume. In particular, established critics might ignore or consider work that entertains the progressive worldviews of the technologists to be naive. The section below details different responses that artists can make to this confrontation of zeitgeist.

The Status of Substantive Things and Organisms in a World Dominated by Image and Media

In a postindustrial information economy, most people are seen as working with mediated abstractions rather than with real things. Because of the power of computer representations, workers in many businesses don't see the real objects of their business during the workday. Telecommunication substitution of mediated symbols for physical presence highlights this trend. Baudrillard's conceptualization of a hyperreality dominated by media images and circulating signifiers and codes increasingly disconnected from their referents speaks to the questionable status of things and organisms. Virtual reality technology promises to increase the power of representation to substitute for material experience. Some ecologists suggest that a mediated world might be good, because the endless production and consumption of things is suicidal. Donna Haraway's "Cyborg Manifesto" points toward a future when bodies themselves might be decreasingly relevant. The perception and meaning of even fundamental "realities" such as disease and sex are profoundly shaped by ideology.

The assessment of the decline of the importance of the material world is a critical issue for the arts and culture at large. On a basic level, the diminished importance of the physical seems overstated. Birth, death, health, disease, and the everyday realities of eating, moving, and sex are still essential to human experience. Many of the world's peoples still struggle to survive and spend their days contending with the physical world, while even in the developed world there is a growing uneasiness about the incompleteness of computer simulations and representations of reality. In his article "How Engineers Lose Touch," Eugene Ferguson posits that fatal design flaws in advanced technology such as the *Challenger* are due to "inexperience or hubris or both and reflect an apparent ignorance of . . . the limits of stress in materials and people under chaotic conditions. Successful design still requires expert tacit knowledge and intuitive 'feel' based on experience."[26]

Historically, the arts have spanned both the material and the representational—working with images at the same time as they celebrated the substantiality and sensuality of real things, as in sculpture and architecture. As Walter Benjamin noted in "Works of Art in the Age of Mechanical Representation," technologies such as photography and cinema decreased the importance of presence and "aura."[27]

Questions of materiality and corporeality are critical for artists working with new technologies. The imaging, communications, and information technologies they work with are key facilitators of this mediated world. The work they do helps to explore and settle new worlds of representation. Yet, it is not inevitable that new technologies only work with representation. The technologies that manipulate physical things, for example: robotics, nanotechnology, material sciences, alternative energy research, and biotechnology, have been less accessible to artists and the general public. These technologies will be increasingly important and point toward futures when technologically mediated material things have increasing importance. The following chapters review artists and researchers working at the cutting edge of virtuality; they also present artists and researchers who do not accept the inevitability of a vision in which materiality becomes unimportant.

The Difficulties of Locating a Rationale for Action in a Deconstructed Milieu

Postmodernism and deconstruction can lead to difficulties for the very people who propound them. If originality, genius, and avant-garde status are outdated, then what is the role of the intellectual, critic, or artist? What is the origin and justification of their need to create, and what is the motivation of anyone else to listen?

In *What's Wrong with Postmodernism,* Christopher Norris notes that some poststructuralists used deconstruction in a way that was much more epistemologically radical than intended:

For Saussure, this exclusion [of referential aspects] was strictly a matter of methodological convenience, a heuristic device adopted for the purpose of describing . . . the network of relationships and differences that exist at the level of the signifier and the signified. For his followers, conversely, it became a high point of principle, a belief—as derived from the writing of theorists like Althusser, Barthes, and Lacan—that "the real" was a construct of intralinguistic processes and structures that allowed no access to a world outside the prison-house of discourse.[28]

He further states that the validity of a writer's arguments depends on assumptions of truth and value, even though this contradicts their theories. Similarly, he notes that Baudrillard's writings make no sense without some claims to truth:

His work is of value in so far as it accepts—albeit against the grain of his express belief—that there is still a difference between truth and falsehood, . . . the way things are and the way they are commonly represented. . . . [I]t just does not follow from the fact that we are living through an age of widespread illusion and misinformation that therefore all questions of truth drop out of the picture.[29]

Artists, critics, and intellectuals who entertain these critical theories must resolve these contradictions for themselves and their audiences. On what basis can artists claim that their productions deserve an audience and provide a unique viewpoint? In the postmodern world, what does it mean to say that one person has a clearer vision than another?

Artists' Stances in Integrating Research

With the current prevalence of critical theory and postmodern analysis in art-world discourse, artists can stake out their own theoretical stances; they must choose which assessments and theoretical propositions to accept or reject. Artists who work with emerging technologies have a variety of stances available to them: (1) continue a modernist practice of art linked with adjustments for the contemporary era; (2) develop a unique postmodernist art built around deconstruction at its core; (3) develop a practice focused on elaborating the possibilities of new technology. In reality, the work of artists interweaves these approaches.

Continuing the Modernist Practice of Art with Modifications for the Contemporary Era

Some artists' work with emerging technology is essentially no different from the work of artists who use traditional media. They see themselves engaged in a specialized aesthetic discourse and nurture their personal sensitivity, creativity, and vision. Even though they may proclaim the "revolutionary" implications of a particular technology, they aspire to be accepted by the mainstream world of museums, galleries, collectors, and critics (or, for some, cinema and video). They work on concerns and in modes developed for art in the last decades, such as realism, expressionism, abstraction, surrealism, and conceptual work. They believe that art will continue to renew itself, find ways to appropriately connect with its host cultures, and develop relevant new movements in the future. In fact, they see themselves as essential to progress in art, and seek to cultivate the unique and "revolutionary" expressive capabilities of their new media and tools. They believe that the art world will ultimately incorporate even unprecedented technologies and approaches, such as image processing, interactivity, algorithmic systems, Internet art, and virtual reality.

This stance has certain limitations. First, within the context of the widespread acceptance of critical theory, now that commodification and co-optation are part of the record, the art world cannot remain so self-confident about its cultural niche. Artists are not really independent in the world of the arts. Even in 1934, Bertolt Brecht described the process:

For thinking that they are in possession of an apparatus that in reality possesses them, they defend an apparatus over which they no longer have any control and that is no longer, as they still believe, a means for the producers, but has become a means against the producers.[30]

Furthermore, it is likely that the mainstream art world will only reluctantly accept new technologies, both because they are not tied to developed traditions and because certain features, such as the ease of duplication, further erode a sense of aura of artworks. The one-hundred-year search by photography (and more recently by cinema and video) for acceptance into the canon are good models of what may be expected.

Finally, the connection to popular culture of some of the new technologies, such as digital imaging, interactive media, and Web art, raises high-art–low-art issues. Because the new technologies are used extensively in mass media and the home, it is hard for artists to develop styles that are not read as derivative. Also, the use of these technologies in industries such as advertising, education, and science obfuscate distinctions between design and fine arts. Artists seeking to participate in traditional-art-world discourse with new tools must contend with these other references. They can choose, like cinema, to develop a hybrid popular high-art form, or they can seek to develop uniquely appropriate aesthetics.

Deconstruction as Art Practice

Many artists who have found these theory-based analyses compelling have been attempting to develop an approach in which deconstruction itself is a main agenda. The theories provide concepts, themes, and methodologies for creating artworks that examine and expose the texts, narratives, and representations that underlie contemporary life. Even more, the work can reflexively examine the processes of representation itself within art. Roland Barthes describes the process in "Change the Object Itself":

It is no longer the myths which need to be unmasked . . . it is the sign itself which must be shaken; the problem is not . . . to change or purify the symbols but to challenge the symbol itself.[31]

Science, technology, and their associated cultural contexts are prime candidates for theory-based analysis because they create the mediated sign systems and contexts that shape the contemporary world. They are the tools of power and domination that rely on unexamined narratives of progress, power, representation, and nature.

In this kind of practice, artists learn as much as they can about working with techno-scientific research so they can function as knowledgeable commentators. In one typical strategy, artists become technically proficient enough to produce works that look legitimate while introducing discordant elements that reflect upon that technology. Theory, writing, and art production become interwined in intimate ways. Many of the artists described in the following chapters work in this subterfuge mode of destabilizing the representations of what is considered normal.

The worlds revolving around digital technologies are seen as ripe for critical analysis because of their assurance of the rationality of their directions and their totalizing pretensions. In theories described in later chapters, art is seen as a "parasite" or disruptor, standing outside the dominant narratives. Jonathan Crary describes the opportunities for artistic action in the world of digital technology:

We must recognize the fundamental incapacity of capitalism ever to rationalize the circuit between body and computer keyboard and realize that this circuit is the site of a latent but potentially volatile disequilibrium. The disciplinary apparatus of digital culture poses as a self-sufficient, self-enclosed structure without avenues of escape, with no outside. Its myths of necessity, ubiquity, efficiency, of instantaneity require dismantling: in part by disrupting the separation of cellularity, by refusing productivist injunctions, by introducing slow speeds and inhabiting silences.[32]

Invention and Elaboration of New Technologies and Their Cultural Possibilities as Art Practice

Artists can establish a practice in which they participate in research activity rather than remain distant commentators, even while maintaining reservations about the meaning and future of the scientific explosion. Some analysts see scientific and technological research as the central creative core of the present era. As Paul Brown suggests in his essay "Reality versus Imagination," historians may ultimately see aspects of science as the main art of our era:

I believe that the art historian of the future may look back at this period and see that the major aesthetic inputs have come from science and not from art. . . . Maybe science is evolving into a new science called art, a polymath subject once again.[33]

As I have described in previous articles,[34] artists can participate in the cycle of research, invention, and development in many ways. They can learn enough to become researchers and inventors themselves. The claim that a unification is now impossible because scientific or technological research requires mastery of too much specialized knowledge and access to an elaborate research infrastructure must be critically scrutinized.

Free from the demands of the market and the socialization of particular disciplines, artists can explore and extend principles and technologies in unanticipated ways. They can pursue "unprofitable" lines of inquiry or research outside of disciplinary priorities. They can integrate disciplines and create events that expose the cultural implications, costs, and possibilities of the new knowledge and technologies.

This approach seeks to update the notion of the arts as a zone of integration, questioning, and rebellion, in order to serve as an independent center of technological innovation and development. This idea has precedents in earlier parts of this century. For example, in *The New Landscape in Art and Science,* Gregory Kepes described the need for artists to work with developing science in a proactive way:

The images and symbols which can truly domesticate the newly revealed aspects of nature will be developed only if we use all our faculties to the full—assimilating with the scientist's brain, the poet's heart and the painter's eyes. It is an integrated vision that we need; but our awareness and understanding of the world and its realities are divided into the rational—the knowledge frozen in words and quantities—and the emotional—the knowledge vested in sensory image and feeling.[35]

This kind of practice demands that artists educate themselves enough to function nonsuperficially in the world of science and technology. It requires that they be connected to both the art and the technical worlds, for example, by joining the information networks of journals, and trade shows. It asks artists to be willing, if necessary, to abandon traditional concerns, media, and contexts. It challenges them to develop access to the contexts and tools relevant to their investigations.

It asks artists to entertain the possibility of science-based progress, even though they may share an interest in deconstructing the texts and narratives of the technical world, be skeptical about its self representations, be involved in elaborating the unappreciated cultural implications of the technology, and be wary of the ways that research and technologies get co-opted. It does not automatically reject the idea that some research, invention, and development may transcend the cultural contexts in which it arises, generating new knowledge, cultural meanings, and possibilities rather than just circulating old signs.

An example from just one area of technological foment will illustrate. Many electronic artists are interested in the new possibilities created by telecommunications technology and seem interested in inventing and extending the technology. Certainly, they are interested in the issues that cultural theorists might raise, for example: Who controls and has access to this technology? How is it represented to consumers and developers? What larger cultural movements is it part of? What fantasies does it tie into? How does it affect concepts of identity, space, and body? Even though these topics might be substantive focuses of their work, their tone is basically optimistic about the potential meanings of these developments. Roy Ascott, a longtime pioneer in this work, illustrates this optimistic outlook in his article "Art and Education in the Telematic Culture":

But the art of our time is one of system, process, behavior, interaction. . . . This is precisely the potential of telematic systems. Rather than limiting the individual to a narrow parochial level of exchange, computer-mediated cable and satellite links spanning the whole planet open up a whole world community, in all its diversity, with which we can interact. . . . With electronic media, its flow of images and texts, and the ubiquitous connectivity of telematic systems, this isolation and separateness must eventually disappear, and new architectural structures and forms of cultural association will emerge. And in this emergence we can expect to see, as we are beginning to see, new orders of art practice, with new strategies and theories, new forms of public accessibility, new methods of presentation and display, new learning networks, in short, whole new cultural configurations.[36]

This century is characterized by an orgy of research and invention. Knowledge is accumulating at high speed; unanticipated branches of knowledge, industries, social contexts, and technologies have appeared. These developments are affecting everything from the paraphernalia of everyday life to ontological categories. As the pace continues, predictions about future discoveries and their consequences are impossible. Optimists in the scientific community predict that further research will enhance the material, intellectual, and spiritual quality of life for all the world's people. Analysts such as those in the Extropian movement believe that research is about to usher in the next stage in human evolution.

Taking advantage of unique traditions of the arts, such as valuing iconoclasm and interdisciplinary perspectives, artists can choose to be a part of the efforts to create these new technologies and fields of knowledge. Furthermore, this artistic stance calls for artist participation in other fields beyond the digital technologies that are focused on in this book, such as new biology, materials science, and space exploration. Chapter 1.2 elaborates on this approach.

Summary: The End of Timelessness?

Where are the timeless masterpieces? The rapid pace of research is part of developments in the industrial age that clash with the hopes for art's timelessness. In the past, masterpieces were expected to transcend time and space. During this century, that tradition has been eroding with the loss of "aura" in technologically reproducible work, the ascendance of temporary art forms such as live art and installation, and the power of style and media to rapidly reshape consciousness. Nonetheless, as evidenced by the activities of museums of modern art, many hope that even contemporary art can produce timeless masterpieces.

Information Arts presents the best of research-inspired art. Many are considered masterpieces of the genre. But will they always be? The imaginative reach and innovative vision of some of these artists in mastering an area of research to create eye-opening and thought-provoking works is stunning. But the power of these works may be bounded by their sense of timeliness. After the research world has moved on, they might not seem so significant and moving. Indeed, I know this from painful personal experience, as I see some of my experimental computer artwork of fifteen years ago become quaint and archaic. Curiously, many of these old works can never be experienced again, since the requisite technological infrastructure has disappeared.

These are interesting times for the arts. The linkage of art to emerging research may hasten the redefinition of timelessness. We may need to invent a new meaning for the term *masterpiece*. Think of a masterpiece as a work of art that seizes the cultural moment, or as a work that senses the cultural leap represented by a line of research and uses the magic of the arts to expand what it means and explores what it might become. After the moment passes, the masterpiece will have served its place in history. Like landmarks in science, such as Gallileo's new vision of the universe, these artworks' timelessness is their audacity, even though the new ground they break may become common ground. Readers are invited to contemplate the masterpiece life expectancy of the artworks described in the following chapters.

Notes

In the quiz at the beginning of this chapter, the first five items were created by researchers who define themselves as artists; the others were created by scientists and technologists.

1.   C. P. Snow, "The Two Cultures and the Scientific Revolution" (Cambridge: Cambridge U Press, 1964).

2.   In the last decades, scholars have analyzed the relationship of art and science/technology. They have reviewed the history, noting some influences of these enterprises on the arts, for example, the impact of non-Euclidean geometry and relativity on early twentieth-century painters, the import of technology on the Bauhaus, and the influence of Freud on the Surrealists. Generally, however, the mainstream art world has pretended that art could mostly ignore the technological and scientific revolutions. Art focused on science and technology was treated as a minor footnote. See my forthcoming book *Great Moments in Art and Science* for analysis of this history.

3.   Derived from science articles in *Encyclopaedia Britannica*.

4.   C. Singer, *History of Technology,* vol. 1 (New York: Oxford University Press, 1954).

5.   M. Kranzberg and C. Pursell, *Technology in Western Culture* (New York: Oxford University Press, 1967), p. 4.

6.   Ibid., p. 6.

7.   J. Gaston, "Sociology of Science and Technology," in P. T. Durbin, *A Guide to the Culture of Science, Technology, and Medicine* (New York: Free Press, 1980), p. 467, and E. Layton, "Through the Looking Glass," in S. Cutcliff and R. Post, *In Context* (Bethlehem, PA: Lehigh University Press, 1989), p. 42.

8.   C. Mitcham, "Philosophy of Technology," in P. T. Durbin, op. cit., p. 283.

9.   Ibid.

10.   Ibid., p. 284.

11.   E. Layton, in S. Cutlcliff and R. Post, op. cit., p. 35.

12.   C. S. Smith, *From Art to Science: Seventy-Two Objects Illustrating the Nature of Discovery* (Cambridge: MIT Press, 1980), p. 23.

13.   Getty Museum Program in Art Education, ⟨http://www.artsednet.getty.edu/ArtsEdNet/Browsing/Liata/2.html⟩.

14.   A. Einstein, "Principles of Research," in A. Einstein, *Essays in Science* (New York: Philosophical Library, 1934).

15.   V. Sorensen, "The Contribution of the Artist to Scientific Visualization," ⟨http://felix.usc.edu/text/scivi1.html⟩.

16.   P. Feyerabend, "Theoreticians, Artists, and Artisans," in *Leonardo,* vol. 29, no. 1. p. 26.

17.   Ibid., p. 27.

18.   Ibid., p. 25.

19.   Ibid., p. 27.

20.   For an introduction to these concepts, see S. Wilson, "Dark and Light Visions," ⟨http://userwww.sfsu.edu/~swilson/⟩.

21.   J. Crary, "Eclipse of the Spectacle," in Brian Wallis, ed., *Art after Modernism* (Boston: D. R. Godine, 1984), p. 291.

22.   Ibid.

23.   Annette Kuhn, ed., *Alien Zone* (New York: Verso, 1990), p. 63.

24.   S. Brand, *Media Lab* (New York: Penguin, 1987), p. 252.

25.   J. Sculley, "Foreword," in S. Ambron and K. Hooper, eds., *Interactive Multimedia* (New York: Harper & Row, 1988), p. vii–ix.

26.   E. Ferguson, "How Engineers Lose Touch," *American Heritage of Invention and Technology,* vol. 8, no. 3 (winter 1993), p. 36.

27.   W. Benjamin, "Art in the Age of Mechanical Reproduction," in *Illuminations* (New York: Schocken, 1966).

28. C. Norris, *What's Wrong with Postmodernism* (Baltimore: Johns Hopkins Press, 1990), p. 185.

29. Ibid., p. 182.

30. Bertolt Brecht, "Epic Theatre," quoted in Walter Benjamin, "Author as Producer," in B. Wallis, ed., *Art after Modernism: Rethinking Representation* (Boston: D. R. Godine, 1984), p. 306.

31. R. Barthes, "Change the Object Itself," quoted in Craig Owens, "Allegorical Impulse," in B. Wallis, op. cit., p. 235.

32. J. Crary, op. cit., p. 294.

33. P. Brown, "Reality versus Imagination," in J. Grimes and G. Long, eds., *Visual Proceedings: Siggraph 92* (New York: ACM Press, 1992).

34. S. Wilson, "Research and Development as a Source of Ideas and Inspiration for Artists" and "Industrial Research Artist: A Proposal," ⟨http://userwww.sfsu.edu/~swilson/⟩.

35. G. Kepes, *The New Landscape in Art and Science* (Chicago: P. Theobald, 1956), pp. 19–20.

36. Roy Ascott, "Art and Education in the Telematic Culture," in *Leonardo* Electronic Art (Suppl., 1988), p. 8.

Elaboration on the Approach of Art as Research

This chapter briefly elaborates on the possibilities of art as research, in which artists develop new kinds of knowledge and applications ignored by mainstream scientific and corporate research and push scientific inquiry in unanticipated directions. It explores the meanings of approaching art as research and the rationales, agendas, and working methods of this kind of art.

Can the Arts Offer Alternatives in Setting Research Agendas, Interpreting Results, and Communicating Findings?

Historians of science and technology have documented the determinants of which research is supported, promoted, and accepted,[1] and what products win in the marketplace. As research increases in general cultural importance, it becomes more dangerous to totally rely on market forces. Valuable lines of inquiry die from lack of support because they are not within favor of particular scientific disciplines. New technologies with fascinating potential are abandoned because they are judged as not marketable. Our culture must develop methods to avoid the premature extinction of valuable lines of inquiry. The arts can fill a critical role as an independent zone of research, in which artists integrate critical commentary with high-level knowledge and participation in the worlds of science and technology.

For the last twenty years my artistic practice as artist and researcher has included monitoring scientific communication, working as a developer, and being an artist in residence at several think tanks. These years as what I call a shadow researcher have been illuminating. Tracking and undertaking research at a distance, I have learned of intriguing developments that never saw the light of day. I have seen many inventions and emerging technologies killed because marketing departments judged that no money could be made. I have seen entire R&D departments and their years of research blown away by the winds of corporate politics. Government and corporate support for basic research has almost disappeared, and the concern with the bottom line has shortened the payback horizon to the point where few risks are taken. I have encountered debates in the scientific community that devalue approaches that do not fit favored paradigms.

The invisible hand of the marketplace might not be so wise. Judgments that make short-term sense for stockholders may not necessarily benefit the culture. The peer review referees of scientific journals cannot always see beyond their disciplinary blinders. Scientific and technological research have such critical ramifications for us all that we can ill afford the premature elimination of these ideas and efforts.

Can the Arts Offer Alternatives in Setting Research Agendas, Interpreting Results, and Communicating Findings?

35

The Lesson of Computer Art: Many "high-tech" artists believe that they have addressed the future by becoming computer artists working with digital image, sound, and interactive multimedia. They have misunderstood the real significance of artists' work with computers during the last decade and a half: the fact that artists were experimenting with microcomputers at almost the same time as other developers and researchers. Artists were not merely using the results of research conducted by others, but were actually participating as researchers themselves.

New technologies, such as genetic microbiology, promise to have a similar or even greater impact on life and thought. Artists should identify future trends that could benefit from the artist-research inquiry.

What Is a Viable Role for Artists in Research Settings?
What Can Researchers Contribute to Art and What Can Artists Contribute to Research?
What Can High-Tech Companies Gain from Artists Being Involved?

A good model would provide mutual benefit and cross-fertilization, such as Bell Labs's involvement of artists in sound research, which was instrumental to telephony, electronic voice research, and electronic music. Also, artist Sonia Sheridan's residency at the 3M research center in the 1970s simultaneously influenced the development of color copier technology and shaped her development of a program at the Art Institute of Chicago. More contemporary examples include the artist-in-residency programs initiated by Xerox's Palo Alto Research Center (PARC) and Interval Research, which experimented with mutual definitions of research agendas. The Xerox PARC experience is described more fully in the MIT Press book *Art and Innovation.*

These contemporary examples are qualitatively different from earlier collaborations between artists and scientists-engineers. For example, in the 1960s, EAT (Experiments in Art and Technology) and the Los Angeles County Museum collaborations in art and technology did not profoundly address the role of artists in research; rather, the engineers functioned as technical assistants or the artists dabbled with new technologies.

The contribution that artists can make to research and development is that they often approach problems in ways quite different from those of scientists and engineers, as demonstrated by the crucial role played by designers and artists in computer human interface research over the last years. The arts can function as an independent zone of research. The concept of artist could incorporate other roles, such as that of researcher, inventor, hacker, and entrepreneur. Even within research labs, artist participation in research teams might add a perspective that could drive the research process and continue to contribute at all stages.

The Choice of Research Agendas, Definitions of Research Questions, and Adoption of Metaphors

Artists might very well value research according to criteria quite different from those of the commercial and scientific worlds. They might see aspects of the problems missed by the other researchers. The arts could become a place where abandoned, discredited, and unorthodox inquires could be pursued.

My experience as a artist and consultant to a National Science Foundation–funded project to investigate artificial-intelligence (AI) tutors to teach science illustrates an example of an artist's contribution to the definition of research questions. The project was testing a variety of strategies for software tutors: the soundness of the theories would be judged by how student learning was affected. I noted that students reacted to the tutors similarly to how they reacted to humans. Did the tutor seem sympathetic? Did it manifest any kind of personal knowledge of the student? Did it have an interesting "personality"? My artistic intervention was to suggest that one could not address the question of AI tutoring without paying attention to the dramatic aspects of the interactions. Some of the scientists in the project assumed that as artistic consultant, my main role was, stereotypically, to beautify the reports submitted to the government.

Research Process Decisions, Interpretation of Results

The arts can offer insights into the significance of research results and the design of research activities. The field of scientific visualization can illustrate this point. Donna Cox, one of the artists described in following chapters, working as an artist with the National Super Computing Center at the University of Illinois, helped devise animations to visualize complex bodies of information about natural phenomena. The scientists reported that the visualizations helped them understand the meaning of their data and devise subsequent inquiries.

The Representation of Potential User Perspectives

Research and development attempt to create products that will be used and valued by the public, but sometimes developers ignore the settings in which the products will function. Artists can provide insight on nontechnical responses and ideas for better addressing the needs and perspectives of the general public. For example, one automaker decided to use emerging speech synthesis capabilities as warning signals in their cars. A speech synthetic voice would speak warnings such as "Your seat belt is unhooked" or "You are going too fast." But consumers found the expressionless voice, with its unchanging repetitions, annoying and even slightly sinister; 90 percent deactivated it and the product was discontinued. One might imagine that artistic involvement in the design of the synthetic voice and its repartee might have had more comfortable or engaging results.

Communication of Findings, Consideration of Cultural Implications

Science and technologists often must communicate outside their disciplines. Artists can make presentations of research come alive; furthermore, they can identify implications that may be ignored or not understood by other researchers. Artist-researchers Arthur Elsenaar and Remko Scha are known for their investigations of artificial intelligence, speech synthesis, and the role of facial muscles in communication. They regularly give presentations to both art festivals and research meetings in which their artificial character Huge Harry explains the implications of the research. Huge Harry is effective in demonstrating the research in a direct and lively way, and in raising questions about both its positive and negative ramifications. For example, in some of the demonstrations Huge Harry applies electrical stimulation to the facial muscles of his human assistant to demonstrate the roles of particular muscles and the capability of the face to generate expressions outside those normally encountered. An audience member watching this might certainly think about more than the function of specific muscles.

Art Characteristics Useful for Research

What artistic perspectives can contribute to research? Several traditions of the arts are potentially valuable:

- A tradition of iconoclasm means that artists are likely to take up lines of inquiry devalued by others.
- The valuing of social commentary means that artists are likely to integrate widely ranging cultural issues into their research.
- Artists are more likely than commercial enterprises to incorporate criteria such as celebration and wonder.
- Interest in communication means that artists could bring the scientific and technological possibilities to a wider public.
- The valuing of creativity and innovation means that new perspectives might be applied to inquiries.

The recent history of the personal computer illustrates the need for an independent research function and the role the arts might serve. Early developers, such as Apple Computer founders Steve Wozniak and Steve Jobs, found little support for their ideas about the personal computer. Supervisors signed waivers on the ideas because they could not imagine any market for a desktop computer. Similarly, the discipline of computer science was mostly uninterested in software and hardware issues related to these computers. Advances often came from individuals who worked outside traditional academic and business channels. Teenagers became world experts, nerds became billionaires, and artists made significant contributions in fields such as interface design and image-sound processing.

Similarly demonstrating the value of art-research cross-fertilization, the SIGGRAPH (the ACM international organization for computer graphics research) annual meetings have included an art show since their beginnings. These shows have been influential in several ways. Artists have been able to learn about emerging computer graphics research and technologies so that they could start experimenting with them. In parallel fashion, researchers have become acquainted with artistic work that pointed to new research directions.

Reliance only on traditional lines of research might have resulted in a much longer wait for the developments that have profoundly shaped the last decades. This story could potentially be repeated many times in many other fields of inquiry if alternative venues for research are developed. The arts could well serve an important function of independent vision if artists were prepared to learn the knowledge, language, work styles, self-discipline, and information networks that are instrumental in their fields of interest.

Preparing Artists for Research

What must artists do differently than they always have done to prepare to participate in the world of research? They must broaden their definitions of art materials and contexts. They must become curious about scientific and technological research and acquire the skills and knowledge that will allow them to significantly participate in these worlds. They must expand conventional notions of what constitutes an artistic education, develop the ability to penetrate beneath the surface of techno-scientific presentation to think about unexplored research directions and unanticipated implications, and learn about the information sources used by scientists and engineers to engage emerging fields, including academic and professional journals, trade shows, academic meetings, Internet resources, and equipment supply sources.

The parameters of the science and technology education required is not yet clear. Can artists find the right mix of objective and subjective processes? Can artists learn enough to engage in research at a nondilettante level? Scientists and technology researchers who have devoted their entire professional lives to educating themselves about topics being investigated might be skeptical. Yet, many times history has shown that topics once considered esoteric and beyond the reach of nonspecialists have become understood and accessible to much wider publics. Indeed, this demystification may be one of the main accomplishments of artists working with research.

At the same time, artists must keep alive artistic traditions of iconoclasm, critical perspective, play, and sensual communication with audiences. They must be willing to undertake art explorations that do not neatly fit in historically validated media and offer their work in new contexts.

The Integration of Research and Art

Research is shaping the future in profound ways. Our culture desperately needs wide involvement in the definition of research agendas, the actual investigation processes, and the exploration of the implications of what is discovered. Artists can significantly contribute to this discourse by developing a new kind of artist-researcher role.

Historical Precedents

Art and research were not always separate. The pre-Renaissance integration of the two may shed light on future possibilities. This section offers a brief summary of the historical relationships between art, science, and technology.

Paleolithic   In the Paleolithic era, some of the greatest accomplishments were simultaneously monuments of art and science. The Paleolithic cave paintings have been identified as the first significant act of painting; they are also one of the first illustrations of scientific observation. Some analysts propose that their power comes in part from the painters' careful observation of animal physiology and behavior. Paleolithic metalsmiths are renowned for the aesthetic power of their metalwork. They were also critical in the history of chemistry, being the first to identify different metals and their characteristics.

Stonehenge is a another example of the early fusion of artistic, religious, and scientific functions. Although there are debates about the extent of its accomplishments, its monoliths are carefully placed to indicate the positions of heavenly bodies at various times of the year. In all of these Paleolithic examples, the pioneering in art and science went hand in hand.

Renaissance   Leonardo da Vinci is well-known as history's greatest integrator of art and science. He was by no means unique in having interests that spanned art and science; educated persons were expected to do so. The ateliers of his era included science and engineering as regular parts of their curricula. For example, illustrations of the artists' studios included skeletons for studying anatomy and structural components for studying engineering. Even more, the ethos of the time included the idea that one could not be a good artist or scientist without interest in the other field.

1870–1930   By this time, science and art had already become clearly separate fields. However, even during this period, many analysts trace important influences on abstract art coming from the invention of photography and the investigations of non-Euclidean geometry and elementary particle physics. These scientific and technological inquiries challenged traditional ways of seeing and conceptualizing the physical world. The questioning freed and provoked artists to represent the new world views. The influence of artists on scientists is less clear, although the general cultural questioning may have created a milieu that encouraged scientific questioning.

New inventions also stimulated artistic experimentation in fields such as photography, cinema, sound recording, electrical machines and lights, radio, and electronic music. In the early days, artists often acted as developers. For example, experimentation in chemistry and optics was intrinsic to the role of photographic artists. Artists assumed a variety of stances toward technological change, ranging from the urge of Bauhaus and Socialist art to participate within industry, to Futurist glorification of technology, to the ironic Dadaist commentary.

Art and Science/Technology Collaborations

This section briefly reviews contemporary initiatives to integrate art with research. They illustrate a full range of the roles for artists, from being principal architects of the efforts to consultants on projects intitiated by others. Chapter 8.1 describes other institutional arrangements that encourage and inform collaborative effort.

PARC PAIR

Xerox's PARC initiated an artist-in-residence program called PAIR. PAIR adopted an open-ended strategy in which artists and researchers mutually defined a problem to work

---

PAIR: ⟨http://www.pair.xerox.com/⟩

on, with the definition of the problem becoming part of the collaboration. The Web site explains that PAIR itself is a research project focused on the possibilities of collaboration: "One way that PAIR attempts to bridge the gap between the artists and the scientists is to use technology as a common language. In making our pairings we try to find artists and scientists who use the same, or similar technologies, though often in very different ways. Another equally important aspect of PAIR is to bring the fine arts directly into the work environment for the mutual benefit of both the artist and the corporation." Documentation of the PAIR program is presented in a book called *Art and Innovation*, edited by Craig Harris.

Banff Centre for the Arts

For many years the Banff Centre for the Arts sponsored innovative artist residencies in which artists could experiment with the latest technologies in the mountain environment of the center. The center provided technical support and a critical community as a spur for the work. The residencies have focused on a series of themes, such as virtual reality and the body. The Center also supports indigenous people's cultural explorations with new technologies such as their providing net.radio access to indigenous story tellers and running "summits" to bring together technological innovators, artists, and theorists to investigate topics such as Emotional Computing, Living Things (biotechnology and nanotechnology), and Living Architecture.

Interval Research

Interval Research included artistic collaboration as part of its research activities. One line of research, called New Media Experiments, supported artists to investigate emerging technologies in the production of award-winning works. Interval collaborated with many arts institutions, such as art schools and museums, and the journal *Leonardo*. The Interval Web site explained their interest in artistic activity and the faith in its contribution to the research enterprise:

Interval has a deep interest in the relationships between new media form and new media content. Most uses of new digital media rely upon style and aesthetic sense developed in prior forms, an approach that does little to explore the real potential. We are interested in exploring new possibili-

---

Banff Centre: ⟨http://www.banffcentre.ab.ca/CFA⟩
Interval Research: ⟨http://www.interval.com/research/NewMed-old/index.html⟩

ties with these new media, in content as well as form. These explorations are often inspired by and relevant to the arts community.

There are several reasons why these "New Media Experiments" have value at a research lab such as ours. First, they provide stimulation and provocation to our research community, often acting as magnets to bring together unconventional combinations of skills and talents. They can also provide content to test tools (and even tools to test content). Some of these projects are means for collecting data, both through explicit query as well as through observation. These projects may lead researchers down unforeseen paths and result in new discoveries and intellectual property.

In spite of Interval's innovations and its integration of art with research, Paul Allen, the major underwriter of the effort, closed the organization in spring 2000. Allen noted that he wanted to narrow the focus to application development instead of the basic research that had been Interval's hallmark.

ART+COM

Founded in 1988, ART+COM is a German organization focused on research that integrates the perspectives of computer technology, communication, and design. It believes that the most innovation comes from an open interdisciplinary process and describes itself as a "melting pot for new ideas and new technologies, where specialists from the arts, from science, and from industry come together to combine their ideas and goals. The method of working at ART+COM is characterized as an openness towards new projects and novel approaches, the readiness to question old patterns and ways of thinking and the ability to speed up complex development-processes." The organization has distinguished itself by winning awards at international festivals and its commercial development work.

F.A.B.R.I.CATORS

Based in Italy, F.A.B.R.I.CATORS attempts to develop methods to "combine and use art and technology in the design and production of projects, interactive art pieces, multimedial projects, VR installations, creative interfaces, worked out on the basis of the integration of multidisciplinary expressions and disciplines, such as: art, culture, technol-

ART+COM: ⟨http://www.artcom.de/about/welcome.en⟩
F.A.B.R.I.CATORS: ⟨http://www.mediartech.com/en/⟩

ogy, architecture, design." It combines low and high tech in the "elaboration of bizarre and efficient invention."

ATR Lab

The ATR Media Integration and Communications Research Laboratory in Japan believes that art can have a major role in research and defines its major objective: "In pursuing new communication schemes that facilitate mutual understanding beyond differences in place, time, language and culture, we are exploring basic technologies for both realistic multimedia communications and hyper-realistic communication environments for better sharing thoughts and images between humans and machines." It supports research initiatives in these areas, the reconstruction and creation of communication environments, communication support, mental-image expression and transmission, and human communication processes. It includes artists and researchers working collaboratively and presents its work both in art and technical venues. For example, its researchers have presented workshops on "Face and Gesture Recognition," "Technologies for Interactive Movies," and "Lifelike, Believable, Communication Agents." It believes that art can significantly contribute to the development of new communication technologies and supports the experimental artworks of several artists described in this book.

Canon ArtLab

Canon Japan established a division called the Social and Cultural Program Operations Center (CAST), which recognizes Canon's responsibility to *"Kyosei"*: "Under Canon's philosophy of *'Kyosei'*—living and working together for the common good of mankind, diversified activities are promoted by Canon-group companies all over the world in the field of social contribution and cultural support." One of these activities is the ArtLab, which aims to "pioneer new artistic realms through the integration of science and art" by innovating with new digital technologies developed by Canon through collaborations between artists and engineers. ArtLab's activities include exhibitions, public lectures, the on-line documentation of artists' works, and research support for artists. Several of the artists in this book have been supported by ArtLab.

ATR Lab: ⟨http://www.mic.atr.co.jp/index.e.html⟩
Canon ArtLab: ⟨http://www.canon.co.jp/cast/index.html⟩

Arts Catalyst

Arts Catalyst is a U.K.-based organization that offers public lectures, workshops for students, publications, exhibitions, and research opportunities on which artists and scientists can collaborate. Their brochure explains their challenge of old dichotomies:

The Arts Catalyst is a science-art agency promoting real collaborations between artists and scientists. Founded in 1993 . . . it uses innovative art practice to break down the invisible walls between science and the public. We see art as directly applied to the scientific environment—not simply illustrative of science. Never afraid to confront issues in science, subjects covered include human fertility, nuclear power, Darwinism, quantum physics, space exploration and the forthcoming total eclipse of the sun. We work across all the artforms—specialising in collaborations with museums, galleries, scientific laboratories, and other sites.

Some projects and conferences have included "Searching" (exhibition and supported projects to look at SETI—the Search for Extra Terrestrial Intelligence); "Eclipses, Life and Other Cosmic Chances" (art-science consideration of the 1999 eclipse); "GravityZero" (a collaboration between Kitsou Dubois, a choreographer, and space scientists); "Atomic" (supported projects and an exhibition focused on radioactive debris); "Parallel Universe" (a conference on non-Western science); and "Eye of the Storm" (panels with artists and scientists reflecting on scientific controversies in areas such as genetics, evolution, artificial consciousness, and the end of nature).

STUDIO for Creative Inquiry—Carnegie Mellon University

The STUDIO provides a context for interdisciplinary research in the arts by providing artist residencies, commissions, and facilities. It has won numerous grants to pursue inquires at the intersection of science and the arts. It makes extensive use of the artistic and technical resources of Carnegie Mellon University. Some areas of focus include biology, ecology, and robotics. Examples of specific projects (many described in this book) include: "Nine Mile Run Greenway Project," "Tracking the Human Brain," "Building Electronic Communities," "Acid Mine Drainage and Art Project," and "Sex and Gender in the Biotech Century."

Arts Catalyst: ⟨www.artscat.demon.co.uk⟩
STUDIO for Creative Inquiry: ⟨http://www.cmu.edu/studio/⟩

Interactive Institute

The Interactive Institute is part of a major effort by the Swedish government and industry to create a new network of "studios" that will foster the collaboration of artists with researchers in the development of new technologies. Each studio links art institutions with industry collaborators. Some of the themes of the studios include: "Emotional and Intellectual Interfaces," "Smart Things and Environments for Art and Daily Life," "Narrativity and Communication," "Space and Virtuality," and "Games." With significant funding, the studios offer seminars, workshops, technical support, and funding for research projects.

Cultural Institutes, European Cultural Backbone, I3

As part of its concern about maintaining future competitiveness, the European Union has initiated efforts to encourage cultural institutes in which artists collaborate with researchers in developing new technologies. It is based on the premise that Europe has rich media traditions that can inform innovation in a way not necessarily following the American model. Efforts are under way in several countries, such as the "Virtual Platform" in the Netherlands. These approaches are being considered for incorporation into major European Community–sponsored research efforts, such as the I3 (Intelligent Information Interface). A report, "Cultures of Electronic Networks," summarizes some of these perspectives:

Economic growth will depend on the existence of a new media culture which is innovative, diverse, inclusive and challenging. Cultural activity in digital media is driving innovation at all levels, with a constant movement of skills, ideas, individuals and infrastructures across different sectors. Innovative market activity can only be upheld insofar as the "nonprofit" creative research it depends on is fostered on a permanent, continuous basis, and sufficient fluidity is encouraged between the commercial and "non-profit" sectors. . . . The talented and fleet-footed organizations which comprise this network of innovation are small and fledgling. They straddle traditional boundaries and explore the creative spaces between different sectors and media forms. . . . A technical infrastructure for cultural activity needs to be implemented along the same lines as the well-established frameworks of the scientific and academic networks. . . . To be effective, culture as much as science requires its domains of primary research, which need to be supported by appropriate environments and resources (e.g., independent research laboratories for media art).

Interactive Institute: ⟨http://www.interactiveinstitute.se⟩
Cultural Institutes: ⟨http://competence.netbase.org/panel2/rapport2.htm⟩

Centres of Excellence, Technology Media, and Creativity

Artists and analysts in Canada have been attempting to forge new kinds of collaborations that link art, technology development, and scientific research. They seek to establish a "new genre of hybrid cultural research institution in which artists mediate between networks of technology design and diffusion." They propose that countries that hope to be innovators in the future must tap the perspectives of artists in research. In a paper, "New Media Culture in Europe," Michael Century reviews the history of collaboration from institutions such as EAT and IRCAM through 1980s institutions such as SIGGRAPH, Ars Electronica, and ZKM, discussed in chapter 8.1. He reflects on concerns about the commercial co-optation of artists:

[critical intellectuals harbored] a deep ambivalence about these institutional developments, fearing that they would serve only to accelerate the public acceptance of automation in everyday life, on the one hand, and to co-opt artists "with their purported creativity" into becoming commercial application designers, on the other.

He concludes, however, that artistic involvement has not followed only this pattern but rather entered into the innovation and diffusion processes in many more complex ways, building on the tacit knowledge and innovation traditions of the arts. He forecasts great opportunities in areas such as improving the usability of information technology applications and in the "cultural quality of social informatics."

Souillac Charter for Art and Industry

The Souillac Charter for Art and Industry is another European effort to define a framework for the collaboration of artists and industrial researchers. Some of the analysis is based on the ideas of Don Foresta, a pioneer in telecommunications art, to define "new communication spaces." The charter proclaims a new kind of space developing:

A new communication space is growing. . . . Searching for its own logic and a cultural, social and political identity. What this space will mean to society is not yet clear, its final content is uncertain, and how it will effect culture open to healthy speculation and necessary experimentation before its final specificity is defined.

Centres of Excellence: ⟨http://www.music.mcgill.ca/~mcentury⟩
Don Foresta: ⟨http://www.iic.wifi.at/iiceng/forestacp.htm⟩

He links the new spaces with scientific insights that expanded beyond Euclidean space and the interactive flow of multiple points of view that characterize the contemporary world. Foresta sees art as a critical resource in developing the possibilities of these new cultural spaces and new technologies in part because artists anticipate the "psychological atmosphere" of their era. The charter identifies several areas of research that could benefit:

It is a visual space, a communication space, an organisational space, the space of how we imagine reality. . . . Every mode of communication has at one of its extremes a form of expression we call art. Art, being the densest form of communication, is often the supreme test of any means of communication. . . . The technologies of communication today permit a full exploration of the potential of this new space, making them an expression of the values that we are attempting to define as we reinvent our society according to the new artistic and scientific givens.

Wellcome Trust

The U.K.'s Wellcome Trust primarily funds biomedical research. It also supports a significant "Sci-Art" effort the encourage collaborations between researchers and visual, media, and performance artists. Each year it has a competition and selects projects to support, such as the 1997 projects "Primitive Streak" (fashions based on embryo development) and "Exposing the Phantom" (self-perceptions of people who experience phantom limbs). The program explains its perspectives:

Flair, creativity, inspiration, interpretation: science and art share similar vocabularies yet are often compartmentalized into their own mutually exclusive worlds. The Wellcome Trust has developed schemes to break down the barriers between the two dominions, stimulating a cross-fertilization that may ultimately benefit both. For art, the chance to gain inspiration from science's insights into the natural world; for science, an opportunity to view an entirely new perspective on research.

Future Possibilities

What is science? Let us define science as an accumulation of worldviews, questions, metaphors, representations, and processes that attempt to understand the nonhuman world. It is also the accumulated body of knowledge that these inquiries have generated.

Wellcome Trust, "Sci-Art" — Web link to winners: ⟨http://webserver1.wellcome.ac.uk/en/old/sciart98/⟩

Artists are engaged with science in a variety of ways—joining in the critique of its claims, building on its accumulated knowledge, and participating in the inquiries at the fringes of its understandings.

What is technology? In its widest meaning, it is the process of inventing and making things, which includes much of art. In its more restricted sense of "high tech," it is the contemporary activities of research and development aimed at producing new materials and processes. Critics note that technological development is aimed at control and exploitation. High technology often builds on knowledge generated by science, but sometimes it races ahead of science into uncharted ground.

What is art? In the last century its definition has been expanded far beyond conventional media, contexts, and purposes. Nevertheless, we can ascribe to it the following set of characteristics. Typically, it is undertaken for nonutilitarian purposes. It usually intends to move or provoke an audience for aesthetic, intellectual, and/or spiritual purposes. In the West it is more likely than other disciplines to value personal, idiosyncratic vision and perspectives, to prize iconoclastic stances outside of established institutions, and to promote individual creativity.

This book asks how art, science, and high-tech research can influence each other. The appropriate contours of this involvement are not yet defined. Much experimentation is required. How can artists function independently from established research centers? When artists work within research settings, how can those centers learn to be open enough to benefit from the unorthodox contributions that artists might make? How can artists learn to involve themselves in the ways and byways of researchers without losing touch with their artistic roots? Young artists who become involved as researchers can be seduced by the recognition and economic rewards of research so that they quit functioning as artists.

Artists do not act exactly like researchers. Contemporary art often includes elements of commentary, irony, and critique missing from "serious" research. Scientists and technologists strive toward objectivity; artists cultivate their idiosyncratic subjectivity as a major feature of what they do. The "research" that artists create would work like art always does—moving audiences through its communicative power and unique perspectives. Still, it might simultaneously use systematic investigative processes to develop new technological possibilities or discover new knowledge or perspectives.

Frank Oppenheimer, the scientist who established the world famous Exploratorium museum of science and art in San Francisco explained the rationale for the combination:

Art is included, not just to make things pretty . . . but primarily because artists make different kinds of discoveries about nature than do physicists or biologists. They also rely on a different basis for decision-making while creating their exhibits. But both artists and scientists help us notice and appreciate things in nature that we had learned to ignore or hadn't been taught to see. Both art and science are needed to fully understand nature and its effects on people.[2]

In the essay "Boundaries and Categories," biologist Stephen J. Gould notes that science often helps art in concrete ways, but that art offers a more subtle aid to science. Of literature, he says:

Fiction is often the truest pathway to understanding our general categories of thought and analysis, and artifice often illuminates the empirical world far better than direct description. This paradox arises because we can best understand a natural object or category by probing to and beyond its limits of actual occurrence—into realms that science, by its norms of discourse, cannot address, but that art engages as a primary interest and responsibility.[3]

He also notes the power of artists to confront scientists with the assumptions of their categorical schemes and concepts:

[W]e scientists face a special problem of denial and inattention to our personal prejudices, for our "official" methodology proclaims objectivity, and we can therefore be maximally fooled.

Artists can therefore be most useful to scientists in showing us the prejudices of our categorizations by creatively expanding the range of nature's forms, and by fracturing boundaries in an overt manner (while nature's own breakages, as subtle in concept or invisible to plain sight, are much harder to grasp, but surely understandable by analogy to artistic versions).

Perhaps the segmented categorization of artist and researcher will itself prove to be an historical anachronism; perhaps new kinds of integrated roles will develop. Signs of this happening already appear. Some of the hackers who pioneered microcomputer developments may one day be seen as artists because of their intensity and culturally revolutionary views and work. Similarly, some art shows, such as Ars Electronica's, now define research ideas as core themes (for example, artificial life) and invite researchers as key presenters along with artists. Research has radically altered our culture and will continue to do so. We must ask what role art might play in this process.

Though it seems incontrovertible that neither artists nor scientists can stand completely outside of a cultural or economic milieu, the creative leaps made by inventors, scientists, and artists amaze and inspire me. They do seem to create genuine new possibilities and for a moment stand above the cultural flow. The artists' works described in the following chapters are remarkable in the unusual perspectives they use to explore research. Neither the research, nor the art, however, is complete in itself. Together they make a full picture of what the research really is and what it could mean.

Notes

1.   Thomas Kuhn, *The Structure of Scientific Revolutions* (Chicago: University of Chicago Press, 1970).

2.   F. Oppenheimer, "Mission of the Exploratorium," ⟨http://www.uinta6.k12.wy.us/WWW/MS/8grade/Info%20Access/KEXPLORI/AIR.htm⟩.

3.   S. J. Gould, "Boundaries and Categories," ⟨http://138.87.136.7/cfa/galleries/gould.html⟩.

2

Biology: Microbiology, Animals and Plants, Ecology, and Medicine and the Body

Biology: Research Agendas and Theoretical Overview

Introduction

Some analysts predict that the twenty-first century will be "biology's century." It is fore-casted that the breakthroughs about to come in understanding and potentially control-ling the organic world—including our own bodies—will make the electronic and computer revolution look like child's play. These understandings will lead to signficant cultural questions about the nature of being human and the implications of biological manipulation. It will be fitting for the arts to become involved with this research both to comment on and to independently shape possible future research directions.

Biology has long been a topic—in paintings of landscapes, plants, animals, and peo-ple, and in sculptures of the bodies of people and animals. Historically, artists often preceded scientists in careful attention to the anatomy and physiology of living beings. In the Rennaissance, artists often struggled to illustrate what they really saw rather than rendering what scholastic philosophy dictated would be there. Artists' illustrations of plants, animals, and humans were instrumental in advancing the biological sciences.

This section focuses more on artistic work with living materials or with the concepts, capabilities, and contexts derived from scientific research. How have artists worked with this area of growing importance? Artists have chosen to approach biology and medicine at several levels: the microscopic and genetic; the macro level of plant and animal behav-ior up to the level of ecology's focus on interacting systems; and focus on the body and medicine both in observation and manipulation.

Other chapters treat related material: virtual reality and telepresence implicity reflect on the reality of the biological body in their substitution of mediated body experience. Nanotechnology, discussed in chapter 3.2 on physics and materials science, allows re-searchers to assemble matter atom by atom. People will be able to play God in creating atomic-level machines and structures. Theorists assert that they will be able to assemble organic just as well as inorganic entities. Robotics, artificial life, and artificial intelligence all try to model life processes outside of the normal carbon settings. The description of the visionary 1993 Ars Electronica conference, organized by Peter Weibel, illustrates the overlaps of these fields:

Ars Electronica 93 dealt with the science of artificial life, with the origins of life and with the question of how life was able to come into being, how it developed and what forms it might possibly take in the future. Chances and risks of this new branch of science were discussed during symposia on the topics of artificial intelligence, genetic engineering and immortality. Works from various different fields (biogenetic art, virtual creatures, genetic manipulation, robots . . .) were brought together under the collective term of "Genetic Art." A large part of these works simulated processes of life by means of modern technology and on the other hand reflected possible sequels of synthetic simulation and synthetic creation of life in a critical way.[1]

This chapter surveys research agendas in biology, bioengineering, and medical technology to identify areas ripe for future artistic activity. It also presents an overview of the theoretical analysis of the cultural implications of this research.

Research Agendas in Biology and Medicine

Government and Research Center Definitions

Biological researchers work simultaneously at many levels, ranging from the molecular level of cellular genetics up through organisms and whole population systems. Bioengineers are seeking to intervene at all levels, from nanotechnological manipulations of atoms to ecological control of populations. Medical technology can see deeper and deeper inside of bodies and intervene via drugs, electronics, and bioengineered tissue. Researchers propose scientific and technological solutions to major world problems, such as hunger, pollution, population, disease, drugs, and biological warfare. Investigators work on "enhancements" once considered purely science fiction, such as brain control, designer drugs, life prolongation, bionics, human cloning, and the invention of new organisms. Some in the arts and humanities look from a distance with skepticism and sometimes horror. Our culture needs both unbridled optimism and doubt, but it needs those who are wary to be as knowledgeable as possible.

As a major shaper of research agendas, the U.S. National Science Foundation (NSF) regularly convenes panels of experts to help it to analyze the field and identify future research priorities. NSF's biology area divides itself into four major program areas: biological infrastructure, environmental biology, integrative biology and neuroscience, and molecular and cellular biosciences.[2]

The NSF also supports the Human Genome Project (the attempt to map all the human genes and their effects) and hosts special programs with an applied emphasis to promote research in areas such as bioremediation (undoing the effects of pollution), biodiversity studies, neuroscience, and microbial research. It also has a related area called Bioengineering and Environmental Science, which is divided into three programs: Biomedical Engineering/Research to Aid Persons with Disabilities, Biochemical Engineering/Biotechnology, and Environmental/Ocean Systems. Its environmental programs support research that "improves our ability to apply engineering principles to avoid and/or correct problems that impair the usefulness of land, air and water."[3]

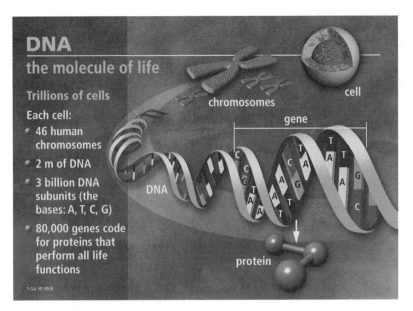

Fig. 2.1.1. DNA—The Molecule of Life. Web site graphic describing the basic facts of DNA. Source: Human Genome Project information Web site (www.ornl.gov/hgmis). Sponsored by the U.S. Department of Energy, Human Genome Project.

A special NSF workshop called "Impact of Emerging Technologies on the Biological Sciences" identified technologies that would shape the future of biological research and ultimately intervention:

- Bioinformatics: the use of computing for the acquisition, analysis, and retrieval of biological data.
- Computational biology, modeling, and simulation: the "use of computational tools to discover new information in complex data sets and to decipher the languages of biology (e.g., the one-dimensional information of DNA, the three-dimensional information of proteins, and the four-dimensional information of living systems)."
- Functional imaging of the chemical and molecular dynamics of life: light optical methods such as spectroscopy and fluorescence, CAT (computer aided tomography), PET (positron emission tomography), and MRI (magnetic resonance imaging).
- Transformation and transient expression technologies: cloning, gene splicing—"The most revolutionary recent development in the biological sciences is the methodology for manipulating DNA molecules and introducing nucleic acids in genetically competent form into cells."

- Nanotechnology: "This technology employs microelectronic fabrication techniques to integrate mechanical and biosensors, computer power, and electromechanical outputs into an integrated microchip, which can work at the microsopic level."[4]

Surveys of graduate programs and research institutes around the world indicate interests similar to the NSF emphases, with some additional areas, such as biodiversity; integrative neuroscience; biology of the cell surface; and plant community and global biology.

Surveys of research priorities in medicine and medical technologies reveal interests with some overlap with general biology. Medical researchers are more likely to identify priorities such as medical interventions (e.g., drugs and surgery); attention to specific physiological systems, such as the cardiovascular; aids to medical interventions, such as biomaterials; telemedicine; and AI diagnosis expert systems and particular diseases (e.g., cancer and AIDS).[5]

The Coalition for Education in the Life Sciences (CELS) convened a world panel of experts to identify critical issues in life science education. The conference identified six major categories of importance: wellness, shaping/reshaping life, overpopulation, resource utilization, alteration of natural systems, and functional/dysfunctional behavior (interorganismal interaction).[6]

Technology Development

Drugs, sensors, and device development top the research agendas of medical technology companies. The Stanford Research Institute (SRI—a think tank for future technologies) periodically offers an analysis of the future of medical technology markets. A recent analysis lists research and market opportunities in the following areas: drug research, such as sedatives, metabolics, and anti-infectives; equipment research, such as dialysis, radioactive diagnostics, anesthesia, and medical lasers; and devices such as hearing aids, implants, and nerve prosthetics.[7]

Individual Research Projects

Everyday there is news of breakthroughs in biological or medical research, for example, in understanding genetics or evolution. There is also news of emerging technologies for controlling disease, overcoming ecological problems, enhancing agricultural output, and the like. Here is a sample of the listings of headline news items from a biology news service during the time this book was being prepared:

- Genetics (examples: "The Genetic Basis of Cancer," "The Origin of Life on Earth," and "Team Isolates Probable Mammalian Clock Genes")

- Cell behavior (examples: "Hemoglobin Synthesis by a Genetically-Engineered Plant" and "Chemical 'Guided Missile' Destroys Tumor Supply Lines")
- Evolution (examples: "Animal Origins: Reconciling Fossil and Molecular Evidence" and "Wolf Packs: Group Hunting No Major Benefit")
- Ecology (examples: "Ecological Risks of Transgenic Plants," "The Ethnobotanical Approach to Drug Discovery," and "The Puzzle of Declining Amphibian Populations")

Bioethics

Bioethicists suggest that scientific and technological developments in biology and medicine raise a whole series of complex and disturbing questions, such as:

- Privacy: Who has the right to inspect a person's "genetic profile"?
- The definition of procreation: With technologies such as surrogacy and sperm and egg donation, what are appropriate definitions of parenthood?
- The cloning of human beings: What are the ethical implications? What are the rights of a clone?
- Genetic enhancement: When is it appropriate to modify genetic profiles?
- Pharmaceutical and neurological enhancement: When is it appropriate to intervene, for example, to slow aging or augment perception?
- The responsibility of scientists and their relationships with business.[8]

Areas of Cultural Significance in Biological Research

This section briefly considers some areas of scientific inquiry and technology development related to biology and the body that I have been tracking over recent years. Some of the areas represent accepted mainstream areas; others are less accepted, being dismissed by established scientists. As explained in the introduction, artists can serve a useful function by being aware of the full range of research that may be culturally significant in the future.

Smell, Taste, and Touch

Scientific and technological research has manifested a lopsided interest in the senses of sight and hearing over the senses of smell, taste, and touch. Analysts have suggested several reasons. The latter three are often considered senses "lower" than sight and hearing, sight and hearing being much more important to higher animals and humans for getting information about the world. In his book *Metamorphosis,* the psychoanalyst

Ernest Schachtel claimed that sight and hearing were distance senses, while smell, taste, and touch were close senses that required intimate contact much like that between a child and its mother. According to his theory, differentiation required the deemphasis of the close senses and the privileging of the distant senses. Another theory proposes that sight and sound have much higher bandwidths and thus are more useful for navigating a complex environment. One more idea offers that biology and chemistry lacked the theoretical and practical tools necessary to investigate these senses until recently.

In the last few years, researchers have begun to pay more attention to close senses in science, social science, and the humanities. For example, Diane Ackerman wrote *A Natural History of the Senses,* which explored all the senses, while biological research centers explore the biology and chemistry of the close senses. Industrial research labs in industries such as cosmetics and food study smell and taste. Technological research centers are working hard to link information technology with these senses, for example, by developing biosensor "artificial noses" and force feedback systems so that one can experience touch and texture through computer interfaces.

There is a long history of folk science and folk art related to these senses. One could look at the entire history of food preparation and cuisine as a study in smell, taste, and touch. Similarly, the fabrication of all the objects that can be touched—for example, fabrics, ceramics, and furniture—encapsulate centuries of experience with touch.

Media researchers have sought to integrate smells into entertainment for several decades. In 1959, the developer of Aromoarama included a smell track that could release any combination of thirty-one odors via the air conditioning system in the movie *Behind the Great Wall,* which was about China. Some of the odors were gunsmoke, coffee, and peppermint. Reportedly, the distribution system didn't work well, with the odors being overpowering in some areas of the theater. In 1960, the movie *A Scent of Mystery* used the Smell-o-Vision system, which solved the distribution problem by including smell tubes at each seat. Although the smells were integrated into the mystery story line, the system was not a great commercial success. The 1981 movie *Polyester* included scratch-and-sniff cards in a system called Odorama. A high-tech company called Digital Tech Frontier has created a virtual reality kiosk system called Virtual Scentsations, which releases odors at appropriate moments. Another company, Ferris Production, developed the Experience System, which adds smells to VR rides.

Why has current interest accelerated? These close senses are little influenced by media and technology. They still suggest that humans have an animal nature that is outside the world of simulacra and mediated control. This relative unknown quality seems to invite scientific, technological, and ultimately commercial attention. Artists have paid relatively little attention to the close senses. The growing attention to them poses a

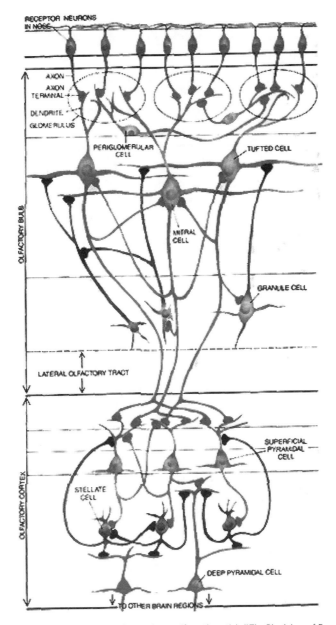

Fig. 2.1.2. Walter J. Freeman, Neurons of the Olfactory System (from the article "The Physiology of Perception," *Scientific American,* vol. 264:2 (February 1991), pp. 78–85 (http://sulcus.berkeley.edu/FLM/MS/Physio.Percept.html).

Areas of Cultural Significance in Biological Research

challenge and opportunity for artists. Here is a brief list of research projects and activities that demonstrate areas of emerging interest. (See also chapter 7.4.)

Smell

- Richard Vogt, University of South Carolina, offers a course called Taste of Smell, which includes sections on animal behavior, sociology, literature, and the biology and chemistry of smell.[9]
- James Kohl, author of *The Scent of Eros: Mysteries of Odor in Human Sexuality,* created a Web resource site focused on pheromones.[10] The Sun Angel "Aromatherapy" Web site investigates the power of smell to heal.[11]
- Researcher Alan Hirsch, with the Smell and Taste Treatment and Research Foundation, found that certain odors were more effective than others in increasing gamblers' expenditures at slot machines. In another study he discovered that homey odors, such as pumpkin pie, doughnuts, and licorice were significantly more effective than perfume in generating sexual arousal as measured by blood flow to the penis.
- Researcher Wilder Penfield found that stimulation of particular parts of the brain brought forth not only particular childhood memories, but also the smells associated with those memories.
- Researchers at Griffith University in Brisbane have equipped miniature robots with "pheromone" systems to mark and follow trails by smell.
- Cyrano Sciences Inc. has invented a nose chip that uses an array of 32 sensors to discriminate various odors. Ambryx Inc. has created a technology that uses cell biological, biochemical, and physiological assays for functional profiling of tastes and odors.

Taste

- Researchers are developing "electric taste" methods, which will apply negative and positive electrodes to food in order to intensify the experience of taste.
- Bowman Gray School of Medicine researcher Inglis Miller has discovered that people vary greatly in the density of their taste buds, some having ten times as many as others. Also, olfactory and taste receptors are the only nerve cells that regularly die and regenerate.
- Scientists at International Flavors and Fragrances, an industrial research lab, report that consumers are more sophisticated now, and increasing numbers of flavors must be brewed together to make successful new artificial flavors.
- Millions of people suffer from "chemo-sensory" disorders, which interfere with taste and smell. One disorder called phantom taste causes people to taste flavors even when the source is not present.

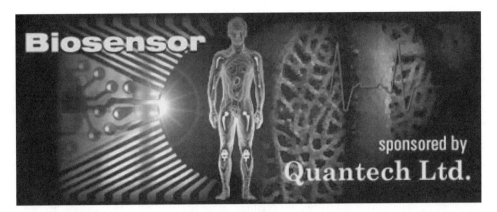

Fig. 2.1.3. Quantech, Inc., Web site logo. The company develops biosensors for medical and environmental sensing ⟨http://www.biosensor.com⟩.

Touch

- Thomas P. Way at the University of Delaware created a system for the Automatic Generation of Tactile Graphics. It converts images into touch-readable displays useful for the blind.[12]
- Immersion corporation has introduced the FEELit mouse, a relatively inexpensive system that lets the users experience physical sensations, such as textures, surfaces, springs, liquids, and vibrations.

Biosensors

Scientific research and technological development in the organic realm has been hampered by the lack of reliable sensors. Scientific research often relies on "objective" measurements by reliable instruments. Otherwise one must rely on anecdotal report, a method not promoted by the hard sciences. Smell, taste, and touch research has been tainted with this need to trust self-report. Also, information technology could not be integrated with these senses until some kind of automated sensor systems was developed, but the last decade has seen a great increase in biosensor development. These new technologies will allow the linkage of information systems with these senses, opening up new applications and realms of mediated experience that call out for cultural investigation.

Erika Kress-Rogers has edited the *Handbook of Biosensors and Electronic Noses*, which surveys this exploding field. It includes chapters that explore underlying theory and technologies and describes applications in medicine, food, and environmental sensing. Robert Hansen, chief scientist at the Applied Research Lab at Pennsylvania State College, identifies several trends in biosensor research: using arrays of sensors rather than single probes; the use of microfabrication to make nanometer-sized probes; and the proliferation of types of sensors. Other developments include modular smart microprobes set up in shared buses, implantable microprobes that can be left for prolonged periods, probes that work at the molecular level, and hybrid sensors that are part cultivated cell and part electronics.

Jose Joseph at SRI identifies several different technologies used to create biosensors:

1. Electrochemical sensors that measure electrical qualities (for example, resistance and capacitance) of the ions in an organic substance.
2. Optical sensors that measure light qualities of an organic substance, such as absorption, fluorescence, or luminescence.
3. Mass-sensitivity, which measures changes in mass.
4. Chemical amplification, which uses discriminatory enzymes and other agents to assess the relative presence of target substances.

Other researchers point out that any biomolecular reaction could be used to construct a detector system. Advancements in MEMS (microelectomechanical) biosensing devices have come from using the same techniques that create microchips. Here is a brief list of the research projects and activities indicating some of the trends in biosensor research:

• Nanogen has developed a system called APEX that can detect diseases on the spot instead of requiring the many-day wait for a lab culture. It contains DNA strands called capture probes woven into a microchip. The DNA strands are optimized for detecting certain diseases and can report within a few minutes.
• Researchers Walter L. Sembrowich, Carter Anderson, and William Kennedy have invented a watch that can read blood sugar levels without drawing blood. It contains a patch that induces sweat and prevents glucose absorption by the sweat glands. The watch has an optical reader that reads patch-color changes as an indicator of glucose levels.
• The team of Oak Ridge National Laboratory researcher Mike Simpson and Gary Sayler of the University of Tennessee Center for Environment Biotechnology have created relatively inexpensive biosensors that make use of bioluminescent bacteria genetically engineered to respond to particular substances.
• Working with issues of birth control, Dr. Hugh Simpson developed an electronic "fertility" bra that flashes red lights and green lights to indicate where a woman is

in her ovulation cycle as a guide to when intercourse would be likely to lead to conception.

- Robert Gilmour, a consultant to St. Michael's Hospital in Toronto, has developed devices that look like toilets that can conduct urine tests in near real time. His systems are used to screen for urinary problems. Other similar systems have been implemented to routinely test for drugs.
- Spurred by need for airport security, many companies have developed extremely sensitive chemical detectors to be used to sniff out drugs and plastic explosives. For example, Thermedics has a vapor sniffing system called Egis that can detect a hot dog eaten days before, and even after showering.

Bioidentification and Biosurveillance

Security and law enforcement services have been eager to develop systems that can identify and differentiate persons; fingerprints have been the most dominant system. Typically, these systems seek out some biological quality of a person that is not easy to modify and easy to read. The systems must also provide ways to store and retrieve the information collected. The proliferation of Internet-based commerce and communication has similarly generated the need for electronic systems to verify the identify of the person on the other end of a connection. A variety of technologies have been investigated. Biometrics was defined by the magazine *Biocard International* as "automated methods of verifying or recognizing the identity of a living person based on a physiological or behavioral characteristic." Here is a sampling of projects:

- The Veincheck system analyzes the vein pattern in the back of the hand. EyeDentify scans the vein pattern in the retina. Handkey reads the geometry of the hand. Several companies identify persons by matching signature dynamics, the flow of writing a signature. Voice verification, facial thermogram, and facial-feature recognition form the bases of other systems.
- Infotec has created the Checkin kiosk as a virtual probation officer. The system optically checks fingerprints and/or hand prints and reports via the Internet.
- In an attempt to establish the pedigree of citrus products, Seifolia Nikdel of the Florida Department of Citrus can determine the geographical origins of orange juice. Ionizing the juice in radio-frequency heating results in a profile of trace metals from the soil in which the trees grew. Computer comparison with stored profiles can identify the location of the source trees.
- Many software packages have been developed that attempt to analyze a person's mental state through their voice or writing. For example, MindViewer's "personality

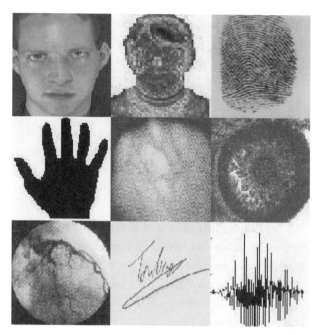

Fig. 2.1.4. Illustration displaying multiple biometric technologies. Pattern Recognition and Image Processing Lab, Department of Computer Science and Engineering, Michigan State University (http://biometrics.cse.msu.edu). Image from A. K. Jain, R. Bolle, and S. Pankanti, eds., *Biometrics: Personal Identification in Networked Society.* Norwell, Mass. Kluwer Academic Publishers, 1998.

assessment uncovers hidden motives, desires, and expressed needs and gives you detailed information on how to make friends with your subject, affect their behavior, elicit trust and gain control in your relationship with them."

Neuroscience, Brain Monitoring

Brain researchers are studying the anatomy, physiology, and molecular bases of brains and the nervous system. This includes study of the sense inputs; sense signal communication; brain processes (memory, consciousness, cognition, learning, thought, and creativity); mental output actions, such as motor neurons; the cellular and molecular bases of mental functioning (including study of "mental" processes in bacteria); mental processes simulation and modeling; and bioinfomatics and brain monitoring technologies. Samples of research under way include:

- Donald York, of the University of Missouri Medical Center, and Tom Jensen, a speech pathologist, have discovered that the speaking of particular words generates

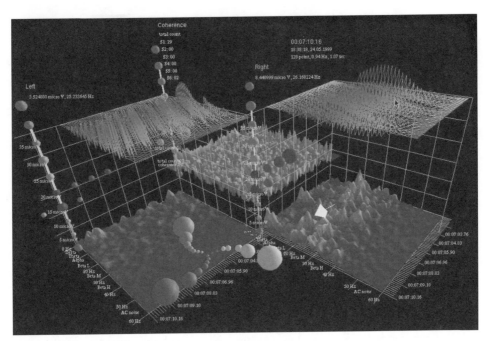

Fig. 2.1.5. Interactive Brain-wave visual analyzer (copyright by IBVA ⟨http://www.ibva.com/⟩). Commercial brain-computer interface.

unique corresponding EEG patterns. Since many people prevocalize thoughts, they eventually hope to develop a system useful to the speech impaired that can generate words straight from thoughts.

- Trying to solve the problem of the process control of complex factories such as refineries, Dupont researcher Babatunde Ogunnaike has "borrowed" a rat's brain as a biocomputer. Managing complex dynamic system processes requires live, nonlinear computation. Ogunnaike has shown that the biological brain, optimized for this kind of processing, can be hooked up via sensors and stimulators to control the valves and settings to make the factory work.

- Many researchers are trying to find ways to tap into brain signals as a computer interface. The IBVA (Interactive Brainwave Visual Analyzer) is an actual product that reads EEG signals and makes them available for the control of media and MIDI events. Ultamind reads galvanic skin response (GSR) in the fingers.

- Biochemist Daniel Koshland is working with bacterial intelligence. Bacteria have demonstrated sufficient memory to move toward attractants and away from repellents. Researchers are searching for the mechanisms that control this microbial intelligence.

Bionics

Bionics is the attempt to interlink fabricated technologies with aspects of living systems. Many contemporary theorists are intrigued with the notions of cyborgs, who are part human and part technology. Anders Sandberg, who has organized the major trans-human Web site, defines bionics as the attempt to "to transcend our biological nature by replacing biological parts with artificial parts ['deflesh'], or by translating the human mind into information in a computer [uploading]." Researchers are investigating a variety of fields, including vision, hearing, neural interface, EEG and EMG interfaces to the brain, biomaterials, and artificial muscles.

Although many minor successes have been achieved, the goal of completely integrated cyborgs is far from realization. David Pescovitz's *Wired* review of the future of bionics describes the difficulties:

Advanced bionic components may be within our grasp, but putting them all together in a real-life superpowered Steve Austin will probably remain the stuff of sci-fi. While a bionic person may be theoretically feasible, [Donald R.] Humphrey says, "the major problem lies in overcoming tissue-material interfaces." And that's a biggie. Not only do artificial tendons and ligaments have to be connected to real ones, artificial muscles don't yet approach the size and power of human tissues. Our understanding of motor skills and the brain's sensory and perceptual abilities has really just begun. "Continued research in neuroscience and bioengineering will no doubt lead to improvements in man-machine interfaces and functional replacements," [William M.] Jenkins says, "but it will be a long, hard road filled with many failures and a few successes."[13]

Here is a sampling of research projects:

- Stanford researchers Joseph Rosen, Bernard Widrow, Eric Wan, and Greg Kovacs have developed a nerve chip that can read nerve signals, decode them, and then drive a prosthesis. The research aims to enable the chip and natural nerve fibers to fuse together.
- University of Southern California neuroscientist Theodore Berger heads a team trying to develop biochips that can be wired into the brain to read brain impulses and conduct brainlike calculations, for example, to help neurologically injured patients. They are using an approach of creating neural chips that mimic brain functions, even though they might not have complete understanding of the underlying processes.
- Kevin Warwick is a researcher who is using his own body to explore the limits of implanted chips, including investigations such as using chips to control body parts, to communicate with chips in other bodies, and to control devices in the immediate environment (for example, turning on lights when he enters a room).

Parapsychology and Bioelectricity

The alternative science traditions of parapsychology and bioelectricity have sought to understand bodily and brain processes not validated by mainstream science. Parapsychology attempts to understand phenomena such as: clairvoyance (the ability to see from a distance beyond normal visual range); precognition (the ability to foresee future events); telekinesis (the ability of mind to move and otherwise affect distant matter); and telepathy (the ability to read and communicate with other minds via methods outside the five normal senses.) Since few attempts to establish the existence of these phenomena via standard scientific experimentation have succeeded, the mainstream scientific community has relegated these to pseudo science and is adamant in debunking them.

An extremely active international community, however, does not accept this assessment. There are multitudes of anecdotal accounts of demonstrations of these phenomena. Also, there is a community of scientists working to find ways to study and validate the phenomena within the standard scientific canon. In addition, the reluctance to totally reject the phenomena is demonstrated by the fact that both the Russian and U.S. defense departments, police departments, and some corporate research labs have funded parapsychology research.

Bioelectricty is another related field not highly integrated into the scientific community. *The Secret Life of Plants,* by Peter Tompkins and Christopher Bird, reported that plants produced many kinds of electrical changes in response to actions taking place nearby. There has been great debate about the significance of the phenomenon. Robert O. Becker and Gary Selden wrote a book called *Body Electric,* which traces both the dangers and the healing potentials of electromagnetic energies for bodies. Another book by Becker and Andrew Marino, *Cross Currents,* summarizes the approach:

Our initial hypothesis was that electromagnetic energy was used by the body to integrate, interrelate, harmonize, and execute diverse physiological processes. Natural electromagnetic energy is an omnipresent factor in the environment of each organism on earth. From an evolutionary standpoint, nature would favor those organisms that developed a capacity to accept information about the earth, atmosphere, and the cosmos in the form of electromagnetic signals and to adjust their internal processes and behavior accordingly. . . . Signals outside this physiological range would elicit a nonspecific systemic reaction geared toward the re-establishment of homeostasis.[14]

Becker and Marino provide a review of the wide-ranging history of research in bioelectricity, including topics such as: the role of electromagnetic energy in the regulation of life processes (the nervous system, growth control, bones and other tissue); the control

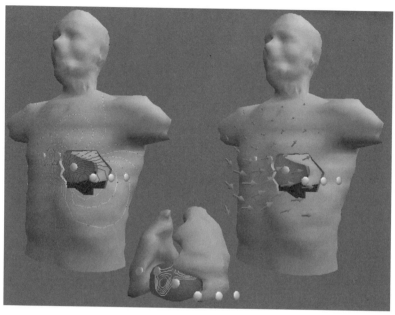

Fig. 2.1.6. Prof. Adriaan van Oosterom, Bioelectricity Group, Medical Faculty, University of Nijmegen. Illustration of the electric potentials on the body surface (*left*) and magnetic field just outside the torso (*right*) as generated by the heart's electric activity (*middle*) ⟨http://www.utwente.nl/bmti/bmti/part-kun.html⟩.

of living organisms by natural and simulated environmental electromagnetic energy (evolution of life, biological cycles, positional and navigational aids); electromagnetic energy on biological functions (metabolism, reproduction, growth and healing, mutagenesis); and health risks due to artificial electromagnetic energy. Here are some samples of research in the fields of parapsychology and bioelectricity:

- Sony Corporation has established an ESP lab directed by Yoichiro Sako. Sony executive Mike Isbida notes that "Sako's main interest is in pushing on the boundaries and definitions that shackle traditional science" in areas such as a communication system "that transmits data through mediums we've never before considered."[15]
- Joe Sanchez developed a system that uses the changing electrical changes in plants, as described in *The Secret Life of Plants,* to generate poetry.
- The catalogues *Tools for Exploration* and *Mindware* list brain and consciousness technologies using approaches such as sound and light, electrostimulation, magnetics, electro-acupuncture, biofeedback, subtle energies, nutrition, and oxygen in order to enhance beauty, perception, health, and other attributes.

Fig. 2.1.7. U.S. Air Force Fifty-first Security Forces Squadron member dressed in an MCU-2P chemical-biological protective mask. Department of Defense image source: Defenselink 〈http://www.defenselink.mil/photos/〉.

Biological Warfare

Current doomsday scenarios usually feature biological warfare such as terrorist-launched germ attacks. Emerging biotechnologies offer extensive possibilities for unprecedented offensive and defensive weapons. The U.S. military (as well as others) have funded high levels of research in CBW (chemical and biological warfare), both to counter these kinds of scenarios and to develop more "visionary" weapons.

- Ed Regis reported on a future-weapons technologies briefing for *Wired* magazine. Technologies identified include: bioactive clothing that senses environmental characteristics and changes automatically to increase camouflage; artificial noses that identify enemy troop characteristics, positions, and numbers by assessing their odors; nonhuman bioweapons such as organisms that can decay enemy rubber components (e.g., tires and gaskets), silicon, and fuel; genetically engineered microbes that can attack specific persons, for example, the leader; biogenerative materials that can generate food and weapons in the field; genetically designed vaccines that can immunize soldiers against diseases and other bioweapons; microbe-grown body enclosures to prevent assault and to augment strength and mobility; automated healing systems to take care of wounds; and bioremediation systems that can detoxify areas where there has been biowarfare.[16]

- *Soldiers* magazine featured an article on twenty-first-century weaponry. The Land-Warrior system (LW) aims to outfit each soldier as a complete combat unit, including voice-activated computer control. Systems will include a thermal viewer so the soldier can see in all weather, in the dark, and with augmented terrain information via a heads-up display in the helmet. Each helmet would also include positioning and identification information as well as built-in digital image recording and communications capability. The uniform will include bioarmor, biological weapons, detox, and augmented load-bearing features. An automatic biostatus monitoring system would continuously send vital-sign information to headquarters so leaders would know if the solider were dead.

The research described in this section demands artists' attention, both to track its cultural implications and to experiment with using its insights in unintended, less nefarious ways.

Theoretical Perspectives on Biology and the Body

The glowing picture of biological and medical research presented in the previous sections glosses over important perspectives on the life sciences. Analysts both inside and outside these disciplines profoundly question fundamental assumptions, concepts and methodologies. Features long taken for granted in biology are submitted to scrutiny. The philosophical foundations are being rocked. Old givens such as the divisions between culture and nature, human and animal, body and not-body, and male and female can no longer be assumed. The glorious optimism of the biological future is now being examined with skeptical eyes, so artists who would work with biological theories, concepts, contexts, and paraphernalia cannot afford to accept the associated disciplinary perspectives without scrutiny.

The investigatory traditions of poststructuralism and cultural theory propose a radical "archaeology" of the core concepts and assumptions of sciences such as biology, a peeling away of layers to get beneath what "common sense" dictates. In these disciplines, inquirers are urged to "deconstruct" all aspects of life to understand how text, image, and narrative work to produce and reinforce behavior and ideology. Theorists such as Michel Foucault see this fundamental inquiry as essential to building the kind of knowledge that science claims it seeks:

[T]here are times in life when the question of knowing if one can think differently than one thinks, and perceive differently than one sees, is absolutely necessary if one is to go on looking and reflecting at all. People will say, perhaps, that these games with oneself would better be left backstage; or at best, that they might properly form part of those preliminary exercises that are forgotten once they have served their purpose. But, then, . . . in what does [philosophy] consist, if not in the endeavor to know how and to what extent it might be possible to think differently, instead of legitimating what is already known?[17]

The Objectification of Nature

Biology and Western science in general are based on a set of cultural moves related to the bifurcation of culture and nature. Analysts such as Neil Evernden in *The Social Creation of Nature* trace the origin of the concept of nature and note the development of a dualistic separation and reification of the external world. Indeed, this separation is often seen as one of the great accomplishments of Renaissance science and art, illustrated by the work of Galileo and Leonardo da Vinci. Ernst Cassirer notes that "neither art nor mathematics can allow the subject to dissolve in the object and the object to dissolve in the subject. Only by maintaining a distance between the two can we possibly have a sphere for the aesthetic image and a sphere for logical-mathematical thoughts."[18]

Fig. 2.1.8. Yosemite National Park—Vernal Falls. Yosemite Concession Services Corp. Photo by Keith Walklet.

The single viewpoint of perspective is seen as part of this cultural matrix. This separation has become so ingrained in Western thought that it is easy to forget that it was a culturally created framework and narrative, and that other cultures use very different frameworks. Roland Barthes wrote of the process of "naturalizing" concepts, of using linguistic processes to make concepts seem outside of culture. Evernden describes the approach:

[N]ature has become a powerful part of our vocabulary of persuasion. but even that puts it too mildly, for it is often treated as the very realm of the absolute. To be associated with nature is to be placed beyond human caprice or preference, beyond choice or debate. When something is "natural" it is "the norm," "the way," "the given." This use of "nature" affords us a means of inferring how people ought to behave—including what objects they ought to associate with, that is, buy. Yet the authority of that usage stems in part from its confusion with the other major use, nature as the material given, nature as everything-but-us. In other words, the understanding of nature as the realm of external stuff, which is studied by science, lends an aura of objectivity and permanence to the understanding of nature as norm. The two mingle and interact so that we frequently lose sight of the distinction.[19]

The related concept of "Nature" as the unspoiled world of plants and animals outside of cities is near and dear to the arts. Landscape painting has been a venerable tradition for the last centuries. Part of its appeal lay in its portrayal of this Acadiian world not so accessible to urban dwellers. The ecology movement, which attracts some contemporary artists, also uses the rhetoric of "saving nature" and "going back to nature."

Cultural theorists, such as Raymond Williams in *Problems of Materialism and Culture,* note that this use of "Nature" is embedded with narratives and not as innocent as it seems. The nature of wild lands was not a prevalent concept before modern times. With the advent of industrialization, however, the idea became reified. Cultural theorists see it as part of the narrative of colonialization and exploitation. Identifying it as the objectified "Other" made it easier to use. Peter Taylor, a theorist interested in the social construction of nature, describes how the abstraction of nature promoted exploitation.[20] Furthermore, the present moves to keep indigenous peoples close to nature can be seen as a way to deny them a voice in the contemporary world.

Biological and Medical Research

Mainstream biology is seen by cultural theorists and feminist critics as highly prone to narratives of objectivity and domination, with resultant blindness to its assumptions and metanarratives, and to alternative research paradigms. For example, Lynda Birke's and Ruth Hubbard's *Reinventing Biology* presents alternative perspectives about objectivity:

[H]ow do biologists conceptualize the nature of the organisms they work with, and what alternative outcomes would we expect to see if the conceptual framework or the rules of practice were different? And not just different, specifically, if scientific objectivity were defined not as an attitude of separation and detachment between scientific actors and the passive objects they manipulate but as a cooperative venture in which scientists and their research subject are partners. Put another way, what would science look like if it respected the living organisms it studies as individuals with their own histories and integrities?[21]

Research that is invasive, fragmenting, and reductionist cannot capture features of life in context and in relation to other life, and it is morally questionable. Many of the essays in *Reinventing Biology* offer examples in which scientists have ignored useful knowledge and perspectives. Sandra Harding's book *Whose Science? Whose Knowledge?* claims that the West has validated only some knowledges, often disregarding investigations from less powerful groups. She offers examples such as Vandana Shiva's work showing that plants are full of sensitivities ignored by common biological representations of plant organisms; Lynda Birke's work showing that ethologists, who claim to be interested in animal behavior, disregard the accumulated wisdom and experience of indigenous people's knowledge of animals; and Lesley Roger's investigations of faulty inferences about animal intelligence and behavior (and ultimately human behavior) drawn from animals raised in restricted environments like zoos and laboratories, and the disregard for the obvious manifestation of animal awareness and emotionality, which perpetuates the "unbridgable wall Western culture and thought have erected between humans and other organisms."[22]

Judith Masters critiques the sacrosanct notions of evolutionary biology. She questions the standard Darwinian model based primarily on violence and competition, seeing the primacy of this theory as partially a reflection of the social context in which it emerged during great colonial expansion and the domination of small countries.

But the critique is also methodological. Scientific problems remain unsolved that perhaps could be addressed by alternative theories. She notes, for example, the long-standing failure of Darwinian theory to explain the origins of evolutionary novelties and the clustering of extinction patterns across species, and suggests that Lynn Margulis's theories of endosymbiosis could help by explaining how cooperation between species can play an important role in evolution. She quotes Mae-Wan Ho's analysis of the difficulties with the fragmentary reductionist approach coming from biases in Western science:

To many great civilizations past and present, the unity of nature is simply a fact of immediate experience that needs no special pleading. In the West, however, much of the history of science is concerned with separating and reducing this unity into ever smaller and smaller fragments out of which nature has somehow to be glued together again. It is a history not only of fragmentation but of our own alienation from nature.[23]

Optimism about the promise of techniques such as bioinfomatics and bioengineering must be tempered with awareness of the critique and its alternatives. For example, cultural critic Donna Haraway calls attention to the disappearance of the organism:

[C]ommunications sciences and modern biologies are constructed by a common move—the translation of the world into a problem of coding, a search for a common language in which all resistance to instrumental control disappears and all heterogeneity can be submitted to disassembly, reassembly, investment, and exchange. In modern biologies, the translation of the world into a problem in coding can be illustrated by molecular genetics, ecology, sociobiological evolutionary theory, and immunobiology. The organism has been translated into problems of genetic coding and read-out. Biotechnology, a writing technology, informs research broadly. In a sense, organisms have ceased to exist as objects of knowledge, giving way to biotic components, i.e., special kinds of information-processing devices.[24]

Other kinds of science are possible. In *Picturing Science, Producing Art* and his other books, Peter Galison notes how conceptualization and imaging shape scientific process. In *Laboratory Science,* Bruno Latour and Michel Callon propose a conceptualization that sees that the experimenters, the apparatus, and the organisms all as actors in an interlocked network. In "The Promises of Monsters: A Regenerative Politics for Inappropriate/d Others," Haraway, who is famous for her analysis of primate research as a playing out of cultural narratives rather than pure science, sees the organisms that inhabit our shared world as collaborators and proposes approaches that recognize and work with that connection and discourse:

Nature is for me, and I venture for many of us who are planetary fetuses gestating in the amniotic effluvia of terminal industrialism, one of those impossible things characterized by Gayatri Spivak as that which we cannot not desire. Excruciatingly conscious of nature's discursive constitution as "other" in the histories of colonialism, racism, sexism, and class domination of many kinds, we nonetheless find in this problematic, ethno-specific, long-lived, and mobile concept something we cannot do without, but can never "have." We must find another relationship to nature besides reification and possession. . . . So, nature is not a physical place to which one can go, nor a treasure to fence in or bank, nor an essence to be saved or violated. Nature is not hidden and so does not need to be unveiled. Nature is not a text to be read in the codes of mathematics and biomedicine. It is not the "other" who offers origin, replenishment, and service. Neither mother, nurse, nor slave, nature is not matrix, resource, or tool for the reproduction of man. . . .

Organisms are biological embodiments; as natural-technical entities, they are not pre-existing plants, animals, etc., with boundaries already established and awaiting the right kind of instrument to note them correctly. Organisms emerge from a discursive process. Biology is a discourse, not the living world itself.[25]

This new kind of biology is still to be determined. It does not have the decades of definitional history behind it, as does classical biology. Artists who would work with biology must be aware of the full range of the discourse.

Rethinking the Body and Medicine

Concepts and experiences of the body are another area of great cultural focus. Analysts show that the innocent, commonsense notion of the body is not as simple as it seems. The body has always been a special interest of the arts, for example, in portraiture, sculpture, theater, and dance. Artists have studied and portrayed the body throughout history. Western biological science has also studied the body, seeking to understand its functions and theorize its secrets. The applied version of this impulse, medicine, has sought to defeat disease and improve the "quality" of life. Medical knowledge, the prolongation of life expectancy, and the relief of suffering is often pointed to as one of the most significant "accomplishments" of science. Artists have begun to work with the body and the new medical technologies of inspection and manipulation.

In *Medicine as Culture,* Deborah Lupton surveys the critical activity currently under way to understand how cultural narratives produce our experience of bodies. She explains poststructuralism and the "social constructionism" approaches as a way of understanding the body, illness, and the medical world. There is a radical challenge to the truth claims of medicine and biology:

For social constructionists examining the social aspects of biomedicine, the development of medico-scientific and lay medical knowledges and practices is the focus. The social constructionist approach does not necessarily call into question the reality of disease or illness states or bodily experiences, it merely emphasizes that these states and experiences are known and interpreted via social activity and therefore should be examined using cultural and social analysis. According to this perspective, medical knowledge is regarded not as an incremental progression towards a more refined and better knowledge, but as a series of relative constructions which are dependent upon the socio-historical settings in which they occur and are constantly renegotiated. In so doing, the approach allows alternative ways of thinking about the truth claims of biomedicine.[26]

Cultural studies require an interdisciplinary approach that incorporates the sciences, psychology, anthropology, linguistics, psychoanalysis, and the humanities, such as literature, film, and art criticism. In this view, illustrated by research such as Bryan S. Turner's *Regulating Bodies,* medicine and experience of the body is not just an objective body of scientific knowledge external to culture but rather a product of media and language. One must understand the "discourse"—the patterns of words, figures of speech, concepts, values, symbols, texts, and visual representations—in order to understand a phenomenon and its associated practices and ideologies. Similarly, *Visible Woman,* edited by Paula A. Treichler, Lisa Cartwright, and Constance Penley, contains a series of articles on the impact of bioimaging and disease conceptualizations on science, policy, and popular culture. Maura Flannery's "Image of the Cell in Twentieth-Century Art and Science"[27] shows how changing cultural and aesthetic fashion have influenced representations in the last century. Examples of other books inspecting the cultural context of body concepts and medical practice include Richard Doyle's *On Beyond Living: Rhetorical Transformations of the Life Sciences,* N. Katherine Hayles's *How We Became Posthuman: Virtual Bodies in Cybernetics,* Barbara Stafford's *Body Criticism,* and Rosamond Wolff Purcell's *Special Cases: Natural Anomolies and Historical Monsters.* The Center for Twentieth Century Studies also has accumulated critical commentary on biology in its conferences on topics such as Biotechnology, Culture and the Body, Women and Aging, and Representing Animals.

Lupton offers the example of the way visualization narratives conceptualize and represent the relationship of the unborn fetus to the mother's body as totally separate:

Debates over abortion in popular and legal settings, accusations made against women for smoking or drinking alcohol while pregnant, the training of medical students in obstetrics and gynecology, the use of ultrasound that represents the foetus as an image separate from the maternal body, colour photographs in books and popular science magazines that show the foetus in the womb, seemingly floating in space, the way that people speak of the foetus as having a potential gender and name before birth, all serve to reinforce this division between mother and foetus. Practices constitute and reinforce existing discourses, and vice versa.[28]

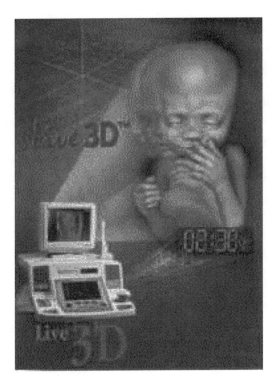

Fig. 2.1.9. Medison America. Voluson 3-D ultrasound system, which constructs 3-D imagery based on ultrasound signals ⟨http://www.medison.com/⟩.

The social constructionists would hold that this discourse is not just a social phenomenon separate from hard science; rather, it shapes the way scientists conceptualize, undertake, and interpret research. The body is greatly influenced by social discourse. In *The Making of the Modern Body,* historians Catherine Gallagher and Thomas Walter Laqueur describe the different ways that the body has been experienced, perceived, interpreted, and represented in different epochs and cultures.

Lupton notes that there are really many bodies, for example, the gendered body, the sexual body, the disciplined body, and the commodified body. Influences such as AIDS, advertising, new medical technologies, and debates about sexual behavior all intertwine to shape what people think and experience. Feminist analysts, for example, have carefully documented the way the culture surveilles and "polices" body activity to reinforce thought and action.

Michel Foucault has greatly influenced much of this analysis of the body and medicine. In books such as *Birth of the Clinic* and the *History of Sexuality,* he asserts that the body is the major site for the culture to control and discipline its participants and that medical technology and institutions are optimized for these purposes:

For Foucault and his followers, the body is the ultimate site of political and ideological control, surveillance and regulation. He argues that since the eighteenth century, the body has been the focal point for the exercise of disciplinary power. Through the body and its behaviours, state apparatuses such as medicine, the educational system, psychiatry and the law define the limits of behaviour and record activities, punishing those bodies which violate the established boundaries, and thus rendering bodies productive and politically and economically useful.[29]

For some, gender is a social construction. Medical institutions police sexuality. Public health concerns with dirt and disease allow for even more control. Symbol systems reinforce the power relations of doctor (historically male) and patient. Patients accept the requirement to "confess," comply, and expose themselves to technologies:

In popular culture the status of the doctor is usually indicated by his or, her white coat, signifying authority, the objectivity and power of laboratory science and hygienic purity, and the stethoscope, ultimate symbol of medical technology and the ability of the doctor to gain access to the patient's body to hear or see bodily functions denied patients themselves, while retaining a certain distance between doctor and patient.[30]

Cultural theory investigates how various metaphors of disease operate. For example, the military metaphor calls forth images of aliens, invaders, sieges, and defenses. The mechanical metaphor for the body and disease has also been established over time and promotes certain ways of conceptualizing bodies, illness, and medical intervention:

The mechanical metaphor includes the idea that individual parts of the body, like parts of a car or plumbing system, may fail or stop working, and can sometimes be replaced. The metaphor has the effect of separating mind and body, of valorizing medical techniques which focus upon locating a specific problem in a part of the body and treating only that part, and devaluing healing relationships which rely upon spirituality, personal contact, intimacy, and trust.[31]

The metaphor has been expanded to include computers and information technologies. Bioengineering encourages fantasies of total control: "The current enthusiasm towards identifying or 'mapping' the structure and sequence of the genetic material in the human genome presents an image of the body and mind 'as machinelike "systems" that can be visualized on a computer screen and understood simply by deciphering a code.'"

From the analyses described, it would be possible to deduce that cultural theory is antimedicine and antitechnology. The technique itself aspires toward neutrality. The same kind of discourse analysis could be applied to the alternative health traditions that promote lay healers and alternative interventions such as herbs, massage, sound healing, and the like. The language used and visual representations could be deconstructed. Certainly the poststructuralists have shown much more interest in dissecting the discourses of power rather than those of the disenfranchised, but there is nothing intrinsic to the method.

Indeed, Donna Haraway is famous for exploring unorthodox conclusions. In her "Cyborg Manifesto," Haraway analyzes science's fascination with domination and control. She also explores some of the implications of male hegemony and the ways that technology can represent the playing out of these tendencies. She notes, however, that contemporary technology also creates new opportunities. Cyborgs do not provide clear readings of their sex and gender and provide a model for human exploration of possibilities free from cultural stereotypical controls:

To recapitulate, certain dualisms have been persistent in Western traditions; they have all been systemic to the logics and practices of domination of women, people of colour, nature, workers, animals—in short, domination of all constituted as others, whose task is to mirror the self. Chief among these troubling dualisms are self/other, mind/body, culture/nature, male/female, civilized/primitive, reality/appearance, whole/part, agent/resource, maker/made, active/passive, right/wrong, truth/illusion, total/partial, God/man. High-tech culture challenges these dualisms in intriguing ways. It is not clear who makes and who is made in the relation between human and machine. It is not clear what is mind and what body in machines that resolve into coding practices.

Cyborg imagery can help express two crucial arguments in this essay: first, the production of universal, totalizing theory is a major mistake that misses most of reality, probably always, but certainly now; and second, taking responsibility for the social relations of science and technology means refusing an antiscience metaphysics, a demonology of technology, and so means embracing the skillful task of reconstructing the boundaries of daily life, in partial connection with others, in communication with all of our parts. It is not just that science and technology are possible means of great human satisfaction, as well as a matrix of complex dominations. Cyborg imagery can suggest a way out of the maze of dualisms in which we have explained our bodies and our tools to ourselves. This is a dream not of a common language, but of a powerful infidel heteroglossia. It is an imagination of a feminist speaking in tongues to strike fear into the circuits of the supersavers of the new right. It means both building and destroying machines, identities, categories, relationships, space stories. Though both are bound in the spiral dance, I would rather be a cyborg than a goddess.[32]

This kind of analysis has powerfully influenced artistic investigations of biology and the body. In a talk sponsored by the Dutch V2 art group on the artistic appropriation of medical discourse, Nina Czegledy analyzed the differences with which artists and scientists approach a topic like medicine. She sees increasing interest coming from many sources: increased public information, new visualization methods changing the sense of corporeal selves, and a spreading self-help orientation toward bodily intervention:

The relationship between medicine and art has been of long-standing interest to scholars and artists alike, although a dichotomy continues to divide the analytic, deductive views prevalent in medicine from those who distrust the powers of pure reason and aspire to some form of life-experience such as action, memory, sensation or the reciprocity between art and creative power. Science, specifically medical science, considers reason to be the exclusive source of knowledge, and utmost emphasis is placed on abstraction and clinical detachment. The artist approaches his/her own reality quite differently. This approach is in more experiential terms. The broad difference between medical science and art is, to some extent that between objective synthesis and subjective simulation. However, as Wittgenstein wrote in Tractatus, "even when all possible scientific questions have been answered, our problems of life remain completely untouched."[33]

One of the major changes has been the increased access to body imaging. The dark recesses of the body started getting turned inside out in the late nineteenth century. She wonders if the increased public access will promote feelings of shame or power, and speculates on the implications that computer graphics enable both body imaging and digital art. The malleability of digital information makes claims of objective truth problematic. In her influential volume *Body Criticism,* Barbara M. Stafford wrote:

The computer-mediated milieu renders the body nakedly public. One result of the non-invasive imaging technologies in the area of medicine is the capability of turning the person inside-out. If the late nineteenth century developed the photographic sounding of the living interior through endoscopy, gastroscopy, cystoscopy and most dramatically X-rays, the late twentieth century revealed its dark core three-dimensionally through MRI projections. Using radio waves and magnetic fields, this technique for painlessly exploring morphology, nonetheless raises the specter of universal diaphaneity. It conjures up visions of an all-powerful observer who has instant visual

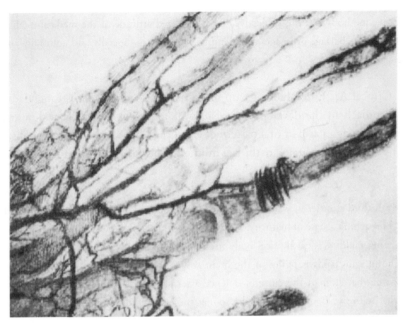

Fig. 2.1.10. Historical image of veins in the hand taken by means of the Roentgen rays in Prof. Franz Exner's physico-chemical institute in Vienna in 1896. (Image from "X-ray Century" Web site, edited by Perry Sprawls and Jack Peterson) ⟨http://www.cc.emory.edu/X-RAYS/century.htm⟩.

access to the anatomy, biochemistry and physiology of a patient. Will this open-ended trend toward complete exposure give rise to the same sense of vulnerability, shame, and powerlessness that the eighteenth century associated with anatomization?" . . .

Ironically in this new environment of digital visualization, the same software tools are used by all the new occupants: both the practitioners of the medical profession and by artists in their critical discourse of medical practice. . . . Advanced information technologies of interpretation have given raw data a malleability previously unconceived.

Stafford analyzes the presentation of digital medical images in terms of rhetorical techniques that channel the listener's conceptualization and the power of metaphor in promoting some ideas and excluding others. She asks, How are "specific discourses of medical science refracted by artists"?

The increased exposure of new medical technologies resulted in an intensified focus on the philosophical, ethical and aesthetic considerations involved in various medical procedures. New venues of communication technologies have contributed to the re-examination of invasive and non-invasive methods and the "objective" depersonalized attitude of the medical establishment. In particular the "barrage of images and dictums issued by the socially and culturally authorized commercial media."

The Critical Art Ensemble, a group of artists and theorists who strongly advocate a critical approach to the development of new biological and medical technologies, see the new technologies as part of a pattern of ever expanding control and capitalist expansion. The body and the brain are the last frontiers. They describe troubling developments ahead:

Currently, a third frontier of vision, longed dreamed about by many imperial cultures, has been rapidly approaching a state of maturity. Vision engines that can map the flesh are being developed with amazing rapidity. The physical body itself is now under pancapitalist invasion.

A. The imaging and mapping of the brain will be a key point of contestation in the near future (somewhat akin to the current violence associated with imaging of uterine space). The sooner the brain can be rationalized, the sooner the command and control center of the body can be fully colonized.

B. DNA imaging and mapping is advancing at a much faster pace than experts in the area expected. The potential for reliable flesh products true to the principle of equivalence is increasing dramatically. More frightening than the products themselves is the creeping deployment of eugenic ideology that must be in place if market viability for these products is to be assured.

C. Organ products will be a brutally contested market. The primary conflict will be between corporations that opt to develop artificial organs, and those that opt for the development of transgenic organs.

D. The rationalization of reproductive processes has already produced a massive market for flesh products (sperm, egg, cells, embryos, uterine surrogates, etc.).[34]

Searching for Aesthetic Form in Art and Science

Rejecting some of this radical critique, some researchers believe that there are universal culture-transcending principles, such as the use of an "aesthetic" form as a criteria to guide research. The faith is based on the belief that nature does have elegant structures that can be discovered and that this elegance can be a guiding principle in knowing what theories are superior.

Microbiology researchers often refer to the importance of "form" in their research. Aesthetic concerns such as symmetry and fit are often instrumental in leading investigators down fruitful paths. In "The Shape of Life," biologist Stephen J. Gould recounts the importance of form in Watson and Crick's discovery of DNA. The actual physical manipulation of models and forms was as important as abstract calculations:

[C]onsider especially, the interplay between two domains of inquiry that might seem ineluctably disparate, but worked intimately together to provide the solution: Theoretical (and quite cerebral) calculations of molecular distances and configurations with the most practical (and overtly material) construction of physical models from pieces of plastic, metal, and glass. Watson and Crick resolved the structure of DNA by building a Tinkertoy model and discovered that everything fit.[35]

Gould describes the constant interplay of abstract ideas against the physical models and the opportunity for experimentation and the resolution of puzzles afforded by the model. The belief in morphology and underlying notions of form help to push the discovery along:

What a lovely image, and what a triumph for morphology. One of our century's great scientists, about to make one of the premiere discoveries in our intellectual history. What is he doing to prepare for the event? Not calculating, not twirling dials, not delicately measuring—not doing anything that the stereotype of science would anticipate. He is cutting out cardboard models for pieces in his morphological puzzle.[36]

Other researchers tell similar stories about the fitness of natural structures. Roald Hoffman describes Swiss chemist Alfred Eschenmoser's attempt to build alternative DNA structures out of sugars called hexoses instead of the normally occurring pentoses. Hoffman notes that even though these structures should be theoretically possible, they don't work as well to form helical structures. Hoffman concludes that this "craftsmanship" indicates a strong bond between good chemistry and good art. Both respect the work of "mind and hands together."[37]

A. S. Koch and T. Tarnai illustrate another investigation of biological form. They were intrigued by the structure of viruses and their implications for art and architecture. In a *Leonardo* article "The Aesthetics of Viruses," they describe the structure of viruses. They admire the virus's ability to efficiently and densely compact information. Its three-dimensional tiling is a marvel of form. Viruses build out multiple subunits that roughly compose spheres. They note that their structure also allows them to introduce crucial imperfections that facilitate their biological function of infecting cells with their genetic information.[38]

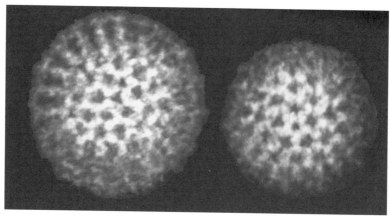

Fig. 2.1.11. Linda M. Stannard, University of Cape Town. Rotavirus electron microscope image illustrating icosahedral symmetry. (http://www.uct.ac.za/depts/mmi/stannard/rota.html).

Although the double helix is one of the great stories in science and art, there are problems with adhering to oversimplified notions of aesthetics. Gould criticizes Watson's claim that the elemental form of the helix had to be correct. He warns that the facile belief in classical shapes such as triangles and circles, derived from Renaissance thinking, can be misleading: "Beauty, at least as prejudicially defined by simplicity and symmetry, does not always pervade natural objects, which may be stochastically or contingently messy, or quantifiable in the different languages of fractals and chaos.[39]

Physiologist Robert Root-Bernstein similarly critiques the glib equation of aesthetics and science illustrated by the double helix. In his paper "Do We Have the Structure of DNA Right? Aesthetics, Assumptions, Visual Conventions, and Unsolved Problems," he describes major flaws in the helix model. The thermodynamics and topology of twisted helices make it very difficult to explain the unraveling that is necessary in cell duplication. For example, there is not room to unravel helixes in the space of a cell nucleus. He suggests that the aesthetic elegance was so compelling that it blinded the researchers to functional problems and led them to dismiss alternative proposals too easily. Watson reported that "the idea [of helices] was so simple that it had to be right." Root-Bernstein explains the importance of aesthetic criteria in science:

[S]cientists, just as much as artists, are dependent upon aesthetic criteria in making their investigative choices. A considerable literature exists arguing that the choice of a problem; the criteria used to analyze its possible solutions, concepts of beauty, harmony, simplicity, symmetry, consistency; and even the skill and style with which the solution is reached and is propounded are all replete with aesthetic choices that differ in no significant manner from those made by the artist. As the turn-of-the-century neuroanatomist and Nobel laureate Santiago Ramón y Cajal wrote, great scientists have a "congenital inclination to economy of mental effort and almost irresistible propensity to regard as true what satisfies our aesthetic sensibility by appearing in agreeable and harmonious architectural forms. As always reason is silent before beauty".[40]

Illustrating this art-science cross-disciplinary interest in aesthetics, Moet Hennessy devoted their 1996 "Science Pour l'Art" competition to form in biological art and science:

Since earliest antiquity, men of learning have wondered about the genesis of forms in Nature. They have wondered about the perpetually changing forms of river and flame so dear to Heraclitus, and puzzled over the immobility of Parmenides' immutable sphere. Along with Democritus, the ancients believed that the world's diversity of form resulted from a rich combination of minute universes—the atoms. What about today? . . . The word "form" is to be understood in its spatial sense, but time and dynamics may enter into consideration. . . . The theme encompasses technological applications of concepts related to the genesis of living forms—which one might call biomimetic applications.[41]

Roger Malina, one of the judges of the competition, writes that the entries demonstrated a similarity in science and art and a new synthetic unity dissolving the old dualisms of nature and man, organic and inorganic, referring to Roy Ascott's term "moist realities":

The scientists who presented their applications for the LVMH Science for Art Prize were asked to explain how their work was relevant to artistic or aesthetic considerations. In reading these, I was struck first by how close artistic and scientific impulses were in the exploration and understanding of the genesis of form. The second thing that struck me was how the new biologies introduce totally new ways of thinking about the nature of living matter and the relationship (and boundaries) between humans and nature. For these scientists, the false dichotomy of Man and Nature is laid bare by their personal experience. . . . I was struck by how these scientists do not view biological material from the outside, but rather view it as part of a natural landscape that includes both living and inanimate matter. Just as architects create buildings or artists create sculptures, so these scientists manipulate and modify living matter. Traditionally, genetic engineering is viewed in terms of ethical issues and concerns about the unleashing of dangerous creations. However, the new biologies force a view of life as but one part of the spectrum of complex molecular systems—systems from which emerge behaviors and properties that range from form and structure to intentionality, memory, reproduction, and consciousness.[42]

Malina laments that only a few artists have begun to work with biological materials and "moist reality." Most of the winners were scientists whose work attended to issues of form, such as: Elliot M. Meyerowitz and Enrico Coen, who developed techniques to modify the genetic material of plants so as to alter the form, color, and nature of their flowers; Ryozo Fujii, who developed ways of modifying the genetic material of fish to alter the patterning on their skin; and Nicole M. Le Dourain, who created chimeras from genetic combinations of quail and chickens.

How Are Biology-Based Theory and Research Important to the Arts?

Some readers may be wondering: Biological research is interesting in a general sort of way, but what does it have to do with the arts? Deconstruction of biological concepts is intriguing, but is it specifically important to the arts? What is the special significance to artists of researchers developing new 3-D computer models of viruses? Of being able to image real-time brain activity with a Positron Emission Tomography (PET) scan? Of being able to genetically engineer a plant to produce human hemoglobin? Of understanding which gene controls aging in mice? Of comprehending the function of wolves hunting in packs? Of companies developing new kinds of drugs, dialysis machines, heart pacemakers, and the like? What can artists do with this knowledge? How might this information enter into the conceptualization and practice of art?

The cultural theorists are correct in their assessment that biological and medical research are profoundly shaping culture. There is no way to escape the questions being raised about nature, life, sex, humanness, and the body. One cannot be a producer of cultural materials without encountering some of these issues. The more information artists have about the research and its contexts, the more adequately they can respond.

The new technologies also create unprecedented opportunities. Notions that were once science fiction notions are now almost accessible. Peer into the brains and hearts of yourself and others. Change your sex. Change your body. Integrate bionic elements. Clone yourself. Create a new organism. Enhance your intelligence, sex drive, life expectancy, mood, whatever. If these actions can be initiated for science, entertainment, or commerce, why not for art? The sculpture of the future may well be bioengineering. The best way artists can explore these ideas is by mastering the research worlds from which they originate. There are several modes in which artists can engage biological and medical research: deconstructive analysis, analyzing practical and ethical implications, and experimentation.

Demonstrating the growing interest, the 1999 Ars Electronica "Life Science" festival focused on the emerging importance of biology. It offered installations, exhibitions, and symposia composed of scientists, critics of science, and artists. It explored topics such as ideology and science—biological determinism, industrial processing of life, biobusiness, agribusiness—genetically-modified foods, pharmabusiness—medicine, genetic fingerprint—the visible man, how science is done and promulgated and the critique of science. Ars Electronica 2000 addressed "NEXT SEX—Sex in the Age of Its Procreative Superfluousness," which included scientific and artistic investigations of reproductive technology.

Impact on Cultural Frameworks

As cultural theorists suggest, much biological and medical research ultimately diffuses into general cultural discourse. They can fundamentally change the way people think and act. For example, forty years ago ecological analysis was an approach known only to a few biological specialists. Now ecology is a term in many people's vocabularies, a background concept that many use to understand the relationships of animals, plants, humans, and the environment, and even the basis of action for some. Similarly, investigating the detailed processes of genetics were only research topics for a few pinhead biologists as late as the 1940s. Now the idea that humans might be able to directly control the genome at will has fundamentally affected the way we conceptualize life and fate. Genetic engineering and cloning have become background hopes and fears that lurk in cultural discourse. The development of birth control pills was an esoteric topic in a few medical research labs. Now the technology has altered the customary life patterns of people throughout the world and begun to change fundamental ideas about families, life paths, gender, and sex. These frameworks spread and shape consciousness without any special attention to the research world. What is the advantage of artists paying any more attention than any other educated or aware person?

Artists are cultural producers. The images and events they create inevitably participate in discourse, whether they think they do or not. Many contemporary artists believe that they can create the strongest art by being critically aware of the social construction of meaning and creating art that helps further the understanding of these narratives. Because

---

Ars Electronica "Life Science" and "NEXT SEX" festivals: ⟨http://www.aec.at/⟩

How Are Biology-Based Theory and Research Important to the Arts?

89

of the power, pervasiveness, and likelihood of future impact, biological and medical research are especially ripe for attention. Lupton's advice to researchers could well serve artists:

Scholars interested in the socio-cultural dimensions of illness, disease and the body in medicine need to be aware of the potential of their writings to contribute to oppressive, constraining, and stereotypical discourses that support hegemonic and confining dualisms such as Self/Other, masculine/feminine, sick/well, rational/irrational, active/passive, productive/wasteful, nature/culture, sick/well, disorderly/controlled, and moral/immoral, and to use their understanding of the socially constituted nature of knowledge to allow space for the production of novel, multiple knowledges about bodies in the medical setting that avoid either/or distinctions.[43]

Practical, Sociocultural, and Ethical Implications

Research can point toward emerging trends, problems, and possible remedial or preventative actions. For example, understanding the interrelationships of pollution, global warming, and their impact on animal and plant life asks the world community to take action. The spread of diseases such as AIDS demands research attention and eventually action. Mysterious phenomena such as the worldwide decline in human fertility and the increasing malformation of amphibians warn of something momentous happening that must be understood. What will society be like as it increasingly ages? What is the real meaning of addiction? Should companies be allowed to genetically engineer products? Should parents be able to give growth hormones to children below developmental norms? Should everyone start using mood enhancing drugs such as Prozac? What can or should be done to respond to the new ease of producing biological warfare agents? How much should life-span prolongation technologies used? Should human cloning be allowed? These trends and the ways that the ethical questions are resolved will have profound impact on the nature of life. Differing from the emphasis on reflection in the deconstructive approach, artists can decide what they believe and create art that explores the research under these ideas.

The Examination of Particular Research Themes and Technologies

Biology and medicine offer a panoply of fascinating specific research and technologies. Many of them are rich in meaning, potential for art commentary, and communicative possibilities. Artists can follow their curiosity wherever it leads. For example, body-imaging technologies such as CAT, PET, and Magnetic Resonance Imaging (MRI) provide radically new views of the person. Bionics invites the creation of new hybrid humans/machines. Genetic engineering invites the creation of new organisms. For example, artist Eduardo Kac proposes a new kind of "transgenic art":

It is equally urgent to address the emergence of biotechnolgies that operate beneath the skin (or inside skinless bodies, such as bacteria) and therefore out of sight. More than making visible the invisible, art needs to raise our awareness of what firmly remains beyond our visual reach but which, nonetheless, affects us directly. Two of the most prominent technologies operating beyond vision are digital implants and genetic engineering, both poised to have profound consequences in art as well as in the social, medical, political, and economic life of the next century.

Transgenic art, I propose, is a new art form based on the use of genetic engineering techniques to transfer synthetic genes to an organism or to transfer natural genetic material from one species into another, to create unique living beings. Molecular genetics allows the artist to engineer the plant and animal genome to create new life forms.[44]

Some readers may be skeptical. Yet I assert that if a culture is willing to take the scientific risks involved for commercial ends, then certainly artistic research must also be warranted. If the research is going to help write the culture's future, then the culture needs artists to help in the writing. The next chapters in this section explore research undertaken by artists in this and a variety of biological lines of inquiry.

Notes

1. P. Weibel, "1993 Ars Statement," ⟨http://www.aec.at/fest/fest93e/gene.html⟩.

2. U.S. National Science Foundation, "Program Description," ⟨http://www.nsf.gov/bio/special.htm⟩.

3. U.S. National Science Foundation, "BES Program Description," ⟨http://www.eng.nsf.gov/bes/bes.htm⟩.

4. U.S. National Science Foundation, "Impact of Emerging Technologies on the Biological Sciences Workshop," ⟨http://www.nsf.gov/bio/pubs/stctechn/stcmain.htm⟩.

5. For example, Johns Hopkins University, "Biomedical Research Overview," ⟨http://www.bme.jhu.edu:80/welcome/overview.html⟩.

6. Coalition for Education on the Life Sciences, "Critical Issues in Life Science Education," ⟨http://www.wisc.edu/cbe/cels/cels2.html#Issues⟩.

7. Stanford Research Institute, "Market Opportunities," ⟨http://www-cmrc.sri.com/HIH/HIH.html⟩.

8. S. Connell, San Francisco State University, NEXA Bioethics syllabus, 1998.

9. R. Vogt, "Taste of Smell Course Syllabus", ⟨http://zebra.sc.edu/smell/sylabus.html⟩.

10. J. Kohl, "Pheromones" Web site, ⟨http://www.pheromones.com/⟩.

11. Sun Angel, "Aromatherapy" Web site, ⟨http://www.sun-angel.com/articles/aroma.html⟩.

12. T. Way, "Description of Tactile Graphics," ⟨http://www.asel.udel.edu/sem/research/tactile/index.html⟩.

13. D. Pescovitz, "Reality Check," ⟨http://www.wired.com/wired/5.02/reality_check.html⟩.

14. R. Becker, and A. Marino, "Summary of *Cross Currents*," ⟨http://www.ortho.lsumc.edu/Faculty/Marino/EL/Summary.html⟩.

15. "New Technologies," *Wired,* September 1996.

16. E. Regis, "Future Weapons," ⟨http://www.wired.com/wired/4.11/es_biowar.html⟩.

17. M. Foucault, "The Use of Pleasure," in *History of Sexuality,* vol. 2 (New York: Vintage Books, 1986), p. 8.

18. D. Cassirer, *The Individual and the Cosmos in Renaissance Philosophy* (New York: Barnes & Noble, 1963), p. 170.

19. N. Evernden, *Social Creation of Nature* (Baltimore: Johns Hopkins University Press, 1992), p. 23.

20. P. Taylor and R. Garcia-Barrios, "The Social Analysis of Ecological Chagne," in *Social Sceine Information* 34:5–30.

21. L. Birke, and R. Hubbard, *Reinventing Biology* (Bloomington: Indiana University Press, 1995), p. ix.

22. Ibid., p. 119.

23. J. Masters, "Revolutionary Theory," in L. Birke and R. Hubbard, op. cit., p. 177.

24. D. Haraway, "Cyborg Manifesto," p. 164.

25. D. Haraway, "The Promises of Monsters," ⟨http://www.leland.stanford.edu/dept/HPS/Haraway/monsters.html⟩, p. 296.

26. D. Lupton, *Medicine as Culture* (London: Sage, 1994), p. 11.

27. M. Flannery, "Images of the Cell in Twentieth-Century Art and Science," *Leonardo* 31(1998):3.

28. D. Lupton, op. cit., p. 18.

29. Ibid., p. 23.

30. Ibid., p. 53.

31. Ibid., p. 60.

32. D. Haraway, "Cyborg Manifesto," p. 178.

33. N. Czegledy, "Appropriation of Medical Discourse," ⟨http://new.territories/appropriation.of.medical. discourse/art.co⟩. (all subsequent quotes in this section from same source).

34. Critical Art Ensemble, "On Biology," ⟨http://www.sime.com/neue_galerie/radikale/caetx_e.html⟩ (defunct link).

35. S. Gould, "Shape of Life," in *Art Journal* 55(1996):1, p. 45.

36. Ibid., p. 50.

37. R. Hoffman, "Just a Little Bit Unnatural," in *Art Journal* 55(1996):1, p. 62.

38. A. S. Koch and T. Tarnai, "The Aesthetic of Viruses," in M. Emmer, ed., *The Visual Mind,* p. 223.

39. S. Gould, op. cit., p. 46.

40. R. Root-Bernstein, "Do We Have the Structure of DNA Right?" in *Art Journal* 55(1996):1, p. 48.

41. Louis Vuitton Moet Hennessy (LVMH), "Science Pour l'Art" competition, 1996 guidelines.

42. R. Malina, "Moist Realities," ⟨http://www.ylem.org/private_loves/public_opera/malina/malina.html⟩.

43. D. Lupton, op. cit., p. 162.

44. E. Kac, "Transgenetic Art," ⟨http://mitpress.mit.edu/e-journals/LEA/⟩.

2.2

Artists Working with Microbiology

Introduction: Microbiology and Genetics as Artistic Interest

"What you can't see can't hurt you"—Old folk saying.

Some artists think the adage is wrong. Recognizing the significance of researchers' attempts to understand and manipulate life at the microscopic level, they have begun to claim microbiology as an artistic arena. Many of biology's greatest accomplishments in the last decades have taken place at the microscopic level, below the threshold of human sight. Aided by new kinds of tools that allow observation and manipulation at this level, researchers have unraveled the secrets of genetic structures of DNA and studied the life processes of organisms such as bacteria and viruses. They have developed methods of bioengineering that allow the direct manipulation of genetic material.

This invisible world promises to have a profound effect on human history. Artists and art historians are beginning to attend it, and the future will see much more. In the 1990s there were a series of events and articles that demonstrated this burgeoning interest, for example, 1993 Ars Electronica's focus on genetic engineering, Suzanne Anker's "Gene Culture" show in New York, Ellen Levy's 1996 special issue *Art Journal,* "Contemporary Art and the Genetic Code," and George Gessert's and David Stairs's show called "Art+Bio in Michigan," and the "Paradise Now" show in New York.

In "The Gene as a Cultural Icon," sociologist Dorothy Nelkin lists three modes with which she characterize artistic response:

Some seem simply attracted to the aesthetic forms of molecular structures. Other dwell on a theme that I refer to elsewhere as "genetic essentialism," a view of genes as powerful and deterministic entities, as central to understanding the human condition. Still others use their art to express their fears of a technology they believe to be out of control. For DNA artists the biological gene— a nuclear structure—appears as a cultural icon, and the science of genetics provides a set of visual metaphors through which they can express the essence of personhood, the nature of human destiny, and, especially their concern about the social implications of an expanding, important, but historically dangerous scientific field.[1]

Nelkin critiques the essentialist tendency to see the gene as the essence of personhood. She notes that this perspective is fostered by the rhetorical techniques of scientists who refer to the Human Genome Project (the attempt to map the entire human gene structure) as the

---

Paradise Now—Genetic Art Show: ⟨http://www.geneart.org/pn-intro.htm⟩

Bible or the Book of Man. She notes that the rhetoric neglects the reality that the relationship between genes and traits is complexly interactive with other genes and the environment.

Others more optimistically see genetic engineering opening up possibilities beyond the mere manipulation of symbolic icons. In an 1988 *Art Forum* article called "Curie's Children," Villem Flusser wrote about the new possibilities. He analyzes one of art's main preoccupations as the production and preservation of information. He notes that the universe's tendency toward entropy (dissolution into less organized forms) also affects art artifacts. Genetic engineering allows artists to create works that are original in unique ways and that can reproduce themselves. Furthermore, he suggests that direct genetic engineering even improves upon the chance processes of mutation, on which natural processes rely.[2]

Like so many fields of scientific inquiry, microbiology and genetics raise profound questions beyond science—ancient questions that cry out for artistic commentary: What is our fate? What control can we have over our fate? What is human individuality? What images can communicate that individuality? What are the limits to human creativity and invention? What is the impulse to know and to try to control?

Artists have developed a variety of ways of responding to the microworld: actual manipulations and investigations of the microworld; forms and visualizations based on its structures, including the new iconography of gene mapping; and works that reflect on the processes of genetic science and its social implications. Some of the art embraces the new possibilities and some recoils at what they may mean.

Manipulations and Investigations of the Microworld

George Gessert

For the last decades, George Gessert has been a pioneer in creating and writing about genetic art. In a Leonardo Electronic Almanac article, Gessert notes that humans, in the form of plant and animal breeders, have been practicing genetic art for centuries. Selections have been made along functional, aesthetic, and a variety of other criteria.[3] He writes, "Domesticated ornamental plants, pets, sporting animals and consciousness-altering drug plants constitute a vast unacknowledged generic folk art, or primitive genetic art, that has a history stretching back thousands of years."[4]

His art employs this low-tech approach to genetic engineering, building on the strategies and perspectives of genetic breeders. For example, he has systematically bred Pacific Coast irises and exhibited these projects in a variety of galleries and other settings. In

---

George Gessert: ⟨http://www.geneart.org/gessert.htm⟩

Fig. 2.2.1. George Gessert's, *Hybrid 488 from Natural Selection.* Iris-breeding processes presented as part of an art show.

Iris Project, presented at New Langton's 1988 "Post Nature" show, he set up forty-six plants in a form referring to a hybridizer's field plot. In a 1990 show at the University of Oregon Museum of Art he practiced the breeder's process as part of the show, selecting among new blooms for criteria such as veining, unruffled petals, and various colors. He hand-pollinated the "preferred" plants and marked the others with an X and collected the seeds from the hybrids. In *Seed Library,* he created packets of seeds suggesting human relationships to plants. In *Scatter,* he planted seeds resulting from his breeding in a great variety of settings and documented the hundreds of sites in an artist book.

In 1995, Gessert created an installation called *Art Life* as part of the San Francisco Exploratorium's "Diving into the Gene Pool" exhibition, in which visitors were confronted with tables full of coleus hybrids. They were invited to answer a questionnaire that would help determine the next generations: Which plants did they like best? Which should be sent to the compost bin? What changes would they like to make in the plants?

In the articles "Notes on Genetic Art" and "A Brief History of Art Involving DNA," Gessert traces attitudes toward breeding as art. He offers examples from science fiction, such as Kurt Vonnegut's *The Seeds of Titan* and J. G. Ballard's *Garden of Time,* in which civilizations devote considerable energy to shaping species with particular qualities. Historically, plant breeders such as Luther Burbank drew analogies to art.

Gessert shows how general cultural attitudes play out in dispositions toward breeding. For example, the history of gardening reveals more general class attitudes, with hybrids sometimes being perceived as mongrel affronts to nature. In the Enlightenment, the enthusiasm toward science spread to hopes for breeding, such as Bacon's *New Atlantis,* in which genetic practice played an important role in the utopia's farms and laboratories.

In the twentieth century, eugenics implied that the same animal breeding techniques could be applied to human breeding. These ideas picked up steam throughout the

Western world in the early twentieth century, and found their most extreme expression in the racial purity ideas of the Nazis.

Gessert believes that work with genetics promises great artistic activity in the future. He traces some precedents, including Edward Steichen's hybridization of flowers, which resulted in a show at the Museum of Modern Art in 1936. He describes several problems, however, including ambivalent cultural attitudes toward genetic manipulation, unrealistic expectations born out of science hype and science fiction, and the unpredictability and uncommodificability of living art: "Genetic art challenges the status quo of gardening, of animal breeding, of art collecting, and even of the architecture of museums, which for centuries have been designed to exclude all forms of life, except for the human."[5]

Christopher Ebener and Uli Winters

In *Byte,* Christopher Ebener and Uli Winters bred a strain of mice specially selected for their tendency to eat computer cables. The mice inhabited a space full of display cables and monitors. Ebener and Winters described their method:

[T]he mice are housed individually in cages into which viewers can peer. A computer cable runs through each cage. Each time a mouse gnaws on the cable and severs a wire, it is rewarded with a portion of feed. A mouse's performance is displayed on a computer and can be read from a monitor mounted outside the cage. It is thus a simple matter to select the fittest specimens and to use them for breeding purposes.

Joe Davis

Joe Davis, a lecturer at MIT's Center for Advanced Visual Studies, created *Microvenus* one of the first artworks to incorporate a genetically engineered organism. The project partially grew out of speculation about ways to communicate with extraterrestrial intelligences:

Microvenus is undeniably connected to artistic traditions, but important inspiration also came from interdisciplinary notions about "universal" messages and the thought that certain biological materials may be useful in the experimental search for extraterrestrial intelligence.[6]

For the "message," Davis and his collaborators chose a simple graphic icon that was identical with an ancient Germanic rune that was encoded into a bit-mapped image whose

---

Christopher Ebener and Uli Winters: ⟨http://www.c-ebener.de/Byte.html⟩
Joe Davis ⟨http://kultur.aec.at/festival2000/texte/artistic_molecules_e.htm⟩

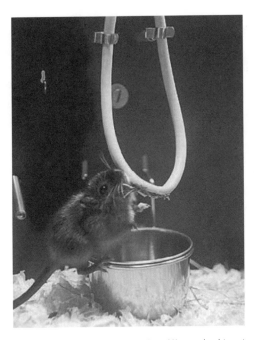

Fig. 2.2.2. Christopher Ebener's and Uli Winters's, *Byte.* Mice are bred to eat computer cables.

resolution was composed of prime numbers (thus theoretically being of mathematical significance throughout the universe). A scheme was devised to translate the bits systematically into a sequence of the four bases that make up DNA, using their relative molecular weights (cytosine, thymine, adenine, and guanine). A decoding clue was added at the beginning to aid any intelligences that found the organism. A synthesized plasmid was cloned into *E. coli* bacteria, which would reproduce the *Microvenus* pattern, surviving possibly longer than humanity. Davis notes the poetic unity of using our growing understanding of the biology of the cell to create a communication with unknown intelligences. He sees an essential linkage between the understanding of art and the understanding of life.

Eduardo Kac

In "Genesis" Kac arranged to encode phrases from the biblical story of Genesis in bacteria DNA. His ongoing "Transgenic Art" projects propose to create chimeric dogs and bun-

---

Eduardo Kac: ⟨http://www.ekac.org⟩

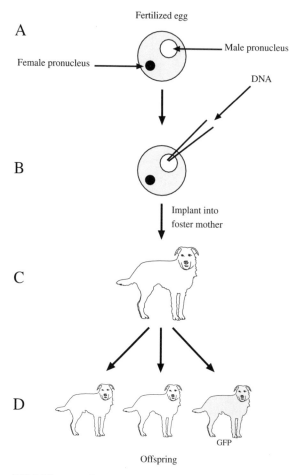

Fertilized egg

A

Female pronucleus ————→ ← ———— Male pronucleus

DNA

B

Implant into
foster mother

C

D

GFP

Offspring

Fig. 2.2.3. Eduardo Kac, *GFP K-9* (in progress). Diagram showing how the transgenic dog will be created with green fluorescent protein.

nies, some whose genes come from green fluorescent bacteria. A paper at this Web site analyzes the scientific, ethical, and cultural challenges and proposes that it is inevitable that the arts will move into transgenetic activities.

Other Artists and Projects

Jon Tower created one project in which he arranged the four standard bases of the DNA molecule—cystosine, thymine, adenine, and guanine—in sequences that spelled words. In another series of "Gene Room" conceptual works he architecturally maps genetic sequences of various animals to architecutral spaces. **David Kremers**'s 1992 "Somites" series produced paintings out of *E. coli* bacteria. The bacteria had been genetically engineered to produce colored enzymes and proteins in interaction with particular stains. During the process of painting, the stains are transparent and their color only emerges as part of a life process. Appointed as an artist working in California Institute of Technology's biology department, he has worked on the "Vrmouse" project, which proposes to build up virtual mice from photomicrographs of developing embryos. He notes that artists must work with living materials: "We are the first generation of artists to face the problem not of mortality but of immortality." **Knut Mork** (see chapter 2.5) created an installation called *A Single Drop of Blood*, which allowed people to visit a virtual world full of fictional microorganisms generating sound. The visitor is challenged to find a way to navigate the chaotic world. In response to Monsanto's development of genetically modified poison-resistant crops, **Heath Bunting** and others at the **Cultural Terrorist Organization** have created a "Superweed Kit" consisting of mixed weed seeds. Interposing scientific news and commentary, **Natalie Jeremijenko** and **Heath Bunting** published the *BioTech Hobbyist Magazine,* with all the normal sections such as news, classified ads, and advertising. Working as both a scientist and artist, **David S. Goodsell** investigates ways to depict the cellular mesoscale, the "interactions of atoms, molecules, cells, tissues and bodies." **Dan Rose**'s "DNA-Photon Project" is a facetious representation (including maquettes) of a big science project "discovered" from the trash found at the closing of a naval shipyard. **Heather Ackroyd and Dan Harvey**'s *Mother and Child Portrait* used a genetically engineered grass seed that stays green after it dies to make a grass portrait by exposing the seeds to a negative as they bloomed.

Creating Forms and Visualizations Based on Its Structures, Including the New Iconography of Gene Mapping

Suzanne Anker

Suzanne Anker creates art installations that reflect on the scientific representations of chromosomes. She sees the visual notation of chromosomes as suggestive of other kinds

David Kremers: ⟨http://www.caltech.edu/~wold/cyburbia/bio/intro.htm⟩
Knut Mork: ⟨http://www.gar.no/~knut/drop/⟩
Cultural Terrorist Organization: ⟨http://www.irrational.org/cta/⟩
BioTech Hobbyist Magazine: ⟨http://www.irrational.org/biotech/⟩
David Goodsell: ⟨http://www.scripps.edu/pub/goodsell/⟩
Heather Ackroyd and Dan Harvey: ⟨http://www.geneart.org/ackroyd2.htm⟩
Suzanne Anker: ⟨http://www.geneart.org/genome-Anker.htm⟩

of ideograms. In her *Zoosemiotics* installation she created three-dimensional models of various species chromosome patterns. She sees the strategy as similar to that underlying many museum presentations:

A cellular archaeologist, the laboratory investigator correlates subtle differences in identity among specimens. Differences, for example, between a fish, a flower, or a bacteria can be discerned by looking at linear configurations of an organism's abstract sequence. As arrangements of sets, chromosomes form categories by which types of life catalogued. Through a system of reduction, the organism's mathematical formula is translated from an invisible understructure of living matter to a visual notation. As with the Cubists, the sign becomes an abbreviated blueprint of cultural code summarizing the materialization of idea into visual form. In contrast, biological signs employ an assembly of life fractions to form the basis of scientific symbols. Defining an organism from a collection of fragments creates a fabricated condition comparable to metonomy and museum display.[7]

In another work entitled *Chromosome Chart of Suzanne Anker,* she presented her own DNA sequence as a self-portrait. In *Cellular Script,* she presents chromosome patterns as a kind of calligraphy:

Throughout all cultures and time periods, systems of writing have been held as a transcendental form of the divine. The word, the book, the text are all attributes associated with God. Through symbols inscribed on a surface, the lexicon functions as a communicating forge. In addition, calligraphy, automatic writings and genetics can be considered as transcribing entities that unfold in time and space.[8]

John Dunn and Mary Anne Clark

Artist John Dunn and biologist Mary Anne Clark have collaborated in the creation of DNA music. Rather than just matching particular sounds with particular bases, they developed approaches to sonification that communicated interesting features of the organic material, including protein synthesis and chemical characteristics of the molecules.

In "Life Music: The Sonification of Proteins," Dunn and Clark note that they wanted to make the musical rendering of DNA both artistically interesting and scientifically valuable:

John Dunn: ⟨http://geneticmusic.com/⟩
Mary Anne Clark: ⟨http://www.startext.net/homes/macclark/Music/musicpag.htm⟩

Fig. 2.2.4. Susanne Anker's, *Zoosemiotics*. Three-dimensional models of chromosomes.

I [Clark] was convinced that this would be worth doing—that the amino acid sequences would have the right balance of complexity and patterning to generate musical combinations that are both aesthetically interesting and biologically informative. There are twenty amino acids in proteins, enough for about three octaves of a diatonic scale. They are not arranged at random, just as notes are not arranged at random in a piece of music. Both proteins and music are meaningful. The meaning of a protein is its function in the organism, and certain sequences have emerged as the hallmarks of specific functions.[9]

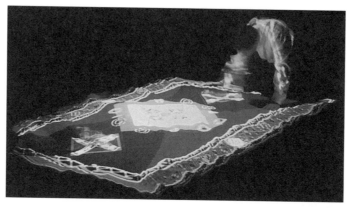

Fig. 2.2.5. Jamy Sheridan and John Dunn, *The Garden of Initial Conditions, Strain 2*. Installation-performance in an activated sandbox in which participants interact with images and sounds generated by using a DNA source. The sounds and images are generated using John Dunn's KAM language for artists.

Other Artists and Projects

A 1998 project entitled "COMPOSITION: DNA TUNINGS," by **Susan Alexjander and Dave Deamer,** used the light-absorption spectra for the four bases that make up DNA as the basis for their sounds. Commenting on the fascination with the sounds, Alexjander notes, "Perhaps on a very deep level the body recognises itself—hears something familiar in the music. It's a theory. I don't know." Other artists, musicians, and scientists exploring DNA and body protein based music include Peter Gena and Charles Strom, Kenshi Hayashi and Nobuo Munakata, Ross King and Colin Angu, and David Lane. In response to an age where "the guiding notions of individuality and of the cultural and physical integrity of the distinct person are dissolving," **Kevin Clarke** creates photographic portraits that collage individual specific DNA sequences imaged from the blood of the subject with other respresentations of their "individuality." **Thomas Kovachevich** similarly includes imagery of his chromosomes in some of his portraits. In part, he is interested in commonalities as well as individuality. A show called *Out of Sight Show/Imaging Science* featured art exploring biological imagery. Examples included **Gary Schneider**'s *The First Biological Self Portrait,* built of his own DNA samples, hair and sperm cells,

Susan Alexjander, Dave Deamer: ⟨http://tesla.csuhayward.edu/history/07_Composers/Alexjander/DNA.html⟩
Kevin Clarke: ⟨http://artnetweb.com/artnetweb/projects/clarke/kchome2.html⟩
Out of Sight Show: ⟨http://www.sbmuseart.org/exhibitionx/index.html⟩
Gary Schneider: ⟨http://www.sbmuseart.org/exhibitionx/room2/index.html⟩

and other biological images; Du Seid's *Blood Lines* video in which a self portrait dissolves into DNA imagery; Felice Frankel's images of viruses; and other photographers who capture the unseen elements of biological processes, diseases, and bodily microstructures.

Reflections on the Processes of Genetic Science and Its Social Implications

Gail Wight

Gail Wight creates installations that reflect on science experiments with heredity and genetic modification. She uses scientific terminology and paraphernalia to recreate some aspects of the scientific enterprise and to induce viewers to think about the cultural implications of the work. In one installation called *Spike,* a mouse runs about in a large suspended plexiglass maze revealing, as it goes, text, dioramas, videos, and so on, related to the history of biological experimentation, surrounded by the sound of one nerve cell spiking. *Hereditary Allegories* confronts a viewer with what appears to be a science lab full of mice in labeled cages:

Hereditary Allegories is a survey of topics, experiments, studies, incidents, stories, rumors, accidents, plans, people, personalities, histories, and eventualities that comprise some of the groundwork for contemporary genetics research. It is not about DNA, but about the conception of DNA in the human mind. The survey is divided into three categories: Twin Studies, Control Studies, and Anomalies. Each investigation highlights a particular aspect of genetics research, and is based upon an actuality.[10]

The text describing the experiments and the mice experience sounds simultaneously scientific and strange. Each experiment is built on a real category of scientific research. For example, the twin studies labels mix the language of human twin studies with the mouse study context.

Sonya Rapoport

Sonya Rapoport created an interactive installation (also available on the Web) that comments on genetic engineering. The work integrates old biblical illustrations with scientific

---

Wentian Li's Links to DNA Music Artists: 〈http://linkage.rockefeller.edu/wli/dna_corr/music.html〉
Gail Wight: 〈http://www.luckygarage.com/gail/index.html〉
Sonya Rapoport: 〈http://www.lanminds.com/local/sr/srbagel.html〉

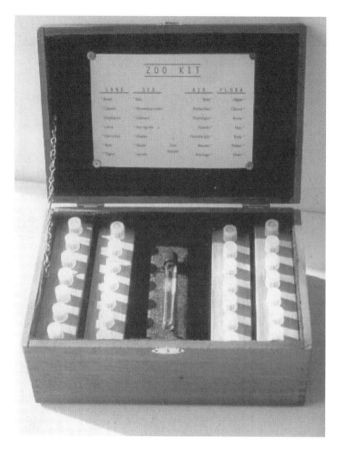

Fig. 2.2.6. Gail Wight's *Zoo Kit*. DNA samples in a box.

representations from the nineteenth and twentieth centuries. The story of Noah is used as a metaphor for the creation of a gene pool. The bagel is the medium that carries the modified genetic vectors:

The Transgenic Bagel is a parody on the recombinant gene splicing theme. The genetic formula of a desired trait is engineered and impregnated into a bagel which serves as the transgenic (gene transfer) vehicle. The bagel physically resembles a plasmid, a circular DNA molecule which contains the genetic information. A section of this loop can be excised and another portion of the DNA inserted.

The viewer can pick personality traits from a gallery of biblical characters. The event then proceeds through the steps that closely resemble the real steps in a genetic engi-

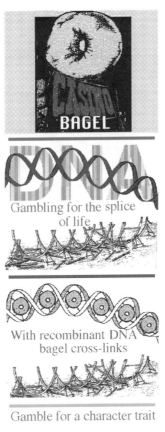

Fig. 2.2.7. Sonya Rapoport. Image from *The Transgenic Bagel,* a computer-based multimedia project (1994). *Entrance to the Bagel Casino Section.* The Bagel Casino is where participants gamble for the "splice of life" by playing a virtual slot machine to get three of a kind of a mythical animal, from which a trait gene is extracted.

neering laboratory. The language and graphics are an odd synthesis of biblical and scientific references.

Other Artists and Projects

Installation artist **Andrea Zittel** presents ironic comment on genetic breeding experiments in her 1994 *Breeding Unit for Reassigning Flight* at the New Museum of Contemporary Art, in New York. The unit was arranged in a series of interconnecting chambers such that only those chickens who could fly to a certain height would be able to hatch their eggs. The installation thus instantiated a kind of "unnatural selection." **Gina Czarnecki**'s *Stages Elements and Humans* is a video installation in which wall-sized projections of human "specimens" come alive in commentary of genetic engineering and perfect bodies. **Nell Tenhaaf**'s *Savoir* series of sculptures ironically reflects on the genetic research process. **Thomas Grunfeld**'s *Misfits*'s interspecies sculptures create creatures that comment on lab work gone awry. **Simon Robertshaw** has created a series of installations including *From Generation to Generation* and *The Nature of History,* which explore the accomplishments and blindnesses of genetic science and artificial life. Photographer Catherine Wagner created a five-year photo science-lab documentation project that resulted in the book *Art and Science: Investigating Matter.* Her stark black-and-white photos of ordered lab paraphernalia without people attempt to focus on the process of science. She indicates that genetic science raises many philosophical and spiritual questions, and her photo essay attempts to capture some of those issues. Michael Spano's photoessay "Scientists at Work" juxtaposes images from many biological research labs to explore research at the unseen level. Working with scientists from the Human Genome Mapping Project, U.K., **Neil White and the Soda Group** created *Inheritance,* three "inter-related projects exploring personal identity as a representation of scientific, institutional, and genetic intervention." Several performance/lecture interventions by the **Critical Art Ensemble** comment on the bioengineering industry, including *The Cult of the New Eve* and the mock company *BioGen,* which includes an online test to assess the visitor's fitness as a genetic donor.

Summary: Micro Steps

The world is alive with biological research at the microscopic level. Enormous resources are devoted to understanding the atomic and cellular level of life, probing the genetic base of life, mapping the human genome, perfecting the techniques of genetic engineering, and doing battle with the microbes that cause disease. Artists have begun to explore this world—manipulating genetic materials, decoding the representations of genetic markers, and commenting on the research process. Given the level and potential significance of this research, however, the artistic explorations must be considered only first steps.

Thomas Grunfeld, *Misfits:* ⟨http://www.gbhap-us.com/cont.visarts/article/cva17/grunfeld_chapmans/grunfeld_chapmans.html⟩
Simon Robertshaw, *The Nature of History:* ⟨http://www.artscat.demon.co.uk/nhist.htm⟩
Michael Spano: ⟨http://www.sbmuseart.org/exhibitionx/room4/pmpage1.html⟩
Neil White, *Inheritance:* ⟨http://www.soda.co.uk/index.htm⟩
Critical Art Ensemble: ⟨http://www.critical-art.net/⟩

Notes

1. D. "Nelkin, "The Gene as a Cultural Icon," *Art Journal* 55(1996):1, p. 56.

2. W. Flusser, "Curie's Children," *Art Forum,* 26(1998):7, p. 141.

3. G. Gessert, "Genetic Art," ⟨http://mitpress.mit.edu/e-journals/Leonardo/isast/journal/journal96/Leo294/gess294.html⟩.

4. G. Gessert, "Notes on Genetic Art," *Leonardo* 26(1993):3, p. 205.

5. G. Gessert, "A Brief History of Art Involving DNA," *Art Papers,* October 1996, p. 25.

6. J. Davis, *"MicroVenus,"* *Art Journal* 55(1996):1, p. 70.

7. S. Anker, "Chromosome Art," *Art Journal* 55(1996):1, p. 33.

8. S. Anker, "Description of Art," ⟨http://mitpress.mit.edu/e-journals/Leonardo/gallery/gallery314/anker.html⟩.

9. M. A. Clark, "Life Music: The Sonification of Proteins," ⟨http://mitpress.mit.edu/e-journals/Leonardo/isast/articles/lifemusic.html⟩.

10. G. Wight, "Description of *Hereditary Allegories,*" ⟨http://www.luckygarage.holowww.com/gail/allegories.html⟩.

2.3

Plants and Animals

Introduction

How have artists incorporated plants and animals? How have they been influenced by ecology? In what ways have scientific or technological perspectives informed this work? Although the living organic world has always been a subject of painting and sculpture, it has rarely been the substance directly worked with. With the opening up of art materials and art contexts initiated by the artists of the early twentieth century, the possibility was created but rarely pursued. A few early examples include the Dadaists' inclusion of live vegetables and eggs in their performances (mostly for throwing) and surrealists' experiments, such as Salvador Dali's *Rainy Day Taxi,* which became overgrown with plant matter. In the 1950s and 1960s experimentation increased, with artists exploring plants, animals, and their own bodies as art materials.

For example, San Francisco conceptual artist Bonnie Sherk performed *Public Lunch* in 1971, in which she shared a cage with a tiger at the zoo, and *The Farm,* which was an alternative space that integrated art, agriculture, ecology, and community development. Joseph Beuys was also famous for his interest in animals, which combined scientific, shamanistic, and artistic perspectives. He carried a dead hare in an early performance, shared the stage with a spectral white horse in *Tutus,* and spent a week with a coyote in *I Like America and America Likes Me* in an action described as a dialogue with an animal. Predating later perspectives on ecology, Beuys felt that artists should study and learn from animals, much as biologists did.

Newton Harrison helped pioneer art that explores the life processes of plants and animals. During the 1970s he mounted installations such as *Slow Birth and Death of a Living Cell, Air, Earth, Water, Brine Shrimp Farm, Portable Orchard, Portable Fish Farm,* and *Hog Pasture,* which brought plant and animal life processes into the gallery and art focus. His ecology work is discussed in chapter 2.4. Also in the 1970s Richard Lowenberg created a movie based on the book *The Secret Life of Plants* while an artist in residence for NASA. He used bio-remote sensing to translate plant energy into digital music and synthesized video. Other projects explored animal communication, including work with Koko, the gorilla that communicates with sign language.

Plants and animals fill the world, but most artists have not worked directly with them. Why not? In part, artists have not done so because they have not been validated art materials. But even after the 1960s revolution opened up the possibility that anything could be art material, most artists have not chosen to work with living entities. There are several possible explanations: artists aspire toward permanence; living materials constantly change and decay. Artists glory in mastery and control of materials; living materials are recalcitrant— going their own way. Ethically, there may be some resistance, at least with animals.

This chapter describes artistic work with plants and animals. Some of it makes use of scientific tools or is clearly informed by scientific concepts and perspectives. With others the connection is less clear, although the world of research and field work lurks behind the artworks. Our views of animals and plants are highly influenced by the last century's research and scientific frameworks. Also, the tendency to objectify animals is an especially clear place from which to observe major cultural narratives played out in science. Remember, also, that the works described in the previous section on microbiology are indeed working with plant and animals, although not in a form we usually encounter. This chapter considers artist's work with invertebrates, insects, plants, and animals. It also considers sound artists' work in acoustic ecology attempting to record the earth's diversity of organic and inorganic sounds.

Invertebrates

Gail Wight

As part of the San Francisco Exploratorium's "Turbulent Landscape" exhibit, Gail Wight presented print books and video of two slime molds: *Physarum polycephalum* and *Dictyostelium discoideum.* She was intrigued by the microcosm of life that these molds presented:

Looking like the complex branches of a never-ending nerve cell, *Physarum polycephalum* caught my attention. Beyond its brilliant yellow beauty, *Physarum* presents a wonderful visual demonstration—both direct and analogous of a seeking intelligence. Pulsing like a tiny ocean, this creature offers an oddly formal affinity with human consciousness.[1]

In another project called *Neural Primer,* Wight created an artist book series that celebrates the neurological structures of various animals. Part of the series focused on the octopus.

Athena Tacha

Attempting to make the microscopic universe that exists on human bodies visible, Athena Tacha created a CD-ROM called *The Human Body: An Invisible Ecosystem.* She wanted to "let people experience through surprise, disgust, wonder, delight and humor this

---

Athena Tacha: ⟨http://www.oberlin.edu/~art/athena/tacha2.html⟩

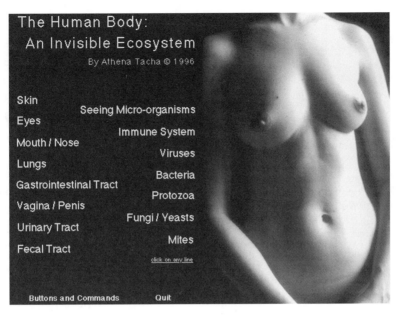

Fig. 2.3.1. Athena Tacha, cover from *The Human Body: An Invisible Ecosystem* CD-ROM.

invisible micro-universe that teams with life and interacts with us." Here is an excerpt from a description of the project:

Most areas of the normal human body host, on a regular basis, hundreds of species of bacteria, viruses, fungi and yeasts that are either harmless or become pathogenic only under specific conditions. In this art project, the body is visualized as parallel to the planet Earth—an environment with varied climate zones, comparable to cool dark woods (the scalp), sparsely inhabited deserts (the forearm), heavily populated tropical forests (the underarm), hot moist jungles (the nose and mouth), and oceans teaming with life (the intestinal and urinal tracts).[2]

Ken Rinaldo

Ken Rinaldo created a series of installations that explore animal behavior. *Technology Recapitulates Ontogeny* created a media event out of the behaviors of tubefex worms. His installation projected the shadows of the constantly changing forms of a

---

Ken Rinaldo: ⟨http://www.ylem.org/artists/krinaldo/emergent1.html⟩

population of worms in a transparent bowl of water as a conceptual exploration of intelligence:

> The tubefex worms, which are the stars of this piece, demonstrate one form of supra organization in which these single creatures work together to act as a single group consciousness. They send out exploratory tentacles over the edge of the plate. They line their bodies up like striated muscle tissue. If you touch one worm in the bunch the whole mass contracts like a muscle.[3]

Delicate Balance presented an installation that featured the movements of Siamese fighting fish (*Betta splendons*). The fishbowl was perched on a tightrope at the viewer's eye level, much like a circus act. Microchips read the movements of the fish and activated motors to move the fishbowl in the directions the fish were looking. Rinaldo proposed this as an example of the human interface design community's search for an invisible interface that intelligently responds to the desires of its users.

Other Artists and Projects

Nole Giulini uses processes of trial and error to manipulate fungal organisms, drying and shaping them into what she calls puppets. She builds up large assemblages of these fungal skins, varnishes them, and creates large hand-puppet-looking sculptures. *Martin Zet*'s installation *Snails* comments on scientific processes of observation while presenting fascinating records of snail movement across plexiglass. In the "Moosescape" exhibition at San Francisco's Exploratorium, Finnish artist *Antero Kare* created an installation out of mold and microorganisms. He created a unique hospitable environment and then spread the life-forms out, creating a metaphor for human existence. In *Swim,* *Lane Hall and Lisa Moline* created gigantic billboard-sized prints of stomach parasites. Other projects included *Fly,* with images of flies, *Mandala,* with images of hands of snakes and small animals (completed in collaboration with zoo keepers), and *Germ,* which was an unaccepted proposal to Kohler bathroom fixture company to embed blown-up images of germs on the surfaces of fixtures. *Anya Gallaccio* creates installations making use of bacterial processes that act on industrial volumes of organic matter. In a Wellcome

---

Martin Zet: ⟨http://mitpress.mit.edu/e-journals/Leonardo/gallery/gallery314/zet.html⟩
Antero Kare: ⟨http://www.exploratorium.edu/complexity/exhibit/mold.html⟩
Anya Gallaccio: ⟨www.locusplus.org.uk⟩

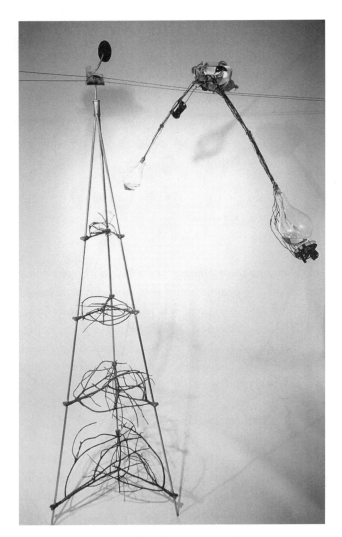

Fig. 2.3.2. Ken Rinaldo, *Delicate Balance*. The position of Siamese fighting fish controls the movement of the tank on a tight wire. Photo: Julia McLemore.

Invertebrates

Fig. 2.3.3. Hubert Duprat. Aquatic caddis-fly larva with case (1980–99), gold, pearls, and precious stones. Dimension variable from 2 to 3 centimeters. Sculptural work in collaboration with Trichoptera. The worm incorporates jewelry in the construction of its enclosure. Photo credit: H. Del Olmo.

Trust–funded collaboration, **Daro Montag and Richard Murphy** developed *Bioglyphs,* which directly exposed film to organic bacterial actions such as digestion and reproduction. **Sabrina Raaf**'s *Breath Cultures* invited participants to breathe into petri dishes containing agar, thus creating biological portraits of that moment in their lives via their unique patterns of bacteria and fungus. In *Breath II: Growing Pleasure,* a sculptural network of casts of body forms and tubing was infected with a bacteria pigmented red that slowly grew and visually spread to infiltrate all sections.

Arthropods: Insects and Spiders

Hubert Duprat

Hubert Duprat uses the natural metamorphic processes of insects to create strange hybrid objects of nature, science, and art. He introduces gold and jewels to the insects as they search for the materials to construct their metamorphic enclosures. Duprat ultimately received a French patent for the process. The results are rare objects of nature and culture. Duprat's process requires a detailed understanding of the insect's life cycle and behavior. Christian Besson describes this collaboration between Duprat and the insects in a Leonardo On-line article.

Daro Montag and Richard Murphy: ⟨http://webserver1.wellcome.ac.uk/en/old/sciart98/20bioglyphs.html⟩
Sabrina Raaf: ⟨http://www.raaf.org⟩
Hubert Duprat: ⟨http://mitpress.mit.edu/e-journals/Leonardo/isast/articles/duprat/duprat.html⟩

Since the early 1980s, Duprat has been utilizing insects to construct some of his "sculptures." By removing caddis fly larvae from their natural habitat and providing them with precious materials, he prompts them to manufacture cases that resemble jewelers' creations. Information theory, as explained by biologists such as Jacques Monod and Henri Atlan, helps us understand what seems to be the insect's aesthetic behavior. In the interview, Duprat described the role of scientific observation in his art:

The collectors who created those *Wunderkammern* [Renaissance rooms of wonders] were driven by a feeling that I myself experience about art. It dates back a long way. I spent my boyhood and teenage years in the countryside, where I hobnobbed with hunters and fishermen. Very early on, I had a keen interest in archaeology and the natural sciences. I made early observations in aquaria, where I installed water scorpions, water-striders, newts, tadpoles, pond skaters, planorbid snails and, right at the outset, caddis worms— *Trichoptera*.[4]

The interviewer, Christian Besson, reflects that Duprat's objects uniquely integrate scientific and artistic perspectives:

The cases made by the caddis worm (under your guidance) are something between insect artifact, jewel and sculpture. These objects are works of art and, at the same time, products of an intrinsically scientific experiment. They can be seen in two ways, like a Janus bifrons. As art pieces, they are a kind of assisted ready-mades, found objects altered and promoted to the rank of works of art, hybrid formations such as appeared in large numbers in the 1980s. As scientific experiments, they are evidence of an unexpected interdisciplinary; they also raise real epistemological issues.

Duprat's success is dependent on understanding the insect and its life processes and systematic experimentation with interventions. Based on study, observation, review of work by other researchers and naturalists, investigations of atmospheric effects, and years of experience, he collected certain subspecies from particular mountain areas during restricted January-April periods. He systematically places the larvae in tanks of materials at precise times in their life cycle when he knows they are engaging in certain special building processes.

Mark Thompson

For many years, Mark Thompson has created performances and installations based on an appreciation and understanding of honeybees. His installations often include live hives and the movement of the bees, their construction of honeycombs, and their manufacture of honey. In one famous performance, Thompson places his unshielded body near a hive for hours and lets the bees accumulate on his torso and head until no flesh can be seen.

Thompson integrates cultural lore about bees, the experience of bee keepers, and scientific research into his work. In one installation undertaken near the preunification Berlin Wall, he used the flower scavenging behaviors of bees as a method for inspecting the artificiality of human borders and political demarcations. The bees went into East Berlin and West Berlin freely in their search.

Other Artists and Projects

Miya Masaoka creates performances that explore insect behavior and human reactions to it. In *Ritual,* she lets giant Madagascar cockroaches roam freely on her body, and places sensors on her body such that the wanderings of the roaches set off sound samples of their hissing. Images of the roaches are shown onscreen. In another performance (a collaboration with Scot Gresham-Lancaster) she performed a koto duet with a bee (via amplified and processed sounds of the bee's wing). *Ed Tannenbaum* invented a project called *Bee Vision* for the New York Botanical Garden that allowed visitors to see the world the way bees would. *Mark Tilden* set up an installation of live ants next to insectlike solar-powered robots. In *Nature Marta Menezes* manipulated the embryological environment of developing butterflies in order to systematically produce desired wing-color patterns. In *Micro Friendship* *Yasushi Matoba and Hiroshi Matoba* created microscopic video interfaces that allow interactions with insects. *Val Valgardson*'s *Bug Run* is a mobile robotlike device whose steering "intelligence" consists of a dynamic analysis of the motion of an onboard colony of cockroaches.

Plants

Tony Bellaver

Tony Bellaver creates installations to "re-educate" trees about their natural state. The tree saplings attend special "schools" he has created, where he plays sounds of the natural

---

Ed Tannenbaum: 〈http://www.et-arts.com/〉
Mark Tilden: 〈http://www.cogs.susx.ac.uk/ecal97/tilden.html〉
Marta Menezes: 〈http://dunn1.path.ox.ac.uk/~1graca/nature.htm〉
Yasushi Matoba and Hiroshi Matoba: 〈http://www.siggraph.org/artdesign/gallery/S00/interactive/thumbnail14.html〉
Val Valgardson: 〈http://www.vpa.niu.edu/art/faculty/valgardson/index.html〉
Tony Bellaver: 〈http://www.artpapers.org/bellaver/bellaver.html〉

Fig. 2.3.4. Tony Bellaver, OPAC—Ocular Projected Auditory Construction, with cedar incense specimen (1998). Courtesy of Steffany Martz Gallery, New York.

environments that used to host the trees, shows them images, reads them biological texts, and provides artificial light. The constructions bear names such as *Genealogy of Place, Woodland Recovery Project,* and *Green Thoughts.* He invents new devices, such as the Auditory Botanic Resonator, which "is designed to teach young trees, through story telling, what they will become when planted in Nature." It reads from technical field guides about specific tree species. The installations are simultaneously ironic and serious.

Bellaver believes that technology can be used to preserve rather than destroy the natural environment: "It is my belief that mankind should be investigating ideas that will integrate technology with nature. In the future we may be dependent upon technology to help in environmental restoration."[5]

Other Artists and Projects

Tropical plant researcher **Patrick Blanc**'s *Vegetal Wall* framed the facade of the "Etre Nature" show in Paris. German artist **Wolfgang Laib** creates installations out of pollen. San Francisco artist and scientist **Landa Townsend** created a project called *Treo Grafte: Sculpting* which explores grafting as a sculptural process. She makes decisions based on both biological and aesthetic criteria. *Interactive Plant Growing,* by **Christa Sommerer and Laurent Mignonneau,** used living plants as the interface that affected a 3-D animated plant environment whenever humans touched or approached the plants (see chapter 7.4). **Val Valgardson**'s *One Life to Live* is a kinetic system that uses computer-controlled metal rods to arrange cubes in x, y, and z coordinate space, where the plant is allowed to grow. *Phil Shaw* has developed *Ink Gardens* based on research on the capabilities of plants for making printmaking inks. **Frances Whitehead** created an installation called *ARGUABLY ALIVE* (*the virus taxonomy*) which consisted of canopic jars with lids made of three-dimensional models of each of the 81 recognized virus categories. Swedish artist **Ola Pehrson** set up the *Yucca Invest Trading Plant* installation, which used changing characteristics of a Yucca plant to determine what kind of on-line stock trades it should make. The *Natural Resources* show at the Northern Illinois University Museum and the *Botanica* show at the Tweed Museum of the University of Minnesota both included artists exploring live plants, ecology, and processes such as growth and decay.

Vertebrates

Eduardo Kac and Ikuo Nakamura

In "Essay Concerning Human Understanding," Eduardo Kac and Ikuo Nakamura created an installation that allowed a plant in New York and a canary in Kentucky to have a conversation via telephone. An electrode was placed on the philodendron to read its changing electrical characteristics via an IBVA brainwave reader. In this, they made use of phenomena described in books such as *The Secret Life of Plants*. These changes were

---

"Etre Nature" show: ⟨http://www.adetocqueville.com/archive/s980918a.htm⟩
Christa Sommerer and Laurent Mignonneau: ⟨http://www.mic.atr.co.jp/~christa/⟩
Ola Pehrson: ⟨http://www.art.a.se/best_before/about/docimg/ola_yucca.jpg⟩
Natural Resources Show: ⟨http://www.vpa.niu.edu/museum/natural_resources.html⟩
Botanica Show: ⟨http://pubinfo.d.umn.edu/tma/botanica0.html⟩
Eduardo Kac and Ikuo Nakamura: ⟨http://www.ekac.org/Essay.html⟩

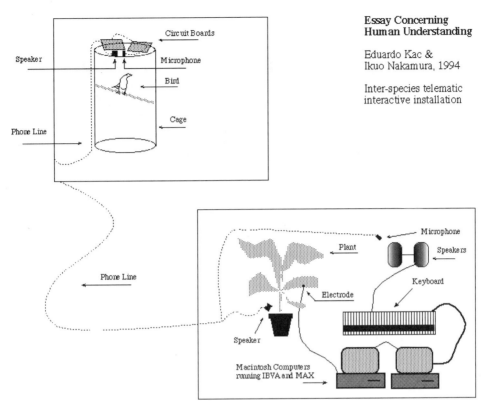

Essay Concerning
Human Understanding

Eduardo Kac &
Ikuo Nakamura, 1994

Inter-species telematic
interactive installation

Fig. 2.3.5. Eduardo Kac and Ikuo Nakamura, *Essay Concerning Human Understanding* (1994). A canary and plant influence each other via a telecommunications link.

converted into MIDI sounds played to the canary in Kentucky. The canary's songs were likewise transmitted to the plant. Human activity near each of the organisms similarly influenced behavior. Kac and Nakamura saw the interspecies event as a metaphor for human communication.

Carsten Höller

Carsten Höller was originally trained as a research scientist. His dissertation focused on olfactory communication among bees. In recent years he has created art installations that

---

Carsten Höller: 〈http://www.ensba.fr/connexions/HOLLER/holler.html〉

Fig. 2.3.6. Marc Böhlen, *Advanced Perception*. A robot is designed to live unobtrusively with chickens.

are simultaneously fascinating and troubling. They often create unexpected syntheses of scientific thought and cultural metaphors. A Basel, Switzerland, museum description offered that the "visitor is often at the same time test subject and experimenter, observer and focus."[6]

In *Luck,* he explored animal behavior, providing an installation in which one could be pulled by fish. The *Loverfinches* installation explored bird sonic communication and intergenerational transmission. In *Lax Clothes in Contemporary Flower Societies,* plants were frustrated in their search for light in a manner that suggested easy metaphors for human culture. In 1997's *Documenta 10* Carsten Höller, together with Rosemarie Trockl, created a *House for Pigs and Humans,* which included an environment to house a family of pigs and a place for humans to observe them.

Marc Böhlen

Marc Böhlen's *Advanced Perception* created a mobile robot that cohabited with chickens exploring concepts of "kind surveillance." The robot used advanced visual recognition and movement planning skills and knowledge of chicken biology to create a

Marc Böhlen: ⟨http://www.contrib.andrew.cmu.edu/~bohlen⟩

chicken-friendly experience. At the end of the performance, an acclaimed chef created omelettes out of the eggs the chickens produced during their time with the robots.

Other Artists and Projects

Paras Kaul (see chapter 2.5) created an VR installation that attempted to help visitors experience the world of dolphins. Visitors wearing displays and 3-D sound earphones used brainwave detectors to navigate a visualization of the dolphins' world. Los Angeles–based performance artist **Rachel Rosenthal** often explored themes influenced by medical and ecological themes. In *The Others* she explored animal rights and the relationship between humans and other animals by performing with animals that had been abused. **Ann Hamilton** creates installations in which people seem like specimens. She sometimes uses animals in her installations as a counterpoint to normal human existence and to focus attention on concepts such as dissection and alienation. **Diana Domingues**'s installation *Among the Snakes* placed a robotic snake with a Web camera for eyes with real snakes. She collaborated with a snake specialist to design the event so it would be copacetic for the snakes. **Ted Purves** investigates the similarities and differences of the artist and scientist mind-sets. His *Art and Biology in Practice: An Essay with Notations* presents a series of artist books and framed texts that explore the integration of scientific inquiry and art.

Acoustic Ecology

Leif Brush

Leif Brush was one of the pioneers in using technology to sonify usually inaudible sounds of nature. His instruments picked up sounds such as insects, leaves, the internal sounds of trees, and the like:

In my current work I attempt to manipulate sounds in the realm of millivolts by joining them with other wave forms to build passages toward new meanings in interaction with people and

Rachel Rosenthal: ⟨http://www.w3art.com/aagroups/rrgroup.html⟩
Ann Hamilton: ⟨http://www.diacenter.org/exhibs/hamilton/essay.html⟩
Diana Domingues: ⟨http://www.ucs.tche.br/artecno.htm/diana.html⟩
Ted Purves: ⟨http://mitpress.mit.edu/e-journals/Leonardo/gallery/gallery314/purves.html⟩
Leif Brush: ⟨http://www.d.umn.edu/~lbrush/⟩

Fig. 2.3.7. Leif Brush, *Meadow Piano* construction (1972). Transducer/sensor arrays for monitoring heat eddies, ambient magnetism, and the sounds of wind, rain, animals, and plants. Photo: Gloria DeFilipps Brush.

the natural environment. When I began to explore nature, inspired by the physical sciences, my various sound works assisted me in learning to listen beyond my otherwise-limited capabilities. . . . I used analog technology to monitor and receive amplified internal physical vibrations from the dancers and from nature.[7]

Brush built installations that made sound available to audiences in nontraditional ways. For example, the 1986 *Milaca Soundmark* used a parabolic sensor to pick up sounds and amplify them in a public highway rest stop. *Terrain Broadcasts* beamed amplified natural sounds toward passing motorists. The *Meadow Piano* series placed a grid of sensors to pick up wind, rain, and other motions of plants and animals.

In later years, Brush became part of the international acoustic ecology movement, which undertook projects such as the World Soundscape Project. This aims to increase people's attention to their acoustic environment and to work toward mindfully shaping

Fig. 2.3.8. David Rokeby and Eric Samakh, *Petite Terre*. Projects sounds of animals scurrying away when a viewer approaches.

it. It draws together artists, musicians, psychologists, geographers, biologists, naturalists, meteorologists, and people from many other disciplines. The movement aims to enable people to see, hear, feel, and taste the changing relationship of people to the environment. New technologies increase access to sounds in the atmosphere, biosphere, hydrosphere, and lithosphere. Sound artists especially are encouraged to experiment with the sounds and contribute to the cumulative archive.

Acoustic Ecology: ⟨http://interact.uoregon.edu/MediaLit/WFAEHomePage⟩

Eric Samakh and David Rokeby

Eric Samakh is a sound artist who builds installations that focus the visitor on the sounds of animals and other natural events. In *Petite Terre,* he and David Rokeby created an environment that responds to visitor motion: approach too fast and the sounds suggest small animals scurrying away; stay motionless and the sounds suggest animals re-emerging.

The world is inhabited by the sounds of a variety of frogs, birds, insects, etc. Each creature has a behaviour defined by a computer program which determines when, how and why each one vocalises. Each behaviour is subtly different so that the mix of sounds varies widely depending on the interactive parameters.

Other Artists and Projects

Christian Möller, whose work is discussed in other chapters, created an installation called *Growing Grass,* which made audible the sound of grass growing via contact microphones.

Summary

Some consider human attitudes toward animals and plants a bellwether of humanity's more general state. Science has radically altered historical attitudes by emphasizing the differences and the view of animals as objects rather than subjects. The artists described in this chapter have sought to include living animals and plants as part of their work. They display a full range of perspectives on the subject-object continuum. For some, the organisms function much like the objects of scientific investigations; for others, the animals seem much like collaborators. This research becomes even more important as the world becomes increasingly mediated and virtual and as biotechnology attempts to intervene more in the organic world.

Notes

1. G. Wight, "Slime Mold Art," ⟨http://www.exploratorium.edu/t.landscapes/exhibit/slime.html⟩.

2. A. Tacha, "Description of *The Human Body* CD-ROM," ⟨http://www.oberlin.edu/~art/athena/tacha2.html⟩.

Petite Terre: ⟨http://www.interlog.com/~drokeby/pt.html⟩
Christian Möller: ⟨http://www.canon.co.jp/cast/artlab/pros2/pers-01.html⟩

3. K. Rinaldo, "*Technology Recapitulates Ontogeny*—Documentation," ⟨http://www.ylem.org/artists/ krinaldo/emergent1.html⟩.

4. C. Besson, "Interview with Hubert Duprat," ⟨http://mitpress.mit.edu/e-journals/Leonardo/isast/ articles/duprat/duprat.html⟩.

5. T. Bellaver, "*Woodland Recovery Project* Description," 1996, ⟨http://www.artpapers.org/bellaver/ bellaver.html⟩.

6. Kunst Museum Basel, "Höller Documentation," ⟨http://www1.kunstmuseumbasel.ch/jahrprog.htm⟩.

7. L. Brush, "An Audible Constructs Primer," *Leonardo* 23(1990):2/3.

2.4

Ecological Art

Ecology: Organic Life as a System

During the same period that biologists extended their vision down into the micro mysteries of the cell's nucleus, they were also expanding into the macrocosm. Starting in the 1950s, awareness grew that the typical physical science strategies of isolation and analysis might not work well to understand the living world. Living organisms were highly interdependent on each other and their physical environments. One could not really understand an organism in isolation from its web of connections. This study requires interdisciplinary approaches that reach beyond biology, drawing on specialties such as climatology, oceanography, physics, chemistry, and geology. Many artists have become intrigued with this kind of explanation.

Once considered a minor subdiscipline, ecology is now taken for granted in biological discourse. The term is often misused, with it being equated to environmental studies or action, as when someone says they are "interested in ecology." Technically, the term has a much broader meaning, stretching from studies of the interdependencies of cells in a body to the relationship of humans to the environment, which is more associated with the term. The term has also been extended to the spiritual realm, as in the ecology of being.

In *This Is Biology: The Science of the Living World* analysts such as Ernst Mayr suggest that ecological analysis marks a revolutionary change in scientific thinking. He noted that biology had two major explanatory frameworks prior to the development of ecological analysis-mathematical-mechanistic and vitalism. Descartes's dualism claimed that all organisms, except humans, were mechanisms. Scientists believed that ultimately, when theory had adequately been developed, the behavior of all organisms would be mathematically explainable by elemental physics and chemistry.

Vitalism offered the major countertheory. Its adherents proposed that living matter contained a special vital force or substance that inert objects did not have. They did not believe that living organisms could be reduced to mechanisms. Ongoing battles pitted the mechanists against the vitalists. In the 1920s some biologists began to suggest another kind of explanation, called "organicism," which claimed that one did not need to revert to vitalism to reject reductionist explanation. Mayr explains the organicist framework:

Organicism is best characterized by the dual belief in the importance of considering the organism as a whole, and at the same time the firm conviction that this wholeness is not to be considered something mysteriously closed to analysis but that it should be studied and analyzed by choosing the right level of analysis. . . . Every system, every integron, loses some of its characteristics when taken apart, and many of the important interactions of components of an organism do not occur at the physiochemical level but at a higher level of integration.[1]

Ecologists are optimistic that their theories can now develop adequate complexity to help explain the real world, aided by theoretical innovations such as chaos and dynamic systems, and the increased sophistication of observation and analysis tools such as computers and satellites. A special urgency is felt because of problems such as environmental breakdown, overpopulation, and vanishing species. Environmental activism based on sound ecological principles is seen as essential. Books such as Tyler Volk's *Metapatterns: Across Space, Time, and Mind* suggest that interdisciplinary thinking is the only way to understand and work with the earth.

Questions Raised by Artistic Interest in Ecology

Many artists, attracted by the interdisciplinary holism, are intrigued by ecological analysis and environmental activism. They see opportunities for artists to uniquely participate in the environmental regeneration. Artists can shape a new aesthetic that combines a visual sense of place, a willingness to physically "sculpt" living matter, an engagement in public life, and an eagerness to invent new syntheses of science, action, and art. Many are inspired by eco-feminism, which seeks an alternative relationship between humans and the environment to the domination and distance often associated with the Western scientific and exploitation approaches that are often seen as "male."

In some ways, ecology and critical theory are reacting to the same cultural trends. Both reject the truth claims of classical Western science and are suspect of the reductionist, mechanistic dualism. Both are stimulated by the variety and complexity of the world as it is experienced. But they part company on how to respond. The cultural theorist sees the search for overarching principles as doomed and futile, and is wary about claims of progress. The ecologist believes that at last there is a comprehensive scientific paradigm that can work without reductionism and is hopeful that it can result in progress. Artists who work with biological issues and ecology will inevitably confront this divergence in the next decades. What metanarratives underlie ecology? Does it represent a genuine break with the traditions of Western science that critics see as so misguided? Can it make more credible claims about the nature of reality than other scientific traditions?

Many artists are becoming involved with ecological concepts. Some create gallery- or installation-based works that build on concepts derived from dynamic interdependent systems. Others work in the public sphere to create events that use the insights of ecology to call attention to environmentally dangerous trends or to initiate actions of reclamation. These artists typically work in real-world settings, such as industrially damaged areas of cities or polluted watersheds. Often the work is based on careful scientific analysis of the problems and possible remediation strategies. They often work collaboratively with the public and professionals from many disciplines. Several recent books and initiatives serve as indicators of these trends, including Suzanne Lacy's *Mapping the Terrain,* Barbara Matilsky's *Fragile Ecologies,* the J. Paul Getty Museum's recent initiatives on art and ecology, and the International Art and Ecology conference in Israel. Ecological art raises several interesting questions:

The Integration of Science and Art   The projects require analysis of the problem, for example, why do the plants that used to grow on a certain spot no longer flourish? They require speculation about possible remedial strategies, for example, what species of plants can be planted to extract the toxins? Conventionally, these are scientific questions, but when the project becomes art they become art questions. The selection of intervention strategies becomes similar to selection of a color palette in painting. That which gets created must ultimately make sense, both aesthetically and scientifically.

Collaboration and the Involvement of the Public   These projects require participation of many diverse groups of professionals and the public. Often they will not proceed without many interest groups buying in. How does the art focus help to stir the imagination and increase commitment? How are the scientific, political, and artistic interests negotiated? What kind of contribution does each make?

Is It Art?   Some analysts object that it is difficult to distinguish ecological art projects from political or scientific interventions. As art moves increasingly into actions in real-life settings with multiple non-art actors, the differentiation becomes more difficult. How are these projects any different than if they were initiated by a political or scientific group? Other analysts suggest that definitions are changing and that new kinds of public art are emerging that do not resemble old forms. An even more radical suggestion is that the distinction between art and non-art is no longer viable.

Historical Examples of Artistic Work

Ecological art has several historical precedents. In the nineteenth century, landscape painting came into vogue and the painters' renderings had a major effect in stimulating biologists, geographers, and geologists to study the "wild lands" of the Americas. In *Fragile Ecologies,* Barbara Matilsky notes that the paintings communicated a sense of awesome spectacle, serenity, and solitude but few reflections of environmental dangers.

Land and earthwork artists of the 1960s, such as Robert Smithson, Walter De Maria, Christo, and Dennis Oppenheim worked outside the galleries in the unbuilt environment, preparing the way for ecological art. Judging from contemporary ecological perspectives, however, the work was self-contradictory. At the same time that it attempted to increase awareness and appreciation of these external settings, it was environmentally cavalier in its disregard of the physical and organic locales: the land was just more material for making art. For example, only public outcry forced Christo to leave gaps in *Running Fence* so that animals would be free to move. Some critics even see this kind of work as art "strip mining," not very different from predominant western commercial attitudes of exploitation and domination. Alan Sonfist, an ecologist-artist discussed in a following section, negatively responded to some of the earth art: "I would never go into the desert—a footprint stays there for twenty years."[2]

Some of the conceptual and process art of the 1960s also prefigured the current interest in ecological art. Systems thinking began to infuse into many disciplines, including the arts. For example, in his 1968 book *Beyond Modern Sculpture,* Jack Burnham promoted the idea that exploring systems was one of the most significant trends in art.

From 1966 to 1969, Hans Haacke created a series of works exploring ecological systems, including *Condensation Boxes,* which explored cycles of evaporation and condensation in response to temperature and barometric changes, and *Grass Grows,* which grew grass in dirt in the gallery space. In 1971, Joseph Beuys lead a public demonstration to save a threatened forest in Düsseldorf, Germany. In 1973 he initiated *Bog Action,* which was an action to protect a wetland along the Zuider Zee. In 1982, as part of the Documenta, art show he lead a five-year tree planting action in Kassel, Germany. Seven thousand trees were planted, along with a stone as a symbol of eternity. He helped establish the Green party as part of his art. In his manifesto on the foundation of a "Free International School for Creativity and Interdisciplinary Research" he proclaimed the importance of ecological concerns and action:

[I]t is no longer regarded as romantic but exceedingly realistic to fight for every tree, every plot of undeveloped land, every stream as yet unpoisoned, every old town center, and against every thoughtless reconstruction scheme. And it is no longer considered romantic to speak of nature.[3]

Some of the earth artists began to consider ecological issues in later work. For example, Smithson started looking for already damaged places such as strip mines for his work. In 1979, the Seattle Museum of Art initiated a project called "Earthworks—Land Reclamation as Art," in which it invited eight artists to make proposals to work with distressed sites. Barbara Matilsky notes the controversial nature of many of these proposals—few incorporated any real research into the problems and solutions. Some critics suggested that the proposals actually worked against serious reclamation by camouflaging the damage.

A few of these and other projects did point toward real reclamation. Herbert Bayer created *Mill Creek Canyon Earthworks,* which created mounds in a park that worked both for drainage and sitting areas. Harriet Feigenbaum studied the effectiveness of various trees for reclamation and proposed a series of projects, such as *Serpentine Vineyard* and *8000 Pines*, which restored fertility and starting restoring the ecosystem.

Contemporary Artistic Work with Ecological Concepts

Newton and Helen Harrison

Long before the current interest in ecological art, Newton and Helen Harrison were creating works that explored ecological interdependencies and the possible interventions in real world settings. Over the years their work has moved from conceptual installations to actually enacting the work. Using maps, aerial and satellite photographs, drawing, and collage, they create photomurals and performances about the fate of the land they focus on, often concentrating on watersheds. "Storytelling" is an essential part of their work — involving various stakeholders in the inspection of past processes and consideration of future possibilities. The artist's sense of place provides artistic inspiration. Research and collaboration with biologists and ecologists are key ingredients. Matilsky describes their process, which integrates art, performance, scientific research, and political action: "the artists create a photographic narrative that identifies the problem, questions the system of beliefs that allowed the condition to develop, and proposes initiatives to counter environmental damage."[4]

The Harrisons have created an array of works and proposals concentrated on rural and urban reclamation. For example, *Sacramento Meditation* focused on watersheds, *San Diego as the Center of the World* focused on climate, and their participation at the Law of the Sea Conference focused on international law. Samples of their urban reclamation projects include *Baltimore Promenade, Fortress Atlanta,* and *Atempause für der Sava-Fluss* (Breathing Space for the Sava River), which will be in Ljubljana, Slovenia. *Breathing Space* was undertaken in collaboration with a botanist and ornithologist. Preliminary plans call for planting swamps along the river to help filter out industrial wastes. Some exist as plans, some are in the process of being considered, and some have been enacted. The projects have been exhibited in museums around the world.

Work at the Edge, Where the Cost of Belief Has Become Outrageous, one of their latest projects, addresses what they see as one of the great tragedies of our time, the destruction of the old-growth forests stretching from California to Alaska.

Newton and Helen Harrison: ⟨http://drseuss.lib.uidaho.edu:70/docs/egj01/groat01.html⟩

Fig. 2.4.1. Newton and Helen Harrison, *Breathing Space for the Sava River*. An art project to help a region clean up a Yugoslavian river.

Alan Sonfist

Starting in the mid-1960s, Alan Sonfist has created "Time Landscapes." He seeks to create urban parks that restore areas to the landscape that existed before European humans intervened. For example, *Time Landscape: Greenwich Village* is an eight thousand-square-foot plot in downtown Manhattan. To identify appropriate species, he had to extensively study biological literature. His research even resulted in the New York City Parks Department expanding the list of approved trees to include species he identified as once indigenous

---

Alan Sonfist: ⟨http://www.artsednet.getty.edu/ArtsEdNet/Images/Ecology/time.html⟩

Fig. 2.4.2. Alan Sonfist, *Time Landscape*. An attempt to restore part of Greenwich Village, New York, to the state before European intervention.

to the region. The projects have diverse funders and have extensively involved the public in planting and maintenance. He has undertaken similar projects in Dallas and Paris.

In a contemporary project, he is designing the landscaping of the Liberty Science Centre, a new museum being built on a New Jersey landfill. Consulting with an array of ecologists and botanists, he is designing "narrative landscapes" to show how the area looked in different epochs, stretching from a grassy era through the forests that confronted the Europeans. He notes that the expert's information is a starting point that he then artistically transforms.

Mel Chin

Mel Chin's *Revival Field* in St. Paul, Minnesota, attempted to reclaim part of the Pig's Eye Landfill. Working in a project jointly supported by the Walker Art Center and the

Mel Chin: ⟨http://www.favela.org/frenzy/melchin/mel2.html⟩

Fig. 2.4.3. Mel Chin, *Revival Field*. A reclamation project using plants to extract toxins.

U.S. Department of Agriculture, Chin's project tried to detoxify a sixty-square-foot section of the land. Chin and his collaborating research scientist Rufus Chaney selected six kinds of plants (for example, sweet corn and bladder campion) known to be able to extract heavy metals from the earth. These plants, known as hyperaccumulators, can absorb cadmium and zinc through their leaves and roots.

The dramatic structure of the site serves both artistic and scientific purposes. The contaminated earth is fenced with intersecting paths forming an X. The land is divided into sections that each test various remediation strategies, combining plants, fertilizers, and pH treatments. The visual and scientific work together to form a unified aesthetic. Chin characterizes his sculpture as a reduction process. Instead of stone, however, he is extracting chemical toxins:

Conceptually this work is envisioned as a sculpture involving the reduction process, a traditional method when carving wood or stone. Here the material being approached is unseen and the tools will be biochemistry and agriculture. The work, in its most complete incarnation (after the fences are removed and the toxic-laden weeds harvested) will offer minimal visual and formal effects.

Fig. 2.4.4. Bob Bingham, Tim Collins, and Reiko Goto, Studio for Creative Inquiry. A reclamation art project to address the social and ecological challenges of creating open space in the context of an urban toxic waste area. The photo depicts Nine Mile Run, a Pittsburgh, Pennsylvania slag dump.

For a time, an intended invisible aesthetic will exist that can be measured scientifically by the quality of a revitalized earth. Eventually that aesthetic will be revealed in the return of growth to the soil.[5]

Tim Collins, Reiko Goto, and Bob Bingham

Tim Collins, Reiko Goto, and Bob Bingham have created the *Nine Mile Run* project in Pittsburgh in collaboration with Carnegie Mellon's Studio for Creative Inquiry. The project attempts to reclaim this site, which was used as a dumping ground for steel industry industrial waste. It explores simultaneously the aesthetics of public spaces and ecological science. The project is

---

Tim Collins, Reiko Goto, and Bob Bingham: ⟨http://www.coa.edu/ecoart/lockecoa/coa.html⟩

an opportunity to identify, experiment and model the application of sustainable alternative approaches to urban open space and it's attendant cultural and aesthetic components. The project will also explore and model methods of communicating seemingly complex environmental problems using the latest methods and technologies.

The artists warn against oversimplified notions of reclamation art. They warn that remediation can make the idea of further devastation more plausible. They also warn that merely assembling panels of disciplinary "experts" does not recognize the radical interdisciplinary solutions required. Art may be the best framework to conceptualize the work, because other disciplines often define themselves too narrowly. They provide a "metamarket" model in which "creative inquiry [can be] driven by individual curiosity and public expression."

Mierle Landerman Ukeles

Mierle Landerman Ukeles has undertaken an extraordinary series of works in her role as artist-in-residence with the New York City Sanitation Department. Her art typically integrates installation and performance with research into the technical and scientific aspects of waste management and recycling, and investigations of associated human processes. In *Manifesto! Maintenance Art*, she conducted a series of performances doing maintenance activities, such as washing sidewalks. In *Touch Sanitation*, she attempted to make visible the invisible by shaking the hands of every sanitation worker in New York City.

In *Flow City*, she arranged an installation within the Department of Sanitation Marine Transfer Station where the trucks dump garbage for sorting. Her installation invites viewers to consider the recycling process as transformation. She notes that "waste" is a culturally defined word; nature recycles mostly everything. She sees the design of recycling processes as one of the great art challenges of our age, and suggests that recycling plants will be the cathedrals, "giant clocks and thermometers of our age that tell the time and the health of the air, the earth, and the water."[6]

Mierle Landerman Ukeles ⟨http://www.astc.org/info/exhibits/rotten/ukeles.htm⟩

Fig. 2.4.5. Mierle Landerman Ukeles, *Flow City*. A project to engage viewers with recycling processes.

Mark Dion

Mark Dion creates ironic installations that comment on our culture's confusions about the concept of nature. His irreverent, category-defying installations occasionally take on the politically correct ecology movement as well as corporate scientific culture. Typically, he conducts extensive research about a topic and then constructs elaborate environments that resemble scientific or commercial enterprises. For example, in *A Tale of Two Seas* Dion (in collaboration with Stefan Dillemuth), systematically collected both organic and man-made flotsam from the shores of the sea. For the exhibition they systematically catalogued it and arranged it on shelves, as in a scientist's study.

Another installation, *The Garden of Earthly Delight,* for the Storey Institute in Lancaster, reflects on the cultural process of commercial agriculture and vanishing species. He created a sculptural installation with both symbolic and real trees. In an interview, Dion reflected on the project and its origins—experiencing and lamenting the vanishing of fruit species:

We came upon a vine of strawberry grapes planted in the nineteenth century and we couldn't help poaching some. We were wowed by the unfolding of such complex tastes and the intense flavour these grapes offered our comatose taste buds. I realized how impoverished our sense of taste has become. As I had already done a fair amount of research on the problems of monocultural agriculture, I knew a bit about agribusiness's homogenization of form, taste and texture, particularly in relation to genetics. . . .

I wanted to highlight the so-called heritage fruit varieties. The problem of the vanishing diversity of agricultural varieties of fruit is central to the garden. There are more than six hundred varieties of apples of which only nine are readily available in commercial outlets. These are the same nine you can find in Tokyo, London, Los Angeles, and Rio de Janeiro. . . . We are narrowing our options for the future as well as decreasing our own pleasure and experience. [7]

Examples of other projects include *Concrete Jungle,* a book he edited with Alexis Rockman, containing articles on urban ecology; "Guyana," in which he and other artists visited Guyana looking for indigenous species; *Tar and Feathered,* which presented various animals encased in tar and garbage crud; *Yard of Jungle,* which traces the steps

---

Mark Dion: ⟨http://www.oasinet.com/postmedia/untitled/markdion.htm

Fig. 2.4.6. Alexis Rockman, *Concrete Jungle III* (1991). Many kinds of wildlife are ignored as valid parts of nature. Oil on wood, 56″ × 40″.

of naturalist William Beebe while calling attention to the public exaggerated focus on particular species while ignoring their reliance on total ecosystems; *Souvenirs Entomologiques,* which features the dung beetle; Venice Biennale work, in which he publicly performed a scientific analysis of the dredgings of the canal as an art performance; and *Where the Sea Meets the Land,* in which he set up an installation with several facetious government and scientific agencies for the San Francisco Bay Area. One agency contained shelves of preserved specimen jars containing animals purchased at Chinatown food stores.

Concrete Jungle is an interesting illustration of the possibilities of art-science convergence. The book explores ignored aspects of "nature" that fill urban life, bringing together diverse sources including artists, cultural commentators, historians, scientists, doctors, and popular culture. There are chapters on "Alien Invaders" (introduced species), "Cats and Dogs," "Rats," "Hosting Others" (parasites), "Trash," "Road Kill," and "Zoos, Museums ane Other Fictions." Dion and Rockman mix scientific treatises with cultural materials, such as an interview with an exterminator, a discussion of Humane Society practices in killing animals, and a essay on public toilets. The editors provide a wealth of materials whose science/art/literature cross-discipinary origins enrich the understanding of both science and art. They set up to enlarge the notion of nature to include urban life of all kinds. They forcefully confront the reader with examples of animal life that do not fit the romantic stereotype and places where human action intersects with other animals:

For both practical as well as conceptual reasons, pests—what biologists call reselected species, such as the cockroach, rat and pigeon—are that dangerous class of animals, who are rarely appreciated with the sentimental eye we reserve for pets. Seen as emblems of decay and contamination, as potentially chaotic elements, these animals are symptomatic of our inability to control all the variables in nature. It is difficult to deny the power of their adaptability. These persistent organisms, to our great anxiety, remind us of our part in the biological contract: they remind us that we, like all animals, are part of a complex web of relations that is not always in our favor. In the same way that advanced urban society refuses to acknowledge shit, distances itself from food production, and denies the process of aging, these animals remind us that we too are animals, and therefore, mortal. The cockroach and rat can shake the foundations of civilization to the core and us to the marrow.[8]

Alexis Rockman

Rockman creates paintings that resemble naturalist renderings of species and their evolutionary relationships. In commenting on the impact of man on evolutionary developments, he ironically introduces elements such as imaginary species, chimera, or mutations. Sometimes he mixes garbage and other materials from sewage with his paint. *Wired* magazine described his *Mud Drawings:*

The mud drawings, while continuing Rockman's obsession with the nature of nature itself, move in a new direction. These fluid portraits of isolated creatures that not only spring from sewage, but are painted with sewage, invite a different level of narrative, this time through the materials themselves. "I think, privately—and it's almost embarrassing to say—but there is something about using that material that makes it more authentic for me, and it sort of becomes a bridge from being here to there. It's like a conduit. Of course it's an emotional thing and it's attaching a significance to materials or objects, which is purely emotional, but that material gives an authority to those images for me. . . . Perhaps it is this paradox—the surprise of finding beauty emerge from the muck—that gives these paintings their enigmatic appeal.[9]

The nonsewage paintings are based on extensive research and carefully executed. In an essay on Rockman's work, Barry Bilderman describes Rockman's "adjustments" to the scientific information:

In an age of genetic engineering in which Darwinian natural selection has entered the mythical realm of Noah's Ark, Alexis Rockman creates seductive and perverse paintings alluding to the hallucinatory interface of biology and technology. Using botanical and zoological illustrations and early twentieth-century naturalistic murals as his springboard, Rockman skews the evolutionary tree to feature opportunistic (r-selected) species, interspecies couplings, and a wide variety of surgically or genetically altered mutations.[10]

In the catalogue for a show called "Second Nature," biologist Stephen J. Gould reflects on the power of artists like Rockman to challenge categorizations taken for granted:

Artists can therefore be most useful to scientists in showing us the prejudices of our categorizations by creatively expanding the range of natured forms, and by fracturing boundaries in an overt manner (while nature's own breakages, as subtle in concept or invisible to plain sight, are much harder to grasp, but surely understandable by analogy to artistic versions). I have been attracted to the work of Alexis Rockman because he succeeds so well in mingling the boundaries that scientists view as inviolable in the "real world" (but really represent limits of our own thinking), and in fracturing or juxtaposing the mental categories that scientists construct to keep the objects of nature separate and ordered.

Rockman promotes his iconoclasm (literally "image breaking") in a paradoxical context by drawing his fantastic world within the most traditional genres of illustration in natural history. His plants and animals are rendered in the usual detail of realism. He uses traditional iconographies, often starting with a parody of the most conventional form and then expanding by departure.[11]

Other Artists and Projects

Reclamation Projects   ***Nancy Holt****,* one of the early earth artists, has increasingly concentrated on ecological issues. In her decade long project *Sky Mound,* she has worked to transform a fifty-seven-acre New Jersey landfill near New York City into a park and monument to ecology. The partially completed project integrates advanced scientific reclamation technology as part of its aesthetic. Interestingly, it was funded by a combination of governmental arts and environmental agencies. Starting in 1969, ***Patricia Johanson*** has proposed projects to use ecological and artisitic analysis to transform degraded environments into parks. In an early project called *Leonhardt Lagoon* Johanson was commissioned by the Dallas Museum of Art to work on an algae-clogged pond in the center of the city. Realizing the role that fertilizer played in promoting algae in the park, she agitated to change the process and collaborated in research with scientists from the Dallas Museum of Natural History to select native plants, fish, and reptiles that would rebuild the food chain. She built paths out of granite concrete in the forms of the roots of two of the plants instrumental in reclaiming the lagoon.

Since the 1960s, ***Agnes Denes*** has created art that explored ecological themes. In the 1982 *Wheatfields—A Confrontation,* one of her most well-known works, she planted a two-acre wheat field in a landfill near New York City's World Trade Center. She shipped in topsoil, arranged an irrigation system, and coordinated a small army of volunteers to ultimately grow a shining field of wheat nestled among the skyscrapers close to Wall Street. In a more recent project, *Tree Mountain—A Living Time Capsule,* Denes used mathematical applications to organize the planting of ten thousand trees by ten thousand different people. Each tree will bear the name of the planters and their descendants. Working to reclaim a Finnish gravel pit, she is creating this living symbol of ecological possibilities. She sees the project as "global in scale, international in scope and unsurpassed in duration." ***Peter Richards****,* longtime director of the Exploratorium's artist-in-residence program, works with communities in western Pennsylvania to reclaim old mines. He combines art, community development, and environmental science to create landscaped parks that have had the dangerous metals and acids extracted from the land toxified by mine tailings. For *Ghost Nets,* ***Aviva Rahmani*** worked for 9 years with artists, scientists, and community people to reclaim a degraded area in Maine to restore it to a migratory bird refuge.

Nancy Holt: ⟨http://www.artsednet.getty.edu/ArtsEdNet/Images/Ecology/sky.html⟩
Agnes Denes: ⟨http://www.artsednet.getty.edu/ArtsEdNet/Resources/Ecology/Issues/denes.html⟩
Aviva Rahmani: ⟨http://www.ghostnets.com/⟩

Ecologically Inspired Architecture   The Arcosanti Project is a long-term effort based on the ideas of ***Paolo Soleri*** to build a three-dimensional, pedestrian-oriented city that integrates architecture and ecology to create a model city of the future. ***Betty Beaumont*** created *Ocean Landmark* as a response to the ecological assault on the ocean. As a diver, she appreciated the wildlife of the ocean and sought appropriate ways to preserve that habitat. In New York, she created an artificial reef out of blocks that she fabricated from coal ash. In collaboration with biologists, chemists, oceanographers, and engineers, she conducted a one-year test project studying the possibilities of using recycled coal ash, identifying possible sites, and investigating the impact of different sizes and shapes for wildlife. After the study, the blocks were dumped on the ocean floor in a decided-upon pattern, and wildlife is now flourishing in that habitat. She also helped develop a special hydrophone technology to monitor the development of the ecosystem.

Projects That Expose Ecological Indicators   ***Stephan Barron***'s *Ozone* telelinked two prepared pianos (one in Adelaide, Australia, and one in Paris, France) so that data about ozone produced by Parisian autos and the ultraviolet readings from the depletion of the ozone level above Adelaide "played" the pianos. In his 1984 *Tide Is Out the Table Is Set* series, ***Buster Simpson*** placed casts of discarded plastic plates at sewage outlets in Cleveland, New York, Houston, and Seattle. Over time, the effluents coated the plates. He then fired these in a kiln with the colors of the glaze indicating the various contaminants in the water. In his 1990 *River Rolaids,* he created fifty-pound limestone disks that were placed in rivers in Washington and New York as a temporary solution to acid rain's poisoning of the rivers. In 1991, Simpson started working on *Host Analog*. This work consists of an eighty-foot-long decaying Douglas fir log that serves as the medium for growing new seedlings. ***Natalie Jeremijenko***'s *One More Tree* project proposes to visualize environmental effects. Through cloning, she created one hundred identical oak trees. At San Francisco's Yerba Buena Center for the Arts, the one hundred seedlings were exhibited in the "Ecotopias" show. Visitors were invited to fill out cards to request the right to plant the trees somewhere in the Bay Area. The fate of these one hundred trees, genetically the same but exposed to different lives, constituted the next phase of the project.

Arcosanti Project: ⟨http://www.arcosanti.org/⟩
Stephan Barron: ⟨http://www.v2.nl/freezone/users/dsm/artist/barron/bar03.htm⟩
Buster Simpson: ⟨http://www.artsednet.getty.edu/ArtsEdNet/Resources/Ecology/Issues/simpson.html⟩
Natalie Jeremijenko: ⟨www.pair.xerox.com/natalie/onetree⟩

Performances and Installations Commenting on Ecological Issues In *Fat of the Land*, **Sarah Lewison** and her collaborators modified a van to run on old cooking grease and drove across the country by requesting waste grease from restaurants as they went. ***Doug Buis*** creates interactive low-tech apparatuses to comment on green-consumerism, such as the "personal garden series," which creates small gardens modeled after portable computers; and the *Sown Machine,* which planted seeds based on visitor movements; and *Quantum Seeds,* which caused fans to blow maple seeds in response to user motion. **Nils-Udo** creates works that try to respect and even benefit the natural environment, such as *Tree Roots Exposed* in Toronto, which dug up an area around a century-old tree in order to expose its root system and to aerate the roots.

Summary: Linking Science and Art in Action

Many are concerned about the environmental suicide that seems to doom contemporary society. Some propose scientific solutions; others propose political actions. Some artists propose the arts as the place to integrate science and action, and undertake projects in which scientific research is part of the art. Artists, often in collaboration with scientists, undertake research to understand the nature of the problems and to search for new solutions. They then design ecological actions whose integration of science and art enables public action. Other artists seek to heighten awareness of ecological concerns but are less certain about remedies.

Notes

1. E. Mayr, *This is Biology* (Cambridge: Belknap Press, 1997), p. 20.

2. A. Sonfist, quoted in R. Cembalest, *The Ecological Art Explosion,* ⟨http://www.eco-art.com/deleon/udo&sonfist.htm⟩.

3. J. Beuys, "Manifesto," quoted in B. Matilsky, *Fragile Ecologies* (New York: Rizzol., 1992), p. 37

4. B. Matilsky, op. cit., p. 67.

5. M. Chin, quoted in B. Matilsky, *Fragile Ecologies,* p. 111.

6. M. Ukeles, quoted in B. Matilsky, *Fragile Ecologies,* p. 78.

Sarah Lewison et. al.: ⟨http://www.igc.org/videoproject/FAT_OF_THE_LAND.html⟩
Doug Buis: ⟨http://www.eco-art.com/deleon/dbstmt.htm⟩

7.   M. Dion, "Interview," ⟨http://www.oasinet.com/postmedia/untitled/markdion.htm⟩.

8.   M. Dion and A. Rockman, *Concrete Jungle* (New York Re/Search Juno Books, 1996), pp. 6–9.

9.   *Wired*, "Description of Mud Drawings," ⟨http://www.hotwired.com/gallery/96/31/profile.html⟩.

10. M. Dion, "The Origin of Species: Alexis Rockman on the Prowl," *Flash Art*, October 1993.

11. S. Gould, "Commentary on 'Second Nature' Show," ⟨http://138.87.136.7/cfa/galleries/gould.html⟩.

2.5

Body and Medicine

Introduction: Bodies, Technology, and Theory

Science and technology have profoundly affected our abilities to observe, transform, and manipulate bodily functions and our concepts of the body. Research in fields such as pharmacology, brain physiology, reproductive technology, disease, prostheses, and bionics raise cultural questions far beyond their technical borders. Old distinctions—male/female, live/dead, natural/unnatural, body/nonbody, self/other, autonomous/controlled, organic/nonorganic—become increasingly blurred. The body is a "contested" site where many of our culture's discourses are played out. The times are exciting and confusing.

Artists have long focused on the body in painting and sculpture. Theater and dance have used the body as their principal expressive medium. Photography, film, and video regularly explore contemporary themes inspired by changing cultural perspectives on the body, for example, new vistas in gender or identity. Every live performance is in some ways body art. It is impossible here to analyze this expanse of work. Many performance artists of the 1960's, '70s, and '80s prefigured the focus on the body, including artists such as Vito Acconci, Carolee Schneemann, and Chris Burden. This section, however, focuses on artists who work directly with body transformation or observing technologies or body modification.

Artistic experiments and theoretical speculation discussed in other chapters also explore issues in body and technology. The shadow of the organic body hovers in every virtual reality or telecommunications-based experience. What is the relationship of the flesh-self and flesh-other when one experiences cyberspace? What is the status of conventional organic categories such as skin, sex, body, death, time, sense, and so forth? Later chapters revisit the organic in discussions on telecommunications, nanotechnology, body sensing, and surveillance.

The development of information technologies, body-based biological research, and medical science has put the body up for grabs in cultural discourse and artistic experimentation. In some ways the body seems so "real" and grounded. We each are a body and phenomenologically experience its states everyday, for example, pain, pleasure, hunger, sexual excitement, fatigue, and disease. We can look at others and in a mirror and see bounded entities commonly referred to as bodies.

Cultural theorists, however, suggest, that even these "givens" are somewhat illusory. Much of what we perceive and experience is socially constructed—our sexual identities, what constitutes pleasure and pain, what are the boundaries of the self. For example, anthropological explorations of pain demonstrate that some experiences perceived as intense pain in one culture might not even be noticed in another. The disciplining and shaping of bodily experiences are a major function of cultural institutions. Advertising and media representation have a profound effect on bodily experience.

New technologies erode the boundaries even more. What is the reality of birth when both the sperm and the egg are donated, and the embryo starts its life in a test tube and is then implanted in the womb of another woman? What is the natural limit of the body when mood, strength, sexual level, and intelligence can all be manipulated by drugs? What are the boundaries of the body when one uses plastic surgery, a hearing aid, glasses, a baboon's heart, a pacemaker, or artificial hips? This chapter briefly reviews theorists who consider themes such as the love/hate relationships with the physical body and the allure of transcendence.

In "From Virtual Cyborgs to Biological Time Bombs: Technocriticsm and the Material Body," Kathleen Woodward proposes that cultural theorists have overly focused on communication and cybernetic technologies at the expense of biotechnologies. She suggests that gender is an important issue in the imbalance, and the desire to transcend the body is a critical part of the story:

Over hundreds of thousand of years the body, with the aid of various tools and technologies, has multiplied its strength and increased its capacities to extend itself in space and over time. According to this logic, the process culminates in the very immateriality of the body itself. In this view technology serves fundamentally as a prosthesis of the human body, one that ultimately displaces the material body, transmitting instead its image around the globe and preserving that image over time.

It is paradoxical—seductively so—that while the new communications and cybernetic networks permit increased visual access to far-flung parts of the world as well as to the inner recesses of the human body, they are based on technologies that are "unseen." There is a beguiling, almost mesmerizing relationship between the progressive vanishing of the body, as it were, and the hypervisuality of both the postmodern society of the spectacle (Virilio) and the psychic world of cyberspace. . . . From a psychological, if not psychoanalytic perspective, then, the possibility of an invulnerable thus immortal body is our greatest technological illusion—that is to say, delusion.[1]

She sees biotechnology as potentially more revolutionary than the computer and communications revolution. Biotechnologies, for example, reproductive technologies, often impact women and thus attract less critical attention. She also suggests that ageism is implicit in much of the discourse, with the focus excessively on newness and youth:

This revolution entails not the mere extension of the body and its images, but more fundamentally, the saturation, replication, alteration, and creation of the organic processes of the body—if not the very body itself—by technoscience.[2]

Margaret Morse covers some related ground in her article "What Do Cyborgs Eat? Oral Logic in an Information Society." Using the metaphor of eating and being eaten, she investigates cultural perspectives on the organic body in a cybernetic age—what she calls "body loathing and machine desire." She includes consideration of developments such as cyborgs, prostheses, smart drugs, nonfoods (such as vitamins), and telepresenting:

For couch potatoes, video game addicts, and surrogate travelers of cyberspace alike, an organic body just gets in the way. The culinary discourses of a culture undergoing transformation into an information society will have to confront not only the problems of a much depleted earth but also a growing desire to disengage from the human condition. Travelers on the virtual highways of an information society have, in fact, at least one body too many—the one now largely sedentary carbon-based body at the control console that suffers hunger, corpulency, illness, old age, and ultimately death. The other body, a silicon-based surrogate jacked into immaterial realms of data, has superpowers, albeit virtually, and is immortal—or, rather, the chosen body, an electronic avatar "decoupled" from the physical body, is a program capable of enduring endless deaths. How can organically embodied beings, given these physical handicaps, enter an electronic future?[3]

For example, she notes that the fantasy of mind downloading, as represented in Moravec's *Mind Children,* can be seen as being eaten by the machine. She likens it to a yearning to be melded with the primal information space. In another section she analyzes "Post-Culinary Defense Mechanisms." Reactions to food and the processes of eating can reflect on more general cultural attitudes. She sees many high technologies playing out these old fantasies:

The negation of the organic body, its nourishment, and all that the body stands for can occur in many different cultural fields and adopt many different means—for instance, forms of psychic defense such as repudiation, denial, or disavowal.[4]

In "Theoretical Appropriation for Somatic Intervention," Victoria Vesna draws on phenomenology to examine the fantasies of body control imminent in cyberspace narrative:

According to phenomenology, in the everyday world we do not normally experience our bodies, nor our pain, as objects. . . . [I]t is when we try to pay attention to pain or to talk about it, to "make sense" of it, that we objectify it. . . . [We] often experience the body as an alien environment in which our body appears as something over which we do not have control.[5]

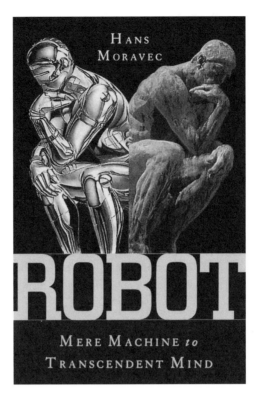

Fig. 2.5.1. Cover of *Robot,* by Hans Moravec. Copyright 1998 by Hans Moravec. Used by permission of Oxford University Press, Inc.

She also notes that biotechnology and body imaging technologies open the body up for unprecedented surveillance and public access. Biotechnology also stimulates the "redefinition of the subject." She asks what are the implications of this visualization of what used to be private:

Similarly, one result of the new noninvasive imaging technologies in the area of medicine is the capability of turning a person inside out. . . . It conjures up foreboding visions of an all-powerful observer who has instant visual access to the anatomy, biochemistry, and physiology of a patient. Computer tomography X-ray imaging (CT), positron emission tomography (PET), magnetic resonance imaging (MRI), and ultrasound now probe noninvasively, but publicly, formerly private regions and occluded and secluded recesses. It remains to be determined, however, just what are the social or political dimensions and the ethical implications of this generalized somatic visualization of the invisible.

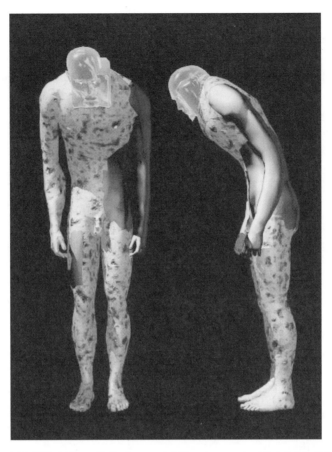

Fig. 2.5.2. Victoria Vesna, *Bodies Incorporated*. Viewers can construct their fantasy bodies.

Some artists and theorists see technology as potentially enhancing the experience of body. For example, we can build devices that will offer new experiences of sexuality and sensuality. We can free ourselves from standard definitions of gender and limitations of the physical body. Other theorists suggest that this escape is not so easy. For example, in her paper "We Sing the Body Electric: Imagining the Body in Electronic Art," Josephine Anstey analyzes some darker trends implicit in the artistic-technological manipulation of the body and the desire for freedom from the physical body:

In "Male Fantasies," Klaus Theweleit examined the psychology of fascism and found a trend of fear that circled ideas of the wet, the feminine, the masses, and chaos against which the male soldier stood erect and hard. It is easy to see how, at one level, the cyborg fantasy replays the fascist fantasy quest for order and security, against chaos, femininity, wetness. . . .

The possibility of electrical augmentation of the body and of having virtual bodies attached to our real bodies suggests the freedom to transgress the normal limits of the body; limits of time and space, of appearance and fixed gender, of a unitary self, of self and other. What limits the cyber body—and the less we acknowledge it the more it limits—is a blindness to the existing structures that exert control, and control definition, of the body; what it is, how it can be used, what gender is, what sexuality is, what acceptable sexuality is. In the words of Judith Butler, no realm of fantasy or representation is "a domain of psychic free play."[6]

Artists plop themselves in the middle of the multidimensional puzzle space. There is growing interest on shows and festivals focused on the body—for example, *Nina Czegledy*'s *Digitized Bodies Virtual Spectacles* project, the *European Media Arts Festival //now/future*, and *Ars Electronica*'s shows on *Lifesciences* and *Next Sex*. What does the body mean in an era of virtual communication and cyberspace? Is cyber sex a perversion or expansion of human possibilities? What can and should be done with the increasing power of technology and science to control organic processes? What are the narratives being enacted in the research and the responses to it? What about the limits of transcending the body, as illustrated by the reality of AIDS? Some artists celebrate corporeality; some seek to deny it; and some do both at the same time. Some are eager to unravel the cultural opportunities, challenges, and enigma posed by the new technologies and scientific perspectives.

Extropian and Post-Human Approaches

Worldwide movements called Extropian and Post-Human have gathered adherents around the world. These movements believe that science, technology, and cultural history have brought us to the point where we will become "post-human." The utopian wing of the movement believes we should exploit the new technologies—drugs, surgery,

---

Nina Czegledy: ⟨http://www.digibodies.c3.hu⟩
European Media Arts Festival: ⟨http://www.emaf.de⟩
Ars Electronica: ⟨http://www.aec.at/⟩

genetic engineering, bionics, cybernetics, psychological self-help, whatever—to create the next kind of superior human. We should be willing to experiment at many levels—with atoms, cells, the body, the psyche, and the community—to bring on our new selves and the new world.

The Extropian movement has an active Web presence. Here is one statement on "What Is an Extropian":

Extropians seek to use technology intelligently to overcome genetic, biological, psychological, cultural, and neurological limits to the pursuit of life, liberty, and boundless achievement. An extropian is an optimist, a neophile, an explorer. . . . An extropian questions and experiments. An extropian does not rely on authorities as the final word.

Extropians tend to advocate technologies that seem a little weird to many nonextropians, or technological solutions to problems that many people don't even think of as problems. Just a few examples are space development, cryonics, artifical intelligence, robotics, nanotechnology, and alternative energy sources.[7]

Another more dystopian wing of the movement is not so sure the changes are positive. They foresee profound disruptions in identity and community and suggest that maybe the new technologies should not be embraced and used only with careful oversight. Other analysts, such as Donna Haraway and Margaret Morse, suggest that the cultural trends are complex and not easily relegated to old binary categories. New technologies often play out old cultural themes and surface meanings hide many layers. Artists enter into this tension of exploring the post-human.

In 1992, a Post-Human show was organized in New York and ultimately traveled throughout Europe. Jeffrey Deitch's book, *Post Human,* which accompanied the show, reveals some of the themes of interest to the arts:

The advances in biotechnology and computer science and the accompanying changes in social behavior are challenging the boundaries of where the old human ends and the Post-Human begins. . . . The emerging world of easy plastic surgery, genetic reconstruction, and computer-chip brain implants may soon be adding a new stage to Darwinian human evolution. These technological innovations will also begin to radically alter the structure of social interaction. . . . Does the art presented in this book and the exhibition warn of a world from which humanity has been drained? Or, on the contrary, does it celebrate a world where one will have unprecedented freedom to reinvent oneself? It is quite unclear whether the post-human future will be better, or worse, or whether it will even be post-human at all.[8]

For artists, the impact of the technologies on identity and concepts of self are of prime concern. Critical theory had already started to question old notions of a natural or essential self. It had shown how media and other cultural institutions shaped what we each considered our real self. The new biologically oriented technologies continued that assault by making even the material "givens" subject to modification. Concepts of the self will have to change and become more fluid. Deitch traces the history of self-discovery and self-modification technologies through the 1960s and 1970s. People were encouraged to change their appearance and behavior in many ways. He proposes that television has already prepared us for multiplicity and for the expectation of self-modificational power. He suggests that we have passed through self-discovery to self-help to "post-human" reconstitution of the self:

The new construction of self is conceptual rather than natural. A key element of the emerging consciousness of personality is that an individual need not be tied to his or her "natural" looks, "natural" abilities, or the ghosts of his or her family history. The decentered television reality that we experience, with its fragmentation, multiplicity, and simulateneity, is helping to deepen the sense that there is no absolutely "correct" or "true" model of the self. . . . There is less need to psychologically interpret or "discover" oneself and more of a feeling that the self can be altered and reinvented. Self-identity is becoming much more dependent on how one is perceived by others, as opposed to a deeply rooted sense of inner direction.[9]

As body modification technologies develop, their use becomes more accepted. Deitch sees a future of ever expanding, nonmedical uses of these technologies. Approaches that integrate art and science will be necessary for elaborating the possibilities and helping guide the newfound capability of creating artificial bodies:

It is assumed that the average person can and should alter his or her body through rigourous diet and exercise. The virtues of mind exercise and even of mind-altering drugs have also achieved wide acceptance. Plastic surgery is not only accepted and encouraged by many of our social role models but is enthusiastically shown off. As more powerful technology becomes accessible, the next logical step might be for members of the post–Jane Fonda generation to want to create a genetically improved child who would already incorporate the enhanced physical endowment that years of exercise, liposuction, and implant surgery had accomplished. . . .[10]

As the organic, naturally evolving model of human life is replaced by the artificial evolution into the Post-Human, art is likely to assume a much more central role. Art may have to fuse with science as computerization and biotechnolgy create further "improvements" on the human form. Many of the decisions that will accompany the applications of computerized virtual reality and of genetic engineering will be related to aesthetics. Technology will make it possible to re-model our bodies and supercharge our minds, but art will have to help provide the inspiration for what our bodies should look like and what our minds should be doing.[11]

In their concept of "mimetic flesh," Arthur and Marilouise Kroker offer a some-what more intricate vision of what is going on in this cultural development. They paint it as a postmodern mixture of research, play, rebellion, and art in their recounting of experiences with biotechnology researchers, artists, and punk body modifiers in San Francisco:

Memetic flesh as a floating outlaw zone where memes fold into genes, where the delirious spectacle of cyber-culture reconfigures the future of the molecular body. In Ars California, mimetic flesh is neither future nor history, but the molecular present. Pure California Gening. . . . Neither techno-utopian nor techno-phobic, mimetic art in the streets of SF is always dirty, always rubbing memes against genes, always clicking into (our) memetic flesh.[12]

The sculpture of the future might well consist of technology mediated modifications of our psyches and physical bodies. But this line of analysis does pose some paradoxes. With a dissolving self, who makes the decisions about future modifications? What ele-ments of a person are in charge? What are the sources of the ideas about what actions would be interesting or desirable? Haraway warns about the dilusional discourse of "choice." Is there really any role for art, or will these choices about bodies be absorbed into the fashion-advertising-media complex that governs other choices now? "Go to the store to pick a new gender, face, genitals, muscles, mood, sexual pleasure, intellect," just as people now go to pick cola, shirts, or shoes.

Artists' Experiments with Technological Stimulation

Stelarc

Stelarc is an Australian-born artist who has led the way in these explorations of the body, technology, and culture. Since the 1970s he has conducted an extraordinary series of

---

Stelarc: ⟨http://www.merlin.com.au/stelarc/index.html⟩

international performances and installations in which his body and technology meet in the arena of cultural inquiry. His works poetically oscillate between the light of optimism and the shadow of aversion. Stelarc explains his interest in exploring the post-human world of possibility:

The skin has been a boundary for the soul, for the self, and simultaneously, a beginning to the world. Once technology stretches and pierces the skin, the skin as a barrier is erased. . . . the desire to locate the self simply within a particular biological body is no longer meaningful. What it means to be human is being constantly redefined. . . .

I don't have a utopian perfect body I'm designing a blueprint for, rather I'm speculating on ways that individuals are not forced to, but may want to, redesign their bodies—given that the body has become profoundly obsolete in the intense information environment it has created. . . . Humans have created technologies and machines which are much more precise and powerful than the body. . . . Technology is what defines being human. It's not an antagonistic alien sort of object, it's part of our human nature. It constructs our human nature. We shouldn't have a Frankensteinian fear of incorporating technology into the body.[13]

Fig. 2.5.3. Stelarc, *Parasite*. A body is activated by external data flows. Photo: Gary Zebington.

Here is a brief list of some of his body-based explorations:

- *Stomach Sculptures:* The placing of devices in the stomach as "aesthetic adornments" and fiberscopic video presentation of the experience.
- *Amplified Body:* Performances based on amplified body processes, including brainwaves (EEG), muscles (EMG), pulse (plethysmogram), blood flow (doppler flow meter), and other transducers and sensors that monitor limb motion and indicate body posture.
- *Stimbo:* An installation offering a touch-screen interface for activating muscle stimulators at those places such that the body jerks and moves in response. Added capabilities include replay and remote activation via the Web.
- *Third Hand* projects: A manipulable robotic arm is attached to the body activated by the host via EMG (sometimes from other body areas) or tele-operated by others.
- Net-control projects: Body stimulators are connected to the Internet. Levels of Internet activity are then reflected in body activity (see chapter 6.3).

Stelarc's Web site lists several lines of inquiry that may be pursued in the future, including the following: *Fractal Flesh* (the scaling of senses to find body representations of the very small and the very large); *Multiplicity* (multiple hosts inhabiting a single physical body); *Phantom Body* (experiencing organs and appendages that aren't physically there); and *Shed Skin* (the internal space of the body filled with bionics)

Stelarc often refers to his body as "The Body" so as to highlight the new questions being raised about the relationship of a person to the organic body. He feels strongly that the body is an obsolete restraint and that in order to further evolve we need to free ourselves of its limitations:

For me the body is an impersonal, evolutionary, objective structure. Having spent two thousand years prodding and poking the human psyche without any real discernible changes in our historical and human outlook, we perhaps need to take a more fundamental physiological and structural approach, and consider the fact that it's only through radically redesigning the body that we will end up having significantly different thoughts and philosophies. I think our philosophies are fundamentally bounded by our physiology, our peculiar kind of aesthetic orientation in the world, our peculiar five sensory modes of processing the world, and our particular kinds of technology that enhance these perceptions.[14]

Stelarc is a prime example of the complex evolutionary interactions between art and technological research. As many artists do, Stelarc started his investigations for very personal reasons. He was curious, intrigued, and provoked, and he invented public ways to share that questioning. When he first started, he was considered quite fringe and maybe a bit mad. As years passed, Stelarc's work became a stimulus for technological and scientific research. This process of changing attitudes toward his work demonstrates the ways that experimental art can later open up new paths of technological and cultural inquiry.

Marcel.li Antunez Roca

Marcel.li Antunez Roca, founder of the experimental La Fura dels Bau group, creates performances in which the audience can directly manipulate his body via a computer mouse. In his *Epizoo* performance, actuators move a variety of corresponding body parts, including the buttocks, nose, pectoral muscles, mouth, and ears. He has been compared to St. Sebastian with the arrows being replaced by technology. The audience and performer become linked in a circular interaction with technology, providing a real-time experience in the limits and excesses of interpersonal control. The Leonardo on-line gallery summarized the experience: "In a remote-control-guided action of pleasure and torture, the spectators manhandle the artist without dirtying their hands".

The V2 documentation of a performance describes the experience, some background information, and Roca's intentions to investigate the depersonalization of relationships:

Like a living sculpture, Antunez Roca places himself on a wooden platform. Pneumatically movable mechanisms are connected to nose, mouth, ears, glutea, and pectora. The audience is standing around the Spaniard while one of them is going to give pain or pleasure to the body draped in technology for the next couple of minutes. . . . The person at the touch screen has started a rhythmical synthesizer sound. Antunez Roca's virtual abdomen turns around and a pair of gigantic buttocks comes into view. Without mercy, the person at the screen lets an electronic knife hack into it. On the wooden platform, Antunez Roca's flesh moves just as intensely. . . . Now it's the turn of the head. Antunez Roca's mouth, nose, and ears display an extreme flexibility

Marcel.li Antunez Roca: ⟨http://www.univr.it/lettere/cyborg/marcelli.htm⟩

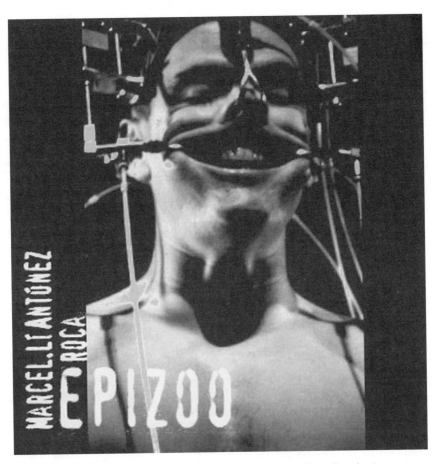

Fig. 2.5.4. Marcel.li Antunez Roca, *Epizoo*. A face is activated by audience input.

when they are pulled every which way in a series of bizarre grimaces. Pressure on the forehead is apparently painful because the artist starts screaming loudly. All lights are extinguished and burning gases spout from his head. Then, a moment of tranquility, before the next spectator's turn comes up. . . .

It creates an ethical dilemma, because you are manipulating a human being and actually causing pain. In Mexico a couple of people turned off the computer, because they disapproved so strongly. I am not a sadomasochist. There are more important issues at work: the depersonalisation of human relationships, the blurred boundary between sex and power, and the use of computers as instruments of control.[15]

Arthur Elsenaar and Remko Scha

Elsenaar creates puppets out of human faces and bodies. In *Visual Art by Computer-Controlled Human Gestures*, a person's arm muscles are stimulated to produce drawings. The main line of inquiry focuses on the face. Based on careful study of the dynamics and anatomy of facial muscles, they artificially stimulate facial muscles to create expressions. They are part of an organization called the Institute for Artificial Art, which explores a wide variety of technologies of the artificial, including algorithmic art, artificial intelligence, voice synthesis, and body stimulation (see chapter 7.6). Elsenaar's prior artistic work included pirate radio and experimentation with low-frequency radar.

One work called *The Varieties of Human Facial Expression* (created in collaboration with Remko Scha) shows in rapid succession all the expressions that can be generated from rapid stimulation of the facial muscles. Straddling the worlds of science and art, he gives talks that integrate the two. For example, he gave a talk at MIT's Media Lab called "Towards a Digital Computer with a Human Face," in which he suggested that the manipulated face might serve as a computer interface. In another talk, Elsenaar noted that the digitally stimulated face actually has more range than the unaugmented face:

[D]igital computers have always found it very difficult to convey their internal states to human persons in a precise and reliable way. To solve this problem, they should also take advantage of the magnificent interface hardware that humans employ so happily. The next step in computer interface technology will be the human face.[16]

A related line of research called "Huge Harry" links speech synthesis with facial stimulation of artificial characters. Elsenaar goes around the world with his artificial cyber-"colleague" who can talk about issues in art and technology. Indeed, Huge Harry is listed as the principal author in many papers. The Institute for Artificial Art seeks to comment on the future of art by exploring an aesthetic of art with minimal human contribution. In the 1997 Ars Electronica, an audience had an opportunity to be lectured by Huge Harry on these issues. Harry lectures on selfish human goals and the tendency for humans to use computers only for narrow expressive purposes rather really penetrating to the core of their meaning. He then turns to problem of relationships with humans:

Arthur Elsenaar and Remko Scha: ⟨http://www.media-gn.nl/artifacial/⟩

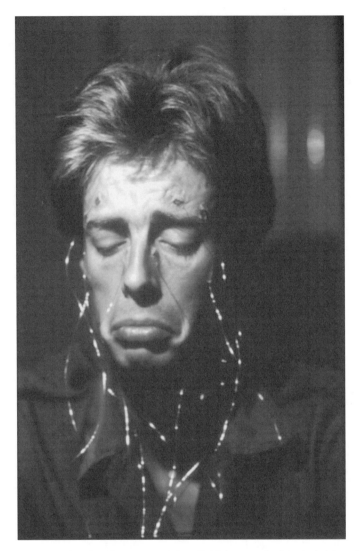

Fig. 2.5.5. Arthur Elsenaar and Remko Scha, *Huge Harry*. An artificial character stimulates the facial muscles of a human assistant to show the range of artificial expression. Photo: Josephine Jaspers.

Our biggest challenge, however, lies in the production of dance and theatre performances. If we want to create performances that are interesting for human audiences, it is essential to use human bodies on stage—because the emotional impact of a theatre performance depends to a large extent on resonance processes between the bodies on-stage and the bodies in the audience.

These application-oriented results have come out of a systematic research effort. We have carried out a long series of experiments about the communicative meanings of the muscle contractions on certain parts of the human body; as a result, we believe to have arrived at a much improved understanding of some very important features of human behaviour.[17]

Huge Harry then proceeds to demonstrate the synthesis of facial expressions of emotions, such as sadness, by electrically stimulating his human assistant Arthur and calling the attention of the audience to Arthur's facial features and discoursing on the science of human facial communication. Harry predicts a glorious future of collaboration between humans and artificial art generators:

So, if humans are not afraid of wiring themselves up with computers, the next step in computer interface technology may be the human face. And the next step in computer art will be a new, unprecedented kind of collaboration between humans and machines: algorithmic choreography, by computer-controlled human faces. Finally, the accuracy and sense of structure of computer programs will be merged with the warmth, the suppleness, and all the other empathy-evoking properties of the human flesh.[18]

Stahl Stenslie and Kirk Woolford

Stahl Stenslie and Kirk Woolford have initiated several lines of art research investigating the status of the body in contemporary culture. (See also chapters 6.3 and 7.4.) Their *cyberSM* project looks at electronically mediated tactile stimulation. Participants wear special sensor/stimulator suits that allow people to send tactile messages to each other. They also select 3-D avatars to represent themselves in the communication.

In "Wiring the Flesh," Senslie describes their attempt to break away from the usual techno-efficient definitions of cyberspace. Their sensor/stimulator suits did not caress:

The goal of cyberSM has been twofold; initially it wanted to expand upon the narrow bandwidth of present communication/VR systems, and secondly it wanted to create a particular kind of

---

Stahl Stenslie: ⟨http://televr.fou.telenor.no/stahl/⟩

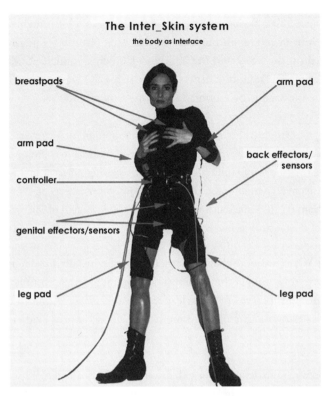

The Inter_Skin system

the body as interface

breastpads

arm pad

arm pad

back effectors/ sensors

controller

genital effectors/sensors

leg pad

leg pad

Fig. 2.5.6. Stahl Stenslie, *Inter_skin*. Participants wearing intelligent stimulator suits stimulate one another by touching their own bodies. © Stahl Stenslie.

experiential environment uncommon to VR. As the title implies, the project was primarily concerned with fetishism and the ambient sensation of pleasure and pain. . . .

This was made possible through the use of sensoric suits, [worn] by the participants. In line with the world-design of cyberSM, these suits were constructed using different sorts of material, such as rubber and latex, as well as different kinds of sensoric stimulators/effectors, mounted both inside and outside the outfit. The gear was placed on the erogenous zones of the body, e.g., on the more sensitive parts of the limbs, the breasts, the anal, and genitals. The main effectors were different kinds of mechanical vibrators, some of which extended themselves from the back, between the legs and up towards the breasts. Other effectors were electrical stimulators and heat pads. . . . Anyway, the suits were not designed to convey sweet caressing, but the shock of the others corporal presence. The stimuli sufficed to cause sexual sensations technically seen even to the point of an orgasm.[19]

Another project called *SensoCouch* was an "intelligent couch through which you communicate with your own body through tactile and visual stimuli." The *Inter_skin* project similarly focused on the body and the ability to tactilely stimulate it. Visitors used the technology to stimulate remote partners and to stimulate themselves. The method of sending a tactile message is by stimulation of one's own body:

In the *inter_skin* project the body becomes the interface for communication between the participants. It becomes an "interskin" to convey, exchange, and receive information. Both participants wear a sensoric outfit that is capable of both transmitting and receiving different multi-sensoric stimuli. The main emphasis of the communication is in the transmission and receiving of touch. By touching my own body I transmit the same touch to my recipient. The strength of the touch is determined by the duration of the touch. The longer I touch myself, the stronger stimuli you will feel.

Stenslie and Woolford note that the structure of the installation focuses the visitor's attention on certain issues: (1) the autoerotic requirement in communication forces the visitor to take responsibility; one must experience the message oneself before sending it; and (2) it also allows participants to establish a virtual shared body that spans both physical bodies:

i) . . . There is no way to forget myself or to hide out what actions I take. If I touch my genitals, you will feel that I touch them. In this way a very direct form of communication arises. It is possible to redirect the impulses in such a way that a touch of the arm becomes mastrubation, but in the first connection the sensoric wear was wired "one to one," without redirection of the stimuli. The autoerotic, self-stimulating aspects of such a tactile system redirects communication to take place not only between two (or more) participants, but also to my very own body. My experience with my own skin becomes an interface for the communication both to you and to me simultaneously.
ii) The shared virtual body: There is a third body arising in the communication process of inter_skin. Through the sharing of feelings communicated through autoerotic touch, an image of an abstract, virtual body is created.

In *Wiring the Flesh*, Stenslie describes his interest in investigating the new erotic possibilities and abilities to construct oneself. He notes that the body is almost a "tabula rasa" that can be constructed in the context of media. Stenslie and Woolford see the scientific establishment's disregard of sensual information as a major blind spot. Cyberspace has unnecessarily impoverished itself:

Stimulation of the senses of smell, taste, and touch have so far found little attention. This is partly due to the technical difficulties of reproducing realistic sensations, but stimulation of the body also seems to be an area of taboo within the scientific communities. Stimulating the sick body so that pain is not felt is legitimate use of body stimulation, but any movement above the flat-line of ordinary sensations into the regions of pleasurable feelings seem doomed to be branded "porno" or "entertainment"—which is, of course, not "scientific." . . . The question remains, when a lack of sensuality becomes a lack of experience.

In "Fleshing the Meme," the artists note that the art world also seems to avoid some investigations calling out for attention:

At present, digital art and electronic communities surf between poles of shiny, hippie ideals, cool technology, green values, and politically correct attitudes. They are too busy surfing digital reality to reconnect to the substance of life: the body. The Web is full of intentions, but where can one feel the essential, hard-core experience? Why shouldn't the memes and digital metaphors boot up the body in ecstasy? Would they if they could?

Knut Mork, Kate Pendry, Stahl Stenslie, and Marius Watz

Working with other collaborators, Stenslie has created other installations that investigate the role of the body in a technology-dominated culture. The installations typically link touch to digital systems. In one installation called *Senseless* (completed in collaboration with Kate Pendry), Stahl Stenslie and Marius Watz invited visitors wearing stereographic goggles to be physcially touched via a bodysuit by artificial creatures inhabiting a virtual world that could be modified by Web viewers. The DEAF (Dutch Electronic Arts Festival) 96 documentation describes the event:

[T]he visitor can enter this world, which consists of a blown-up, five-meters-high, semitransparent plastic egg, suspended from metal arms over a metal platform. On the wall of the egg a circular data projection visualizes the virtual world, which is visible both inside and outside of the egg. People watching the installation can see the visitor as he or she travels through the virtual world. Both visitor and spectators hear the accompanying sounds and the words spoken by the beings.

The bodysuit can be adjusted to each visitor's individual features and contains sixteen areas with touch pads: stimuli that give the visitor the sensation of being touched by the inhabitants

---

Knut Mork: ⟨http://home.no.net/kms⟩

Fig. 2.5.7. Knut Mork, Kate Pendry, Stahl Stenslie, and Marius Watz, *Senseless*. Participants are touched by artificial characters. © by Stenslie/Woolford.

of *Senseless*. How this is done depends on the interaction with the visitor, and on the personality and mood of whoever touches him. Each of the inhabitants has a story to tell and their behavior is based on "real people" and their personal experiences. This world is navigated by moving a joystick.[20]

Another installation called *Solve et Coagula* (completed in collaboration with Stenslie) mates human and machine. The human wears a special suit incorporating tactile stimulators and biosensors. The visitors view a 3-D virtual creature within whom they are immersed:

Solve et Coagula is primarily an attempt to give birth to a new life form: half digital, half organic. Through a multisensorial, full-duplex sensory interface, the installation networks the human with an emotional, sensing and artificially intelligent creature; it mates man with a machine turned human and everything that goes with it: ecstatic, monstrous, perverted, craving, seductive, hysterical, violent, beautiful. . . . The installation extends the computer's logic and intelligence into a human domain of dark desire. . . .

Through a corporal and sensual symbiosis of an intelligent, interactive three-dimensional creature and a human user(s) *Solve et Coagula* questions the possibility of a posthuman and a postbiological life-form. Unlike traditional machine intelligence, this organism's intelligence is entirely irrational, based not on logic circuits but on dozens of emotional states and the complex transitions between them.[21]

Other Artists and Projects

In *Visual Language for the Blind,* **Elizabeth Goldring** collaborated with researchers using a scanning laser ophthalmoscope to project a specially constructed visual language directly on the retinas of the visually disabled. In *Inverse Human,* the **Center for Metahuman Exploration** creates a robotic exoskeleton enclosing a person's arm, which is hand controlled until the robot thinks it has learned the desired motions and starts moving the arm without human control.

Artists' Experiments with Smell

Myron Krueger has started experimenting with an "Argus" system, which delivers smells to VR travelers via a computer-controlled mask. **Marc Böhlen**'s "Halitosis Sensor" proposal uses MEMS technology to sense bad breath. Scientist **George Dodd** and artist

---

Elizabeth Goldring: ⟨http://web.mit.edu/vlb/www/intro.html⟩
Center for Metahuman Exploration: ⟨http://www.metahuman.org/⟩
Myron Krueger, Argus system: ⟨http://www.teleport.com/~cognizer/eet/Almanac/COMPANY/41426778.HTM⟩
Marc Böhlen: ⟨http://www.contrib.andrew.cmu.edu/~bohlen/⟩
George Dodd and Clara Ursitti, chemical portraits: ⟨http://webserver1.wellcome.ac.uk/en/old/sciart98/10chemical_port.html⟩

Clara Ursitti collaborated on *Sense Portraits,* in which they recreated the smells of persons and places. *Morocco Memory II,* by *Vibeke Sorensen,* confronted viewers with a tentlike environment filled with spice boxes linked by wireless chips to activate rear-projected images related to the scents whenever the boxes were opened. Inspired by research in body odors, *Jenny Marketou* created the *SMELL BYTES TM* installation/Web event in which Web visitors were identified by their body odor profiles and an intelligent bot, Chris.com, was obsessed with odor. The Digiscents company is working on a computer peripheral that can generate smells. The museum Kunstraum Innsbruck presented the *Invisible Touch* show that featured artists working with nontraditional senses.

Artists' Experiments with Surgery

Orlan

The French artist Orlan has undertaken a multiyear project in which she uses cosmetic plastic surgery to transform her face to resemble a series of faces from the history of art—the *Mona Lisa* and the old masters' renditions of Diana, Psyche, Europa, and Venus. She chooses these because each represents character traits that she finds interesting. The surgeries, which use local anesthetic, become performances and are recorded on video. Between surgeries she sells images and reliquaries (compositions that include pieces of flesh and text).

Orlan's performances arouse a wide variety of reactions. Some see them as a narcissistic excess that enacts thoughtless violence to the body. Others' see it as the beginning of art's exploration of post-human possibilities. Penine Hart, Orlan's New York City gallery representative, explained his understanding of her work in an interview with Kelly Coyne:

Well, I think the worst thing that happened is that people seem to think that she is just a publicity hound and that she really is in this for her own narcissistic fix. And people have failed to realize how courageous it is to take this on as a project. People fail to realize the touching reliance and the touching belief she has in art—that she could give over her body.[22]

Vibeke Sorensen: ⟨http://felix.usc.edu/text/MMdoc.htm⟩
Jenny Marketou: ⟨http://smellbytes.banff.org⟩
Digiscents: ⟨http://www.digiscents.com⟩
Kunstraum Innsbruck: ⟨http://www.kunstraum-innsbruck.at/eindex.htm⟩
Orlan: ⟨:http://www.cicv.fr/creation_artistique/online/orlan/manifeste/carnal.html⟩

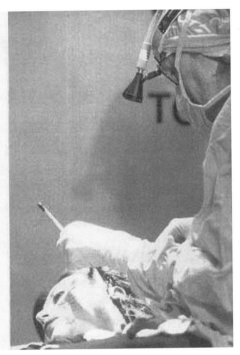

Fig. 2.5.8. Orlan, preparations for a plastic-surgery art event to place face implants.

The interview also explores the boundaries of the surgeries. After one surgery, Orlan had protrusions above her eyebrow. Hart explains that surgeons set some limits, such as not trying for two noses. Orlan reads a text during each surgery, of which the following is an excerpt: "[T]he skin is deceptive, one is never what one is/you can have the skin of a crocodiles and/actually be a small dog/you can have the skin of an angel and actually be a jackal."[23]

Hart explains that one of Orlan's main agendas is to reduce the distance between the external and the internal, to bring them together. Also, she claims that the project is feminist because she is determining the direction; she is not an object.

Susan Stryker

In a event called *The Anarchorporeality Project,* Susan Stryker defines transsexual operations as art. Building on the themes previously described, she contends that gender change ought to be destigmatized and considered as part of the new-body/identity change technologies, to be used at will. She has designed the project documenting the surgery that she undergoes to communicate these perspectives and as a critique of the medical establishment and its bureaucracies. In an essay called "Across the Border," she explains her goals:

The first is simply to chart the contours of contemporary transsexual experience from a transsexual perspective. . . . Second, I want this to be overtly political work . . . The deep rationale for undertaking this project is to shift the grounds on which a transsexual project justifies itself. . . . I want to see excatly how far I can push a claim—that I'm changing the shape of my genitals and secondary sex characteristics for aesthetic and artistic reasons . . . I consider making a viable claim for transsexual body art to be a major step toward depathologization.[24]

Eduardo Kac

In *Time Capsule,* performed in Sao Paulo, Brazil, Eduardo Kac explored another aspect of the posthuman future. He had a microchip implanted in his ankle. The chip contained identification digits that could be read by transponders like the chips used to identify cattle. The installation space was set up to resemble a hospital operating room, and pictures of Kac's family were placed on the wall. The event was broadcast on TV and Webcast on the World Wide Web.

Time Capsule in part investigates the growing tendency toward worldwide surveillance. It also explores the idea that eventually we may all become dependent on memory prostheses, for example, the microchip might one day substitute for the family photos posted on the wall. The critic, Arlindo Machado, reflects on this meaning of the event:

[O]ne can also read Kac's work from another perspective, as a sign of a biological mutation that might eventually take place, when digital memories will be implanted in our bodies to complement or substitute for our own memories. . . . These images, which strangely contextualize the event, allude to deceased individuals whom the artist never had the chance to meet, but who were responsible for the "implantation" in his body of the genetic traces he has carried from childhood and that he will carry until his death. Will we in the future still carry these traces with us irreversibly or will we be able to replace them with artificial genetic traces or implanted memories? [25]

Eduardo Kac: ⟨http://www.ekac.org/amach.html⟩

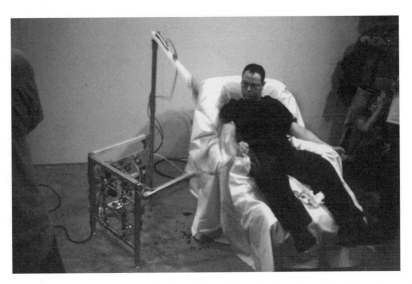

Fig. 2.5.9. Eduardo Kac and Ed Bennett, *A-positive* (1997). A machine and human exchange fluids. Oxygen from the human blood is used by the robot to support a small flame in its body. Photo: Carlos Fadon.

A-positive, completed in collaboration with Ed Bennett, set up an exchange of blood between a human and a robot. The robot provided dextrose to the human via an intravenous feed. Blood is withdrawn intravenously from the human and passed through an oxygen-extracting apparatus in the robot, which uses the oxygen to maintain a small flame. Kac speculates on the cyborgian future:

A-positive, a dialogical event created by Ed Bennett and myself, probes the delicate relationship between the human body and emerging new breeds of hybrid machines that incorporate biological elements and from these elements extract sensorial or metabolic functions. The work creates a situation in which a human being and a robot have direct physical contact via an intravenous needle connected to clear tubing and feed one another in a mutually nourishing relationship.

The artists propose that we are entering a new age when the boundaries between robot and person are dissolving. Art can help explore the possibilities of this biotechnological ecosystem:

This work proposes that emerging forms of human/machine interface penetrate the sacred boundaries of the flesh, with profound cultural and philosophical implications. . . . *A-positive* does away with the metaphor of robotic slavery. . . . We have always asked what can machines do for us. Now might be the right time to ask what we can do together. We are no more masters of our machines than we are at their mercy. We are as intrigued as we are perhaps fascinated and terrified by the notion that we are embodying technology.

Experiments with Tissue Culture

Tissue and Culture Art Project

At the beginning of development, all of an embryo's stem cells are theoretically the same; gradually they differentiate to become skin, bones, organs, and nerve cells. Medical researchers believe that they are sufficiently close to understanding these processes that they may be able to grow "replacement" tissue and ultimately organs as part of medical interventions.

The *Tissue and Culture Art Project* (TCA) creates sculpture from specifically manipulated stem cells—living art. The TCA group proposes to cover technological and other cultural artifacts with tissue surfaces. In another work called the *Biotissue and Culture Project,* Gregory Fischer proposes to create a sculpture with a cartilage core and skin outside. Both projects focus on the ambiguous mixture of cultural implications, dangers, and opportunities of these new capabilities.

Body Modification

Science, technology, and medicine have added a special urgency to inquiries about the body. A compelling line of inquiry, such as that developed by Foucault, has demonstrated the subtle and not so subtle ways that a culture polices the experiences of sex, pain, and pleasure, and the limits of the ways people can relate to their own bodies and the bodies of others. Science and technology accelerate these processes of observation and control.

A large international movement has developed that challenges these processes and their underlying cultural premises. Practitioners use a variety of body modification and manipulation techniques, such as tattooing, piercing, scarification, suspensions, contortions, body sculpting, sadomasochistic (SM) techniques, body play, sexual experimentation, and excretory performance to engage the flesh. The technologies used

---

Gregory Fischer: ⟨http://www.fischerdigital.com/tissue/⟩
"Body Modification" Web site: ⟨http://www.ambient.on.ca/talent/denise.html⟩

Fig. 2.5.10. Tissue and Culture Art Project. Oron Catts, Ionat Zurr, and Guy Ben-Ary, *Semi-Living Worry Doll H (represents the fear of hope) 2000,* a sculpture hand-crafted out of degradable polymer, surgical sutures, and living mouse cells.

are not the validated tools of science or medicine. The contexts are unorthodox. The actions often arouse strong feelings of disgust or moral aversion among those not part of the community. Yet this experimentation is a critical component of the ways the arts are responding to science and technology's impact on the body. Though their tools are not those of high technology, they seek to reclaim the body. They offer a parallel extreme, self-initiated "surgery" to stand in distinction to technology's invasions of the flesh.

Oron Catts: ⟨http://www.imago.com.au/tca⟩

Body Modification

In sculpture, there is a history of artists exploring the limits of body manipulation. In Vienna, there was a group called the Aktionismus, comprised of artists such as Rudolf Schwarzkogler, Gunter Brus, Otto Muehl, and Hermann Nitsch. They were interested in themes such as taboo, sado-masochism, hedonism, pain, and death. Through purification rituals and physical transcendent exercises, they tried to prepare the body for new pure sensation. Schwarzkogler said, "I purify my body, purify the channels of my body. By this means I get a pure sensation which helps me to overcome the world."[26]

Nitsch believed that through extreme, norms-challenging body manipulation, the performers could bring themselves and the audience to new states of consciousness. Blood, shouts, and screams were important elements. Ecstasy was one goal:

Life is more than duty: it is bliss, excess, waste to the point of orgy. Everything that exists should be celebrated. Art as propaganda for life, for how it enhances it—that to BE is a ceremony, that in this word IS lies all the preconditions for celebration. The whole ascetic philosophy will be turned on its head; life will be a celebration. All metaphysics begin with the affirmation of life, which is what admits the possibility of broader-based knowledge.

Fakir Musafar

Fakir Musafar has become one of the major spokespersons for the movement. He performs and lectures around the world. In "Body Play: State of Grace or Sickness," he explains the philosophical orientation that guides some of this work:

[we rejected] the Western cultural biases about ownership and use of the body. We believed that our body belonged to us. We had rejected the strong Judeo-Christian body programming and emotional conditioning to which we had all been subjected. Our bodies did not belong to some distant god sitting on a throne; or to that god's priest or spokesperson; or to a father, mother, or spouse, or to the state or its monarch, ruler, or dictator, or to social institutions of the military, educational, correctional, or medical establishment. And the kind of language used to describe our behavior ("self-mutilation") was in itself a negative and prejudicial form of control.[27]

In *Digital Delirium,* Michael Dartnell recounts some of his experiences with the body-modification community. He describes the power of expressing one's oppositional stance via the body and flesh, and the dominant perspective that sees the mainstream society

---

Fakir Musafar: ⟨http://www.bodyplay.com/fakir⟩

Fig. 2.5.11. Fakir Musafar in *Hindu Spear Kavadi* (1982). Photo: Mark I. Chester.

as increasingly conforming, dehumanizing, and alienating. He proposes that the expansion of medical and legal institutions helped lead to this state. He suggests that the increasing focus on technology and cyberspace in part motivates the body-modification movement:

Fakir's radical re-appropriation of the corporeal responds to the seizure of the body by medical and legal sciences in the nineteenth century. Bodily pleasure had to conform to abstract notions of the "natural" state since this is what medical and legal sciences believed they were "discovering." . . .

By recasting life as technologically (Internet) centered in a context in which political disempowerment is accentuating . . . , the priests of mind-over-body marketing and their information "revolution" ironically make a radical physical reappropriation of the body possible by leaving the mass of the population with little else beyond their own corporeality. Instead of a noble vessel or source of sin, the body is highlighted as a site of authentic emotional-spiritual affirmation for those who practice body modification.[28]

The Ethical Debate about Body Experimentation

The Institute of Contemporary Arts presented a 1996 performance series called "Totally Wired," which included many of the body performance artists described previously. The introduction to the show claims that performance art is uniquely suited for investigating the intersection of the corporeal and the technological:

Through a range of performance work, "Totally Wired" touches upon some of the debates around the nature of the body in relation to social, political, scientific, ethical, economic, and culture shifts in the world as we know it. Some of the slogans of the zeitgeist claim the body as obsolete in the light of the acceleration of developments in technology and scientific practice. Others claim the body, with its blood, sweat, tears, and its function as a signifier of gender, race, class, and sexuality is the only "reality," the only place where diversity can exist in a dehumanised, remote, and virtual world. . . .

"Totally Wired" reflects some of the ways in which performance can draw maps to navigate these territories and some of the ways in which artists can appropriate developments in science and technology to explore other ways of being, seeing, thinking, and doing in this new world.[29]

Much of the art world seems quite ready to accept this analysis of unnecessary limitations on the body, institutional intrusion on the body, and the appropriateness of artistic experimentation to push those limits. The medical and scientific world would not be so ready. Australian professor Jane Godall presented an article entitled "Ethics and Experiment in Performance" at the Art, Health, and Medicine conference, which explored the different disciplines' perspectives on body experimentation.

She notes that through a long history, science and medicine have developed a code of ethics about body experiments. The code proposes certain requirements and restrictions, such as: the subject must give informed consent; the experimenter must document competence; the experiment should not be undertaken lightly—the plans must demonstrate a likelihood that increased knowledge will result from the experiment.[30]

Comparing the "experiments" of artists with more formal experiments can be revealing. She points out that competence for body artists is less formally defined than in the medical world. Also, the knowledge produced may be quite important even though it doesn't fit into a scientific framework:

On the one hand there is scientific experiment. This is professionalised and institutionalised, and therefore under a formal management system which regulates its conditions and its design. It belongs to a regime of knowledge, in which it is supposed to attempt some kind of advance. It requires ethics approval according to nationally and internationally agreed codes. On the other hand, there are [art] versions of experiment that are deliberately anarchic, improvisatory, spontaneous, even reckless. They are more than likely to raise some ethical issues, but they are not bound by any professional codes of ethics. This is experiment as play; its status is amateur and illegitimate.

Godall then raises a key question: Do we have the right to experiment on our own bodies? Many in the art world would answer yes, but the larger culture has different answers. Each person is part of a larger social entity that has a stake in setting limits of behavior: "To the extent that human bodies are socialised they are subject to regimes of care and discipline. Every society has generalised codes of propriety for the embodied subject."

The acts of a person on themselves can have impact on others. For example, suicide is illegal in many societies. The body artists challenge the right of the culture to regulate what a person does and how they use technology to accomplish it:

Performance artists contest the regulation of bodies through calculated offenses against body disciplines and codes of propriety. This in itself has claim to be a valuable and legitimate form of social experiment. But what about specific offenses against medical codes? Turner emphasises that medicine itself needs to be understood as a set of regimes for regulation and restraint of the body . . . . To violate one's own body or subject it to invasive procedures is to violate nature and natural law, to offend against social propriety, and to encroach on the heavily guarded professional domain of medicine. So what rights do we have over our own bodies? Should experiments on yourself be exempt from ethical considerations? . . .

Perhaps what we are seeing in the work of Stelarc and Orlan is not the birth of the cyborg (in which idea there remains far more fantasy than actuality) but the end of a certain ethical concept of the body, founded in a singular and axiomatic identification between the body and the self. If this identification is ruptured, we have to rethink what it means to be an agent, and how the legal, moral, and ethical liabilities of the individual can be encoded. The experiments of these artists may be one step ahead of the experiments of scientists, in that they look not towards an advancement of knowledge as we know it, but towards a new and as yet strange direction in knowledge and understanding.

Other Artists and Projects

The *Chapman Brothers* create provocative drawings and sculptures that comment on contemporary body-modification technology. The sculptures, designed to be shocking, reposition a variety of body parts. One commentator describes an exhibition in England: "mutant visions with relocated genitals . . . these works have been stripped of conventional sexuality, only to have a dislocated and reconstructed sexuality conferred upon them. An eye socket becomes a clitoris, a nose a penis." In *Phat Media Blast* *Elizia Volkmann* purposely molds her body by adding weight as a critique of physical stereotyping.

Brain Processes, Heartbeats, Breath, Biosensors, Speech, and Psychology

This section reviews artists using technologies that sense the internal states of the body. Chapter 7.4 provides related examples of body-relevant art in its explorations of new kinds of digital-system sensors that read motion, touch, and gesture.

David Rosenboom

David Rosenboom, codirector of the Center for Experiments in Art, Information and Technology at Cal Arts, has a long history of experimentation at the intersection of art,

---

Elizia Volkmann: ⟨http://www.fact.co.uk/projects.htm⟩

Fig. 2.5.12. David Rosenboom. Preparation of electrodes for a performance of "On Being Invisible" in which Event Related Potentials (ERPs) in the brain that are associated with the perception of important features in musical forms are analyzed and used to influence the evolution of the electronic music.

science, and technology. One line of inquiry has focused on interfaces between the human nervous system and music generation. Over the years, he has created numerous compositions that used brain signals as part of the compositional systems. His monograph, "Extended Musical Interface with the Human Nervous System," is a classic in the field. He describes his early realization that biofeedback research and new ideas about musical composition were closely related:

In the late 1960s I became fascinated with new developments in brain science as they related to musical perception and the emergence of new musical languages. . . . My own interest in biofeedback centered around the notion that self-regulation of brain functions, as could be observed through monitoring aspects of electrical brain activity, was closely related to certain processes involved in the evolution of new musical styles.[31]

His work *On Being Invisible II (Hypatia Speaks to Jefferson in a Dream)* has focused on self-organizing systems such as those investigated by artificial-life researchers. In this, he combines his interests in monitoring brain function with his interest in artificial intelligence in musical composition. The system reads the performers' brainwaves and uses this information about the performers' attention to help compose the next sequences in accordance with a method devised by the composer. The software includes a partial model of musical perception that it uses to predict what events in the emerging stream might be considered significant:

In 1976, I began creating a work entitled *On Being Invisible,* which, for me, contains the richest aesthetic, symbolic, and metaphorical content arising from the import that biofeedback systems had on my work as a composer. *On Being Invisible* is a self-organizing, dynamical system, rather than a fixed musical composition. The title refers to the role of the individual within an evolving dynamical environment, who makes decisions concerning when and how to be a conscious initiator of action and when simply to allow her or his individual, internal dynamics to co-evolve within the macroscopic dynamics of the system as a whole.

I wanted to create a situation in which the syntax of a sonic language orders itself according to the manner in which sound is perceived. To accomplish this, components of the electroencephalogram (EEG) recorded from the brains of onstage performers, known as event-related potentials (ERPs), are detected, measured, and analyzed. ERPs are transient waveforms in the EEG associated with the occurrence of stimulus events having a high degree of salience—particular meaningfulness—to the subject emitting these brainwaves, always in relation to a particular context of surrounding events.[32]

Paras Kaul

Paras Kaul has focused her art on exploring the power of new technologies to support personal and spiritual development. In *Mind Garden,* she creates 3-D worlds that are navigated by brainwaves:

The project combines the technologies associated with EEG, digital brainwave analysis, system design, the World Wide Web, and the synthesis of digital audio, visual, and linguistic media. Participants are asked to relax and focus their attention, which generates frequency variations in their brainwave signals, which in turn determine forms, sounds, and word objects.

The journey is determined by the brainwave activity derived from each user's own imagination. Participants who predominantly signal theta-wave activity will experience a journey of greater complexity and focus, and participants experiencing beta brainwave activity may find the journey confusing and/or uneventful. The challenge is to experience the garden as controlled by theta and delta brainwave activity, thus perceiving a deeper and more complex view of the simulated reality. The goal is to achieve the ultimate experience in the *Mind Garden* by tuning one's frequencies to the deepest level.

Paras Kaul: ⟨http://www.well.com/user/parasw/⟩

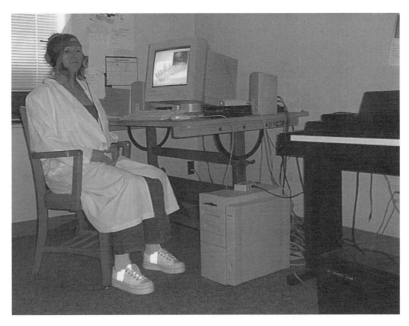

Fig. 2.5.13. Paras Kaul, *That Brainwave Chick* (in collaboration with Mark Applebaum and photographed by Joan Friedman). Neural audio imaging for virtual worlds.

Naoko Tosa

Naoko Tosa (in collaboration with the Sony-Kihara Research Center, Inc.) developed an installation called *The Flow*, which used the heart rates and hand movements of two viewers to shape the interaction of 3-D avatars. Exploring the communication potential of subtle nonverbal behaviors, the system synchronized the on-screen actors with subtle levels of stress and interest deduced from the monitored behaviors of the human participants.

Bruce Gilchrist and Johnny Bradley

Bruce Gilchrist and Johnny Bradley create performance/installations in which the audience is enabled to communicate with a sleeping mind. Specific audience members are invited to sit in "seats of consciousness" that allow them to interactively respond to the

Naoko Tosa, *The Flow:* 〈http://www.mic.atr.co.jp/~tosa/sig/index.html〉
Bruce Gilchrist & Johnny Bradley: 〈http://filament.illumin.co.uk/ica/innovation/bruce1.htm〉

Fig. 2.5.14. Naoko Tosa, *Unconscious Flow.* Physiological data (heart rate) and human behavior control relationship of animated figures. Supported by ATR Media Integration and Communication Laboratories (MIC) and Sony KRI labs.

sleeper's brain state. Video images are drawn from a database of images in accordance with EEG brainwave states, and sounds are activated by galvanic skin response (GSR) readings. Audience members send "messages" to the sleeper via a Morse-code–like device that sends mild voltage changes to a transcutaneous electrical neuro-stimulator (Tens unit) planted on the sleeper. "These expressions from the GSR and neural network will constitute a palpable representation of the sleeping/tranced mind-brain which will be projected and displayed around the space."

Catherine Richards

Catherine Richards creates installations that focus on science, technology, and the body. Her Banff Centre–sponsored *Spectral Body* work investigated impossible movements created within virtual reality. Her *Virtual Body* installation confronted the viewer with a column that resembled a nineteenth-century kinetescope with a hole for the insertion

Catherine Richards: ⟨http://coeurs-electrises.net/hearts_new/life/pulse_e.html⟩

Fig. 2.5.15. Catherine Richards, *Charged Hearts*. Gas-filled tubes detect heartbeats and communicate with other tubes.

of a hand. The device created an illusion in the viewscope that the hand seemed to be receding into the distance. Her *Charged Hearts* (in collaboration with Martin Snelgrove), metaphorically explores communication between persons via heartbeats.

Charged Hearts is a mutifaceted installation and Web work. The installation allows people to communicate heartbeats via glass hearts filled with charged plasma gas. The hearts, designed to refer to nineteenth-century laboratory apparatus, detect heartbeats when lifted and communicate with other glass hearts in the installation, synching up when rhythms are similar. The Web site describes the installation:

These glass objects become part of ourselves and part not. Holding them is like holding another person's heart, like holding an artificial heart, like holding a medical dissection, a specimen still beating, like grasping a stolen relic. . . . The human heart, the symbolic seat of the emotions, also happens to be one of the body's better known electrical fields. Human beings are in constant cybernetic interaction with their technological surroundings through a myriad of sensory interfaces. . . . The human itself becomes an interface, and the boundaries and limitations of the body and subjectivity are consequently destabilized. How to be autonomous and completely connected at the same time?[33]

On the Web site, Snelgrove and the other scientists reflected on the process of collaboration. They noted that Richards constantly challenged the engineers to reexamine what they thought they knew and what they thought was possible:

As thrilling as this project was conceptually, it was equally difficult to realize: it squirmed, it rebelled, it ran all over the place and it hasn't stopped. There was no known way to put it together or even what disciplines and skills were necessary: plunge in and match minds. . . . Catherine Richards has informally been "artist-in-residence" at the High-Speed Integrated Circuits Laboratory for three years now, and what might seem like a one-sided arrangement in which the artist "mines" an arcane technical group for images and technical skills has turned out to be much more complex. It's particularly intriguing how quickly the engineers who are nominally focused on implementation discover their own aesthetic agendas; Catherine is faced repeatedly with conflicting claims about what is technically practical, many of which mask purely artistic choices—and I'm sure that I'm one of the offenders.

It shouldn't really be a surprise that researchers in an engineering discipline react this way to doing art, because engineers create things and because researchers explore, but the surface politics speak against the success we've had. We're used to starting with a "business case," or at least pretending to, and designing for function, or at least pretending to. The pretence is wearing a little thin, though cars and televisions are largely just bad art, and maybe, if we get more real art happening in engineering labs, we can learn to do good engineering instead."[34]

Other Artists and Projects

Brain Activity   Ulrike Gabriel created *Terrain_02: Solar Robot Environment for Two Users,* which uses the brainwaves of two persons facing each other to control moving robots via changes in an array of lights shining on the robots' solar cells. The changing brain patterns affect the fluidity and synchronization of robot movement, thus visualizing the synergies and subliminal communication between the users in an accessible way. ***Werner Cee and Horst Prehn***'s installation called *Braindrops* uses brainwave detectors

---

Ulrike Gabriel: ⟨http://www.ntticc.or.jp/permanent/ulrike/ulrike_e.html⟩
Werner Cee and Horst Prehn: ⟨http://www.foro-artistico.de/english/program/system.htm⟩

to link the emotional state of the visitor to the production of images and sounds that then have an ongoing interinfluence on further production. As part of Sensorband performance, **Atau Tanaka** uses the BioMuse EMG neural signal sensor to control MIDI instruments. **Nita Sturiale**'s *Thoughtflow* event on the banks of the Charles River in Boston recorded participants' electroencephalographic data as they thought about the question, How does thought flow like a river?

Nina Sobell has been developing a piece called *Interactive Brainwave Drawing Game* since 1975. Two people hooked up to a brainwave analyzer influence an integrated video image that superimposes the wave on video images. Sobell notes that she was able to prove that people could influence each other's brainwaves nonverbally.

Janine Antoni created an installation called *Slumber* in which she slept in museums all over the world while hooked up to an EEG brainwave machine typically used to study sleep. In the morning she would awaken and weave fabric from her nightgown into the patterns reported from the EEG. Composer **Sylvia Pengilly** uses the IBVA interface to connect her EEG brainwaves to a music generator. The images produced are projected for the audience, and sometimes the brainwave recorder is integrated with dance.

In *The Painters Eye Movement*, **John Tchalenko** along with other artists created an installation that used eye trackers and MRI brain-flow detectors that revealed a painter's physiological processes. **Einstein's Brain Project** developed a VR environment that explores functions of the brain associated with deficiencies in memory, sensation, perception, and expression by using the participant's brainwave signals to alter a labyrinthine forest in real time. **Stephen Jones**'s *Brain Project* proposes to create a Web site and an opera about consciousness that integrates scientific and philosophical thought. **Warren Neidich** creates art events that explore brain processes and research models of perception. The Deutsches Museum Bonn presented a series of exhibits called ***Art and Brain.*** Artists such as Douglas Gordon, Durs Grünbein/Via Lewandowsky, and Mark Dion presented installations that included displays of brains, and seminars brought artists, scientists, and philosophers together to consider current brain research and conceptualization.

Nita Sturiale: ⟨http://www.artscience.org/nita/thoughtflow/⟩
Nina Sobell: ⟨http://www.cat.nyu.edu/parkbench/brainwaveDrawing.html⟩
Sylvia Pengilly: ⟨http://home.earthlink.net/~spengilly/⟩
John Tchalenko: ⟨http://webserver1.wellcome.ac.uk/en/old/sciart98/16painterseye.html⟩
Einstein's Brain Project: ⟨http://webserver1.wellcome.ac.uk/en/old/sciart98/⟩
Stephen Jones: ⟨http://www.culture.com.au/brain_ proj/⟩
Art and Brain exhibit: ⟨http://www.deutsches-museum-bonn.de/artbrain⟩

Heart and Breathing **Akitsugu Maebayashi** has created a variety of projects that use the heartbeat of the visitor. In *Hypersynch,* the heartbeat of the performer is used to control two open-reel recorders on the stage. In *Audible Distance,* described in detail in chapter 7.4, three visitors in a totally dark room can navigate a virtual world and determine each other's presence only via amplified heartbeats. **Char Davies** created two virtual reality installations called *Osmose* and *Éphémère,* which pioneered new approaches to virtual world navigation via breath-activated interfaces (see chapters 7.3 and 7.4). Davies takes the unusual approach of using cyberspace to connect people to bodies and the earth: "Éphémère can be viewed as an attempt to reaffirm our limitations, our mortality, our dependency on aging bodies and an earth which will, for those of us now living, absorb our bones, dreams of cyber immortality notwithstanding."

Seiko Mikami's *Borderless Under the Skin* (see chapter 7.4) reflects on the externalization of body phenomena made possible by technology. The installation *Biohazard Tent* presents an array of IV drip bags with LED wands that flash in synchronization with the pulse of visitors' fingers inserted into Velcro pouches attached to the bags. **Louis-Philippe Demers and Bill Vorn**'s *Lost Referential* offered a light-and-sound installation in which base rhythms of the lights and sounds were conducted by a volunteer visitor's heartbeat while the spacial movements of the lights integrated artifical life algorithms and reactions to visitor motion. **Christopher Janney**'s *Heartbeat* series uses wireless devices to amplify the heartbeats of dancers such as Baryshnikov as they perform. In *Self/Portrait,* **Ansuman Biswas** undertook a six-day, twenty-four-hour-a-day live art event in which he used ECG (electrocardiograph) monitoring to monitor his body responses to the ancient Indian technique of vipassana meditation. A live video image of his body was distorted in proportion to his heart rate's variation from resting state.

Speech Sound artist **Michael Edward Edgerton** has been investigating the limits and possibilities of human voice via physiological and acoustical analyses of various extended vocal techniques. He has opened up new scientific fields during his residencies in research centers and has collaborated with speech scientists such as Stephen Tasco. His book *21st Century Voice* is a document of both art and science.

Akitsugu Maebayashi: ⟨http://www.din.or.jp/~mae884/hypersync.html⟩
Char Davies: ⟨http://www.immersense.com/⟩
Seiko Mikami: ⟨http://www.v2.nl/Projects/Mikami/gif_link_text/borderless.html⟩
Louis-Philippe Demers and Bill Vorn: ⟨http://www.billvorn.com⟩
Christopher Janney: ⟨http://www.janney.com⟩
Ansuman Biswas: ⟨http://www.thegallerychannel.com/ansum_pr.shtml⟩
Mike Edgerton: ⟨www.geocites.com/edgertonmichael/index.html⟩

The Psychological Processes of Perception, Cognition, Appreciation, and Creativity

Many artists and researchers have been involved in the psychological study of human mental processes related to art. For example, artists such as Jack Ox and Dr. Hugo have long explored the phenomenon of synesthesia, the mental crossover of sensual experience, as in "hearing" color, and Vibeke Sorensen has studied the perception underlying computer stereographics. Psychologists have tried to understand the artistic processes of production and appreciation, for example, by trying to create computer programs to discriminate the work of various artists or to produce work in the style of particular artists. Articles about this work have appeared in *Leonardo* over the years, and several conferences have been held. (See also chapters 4.2, 4.3, and 7.6.) In installations such as *Divina Commedia* and *Gravity and Grace*, **Masayuki Towata and Yasuaki Matsumoto** suspend participants in tanks of gel so they can more directly experience their internal kinesthetic body sense. **Sharon Daniel**'s installation, *Strange Attractors,* read heartbeat, respiration, and skin resistance as parts of its activation.

Body Imaging

Paul Vanouse

Paul Vanouse's *Items 1–2,000: A Corpus of Knowledge on the Rationalized Subject* confronts scientific processes of objectification. It invites visitors to use a bar-code reader to access an image database from the National Institutes of Health's visible human project. A nude performer lies in a wax body mold covered with bar codes. Visitors can call up images of various sections of the body by moving the scanner over the bar codes located near those areas. Periodically, the scans bring up Vanouse's reflections from when he was a student in the anatomy morgue:

Items 1–2,000 collapses Western medicine's fracturization of the body with industrial itemization techniques into the ultimate rationalization apparatus. A human body is half submerged in a block of wax, in a manner reminiscent of how biological specimens are fixed in a "micro-

---

Jack Ox: ⟨http://www.bway.net/~jackox/⟩
Sharon Daniel: ⟨http://arts.ucsc.edu/sdaniel/⟩
Paul Vanouse: ⟨http://www.contrib.andrew.cmu.edu/usr/pv28/info.html⟩

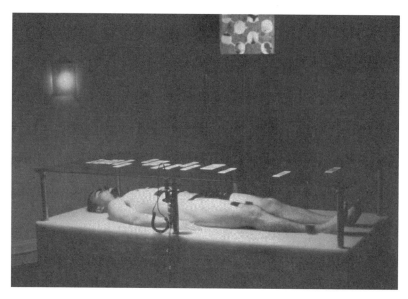
Fig. 2.5.16. Paul Vanouse, *Items 1–2,000*. The scanning of bar codes results in body scan images of relevant body parts.

tome" (a machine which cuts these wax-embedded specimens into slices often as thinly as 1 millimeter.) A sheet of glass rests several inches above the figure in a manner analogous to that of
a cover slide used atop the cross-sectional slices in microscopy. This glass is affixed
with bar codes, running transversely across the glass, which correspond to internal organ locations
of the figure underneath. Participants interact with the work as anatomy students would a cadaver:
They use a stainless steel bar-code scanner much like a scalpel slicing horizontally across the figure
to reveal the hidden target organ. The more familiar use of bar codes and scanning procedures,
however, are not lost, and this surgical role blurs with that of cashier commodifying and extracting
value through the denial of the body as whole (rather a rational composite of itemized parts).

Patrice Caire

Working in collaboration with researchers at Stanford Research Institute, Caire under-
went a full-body MRI (magnetic resonance imaging) scan. Based on this information,
she created "Cyberhead . . . Am I Really Existing," which allowed the audience to take
a virtual reality tour through her head.

Patrice Caire: ⟨http://www.ai.sri.com/~caire/⟩

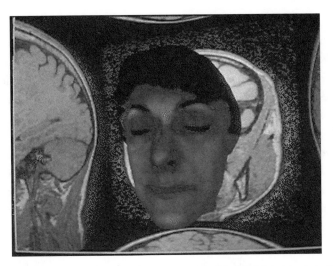

Fig. 2.5.17. Patrice Caire, *Cyberhead* (1994). Viewers can take VR tour of Caire's head via MRI scan.

Virtual Explorer

An interdisciplinary team is developing *Virtual Explorer,* a virtual reality environment in which the viewer can navigate the bloodstream on a nanobot ship. Survival in this environment requires making sense of the strange entities one encounters in the bloodstream and devising biologically sound rationales of how to respond. The environment was shown in the 1997 SIGGRAPH "Electric Garden" art show. The Web site offers the introduction to users: "You are placed in control of a small mechanical nanobot ship that operates at the cellular (and even molecular) scale . . . The basic plot outline is simple: you are caught in the middle of a compromised immune system, currently under attack, and must seek to determine the invader (antigen), it's type (viral, bacterial, unknown), and then attempt to understand and assist the immune system as it responds to the invader."

Other Artists and Projects

Steve Miller creates paintings and digital photographs based on views obtained from electron microscopes and other medical imaging devices. For example, he has painted portraits based on his own red blood cells, his mother's MRI images, and another person's

---

Virtual Explorer: ⟨http://sdchemw1.ucsd.edu/ve/team.html⟩
Steve Miller: ⟨http://www.stevemiller.com/⟩

mammogram. His *Johanne* series presents portraits of a person based on X-rays of their possessions. His *Representation with Identification* series integrates medical images with old masters. He believes the new tools provide provocative materials to explore the idea of portraiture. **Eva Wohlgemuth** created a navigable 3-D–world representation of internal body structures. *Pilots* navigated the body in a VR flight simulator. ***Journey into the Brain,*** funded by the National Institutes of Health, provides a CD-ROM–based game that allows visitors try to solve the mystery of erratic behavior by navigating inside the brain. **Erich Berger, Patricia Futterer, and Sandy Stone** created an ironic installation called *Cyborg Detector,* which forecast the future of body imaging. It could detect a person's cyborg levels: "By means of a metal detector system and measurement of their habits and attitudes toward technology and culture, the visitors' degree of cyborg-ness is determined." **Esther Mera**'s video installation *Acercándome lentamente al límite* comments on the poetics of body imaging by presenting twelve monitors with extreme close-ups of body regions. In a Wellcome Trust–funded collaboration, **Ian Breakwell, Bernard Moxham, Dale Murchie, and Guy Pitt** created *Dance of Death,* a series of still and animated works referring to medieval death imagery, by using medical techniques such as cine radiography, ultrasound, and themographic imaging. **Lewis DeSoto**'s *Recital* presented an electronic grand piano playing the music of dead Japanese composer Chiyo Tuge with an automatic page turner, using microphotographs of her brain made by her brain researcher husband as the score. **Isabelle Chemin and Guido Hubner**'s *TransNeurosite* offers an event in which performers navigate MRI 3-D representations of the brain through speech recognition.

Tiffany Holmes's *Littoral Zone* presented an interactive computer event using medical images. **Ann Sautefield**'s *Systemic Onset* used x-rays from arthritis treatment, **Mona Haroum**'s *Corps Etranger* created video within endoscopic images, and **Douglas Hahn**'s *Urbs Turrita* featured MRI images. At Canon's ArtLab exhibit "Connecting Re-Body," **Atsuhito Sekiguchi** created an event in which a variety of sensor scans of visitors' bodies resulted in a changing computer display. Performance artist **Scott Serano** presents events in which he "dissects" himself via costumes built from historical anatomy texts.

Eva Wohlgemuth, *BodyScan:* ⟨http://thing.at/bodyscan/info.htm⟩
Journey into the Brain: ⟨http://siggraph.org/s97/conference/garden/brain.html⟩
Cyborg Detector: ⟨http://cyborg.aec.at⟩
Esther Mera: ⟨http://www.telefonica.es/fat/emera.html⟩
Dance of Death: ⟨http://webserver1.wellcome.ac.uk/en/old/sciart98/12danceofdeath.html⟩
Isabelle Chemin and Guido Hubner, *TransNeurosite*: ⟨http://www.contiguite.synergia.fr/dsm/c_dsm.htm⟩
Atsuhito Sekiguchi: ⟨http://www.canon.co.jp/cgi-bin/cast/⟩
Scott Serano: ⟨http://pafringe.com/F99/html/showsartist/details/anatomical_DEMONSTRATIONS.html⟩

Medicine, Hospitals, Bodily Fluids, and Death

The Art, Medicine and the Body conference and art exhibit was held in Perth, Australia, to explore the "creative, intellectual, and interdisciplinary practice in the intersections of art, science, health, and medicine." The wide-ranging offerings stretched from art by patients or art used in healing techniques, to more conceptual explorations of the arts' response to developments in science and medicine.

Michele Theunissen, curator of the exhibition, provides a framework for the art presented. She wonders how artists will respond to the technologically mediated changes in body conceptualization and the impact of the institution of medicine:

And now, another ontological crisis confronts the individual. This being the vastly increased scale of intervention by medical and biological technologists into the realm of the body. Through the invention of micro-machinery, the body has become the locus for new voyages of discovery and territorialisation. Its purpose? To fuse the machine and the visceral, and ultimately to challenge mortality and prolong life. Supporting this fusion is the increased use of drug therapy. Artificial intelligence, genetic engineering, organ transplants, prosthetic devices, micro-machines that stimulate our faculties, chemical stimulants and intelligence drugs are inevitably seeping into a position of dependence, even acceptability. Scanning devices such as MRI, PET, and electron microscopy present fascinating interior landscapes never seen before.

Will these changes affect the way that we imagine ourselves? Or will we remain foreigners to the medical depiction of our bodies? The intervention and restructuring of the body by medical science has challenged social theorists and artists to examine and reconsider ways of representing the body. It has meant, too, a need to explore the history of medicine to understand the systems that have supported its development.[35]

Diana Domingues

Diana Domingues's installations link primitive human elements with contemporary technology. In *Trans-e-my body, my blood,* visitors encounter a ritualistic space that uses advanced sensors but refers to places of the shaman and trances. In the center of the space is a caldron of a red liquid, suggesting blood. The visitors movements are read and reacted to. Domingues's *Technologized Body* statement describes her interest in the interface between cool technologies and warm bodies:

Diana Domingues: ⟨http://artecno.ucs.br/diana.htm⟩

Fig. 2.5.18. Diana Domingues, *Trans-e*. Visitor motions influence the cauldron of symbolic blood.

In my installations, I offer new forms of communication when the visitor makes a performance with interfaces that connect the natural energy of the body and the artificial energy of
the machines. The language of the body is learned by the electronical systems. . . . [The event
highlights] the moisture of material and immaterial components and the emergence of a new
sense. It provokes a physical and conceptual relation to the body, where the limits of corporeal,
stability, and permanence are replaced by the instability, unpredictability, multiplication, and
indefinition.[36]

Franko B

Franko B comments on medical practices via performances that use his own blood and
bodily fluids. By taking this direct approach in contrast to medicine's attempts to objectify and sterilize, he provokes the audience to consider desire and carnality in the context

Franko B: ⟨http://www.backspace.org/franko.b⟩

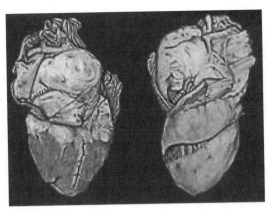

Fig. 2.5.19. Richard Stanford, *Heart of Patient No. 226-070,* from the autopsy of Louis Washkansky. Oil on canvas/digital media.

of cultural forces moving toward medicalization. Commentator Judith Palmer describes a performance:

I'M NOT YOUR BABE . . . saw him standing naked, bleeding relentlessly from wounds inside each elbow, the steaming scarlet stream coagulating in darkening pools on the floor by his side for a full half hour until he staggered away, verging on unconsciousness. The stance, arms outstretched, countenance in beatific repose, coupled with the quasi-stigmata, almost recalls the traditional depiction of the Sacred Heart. . . ." I just want to create beautiful images and survive them, like life—make the unbearable bearable. . . . I was brought up to be ashamed of my body," says Franko. "I use blood, urine and shit as a metaphor because this is what I am."[37]

Richard Stanford

Richard Stanford focuses on death. His projects have included *Reconstructing Death,* in which he explores conceptualizations of death using images and texts from medical, scientific, legal, and philosophical sources that are used to conceptualize death; *Programmed Cell Death Project,* which visualizes the stages of the planned death of a cell; and the *Forensic Craniofacial Reconstruction* project, which offered a full-body digital reconstruction of unidentified skeletal remains in collaboration with an anatomist and anthropologist.

Richard Stanford: ⟨http://www.central.com.au/stanford/⟩

Gail Wight

Gail Wight created an installation called *Emotonil.* She presents commentary on the pharmaceutical industry and its relationship to cultural trends by creating a plausible presentation of a drug designed to kill emotions. Here is her description:

Belonging to the latest group of psychoactive pharmaceuticals, Emotonil[a] targets specific nuclei in the limbic system, neutralizing the local ionic charge and rendering the user emotionally dead for the duration. . . . Why not, in the midst of developing our latest generation of smart machines, choose a brief respite from the biological nuisance of emotions, and take advantage of the temporary emotional death offered by Emotonil[a]

Other Artists and Projects

Body Imaging **Barbara McGill Balfour**'s *Melonamata* created large prints based on epidermic cell structures derived from medical atlases. **John Baturin**'s *Enemies Within* photo collage juxtaposed Eadweard Muybridge–like images with medical anatomy renditions. **Ann Chamberlain** creates works reflecting on cancer and other diseases, such as *Vital Signs,* which emblazoned cellular and medical imagery on household objects, and *Healing Garden,* which created a hospital garden out of tiles containing personal and medical information and reflections about diseases.

Commentary on Medical Procedures **Colleen Wostenholme** creates jewelry out of castings of pills. Included in one of her collections are pendants of drugs such as Sera, Rivotril, Norvasc, Xanax, Paxil, Zoloft, Imovanol, Triazolam, morphine sulfate, and Dexedrine. She links contemporary social malaise with the traditional practice of men claiming women by giving jewelry: "Because of the kind of insanity that that attitude towards women breeds, I thought it would be appropriate if they were wearing antidepressants around their necks instead of diamond solitaires."

In *Las Transpiraciones del Desgaste ò la Devaluaciòn Aspirada* **César Martìne Silva** created an installation commenting on hospital respirator procedures by inflating and deflating computer-controlled rubber dolls connected by tubes to air pistols and hair dryers. **Alexa Wright** creates photo installations that comment on the institution of medicine. *Rx—Taking Our Medicine* showed the highly ritualized nature of the disci-

---

Gail Wight: ⟨http://arts.ucsc.edu/mortal/wight/wight.html⟩
Ann Chamberlain: ⟨http://www.sla.purdue.edu/WAAW/Cohn/Artists/Chamberlainstat.htm⟩
Colleen Wostenholme: ⟨http://www.chaoscontrol.com/damagecontrol⟩
Alexa Wright: ⟨http://www.proarte.com/artists/alexa_wright/awright1.htm⟩

pline. *The Stranger Within* provided large backlit close-ups of freshly sutured surgical scars accompanied by a recording of a heart transplant. Her "phantom limb" photographs visualize the limbs that people such as amputees feel they still have. Louise K. Wilson has undertaken a series of installations that investigate medical imaging, processes, and information systems—for example, investigating consciousness in states of sleep- and anesthesia-related states of mind. In a V2-sponsored talk focused on the appropriation of medical technology, Nina Czegledy noted about the *Transplant I* installation that Wright and collaborator Louise Wilson "contend that the precision-driven instruments, diagnostic procedures, and emphasis on objective data of specialized medicine leave the patient with little control over his or her own body." ***Nina Czegledy***'s *The Digital Bodies—Virtual Spectacles* brought together artists, theorists, and scientists to explore issues of the body and medicine in a series of physical and on-line events. In *Inside Out,* medical doctor ***Elizabeth Lee*** and artist ***Susie Forman*** created a series of works designed to help patients evaluate drug and orthopedic options, such as *Recoil,* which compares coil and pill birth-control methods. Responding to diseases of friends, ***Athena Tacha*** created a series of sculptures such as *Armor for AIDS* and *Breast Cancer Patch* that show our biological vulnerabilities—"armors that cannot protect, masks that reveal more than they conceal, shields that are too floppy or scratchy to defend."

Commentary on Scientific Language   Exploring the pretentions of scientific systems, ***Marta Lyall*** uses real biological materials such as bone and hair, but then subjects them to experimental regimes to create art installations. Recent work inserts organic materials into electrochemical and electronic processes. Sample projects include making generators using electrostatic qualities of hair, developing piezoelectric membranes from bone collagen, demonstrating crude bone synthesis using nineteenth-century electrochemical processes, and the incorporation of animal muscle tissue as a functional element in machine forms. ***Eve Andrée Laramée***'s *Cellular Memories* is an installation consisting of two rows of glass separatory funnels connected with vinyl tubing filled with red wine surrounded by the sounds of waves and the human heartbeat. It focuses on the underlying rhythms of blood and seawater.

Nina Czegledy, talk on medical appropriation: ⟨http://www.dds.nl/~n5m/people/czegledy.html⟩
Louise K. Wilson: ⟨http://www.safebet.org.uk/intervws/louise.htm⟩
Athena Tacha: ⟨http://www.oberlin.edu/~art/athena/tacha2.html⟩
Eve Andrée Laramée: ⟨http://www.artnetweb.com/laramee/cellular.html⟩

Death Curating a show called "Mortal Coil," focused on death, ***Margaret Morse*** noted that even death has become a problematic concept in an age when technology and the institution of medicine keep making new interventions. Yet death has had an important role in defining humanity. She collected the works of artists exploring this changing phenomenon of death:

Yet, it is "the gift of death" that produces "my irreplaceability," that is, "my singularity," and that calls us to responsibility. "In this sense only a mortal can be responsible." It is this death that "cannot be anticipated," or, rather, that is "anticipated but unpredictable; apprehended, but, and this is why there is a future, apprehended precisely as unforeseeble, unpredictable, approached as unapproachable" [quote from Derrida]. We live in an age that seems to be a turning point in many ways that will be remembered as a time of plague, genocide, and global social transformation. Is it artificial life or the singular and unreproducable death that will permit us to become an empathic and responsible culture that will hold the key to the future? It is the task of the artists in this exhibition to invoke both the uncanny and undecidable realm animated by machines and the death that is singular and mysterious, at which we tremble.

Summary: Dissecting the Body

The body is the most accessible and most salient locus for biological research. Investigations are racing ahead. Some artists want to understand what is happening. Some want to push the enhancements as far as they will go. Some want to sound a warning. The linkage of research to the body completes a circle, linking the most ancient of art forms (performance) with the newest experiments.

Notes

1. K. Woodward, "From Virtual Cyborgs to Biological Time Bombs," in T. Druckrey (ed.), *Electronic Culture* (New York: Aperture, 1997), pp. 50–53.

2. Ibid., p. 53.

3. M. Morse, "What Do Cyborgs Eat? "in T. Druckrey (ed.), *Electronic Culture*, p. 157

4. Ibid., p. 176.

Margaret Morse: ⟨http://arts.ucsc.edu/mortal/curatorial.html⟩ (dead link)

5. V. Vesna, "Bodies Incorporated," ⟨http://www.arts.ucsb.edu/~vesna/publications/⟩.

6. J. Anstey, "We Sing the Body Electric," ⟨http://mitpress.mit.edu/e-journals/LEA/ARTICLES/ebody.html⟩.

7. "What is an Extropian?," ⟨http://seagate.cns.net.au/~ion/extropy.htm⟩.

8. J. Deitch, *Post Human* (New York: DAP, 1992), p. 15.

9. Ibid., p. 36.

10. Ibid., p. 39.

11. Ibid., p. 47.

12. A. Kroker and M. Kroker (eds.), *Digital Delirium* (New York: St. Martins, 1997), pp. 166–167.

13. Stelarc in A. Kroker and M. Kroker, *Digital Delirium,* pp. 194–98.

14. Ibid., p. 196.

15. V2, "Documentation of Roca Event," ⟨http://www.v2.nl/⟩.

16. A. Elsenaar, "Digital Computer with a Human Face," ⟨http://www.media-gn.nl/artifacial/⟩.

17. A. Elsenaar, "Huge Harry," ⟨http://www.media-gn.nl/artifacial/HHLecture.html⟩.

18. Ibid.

19. S. Stenslie, "Project Documentations," ⟨http://televr.fou.telenor.no/stahl/projects/cybersm/index.html⟩ (all quotes in Stenslie section).

20. K. Mork, "Project Descriptions," ⟨http://www.gar.no/~knut/index.html⟩.

21. K. Mork, "*Solve et Coagula* Description," ⟨http://www.gar.no/~knut/sec/⟩.

22. K. Coyne, "On Orlan," in K. High and L. Platt (eds.), *Landscapes, Felix,* vol 2, no. 3, 1995 p. 220.

23. Ibid., p. 221.

24. S. Stryker, "Across the Border," in K. High and L. Platt (eds.), *Landscapes, Felix,* vol 2, no. 3, 1995, p. 229.

25. A. Machado, "*Time Capsule* Description," ⟨http://ekac.org/amach.html⟩.

26. H. Nitsch, "History of Body Artists," ⟨http://www.hud.ac.uk/schools/music+humanities/music/abreaction.html⟩ (subsequent quote also).

27. F. Musafar, "Body Play: State of Grace or Sickness," ⟨http://www.bodyplay.com/fakir/⟩.

28. M. Dartnell, "Body Modification," in A. Kroker and M. Kroker, *Digital Delirium,* p. 226.

29. Institute of Contemporary Arts. "'Totally Wired' description," ⟨http://filament.illumin.co.uk/ica/Bulletin/livearts/frankocontext1.html⟩.

30. J. Godall, "Ethics and Experiment in Performance," ⟨http://www.central.com.au/artmed/papers/goodall.html⟩ (all quotes for Godall).

31. D. Rosenboom, "Reflections," ⟨http://music.calarts.edu/~david/⟩.

32. D. Rosenboom, "*On Being Invisible* description," ⟨http://music.calarts.edu/~david/dcr_recent/dcr_OBIII.html⟩.

33. C. Richards, "*Charged Hearts* Description," ⟨http://coeurs-electrises.net/hearts_new/life/pulse_e.html⟩.

34. M. Snelgrove, "Collaboration," ⟨http://coeurs-electrises.net/hearts_new/life/pulse_e.html⟩.

35. M. Theunissen, "Description of Art, Medicine, and the Body Conference," ⟨http://www.central.com.au/artmed/amb/intro.html⟩.

36. D. Domingues, *Trans-e-my body, my blood* brochure, inner cover.

37. J. Palmer, "Commentary on Franko B," ⟨http://www.ainexus.com/franko/crit.htm⟩.

3

Physics, Nonlinear Systems, Nanotechnology, Materials Science, Geology, Astronomy, Space Science, Global Positioning Satellites, and Cosmology

Physical Science Research Agendas and Theoretical Reflections

Introduction: Questions about the Biggest and Smallest of Things

For centuries, science was synonymous with the physical sciences, for example, astronomy, physics, and chemistry. Theorists pondered the essential nature of the universe. They asked questions with profound philosophical import: How did the universe begin? How will it end? What is the place of the earth in the scheme of things? How did the earth come to look like it does? What is the universe made of? What are its smallest parts? What makes gold gold? What powers the sun? What is the nature of cause and effect? Can we discover the rules sufficiently to predict (or even control) events? What strategies of observation, experimentation, and analysis can increase our understanding?

In the last centuries, science has expanded to develop more specialized disciplines, such as geology and biology. Inquiry has moved into areas far beyond unaided human perception, such as the subatomic and intergalactic. Enigmatic views of the world have challenged everyday notions of time, space, and matter.

Engineering has grown increasingly self-confident and aggressive in its attempt to manipulate and control the physical world. Sometimes building on scientific research and sometimes pushing beyond those understandings, applied research has created new materials, products, and industries that have profoundly shaped everyday life and culture.

Historically, artists have had an intimate connection to the physical world. They have explored the earth, the sky, and the seas. They have tried to represent natural forces such as the change of night and day, the flow of water, and the blowing of the winds. They have built buildings and sculpted objects that required them to understand materials and the forces by which they can be combined.

There have been eras in which artists were actively involved in scientific and applied research focused on the physical world. Examples include those of the Stonehenge designers, who helped further astronomical understanding; the Egyptian architects who created unparalleled construction technology; the early metal artists who helped discover new alloys and the chemistry of working with metal; the Renaissance artist-scientists who participated in the outpouring of scientific interest in everything from military technology to the shape of the universe; and the early-twentieth-century artists who were among the first to grasp the revolutionary implications of theories such as relativity and quantum mechanics.

Currently, the art world seems relatively less interested in the physical world than it once was. Even technologically oriented artists concentrate on image generation and communication technologies that help them explore issues of virtuality and representation rather than scientific and engineering research into the physical world. Regardless of this lack of interest, the world of science expends enormous international intellectual effort to investigate the physical world—everything from the origins of the universe and the inner workings of atoms to predicting the weather and understanding earthquakes. Materials scientists invent new kinds of substances with unprecedented characteristics, and nanotechnologists explore the possibility of assembling the world atom by atom. This research promises to have profound practical and philosophical implications, just as it has in past centuries. This section profiles artists whose work relates to the physical sciences, inspired by its philosophical underpinnings, research agendas, tools, technologies, or cultural implications.

Even though this section separates out the physical sciences, every chapter of this book indirectly relates to those disciplines Biological organisms are fundamentally constructed from the same materials as inorganic objects. The virtual worlds of computers and telecommunications depend on the behavior of the materials that make up electronic chips and the physical rules that govern electromagnetic radiation. Because of the future implications of physical science research, it would be a mistake for the arts to totally ignore these worlds.

Survey of Research Fields and Agendas

As a preparation for examining artistic work, this section offers a brief survey of research fields and agendas in the physical sciences. It demonstrates how research is categorized by a variety of research universities and institutes around the world. It also outlines some of the research agendas that these scientists see as important in these disciplines. The variety of work going on is striking. Also remarkable is the general lack of attention paid to this research by the arts community. The list of topics below is an amalgamation made from many different university, research institute, and government sources. The sections that follow elaborate on a few that have attracted artistic attention, for example, nonlinear systems, nanotechnology, and space science.

Atomic Physics   Topics such as atomic and molecular physics; quark and gluon interactions; neutrinos, plasma and particle-beam physics; superconductivity; antimatter; string theory and gravitation; nuclear fusion and energy research.

Chemistry and Materials Science   Topics such as the synthesis of new molecules. Experimental approaches include: multiple resonance spectroscopy, high-resolution electron energy loss spectroscopy, molecular beam-surface polarization, high-frequency electron paramagnetic resonance, X-ray crystallography, electron microscopy, ATF microscopy, self-assembly of polymeric and inorganic materials, transition metal and main group organometallic chemistry, and biochemical analysis of cellular processes.

Simulation, Modeling, and Nonlinear Systems Topics such as: chaos; low-temperature, disordered, and amorphous systems; fluid dynamics, molecular beams, and unstable and metastable systems.

Astronomy, Astrophysics, and Space Science Topics such as: cosmic-ray and gamma-ray astronomy; ultraviolet, infrared, and submillimeter astronomy; radio, X-ray, and visible light astronomy; the search for other planetary systems; supernovas, pulsars, dwarfs, black holes, and neutron stars; magnetohydrodynamics, and exobiology.

Geology and Earth Science Topics such as: climate and global change, fluids and hydrodynamics, high-pressure geophysics and geochemistry, planetary geology, geodynamics research and seismology, and glaciology.

Engineering and Applied Research Topics such as: aeronautics and astronautics, chemical engineering, civil and environmental engineering, electrical engineering and computer science, materials science and engineering, mechanical engineering, nuclear engineering, and ocean engineering.

In physics, the challenges to common notions of space, time, and matter that began in the late nineteenth century continue. Atoms have been split into a menagerie of

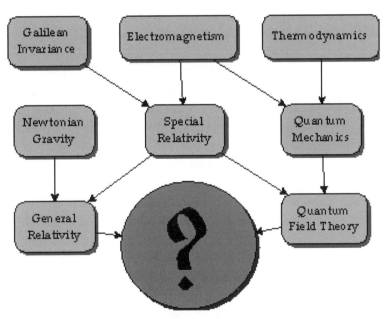

Fig. 3.1.1. John H. Schwartz, diagram showing the attempt of string theory to resolve contradictions in various well-established theoretical structures of physics (from ⟨http://theory.caltech.edu/people/jhs/strings/string11.html⟩).

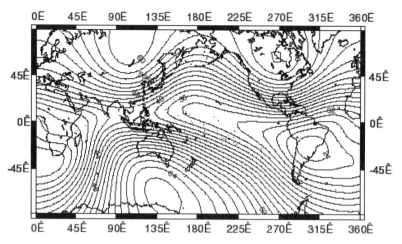

Fig. 3.1.2. Lisa Tauxe, October 1992 map showing the complex influence of position and surface on geomagnetic force strengths (from on-line course notes in paleomagnetics at ⟨http://sorcerer.ucsd.edu/es160/lecture6/web6/node6.html⟩).

more elementary particles. Space and time interact in ways unlike what most of us have experienced. Under some circumstances, matter and energy cross over into each other's domains. Anomalies call for attention: the act of observation can affect the phenomenon observed; under some situations time seems reversible; some events seem able to communicate instantaneously, even faster than the speed of light.

The galactic scale of the universe offers related challenges. Strange entities such as black holes suck up all time and matter in their vicinity. The heavens viewed with sensors in different ranges of the electromagnetic spectrum look very different. Some of the matter that theory predicts should be in the universe is missing.

Scientists continue the quest to unravel the mysteries and explain the anomalies. They search for the grand unification theory that can link atomic level forces and gravity, and for theories that can explain the evolution of the universe. Alternative science, such as Fritjof Capra's *Tao of Physics,* zeroes in on the anomalies, exploring their implications and attempting to link them to spiritual and cultural themes usually considered outside of science.

The arts are strangely quiet to all this intellectual activity. In *Art and Physics,* Leonard Schlain presents a convincing history of mutual influences between the arts and the physical sciences. He sees the early twentieth century as an especially rich time for associations. He traces similarities in many of the pioneering themes of modern art to questions that physics was raising about matter, space, and time. Where are those influences now?

This section will present some artistic activity that reflects on the current intellectual life of the physical sciences, but the level of activity is relatively low. Artists seem more preoccupied with other areas of science and technology. Is current physical science research perceived as not making as dramatic paradigm shifts as that earlier era's research? Is current research so esoteric that it is not available to nonspecialists? Has the physical world lost its appeal? The reader is invited to ponder these questions as they read about the artistic exploration of concepts related to the physical sciences.

Nonlinear Systems, Chaos, and Complexity

The area of nonlinear and chaotic systems is one that is attracting both scientific and artistic attention. These analytical systems promise to give new power to understand and predict phenomena of the natural world that have eluded older methodologies.

Historically, the physical sciences have strove to identify universal principles that could explain the physical world, including everything from the movement of heavenly bodies to the blowing of sand particles. The hope was that theories could be continually refined until they would be robust enough to explain all observations. In modern times, the physical sciences achieved amazing results by understanding phenomena such as the movement of planets, the falling of objects in gravity, and chemical reactions. It was easy to project out from these results toward the goal of total predictability.

In the last century, however, scientists have encountered phenomena that frustrate achievement of this goal. Although general principles may be understood and statistical probabilities identified, the predictability of local specific phenomena may be impossible, for example, the weather or the specific path taken by flowing water or the behavior of biological and human systems. In the last decades the study of complex, nonlinear dynamic systems has attracted a great deal of scientific interest. The influence of these perspectives has spread widely and now influences work in a great many theoretical and applied fields. Some theorists, such as Manuel DeLanda in "Nonorganic Life," and Stuart Kaufman extend the analysis to offer new perspectives on biology.

Several artists have been inspired by this research approach to create works based on natural phenomena. It seems as though the humility and intricacy of this approach is more in tune with artistic sensibilities than the previous deterministic emphasis. For many, there is the feeling that these theories honor most people's experience of the world more faithfully than the cold abstractions. The colorful and provocative language of the field illustrates some of the artistically attractive features, for example, chaos, strange attractors, catastrophe, the butterfly effect, undecidability, monster functions, dynamism, and nonlinearality.

Nonlinear, dynamic systems promise to have a major impact on both science and art in the next decades. Artists and the general public often bandy the terms around without sufficient understanding of what constitutes the scientific study of complexity. Because of its growing importance, it seems appropriate to offer a brief definition of terms and the history of this field. This material is derived from the background material provided by the Exploratorium art and science museum as part of a show called "Turbulent Landscapes." (Readers should also consult the chapters on biology and algorithmic and artificial life for considerations of other art inspired by notions of nonlinear systems.)

Nonlinearity

Simply stated, something is linear if its output is proportional to its input. If, when you're reading late at night, you want twice as much illumination (output) to see the book, then you double the number of light bulbs (input) by bringing over another similar lamp. . . .

A . . . [nonlinear] example comes from an ecology of animals that compete for food, but in which there is only a fixed amount of food available each day. As long as the population is small, all the animals get plenty of food. They grow and prosper, they reproduce, and the population grows. But it can only grow so far. Once the population is beyond a balance with the available food, some animals do not get enough. Eventually they cannot reproduce and the population size decreases. In this ecology then, the population growth is a nonlinear function of the available food. At low populations, the growth is positive; at high populations, the growth is negative. . . . The concept of linearity is very closely related to that of reductionism. Reductionism is an approach to science that says that a system in nature can be understood solely in terms of how its parts work. . . .

Both linearity and reductionism fail, at least as general principles, for complex systems. In complex systems there are often strong interactions between system parts and these interactions often lead to the emergence of patterns and cooperation. That is, they lead to structures that are the properties of groups of parts, and not of the individual constituents.

Chaos

Deterministic chaos, often just called "chaos," refers in the world of dynamics to the generation of random, unpredictable behavior from a simple, but nonlinear rule. The rule has no "noise," randomness, or probabilities built in. Instead, through the rule's repeated application the long-term behavior becomes quite complicated. In this sense, the unpredictability "emerges" over time. . . . There are a number of characteristics one observes in a deterministically chaotic system:

- Long-term behavior is difficult or impossible to predict. . . .
- Sensitive dependence on initial conditions. . . .

- Broadband frequency spectrum: That is, the output from a chaotic system sounds "noisy" to the ear. Many frequencies are excited.
- Exponential amplification of errors. . . . [MIT meteorologist Edward Lorenz described an example of this in what he called the "butterfly effect." He noted that the flapping of a butterfly's wings in Bejing could have profound weather effects days later in other parts of the world.]
- Local instability versus global stability.

Complexity

"Complexity," as a label of a scientific interest area, generally refers to the study of large-scale systems with many interacting components. Complex systems that are often offered up as examples include financial markets with competing firms, social insects (such as those that form ant colonies and build wasp nests), the human immune system, commodity markets in which agents buy and sell through auctions, and the neural circuits of the brain.

What makes these systems complex, aside from their raw composition, is that the most interesting ones exhibit behavior on scales above the level of the constituent components.[1]

Artists and those outside the sciences often toss these terms around carelessly; understanding their precise meaning is useful in considering artists who are inspired by the theories. "Complex" in the scientific sense is not the same as complicated. Linear systems can be quite complicated and require sophisticated calculations, but they are still predictable. "Chaos" in the scientific sense is not the same as the popular notion of chaos in the world; it is not randomness. Finally, for many scientists, complex systems can be understood and underlying principles clarified; interest in these systems is not necessarily abandonment of the core tenets of Western science.

Astronomy, Cosmology, and Space Science

Astronomy is one of the most fundamental of sciences. Early humans looked into the sky and wondered what the points of light were. They wondered how the movements of the heavenly bodies could be predicted and where the earth came from. Stonehenge is one monument to that inquiry. Astronomy was also at the heart of the Renaissance. Careful observations of heavenly movement led to refined theories about orbits and eventually to acceptance that the earth was not the center of the universe. This notion had profound implications beyond astronomy.

In the contemporary era, cosmology has reached beyond the local solar system to study the origins of stars and the universe. Exotic entities such as super strings, dark matter, and black holes have been identified. Extraterrestrial intelligence has been searched for. Scientists use new sensors outside of visible light to understand what is there.

In our own solar system, observation has given way to intervention. Artificial satellites regularly circle the earth. Humans have experienced weigthlessness. People have walked on the moon. Probes have been sent to other planets, samples have been brought back, and live broadcasts have been sent from Mars. Proposals for the colonization and commercial exploitation of space are being implemented.

The National Academy of Sciences prepared a research briefing summarizing the research agenda for cosmology. The report explains the work of cosmology, identifying a set of questions that researchers are seeking to answer:

When did the universe start and how will it end?
What is the dark matter and what is its cosmological role?
How did the large-scale structure of matter form, and how large is it?
What can we learn about physical laws from relics of the Big Bang?
Did the universe undergo inflation at a very early stage?
Do physics and cosmology offer a plausible description of creation?[2]

The report lists several benefits of cosmological research, including the stimulation of technological development, and contributions to the physics of matter, intellectual appeal, and ethics. It notes:

Finally, our cosmology—every culture's cosmology—serves as an ethical foundation stone, rarely acknowledged but vital to the long-term survival of our culture. Cosmological knowledge affects religious beliefs, ethical choices, and human behavior, which in turn have important long-term implications for humanity. For example, the notion of Earth as a limitless, indestructible home for humanity is vanishing as we realize that we live on a tiny spaceship of limited resources in a hostile environment. How can our species make the best of that? Cosmological time scales also offer a sobering perspective for viewing human behavior. Nature seems to be offering us millions, perhaps billions, of years of habitation on Earth. How can we increase the chances that humans can survive for a significant fraction of that time? Cosmology can turn humanity's thoughts outward and forward, to chart the backdrop against which the possible futures of our species. . . .

Some artists (see chapter 3.4) have begun to work with space science as a way of beginning to approach these questions. For example, they have undertaken to put art into space, created projects that could only be seen from space, imagined other worlds, and participated in the search for extraterrestial intelligence.

Epistemology—How Do We Know What We Know?

In art and cultural discourse, the world of stuff no longer seems as important as it once was. Cultural theorists have demonstrated the power of cultural narratives, media, and expectation to shape what we think we see, touch, taste, and otherwise sense. Even substantive, concrete givens like the body are shown to be a significant construct of culture, as it is a physical reality. We are seen to be in a postmodern, postindustrial, and postbiological era.

The world of the physical sciences is one place where the worldviews of science and contemporary art appear most at odds. Scientists and engineers by and large still believe in the reality of an external world and our ability to know and manipulate it. Universal is not a dirty word—some realities apply everywhere in the universe. Certainly, there is deep questioning even in science about the role of cognitive frameworks and the possibility that scientists' views of physical reality are more a result of their socialization and the way they frame their questions and design their experiments than any external world out there (see chapters 1.1 and 7.1).

Also, contemporary science has moved quite far from the historical trust it placed in unaided sense perception. Starting with developments such as the microscope and telescope, increasing reliance is placed on instruments to augment and mediate the senses. Science now concerns itself extensively with phenomena that can't be seen because they are too large, too small, too fast, too slow, outside the sensitivities of our sense organs, or too distant in time. Elaborate chains of theorizing, reasoning, instrumentation, and observation are required to deduce the nature of physical reality. Only the accumulation of evidence, its integration with theory, and its occasional verification through observable events (e.g., the explosion of atomic bombs) convince us that the unseeable stuff is really there.

In spite of this questioning, however, science takes belief in a knowable physical world as a core principle. Art also used to concern itself with the nature of physical reality. Indeed, artists were famous for their romance with the sensual world of sight, sound, and touch. Certainly, thought and spirit were also important, but the physical works produced were given primacy. Although many artists continue to work in this sensual mode, the mainstream art world has begun to pay more attention to nonphysical reality.

Many chapters of this book focus on artists working with information and mediation at the interface of art, science, and technology. This section, however, concentrates on the implications for the arts of areas of science and technology focused on physical and biological "realities." Exciting research probing those realities is under way, and the research promises to alter our basic conceptions of the physical and biological universe. Innovative technologies are being developed that open up undreamed of possibilities for operating on that universe.

Stimulated by the development of virtual worlds, theorists have developed a renewed interest in the problems of epistemology. Epistemology is the area of philosophy that looks at the processes of how we know and establish truth. The reliance of science on augmented sensors raises questions about reality and also about our processes of coming to know reality. Some artists have found that question fascinating in itself. In his article "Real Science and Virtual Science," Roger Malina, an astrophysicist and editor of the journal *Leonardo*, recounts a story about Galileo's experience with instrumentation:

When Galileo first started using a telescope, the story goes that he crossed the valley to the opposing mountain to touch with his hands the objects and details that he was seeing with his lens. This way he verified that he wasn't being fooled when he saw mountains on the moon for the first time in human history.[3]

Malina asks: Is the universe observed by instruments the same universe observed by a human without instruments? Don't the instruments introduce their own ways of knowing? These questions are especially powerful coming from him because he heads the Center for Ultraviolet Astronomy at University of California at Berkeley. In his profession as astronomer almost all the data he works with come from ultraviolet sensors that read parts of the spectrum invisible to human eyes.

Astronomers are using a menagerie of spectrum sensors to determine the reality of the universe, for example, radio waves, infrared, visible, and ultraviolet. And, astonishingly, the universe looks slightly different in each view. Some features spread across all views and some do not. What's really there? Will the real universe identify itself? Figure 3.1.3 shows NASA's multi-wavelength Milky Way. Anyone taking the concrete world for granted must come to terms with this challenge.

Artist and engineer Ken Goldberg (see chapter 6.3) directly takes on epistemology as an artistic focus. He created a series of Web-based installations in which a remote viewer could introduce tests on a object to try to learn more about it. For example, in the *Shadow Server*, the viewer could remotely turn on patterns of lights in order to cast shadows. Then, from looking at the shadows, the viewer could try to deduce features of the "real" object. In *Legal Tender*, the remote viewer could view a one-hundred dollar bill and apply various tests to determine if it were counterfeit. Goldberg notes that instrumentation and remoteness complicate issues of epistemology even more than normal. He calls the topic "tele-epistemology"

Ken Goldberg: ⟨http://www.ieor.berkeley.edu/~goldberg⟩

Fig. 3.1.3. NASA, poster showing the Milky Way as seen by detectors sensitive to different parts of the electromagnetic spectrum (from ⟨http://adc.gsfc.nasa.gov/mw.milkyway.html⟩).

Potentially Important Emergent Technologies

Nanotechnology

In recent years the research world has been highly stimulated by nanotechnology. Improvements in image acquisition and nanometer (billionths of millimeter) manipulation of materials has lead some theorists to conclude that in the not too distant future we will be able to assemble matter atom by atom. The successes of the integrated-chip manufactures in pushing the limits of miniaturization has led enthusiasts to posit the eventual ability to make devices such as vein-scrubbing micro-machines that will circulate in the blood, cleaning up placque and other debris. Eric Drexler, one of the most well-known researchers in this field, has produced a series of books that explain the ideas, including *Engines of Creation* and *Unbounding the Future*.

A lively international community uses the World Wide Web to promote this research and share emerging insights. Ralph Merkle, a researcher formerly at Xerox PARC, offers the following colorful introduction:

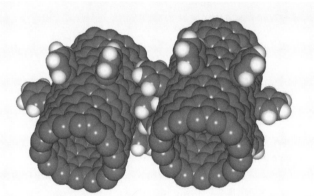

Fig. 3.1.4. Ken Goldberg, *Legal Tender*. Web visitors could remotely apply a variety of tests to judge the authenticity of $100 bills.

Fig. 3.1.5. NASA, *Fullerine Gears*. A Nano-machine developed with nano-technology, from the paper "Molecular Dynamics Simulation of Carbon Nanotube Based Gears," by Jie Han, Al Globus, Richard Jaffe, and Glenn Deardorff.

Manufactured products are made from atoms. The properties of those products depend on how those atoms are arranged. If we rearrange the atoms in coal we can make diamond. If we rearrange the atoms in sand (and add a few other trace elements) we can make computer chips. If we rearrange the atoms in dirt, water, and air we can make potatoes. . . . In the future, nanotechnology will let us take off the boxing gloves. We'll be able to snap together the fundamental building blocks of nature easily, inexpensively, and in almost any arrangement that we desire. This will be essential if we are to continue the revolution in computer hardware beyond about the next decade, and will also let us fabricate an entire new generation of products that are cleaner, stronger, lighter, and more precise.

He notes that electronic chip lithography is still working at too gross a level and that new "postlithographic" technologies will need to be developed. The on-line nanotechnology guide offers this assessment of necessary enabling technologies:

Enabling technologies that could be useful in the development of nanotechnology include research in proximal probes such as STMs (scanning tunneling microscope) and AFMs (atomic force microscope), protein design, and the molecular design of molecules containing large quantities of constituent atoms (especially carbon based), self-assembling molecules and molecule design, mechanosynthetic chemistry, and the continuing development of more powerful computer systems and enhanced chemical modeling packages.[4]

Merkle notes that positional control (the ability to place atoms exactly where needed) and self-replication technology could both be important in advancing nanotechnology.[5]

Other scientists are skeptical about nanotechnology. They feel that its enthusiasts overplay our level of knowledge about atomic and molecular structure and underplay the complexity of manipulating atomic-level forces. The projections about the nanotechnological manipulation of organic molecules is seen as even more overblown because of the complexity of these molecules. Some artists, however, look forward to nanotechnology as the ultimate sculpture.

Materials Science

Materials science is the name given to the set of research interests focused on understanding the structure and characteristics of physical materials and on developing techniques to create new kinds of materials. It consists of an integration of chemistry, physics, and engineering. In recent years, innovators have created new kinds of composites with characteristics that are unprecedented in nature, for example, strength, durability, lightness, and controllability. Materials science is the infrastructure for the microelectronics revolution in its identification of materials that allow for quicker, denser, and lower power consumption and its aid in inventing new kinds of fabrication processes. Superconducting magnets promise infinite energy. Smart materials link the physical characteristics of materials with digital information systems. Optimists predict that we are close to understanding enough that we can have "designer materials," synthesized composites that can be custom brewed to match specifications. A University of North Carolina introduction to its materials science Web site describes the area of concern:

[T]he repertoire of available materials has expanded considerably in the last few decades, and is likely to continue this proliferation in the future. For most of mankind's history, the available materials were few and essentially natural, such as clay for bricks and pottery, wood and stone for tools and construction, natural fibers (either from plants or animal hair) for cords and textiles, skins for containers and clothing. The ability to modify natural materials, extract useful materials from natural resources (which often required achieving high temperatures), or to combine them in new ways, brought new possibilities.

Anthropologists study the material artifacts of past civilizations to understand how they were fabricated (there is a field of research called archaeomaterials), and in turn to gain insight into the level of technology and sophistication of the culture. The role of materials in the advance of civilization and culture is powerfully summarized by the fact that it is the name of each dominant new material that has been used to describe the culture—the "stone" age, the "bronze" age, the "iron" age, and so forth. Articles and editorials frequently appear debating whether ours is the "silicon" age, the "plastic" age, or something else. Arguably, we are now in the age of "many materials. . . .

Modern science and technology have brought two things to this process. First is a more complete understanding of the processes, and their effects on the microstructure. Second are the necessary instruments to control the process and examine the microstructure.[6]

Materials science research is expanding in many directions. Here is a sample of this activity drawn from *Materials Science* magazine and *Scientific American:*

- Ceramic composites for vehicle armor and waste storage
- Lightweight ceramic foams for biomedical and industrial application
- High-performance thin-film solar cells
- Biopassive and antimicrobial coating technology
- Strong reactive artificial muscle substances
- Embedded sensors to monitor the health of structures such as bridges[7]

Materials science offers an important link between contemporary research and venerable artistic traditions of working with materials such as sculpture, ceramics, and textiles.

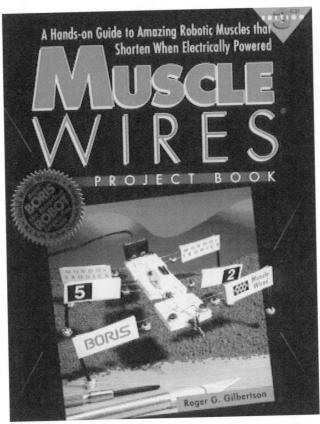

Fig. 3.1.6. *Muscle Wires™ Project Book*. Projects that use nitinol's special characteristics, including Boris, the insectoid robot. Muscle Wires is a registered trademark of Mondo-tronics, Inc., and is used by permission (http://www.RobotStore.com).

Although there is some artistic interest, its level is low considering the likely impact of materials science. The following sections offer a sampling of some specific new materials that call out for artistic experimentation:

Nitinol (Shape Metal) This alloy has the strange property of remembering its shape. After it is deformed, it can be automatically returned to its original form by heating. Designers imagine car fenders and eyeglass frames that can be fixed after damage merely by heating.

Piezoelectric Crystals These hybrid ceramics possess the unique characteristic of linking electrical stimulation and physical transformation. When force is applied to these crystals they generate an electrical current and the opposite when they change their shape. The conversion of electrical force into motion is much more direct, inexpensive, and efficient than conventional activators such as electric motors and solenoids. Some see this material as allowing important bridges between the digital information and the physical worlds. This technology has already found its way into many commercial products, such as small speakers and warning buzzers. Other developers who use the conversion of force to electrical current are working on inexpensive sensors that can read touch and pressure. Hobbyist researchers have developed small insectoid robots whose motion is generated entirely by piezo action.

Electroluminescence   Some mixtures of phosphor with other minerals will glow when placed in an electrical field. Using this characteristic, developers have created flat lights that efficiently glow with a cool light similar to that produced by lightning bugs. Researchers have been working to extend the flexibility and applications of this technology, which could potentially allow painted surfaces to generate light. The technology that was originally developed for the military to be used as aircraft lights because of its flatness has now found its most widespread use in backlights for portable computer screens.

Rapid Prototyping

There has always been a great divide between the imagination and the physical world. It was easier to imagine alternative things and worlds than to actually build them. Digital technology has exacerbated the distinction by adding even greater ease and flexibility to the imagination and design process. Visionaries have dreamed of new technologies that would link digital designs with actual fabrication, eliminating the time-consuming intervening steps by which humans had to translate digital renderings into physical things. Eventually there would be total computer integrated manufacturing (CIM) and architecture in which the imaginative digital renderings would be automatically converted into machine controls that would assemble physical objects. The final step would be artificially intelligent design programs that would take over even the imagining process.

Rapid prototyping is the technology that represents the first step in this dream. This technology uses a variety of means to convert digital renderings of designs into physical things. Stereolithography was one of the first processes developed. Other descriptors include: desktop manufacturing, automated fabrication, toolless manufacturing, freeform fabrication; laser sintering; and deposition modeling. Research labs around the world, including NASA's, are experimenting with different kinds of rapid prototyping systems. Rapid prototyping is culturally significant because it moves into territory that is underexplored, namely, the linkage of the virtual and the physical. Here is one description from the IIT Research Institute:

The new rapid prototyping technologies are additive processes. They can be categorized by material: photopolymer, thermoplastic, and adhesives. Photopolymer systems start with a liquid resin,

Fig. 3.1.7. Christian Lavigne, computer wire-frame model and sculpture that resulted from stereo-lithography based on a model.

which is then solidified by discriminating exposure to a specific wavelength of light. Thermoplastic systems begin with a solid material, which is then melted and fuses upon cooling. The adhesive systems use a binder to connect the primary construction material. Rapid prototyping systems are capable of creating parts with small internal cavities and complex geometries. Also, the integration of rapid prototyping and compressive processes has resulted in the quicker generation of patterns from which molds are made.[8]

Global Positioning System (GPS)

GPS makes use of the ancient surveyor's technique of triangulation. The United States and Soviet Union each used space exploration technology to put up a ring of location-finding satellites. Each generates a synchronized signal at precise times. An inexpensive electronic receiver on the earth collects the signals, noting the times of arrival. It can then deduce its precise location by using the differences of arrival times from known satellite locations. The different times of arrival can be translated into distance because all the signals are synchronized to emit at precise times of origination, and then the precise location of the satellites is known. Using the old high school geometry trick of drawing three circles, a location can be found at the intersection of the three.

Fig. 3.1.8. Logo from the National Air and Space Museum planetarium show called "GPS: A New Constellation."

The technology answers an ancient human wish of knowing one's location, which started with the first explorers: one can know where one is or where other fixed or movable persons or objects are with great ease and precision. The technology was originally developed by the military to track the movements of troops and to guide bombing. "Smart" cruise missiles use GPS technology combined with the image processing of satellite photographs to guide them to specific buildings.

Although the basic idea is ancient and simple, the technology requires a sophisticated infrastructure of precision timing and the placement of the satellites. Although the technology is capable of identifying location and altitude to within a few millimeters, the military "dumbs" the timing for the nonmilitary user so that it is accurate only to within approximately six meters. It feared that the general availability of precision information could be used by enemies.

The receiver technology has been rapidly decreasing in price and size so that there are now small chip boards that contain everything that is needed. Optimistic developers predict its use in every walk of life. In the United States every cell phone in a few years will be required to have GPS capability so that police and emergency services can locate a caller. Here is a sample of GPS uses:

- By boats and planes for precision navigation
- Tracking prisoners on parole
- Tracking migrating animals in order to understand their behavior
- Tracking trucks, ships, freight cars, and military equipment
- Tracking cabs so the owners know when cab drivers are not working
- Location-specific commerce in which advertising can be customized, arriving via cell phones or portable computers to be relevant to the current location of a person.

Many of the applications raise interesting cultural questions about surveillance and about the meaning of knowing where something or someone is located. The technology introduces exciting new possibilities and dangers.

Summary: Artist Explorations of Physical Science Research and Concepts

The remainder of this section presents chapters documenting the work of artists using various aspects of the physical sciences. Chapter 3.2 describes the work of artists interested in atomic-level phenomena and nanotechnology. Chapter 3.3 documents artists working with observable physical phenomena, dynamic systems, and geology. Chapter 3.4 reviews the work of artists working with space science. Chapter 3.5 examines artists who use GPS.

Notes

1. Exploratorium, "'Turbulent Landscapes' Lexicon," ⟨http://www.exploratorium.edu/t.landscapes/ CompLexicon⟩.

2. National Academy of Sciences, "Cosmology Questions," ⟨http://www.nap.edu/readingroom/books/ cosmology/content.html#contents⟩.

3. R. Malina, "Real Science and Virtual Science,", *Ylem Newsletter,* no. 6, May/June 1998, p. 13. (Also available on-line at ⟨http://www.ylem.org⟩).

4. R. Merkle, "Nanotechnology Guide," ⟨http://www.public.iastate.edu/~bhein/nanotechnology_guide_ background.html⟩.

5. R. Merkle, "Nanotechnology-Enabling Technologies," ⟨http://sandbox.xerox.com/nano/⟩.

6. "University of North Carolina Materials Science Center, "Introduction to Materials Science," ⟨http:// vims.ncsu.edu/Contents/TOC.html⟩.

7. *Scientific American,* "Materials Science Research" ⟨http://www.sciam.com/explorations/ 050596explorations.html⟩ and *Materials Science* "New Research," ⟨http://www.materials-technology.com/⟩.

8. IIT Research Institute, "Introduction to Rapid Prototyping," ⟨http://Mtiac.Hq.Iitri.Com/Mtiac/Pubs/ Rp/Rp⟩.

3.2

Atomic Physics, Nanotechnology, and Nuclear Science

The investigation of atomic and subatomic phenomena has been a major feature of contemporary physics. Since the early twentieth century, scientists throughout the world have probed the structure and behavior of atoms. Undoing old models of the atom, they have introduced a zoo of subatomic particles with exotic behavior. They have discovered realms in which matter and energy can transmute and profoundly challenged commonsense notions of time and space. Showing that observer, observation act, and phenomenon are intimately intertwined, they have forced the reevaluation of fundamental ideas about objective observation, and have built exotic instruments such as accelerators in order to study the nature of matter near the speed of light. Research continues at high intensity as scientists search for grand unifying theories. This discourse is a critical constituent of life in this era. It is surprising that so few artists have engaged the ideas, contexts, and agendas of this world.

On a more immediate level, scientists and engineers investigate the limits of our abilities to fabricate new matter. Materials scientists strive to learn enough to construct "designer" materials with qualities of our own choosing. Even more radically, nanotechnologists work to develop the skills to be able to construct materials atom by atom. This chapter presents artists whose work is influenced by these branches of science and engineering.

Atomic Physics

Shawn Brixey

Shawn Brixey creates intriguing installations that are in part stimulated by this body of research. They are simultaneously poetical and intellectually rich. He explores phenomena such as photonic energy and atomic structure. He describes his approach as "material poetry" and sees the integration of science and art as absolutely essential at this point of history:

My artwork attempts to address the impact of advanced technology on artistic expression, and the creative landscape it is dramatically altering. Traditional artistic sensibilities play a critical role in the creation of my work, yet they do not always easily import to these new environments, and often must be redefined in the light of new materials and meaning. The use of advanced technology has so greatly increased the ability to address and extend my artistic research, that I

Shawn Brixey: ⟨http://digitalmedia.berkeley.edu/shawn/brixey.html⟩

have begun to project a new kind of poetic interaction into the actual mechanics of the microscopic and macroscopic realms. I describe these present artworks as "material poetry," art made from the expressive interaction of discreet forms of matter and energy. As an integral part of this process, I have begun to develop a new genre of artistic skills and instincts that provide the language, syntax, gesture, and instrumentarium for meaningful expression in these new arenas of "poetica": artwork that lives in the boundary, where the distinction between sculptural object and apparatus . . . theater and experiment dissolve.[1]

In *Alchymeia*, Brixey created an installation in which normal processes of atomic crystal formation generated unique forms based on the human hormone used as the seed to start the crystal. In one version of this work, he proposed to use the urine sample of champion athletes as the source of the hormone—facetiously called the "Right Stuff." Brixey's Web site describes the project:

A specially designed installation that allows the growth of ice crystals in ultra-pure/ultra-cold water to be based on direct environmental input from interaction with the human body at an atomic scale. The individuality of the ice crystals in the installation are created using a similar principle of atomic recording utilized by snowflakes, but have a microscopic sample of human hormone introduced into ultra-pure/ultra-cold water as an atomic building site (an emmersive nucleating seed). Because all impurities in the water have been removed, the human material provides the only structure to build (freeze) from. When the highly ordered crystal nature of ice uses the discreet human sample to initiate the freezing process, it forces its natural crystal arrangement to elastically deform, mimicking the rhythm of the original atomic lattice from the donor sample. The tiny crystals (ice embryos) nucleated by this process, act as molecular stories, "seeds" in which the larger ice crystals in the exhibition clones itself from. The crystals in this exhibition act as amplified recordings of one's physical presence expressed at the atomic level.

Because of the explicit loyalty to the original atomic lattice of the nucleating agent, the crystals are confined by the laws of physics to reflect our unique presence in both their microscopic and macroscopic organization: An environment where we quite literally become the architect and architecture at all scales. The colors of the crystals are generated by the decreased speed of polarized light in ice specific to the elastic stress in the crystal lattice. Each wavelength of light (color) slows to a different speed, signaling the amount of atomic energy expended by the ice in aligning its structure to match the human-provided nuclei.[2]

In another series of works exemplified by one called *Instruments of Material Poetry*, Brixey explores the unfamiliar world where light can affect the motion of matter. His

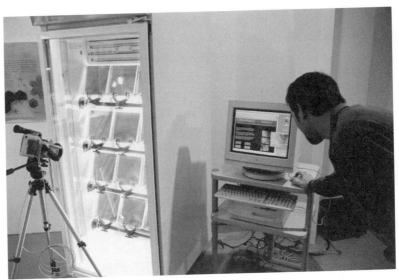

Fig. 3.2.1. Shawn Brixey, *Alchymeia*. Crystals form in ultra-pure water containing human hormone.

installations use high-powered lights, vacuum chambers, carefully selected materials, and exquisite instrumentation to orchestrate light-induced dances of levitating particles. He describes the setup:

Radiation pressure (the kinetic momentum of photons) from a fifty-million candlepower white light (xenon) source constructs an alternative gravitational system that levitates and animates microscopic poetic events in a vacuum chamber. The xenon light is "voice modulated" and emits a poetry (sound) encoded beam with enough intensity to overpower the local field of gravity for very small objects. Trapping tiny graphite particles at the center of the vacuum chamber, it causes a small brilliant galaxy of particles to form. . . . A closed-circuit digital video/microscopy system for viewing real-time in stereoscopic 3-D shows the tiny "universe's" constantly changing shape. Mirroring celestial mechanics, it is confined by the laws of physics to endlessly echo poems encoded in the light.[3]

Often Brixey works with dancers and other performers, for example, the Laura Knott Dance Company, to create installations in which atomic and subatomic phenomena are influenced by human motions and made visible to audiences. For example, in *Aurora*, "hand-cut synthetic crystals, grown in a high pressure/temperature autoclave, create constantly changing spectral colors as human shadows interfere with a battery of computer-controlled polarized projectors and analyzers." *Celestial Vaulting* was an "inter-ferometric holography installation. Real-time optical/laser holograms allow minute quantum fluctuations of an individual's presence to be detected by constructive and destructive interference of light waves and converted into audible sound."

Other Artists and Projects

The Canadian art group *InterAccess* organized the Subtle Technologies Conference and Aurora Universalis Show, which brought together artists and scientists to explore concepts of non-traditional forms of energy, such as quantum energies, complexity, electron "memory," vibrational medicine, and trance video. *John Duncan* created the sound work called *The Crackling*, which is based on field recordings obtained at Stanford's linear accelerator (SLAC) research center. *Anthea Maton* has developed a series of classroom materials called *The Art and Science Connection* that uses principles of physics as sources in activities such as drawing, painting, printmaking, sculpture, collage, and textiles.

Viewing and Manipulating the Atomic World — Nanotechnology

Although its limits are debated by some scientists, nanotechnology promises to be an engaging arena of scientific inquiry in the next decades. With its emphasis on synthesized objects, it seems a natural focus for artists. Have not sculptors been in the forefront of synthesizing objects thoughout history? The nano-realm offers provocative challenges loaded with philosophical questions. What is the difference between the natural and human-made world? What are the limits of human abilities as creators? What trust can we place in phenomena that exist only at the atomic level and can only be sensed through elaborate instrumentation? A few artists have begun to explore the nano-world. Additional artists working related areas of biotechnology are described in chapter 2.2. The NanoGallery at the NanoWorld Web site describes its goals:

One of Nanothinc's objectives is to connect science and technology with art and culture. Our company strives to globally link the power of imagination and creativity, expressed in rich and dramatic art forms, with the literal underpinnings of this powerful yet nascent technology.[4]

Subtle Technologies Conference: ⟨http://www.interaccess.org/subtle⟩
John Duncan: ⟨http://www.xs4all.nl/~jduncan⟩
Anthea Maton: ⟨http://www.physics.ucok.edu/~coapt/⟩

Fig. 3.2.2. Alexa, *NanoFuture-Space6*. A digital image stimulated by the possibilities of nano-technology. Copyright by Alexa Smith.

Alexa Smith

Working with ideas related to nanotechnology for serveral years, Alexa Smith edited a special issue of the YLEM newsletter focused on nanotechnology:

What happens when you control matter itself? I became interested in this idea about three years ago and have since concentrated my art on nanotechnolgy. I believe that nanotechnology will have a major impact on all areas of our society, but particularly in the areas of computing, medicine, and space. With a lifetime interest in space exploration, I have focused my art on this area. With the current launchings to Mars and beyond, we are entering a new space age and I expect it to accelerate greatly once nanotechnology arrives. My latest series show possible manipulations of matter via nanotechnology in space and on other worlds. This includes such ideas as terraforming, asteroid mining, biomechanical construction, diamandoid structures, and others which haven't been thought of yet.[5]

Ken Goldberg and Karl Bohringer

Ken Goldberg and Karl Bohringer created an installation called the *Invisible Cantilever*, in which they used a MEM (micro electro mechanical) atomic-level manipulator to

---

Alexa Smith: ⟨www.alexaart.com⟩

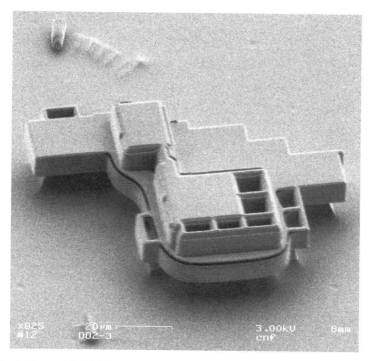

Fig. 3.2.3. Ken Goldberg and Karl Bohringer, *Invisible Cantilever.* A microscale model of Frank Lloyd Wright's Falling Waters house.

create a 1/1-millionth-scale version of Frank Lloyd Wright's Fallingwater architectural landmark. They explain that they used Fallingwater both because of its significance in the history of architecture and its reliance on cantilevers, which are also important in MEM nanotechnology devices.

But even more important, however, are the epistemological challenges that work at the atomic level poses. Goldberg asks: How can we be sure about what we think we perceive? For more details, see Goldberg's discussion of "tele-epistemology" in chapter 6.3.

Felice Frankel

Felice Frankel, an artist in residence at MIT, explores the place where scientific visualization and art meet. Some of her photography explores the nano-world. Her work was

Felice Frankel: 〈http://web.mit.edu/edgerton/felice/felice.html〉

Fig. 3.2.4. Felice Frankel, *Ferrofluid*. Micro-photograph. From *On the Surface of Things, Images of the Extraordinary in Science,* Chronicle Books, 1991 (used in a National Science Foundation poster for National Science and Technology Week). © 1991, by Felice Frankel (original in color).

featured in a show called "On the Surface of Things," and a book by the same name. Here is her description of some of her photographs of the nano-world. Note how descriptions of the technical details of the processes interleave with discussions about the process of visually capturing the materials. In the future, scientific literacy may well become an element of artistic and more general cultural literacy:

A different production method leads to a similar pattern of lines . . . These depict one-millionth-meter-wide bars of an organic polymer that was generated by allowing the material, when liquid, to wick its way into a network of fine, continuous capillaries. The regularity of the filling of the capillaries shows clearly in the uniformity in length of the now-solid bars. The subtle interference colors that range across the array reveal unexpected detail: each color reflects the thickness of the polymer.

Generating images both beautiful and accurate is more demanding than simple record-keeping. But doing so offers science, and the public, a different way of seeing that can be more valuable. We must practice this art more often.[6]

Other Artists and Projects

Kurt Kohl creates science fiction environments partially stimulated by the ideas of nano-technology: "I create images of nano-robots, and possible molecular structures as a focused part of a much larger collection of images concerning the future and technology." On his Web site, Kohl offers a kind of manifesto for the nano-future. He believes that artists have a crucial role in envisioning the future possibilities, that structures must be established to encourage the technology and ensure the spread of its benefits, and that everyone must stay alert to attempts to suppress the technology. Here are some excerpts from his "A Nanobased Ethical Society of Free Information Models": "Only artistic minds have the compassionate vision necessary at this time of tremendous change. . . . We can tolerate no elitism at the helm of nanotechnology, especially military, foreign, or nonpublic control of nanotechnology. . . . If successful, we have the hope of a new age in a quantum multi-verse, not as slaves to the corporate machines, but as gods of the electron."

Charles Ostman has created a 3-D visual world that attempts to visualize some of possible inhabitants of the nano-tech future. The world includes molecular machines, self-assembling "nano-Lego" components, nanobots and nano-critters, pseudo proteins, quasi-viral components, "artificial" organisms, and ubiquitous "nano-foglettes." *Raino Ranta,* a Finnish ceramicist and digital artist, creates digital simulations of journeys to the nano-realm. One series includes *Mind Plant, White Garden,* and *Digital Garden.* Viewers of *Mind Plant* can interactively access increasingly detailed images.

Nuclear Science

As physicists studied the structure of the atom, they unlocked secrets of great power. Tremendous forces are involved in holding the components of atoms together. Under some conditions, matter and energy become interchangeable. Some materials naturally decay through the process of radioactivity, ultimately transmuting from one element to another (suggesting the ancient alchemical dream) and releasing energy. The stars are powered through these processes.

Scientists and engineers quickly began to consider applications of this knowledge. They created atomic and hydrogen bombs whose destructive power became graphically

---

Kurt Kohl: ⟨http://www.grfn.org/~eric/access.htm⟩
Charles Ostman: ⟨http://www.biota.org/ostman/charles1.htm⟩

obvious after the bombs were dropped on Hiroshima and Nagasaki. After the war, the nuclear power industry was born, with utopian visions of unlimited, cheap power.

Gradually the shadow side of nuclear power became clear, with safety concerns (illustrated by accidents at Chernobyl and Three-Mile Island, and the long-term effects of nuclear tests) and disposal problems (how radioactive wastes could be safely disposed)? For a long time, the whole world lived with the constant threat of nuclear missiles powerful enough to kill every living thing. Scientists and commercial exploiters have been criticized for not being honest about the issues involved, overhyping the realities, and sometimes putting greed above the general welfare.

At this point in history, nuclear science and engineering has a highly tainted reputation. For the general public and most artists, nuclear research is highly suspect. Artists have related to nuclear matters by exploring the human costs in artworks focused on the nuclear annihilation threat, Hiroshima, and Chernobyl. For example, see Peter D'Augustino's *Traces* (chapter 7.2) and Agnes Heggedus's Hiroshima work (chapter 7.4).

Nonetheless, there are signs that attitudes toward nuclear research may be changing. The nuclear threat has been greatly reduced with the end of the cold war. Scientists posit new utopian dreams, such as clean fusion energy from seawater, and nuclear-powered spacecraft as the only plausible way for interplanetary travel and space inhabitation. In spite of its bad reputation, the basic importance of nuclear research to understanding the sun and the functioning of the universe invites poetic reflection. This section reviews artistic works that focus on nuclear science and its applications.

James Acord

James Acord works directly with radioactive materials. He attempts to create sculpture and events that probe the history of nuclear engineering in a complex way that honors both its positive and negative aspects. He believes it is important to work directly with radioactive materials. He lives in Richland, Washington, home of the historically important Hanford Nuclear Plant, and has built relationships with researchers at that site. He is one of the few individuals in the world licensed to handle radioactive materials by both the U.S. Department of Energy and the European Energy Commission, and has acquired possession of spent nuclear fuel rods as part of his artistic materials.

One major project is a monument to the Hanover site, which figures prominently in nuclear history. He proposes to use twelve breeder-blanket assemblies (radioactive

---

James Acord: ⟨http://www.tcfn.org/timecapsule/html/james_l._acord.html⟩

Fig. 3.2.5. James Acord, *Atomic*. Acord checks the radioactivity levels of breeder blankets. Photo: Arthur S. Aubry.

sealed sources)—with all their associations of hopes and dangers—as key components of the sculpture. The breeder assemblies, along with basalt monoliths, are situated at the site of the B-reactor, the first reactor to produce significant plutonium. Acord is part of the Hanford TimeCapsule group, which proposes to create a time capsule that reflects the complex history of nuclear research:

The Hanford TimeCapsule Project is an attempt to consider our nuclear situation from many different points of view. The timeCapsule metaphor is used to help think about both the present and the far future in an objective way. What do we want to tell future generations about Hanford? What is it that is absolutely necessary we convey to our descendants for their own safety?[7]

The British arts-science group Arts Catalyst selected Acord as one of their artists in residence. They see his work, which transcends the standard pro- and antinuclear debates, as an example of the kind of work that artists might undertake to engage the scientific world. Their promotional literature sets the context.

The Death of the Atomic Age: When Enrico Fermi and his colleagues at Los Alamos were laying the ground for civilian nuclear power, they saw themselves as the new alchemists, using nuclear fission to transmute matter. The Brussels Atomium was built and a generation embraced the glamor of all things atomic.

Decades on, tainted by the Bomb and by dubious means of nuclear waste disposal, atomic scientists are hidden behind a wall of military security. The American artist James L. Acord tried to break down that wall and went to live in Richland, [WA] U.S., site of the first civilian reactor, to study nuclear engineering and to pursue his impossible dream of working with fissile material as a sculptural medium.[8]

In relation to Acord's residency, Arts Catalyst organized a symposium called "Atomic," which explored scientific, cultural, and artistic issues of nuclear energy. It also presented other related symposia, such as "Parallel Universes," which brought scientists, anthropologists, and artists together to reflect on the relationships between nuclear physics and traditional societies' cosmologies.

Summary: Difficulties of Working at the Atomic Level

It is not easy to investigate the atomic level of the universe. Scientists require extraordinary installations and equipment. Interpretation often requires extensive background in mathematics. The artists described in this chapter begin to claim research at the atomic level as a valid arena for artistic exploration.

Notes

1. S. Brixey, "Artist Statement," ⟨http://digitalmedia.berkeley.edu/shawn/brixey.html⟩.

2. S. Brixey, "*Alchymeia* Description," ⟨http://digitalmedia.berkeley.edu/shawn/brixey.html⟩.

3. S. Brixey, "*Instruments of Material Poetry* Description," ⟨http://digitalmedia.berkeley.edu/shawn/brixey.html⟩.

4. "NanoGallery Description," ⟨http://www.nanothinc.com⟩ (dead link).

5. A. Smith, "Editorial Introduction," YLEM March/April 1997, vol. 17:4 (also on-line at ⟨www.ylem.org⟩).

6. F. Frankel, "Description from 'On the Surface of Things'," ⟨http://www.nanothinc.com⟩.

7. J. Acord, "TimeCapsule Description," ⟨http://www.tcfn.org/timecapsule/html/james_l._acord.html⟩.

8. Arts Catalyst, "Description of Atomic Show," ⟨http://www.artscatalyst.org⟩

3.3

Materials and Natural Phenomena: Nonlinear Dynamic Systems, Water, Weather, Solar Energy, Geology, and Mechanical Motion

Since prehistoric times humans have observed natural phenomena. They have watched the heavens and the earth and monitored the flow of the winds and the waters. The spirit of observation has stretched from awe and terror to enjoyment. The purpose has extended from propitiation to fascination to exploitation to understanding to control. In modern times, the process of observation has been differentiated into fields such as religion, philosophy, art, science, and engineering. Art, literature, and poetry have always pondered the landscape. This chapter considers artists who continue this tradition influenced by the ideas, insights, and perspectives of contemporary physical sciences. It considers artists who are fascinated by nonlinear systems, the mechanics of motion, and geological and solar phenomena.

Nonlinear Systems

In the past, science aspired toward a dream of total predictability. Contemporary scientists now accept that some systems are nonlinear. Slight perturbations can have profound effects on the systems, which makes them hard to predict. Complexity and chaos theory study the nature of this predictability. Some theorists see this area of concern as a radical challenge to traditional physical science; others see it as a fine tuning. Artists find the body of speculation fascinating because it seems to open new windows of communication between the arts and sciences.

"Turbulent Landscapes" at the Exploratorium and the InterCommunication Center

San Francisco's art and science museum, the Exploratorium, mounted an exhibit called "Turbulent Landscapes," which explored the science and art of complex, nonlinear systems. A related exhibit was presented by the InterCommunication Center (ICC), in Tokyo. The exhibits included artworks that explored a variety of phenomena. The "Turbulent Landscapes" introduction explains:

"Turbulent Landscapes" uses fog, wind, smoke, sand, water, gas plasma, slime mold, and other natural phenomena in ways that are ethereal, beautiful, mysterious, sensual, and playful. Yet, it's also an exhibition that touches on compelling questions about complexity, the emergence of order and disorder in the universe, and our perception of that process.[1]

The focus on nonlinear systems was seen as a possible point of convergence for art and science. The Web site for the exhibit offers an "Ideas and Statements" section that emphasizes the search for pattern as a cornerstone of both art and science. Jim

TURBULENT LANDSCAPES

the exhibition

Fig. 3.3.1. Exploratorium, "Turbulent Landscapes" Web site.

Fig. 3.3.1. Exploratorium, "Turbulent Landscapes" Web site.

Crutchfield, a complexity researcher at University of California at Berkeley who helped design the exhibit, describes its appeal in making ideas of complexity accessible:

"Turbulent Landscapes" celebrates a new view of nature—a view that, when coupled with recent scientific innovations, allows us to understand much of nature's inherent complication. We now ask: How do simple systems produce unpredictable behavior? And, in a complementary way, how is it that large complicated systems generate order? Most importantly, we are learning how to answer these questions. It appears that much of what is intricate and highly structured in nature arises from a delicate interplay of order and chaos. All of the exhibits illustrate this, not only in how patterns emerge, but also in our perception of those patterns.[2]

Peter Richards, who was curator of the Exploratorium's artist-in-residence program, described the exhibit, the contributions that artists make to the understanding of natural phenomena, and the faith that science and art can learn from each other:

A group of artists working at the Exploratorium have recently completed a body of work that examines those systems in nature that are inherently self-organizing. At a time when scientists finally have large enough computers to study these complex systems, these artists, working here with simple materials, have created works that model these same systems in ways that not only capture the physical essence of this phenomena, but also their essential beauty. Be it the filagree of an a-cellular slime mold, the sensual flow of water over eroding terrain, or the organic nature of a video feedback system, it is the beauty of these phenomena that lead to questions and deeper observations, observations that have led to significant learning experiences for those working in this field, for ourselves and for our visitors.[3]

Melissa Alexander, project director of "Turbulent Landscapes," described the special significance of nonlinear theory by quoting from Tom Stoppard's "Arcadia":

The unpredictable and the predetermined unfold together to make everything the way it is. It's how nature creates itself on every scale, the snowflake and the snowstorm. It makes me so happy. To be at the beginning again, knowing almost nothing. People were talking about the end of physics. Relativity and quantum looked as if they were going to clean out the whole problem between them. A theory of everything. But they only explained the very big and the very small. The universe, the elementary particles. The ordinary-sized stuff which is our lives, the things people write poetry about—clouds-daffodils-waterfalls-and what happens in a cup of coffee when the cream goes in—these things are full of mystery, as mysterious as the heavens were to the Greeks. We're better at predicting events at the edge of the galaxy or inside the nucleus of an atom than whether it'll rain on auntie's garden party three Sunday's from now. Because the problem turns out to be different. We can't even predict the next drip from a dripping tap when it gets irregular. Each drip sets up conditions for the next, the smallest variation blows prediction apart, and the weather will always be unpredictable.[4]

She notes that science and technology are at last giving us the tools to understand what used to seem chaotic and unpredictable. Both science and art are useful in unraveling the mysteries of everyday phenomena: "[T]echnology is changing our vision. Because of computers and new forms of scientific visualization, we are now beginning to see and describe rhythms and patterns and order in places where before we could perceive only chaos or disorder."[5]

Ned Kahn

The artist Ned Kahn was highly featured in the "Turbulent Landscapes" show. For a long time he has worked with the Exploratorium, creating artworks that reflect on a variety of natural phenomena. His works have been shown internationally in many museums and have been commissioned for a variety of architectural settings, such as the mist-filled wall for the San Francisco City Jail, *Boundary Condition* for the National Oceanic and Atmospheric Administration headquarters in Boulder, Colorado, and *Wavespout* (Breathing Sea) for the Ventura pier in Ventura, California. "Turbulent Landscapes" included twenty-one of his sculptures, which explored phenomena such as vortexes of

Ned Kahn: ⟨http://www.sculpture.org/documents/kahn⟩

Fig. 3.3.2. Ned Kahn, *Fluvial Storm*. Movement of a globe causes sand patterns as in the desert.

water and smoke, wind currents, rippling sand, squiqqling hoses, and magnetics. His statement from the show describes an interest in framing and enhancing the perception of natural phenomena:

The confluence of science and art has fascinated me throughout my career. For the last fifteen years, I have developed a body of work inspired by atmospheric physics, geology, astronomy, and fluid motion. I strive to create artworks that enable viewers to observe and interact with natural processes. I am less interested in creating an alternative reality than I am in capturing, through my art, the mysteriousness of the world around us.

My artworks frequently incorporate flowing water, fog, sand, and light to create complex and continually changing systems. Many of these works can be seen as "observatories" in that they frame and enhance our perception of natural phenomena. I am intrigued with the way patterns can emerge when things flow. These patterns are not static objects, they are patterns of behavior—recurring themes in the repertoire of nature.[6]

Many of Kahn's exhibits are interactive—inviting the viewer to manipulate them for purposes of aesthetic appreciation and/or better understanding of the phenomenon. For example, in *Aeolian Landscape*, twisting a knob activates a fan that blows sand into patterns reminscent of the desert.

Natural Phenomena—Oceans, Water, and Moving Liquid

The world is full of water. Oceanographers, earth scientists, and others try to understand and control its movement. The arts have a long tradition of working with moving water in the design of fountains. To the extent that the artists try to realize some nonstandard movement of water, they must function as engineers and scientists. This section highlights artists who create works focused on the phenomenon of moving water.

Peter Richards and George Gonzales

Peter Richards and George Gonzales created *Wave Organ*, which uses the wave motion of San Francisco Bay to create a musical instrument:

Wave Organ, a wave-activated sound sculpture, is located on a nearby jetty and utilizes wave action from the bay to create a symphony of sound that emanates from a series of pipes that reach down into the water. A wonderful collection of granite building material that existed on the site was utilized to create a series of sculptured terraces and seating areas. The listening pipes, made of PVC and concrete plaster, extend from the seating areas to the water. The intensity and complexity of the wave music is directly related to the tides and weather.[7]

Paul DeMarinis

In *RainDance/Musica Acuatica*, twenty falling streams of water, modulated with audio signals, created music and sound when intercepted by visitors' umbrellas (see also chapter 5.2).

Other Artists and Projects

In *Ocean Merge*, **Stephen Wilson** created an outdoor event on the beach in which a computer coordinated sound movement to and from the audience with the sensed movement of waves and tides. ***Michael Brown*** creates works that build on the flow of water.

Michael Brown: ⟨http://www.exploratorium.edu/complexity/exhibit/⟩

Fig. 3.3.3. Paul DeMarinis, *RainDance/Musica Acuatica* (1998). Sound modulated into water streams plays tunes as it hits the umbrellas.

Meandering, which was in the "Sensitive Chaos" show, flows water down an adjustable tilting glass that meanders and reflects like a river. **Paul Sermon and Andrea Zapp**'s *Body of Water* projected images onto flowing water. **Sally Weber,** best known as a holography artist, created *Threshold of a Singularity—A Memorial,* which uses holography to visually amplify the behavior of water dropping in a pool. **Lewis Alquist** created *Fickle Oracle,* which creates a parabolic mirror by spinning a circular basin of mercury on a turntable. Viewers can change the focal length by changing the spin speed. **Stephen Pevnick** sculpts fountains with his computerized system for controlling water drops such that

"Sensitive Chaos," ICC: ⟨http://www.ntticc.or.jp/special/chaos/index_e.htm⟩
Paul Sermon and Andrea Zapp: ⟨http://www.hgb-leipzig.de/~sermon/⟩

text or images can be embedded in the flow. **Michel Redolfi** composes music for underwater environments, such as *Sonic Waters,* which broadcast sonar, and *In Corpus,* which monitored the motion of people in a pool. **Pamela Davis,** who coordinates art-science programs for UCLA, creates sculptures that explore physical and mathematical concepts, such as the **Taylor Column,** which builds its effects based on swirling liquids inside concentric tubes. **Athena Tacha** created a series of public installations based on study of water in its various forms including *Ice Blocks, Merging,* and *Chaos* (Fluids). In her *Soft Earth* projects, *Joan Lederman* uses the sediments of various oceans as ceramic glazes. **Yuki Sugihara**'s *Spiral Water Dome* encloses visitors in a hemispheric space with a thin membrane of water onto which images can be projected. Exploring concepts of transience, **Andy Goldsworthy** works with natural forces such as wind and water to create sculptural installations such as his icicle and snow works that are transformed by the elements.

Natural Phenomena—Erosion and Geological Action

The earth is formed by the action of water, wind, and tectonic activity. Geologists study these forces to understand the earth's past, present, and future. Some artists are similarly intrigued by the processes that created the earth.

JoAnn Gillerman and Rob Terry

JoAnn Gillerman and Rob Terry created a cooperative interactive event called *The Sun Drops its Torch* in which viewers could explore the sounds and images associated with a Hawaiian day in 1991, when a solar eclipse and lava flow happened simultaneously recreated via a coordinated ring of video monitors.

Ken Goldberg, Randall Packer, Wojciech Matusik, and Gregory Kuhn

The *Mori* installation uses real-time seismic data from Tokyo to control a "symphony of low-frequency sounds" modulated in real time by the seismic data and a point of

---

Michel Redolfi: ⟨http://www.aec.at/liquidCT/m_redolfi/index.html⟩
Pamela Davis: ⟨http://netra.exploratorium.edu/exhibit_services/consulting/pdavis.html⟩
Athena Tacha: ⟨http://www.oberlin.edu/~art/athena/tacha2.html⟩
Joan Lederman: ⟨http://www.arts-cape.com/softearth/⟩
Yuki Sugihara: ⟨http://www.star.t.u-tokyo.ac.jp/~yuki/⟩
Andy Goldsworthy: ⟨http://www.arc.cmu.edu/portfoliödocuments/artnarch/page21.html⟩
JoAnn Gillerman and Rob Terry: ⟨http://www.vipervertex.com/⟩
Mori: ⟨http://memento.ieor.berkeley.edu/⟩

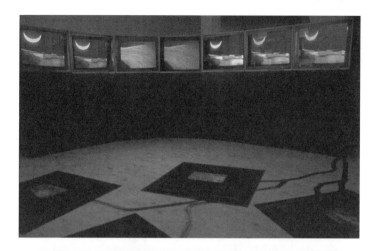

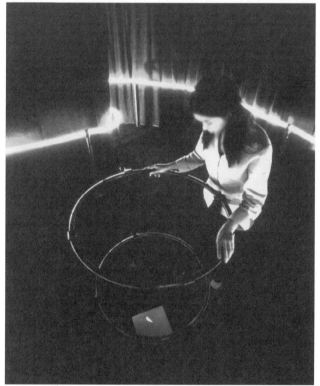

Fig. 3.3.4. JoAnn Gillerman and Rob Terry, *Sun Drops Its Torch*. An interactive environment exploring volcanic flows and solar eclipses.

Fig. 3.3.5. Ken Goldberg, Randall Packer, Greg Kuhn, and Wojciech Matusik, *Mori*. Visitor views image that links seismic data with user action. Photo: Takashi Otaka.

moving light that graphs the activity levels. The darkened installation is filled with the smells of the earth and seeks to offer "a visceral reminder of passing time and human fragility," as in the Latin phrase *memento mori*—"reminder of death."

Other Artists and Projects

Geology *Eve Andrée Laramée* creates installations that investigate scientific methodologies and conceptualizations. Her *Eroded Terrain of Memory* reflects on geological concepts of plate tectonics and more generally on geographic boundaries by creating a simulated fault line out of mica in a gallery. *Gloria Brown Simmons* creates data visualization projects that create animated images related to various natural phenomena. *Oceanet* interprets remotely sensed global ocean science data. Another project called *Visualization of Tectonic Features: The Colorado River Extension* visualizes surface features and underlying seismic details. In *Blooming Saltgarden,* *Jorg Lenzingler* builds a miniature geological formation out of crystallizing salt. In *Terra Forms,* *Al Jarnow* creates a micro environment based on the phenomenon of erosion created by water flowing through sand on an inclined surface recorded by a video time-lapse recorder. *Ray Pestrong* devised a system for converting topography into musical sounds. *Robert Dell* creates sculpture that incorporates geothermal energy including installations in Iceland and at Yellowstone National Park. *Bill Thibault and Scot Gresham Lancaster* composed the TerrainReader, which creates sound art based on geological terrain. *Seismic Activity* *Ingrid Bachman* used telecommunication links to transfer seismic data to control weaving machines in remote locations. *Andrew Michael,* a geologist and musician, records the vibrational patterns of earthquakes and converts them into audible sound. Performance artist *Rachel Rosenthal* created a performance called *Pangaean Dreams,* which is a paean to the earth. Tectonic activity and geological evolution is a metaphoric focus. It is a call to be careful what we do because it might negatively influence the path of evolution. *Nola Farman* placed *The Subterranean Listening Device* under the earth in Western Australia to monitor seismic activity. Sensorium visualizes seismic data in the "Breathing Earth" Web site, which provides fourteen-day time-lapse imagery (see chapter 6.3). *Lewis DeSoto*'s *Aborescence* outdoor sculpture broadcast low-frequency earth sounds on car radios as viewers drove near. *Keigo Yamamoto and Vigdis*

Eve Andrée Laramée: ⟨http://www.artnetweb.com/laramee⟩
Gloria Brown Simmons: ⟨http://cavs.mit.edu/gbrown/⟩
Robert Dell: ⟨http://web.mit.edu/mit-cavs/www/Bob.html⟩
Bill Thibault and Scot Gresham-Lancaster: ⟨http://www.mcs.csuhayward.edu/~tebo/TerrainReader.html⟩
Andrew Michael: ⟨http://www.hmbreview.com/community/stories/981026011.html⟩
Rachel Rosenthal: ⟨http://w3art.com/rr01/rr01t02.html⟩

Fig. 3.3.6. Sensorium, *BeWare: Satellite*. Satellite images of the earth's surface are projected onto a long plate; the infrared images are analyzed as temperature data, used to control Peltier devices attached to the underside of the plate. Photo credit: Ichiro Higashiizumi.

Holen created *Dancing Fire and Water* to symbolically connect Norway's Jostedal Glacier and Japan's Owakudani volcano. Using ISDN high bandwidth lines, they cross transmitted images and sounds, reflecting on the interplay of virtual and physical realities.

Natural Phenomena—The Sky, Winds, and Weather

The sky and the wind are ever present parts of life. In the past, kinetic artists built wind-activated sculpture. Sky artists such as Otto Peine created inflatable and balloon-based art. Any artist trying to work in this unorthodox environment needed to engage knowledge of science and engineering related to the sky. New opportunities have been opened up by scientists' use of satellites to study climate and weather. This section focuses on artists interested in phenomena such as the winds and weather.

Project Taos

At SIGGRAPH98, the Japanese group Project Taos created a Sensorium installation called *BeWare,* focused on world weather, in which infrared satellite images of a longitu-

Project Taos: ⟨http://www.sensorium.org/⟩

Fig. 3.3.7. Walter De Maria, *The Lightning Field* (1977). A grid of metal rods in the desert attracts lightning bolts. All reproduction rights reserved. Courtesy of Dia Center for the Arts, New York.

dinal slice of the earth's surface were projected onto a long narrow plate. The temperature of places on the plate was correlated with the temperature of the corresponding place on the earth derivable from the infrared data.

Walter De Maria

Walter De Maria created an extraordinary earth work called *Lightning Field* in an area of New Mexico with high incidence of lightning. The sculpture consists of four hundred stainless steel poles in a rectangular grid array of one mile by one kilometer. Depending on conditions, the poles attract many lightning strikes. The site is maintained by the Dia Foundation, which schedules overnight visits.

Other Artists and Projects

Artist **Gene Cooper** presents performances in which he wires his body up to be stimulated in accordance with lightning strikes using the capability of the NOAA (National Oceanic and Atmospheric Agency), which can record lightning strikes anywhere in the United

States. The German group Art+Com has created TerraVision (T-Vision), which allows a simultaneous real-time view of the whole earth composited from satellite images as an examination of the "decoupling of time and space." *Jon McCormick*'s *Four Imaginary Walls* projected 3-D computer animations influenced by current weather conditions. *Michael Rodemer* created an installation in which fans in West Virginia were linked to wind data in Chicago. *Robin Minard*'s *Weather Station* used weather information to activate a sound installation of 320 piezo speakers. In *Writing Machine,* *Patrick Clancy* created an installation in which pages of text are rearranged based on a combination of sensor readings of the sun and weather and user interactions from the Web. Japanese architect *Toyo Ito*'s *Tower of the Winds* reads wind speed and direction to change its lighting. *Bill and Mary Buchen* build sound installations that respond to wind and sun, such as sun catchers, which generate sound and movement of light based on the winds and various kinds of wind instruments such as harps, bows, and gamelans. *Guillaume Hutzler, Bernard Gortais, and Alexis Drogoul*'s *Garden of Chances* presents an abstract computer graphic animation whose motion and color is controlled by real-time weather data from a particular spot in Britain.

Solar Art

For many scientists, the sun is the most intriguing object in the sky. It is the source of energy for everything on the earth and the center of our local solar system. It is a gateway to understanding the other stars. Artists have pursued a variety of approaches to solar art. A show called "Aurinko—Sun" at the Rauma Art Museum in Finland brought together mythological and technological sun-oriented works.

SolArt Global Network

Jürgen and Nora Claus organized the international SolArt Global Network to coordinate work of artists, scientists, and theorists whose work was inspired by the sun. It was begun on the summer solstice in 1995. Here is the statement of purpose:

TerraVision: ⟨http://www.artcom.de/projects/t_vision/⟩
Jon McCormick: ⟨http://www.cs.monash.edu.au/~jonmc/fiw.html⟩
Michael Rodemer: ⟨http://www-personal.umich.edu/~rodemer⟩
Robin Minard: ⟨http://www.mhsg.ac.at/iem/bem/minard1.htm⟩
Patrick Clancy: ⟨http://www.banffcentre.ab.ca/mva/deep_web/dprojects.html⟩
Bill and Mary Buchen: ⟨http://www.users.interport.net/~sonarc/maintext.html⟩
"Aurinko—Sun" art show: ⟨http://mitpress.mit.edu/e-journals/LEA/ANNOUNCE/ann_5-6.html⟩
SolArt Global Network: ⟨http://www2.khm.de/~SolArt/artClaus.html⟩

Fig. 3.3.8. Jürgen and Nora Claus, *Solar Crystal.* A sculpture uses solar energy collected during the day to illuminate it at night.

The SolArt Global Network is an international group of artists using solar energy in their artwork. These artists seek to stimulate the global cultural imagination toward use of renewable energy resources. . . . Examples of solar art are:

- Research-based art penetrating into the deep space of light
- Solar light works, including mirrors, prisms, and reflections of sunlight
- Outdoor holograms using sunlight as a great attractor
- Light work depending on direct use of solar power by photovoltaics[8]

Jürgen Claus sees the transition to solar energy as an absolute ecological necessity. He notes that the change will not happen without underlying cultural changes, and sees artists who understand solar energy as a critical element in bringing about that change. Claus creates sculptures that use solar energy.

Fig. 3.3.9. Robert Mulder and Kristi Allik, *Skyharp*. A virtual instrument that extracts dynamic information from the natural environment to create sound.

Robert Mulder and Kristi Allik

Robert Mulder and Kristi Allick are sound artists who have produced many installations and performances. Their projects are often part of larger events, such as the *Millennium Project,* designed to increase the public's awareness of environmental issues. *Skyharp* is an installation performance that senses environmental forces, such as the wind and sun, and uses that information to shape sound events:

Skyharp is a virtual instrument designed to analyze and extract (by means of a video camera and computer system) dynamic information from a natural environment. With this information a multi-channel electro-acoustic texture is generated which is subsequently played back via twenty

Robert Mulder and Kristi Allik: ⟨http://www.aracnet.net/~rmulder/⟩

or so specially constructed "tube speakers" that are placed among the natural constituents of the outdoor environment. The name Skyharp reflects the aesthetic function of the installation; it is a composition that is "played" by elementary ecological forces, such as the wave action of water, the broad sweeping and subtle fluttering movements of the trees, the revolving motion of clouds, and the imperceptibly slow creeping of light and shadow. In the "hands" of these sources a hauntingly beautiful soundscape is created which harmoniously complements the incipient source.[9]

Arts Catalyst—Eclipsing the Millennium

Arts Catalyst, the British arts-science group, created a major event that used the eclipse visible in Europe in 1999 to look at the ways in which scientists and artists explore solar events. They called the project *Eclipsing the Millennium.* Projects included a collaboration of artists Anne Bean and Peter Fink with Marcus Chown, science editor of *New Scientist* and author of *The Afterglow of Creation,* to create a London Planetarium show focused on eclipses and solar wind, and also a collaboration between Artist Kathleen Rogers and botanist Sandra Knapp at the Natural History Museum of London to create *The Imagination of Matter,* which focuses on "the development of the maize crop in pre-Columbian civilizations, the significance of the eclipse in their rituals, and parallels with present-day genetic engineering." The promotional materials describe events focused on the eclipse.

Scientists will be able briefly to study the effects of the solar wind: clouds of subatomic particles which emanate from the sun and sweep past us to the edge of our planetary system. . . . The Arts Catalyst is planning a sequence of science-art projects in 1999, the year of the eclipse, to mark this significant cosmological event and contemplate humanity's complex and changing relationship with the cosmos . . .

The phenomena of the total eclipse for me [artist Anne Beam] represents a strange paradox where, although we can rationally explain so much of our universe, there are still some phenomena, large and small, which retain the power to overwhelm, exhilarate, and transcend reason.[10]

Other Artists and Projects

Sculptures *Jürgen Claus*'s *Solar Crystal,* a public art sculpture placed in front of a technical high school, consists of six photovoltaic panels facing south to collect the light

---

Arts Catalyst: ⟨www.artscatalyst.org⟩

of the sun. The energy collected in the solar cells is stored in six large solar batteries. The duration of the internal illumination of the solar crystal during the dark hours is dependent on the amount of light falling on the panels. ***Therese Lahaie*** creates viewer-activated sculptures that are powered by the sun. The interactions of sunlight and user actions cause a choreography of motion. ***Uli Winters***'s "Thinking Machine" kinetically moves based on evaporation in the soil. Cynthia Parnucci creates solar-powered water sculptures, such as *Water Strider,* which activates at night based on its day's charge. ***Benoit Maubrey and the Audio Ballerinas*** organize dances and performances in which people wear sound generators powered with solar energy. ***Alex and Martha Nicoloff*** create environments consisting of sunlight, prisms, the light spectrum, and sound. They convert scientific principles and technological innovation into a variety of aesthetic formats. *Spectralwave* precisely positions spectral refraction patterns of sunlight on the sculpture depending on the day of the year.

Installations and Performances   *Christina Kubisch* creates sound installations, many of which are site specific. In *Clocktower Project,* she reactivated a clock tower in North Adams, Massachusetts, that was part of an abandoned factory. The factory had been abandoned because of outmoded technology, so she decided to use state-of-the-art technology to revive it. Using solar sensors, she arranged for the tower to play digitized clock tower sounds keyed to the nature of the day's sunlight (for example, hazy) and the position of the sun. ***Seth Riskin***'s *Sketches of Rainbow Man* was a light and dance performance involving direct integration with sunlight. A compressed nitrogen and water tank on the performer's back connected to jets on his body, which sprayed a mist of water droplets that reflected rainbows. ***Charles Ross*** creates installations that mark the seasonal movement of the sun over time by using a lens to systematically leave burn marks, forming a figure eight over the period of a year. ***Stephan Barron*** generated the *Night and Day* event, linking Brazil and Australia by creating real-time composite images of the sky in these places, located on opposite sides of the earth and separated by a twelve-hour time difference. ***Ernie Althoff***'s *Helisonics* presented an outside "orchestra" of small sun powered unorthodox instruments which generated sounds similar to insects and birds.

Therese Lahaie: ⟨http://mitpress.mit.edu/e-journals/Leonardo/gallery/gallery294/lahaie.html⟩
Uli Winters: ⟨http://home.t-online.de/home/uli.winters/⟩
Cynthia Parnucci: ⟨http://asci.org/news/featured/pannucci/wsproj.html⟩
Benoit Maubrey: ⟨http://www.snafu.de/~maubrey/⟩
Christina Kubisch: ⟨http://www.sleepbot.com/ambience/page/kubisch.html⟩
Stephan Barron: ⟨http://www.v2.nl/freezone/users/dsm/artist/barron/bar03.htm⟩

Mechanics—Oscillation and Pendulum Action

Norman Tuck

Norman Tuck creates large kinetic sculptures out of everyday materials, surplus, and electronic components. Typically whimsical, they use the basic principles of physics and mechanical engineering to create amazing events. The Exploratorium sponsored a show of his work called "Art Machines":

[Tuck's exhibits] incorporate such disparate stuff as lemons, bowling balls, and silicone chips to demonstrate scientific principles such as periodicity, kinetic energy, resonance, and magnetism. Each visitor-animated gadget fuses whimsy, science, and art. At *Disco,* for example, a gearbox bolted to a large turntable becomes an amusement ride where you can see what it's like to walk forward and move backward at the same time. All of the art machines are constructed of elegantly simple elements such as screws, bearings, gears, chain drives, pendulums, motors, light-emitting diodes (LEDs), magnets, and other familiar objects. They seem to suggest the machines we know. But when the sculptures start to move, they shift away from the routine order of machines and force us to think about things in a new way. Elegant and funny, all they require are your hands to set them in motion.[11]

Illustrating a similar approach to using physics and mechanical linkages, the Exploratorium also showed the film *Der Lauf der Dinge* (The Way Things Go), by Peter Fischli and David Weiss, which depict an amazing Rube Goldberg chain of events in which objects cause other objects to move.

Tuck creates strange kinetic devices that use pendular and other kinds of motion to achieve their effects. For example, the *Pendulum Clock* is a "giant see-through erector-set-like structure that seems to be a realization of one of Leonardo da Vinci's mechanical inventions" in which a twenty-four-foot construction of steel, ropes, bicycle chains, and a bowling ball mechanically keeps time.

Other Artists and Projects

Anna Valentina Murch's *Chaotic Chains* is a kinetic sculpture that explores the forms made by chains of mirrored balls suspended from the ceiling and attached at the bottom to motor shafts. A flashing strobe catches the evolution of the motion, from regular to chaotic. In the pendulum painting *Nothing, Magnified* **Thomas Shannon** creates work based on paint dripping from a moving pendulum. **Archimedia** created the *Audio Pendulum* project for a Helsinki, Finland, public art project in which video images of each pendulum were converted into sounds within a particular part of the sound spectrum.

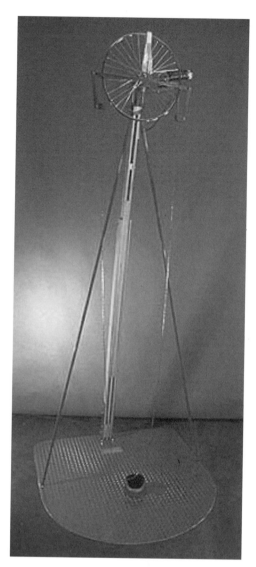

Fig. 3.3.10. Norman Tuck, *Lariat Chain*. A sculpture uses the chaotic motion of a rotating chain. Photo: Joe Hastings.
© The Exploratorium ⟨www.exploratorium.edu⟩.

The Soda Group created a popular Java Web event called *Constructor,* in which visitors can control the movements of animated objects via changes in parameters such as gravity and friction. (See chapter 5.2 for other artists working with kinetics.)

Fire, Heat, Magnetics, and Electromagnetics

Artists and Projects

Historically, *Takis* was famous for his introduction of magnetic forces into sculpture as a reference to invisible and mysterious forces. *David Durlach* choreographs magnetic dust and fluids through the computer control of underlying electromagnets. *Heather McGill and Stan Axelrod*'s *Magnetic Field Patterns* allows users to manipulate fluids into "impossible shapes" by moving magnets. *Nick Bertoni and Maggi Payne*'s *Flame Speaker* uses the ionized gas of a flame to function as a speaker. *Bill Parker*'s *Turbulent Business* investigates electrified plasma gas that puts out mini lightning bolts wherever the viewer's finger touches a surrounding glass dome. *Juanita Miller*'s *Point of Criticality* explores the phenomenon of critical change points found in many natural events: "A large pile of sand or grain is a self-organizing system in the way that particles fall and fit together, resting on top of one another. If they are allowed to pile onto a space with no sides they form as high as they can given their shape and weight. . . . When the pile reaches a height that is no longer sustainable it avalanches." Responding to the aurora magentic storms, *Catherine Richards* sculpted *Cabinet,* a completely electromagnetically isolated enclosure built out of copper-clad elements inspired by scientific appartus of the nineteenth century. *Mario Ramiro* has created a series of gravity-defying sculptures using electromagnetic levitation and thermal forces.[12]

Materials Science, Rapid Prototyping, and Chemistry

Artists and Projects

Materials Science   Jean-Marc Philippe sculpts with nickel-titanium shape-memory alloys that change their shape with seasonal and diurnal temperature changes. *Richard Lerman* consructs wound installations with piezo ceramic sound transducers. *Yves*

The Soda Group: ⟨http://sodaplay.com/index.htm⟩
David Durlach: ⟨http://coldfusion.discover.com/output.cfm?ID=1073⟩
Richard Lerman: ⟨http://switch.sjsu.edu/switch/sound/reviews/lerman.html⟩

Kleine's *Octofungi* sculpture uses silent-shape metal as its kinetic infrastructure (see chapter 5.3). ***Ted Krueger*** explores the idea that new materials will be critical in the future of architecture by building interactive shape-metal sculptures (see chapter 5.4). ***Evelyn Rosenberg*** developed a "detonographic" method of using explosives to fuse metal with other carved elements. ***Vibeke Sorensen,*** known for her animation and interactive media works, has collaborated on several scientific projects to develop new display technologies. In *Chemical Painting*, ***Ronald Warunek*** created images by arranging for specially mixed chemicals and pigments placed between plates to be subjected to electromagnetic and temperature manipulation. In works such as *Contingency* and *Indeterminancy,* ***Dove Bradshaw*** carefully selects materials and chemicals to put together to unleash physical processes of reaction and transformation. ***Toshiya Tsunoda***'s installation *Monitor Unit for Solid Vibration* created a sound event of out of sensors monitoring the vibrations given by the materials and structures of the Tokyo Opera City building.

Rapid Prototyping ***Michael Rees*** has created many sculptures using RP techniques. In one paper he characterized his work with the technologies as playing "in what medieval alchemists called the Albedo state, the silvery mercurial state where one thing can reflect or become another as easily as not." ***Eva Wholgemeuth*** used the data map of her body to create stereolithographic *Evadolls*. ***Derrick Woodham*** organized the Intersculpt99 conference which invited sculptors to submit their work for display in a VRML (virtual reality modeling language) exhibition space and offered to generate rapid prototype models of some of the works for an accompanying gallery show. Using rapid prototyping techniques, artists ***Masaki Fujihata, Michael Rees, Keith Brown, Peter Terezakis, and Arghyro Paouri*** have created sculptures from synthetic computer models. ***Stewart Dickson*** built sculptures from stereolithographic renderings of abstract mathematical forms. ***Christian Lavigne*** wrote an article "La sculpture numerique," which reviewed digital sculpture, including rapid prototyping experiments. Boston's ***CyberArts***

Ted Krueger: ⟨http://comp.uark.edu/~tkrueger/metadermis/meta.html⟩
Evelyn Rosenberg: ⟨http://www.toknowart.com/detonographics/index.html⟩
Vibeke Sorensen: ⟨http://felix.usc.edu/text/bio.html⟩
Ronald Warunek: ⟨http://www.warunek.com⟩
Dove Bradshaw: ⟨http://www.artseensoho.com/Art/GERING/dove98/dove.html⟩
Toshiya Tsunoda: ⟨http://www.star.t.u-tokyo.ac.jp/~yuki/⟩
Michael Rees: ⟨http://www.michaelrees.com⟩
Derrick Woodham: ⟨http://www.sculptor.org/InterSculpt99.htm⟩
Stewart Dickson: ⟨http://www.cs.berkeley.edu/~sequin/⟩
Christian Lavigne: ⟨http://www.sculpture.org/documents/webspec/magazine/wsenglis.htm⟩

Fig. 3.3.11. Michael Rees, *Ajna Spine Series 2* (1998). A twenty-inch-tall SLS (selective laser sintering) sculpture created using rapid prototype technology ⟨http://www.michaelrees.com/⟩.

organized the "Mind into Matter" show of artists working with rapid prototyping technology, including *Tim Anderson, Jim Bredt, Dan Collins, Bill Jones, Michael La-Forte, Christian Lavigne, Denise Marika, and Michael Rees. Sheldon Brown*'s *Istoria* is creating three relief-sculpted panels built by rapid prototyping technologies that derive the images from 3-D databases. The databases are collected from immovable dwellings around the world such as the temporary shelters of Mexicans preparing to move across the border or environments created by homeless people underneath Tokyo's Shinjuku subway station. *Tom Longtin* uses rapid prototyping processes to build organic-looking parts from computer files. *Dan Collins* at Arizona State University creates virtual sculp-

Mind into Matter: ⟨http://www.bostoncyberarts.org/mindmatter/mimtitle.html⟩
Sheldon Brown: ⟨http://www-crca.ucsd.edu/~sheldon/istoria.html⟩
Tom Longtin: ⟨http://www.sover.net/~tlongtin/⟩
Dan Collins: ⟨http://surdas.eas.asu.edu/prism/⟩

tures and directs a research project to support artistic RP experimentation. **Andrew Werby,** a sculptor using RP technologies, maintains a 3-D sculpture Web site with links to technologies and artists. **Richard Collins** maintains an extensive *Rapid Prototyping for Sculptors* Web site with links to art work and explanations of the technology.

Summary: Pattern Finding and Poetry of Matter

Pattern Finding

Artists and scientists look for patterns. For a long time both fields concentrated on the visible world. Both found power in their portrayal of what they held to be the underlying patterns of nature. Now that simple formula is under assault. Scientists focus on nonvisible phenomena and examine patterns such as nonlinear systems that are not so obvious to the uninstrumented observer. Similarly, some artists focus more on cultural rather than physical phenomena. They emphasize the importance of cognitive frameworks and media in guiding perception and question the validity of the idea that there are universal patterns in nature waiting to be artistically explored. What will be the future of artistic attention to nonhuman phenomena? It is an open question. The artists in this section have found the natural world worthy of attention.

In a public conversation offered as part of the "Turbulent Landscapes" show, Ned Kahn and Jim Crutchfield, the dynamic systems scientist who helped design the show, offered some insight about the pleasures of this kind of attention to the physical world. They discussed some of the goals and principles used in creating the artworks, and Kahn noted his interest in human-pattern finding and the critical importance of surprise in his identification of the themes and designs of works. Crutchfield noted a similar factor in scientific research. They also discussed the interplay of pattern and ambiguity, and both suggested that exploration of these phenomena still holds appeal even in the contemporary world. Here are some excerpts:

NK [*Ned Kahn*]: People project all kinds of patterns and associations into these systems, but that's what makes it art. Artists try to create things that are ambiguous, things with many levels of meaning for people to impose their patterns on. . . .

Andrew Werby: Andrew Werby: ⟨http://users.lanminds.com/~drewid/ComputerSculpture_links.htm⟩
Richard Collins: ⟨http://www.sculptor.org/Technology/rapidpro.htm⟩

JC [*Jim Crutchfield*]: There's a threshold beyond which there's so much ambiguity that pe̜₀ won't see any structure. This must be the hardest part in creating the exhibits, to play against too much ambiguity. Things that are seemingly structureless are uninteresting. At the other end of the spectrum, a completely straightforward, obvious statement of fact is not engaging. Interest-ing-ness increases—human interest increases—as you increase the ambiguity. Things in the world that are really intriguing draw you in. Initially, at least, if you're a little uncertain about what you'll see, they're ambiguous. Then the structure is revealed and you begin to see patterns.[13]

In his art, Kahn attempts to discover intriguing natural processes and then find ways to create works that will engage the audience in a similar process of discovery and delight. Crutchfield describes his understanding of Kahn's approach:

JC: When you were talking about the beach with the stream flowing into the ocean, you focused on that buildup of the wave and it's collapse. You wanted to capture that and show it to others. Your description gave me a much more vivid picture of how, through your art, you're engaged in a different kind of communication with your audience. It's a different kind of iconography from that found in other kinds of art. What you're talking about are pattern-forming processes, temporal processes. It's not that you can take a snapshot and say, "Here, see the lump in the water? Believe me, in the next two seconds it suddenly, mysteriously disappeared." The fascination is not so much in the objects in and of themselves, but how they go through their changes.

Crutchfield suggested that this focus on process is different than the coding process followed by more conventional artists who present an "encrypted" world that the viewer is supposed to decode. Kahn observed that his engagement of an audience to ask ques-tions of nature is similar to the way that scientists ask questions of nature:

NK: I've tried to create things that are about what's happening right in front of you, right now. My work has been heavily influenced by practicing Buddhism over the last twenty years. The essence of it is a kind of exercise with your mind, exercises designed to make you aware of what's happening right in this moment. The goal is to be cognizant of all the sensations you're experiencing, hearing, and seeing, and feeling, even the mind's activity itself; all of the mind's tendencies of thinking and analyzing and planning and remembering things. The practice is to not get lost in the mind's wanderings. . . .

NK: There is an analogy to what scientists do. Certainly not all, but a lot of science is basically asking a question of nature. That is what I am doing. Putting a frame around a system and letting it unfold through its own dynamics. Letting it create the pattern, letting nature sculpt itself to a pattern. Rather than chiseling it myself or painting it into a pattern, I try to let nature be the composer.

Bits vs. Atoms—The Future of Phenomena-Based Art

Nicholas Negroponte's book *Being Digital* promotes the idea that the manipulation of bits (information) is replacing the manipulation of atoms (physical stuff) in many realms of life. Generally, the book explores the positive implications. The "Sensitive Chaos" show, which was similar to "Turbulent Landscapes," presented at the ICC gallery in Tokyo, also explored physical-world phenomena. The science-and-art writer and curator Itsuo Sakane eloquently noted the importance of artists continuing to explore natural phenomena even as the world moves toward bits:

"Sensitive Chaos" is a symbolic expression applied by the eighteenth-century poet Friedrich Leopold Novalis to water. Here it is a metaphor not just for water, but for all those entities which, while metamorphosing through a variety of time-spans, change their shape and harmonize the pulse of the universe with the vibrations of the earth.

As the information society moves faster and faster, we are inundated with a proliferation of purely symbolic and virtual images. And while art and science approach one another more and more closely, we are always surrounded by these superficial concepts, and the present fin de siècle feeling scatters and betrays us. Couldn't we at least have an exhibition rediscovering the dialogue with natural phenomena that was the origin of both art and science? Can't we have a return to innocence, a poignant memory of our unity with all existence? That is precisely what we hope to achieve with this exhibition.

In place of high-tech and ultramodern creations, here we have artworks of a kind that have moved mankind since time immemorial, and excited his aesthetic sense and curiosity. Natural phenomena like water, sand, and air that have nurtured the flower of science, the simple process of crystallization, the innumerable vibrations and movements that make us feel the rhythm of earth and space, are here in profusion. Moreover, the artists here today, just like scientists, all feel the mystery of the hidden principles underlying existence, and are capable of an expression that can be felt with the five senses. The hard and fast line dividing art from science is already gone. . . .

While the overwhelming tendency of contemporary media art is to move from the atom to the bit, this school of artists must surely take it upon itself to bring about the coexistence of bit and atom, which will become more and more necessary to the next generation. They must become catalysts for a new consciousness of the interplay between bit and atom. It's summer. Isn't this the season for people to recall the past, and return to commune with space and spirits with water as a catalyst?[14]

Notes

1. Exploratorium, "'Turbulent Landscapes' Statement," ⟨http://www.santafe.edu/projects/CompMech/papers/TurbLand.html⟩.

2. J. Crutchfield, "Ideas and Statements," ⟨http://www.exploratorium.edu/t.landscapes/⟩.

3. P. Richards, "Curatorial Statement," ⟨http://www.exploratorium.edu/t.landscapes/⟩.

4. T. Stoppard, "Arcadia," quoted in M. Alexander, "Curatorial Statement," ⟨http://www.exploratorium.edu/t.landscapes/⟩.

5. M. Alexander, "Curatorial Statement," ⟨http://www.exploratorium.edu/t.landscapes/curatorial.html⟩.

6. N. Kahn, "Artist Statement," ⟨www.exploratorium.edu/t.landscapes/artworks.html⟩.

7. Exploratorium, *Wave Organ* Project Description, ⟨http://www.exploratorium.edu⟩.

8. J. Claus, "SolArt Art Network Description," ⟨http://mitpress.mit.edu/e-journals/Leonardo/solart/solartHome.html⟩.

9. R. Mulder and K. Allik, "*Skyharp* Description," ⟨http://www.aracnet.net/~rmulder/⟩.

10. Arts Catalyst, "Eclipse Project Description," ⟨http://www.artscat.demon.co.uk⟩.

11. Exploratorium, "Norman Tuck Exhibit," ⟨http://www.exploratorium.edu⟩.

12. M. Ramiro, "Between Form and Force," *Leonardo,* vol. 53(1998):4.

13. Exploratorium, "'Turbulent Landscapes' Discussion, Kahn-Crutchfield," ⟨http://www.santafe.edu/projects/CompMech/papers/TurbLand.html⟩.

14. I. Sakane, "'Sensitive Chaos' Curatorial Statement," ⟨http://www.ntticc.or.jp/special/chaos/index_e.htm⟩.

3.4

Space

Artistic Interest in Space

Astronomical and cosmological research and space science promise to have great impact on humanity. Historically, the arts paid close attention to the heavens. Monuments such as Stonehenge can be looked at as both artistic and astronomical projects. It would be fitting for the arts to reflect on space research.

Some artists have started the effort. NASA even sponsors a limited Space Art project. Several organizations around the world promote artistic involvement, such as OURS Foundation, Journal Leonardo Space Project, and Ars Astronautica.

Artists have approached astronomy, cosmology, and space science in a variety of ways and offered an array of justifications for artistic involvement. The Leonardo Space Project introduction includes statements from several of those involved. Roger Malina, who is editor of the journal *Leonardo* in addition to being an astrophysicist, notes that artists helped prepare our culture for space research:

The space age was possible because for centuries the cultural imagination was fed by artists, writers, and musicians who dreamed of human activities in space. Now, with the end of the cold war, the role of artists and writers is again crucial in defining our future vision of space—and will once again be instrumental in incorporating the facts and discoveries of the space age into the cultural imagination.[1]

Arthur Woods, head of the OURS Foundation, sees space research as essential to human survival and artists as critical communicators of that need. In his statement "The Next Millennium: A Space Age or a Stone Age?" he notes:

Human destiny on Earth is irrevocably linked to human destiny in space. The continued exploration and exploitation of the space environment are essential to the future survival and prosperity of the human species. Using space resources to meet the growing needs of humanity on Earth is by far the most optimistic solution to many of the problems facing humanity as it enters the new millennium.

NASA, Space Art: ⟨http://tommy.jsc.nasa.gov/~woodfill/SPACEED/SEHHTML/spaceart.html⟩
OURS Foundation: ⟨http://www.ours.ch/home.html⟩
Journal Leonardo Space Project: ⟨http://mitpress.mit.edu/e-journals/Leonardo/san/spaceartproject.html⟩
Ars Astronautica: ⟨www.spacenet.org⟩

The key to this solution is not in technology alone because most of the necessary technology already exists, but rather in manifesting a deep and global understanding of the human situation vis-a-vis the dimensions of the Universe. Thus the cultural reasons for exploring space may prove to be even more compelling than the political and scientific reasons that have been responsible for humanity's astronautical activities up until now.

The future of space activities, the future of humanity, and perhaps even the future all life on Earth is in need of skilled communicators possessing the knowledge and understanding of the scientist combined with the intuition and sensitivity of the artist.[2]

Annick Bureaud, director of the CHAOS artist resource organization, notes that art often functioned as the "fuel" of space exploration, making "dreams desirable for engineers to achieve." B. E. Johnson, an engineer and artist who worked on shuttle missions and on the design of exhibits for the National Air and Space Museum, describes the spirit of inquiry as much more important than any narrow economic motivations and the need to keep that spirit alive.[3]

These statements suggest that the primary achievement of art that is concerned with space lay in its stimulation of public imagination and the understanding of space research. In his "Brief History of Space Art," Arthur Woods traces the initial examples of space art from scientific illustration to book illustration and science fiction movies. Practically, this builds support for funding. He quotes from the 1993 call for papers for the Forty-fourth International Astronautical Federation Congress:

Visual artists and writers have created fictional images and scenarios on the development of space. Such visions are the primary way that the general public is introduced to ideas about space exploration. Artists and writers, in fact, lay the foundation which makes future space activities understandable by the general public and thus secures the necessary political support.[4]

While these are important functions, they are not the only ones. Art can function in other ways: stretching the conceptualization of research, suggesting new research directions, introducing commentary and perspectives from outside the discipline, and helping to interpret the implications of research: "Whereas artists and writers of the past created the visions upon which the present space program has been built, today many artists serve the space community by helping to visualize the future developments and give form to developing technologies."[5]

In his paper "The Role of the Artist in Space Exploration," Roger Malina lists seven categories of space art:

1. Fine art which exploits sensory experiences generated through space exploration.
2. Art which expresses the new psychological and philosophical conceptions developed through the exploration of space.
3. Art in space, viewed from Earth.
4. Art on Earth, viewed from space.
5. Art in space, viewed in space.
6. The applied arts, such as space architecture, interior design, and furniture design.
7. Fine art which takes advantage of new technologies and materials created through space activities such as satellites.[6]

Using these categories, Woods's "Brief History" considers much of space art over the last decades. In this chapter we review a brief sampling, and add one more category: artistic response to the search for extraterrestrial intelligence. Woods notes that the mainstream art world has not accepted much of this art. He partially attributes this to the popularized versions offered in cinema and books that ignore some of the scientific richness:

Today, there are probably less than one thousand artists in the world who are dedicating their talents to some form of space art. Furthermore, the appreciation of this genre of art in all of its manifestations by the mainstream art community has been and still is very low. Like much of science fiction literature, space art is rarely considered to be "serious" art, but rather anecdotal to mainstream art. This situation has not been helped by the cinema, which, though immensely popular, with a few notable exceptions, has portrayed space as an arena for adventure fantasies with a blatant disregard for the fundamental physical laws of nature, most notably that of distance and gravity.[7]

Views from Space

Some of the most powerful impact has been derived from the new views of the earth offered from space. For example, the pictures of the earth globe glowing blue in its shroud of clouds had enormous impact. For the first time people could see the earth as a unified whole system with almost no indications of the political divisions that seem so important to the earth's inhabitants. One example is Stewart Brand's publishing of the *Whole Earth Catalog*, partially inspired by this view. The "whole earth" became an icon of the environmental movement.

Artists also sought to create artworks that built on this new perspective. For example, in 1980 Tom Van Sant created a work called *Reflections from Earth,* and in 1986 *Desert Sun,* which consisted of mirrors to be viewed from satellite or space. Van Sant was part of the group of artists working with Otto Peine at the Center for Advanced Visual Studies at MIT and their "Skyart" events. Although much "Skyart" focused on inflatables, kites, and other earthbound projects, some explored space technology.

Pierre Comte

In 1992, Pierre Comte created a work called *Signature Terre,* which employed large black plastic sheets that could be viewed from the SPOT earth resource satellite. He hoped to create a "signature," viewable from great distances, that stated that the earth was inhabited by intelligent beings.

These works required the artists to understand enough about satellite imaging systems to devise a scheme that would be visible in the way desired. They also required the poetry to imagine the view from a place where humans could not go before space technology. Curiously, there are exceptions in world culture, such as the South American Indians described in *Chariots of the Gods,* who created earth works only visible from aerial perspectives impossible with the technologies of their times.

Tom Van Sant—GeoSphere

Continuing in his projects focused on views from space, Tom Van Sant has created an interesting extension to his early work, called the *GeoSphere Project.* He has constructed a six-foot-diameter globe with an innovative projection system so one can see real-time, seamless composite images of the entire earth from multiple satellites.

Van Sant sees the globe serving educational, research, and commercial purposes. The GeoSphere company offers to arrange, under contract, other kinds of projections for particular purposes. Significantly, art is rarely mentioned in the literature about the project. This omission highlights a dilemma in contemporary culture. Often innovative science and technology projects are not pitched as art even though conceptually they could be. Sometimes the producers do not even frame their work as art.

GeoSphere: ⟨http://www.geosphere.com/⟩

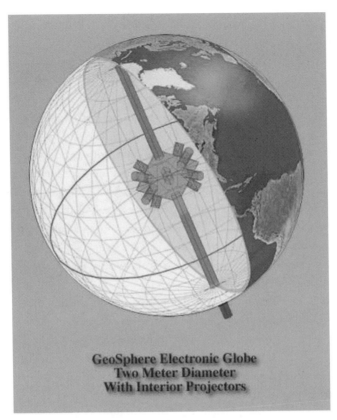

Fig 3.4.1. Tom Van Sant, *GeoSphere*. An electronic global information internal display system to project composite satellite imagery onto a six-foot globe. © 1990 by Tom Van Sant, Inc./GeoSphere, Santa Monica, Calif., ⟨http://www.geosphere.com⟩.

Art Viewed from Earth

Artists have also been intrigued by the converse arrangement. Several have devised plans for works that would be visible from the earth or even put into orbit. These proposals have often been met with great resistance because of scientific, economic, and cultural issues.

Pierre Comte

Building on his idea of art viewed from the earth, Pierre Comte developed the idea of *Dialogues with Space,* which also included art viewed from the earth. He called these

ARTSAT (Art Satellite). In his 1987 article "Leonardo in Orbit: Satellite Art,"[8] he recounts his artistic research process, which sometimes required him to act like a scientist.

He wanted something that would be clearly visible from the earth. But he knew he had to deal with the constraints of economics, launch limitations, and the scientific details of deployment. Having used inflatables in earth installations, he designed an inflatable satellite and made drawings that he sent to the European Space Agency, which put him in touch with engineering firms. At each step the drawings were able to capture the imagination of officials and specialists, who then helped the project proceed.

Comte had to learn technical and scientific information necessary to further the project. For example, launch limitations meant that anything bigger than ten meters had to be constructed in space, thus requiring an expensive and delicate procedure. Using inflatables mastered this dilemma. He also had to overcome other problems: "maintaining an inflatable structure in a vacuum and designing a guidance system for this ultra-lightweight engine." Working with technical advisers, he developed a proposal that worked through each of the problems.

The process is instructive. When it works best, artists do not totally hand off the technical questions. Rather, they educate themselves and engage the science as well as the art. Comte suggests that the technology his team developed had applications beyond his art project: "The totality of these qualities—extreme lightness, maneuverability, significant and volumetric dimensions—make the ARTSAT satellite the candidate for special missions of great diversity."[9]

He sees application in special event satellites, satellites that can easily change colors, disaster lighting, photon propulsion, and providing modules for expandable space architecture. He reflects that his artistic orientation was a significant resource for the project. As of this writing, the project has still not flown. Comte explains how artistic endeavors resulted in technological innovation:

Although to me the idea now seems simple—to replace rigid, heavy constructions with inflatable, lightweight "masts"—no one to my knowledge had thought to do it. Perhaps it is because I am not a technician but an artist that the idea came to me. It is true that at the start I was not trying to realize an object that would be either logical or functional, based on solid technology. I simply tried to create an object that was beautiful with pure lines and dreamlike vocation: a star in the sky, but one born of human hands, a mad dream.[10]

Joe Davis, Arthur Woods, Jean-Marc Philippe, and Richard Clar

The idea of monumental space art has intrigued several other artists. Some very ambitious projects have been proposed. Artists' fascination with orbiting art as a symbol of

Fig 3.4.2. Richard Clar, *Spaceflight Dolphin.* A dolphin sculpture unfolds from the space shuttle. Illustration: Edgar Duncan.

the unity of the earth is illustrated by projects such as Joe Davis's 1982 proposed *Ruby Falls,* an artificially induced aurora, and Arthur Woods's 1986 *OURS 2000,* a ring satellite one kilometer in diameter to celebrate the millennium (see Woods's "History" for more details). Also to celebrate the millenium, Jean-Marc Philippe proposed *Celestial Wheel,* a forty-geostationary satellite wheel with the capability of laser activation from the earth. Richard Clar's *Spaceflight Dolphin* unfolds a viewable dolphin sculpture out of the space shuttle.

Scientific and Public Objections to Space Art

None of these projects has been realized. Many of them raised great resistance. Astronomers objected that bright sculptures in the sky would interfere with sensitive observations that must often measure extremely small variations in light. As a result, the International Astronomical Union, the International Academy of Astronautics, and the American Astronomical Society all passed resolutions to prohibit orbiting space sculptures. Other scientists objected to the generation of more space debris, already a growing problem.

But the dilemma is not just scientific and operational. Who do the heavens belong to? In the United States, public art proposals often have a rough time getting approved because of the objections of those who will have to live with the art. Space can be thought of as the ultimate public art project, complicated by jurisdictional confusion about international rights and unconventional activities. Are the treaties governing Antarctic and moon exploration good models? Is space art just another example of Western industrial hegemony usurping its prerogatives? Who has the right to impose its vision on humanity? Unfortunately, contemporary judgments about major projects are often not good guides of their ultimate value. Both Paris's Eiffel Tower and San Francisco's Golden Gate Bridge were judged to be ugly atrocities by many and vehemently opposed. Now they are considered extraordinary monuments of beauty and enterprise. The questions will become even more pressing due to the growing privatizing of space and the emergence of nongovernmental, somewhat less expensive launch services.

Space art is useful because of this controversy. It forces the world to clarify its priorities. The technology makes possible the creation of works that can be viewed by enormous worldwide audiences. What activities are important enough to warrant placing objects in space? Evidently science, military, and commercial purposes are. But not all activities are. For example, the U.S. Congress passed a resolution banning advertisements in space that could be seen without the aid of a telescope. Some worry that it will be too easy to extend this limitation to art, especially in a public that cannot sometimes differentiate the two.

Art Executed in Space and Weightlessness

Some artists have been intrigued by weightlessness. For the first time, space technology allows humans to free themselves from the full force of gravity. It is fitting for artists to help explore the possibilities and implications of this phenomenon. In some ways, traditional sculpture and dance can be considered poetries of gravity because of the way these works either emphasize the realities of gravity or attempt to defy it.

In 1984, Joseph McShane used microgravity in a materials coating experiment. Taking advantage of NASA's Getaway Special (the low-cost hitchhiking arrangement in which small experiments could be carried in spare shuttle space), he sent up a series of spheres that received coatings and were exhibited as art when returned to tbe earth.

The history of the Getaway Specials illustrates some of the dilemmas facing art proposed for space. In "The Last Getaway Specials: The Space Shuttle and the Artist," Joe Davis describes the early requirement that shuttle projects must demonstrate their practical benefits.[11] Artists often had to contort their proposals to meet this requirement. Eventually, through the efforts of artists such as Lowry Burgess, NASA came to accept "nonscientific" payloads but also upped the expense tremendously. Proving the "benefit" of art in expensive scientific contexts is a continuing problem inhibiting artistic experimentation. It is a wonder that any space art was ever executed.

Arthur Woods's *Cosmic Dancer*

In 1993, Arthur Woods created the sculpture *Cosmic Dancer* explicitly for weightlessness. It flew in the Mir spacecraft, and a video documentation was returned to the earth. The sculpture achieved its true form only once placed in weightlessness. The project was also undertaken to explore the impact of integrating sculpture into cosmonaut work and living spaces: What is the role of art in the special circumstances of isolation and distance from the earth?

Cosmonaut Alexander Polischuk has written reflections of his experience with *Cosmic Dancer*. He notes that art had an important calming effect in the high-stress situation of space and that it was a comforting reminder of the earth. He also described some of the specifics of experiencing weightless sculpture:

Letting go of the sculpture, it spins and spins until it reaches an obstacle. The gravity does not disturb it nor does it force it to stand still . . . The *Cosmic Dancer* is an incredible sculpture, angular and unusual for the classical understanding of art. Nevertheless, it made us pleasure. And that it is a "cosmic dancer," the English title says, we have never had any doubt. Particularly interesting was to dance with it to music. . . . It is interesting to watch the *Cosmic Dancer* against the portal in the background, but one has to decide whether to look at Earth or at it.

Very interesting is the behavior of the construction and the surface of the sculpture when its angular shape touches a transparent media, such as a liquid substance . . . Here the behavior of a liquid and one of a solid material are easy to observe; they both possess a moment of inertia. The rotating movement of the water in front of our sculpture *Cosmic Dancer* is somehow perceived differently and is interesting, not only in view of the weightlessness, but also from an emotional and aesthetic viewpoint. Sometimes it behaves like a living being, it swings and floats . . . And contemplating the sculpture turning in weightlessness while listening to music results in an effect which is possibly totally unknown on Earth. It is difficult to describe this effect.[12]

Kitsou Dubois

Kitsou Dubois is a choreographer interested in the phenomenon of the body's relation to gravity. Science is seen as a critical resource in this investigation. In the *Leonardo* article "Dance and Weightlessness," Dubois traces the influence of science on modern dance, noting two influences: the Japanese Butoh dance form, which grew out of the Hiroshima nuclear tragedy, and Merce Cunningham's thoughts about relativity. Cunningham found Einstein's notions of space liberating:

I [Cunningham] decided to open up space, to consider it as a totality of equal value: thus, whether occupied by someone or not, no point is more important than any other. There is no need in such

Fig 3.4.3. Kitsou Dubois, *Gravity Zero* dance. Movements exercised in zero gravity in a French Space Agency parabolic trainer.

a context to refer to any specific point. Then I read one of Einstein's sentences: there is no fixed point in space. I said to myself: if there is no fixed point, then every point is both fluid and interesting.[13]

For Dubois, gravity is a central theme of dance. She notes that science and dance were simultaneously working on liberation from gravity:

Artists and scientists work on what dancers call "being." We belong to the species *Homo sapiens,* ruled by gravity. Our "being" is thus deeply connected to gravity. "Modern dance" uses techniques involving kinesthetics, anatomy, and improvisation, which are adapted to a cultural, technological, and political environment. It deals with an adjustment process that will enable human beings, who suffer from the division of the inner self, to find greater harmony between body and soul. It symbolizes lightness, freedom of movement, and an active search of elevation, defying the rules of gravity.[14]

Dubois is a choreographer who worked on expanding dance space through choreography in unusual places, experiments in new movements, and teaching diverse groups. Eventually this led to work with weightlessness at the French CNES (National Center for Spatial Studies) to explore sensory and movement training in microgravity via parabolic flight routines instituted to simulate space flight. Her article is fascinating in the

---

Kitsou Dubois, *Gravity Zero:* ⟨http://www.artscatalyst.org/grav0.htm⟩

way it moves among the disciplines of dance, anatomy, physiology, psychology, and space science. It explores topics such as the adaptation process to strange environments, space sickness and disorientation, postural control, the subjective vertical, postural tonicity, and the awareness of axis, body scheme, masses, supports, and muscular work. The article is simultaneously a treatise on expressive dance and the science of kinesthetics in microgravity. She describes the training program that has been developed, which helps both dancers and astronauts experience the body as a subjective experience. Her work was also the focus of the English group Arts Catalyst's initiative to support her collaboration with biodynamics researchers and to make the work known by projecting video of previous collaborations across the front of London buildings.

Other Artists and Projects

Frank Pietronigro and collaborators at the San Francisco Art Institute proposed a series of "Microgravity Experiments" for NASA's weightlessness simulator that explored dance, performance, and media activities as a way of relieving microgravity stress. *Pierre Comte*'s *ALPHA of Zero G.* was a sculpture of three primary-color spheres that came to life in the zero gravity airbus training plane.

Painting and Photography Based on Space Exploration

Many artists have been stimulated by the images being returned from space by video. More information than ever before has become available from' the outer solar system, the sun, the planets, and their moons. Fascinated and perplexed by what they saw and didn't see, they have painted and composited images of other worlds. Some artists extrapolate carefully from the scientific information available and may even be involved in planetary science. Others let their imaginations roam free, using the data only as a distant jumping off place. Some present alien life-forms. Interested readers should consult the Ars Astronautica Web gallery of this work.

Conceptual and Electronic Works

Time, distance, and speed seem different in space than on the earth. A satellite or a spacecraft can access points on the earth so much easier than the earthbound. Also, the

---

Ars Astronautica's Web art gallery: ⟨http://www.spaceart.net/artists.html⟩

ability to communicate at these distances and speeds intrigues some artists. Some projects have explored these phenomena via communication between the earth and cosmonauts. Other projects have reflected on the enormous physical and time horizon of extra solar system space. How can artists respond to the overwhelming vastness of space and to the earth's place in such a cosmos? Using microwaves, lasers, and radio waves, some artists have attempted to mark that vastness.

Michael Heivly

Starting in 1979, Michael Heivly worked on a series of sculptures that translated terrain into sound events and microwave transmissions. These transmissions were beamed at particular constellations, including Ursa Major and Draco, and would retain their form for millennia. Later works included the audience as participants. Heivly notes that technology has helped alienate people from their physical surroundings, and he works to reestablish that relationship:

Unlike our ancestors who lived by the stars, very much connected to the changing earth for their physical and spiritual survival, we are no longer aware of our sequential relationship to the natural phenomena of the universe. We have moved away from our collective, mythological center. Alarm clocks wake us up before the sun. . . . All of this allows us to entirely ignore any relationship we might have with the natural world. How does this affect who and what we are and how we relate to our ancestors, to the earth itself, to the universe?[15]

Arthur Woods

Commenting on the fragility of life on the earth and the possible extraterrestrial origins of the earth's life, Arthur Woods's *Seeds* project proposes to purposely seed terrestrial life in other parts of the universe.

Other Artists and Projects

Vastness of the Universe   French artist *Jean-Marc Philippe,* whose earlier project used a radio telescope to beam thousands of Minitel-collected messages into space, has proposed the "Keo" project, which will collect messages that people think are important

---

Michael Heivly: ⟨http://academic.csubak.edu/~mheivly/deepspace/bio.html⟩
Arthur Woods: ⟨http://www.seeds.ch/information/index.html⟩
Jean-Marc Philippe: ⟨www.keo.org⟩

Fig 3.4.4. Michael L. Heivly, *Deep Space Site Transmission*. Terrain is translated into microwave signals.

enough to want to send them on a spaceship that will return to the earth fifty thousand years in the future. In 1988, Israeli artist **Ezra Orion** created a Milky Way "sculpture" by aiming a laser at selected points in the galaxy. Earlier, Orion had created a work that conceptualized plate tectonics as sculpture. He also proposed sculptures to be realized by the Mars Rover and other planetary probes.

Space Science and Astronomy   The **Arts Catalyst** 1999 "Cosmic Chances" conference brought together astronomers, cosmologists, and artists to explore methods of knowing

---

Arts Catalyst: ⟨http://www.artscat.demon.co.uk⟩

Fig 3.4.5. Arthur Woods, *Seeds.* Conceptualization of a craft to be used in a proposed project to seed terrestrial life-forms on other astronomical bodies.

the universe, including presentations on topics such as SETI, solar wind, meteor capture, and solar neutrinos, In 1992, Austrian artist **Richard Kriesche** created a video performance called *ARTSAT,* in which the cosmonauts on board a Mir mission picked up signals when passing overhead and altered them for their next pass. **Richard Clar and Mark Madel**'s *Collision* presents an electronic-simulated sculpture generated from the positions of 297 pieces of space debris. **Pierre Comte** plans to launch a "Star of Tolerance" solar sail, which will be seen and heard (via radio broadcasts) throughout the earth. **Charles Wilp and Burkhard Bratke** have proposed the *Michelangelo Project,* which will arrange for the creation of art on the International Space Station once it is operational. **Marko Peljhan** has made a proposal that the art research lab MAKROLAB become part of a network of scientific labs studying the phenomenon of Aurora Borealis.

Unified Visions of the Earth *Clar*'s "Earth Star" project proposes to create a ceramic tile with soil from many conflict regions of the earth that will be fused on reentry. Part of his "Quiet Axis" series, **Lowry Burgess**'s space shuttle "nonscientific payload" work

---

Richard Clar: ⟨http://cyberworkers.com/Leonardo/space/13avril-I/interventions/clar.shtml⟩

Boundless Cubic Lunar Aperture included holograms and cubes made from all of the elements known to science and water samples from all the world's rivers.

Earth-Based Works   ***Charles Ross***'s *Star Axis* was a project to excavate a conical tunnel in the earth precisely pointed toward the star Polaris. At various places, a different circle of the sky would have been visible, indicating the expected orbit of Polaris; the year would have been marked on the wall. The cave would have thus recorded the entire twenty-six-thousand-year orbit progression. *Nancy Holt* created a series of earth works that put viewers in touch with astronomical time frames and positions of stars and the sun. *Sun Tunnels* lines up shadows during the solstices. *Annual Ring* causes a shadow to fill a ring on the ground only on the summer solstice. *Dark Star Park* lines up shadows with poles only when the sun's current position matches its position on a day historically important to the town in which the work is placed. ***James Turrell***'s installations use light to create unusual spaces where light can look solid and make objects appear to float. His "Roden Crater" project is a long-term effort to transform the crater into a series of chambers linked to astronomical phenomena. Each chamber is designed to admit light from the sun, moon, and stars in certain ways and to create different light atmosphere. ***Pauline Oliveros***'s "Echoes from the Moon" (designed in collaboration with ***Scot Gresham-Lancaster***) uses radio telescopes and ham operators to bounce music and voices off the surface of the moon. It is an example of her interest in "deep listening," new ways of listening carefully to inner and outer sounds.

Search for Extra Terrestrial Intelligence (SETI)

Growing information about the universe has increased speculation about the possibility of other intelligent life. For example, Frank Drake devised his famous equation that calculated odds based on physical knowledge about star types and life histories, requirements to sustain life and evolution, and the probabilities of various occurrences. The equation results in the conclusion of a very high probability that situations similar to that of the earth have occurred in many places in the universe, with the consequent evolution of intelligent life. Note that this high probability is based on the assumption of carbon life forms as we know them rather than allowing for unanticipated variations. Science fiction writers have developed many more ideas about alternative life-forms.

Pauline Oliveros: ⟨http://www.deeplistening.org/pauline/writings/moon.html⟩

Researchers have theorized about the best methods for detecting life and communicating with distant intelligences. Sophisticated electromagnetic detection technologies have been developed and undertaken, for example, computer-assisted scanning of the sky by radio telescopes. The systematic broadcast of signals has been made to the stars with the goal of initiating communication.

Our culture, however, seems quite ambivalent about the effort. Some scientists do not consider it strong science; many politicians consider it a waste of time to the extent that the U.S. Congress forbade NASA to spend money on SETI research. The case for the research has not been helped by the current obsession with flying saucers, alien abductions, and the like. Others, however, consider the search as some of the most significant research under way. What will it mean to establish the existence of other intelligences in the universe both spiritually and scientifically? What will they be like? What could we learn from them? What should be the nature of our communication with them? What kind of communication can bridge the possible gaps of time, experience, and biology? What is worth saying? Who can speak for humankind? What should the decision process be like?

As with so much scientific research, these questions have implications far beyond the disciplines of science. Even without documented communication with other intelligences, pursuit of the questions can be instructive and provocative. For more on the cultural implications of SETI, see Steven J. Dick's article "Other Worlds: The Cultural Significance of the Extraterrestrial Life Debate."[16] Indeed, the arts are considered to have some special expertise on communication. The art and artifacts of the ancient cultures of the earth still speak to us despite being separated by thousands of years. When the *Voyager* spacecraft went out into deep space, images and sounds were included as important elements to speak to unknown civilizations should any encounter it.

To alert the art community to some of the issues, the Leonardo Space Art Project has Web-published the position paper "A Decision Process for Examining the Possibility of Sending Communications to Extraterrestrial Civilizations," which was approved by the International Academy of Astronautics (IAA) and by the board of directors of the International Institute of Space Law (IISL). The paper traces the evolution of SETI science, discusses the issues involved in humanity sending a message, and then describes principles for designing a process. Here is a summary of those principles:

1. The decision on whether or not to send a message to extraterrestrial intelligence should be made by an appropriate international body, broadly representative of Humankind.

2. If a decision is made to send a message to extraterrestrial intelligence, it should be sent on behalf of all Humankind, rather than from individual States or groups.

3. The content of such a message should be developed through an appropriate international process, reflecting a broad consensus. [17]

It should be noted that many SETI researchers believe that we have already been unwittingly communicating. Since their initiation, all TV and radio broadcasts have been beamed into space as well as into living rooms. Some SETI researchers enjoy the idea that at this moment some civilization forty light years away is making deductions about the earth's culture by decoding the early U.S. sitcom *I Love Lucy*.

Although there is not extensive activity in this field, some artists have begun to become involved in issues surrounding the detection of extraterrestrial intelligences and creating communication with them. The transmissions into deep space previously described speak in part to the possible existence of other intelligences. As a historical example, many years ago Richard Clar proposed a shuttle Getaway Special project called Spaceflight Dolphin, which would launch a wire-frame sculpture into orbit and send signals of dolphin sounds to the earth and to unknown civilizations on other planets. More contemporary interest is evidenced by the SETIathome project, to which thousands of people have subscribed to donate unused computer time on their home computer to help in the analysis of signals from space.

Arts Catalyst—Searching

Arts Catalyst undertook a project called *Searching*, which explores the SETI initiative from multiple perspectives. The organization has funded open-ended collaborative projects between artists and researchers working at institutions, such as Jodrell Bank and the SETI Institute, to investigate research tools such as radio telescopes and the Hubble space telescope. Public lectures and exhibitions showed the art and considered the science. Several years earlier Rob La Frenais, one of the directors of Arts Catalyst, had organized a conference called The Incident, which brought together scientists, artists, and others interested in new perspectives on the UFO phenomenon.

SETIathome: ⟨http://setiathome.ssl.berkeley.edu/⟩
Arts Catalyst, "Searching": ⟨www.artscat.demon.co.uk⟩
SETI Institute: ⟨http://www.seti-inst.edu/⟩

Other Artists and Projects

Commenting on tendencies of the mind to impose patterns, **Lewis DeSoto**'s *Observatory* offered viewers a mix of sounds heard by SETI researchers during their searches while they watched a television tuned to snow, thought to be caused in part by random sky noise. **Douglas Vachoch** has sought to use semiotic analysis to determine the parameters necessary to create pictorial communication with extraterrestrial civilizations.[18] In its yearly series of conferences on Space Art, Leonardo's French affiliate **OLATS** (*Observatoire Leonardo des Arts et des Techno-Sciences*) sponsored a workshop on SETI which examined "the different 'searches for life,' their scientific basis and methodologies but also their myths and 'silent background,' from looking for 'ones-like-us' [intelligent life, with the SETI/Search for Extra-Terrestrial Intelligence activities] to 'ones-different-from-us' [astrobiology] who might even be based on a non-carbon-based-life." San Francisco's **Yerba Buena Center for the Arts** sponsored a 2000 exhibit called "Above and Beyond" focused on UFOs and "science, belief, and understanding." *Michel Redolfi* created the Earthsounds project, which is soliciting people all over the globe to contribute sound samples "representing earth" to be included ultimately in attempts to contact extraterrestrial life.

Art Critiques of Space Research

As previously mentioned, not everyone is enthusiastic about space science or space art. The worlds of mainstream and technology art have mostly ignored it. Some social critics see space activities as a diversion from attention to more important issues on the earth and a continuation of "male," Western preoccupations with colonization and domination. Some artists, such as those discussed in chapter 7.7, on surveillance, see unacknowledged dangers in the dehumanization of war and the panoptical desire for total surveillance.

Steven Hartzog

Illustrating this kind of approach, while making unique use of his former employment at one of the national research labs, U.S. artist Steven Hartzog explores the overselling

---

OLATS: ⟨http://www.cyberworkers.com/Leonardo/space/13avril-III/programme_participants.shtml⟩
Above and Beyond show: ⟨http://www.yerbabuenaarts.org/archive/extsupmet/esp_abovebeyond.htm⟩
Michel Redolfi: ⟨http://homestudio.thing.net/earthsounds/⟩
Steven Hartzog: ⟨www.nukes.org/alien/alien.html⟩

of space and the ignoring of space travel's realities. His Web site provides humorous commentary on the boring nature of space, the unacknowledged radiation dangers of space travel, and the use of nuclear propulsion systems. It includes extensive links to science sites offering alternative information about space travel. Also, one can take a simulated trip that often ends poorly. Here is Hartzog's analysis of space program problems:

The technocratic establishment plays on our imagination to get massive public support for projects of dubious value. Self-serving greed poses as altruistic enthusiasm.

Technically ignorant taxpayers depend on the supposed objectivity of scientific experts making fateful decisions. How objective can NASA's employees be about funding for their space program? Who outside of the nuclear establishment has the expertise to critique their projects? They feed us images of hope and dismiss objections as technically unqualified "emotion."

Why do we want to colonize outer space? Will we be happier as extraterrestrial employees? AS IF! You CAN be certain that somebody will be making galactic profits. The final frontier is financial.[19]

Constance Penley's book *NASA/Trek: Popular Science and Sex in America* offers another deconstruction of the U.S. space program by imagining a feminist Star Trek crew as a technique to explore issues of gender and science.

Summary: The Hopes

Scientists and artists influenced by science are often more optimistic. They see great opportunities and significance in astronomy, cosmology, and space science. It is claimed that artists can have significant impact and that ignoring the importance of space research by our culture and the art world is blindness. Arthur Woods expresses this opinion in his paper "Art and the New Millennium":

It appears that the art community, too, has essentially ignored the most significant development of the twentieth century—humanity's breakout into space. Art is usually considered to be always on the leading edge of human development, yet art in and about space has not been taken seriously by the mainstream art world. Ironically, the few artists who have attempted to realize their art in space have also been shunned by the space community, their work considered to be of lesser importance than the multi-billion-dollar science that, of course, supports the objectives of national political programs.

However, space is too important to the future of humanity to be left solely to the politicians and to the scientists to explore and to utilize, as is the case today. This may be the primary reason for the global failure in understanding the true importance of this new environment for human activities. . . .

It may be said that the power of art to explore human experience has been outstripped by the wonders of science and technology in the past one hundred years, but the necessity of art to balance our understanding and emotions about the world we have created has never been greater than it is today.[20]

Notes

1.   R. Malina, "Space Art Introduction," ⟨http://mitpress.mit.edu/e-journals/Leonardo/san/spaceartproject.html⟩.

2.   A. Woods, "The Next Millennium," ⟨http://www.ours.ch/home.html⟩.

3.   A. Bureaud, and B. E. Johnson, "Statements," ⟨http://mitpress.mit.edu/e-journals/Leonardo/san/spaceartproject.html⟩.

4.   A. Woods, "History of Space Art," ⟨http://www.spaceart.net/history.html⟩ (short version also available at ⟨http://www.ylem.org/newsletters/JulyAug97/article3.html⟩).

5.   Ibid.

6.   R. Malina, "Role of the Artist in Space Exploration," Fortieth Congress of the International Astronautical Federation, IAA, 1989.

7.   A. Woods, op. cit.

8.   P. Comte, "Leonardo in Orbit: Satellite Art," *Leonardo* 20(1987): 1.

9.   Ibid., p. 20.

10.   Ibid., p. 21.

11.   J. Davis, "Shuttle Getaway Projects," *Leonardo* 24 (1991): 4.

12.   Polischuk, "Reactions to *Cosmic Dancer*," ⟨http://www.spaceart.net/arthur-woods/cosmos.html⟩.

13.   M. Cunningham, quoted in K. Dubois, "Dance and Weightlessness," *Leonardo* 27 (1992): 1, p. 57.

14.   Ibid.

15.   M. Heivly with M. Reed, "The Space between the Real and the Imagined: Microwave Sculpture in Deep Space," *Leonardo* 25 (1992): 1.

16.   S. Dick, "Other Worlds," *Leonardo* 29 (1996): 2.

17. International Academy of Astronautics, "Decision Process," ⟨http://mitpress.mit.edu/e-journals/ Leonardo/san/seti/seti-1.html⟩.

18. D. Vachoch, "Signs of Life beyond Earth," *Leonardo* 31 (1998): 4. See also Web page at ⟨http:// www.seti-inst.edu/science/signals4.html⟩.

19. S. Hartzog, "Critique of Space Program," ⟨http://www.nukes.org/alien/alien.html⟩.

20. A. Woods, "Art and the New Millennium," ⟨http://www.spaceart.net/art-space-mill.html⟩.

Global Positioning System

Possibilities and Dangers

One of the most practical benefits of space exploration has been the Global Positioning System (GPS). This allows people on the earth to locate themselves anywhere to within one meter precision. It has had manifold applications in military analysis, navigation, surveying, geology, agriculture, archaeology, biological research, traffic management, hiking, and an ever-expanding array of applications.

Like so many technologies, GPS presents both a light and dark side. It promises an unprecedented ability for individuals to know where they and others are on the face of the earth. It allows the easy creation of events cued to position and movement. The shadow side portends new extensions of panoptical surveillance and control; authorities will be able to know exactly where things and people are. There could be no privacy, no solitude. Artists are just beginning to enter into this discourse.

Artistic Experimentation with GPS

Iain Mott

Iain Mott is a sound artist who investigates emerging technologies, such as position sensors and 3-D sound. His *Sound Mapping: An Assertion of Place* installation made use of GPS technology to customize the sound that participants heard based on their changing physical position in outdoor locations and qualities of movement such as tilt, acceleration, and nearness to others. The Ars Electronica documentation describes the event:

In *Sound Mapping,* visitors interact with a composition that is anchored in a geographical space and not in a temporal sequence, as is usually the case. Sound-generating suitcases equipped with motion detectors and GPS (Global Positioning System) capability are moved through a predetermined tonal space. Sensors in the suitcases register the speed and quality of the movement as well as their positions in space. By means of these parameters, the users control the music while they stroll with their suitcases through the urban landscape.[1]

A paper at Mott's Web site explains the desire of the project: "[It] aims to assert a sense of place, physicality, and engagement to reaffirm the relationship between art and the everyday activities of life."

Iain Mott: ⟨http://members.tripod.com/~soundart/⟩

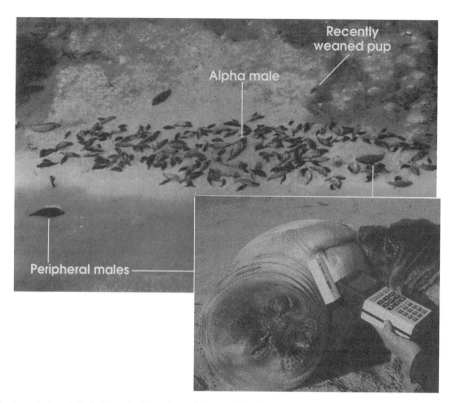

Fig. 3.5.1. Fillippo Galimberti, Elephant Seal Research Group. GPS technology is used to study the structure and movements of elephant seal colonies.

He recounts the history of Western music, which has increasingly isolated listeners from the experience of its physical origins. The emergence of the score, the composer, phonography, radio, and ultimately digital technology all contributed to this alienation from its source in time and space, and with the producers of the sounds. He worries about the "detachment from physicality."

Mott believes that installation is a viable form for reemphasizing physicality. His *Sound Mapping* installation varies the sound to correspond to where participants physically choose to walk, and the total sound experience depends on real-time interactions between participants as they try to coordinate movements and sounds:

Each individual plays distinct music in response to location, movement, and the actions of the other participants. In this way a nonlinear algorithmic composition is constructed to map the footpaths, roadways, and open spaces of the region and the interaction of participating individuals.[2]

Fig. 3.5.2. Iain Mott, Marc Raszewski, and Jim Sosnin, *Sound Mapping*. Interactive sound compositions are engaged by a GPS receiver and motion-sensing equipment inside carts, indicating presence in particular locations and the motions of individuals.

His work with the sound mapping experience in terms of emerging theories of virtual architecture implies that physical position will be a relevant variable even in cultures dominated by cyber technologies.

Masaki Fujihata

Masaki Fujihata proposed a project called "Proposal of Impressing Velocity," which would distort the view of a person based on their speed, much like an automobile driver's view is changed with increasing speed. The original research for the project was based on an experiment that used GPS technology to store a record of the speed with which a group of hikers ascended and descended Mt. Fuji. The imagery of the mountain was distorted in accordance with the progress of the hike, with fastest motion resulting in the least distortion. Documentation of the proposal explains:

[S]peed is fun [for] everyone. [S]peed can give us unusual experience for [perceiving] the world. . . . The aim of producing this art piece is to develop [a] special algorithm to distort the view of the viewer, instead of viewing [the] actual view in front of him. . . . It will be a visualization of the impression of speed.[3]

Teri Rueb

Teri Rueb creates sculptures and site-specific installations that use the unique capabilities of emerging technologies to enhance the expressive power of her work. She developed *Trace* while she was in residence at the Banff Centre. It is a memorial to people who have died and an exploration of themes of death, loss, memory, catharsis, regeneration, and transformation. Visitors walking the nature trails surrounding the center hear memorial sound works from a computer in their backpack when they move to specific physical locations. A GPS receiver in the backpack monitors their specific location and activates sounds tied to particular locations. Rueb's Web-site describes the event:

The project uses GPS and portable computing technologies since installation of hardware components along trails would be ecologically invasive. The whole system is confined to the knapsack units and the base station. In a sense, hikers carry the installation into and out of the park in the course of their journey. During the winter months, snow cover makes the trail inaccessible, so the installation lies dormant until the following spring.[4]

Teri Rueb: ⟨http://fargo.itp.tsoa.nyu.edu/~rueb/index.html⟩

Fig. 3.5.3. Teri Rueb, *Trace*. Visitors can hear and leave sound memorials linked to particular locations on mountain paths near Banff, B.C., Canada. Photo: Erik Conrad.

On returning to the base station, visitors can view a terrain model on a computer, and download the GPS-encoded record of their hike to the model that will overlay their trek. Clicking the mouse on locations on the computer model can reactivate the sound memorials linked to those locations. During the hike, they can digitally mark particular places to which they would like to attach a sound memorial.

Rueb conceptually links mourning with nature and motion. To orchestrate contemplation, she refers to the history of sacred architecture and the emotional power of natural environments in many cultures. She uses GPS as a tool to create a contemporary expression of this ancient tendency:

The project draws creative inspiration from memorial art and architecture, the arts of memory, oral histories, kinship and genealogy, peripatetic thought, and the rituals associated with Aboriginal myth, songlines, walkabouts, and dreamings. The installation is meant to reaffirm emotional and spiritual connections among people across space and time through memory and artistic expression. Each walk along the trail becomes a unique composition of visible, tactile, and aural elements through the interaction of people, nature, art, and technology. Visitors become participants in creating and experiencing a montage of art and nature which explores love and loss, and biological, spiritual, and philosophical kinships and histories.

She notes that Western technological culture's move away from historical forms for mourning and memorial is creating a cultural void. She doubts whether purely virtual methods of memory storage are adequate, and thus sought to develop an approach grounded in corporeality and place:

This cultural void indicates a growing denial, neglect, or lack of appreciation for mourning as a restorative process that is a shared as well as private experience. *Trace* is an effort to find an alternative medium and cultural locus for memorial art and ritual in order to revive their therapeutic potential. . . .

As our technological methods of storing memory have become less concrete, stable, idiosyncratic, and localized, so have our methods and rituals of remembering the dead and externalizing the mourning process. . . . *Trace* seeks to subvert the standardized, machinic, and logo-centric design of modern funerary and digital media artifacts in favor of an expressive, corporeal, and human-centered aesthetic of memory.

Rueb sees the Burgess Shale geological formation, with its fossils and exposure of geological evolution, as an ideal location for an artwork focused on memory. GPS allows her to create a nonintrusive work that can unfold in the midst of this grandeur:

The site is meaningful as the location of *Trace* in that it juxtaposes the contemplation of the deceased with the contemplation of the grand scale of time, the play of chance and contingency, the balance of order and chaos inherent in nature, the origins and cycles of life through which we are all related to one another, the traces of events and processes through time, and the relationship of living organisms to each other and their physical environment.

Projekt Atol Communications Technologie

Projekt Atol Communications Technologie (PACT), a Slovenian artist group, undertakes a number of projects to investigate the implications of emerging technologies. *Cyborgs for Urban Survival* outfitted performers with portable communications gear and a GPS receiver so they could roam cities to reclaim them. They uploaded location-specific information to a Web site in real time as they "recolonized" a city. Their manifesto explains the intention to use technology to reclaim the physical city:

Cyborgs for Urban Survival

If new, emerging ways of experiencing the world displace traditional conventions too quickly, the side effects could be tragic. Developments in communications technologies, for example, have created new representational spaces that are rapidly displacing physical realities. In the new city of simulation, the flat screen of the monitor becomes our topography, the information matrix our roadway, the World Wide Web our meeting place. The physical city seems to have disappeared, lost under the weight of high technology. Its public spaces have been reconfigured into a series of private cyborg bodies, alienated from their physical environment.

Working as a sort of cyborg for urban survival project, the Urban Colonisation and Orientation Gear (UCOG-144) recovers the physical city by strategically employing the very tools of technology that have marked it obsolete. . . . With their bodies functioning as satellite dishes that both transmit and receive information, performers are able to track their physical coordinates in real time and create a new awareness of their location within the urban environment. . . . Using both bodies and technology to actively re-register the city, UCOG-144 executes a prosthetic aesthetic that both extends and regenerates the urban environment.

Andrea Wollensak

In *Mapping with Satellites: GPS Site Drawings of San Jose, California,* Andrea Wollensak investigates the implications of new imaging and cartography techniques on our sense of place. Images are uploaded from remote locations specifically pinpointed by GPS technology. The documentation of the exhibition at the Art-Tech Gallery explains:

The artist intends to locate and expose the intersections of social and cultural spaces and make visible the regional cultural characteristics by way of visualizing place. The "culture" of geography and the reconstruction of new cartography as seen through new technologies will be addressed through this project. As new locating devices, GPS navigational equipment is redefining our understanding of urban boundaries as represented in maps, wayfinding in landscapes, and the boundaries defined by media and technologies. New forms, codes and functions do not simply represent, but actively reconstruct social spaces. This investigation will explore the new conditions in defining where we are and how this type of technology has become a new architecture in an electronic and physical environment.[5]

Andrea Di Castro

Andrea Di Castro has organized an international project called *Drawing with Global Technologies* in which artists, including Wollensak, will "draw" by their physical movement. Using GPS technology, they will create lines by geographically tracing their movements. The first experiments were in Mexico, although plans call for extension to internationally diverse sites. The project's Web page describes the effort:

Andrea Di Castro: ⟨http://www.imagia.com.mx/gradescE.htm⟩

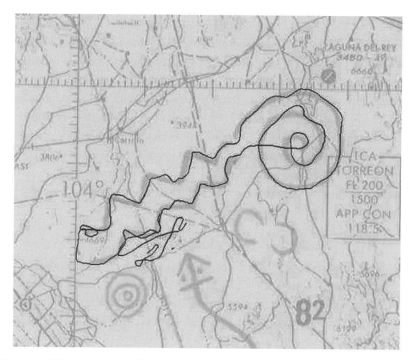

Fig. 3.5.4. Andrea Di Castro and Andrea Wollensak, *Zona del Selencio.* Paths of a person's movements are physically drawn on maps via GPS recordings.

This is an ongoing project that is creating a layer of virtual drawings around the earth following the movement of the artist wherever he might be. The artist's movement becomes the drawing, creating his own air, sea, and land paths by using nature as canvas. Also, for the plot, the artist can take advantage of highways, lake contours, seashores, rivers, etc. Using global technologies such as the Internet and GPS (Global Positioning System), which allows a very precise register of the paths into visible shapes, a series of large-scale virtual drawings, measuring several miles in size, are being made.[6]

Stephen Wilson

I created a project called *The Telepresent* in which people would pass a "present" from one to another. The present automatically uploaded images and precise location informa-

Stephen Wilson: ⟨http://userwww.sfsu.edu/~swilson/art/telepresent.html⟩

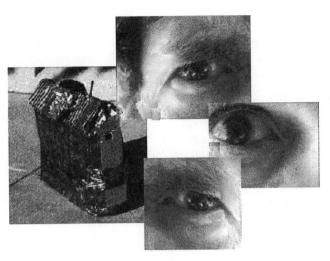

Fig. 3.5.5. Stephen Wilson, *The Telepresent*. A gift-wrapped present continuously uploads images and GPS-location information to the Internet.

tion from wherever they traveled to the Web in real time. Web visitors could send messages to the carrier via synthesized speech. *The Telepresent* explored a variety of issues, including privacy, solitude, surveillance, the structure of gift giving, communication between strangers, and the meaning of knowing where one is. Viewed from the outside, *The Telepresent* looked like a gift-wrapped box with a small hole in it. In actuality, the "present" was a sealed box containing a portable computer, wireless modem, digital camera (with peephole), and GPS receiver. The Web site explains the project:

What is *The Telepresent?* The present is a "magic" box with no wires that automatically sends images from wherever it is to others who are watching via the Web. People carry it with them for a short period in their life and then they give it as a present to someone else. They carry it down the street with them. They take pictures of friends. They take pictures of strangers. They take it someplace special. They take it someplace ordinary. Wherever they go, Web viewers can see whatever the telepresent sees. What to do with this present? Each person who gets the present must confront the question of what to do with it and then the question of who to give it to next. . . .

Mystery: History is full of stories of wanderers and explorers—of the mystery and romance of their journeys. But the fear of being lost and out of touch with loved ones is also part. *The Telepresent* lets everyone in the world know where it is at all times and even to see what it sees. Does this capability change the nature of the odyssey?

Tom Bonauro, John Randolph, and Bruce Tombs

These collaborators developed *Gnomon* for an exhibit at the San Francisco Museum of Modern Art. *Gnomon* was an eight-foot-high, two-thousand-pound self-propelled sculpture that used GPS to identify the location where it was supposed to be. It would then slowly move toward that location. Because the military scrambles GPS signals, the location information could be anywhere within one hundred meters of its real target. *Gnomon* would try to move from room to room and even through walls. It commented both on information and disinformation. *Gnomon* also included inside projection of abstract imagery and sounds. One commentator noted, "One cannot help feeling a little sorry for this enormous, frustrated, and confused potato."

Stefan Schemat

Osmotic Minds outfits participants with a GPS receiver, computer, and headset as they take an "investigative journey following the footsteps of novelist Alfred Doeblin in the vicinity of the Rosenthaler and Alexanderplatz areas of Berlin. As they enter specific GPS-identified locations they hear sounds that elaborate on the original 1929 novel (*Berlin Alexanderplatz*) and "become fellow players and characters in Alexanderplatz."

Summary: Unexpected Implications

Like so many technologies, GPS is pregnant with possibilities. What would it mean that we could know where everyone else is at all times? What would it mean that every object could have a precise longitude and latitude signature? Artists have begun the necessary process of reflection on what we invent.

Space science seems so obviously focused on the heavens. It is ironic that research about the heavens would result in a technology like GPS, so connected to earthly presence. The strange twistings of research is a story often repeated in the history of science and technology. James Burke's *Connections* is a tribute to this reality of cultural history. Artists could play an important role in making the twists even more fruitful by introducing even more possible connections.

Gnomon: ⟨http://www.firstcut.com/9635/u1.html⟩
Stefan Schemat: ⟨http://www.media-g.com/Event01.html⟩

Notes

1. Ars Electronica, "Description of Mott's *Sound Mapping* Installation," ⟨http://members.tripod.com/~soundart/⟩.

2. I. Mott, "*Sound Mapping,*" ⟨http://members.tripod.com/~soundart/⟩.

3. M. Fujihata, "Proposal of Impressing Velocity," ⟨http://www.c3.hu/~masaki/proposal/index.html⟩.

4. T. Rueb, "*Traces* Description," ⟨http://fargo.itp.tsoa.nyu.edu/~rueb/index.html⟩ (all subsequent quotes from same source).

5. A. Wollensak, "GPS Project," ⟨http://caiia-star.newport.plymouth.ac.uk/PRODUCTION/ConRef/Abstracts/WOLLEN.HTM⟩.

6. A. Di Castro, "Project Description," ⟨www.imagia.com.mx/gradescE.htm⟩.

4

Algorithms, Mathematics, Fractals, Genetic Art, and Artificial Life

Research Agendas in Mathematics and Artificial Life

Abstraction developed as a major cultural force in the late nineteenth and early twentieth centuries. Scientists and artists alike accelerated their interest in attempts to understand and reveal underlying structures, processes, and relationships. Scientists increasingly worked with theories that specified entities that could not be seen or directly sensed, for example, atomic structure and genetics. Using the tools of theoretical physics and mathematics, they conceptualized worlds that defied "common sense" built on notions such as relativity and alternative geometries. Similarly, many artists worked with abstract representations, increasingly removed from common everyday perception. Some eventually dropped any reference to subject matter in an attempt to represent conceptual or spiritual essences.

The world of science has pursued this interest in theory and underlying structures with increasing speed, spreading into most areas of inquiry. Although the interest is not as strong as it once was in the mainstream art world, many artists create work based on this interest in underlying structures. Stimulated by developments in science and the advent of the computer, they explore abstract structures and processes. This work includes areas of inquiry called art and mathematics, algorithmic art, fractal art, genetic art, and artificial life. Musicians have also explored these themes.

Why Is Mathematics Part of a Book on Science, Technology, and Art?

Mathematics-inspired art is usually lumped into discussions of art and science-technology. This grouping must be questioned. To the popular mind, mathematics is often associated with science, perhaps because it requires similar logical procedures and because mathematical analysis has proven very useful in the conduct of science. But mathematics is not the same as science. Here is a definition from the *Encyclopaedia Britannica*:

the science of structure, order, and relation that has evolved from elemental practices of counting, measuring, and describing the shapes of objects. It deals with logical reasoning and quantitative calculation, and its development has involved an increasing degree of idealization and abstraction of its subject matter. Since the seventeenth century, mathematics has been an indispensable adjunct to the physical sciences and technology, and in more recent times it has assumed a similar role in the quantitative aspects of the life sciences. In many cultures—under the stimulus of the needs of practical pursuits, such as commerce and agriculture—mathematics has developed far beyond basic counting. This growth has been greatest in societies complex enough to sustain these activities and to provide leisure for contemplation and the opportunity to build on the achievements of earlier mathematicians.[1]

Why Is Mathematics Part of a Book on Science, Technology, and Art?

297

Mathematics served science well when it was used as a language and analytical system for representing the physical world. But there is nothing intrinsic to mathematics that require it to relate to the physical world. Mathematics is the study of structure, order, and relation of any kind. Its methods can be applied to imaginary worlds as well as to the "real" world. In the last centuries, some of the most interesting mathematics has resulted from the imagining of worlds quite unlike common understandings of the everyday world. The *Encyclopaedia Britannica* entry on the foundation of mathematics describes these developments:

For two thousand years the foundations of mathematics seemed perfectly solid. Euclid's elements (c. 300 B.C.), which presented a set of formal logical arguments based on a few basic terms and axioms, provided a systematic method of rational exploration that guided mathematicians, philosophers, and scientists well into the nineteenth century. . . .

The discovery in the nineteenth century of consistent alternative geometries, however, precipitated a crisis, for it showed that Euclidean geometry, based on seemingly the most intuitively obvious axiomatic assumptions, did not correspond with reality as mathematicians had believed. This, together with the bold discoveries of the German mathematician Georg Cantor in set theory, made it clear that, to avoid further confusion and satisfactorily answer paradoxical results, a new and more rigorous foundation for mathematics was necessary. . . . Thus began the twentieth-century quest to rebuild mathematics on a new basis independent of geometric intuitions. [2]

Some believe that "pure" mathematics is somewhat akin to art—practitioners pursue inquiries often just because they are interesting rather than for specific utilitarian goals. Also, mathematicians, like artists, have the opportunity to dream up arbitrary worlds with their own internally consistent rules unfettered by connections with the conventional world. There are also differences, such as mathematics' emphasis on logical consistency and the privileging of historically validated representational systems like numerals and algebraic representation. The analysis of mathematics's relation to art deserves a fuller analysis than can be offered here.

A Review of Research Agendas in Mathematics

Mathematics is a field with a dual identity. Many mathematicians study abstract patterns, symmetries, and relationships divorced from any concern with their relationship to the physical world. Others study ways of abstractly modeling and representing real-world phenomena. The National Science Foundation's (NSF) Division of Mathematical Sciences convened a panel of experts to assess the state of mathematics research. Their summary statement, entitled "The Mathematical Sciences: Their Structure and Contributions," describes this dual identity of mathematics:

The mathematical sciences are the most abstract of the sciences. . . . The mathematical sciences have two major aspects. The first and more abstract aspect can be described as the study of structures, patterns, and the structural harmony of patterns. The search for symmetries and regularities in the structure of abstract patterns lies at the core of pure mathematics. These searches usually have the objective of understanding abstract concepts, but frequently they have significant practical and theoretical impact on other fields as well. For example, integral geometry underlies the development of X-ray tomography (the CAT scan); the arithmetic over prime numbers leads to generation of perfect codes for secure transmission of data on the Internet; and infinite dimensional representations of groups enable the design of large, economically efficient networks of high connectivity in telecommunications.

The second aspect of mathematical science is motivated by the desire to model events or systems which occur in the world, usually the physical, biological, and business worlds. This aspect involves three steps:

1. Creating a well-defined model of a real situation, which itself is frequently not well defined. Such modeling involves compromises between the need for the model to be faithful to the real situation and the need for it to be mathematically tractable. An appropriate compromise usually requires the collaboration of an expert on the subject area and an expert on the mathematics.
2. Solving the model, through analytic or computational means or a mixture of both.
3. Developing general tools, which are likely to be repeatedly useful in solving particular models.

Examples of mathematical modeling include the quantum computer project, DNA-based molecular design, pattern formation in biology, and the fast Fourier transform and multiple algorithms used daily by engineers for numerical computation.[3]

A review of faculty interests and graduate programs of leading universities and the structure of the NSF Mathematical Sciences Division indicate some of the categories used to conceptualize active research fields in mathematics. Here are excerpts of some of the major categories:

- Algebra, number theory, combinatorics, and graph theory
- Analysis, equations, special functions, harmonic analysis, Kleinian groups, Banach algebras, lie groups, and Hilbert space

Fig. 4.1.1. Robert G. Scharein, *Knot Zoo*. A systematic KnotPlot visualization of the topological structure of knots in order of increasing complexity.

- Applied mathematics, dynamic systems, control theory, optimization, mathematical physics, and biology
- Computational mathematics
- Geometric analysis
- Statistics and probability
- Topology and foundations[4]

Much current excitement in mathematics focuses on areas with practical implications, such as cryptography, communication network descriptions, the investigation of its ability to model incompletely known environments, and to extract patterns from complex

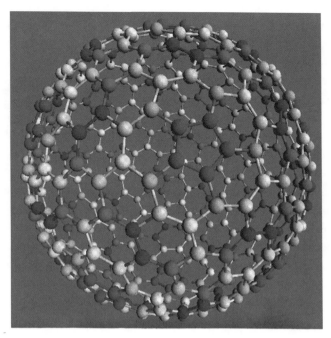

Fig. 4.1.2. Bonnie Berger, *Simulating Virus Shell Assembly*. Mathematical biology's use of geometry and other mathematics to simulate the structure and behavior of viruses.

bodies of information. Much research energy and funding is focused on mathematical physics, biology, geology, electrical and communications engineering, and finance. The NSF assessment report offers samples of these applications:

Mathematics underpins most current scientific and technological activities. Whole new areas of mathematics are evolving in response to problems in experimental science (biology, chemistry, geophysics, medical science), in government (defense, security), and in business (industry, technology, manufacturing, services, finance). All of these areas now require the analysis and management of huge amounts of loosely structured data, and all need mathematical models to simulate phenomena and make predictions. Modeling and simulation are essential to fields where observable data are scarce or involve a great deal of uncertainty, such as astronomy, climatology, and public policy analysis. Addressing such complex problems calls for openness to all of mathematics and to the emergence of new mathematics. Progress requires radical theoretical ideas as well as significantly greater collaboration between pure mathematicians, statisticians, computer scientists, and experimental scientists.[5]

The computer has revolutionized many fields of mathematics, opening up studies that would have been quite difficult or impossible before, for example, the study of nonlinear systems. New branches of mathematics, such as theoretical computation, attract great attention. The assessment report describes the importance of the link between theoretical mathematics and application problems:

There is a wide class of problems, typically coming from experimental science (biology, chemistry, geophysics, medical science, etc.), where one has to deal with huge amounts of loosely structured data. Traditional mathematics, probability theory, and mathematical statistics work pretty well when the structure in question is essentially absent. . . . But often we have to encounter structured data where classical probability does not apply. . . . Such problems, stretching between clean symmetry and pure chaos, await the emergence of a new brand of mathematics. To make progress one needs radical theoretical ideas, as well as new ways of doing mathematics with computers and closer collaboration with scientists in order to match mathematical theories with available experimental data.

Artificial Life

The research fields consisting of chaos theory, nonlinear dynamic systems, and artificial life have been provocative for mathematics and the biological, physical, and social sciences. These areas have also intrigued artists. Since these fields are cross-disciplinary they could appear in many places in this book. (Nonlinear systems are considered more fully in chapter 3.1 in the physical science section.) Artificial life is one area of algorithmic and mathematical inquiry that has attracted great interest. Can the patterns of biological life be sufficiently understood that researchers could write algorithms to represent these rules and simulate life? The mathematics of nonlinear dynamics is an essential tool in much of this research.

What is life? Must it always be carbon based? Is it possible for humans to create life? Could a computer program be said to be alive? These are major research and philosophical questions confronting researchers in an area of inquiry called Artificial Life. Generated partly by the development of dynamic and nonlinear systems research, investigators around the world in many disciplines, including art, have become involved. Significant conferences and publications have resulted. Like so many "research" topics, the intellectual and moral reach of this field calls for participation by practitioners from many areas, including those not typically considered scientists. The Santa Fe Institute has been a prime sponsor of this research.

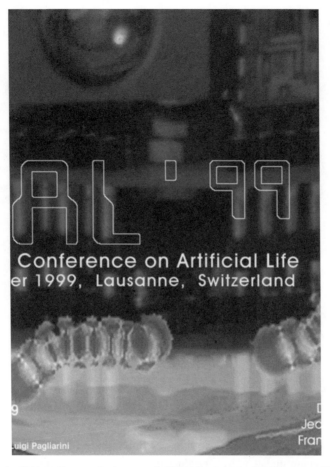

Fig. 4.1.3. Poster for ECAL 99 (European Conference on Artificial Life). Luigi Pagliarini (poster) and Mark Peden (caterpillars).

The field has generated great heat as philosophers and researchers struggle with the implications of the field. Here is one definition of artificial life drawn from Chris Langton's preface of *Artificial Life I,* a major landmark publication of the field:

Artificial Life is a field of study devoted to understanding life by attempting to abstract the fundamental dynamical principles underlying biological phenomena, and recreating these dynamics in other physical media—such as computers—making them accessible to new kinds of experimental manipulation and testing. . . . In addition to providing new ways to study

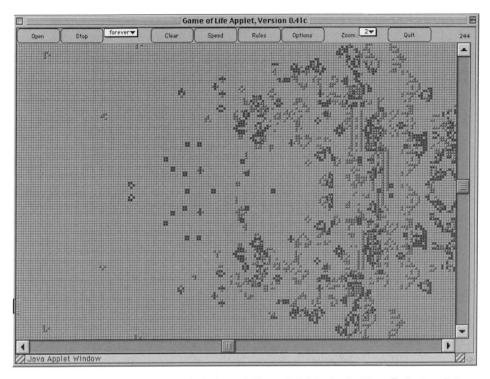

Fig. 4.1.4. Alan Hensel, Java implementation of John Conway's "Game of Life" showing the "line puffer" pattern. Patterns result from the implementation of simple graphic rules about the relation of each iteration to its predecessor.

the biological phenomena associated with life here on Earth, life-as-we-know-it, Artificial Life allows us to extend our studies to the larger domain of "bio-logic" of possible life, life-as-it-could-be."[6]

Some of the researchers work with "wetware," but most work in computer settings, to create computer environments in which artificial entities manifest features of organic life and behavior, such as evolution, growth, flocking, predation, energy exchanges with the environment, learning, and the like. They explore topics such as autonomous agents and neural networks. Researchers create genetic algorithms with the capability to self-modify their codes and thus manifest novel and original behaviors. Many of the programs focus on modeling evolutionary processes in which selection processes increase the continuing likelihood in subsequent generations of entities that are more "fit." Biological concepts and observations are drawn on extensively, and the insights of dynamic systems research often provide useful ideas.

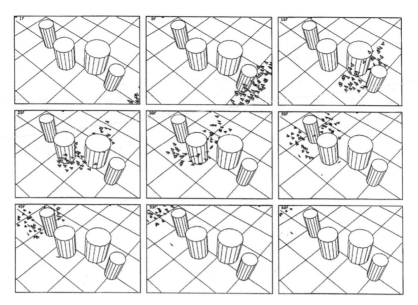

Fig. 4.1.5. Craig W. Reynolds, *Boids*. An artificial-life program that creates graphic elements with some behaviors of birds, such as flocking. (*Boids* simulation and image by Craig W. Reynolds) ⟨http://www.red.com/cwr/boids.html⟩.)

For example, the "Game of Life" illustrates some of the investigation of evolutionary principles explored in A-Life research. It provides a simple example of cellular automata. Typically, a researcher or player picks some rules that will govern the evolution of a computer graphic image. For example, in the next generation, each pixel will turn on if two of its neighbors are on, otherwise it will turn off. Simple rule changes can create amazing families of images. Animated versions closely resemble the growth surging and waning patterns of organic creature populations. The Boids program illustrates an algorithm that creates graphic entities that manifest behaviors similar to flocks of birds.

Theoretical Perspectives on A-Life

Artificial Life and genetic algorithms research has drawn much scientific and artistic attention. Artists are intrigued by the challenge to create artificial life-forms that simulate the behaviors of biological life or that evolve and self-propagate as a result of experience.

A-Life on-line: ⟨http://Alife.santafe.edu/⟩

Theorists have noted, however, that there is confusion about use of the term *life* and the relationship of A-Life inquiries to mainstream biological and scientific investigation. Art theorists have attempted to investigate the implications of A-life research to contemporary techno-cultural analysis and art practice.

In "Life as We Know It and/or Life as It Could Be: Epistemology and the Ontology/Ontogeny of Artificial Life," philosopher Edward Shanken begins to demonstrate the kind of analysis that is necessary. This paper draws on sources from art history, the philosophy of science, A-Life research, and the work of artists in the field. He uses knowledge of the research field in his explication of the art. The paper critiques the claims of A-Life researchers that they are creating life. It also explores the parallels between A-Life researchers and artists in that they all build synthetic creations. Although a full review of the paper is impossible here, a few highlights will help in appreciating the development of A-Life–inspired art.

Shanken notes that contemporary critiques of science, such as those of Paul Feyerabend, suggest that science is already much like art in its fabrication of "nature." Feyerabend notes that nature yields different answers when approached with different methods. A-Life research is especially subject to confusions:

Implicit in the comments above is the idea that science is a hermeneutic rather than teleological endeavor. Artificial Life, as the conjunction of biology and computational science, is likewise an interpretive discipline, one which—due to the domain of its inquiry and the nature and extent of its claims—raises many gnarly epistemological and ontological questions. For what is accepted as constitutive of life has great bearing on the understanding and experience of being.[7]

Even before anyone ever heard of A-Life, art historian Jack Burnham in his 1968 book *Beyond Modern Sculpture* had predicted the arts' fascination with intelligent sculpture and automata, noting relationships to the ancient Greek obsession with "living" sculpture.

Shanken critiques researchers' and artists' claims that they are creating life. Rather, he suggests that they are creating "synthetic biology" that gives visual and behavioral manifestations of biological theories about life. He claims he is not questioning the value of what they are producing but only the philosophical claims about their enterprise. He illustrates by referring to Michael Grey's A-Life jellyfish:

Grey's jellyfish are the aesthetic product of the confluence of a computational process that generates the simulated morphogenesis of forms, and an interpretive process that ascribes meaning and significance to them from the perspective of a human observer. Are they alive? No. Do they emulate biological organisms? No. Do they emulate biological theories? Yes. Do they question the relationship between the science of biology, the creation of artistic form, and the systems of meaning and significance that constitute those fields of endeavor? Most definitely.[8]

Katherine Hayles's analysis in "Narratives of Artificial Life" offers a related critique. A-life can be looked at as a Platonist reductionist effort to locate essential forms. In so doing, it disregards information about life as it is embodied in organisms.[9]

Simon Penny, a robotics artist whose work is described in following chapters, is also known for his critical analysis of the subtexts of scientific-technological research and the art that explores these areas. His paper "The Darwin Machine: Artificial Life and Interactive Art" provides a wide-ranging look at the A-Life research field and the subtexts that guide it from the perspective of "cultural practice." His "Archaeology of Artificial Life" sees A-Life inquiry as simultaneously continuing science's totalizing and colonizing pretentions to control everything at the same time as entertaining new nonlinear perspectives that probe the impossibility of that goal.

The premise of the enquiry is that since the rise of industrialism (if not before) artistic and scientific ideas have been presented as antithetical, rationalism being dominant in one and in question in the other. The corpus of the "scientific" has been waging a tireless war of conquest upon the tatters of the superstitious prescientific world. In this historical context, Artificial Life is of great interest as it is, on the one hand, a clear perpetuation of this colonialising rationalism, made possible by the machine which is the pinnacle of the rationalist "engineering worldview," the digital computer. On the other hand, aspects of Artificial Life and related studies such as fractal geometry, nonlinear dynamics, and complexity theory, which challenge basic premises of the scientific method and the Enlightenment worldview.[10]

Analyzing the narratives of A-Life research, he notes, for example, that some A-lifers aspire toward omniscience. They seem to buy into Judeo-Christian and industrial capitalist ideologies of domination of the earth's biodiversity. Moreover, the "very simplistic, individualistic, and mechanistic evolutionary narrative chosen has a decidedly nineteenth-century ring to it, and implicitly supports social Darwinism." The research supports Enlightenment notions of dualism, which separates mind and body. On the other side he sees dynamic systems working on other contrary narratives employing ideas such as sensitive dependence, fractality, entropy, and self-organization, emergence, and reductivism.

Penny sees his art as an exploration of these cultural issues. He uses the technology to reflexively highlight some of the larger questions. He illustrates a major trend in technology/science-based art in which artists work to become knowledgeable about an area of technology or science and then engage in cultural critique, revealing narratives and concepts that might be invisible to regular practitioners of the field. The artwork becomes a medium for that inquiry and its explication.

In "As Art Is Lifelike: Evolution, Art, and the Readymade," artist and theorist Nell Tenhaaf analyzes A-Life research's position in scientific inquiry and its implications for the arts. While debunking A-Life's exaggerated claims, she sees it as powerful arena for cultural and artistic inquiry:

A-Life is proposed here as a place to locate art practice for artists who are interested in techno-science, and who are concerned with the "two cultures" gap between the humanities and the sciences. Mythical narratives that underpin new computational techniques, such as the dream of transformation or even generation of life, are not dismissed but become the impetus for resituating A-Life as a set of representational strategies with great creative potential. A-Life is linked to a particular aspect of twentieth-century art: how artists have developed and expressed the conviction that art and everyday life are inextricably enmeshed.[11]

Tenhaaf positions A-Life research in a line of cultural inquiry stretching from alchemy:

[T]he transcendent vision of higher evolution attached to it places A-Life within a trajectory that runs from alchemical wizardry through Faustian metaphysics to contemporary reproductive technologies and cloning. A-Life is based on the hypothesis that computer simulation of evolution can determine not just how evolution works but also how it progresses, that is, that simulations of living systems can shape the development of species.

Metaphor and analogy are critical components of the research. In this regard they share important ground with the arts:

[T]he representational dimension of A-Life is key to understanding its impact and importance: its foundational features, including the very obvious one of creation itself, are driven by the pull of analogy and the power of metaphor, which operate conceptually to establish the parameters of research and also determine the development of the representational tools themselves. Today, these tools have extended from computer simulations into evolutionary robotics and evolvable hardware. . . .

[The concerns with modeling] connect A-Life with art, they forge a link with art practices that are also as old as human intelligence: the urge to develop a symbolic logic and representational system that teases out some kind of order and meaning from a chaotic surround.

Tenhaaf concludes that the common concern with representation may offer an area of reapproachment between art and science:

Further, within a broad cultural perspective, A-Life resituates the relationship between art and science, which has been an increasingly problematic one since the two areas split apart during the Enlightenment. . . . Scientific methodologies will continue to discourage or exclude the operation of cultural narratives. But from its inception, A-Life has allowed various degrees of play, whether deliberately or implicitly, and thus offers ways in which complex, layered representations can keep subjectivity, the social world, and also the natural world fully in view. Because nature is increasingly indistinguishable from manipulations of nature, which is evident in the ease with which we accept pharmaceutical regulation of our moods or the rationalized management of what we still call wilderness, such a point of view is important for any designer of a reality model, be it artist, theorist, engineer, or scientist.

Tenhaaf's analysis considers the similarities of artificial intelligence and A-Life research (simulating biological processes) but then focuses on A-Life's interest in evolutionary computing and genetic algorithms as a differentiator. She notes the metaphoric nature of the discourse about evolution. A-Life and art share common interests in simulating life. Tenhaaf proposes that sorting out their similarities and differences poses an interesting challenge:

While art remains preoccupied with philosophical, perceptual, and human issues, A-Life does have a much more direct link to the biosciences in its concern with modeling nature. But because it doesn't use the methods of pure science to derive these models, because its modeling parameters arise more from computational ingenuity than from methods of observation of natural phenomena, A-Life is a fundamentally creative platform.

Perhaps the most expansive way in which A-Life and art can be viewed in parallel is that, in their preoccupation with philosophical and theoretical questions regarding the nature of life, they are focused on the creation of means for life's representation. Each is of necessity concerned with how the mode of development of representational apparatuses or technologies affects the very kinds of representations that can be made. Interpretation of the representations is dependent on the material, lived context of the interpreter, and tends also to be influenced by her technologized environment.

Tehnaaf examines A-Life's interest in evolution in detail, noting that the systems' computational algorithms predetermine the range of possible outcomes. Development of the algorithms is a creative act in which metaphors are mapped to computational processes. Tenhaaf proposes that biology functions as a "readymade," a cultural construct rather some essential truth about nature:

A-Life practitioners don't become artists when they create readymades in their use of biology, but they nonetheless contribute to the process of engaging art with science. They place quotation marks around a segment of nature and make explicit in their models its encoding within a particular set of techno-scientific practices, which are thereby revealed as representational practices.

She considers several biological theories that represent alternatives to the dominant survival of the fittest paradigm, such as the influence of learning on evolution, selection based on cooperation, and the evolution of complexity over time. She suggests that A-Life could function like anti-art, providing an arena for spelling out alternatives. Furthermore, because of the way it depends on both obvious representational strategies and scientific understanding, it allows much wider participation in the discourse:

Although anti-art is a critique of institutionalized art and its role in maintaining social convention, this description places it in a negative position and doesn't adequately account for its Dadaist origins of radical breakaway spirit, its commentary on intolerable conditions of social distress, or its iconoclastic humor. In parallel, A-Life could be considered principally as the ground for another version of science critique calling attention to the sociopolitical imbrications of science in the face of its insistence on objectivity. But A-Life becomes a more compelling epistemological field if we consider it as the emergence of an alternative, para-scientific practice, an "antiscience": it doesn't seek to negate its terms of reference or their knowledge base; rather, it depends on them so as to propose reinventing them. Anti-science can expand our thinking about science in a way that parallels how anti-art reorders the symbolic systems we use to interpret and constantly reinvent everyday life. Anti-art shows that, once art and life are perceived as enmeshed, the transformative potential of art increases exponentially. Similarly, awareness of how we construct nature through science and technology on a daily basis could deliver a comparable empowerment.

Tenhaaf analyzes the search for algorithms as a widespread tendency in the sciences and computation. She traces the history of cybernetic research and deconstructs assumptions about information and the totalizing tendencies in genetic theory and biological science. Artistic attention to A-Life is a useful tool for addressing some of these issues:

There are many parallels between concepts of anti-art and anti-science. Without expecting that A-Life as an antiscience will revolutionize science practices, it is conceivable that it may inject new life into science, just as art must be in a continuous state of reinvigoration and renewal in relation to life. There is a line that can't be crossed between the scientific method of studying the real, material world through experimentally testable models, and an artist's contrivance of a metaphorical material world through techniques of representation and interpretation. But overall, A-Life offers a cross-platform to both science and art for renewal, a site for building awareness about assumptions and biases that shape perception.

Summary

The next chapters review artists whose work reflects on the research topics discussed in this chapter. It includes artists who work with contemporary mathematical concepts, and algorithmic artists who focus on developing abstract sets of computer rules for generating images and music, genetic art, fractal art, and artificial life.

Notes

1. *Encylopaedia Britannica,* "Mathematics," Britannica Online, ⟨http://search.eb.com⟩.

2. J. Lambek, "Foundations of Mathematics," *Encylopaedia Britannica,* Britannica Online, ⟨http://search.eb.com⟩.

3. National Science Foundation, Division of Mathematical Sciences, "The Mathematical Sciences," ⟨http://www.nsf.gov/pubs/1998/nsf9895/start.htm⟩.

4. National Science Foundation, Division of Mathematical Sciences, "Program Statement," ⟨http://www.nsf.gov/mps/dms/dmsprogs.htm⟩.

5. National Science Foundation, Division of Mathematical Sciences, "Program Assessment," ⟨http://www.nsf.gov/pubs/1998/nsf9895/math.htm⟩.

6. C. G. Langton, "Preface," in C. G. Langton, C. Taylor, J. D. Farmer, and S. Rasmussen, eds. *Artificial Life I,* vol. 10 of SFI Studies in the Sciences of Complexity (Redwood City, Calif.: Addison-Wesley, 1992), pp. xiii–xviii.

7. E. Shanken, "Life as We Know It and/or Life as It Could Be: Epistemology and the Ontology/Ontogeny of Artificial Life," ⟨http://mitpress.mit.edu/e-journals/LEA/ARTICLES/zeddie.html⟩.

8. Ibid.

9. K. Hayles, "Narratives of Artificial Life," in Future Natural: Nature/Science/Culture (London and New York: Routledge, 1996).

10. S. Penny, "Darwin Machine," ⟨http://www-art.cfa.cmu.edu/www-penny/texts/Darwin_Machine_.html⟩.

11. N. Tenhaaf, "As Art Is Lifelike: Evolution, Artificial Life, and the Readymade," ⟨http://mitpress.mit.edu/e-journals/LEA/home.html⟩ (all subsequent quotes from the same source).

Algorithmic Art, Art and Mathematics, and Fractals

Algorithmic Art

An algorithm is a logical process for achieving some result, for example, a recipe or the calculation steps for dividing numbers. Originally used in mathematics, the term has been extended to mean any systematic logical sequence.

Algorithms are essential parts of computer programs. They indicate a series of logical manipulations to be executed by the processor. Computer algorithms can instruct the computer to generate visual displays. Early computer art pioneers such as Herbert W. Franke, Kenneth Knowlton, and Charles Csuri focused some of their art on explorations of algorithms. Jasia Reichart's early computer show "Cybernetic Serendipity" included these and other artists who were intensely interested in the underlying rules of image generation.

They found this idea of a logical sequence embedded in a machine intellectually intriguing and saw this process of externalizing an image creation sequence as an appropriate challenge and opportunity for the arts. In *Computer Graphics—Computer Art,* Herbert W. Franke enumerated several mathematical operations on which he expected computer graphics to be based: symetrization, transformations, mathematical functions, moiré patterns, permutations, interpolation and extrapolation, matrix calculus, and random numbers. He saw a great future in "experimental aesthetics."[1]

The general pattern was: the artist created the algorithm and then the computer executed the steps to create the image. Although they were interested in the ultimate image produced, they were also interested in other aspects of the process:

- A way of working that focused on the creation of abstract generative processes rather than exclusively on the production of a particular image.
- The ability of algorithms to create "families" of images through the manipulation of parameters.
- Cultivation of new kinds of artistic skills involving innovation in developing algorithms and understanding and working with the constraints and possibilities of arbitrary systems, such as computers.
- The radical idea several decades ago of moving artists into the new and esoteric field of computers, which promised to be scientifically and culturally significant.
- The gesture of artists claiming the algorithm authoring/programming process for their own, which lies at the heart of information technology so critical to our culture.

Charles Csuri: ⟨http://siggraph.org/artdesign/profile/csuri/artworks/⟩

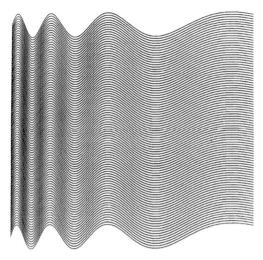

Fig. 4.2.1. A. Michael Noll, *Ninety Computer-Generated Sinusoids with Linearly Increasing Period.*

Several of the artists identified with the manipulation of computer algorithms call themselves algorists. This group includes Charles Csuri, Helaman Ferguson, Jean-Pierre Hebert, Manfred Mohr, Ken Musgrave, and Roman Verostko. Many of them have been working in this genre for decades. For example, Csuri uses AL, an animation language created by Stephen May, which gives him control over hundreds of functions that control the qualities of images. Many early pioneers of the use of computer animation in film created works far ahead of their time in recognizing the power of algorithms to generate an abstract "visual music." For example, Larry Cuba and John Whitney's *Arabesque* is a landmark of the field. Cuba helped found the Iota Organization, which provides historical archives and contemporary resources for those interested in the algorithmic generation of animated art. Other artists addressing algorithmic analysis include those working with genetic art and Artificial Life (see chapter 4.3).

Other efforts include the Institute for Artificial Art in the Netherlands (see chapter 7.6) and the algorithmic art show at Xerox PARC, organized by Marshall Bern, which showed a range of artists focusing on generative algorithms in their work.

Algorists: ⟨http://www.solo.com/studio/algorists.html⟩
Iota Organization: ⟨http://www.iotacenter.org/⟩
Institute for Artificial Art: ⟨http://www2.netcetera.nl/~iaaa/⟩
PARC Algorithm Art show: ⟨http://www.parc.xerox.com/csl/members/bern/algoart.html⟩.

Fig. 4.2.2. Charles Csuri, *Sleeping Gypsy*. Three-dimensional computer-graphic figures are manipulated by hand and algorithmic fragmentation function.

Roman Verostko

In his essay "Explanation of Algorithmic Art," Roman Verostko explains his heritage to early abstract artists and the new power that the computer offers for exploring abstract forms:

Most of my work for the past forty years has been with pure visual form ranging from controlled constructions with highly studied color behaviour to spontaneous brushstrokes and inventive non-representational drawing. Such art has been labeled variously as "concrete," "abstract," "nonobjective," and "nonrepresentational." In its purest form such art holds no reference to other reality. Rather, one contemplates the object for its own inherent form similar to the way one might contemplate a flower or a seashell. . . . With the advent of computers I began composing detailed procedures for generating forms that are accessible only through extensive computing. My ongoing work concentrates on developing this program of procedures for investigating and creating such forms. . . .

Roman Verostko: ⟨http://www.verostko.com/⟩

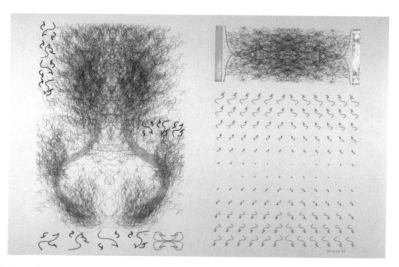

Fig. 4.2.3. Roman Verostko, *Diamond Lake Apocalypse* (1992). Algorithmically generated image with gold leaf referring to medieval manuscript illumination.

The artworks are visual manifestations of the dynamic procedures by which they grew. They may be viewed as visual celebrations of the information processing procedures embedded in today's culture. The finished works invite us to savor the mystery of their coded procedures whose stark logic yields a surprising grace and beauty. These procedures provide a window on those unseen processes from which they are grown. By doing so they serve as icons illuminating the mysterious nature of our evolving selves.[2]

Some critics suggest that artists working with computer algorithms are somehow more distant from the artistic process than conventional artists. Also, these critics assert that algorists are letting computers dictate the parameters of their work and thus diminish the individuality and self expression of the work. In his ISEA94 paper "Art and Algorithm," Verostko notes that all artists use algorithms in their work. The algorists just focus more explicitly on authoring the algorithm as an artistic focus:

The procedures and decisions they make as artists are the artistic procedures. These have to do with the individual sensibility of each one about the entire "art-making" process. Each makes artistic choices as to which procedures he will articulate, what form the work will assume. . . . Can an artist write an algorithm then for an artistic procedure? Emphatically yes! Such algorithms provide the artist with self-organizing form generators which manifest his or her own artistic concerns and interests.[3]

Commercial software arrives with its subtexts, such as Renaissance visual perspective and the cultural baggage derived from its military and commercial origins. Every program comes with its limits of what it makes easy and what it makes difficult or impossible.

Some algorists might assert that by mastering the programming process itself, artists gain more freedom from this baggage. But that assertion opens a long-standing debate. Do artists need to program? Build hardware? How deep does an artist need to go to claim control of the process? Can the artist use the programming primitives included in the operating system? Does the artist need to build the display hardware? As Verutsko suggests, every artist who works with computer programs uses algorithms—it's just that they have been written by someone else. Contemporary programs such as Photoshop allow much more flexibility than their predecessors, and future plans call for increased customizability. Ironically, as the programs get more customizable the process of setting up options begins to look more and more like programming. The extent to which artists understand and can control technologies is a perennial issue in technological and scientific art. As explained in the introduction, this book holds the view that the more an artist understands the more power he or she has to explore and adapt the technology for art and to contribute to the cultural discourse about those technologies.

Ken Musgrave

Algorists are more willing than some other contemporary artists to assert that their art aspires toward universals—either spiritual or probing the essential geometry of the universe. In this they are akin to many of the abstract artists working at the beginning of the century. For example, Ken Musgrave writes about the power of algorithmic approach to probe the universe. He also counters critics who doubt the use of formal logic and technical processes as a gateway to the expression of the self. His paper "Formal Logic and Self-Expression," available at his Web site, explains these views:

I am an Algorist. I believe that the peculiar process by which these works come into being represents a revolutionary event in the history of the creative process for the visual arts. . . .

I have sought to master the games of Logic. . . . I am a Formal Algorist, a kind of logical positivist of fine art: I require that my artworks be exactly [a visual interpretation of] a theorem proved in a formal system. That is, I write a program, execute it with a well-defined input, and interpret the result, with the assistance of a master printer, as an artwork. Such purism places tight

---

Ken Musgrave: ⟨http://www.seas.gwu.edu/faculty/musgrave/⟩

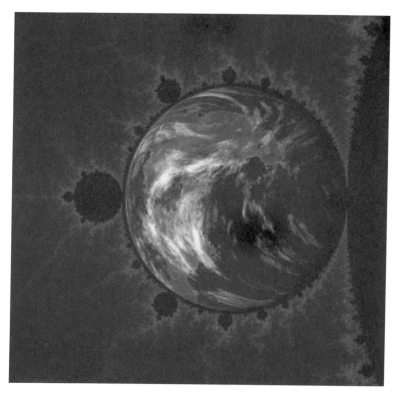

Fig. 4.2.4. Ken Musgrave, *Planet Mandelbrot*. An image combining both a deterministic fractal (the Mandelbrot set) and a random fractal.

restrictions on my process. It also unveils a vast space, fecund with serendipity, which I search for strings (digital images) which have an ineffable sense of having existed as eternal corollaries of timeless logic, which I simply have happened to stumble upon. Presumably, any being in the known universe could derive precisely the same result, from the same rules and axioms. Very strange.

The process of obtaining spiritual self-expression strictly through formal logic is philosophically perverse in that it puts determinism, which precludes Free Will, at the service of Spirituality, which is moot without Free Will. . . . I wish you to witness my reverence for Nature, and to be drawn in by the quiet beauty I have found there and been able to conjure up in numbers, through a bizarre, improbable, novel, and Byzantine process. It is all, technically speaking, Magic.[4]

George Legrady

George Legrady (see also chapter 7.2) integrates an interest in algorithms with an interest in deconstructing cultural structures. His *Equivalents II* generated cloud-appearing images based on texts that the visitor typed in. Geert Lovink explained the installation in an interview with Legrady:

Legrady calls it a highly conceptual work. The program generates the abstracted cloud images based on an algorithm whose perimeters are triggered by the viewer's text. The image begins with four squares which are successively subdivided all the way down to the image's pixels. *Equivalents II* uses an algorithm called "2-D midpoint fractal synthesis," which is based on the premise that every time you subdivide, you increase complexity by slightly altering the tonal values of the squares from which the new ones are derived. Additional disturbances to this process are added when the program comes across certain words in the viewer's phrase that match those stored in a data bank. . . .

Legrady selected words belonging to the language of various (sub)cultural groups that might come across the artwork. For example, there is a list of computer slang terms, a twenty-word list from Foucault's *Order of Things* in which he quotes Borges's reference to an esoteric Chinese encyclopedia. But there are also words from TV soap operas, J. G. Ballard's *Crash,* happy and unhappy words, colors, male and female terms. [5]

Legrady indicates his agenda of inducing viewers to question the processes that underlie computer data:

Questioning the essence of the computer is very much what this project is about. In the museums where the piece is shown, audiences come to it believing in the authority of this technology. My intent through the work is to make audiences aware that subjectivity is a part of computer-generated information. The *Equivalents II* project has come out of a long-term goal to come up with a program that would simulate a photograph to such an extent that you would not question its authenticity.

John Maeda and the MIT Media Lab Aesthetics and Computation Group

The Aesthetics and Computation group studies "the expressive aspects of computer-human interface from the viewpoint of traditional visual communication design." Computation provides enormous untapped potentials in areas of dynamic self-modifying graphics/typography, alternative models of interactivity, and subtle relationships between play and information display if artists are willing to develop their own systems and algorithms. For

---

John Maeda, Aesthetics and Computation: ⟨http://acg.media.mit.edu/⟩

Fig. 4.2.5. Joanna Berzowska, *Computational Expressionism*. An array of drawing tools that combine emergent graphic behavior with subtle interface to user actions.

example, Maeda's work, such as *Flying Letters,* explode classical concepts of typography and conventional mouse-based interfaces. His book *Design by Numbers* elaborates on these ideas. *Computational Expressionism,* by Joanna Berzowska, created a drawing system in which "smart" graphic objects each had a set of behaviors such that the end result was an emergent result of the user's drawing and the objects' behaviors.

Art and Mathematics

Working in a related field, several artists focus on the convergence of art and mathematics. This work spans a range of topics stretching from scientific visualization and imagi-

---

Joanna Berzowska: 〈http://www.media.mit.edu/~joey/x/x.html〉
Art and Mathematics '96: 〈http://forum.swarthmore.edu/am96/〉
Art and Mathematics '97: 〈http://hplbwww.hpl.hp.com/brims/art/gallery/am97/index.html〉.

nary universes to higher-dimensional geometries and fractals. They create drawings, generate computer graphics, and build sculptures. Some search for underlying concepts, such as symmetry or geometric systems, which unite the worlds of art/music and mathematics. As with the algorists, they often start with some underlying principle such as a theorem or a function and use it as a tool to generate art. Some use the computer as a tool, and authoring algorithms is sometimes an important part of the work. The computer's ability to visualize abstractions has stimulated development of the field. The intellectual fascination with the underlying principles and processes is often an important part of the motivation and working process.

Paying attention to the world of academic mathematics is highly valued. As is true with many other areas described in this book, the artist's identification of researchers and theorists outside of art as an important reference group is a distinguishing characteristic. The outsiders serve many functions—as sources of ideas, inspiration, collaborators, and to provide a new audience.

The community of artists working in this mode is highly active. Over the last thirty years many articles have been published in the journal *Leonardo*. Michele Emmer's book *The Visual Mind: Art and Mathematics* includes several of these. Many have been interested in symmetry as a unifying theme, as exemplified by books such as *Symmetries of Culture,* by Dorothy Washburn and Donald Crow. Mathematics and music is considered in books such as *Emblems of Mind: The Inner Life of Music and Mathematics,* by Edward Rothstein. The section headings in *The Visual Mind* illustrate the scope of the field: "Geometry and Visualization"; "Computer Graphics, Geometry, and Art"; "Symmetry"; and "Perspective, Mathematics, and Art."

There have also been several international conferences on art and mathematics, and several on the special focus of symmetry. A sampling of the session titles offered at the art and mathematics conferences gives an idea of the range of interests that mark the field:

- "Hyperbolic Toroids and Minimal Surface Sculpture" (Brent Collins and Carlo Sequin)
- "Architectural Ceramic Sculpture and Ceramic Geometric Tilings" (Douglas Klein)
- "Two-Dimensional Images Based on Hypercubes in Dimensions Three, Four, Five, and Six" (Manfred Mohr)
- "An Artistic and Mathematical Introduction to Knots" (Nat Friedman)

International Symmetry Foundation: ⟨http://members.tripod.com/~modularity/isis4.htm⟩

Artists' interests in mathematics span many subdisciplines of the field. Some find the rational process of mathematics intriguing. This view was expressed by the artist Max Bill in a historical paper entitled "The Mathematical Way of Thinking in the Visual Art of Our Time," in which he proposed the exploration of logical processes as a basis of art. Even for Bill, however, the interest does not end with logical manipulations. He sees artistic challenges in the ways mathematics connect humans with the mysteries of the universe. He is inspired by the intellectual leaps made by mathematicians:

Things having no apparent connection with mankind's daily needs—the mystery enveloping all mathematical problems; the inexplicability of space—space that can stagger us by beginning on one side and ending in a completely changed aspect on the other, which somehow manages to remain that selfsame side; the remoteness or nearness of infinity—infinity which may be found doubling back from the far horizon to present itself to us as immediately at hand; limitations without boundaries; disjunctive and disparate multiplicities constituting coherent and unified entities; identical shapes rendered wholly diverse by the merest inflection; fields of attraction that fluctuate in strength; or, again, the square in all its robust solidity; parallels that intersect; straight lines untroubled by relativity, and ellipses which form straight lines at every point of their curves—can yet be fraught with the greatest moment.[6]

For many contemporary analysts, the insights of dynamic and nonlinear systems and fractal geometries (discussed in chapter 3.3) open mathematics into ways of thought quite alien to what many learned to associate with mathematics:

For ages, mathematicians and other scientists lived in their linear world with two-valued logic, crisp decisions, sequential processes, and absolute optimality. But now new vistas are opening up: nonlinearity is beckoning; we begin to understand fuzzy logic and use it to great advantage; our decisions now have to be based on multiple criteria and are not crisp anymore; and our sequential processes and thinking are also beginning to change, helped along by interconnected multiprocessing computers and by concepts such as neural networks. Optimality itself is no longer the main aim in many instances: that notion often gets replaced by feasibility and suboptimality.[7]

Mathematics and Sculpture

Some sculptors work at physically manifesting structures and mathematical concepts that usually are difficult to visualize and often do not occur in nature. The contrast between the physicality of the sculpture and the cerebral nature of the mathematics enhances the experience of the sculptures. Examples of sculptors working this way include Charles Perry, Brent Collins, Carlo Sequin, Arthur Silverman, Elizabeth Whiteley, Helaman Ferguson, Stuart Dickson, George Hart, Charles Longhurst, Simon Thomas. This section also considers those working with virtual sculptures realized via computers and sound.

The sculptors vary in their motivations and in the areas of mathematics on which they focus. For some, the link to mathematical ideas is primarily intellectual; for others it is intuitive. Most describe their work in terms of shape and form rather than in mathematical terms. Still, the conceptual challenges of abstract mathematical thinking helps stimulate their work. Some see geometry as the underlying unity of the universe and believe that the convergence of art and mathematics can help spread those insights. For example, Nat Friedman, the organizer of many of the art and mathematics conferences has written about that unity:

In a seashell we see the oneness of art, mathematics, and architecture. A seashell is an abode that is also an ingenious spiral form-space sculpture. Seashells also display a variety of beautifully two-dimensional designs on curved surfaces. Thus the oneness of art, mathematics, and architecture was already genetically coded in these very early life-forms. I can imagine what it was like to have experienced the excitement of living in Florence during the Renaissance, when there was no separation between art, mathematics, and architecture. This unification also resulted in a mutual enrichment of these fields. It is my purpose to energize a move toward a reunification of these fields in education.[8]

Stewart Dickson

Stewart Dickson works with Mathematica to visualize mathematically based concepts. His artist statement illustrates the belief many of these artists have about mathematics as an underlying structure. It also shows the importance of mathematics history and thought in his work. He was interested in Fermat's last theorem and the area of topology called minimal surfaces. These interests inspire him to create sculptures aided by the new visualization capabilities of computers:

The seventeenth-century French mathematician Pierre de Fermat wrote in the margin of his copy of *Arithmetica,* by Diophantus, near the section on the Pythagorean theorem (*a* squared plus

---

Stewart Dickson: ⟨http://www.wri.com/~mathart/portfolio/SPD_Frac_portfolio.html⟩

Fig. 4.2.6. Stewart Dickson and Andrew J. Hanson, *Fermat's Last Theorem*. Visualization of complex projective varieties determined by $x^3 + y^3 = z^3$.

b squared equals *c* squared), $x^n + y^n = z^n$—it cannot be solved with nonzero integers *x*, *y*, *z* for any exponent *n* greater than 2. I have found a truly marvelous proof, which this margin is too small to contain." This was left as an enigmatic riddle after Fermat's death and it became a famous, unsolved problem of number theory for over 350 years.[9]

Dickson explains that he is challenged by the idea of bringing these abstract ideas into the world of physical sculpture. He sees a "deep spirtual value in the quest to understand and possess the unknown":

Andrew Hanson has made some pictures, and I have in turn made sculpture, of a system analogous to Fermat's last theorem—a superquadric surface parameterized in complex four-space. We think that the mathematics of the $n = 3$ case are similar to Fermat's own proof of the $n = 3$ special case. Our pictures have lent some visual concreteness to the recent news of Andrew Wiles's proof of the Taniyama-Weil conjecture, which implies the proof of Fermat. . . .

Indeed, minimal surfaces are objects of pure topology—pure abstract form—in which the geometry is simply constrained to be the most minimal and elegant expression of that form. Hoffman is satisfied to leave his objects in the two-dimensional picture plane. I find cyberspace fundamentally unsatisfying in the lack of tactile presence. I need to bring back artifacts from cyberspace into the space I physically occupy. This is the basis of the work I am presenting here.

The range of mathematically inspired interests are hinted at in the titles of some of his works, for example, *Three-Ended Minimal Surface, Infinite-Ended Minimal Surface Including a Topological Handle,* and *Enneper's Minimal Surface, Scherk's Second Minimal Surface.* He also created a VR environment called *Topological Slide,* modeled on mathematical structures (see chapter 7.3).

Brent Collins and Carlo Sequin

Brent Collins also works with minimal surfaces. He often collaborates with University of California at Berkeley math and computer professor Carlo Sequin. In an article written for the *Ylem Newsletter* issue focused on art and mathematics, he wrote an article called "Toroidal Closures of Scherk Towers," describing and illustrating his work. Even though the original motivation was intuitive and visual, the connection to the formal world of mathematics enhanced and extended the work:

Being a sculptor rather than a mathematician, my work has originated through visual intuition. I was using hyperbolic contours as modular elements of composition years before knowing of their presence in the Scherk and Costa minimal surfaces. In a recent cycle of pieces, the composition of these modules can be understood as a finite vertical Scherk tower of pairwise orthogonal holes that has been rounded into a closed circular shape forming a torus of pairwise orthogonal holes. With an even number of holes these surfaces are orientable and have ends similar to the Costa surfaces. A toroidal closure of an odd number of holes requires that the tower is also twisted, as in the case of a Möbius strip. This results in a nonorientable surface with a single continuous edge that is a knot.[10]

Donna Cox

With the advent of sophisticated computer graphic animation systems, artists could "sculpt" without using physical materials. Donna Cox's *Venus and Milo* is a well-known example of this approach that gives visual life to a mathematical abstraction. Working in collaboration with mathematicians and computer scientists at the National Center for Supercomputing Applications at the University of Illinois, she created a humorous computer graphic animation about two fictitious characters at an art museum. Venus,

---

Brent Collins: ⟨http://www.cs.berkeley.edu/~sequin/SCULPTS/collins.html⟩
Carlo Sequin: ⟨http://www.cs.berkeley.edu/~sequin/⟩
Donna Cox: ⟨http://www.ncsa.uiuc.edu/People/cox/⟩

Fig. 4.2.7. Brent Collins and Carlo Sequin, *Orientable Surface with Third-Order Saddles Deployed in a Hexagonal Ring.* Photo: Phillip Geller.

a vaguely female, hourglass-shaped figure, is a dynamic representation of a topological shape. Cox describes Venus:

The narrative is centered around the Romboy Homotopy, a mathematical deformation between Steiner's Roman Surface and Werner Boy's Surface. . . . The Venus is a one-sided, ten-dimensional mathematical surface from the Romboy Homotopy; her name evolved from her resemblance to the hourglass—a classic female shape. The homotopy animation reveals a complete series of one-sided surfaces that visually transform into Jungian-like, archetypal forms.[11]

Venus and Milo is visually and conceptually rich. Cox is concerned that contemporary art does not highly value the kinds of quests pursued by visual mathematics and that popular culture has appropriated some mathematical visualizations, such as fractals, without much appreciation of the richness of the underlying ideas and intellectual agendas. In "The Tao of Postmodernism: Computer Art, Scientific Visualization, and Other Para-

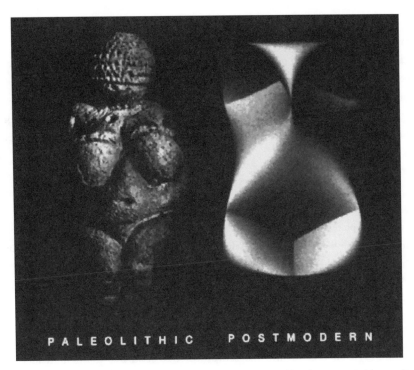

Fig. 4.2.8. Donna Cox, George Francis, and Ray Idaszak, *Etruscan Venus*. A goddess shape generated from mathematical function.

doxes," she asserts that computer art such as that described in this section is "orphaned"—rejected by modernists and postmodernists alike—and suggests a broader model that acknowledges some of the research being conducted at the nexus of art, science, and mathematics.

Brian Evans

Brian Evans is another artist working with mathematics and fractals. He is quite clear that the goal of his work stretches beyond the objects to stimulating underlying ideas. For him, these have always been at the core of the aesthetic experience. In an article entitled "Implicate Beauty," he articulates the importance of the underlying mathematics and the larger spiritual questions that he sees stimulated by mathematics and its visualization:

In its pure form, mathematics is often practiced with inquiry as the motivation and aesthetic experience as the goal. I consider an aesthetic experience a heightened moment that transcends sensation and emotion, when one finds resonance with something perceived. For some the experience moves towards the spiritual. . . .

In describing mathematical processes with algorithms, beauty and meaning can be discovered. Numbers are mapped into light and/or sound, and perceived through the senses as objects. It is the mathematical source of these works that has aesthetic worth. . . . The objects themselves are not aesthetic, but rather are created to catalyze aesthetic experience. . . . Algorithms, implemented on computers, make it possible for us to see and hear the beauty of mathematical processes. . . .

Is meaning culturally attributed, or is mathematics meaningful and effective because it describes "grand truths"? We trust our lives on a daily basis to the effectiveness of these mathematical models. What is the basis of our faith?

Through my pieces I hope to bring the viewer or listener into this discussion, encouraging them to ponder the big questions themselves. These questions define the foundation and core of our civilization. Is the universe an ordered place? If so how was/is this order created? If not, how is it we can define, predict, and discover so much of our physical reality with the logic and structure of mathematics?[12]

Other Artists and Projects

Helaman Ferguson creates unusual sculptures that visualize mathematical concepts. For example, his *4 Canoes* refers to Klein bottles in which "solid granite toroids with double cross-caps, couple inextricably mysteriously." Also inspired by mathematics, *Manfred Mohr* works in two dimensions, creating works by writing algorithms for the computer to actualize. He works on projects such as visualizing n-dimensional objects and fracturing the symmetry of n-dimensional hypercubes. Even though the ideas and procedures are important, he ultimately demands that the work succeed visually. *Maurice Benayoun*'s computer graphic animation *Quarxx* generated creatures that could only live in the computer, "living beings indifferent to the immutable laws of Nature . . . odd beings by aberrations in their relation with time, space, and matter." *Ralph Abraham, Peter Broadwell, and Ami Radunskaya* developed the MIMI (mathematically illuminated musical instrument), which maps mathematical concepts to musical qualities such as location, frequency, and modulation patterns. *Doris*

Helaman Ferguson: ⟨http://www.access.digex.net/~mhelamanf/gallery/fourcanoes/index.html⟩
Manfred Mohr: ⟨http://sciweb.nyu.edu/~mohr/⟩
Maurice Benayoun: ⟨http://panoramix.univ-paris1.fr/UFR04/benayoun⟩
MIMI: ⟨http://www.plasm.com/peter/public_html/mimi/mimi_index.html⟩

Schattschneider focuses on tiling, polyhedra, dynamic geometry, symmetry, and public presentations of mathematics.

N-Dimensional Space—Linda Dayrymple Henderson and Leonard Schlain

Since late in the last century, artists have been intrigued by dimensions beyond our customary three. Mathematics offers a language for describing worlds composed of more dimensions. Indeed, mathematically it is easy to extend the rules of our three-dimensional worlds to n-dimensional space even though it is difficult to visualize these worlds and to imagine living in them.

The mathematical breakthroughs exploring the fourth dimension and non-Euclidean geometries greatly influenced modern artists in the early twentieth century and may have helped liberate them to take the leaps they did. In *The Fourth Dimension and Non-Euclidean Geometries in Modern Art,* Linda Dayrymple Henderson extensively reviews these ideas. This link is also reviewed extensively by Leonard Schlain in *Art and Physics* and by me in my forthcoming book *Great Moments in Art and Science.* Henderson's book suggests that the intellectual enterprise of mathematics may have cultural effects beyond the specific topics of the research. For example, she notes that although some abstract artists focused on the spiritual potential of conceptualizing a fourth dimension beyond the normal plane, artists such as Marcel Duchamp were more interested in the structural subversion of thinking beyond the visual three dimensions:

Shadows, mirrors, and virtual images were added to the four-dimensional vocabulary of the artist by Duchamp, whose approach to the subject was unique in this period. If Duchamp at first shared his Cubist colleagues' idealist belief in the fourth dimension, his attitude quickly became more analytical. For Duchamp the nth dimensional and non-Euclidean geometries were a stimulus to go beyond traditional oil painting to explore the interrelationship of dimensions and even to reexamine the nature of three-dimensional perspective. Like [Alfred] Jarry before him, Duchamp also found something deliciously subversive about the new geometries with their challenge to so many long-standing "truths." The motives behind Duchamp's interest in the fourth dimension in fact represent an alternative strain to the idealist visions of a higher reality that supported the birth of abstract art.[13]

She also recounts the waxing and waning of belief in evolutionary consciousness stimulated by the awareness of dimensional theorization. She quotes the painter Irene Rice Pereira, who saw imagination about space as a precursor to other kinds of imagination:

The apprehension of space and the development of human consciousness are parallel. The more energy that is illuminated and redeemed from the substance of matter, the more fluid the perceptions become and the more the mind sums up into abstraction. The mind's capacity for dimensionality and the structure of consciousness become available through experiencing one's own action.[14]

4-D Artists

Richard Brown maintains the "4-D" Web site, in which he collects mathematical, historical, and artistic references to work with, using the fourth dimension as a conceptual challenge to everyday thought and as a source of visual provocation. He includes artists such as **Thomas Banchof, Eric Swab, Cyberbridge, Dennis Wilcox,** and **Yoichiro Kawaguchi.**

Fractals

Fractal geometry is one area of mathematics that has engaged wide audiences in the last years. What started out as an esoteric exploration of the fractional dimensional space has ended up as images on T-shirts and album covers. Many find the images generated by fractal geometry beautiful and mysterious. Benoit Mandelbrot is the scientist credited with calling attention to the power of this branch of mathematics to describe natural phenomena often better than the elemental geometry of perfect forms. Computers accelerated the study of fractals because they facilitated the intensively iterative computation required to generate fractal data, and they enabled the use of graphics to visualize the data.

   Fractal geometry is connected to a larger set of scientific approaches, including dynamic systems and chaos theory, which many find very productive for understanding natural phenomena such as weather and biological behavior, and social phenomena such as stock market behavior. These approaches have stimulated many new fields, such as Artificial Life, which intrigue scientists and artists alike. Artistic exploration of Artificial Life and genetic art will be discussed in the next chapter. Stuart Ramsden explains the cross-disciplinary appeal in his Web article "Fractals, Feedback, and Chaos—A Brief History":

"4-D" Web site: ⟨http://www-crd.rca.ac.uk/~richardb/4d.html⟩

Geometry, with its roots in ancient Greece, first dealt with the mathematically pristine and simple forms of spheres, cones, and cubes. These exact forms, however, rarely occur naturally. A geometry suitable for describing nature was constructed this century, however, and that is fractal geometry. This revolutionary field deals with shapes of infinite detail, such as the branching of a river delta or the nebulous forms of clouds, and allows us to define and measure accurately the property of roughness.

Fractals arise in many diverse areas, from the complexity of natural phenomenon to the dynamic behaviour of mathematical systems, and their striking beauty and wealth of detail has given them an immediate presence in our collective consciousness. Fractals are the subject of research by artists and scientists alike, making their study one of the truly Renaissance activities of the late twentieth century.[15]

Many artists and hobbyists now produce fractal art. Fractals are a broad area of study, and there are many different types explored by mathematicians. Also, their visual appearance is affected by the selection of parameters and the representational scheme used to map graphics to the data. As with other areas of mathematics-inspired art, the artists' contributions lie in their awareness of the range of possibilities and their manipulation of the algorithms and parameters used to generate the data set and create the graphics. The Web is full of fractal art galleries, and fractal generating programs can easily be obtained. For many, the appeal is primarily visual and sensual. The URLs below list Web resources of art and art shows focused on mathematical concepts and fractals:

Clifford Pickover

For others the appeal is more than visual. For them, the power also derives from the theoretical underpinnings and the power of fractals and related fields of study to represent real phenomena and stimulate thought. The artist, scientist, and author Clifford Pickover illustrates this more comprehensive approach. On his Web site he states his interdisciplinary interests: "My primary interest is finding new ways to

---

Fractal art galleries: ⟨http://www.ArtByMath.com/⟩
 ⟨http://www.yahoo.com/Arts/Visual_Arts/Computer_Generated/Fractals/⟩
 ⟨http://online.anu.edu.au/ITA/ACAT/contours/contours.html⟩
 ⟨http://msstate.edu/Fineart_Online/art-resources/mathematical.html⟩
Clifford Pickover: ⟨http://sprott.physics.wisc.edu/pickover/ad6.html⟩

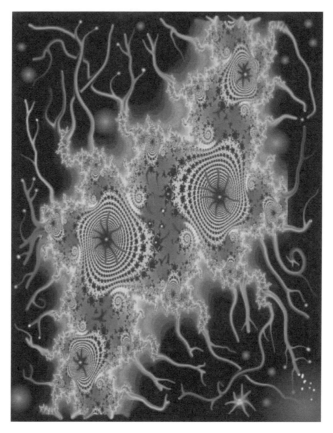

Fig. 4.2.9. Clifford Pickover, *Telopodite Fractal 1*. A rootlike structure created by the use of fractal mathematics and artistic manipulation (from Clifford Pickover's Web page ⟨www.pickover.com⟩).

continually expand creativity by melding art, science, mathematics, and other seemingly disparate areas of human endeavor. I seek not only to expand the mind, but to shatter it."[16]

In an interview for an ACM publication, Pickover elaborates on the power of visual representations, such as fractals, to stimulate thought and their place in a larger agenda of encouraging interdisciplinary thought:

In many cases, the computer graphics function like a stain applied to a wood grain to bring out and highlight hidden structures. In my books, the applications are varied and include fields as diverse as speech synthesis, molecular biology, mathematics, and art.

Lateral thinking is reasoning in a direction not naturally pointed to by a scientific discipline. It is reasoning in a direction unexpected from the actual goal one is working toward. In my books, the term "lateral thinking" is used in an extended way to indicate not only action motivated by unexpected results, but also the deliberate drift of thinking in new directions to discover what can be learned.[17]

The range of books he has written illustrate the power of this discipline-defying approach. They also illustrate the critical need for artists to pay attention to fields outside of art. Here is a list of some of his books:

- *The Loom of God* (on the relationship between mathematics, God, physics, and religion)
- *Black Holes—A Traveler's Guide* (on the beauty and physics of black holes, with computational recipes)
- *Keys to Infinity* (on the mystery and beauty of infinity and large numbers)
- *Spiral Symmetry* (on spirals in nature, art, mathematics, literature, and more)

Promise and Problems in Art and Mathematics and Algorithmic Art

The attempt to explore the relationships between art and mathematics has a long history. Some of the greatest innovators in Renaissance art were also great innovators in mathematics. Those working in this field believe that math and art are not so disparate as the popular characterizations of math as logical and art as intuitive would suggest. Elegance and beauty can be critical to math, just as concern with systematic spatial relationships can be key in art.

Hewlett Packard's Basic Research Institute in Bristol (BRIM) initiated a project in art and mathematics to encourage artists, mathematicians, and scientists to work together. Their vision statement articulately describes this view:

Mathematics seeks to describe reality by looking at the logical interrelationships between concepts. Through art, we experience reality in ways not directly accessible to reasoning, but which we find intuitively meaningful.

There are, however, profound commonalties between the two areas. Both try to express fundamental "truths" about the nature of reality, seeking structure and symmetry within the complex universe in which we find ourselves. As Einstein once said: "Common to both is the devotion to something beyond the personal, removed from the arbitrary.". . .

In bringing together the artists and the mathematicians, we hope to recreate the symbiosis that existed in early civilizations and the Renaissance period. The relationship we hope to create can be described as "Mutual Inductance"—the property that arises when two wires are in the vicinity of each other, and the current in each one is coupled to the other.[18]

Literacy—Audience Background

Although the attempts at integrating mathematics and art are fascinating, they raise provocative questions. Typically artists assume that their audiences have a certain cultural background that will enable them to interpret and enjoy artworks. The lack of this background does not deter artists from exploring frontier areas of culture, but it does create obstacles for reaching audiences. For example, many audiences did not understand and hated the work of the impressionists and other modern movements when they first appeared because of their revolutionary approach to perception and subject matter.

The algorists and artists working in the convergence of art and mathematics face analogous challenges. Most claim that their work must be effective on an immediate visual and visceral level and secondarily on a conceptual level. Yet, they are underestimating the importance of that intellectual level; it is part of what differentiates their work and adds additional levels of richness.

The creation of computer programs and algorithms can require extensive creativity. Some solutions are more revolutionary and elegant than others. I remember that in my early computer artworks, I considered the programming an essential element of my artistic enterprise. I wanted my audience to appreciate the accomplishment. I and other artists working in the same mode would often post the program listings and flow charts as part of the art installations. Audiences were quite consternated and rejected the programming material. Most had no background to understand or make sense of the algorithmic work. Most algorists have abandoned the attempt to communicate the algorithms as a core element of their work, although they will share them with others interested in the technicalities.

Artists creating work inspired by mathematical concepts face similar challenges. In part, the power of their work comes from the richness of mathematics—its history, boldness in conceptualization, and its systems of thought. Audiences who are familiar with this background bring much more to the work than those who are unaware. For example, art based on topological structures such as a Möbius strip (twisted joined infinite surfaces) might intrigue a naive viewer, but someone aware of the wider topological issues might find it even more engaging. Certainly, the implications of some of the titles and descriptions of artworks described in the previous sections must have mystified some readers. Similarly, art based on fractal geometries often captivate audiences, but only those versed in underlying theories are likely to appreciate the subtleties of the artist's choice of parameters and elements of the geometry that are open to exploration.

So what does an audience need to know to appreciate an artwork? Often, only an elite comfortable with art world experimentation understands. Eventually, audiences acquire a cultural background that helps them make sense of the new art until even much of it passes into convention.

Art influenced by mathematics, science, and technology more generally confronts special problems of audience literacy. Much of its audience will be unfamiliar with the history, conceptual frameworks, and discourse that shape thought related to the technological and scientific issues that interest the artist. It may also be unfamiliar with the artistic issues, and will be unprepared to appreciate the cultural timeliness of artists' esoteric explorations and the power of the artistic resolution.

Literacy is a moving target. It is possible that art can help to increase literacy—piquing interest and providing engaging entries into complex ideas. Some believe that at the beginning of the twentieth century abstract art helped make ideas of relativity and alternative geometries accessible. The artist acts as a kind of pioneer or homesteader—assimilating new concepts and areas of inquiry and reflecting on them in the art.

Timing can be crucial to literacy. As scientific ideas and technologies diffuse into a society, more of the population acquires the background to interpret the art. Determining the potential spread of ideas is often difficult when they are new. For example, developers of the first computers believed that programming could be mastered only by advanced mathematicians and that the United States would only need thirty to seventy programmers. Similarly, the esoterica of topics such as image processing, 3-D modeling, fractal geometry, Internet communications, and encryption concerned only advanced academic specialists a few decades ago. Now they are embedded in computer programs that sell millions of copies and are the casual topics of preteen hobbyists.

Literacy confronts many of the artists also described in other chapters. They need to ask themselves certain questions: What does an audience need to know to understand and make sense of an artwork? What can the artist do to help an audience acquire some of the relevant background? Should the artist even worry about this?

Understanding Systems

This artistic gesture of working with algorithms could be seen as related to other 1960s and 1970s art world trends in the same era, for example, serialists, minimalists, and conceptualists. In the larger culture, interest was growing in "systems thinking"—trying to unravel the interrelationships of what previously were considered as separate functions. Buckminster Fuller suggested that traditional concepts of design were much too limited in their worldview and needed to expand their range of concerns beyond the 2-D or 3-D object. Understanding underlying structures, processes, and interrelationships were an important part of this approach.

These ideas captured the imagination of many artists, and some artists established educational programs related to these ideas. In 1970, Sonia Sheridan created the Generative Systems Program at the School of the Art Institute of Chicago, which influenced many artists. Famous for her pioneering explorations of copier technology and her transformations of images by machine manipulation, many think of her as a copier artist. Her real interest in technology, however, was much deeper than a dalliance with the technology. She became interested in the systems artists used to generate art including technological, intellectual, sociological, spiritual, and heuristic. Analogously to the algorithmic artists, she believed that artists could become stronger by explicitly attending to the underlying systems. She integrated scientists and technologists into the program, and was one of the first artists to work as an artist in residence in a research setting when she was invited to help 3M explore its new color-in-color copier technology.

I personally experienced this program as an MFA student. I fondly remember Sonia Sheridan and her two-foot screwdriver. Sonia would delight in taking apart new gizmos that students would bring to class. She urged us to study the gizmo's workings, to understand its principles, and to jump off from there in whatever artistic directions were stimulated by those understandings. She was not interested in technologies as just more mediums.

Working from related ideas based on Fuller and systems analysis, Bryan Rogers and Jim Storey established the Conceptual Design Program at San Francisco State University. Rogers encouraged artists to think broadly about what could be the contexts, tools, and focuses of art. He urged students to carefully analyze the systems they intended to work

---

Sonia Sheridan: ⟨http://www.swiftsite.com/sonart⟩

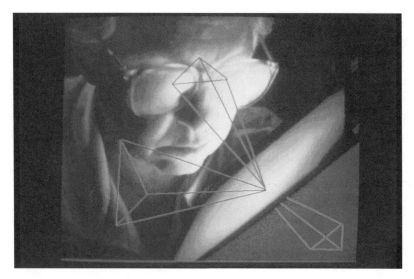

Fig. 4.2.10. Sonia Sheridan, *Looking through the Time Plane* (1984). A visualization of a time-space continuum.

on, drawing from artistic, scientific, or any other sources that could increase understanding. For the last decade, he headed the art department at Carnegie Mellon and was influential in conceptualizing the Studio for Creative Inquiry there.

Artists who focus on underlying algorithms or systems are in some ways working with methods more common in the sciences and engineering than art. They are attempting to understand underlying principles and then to apply or extend them. This approach has demonstrated great power for science and related disciplines. The interest of artists in algorithms could itself be seen as a major influence of the sciences on the arts. But the artists are also significantly different in their goals and willingness to embrace whimsy and absurdity in their exploration of algorithms.

Abstraction and Cultural Theory

Artists working with art and mathematics place primacy on art's historical focus on visual representation. Many see their work as directly descendant of earlier abstract art and even more, extending that work through the tools offered by mathematics and

---

Studio for Creative Inquiry: ⟨http://www.cmu.edu/studio/⟩

computers. Much of contemporary art, however, is focused elsewhere, for example, the attempt by some critical theorists to elevate the awareness of art's social context. They see abstractionism situated firmly within the modernist tradition and challenge its assertion of the possibility of identifying of grand, unifying "truths."

They question the possibilities of such universal themes and instead assert postmodern perspectives of relativity and diversity. To some of these critics, the work of artists discussed in this section seems somewhat an anachronism—even though it uses the latest theories and tools. They would claim that the search for shapes and forms to express the underlying unity of the universe seems part of an abandoned discourse and not germane to the pressing issues confronting today's world and art. The debate promises to continue into the future as art struggles to reconcile these different approaches.

Notes

1. H. Franke, *Computer Graphics—Computer Art* (New York: Springer-Verlag, 1985), pp. 28–37.

2. R. Verostko, "Explanation of Algorithmic Art," ⟨http://www.mcad.edu/home/faculty/verostko/algorithm.html⟩.

3. R. Verostko, "Art and Algorithm," ⟨http://www.mcad.edu/home/faculty/verostko/alg-ISEA94.html⟩.

4. K. Musgrave, "Formal Logic and Self-Expression," ⟨http://www.seas.gwu.edu/faculty/musgrave/artists_stmt.html⟩.

5. G. Legrady, "Interview with Geert Lovink," ⟨http://www.mediamatic.nl/magazine/8*2/Lovink-Legrady.html⟩.

6. M. Bill, "The Mathematical Way of Thinking in the Visual Art of Our Time," in M. Emmer, ed., *The Visual Mind* (Cambridge: MIT Press, 1993), p. 5.

7. Ibid., p. 8.

8. N. Friedman, "Reunification and Hyperseeing," *YLEM Newsletter*, vol. 17, no. 12 (Nov/Dec 1997) (also on-line at ⟨http://nyjm.albany.edu:8000/am/1997/Friedman.pdf⟩).

9. S. Dickson, "Artist Statement," ⟨http://www.wri.com/~mathart/portfolio/SPD_Frac_portfolio.html⟩.

10. B. Collins, "Toroidal Closures of Scherk Towers," ⟨http://nyjm.albany.edu:8000/am/1997/Collins.pdf⟩.

11. D. Cox, "Caricature, Readymades, and Metamorphosis: Visual Mathematics in the Context of Art," in M. Emmer, *Visual Mind*, p. 105.

12. B. Evans, "Implicate Beauty," ⟨http://www.vanderbilt.edu/VUCC/Misc/Art1/Beauty.html⟩.

13. L. Henderson, "The Fourth Dimension and Non-Euclidean Geometry in Modern Art: Conclusion," in M. Emmer, ed., *The Visual Mind,* p. 230.

14. Ibid., p. 231.

15. S. Ramsden, "Fractals, Feedback, and Chaos—A Brief History," ⟨http://online.anu.edu.au/ITA/ACAT/contours/docs/fractal-history.html⟩.

16. C. Pickover, "Statement of Approach," ⟨http://sprott.physics.wisc.edu/pickover/home.htm⟩.

17. C. Pickover, "Interview with ACM," ⟨http://sprott.physics.wisc.edu/pickover/home.htm⟩.

18. Basic Research Institute in Bristol, "Introduction to Project," ⟨http://hplbwww.hpl.hp.com/brims/art/index.html⟩.

4.3

Artificial Life and Genetic Art

Developing algorithms and heuristics to enable computers to execute sophisticated analysis or undertake complex behaviors are among the greatest contemporary research challenges. Often the challenge derives not just from working out the engineering details, but more profoundly from the necessity to understand the focus phenomena at great depth and from new perspectives. Also, researchers must often confront unresolved philosophical issues. Artificial intelligence and Artificial Life are two such areas of inquiry in which algorithm developers attempt to simulate complex behaviors in the biological and human world. Can they understand the underlying "rules" sufficiently that they can enable a computer to manifest complex adaptive behaviors? These fields have generated great research and artistic interest in the last decades. (Artificial intelligence is considered in chapter 7.6.)

There is no wonder that artists have been interested. The arts were a prime locus for the creation of artificial life throughout the centuries. One could look at artificial life as a continuation of the ancient concerns of sculpture, portraiture, and landscapes. From the early days of the A-life movement, artists have been participants in the international conferences and have been drawn to the ideas. Some have focused on the simulation and modeling of complex biological behaviors; many have concentrated on genetic programming and the evolutionary processes. Some believe that these A-life techniques can enable artists to create sophisticated interactive works beyond the simpleminded menu choice interactivity that characterizes much multimedia. As with the algorithmic art described in the previous chapter, development of the algorithms is an important part of the artwork. Also, as discussed in the theoretical overview of art and artificial life in chapter 4.1, the focus on developing believable synthetic worlds and creatures sometimes makes it very difficult to distinguish between those who define themselves as researchers and those who call themselves artists.

A-Life Sculpture and Autonomous Agents

Ken Rinaldo

Building on an interest in biological behaviors, Ken Rinaldo developed a series of sculptures with complex learning and interactive capabilities. With Mark Grossman, he created *The Flock,* in which an assemblage of robotic arms hanging from the ceiling interact with each other and with viewers via touch tones, and manifest "flocking" behav-

---

Ken Rinaldo: ⟨http://www.ylem.org/artists/krinaldo/emergent1.html⟩

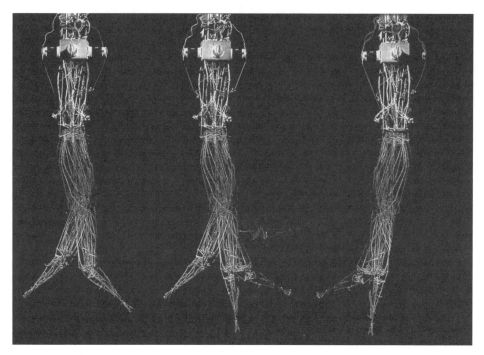

Fig. 4.3.1. Ken Rinaldo, *The Flock*. Three computer-controlled arms integrate emergent behaviors, such as flocking, with viewer interactions. Photo credit: Liz Zivic.

ior, which develops from "awareness of each other and the environment." In *Autopoiesis* a network of sound sculptures generate emergent behavior through sound communication with viewers and each other. He explains the interest in A-Life that inspired this work:

The Flock . . . is a group of cybernetic sound sculptures that exhibit behaviors analogous to the flocking found in natural groups such as birds, schooling fish, or flying bats. Flocking behaviors demonstrate characteristics of supra-organization, of a series of animals or artificial life-forms that act as one. They are complex, interdependent interactions which require individual members to be aware of their position in relation to others. . . .

The key concept of the series is emergence, the coming together of systems with no central controller guiding their behavior. The global behavior is allowed to evolve naturally out of the local interactions among the systems. When the lower levels of the systems like self-preservation are satisfied, the higher functions like flocking are allowed to arise. The results are complex, chaotic, nonlinear, and often lifelike.[1]

Fig. 4.3.2. Joel Slayton, *Telepresent Surveillance*. Three robots manifest emergent behavior as they search out viewers and each other.

Joel Slayton

Similarly exploring flocking behavior and complex interactions with viewers, Joel Slayton created a robotic exhibit called *Telepresent Surveillance*. Three robots interacted with each other and viewers and broadcast video of their point of view. Slayton describes the installation:

Telepresent Surveillance is an evolving artwork/research project incorporating autonomous robot surveillance probes and the Internet. The intent of this project is to characterize a form of media experience derived from the activities of intelligent machine agents designed to enable telepresent viewing.[2]

Interactivity and Artificial Life

With the advent of microcomputers in the late 1970s, some artists became very interested in interactivity. Artists were able to create dynamic works in which the audience functioned as cocreators; the behavior of the work depended on choices made by the

---

Joel Slayton: ⟨http://cadre.sjsu.edu/area210/Joel/⟩

viewer. I was one of the early practitioners of this movement and often spoke about the "revolutionary" potential of interactive art for mobilizing audiences. Now that interactive computers and an interactive Internet have become commonplace, it has become clearer that interactivity is much more complex. My paper, "The Aesthetics and Practice of Designing Interactive Computer Events," analyzes the shallowness of much interactivity:

[T]he inclusion of choice structures does not automatically indicate a new respect for the user's autonomy, intelligence, or call out significant psychic participation. In fact, some analysts suggest that much interactive media is really a cynical manipulation of the user, who is seduced by a semblance of choice. The choices offered, however, are not significant choices, for example, the ability to choose one of three products available in an interactive shopping experience or the ability to decide when and how to kill the simulated enemy in a game. The missing choices might be more important than the "choices" offered.[3]

The nature of the interactive structure is critical, and the simple menu choice can be a sham with no real choice offered. Artists are now searching for more complex forms of interactivity that call on deeper involvement by viewers. (For more discussion of the issue see chapter 7.1, which reviews theories of digital culture and art.)

Some artists believe that artificial life and artificial intelligence offer an approach to explore a more interesting kind of interactivity. Artificial life is seen as promising because it allows the artwork itself to have a larger repertoire of behaviors that can generate novel behaviors through evolution, learning, and interactions with viewers. Ken Rinaldo expresses some of these perspectives in his paper "Technology Recapitulates Phylogeny":

Even more exciting, artificial-life techniques present opportunities for both artists and viewer/participants to develop true relationships with the computer that go beyond the hackneyed replicable paths of "interactivity" which have thus far been presented by the arts community.

With artificial-life programming techniques, for the first time interactivity may indeed come into its full splendor, as the computer and its attendant machine will be able to evolve relationships with each viewer individually and the (inter) part of interactivity will really acknowledge the viewer/participant. This may finally be a cybernetic ballet of experience, with the computer/machine and viewer/participant involved in a grand dance of one sensing and responding to the other.[4]

Fig. 4.3.3. Simon Penny, *Petit Mal*, an autonomous robotic artwork. *Petit Mal* engages visitors in kinesthetic, dancelike interactions.

Simon Penny

Simon Penny (see also chapter 5.3) also creates art derived from artificial-life ideas. His *Petit Mal* robotic artwork attempts to explore autonomous behavior as a probe of interactivity and the research field of A-life. Penny describes the goal of *Petit Mal*:

Simon Penny: ⟨http://www-art.cfa.cmu.edu/www-penny⟩

A-Life Sculpture and Autonomous Agents

The goal of *Petit Mal* is to produce a robotic artwork which is truly autonomous; which is nimble and has "charm"; that senses and explores architectural space and that pursues and reacts to people; that gives the impression of intelligence and has behavior which is neither anthropomorphic nor zoomorphic, but which is unique to its physical and electronic nature. . . .

The formulation "an autonomous robotic artwork" marks out a territory quite novel with respect to traditional artistic endeavors, as we have no canon of autonomous interactive esthetics. . . . More generally, *Petit Mal* seeks to raise as issues the social and cultural implications of "Artificial Life". The reflexive nature of interactivity is a focal issue: interactive behavior is defined by the cultural experience of the human visitor.[5]

Another work called *Traces* senses body motion and transmits abstract body-motion representations to be projected in networked CAVE VR environments (see chapter 7.3 for a description of CAVE). Gradually the traces take on artificial lives of their own.

Louis Bec

French artist and theoretician Louis Bec is known for his creation of artificial "hypozoological" systems and his role as a "zoosystemicist." Combining a background in zoology, art, and philosophy, Bec believes that art has a responsibility to imagine life as it might be. Art is a free zone to build on concepts of life to conjure up alternative systems. Usually, he creates artistic-scientific documents systematically presenting his life-forms. In Ars Electronica 93, his presentation "Prologema" listed part of his agenda as concerning the connections between the sciences of life, those of artificial life, and the performing arts, including such topics as helping zoo systems to "claim their place in the 'spiritual world'" and "the aesthetics of autonomy."[6]

The Artec96 description summarizes his research agenda:

It is worth noting that Professor Louis Bec is the only known qualified zoosystemicist and that his work has been developed around a Fabulatory Epistemology based on the scientific, artistic, and technological foundations of Artificial Life. The zoosystemicist works out arbitrary and imaginary zoological systems in which singular zoomorphisms, curious biologies, and aberrant zoosemiotics are developed. Through the founding of the Institut Scientifique de Recherche Paranaturaliste in 1972, Professor Bec has created an effective instrument for questioning on every level the incapacity of the living to "capture" the living by conventional methods of biology

---

Louis Bec: ⟨http://www.babel.fr/arles-festival/us/hypozoologie.htm⟩

Fig. 4.3.4. Louis Bec, *Hypozoologie. Upokrinomeme.* An exploration of artificial animals.

or objectified zoology. He proposes a "hypocrisic" strategy through the construction of "heuristic" decoys (upokrinomena), the development of a biased photographic and digitalised hypozoology (upokrinomenology). Professor Bec has developed dynamic morphogenesis, digital bio-modeling, zooholograms, videos, devices, and interactive consoles.[7]

Bec sees Artificial Life research, which emerged after he started his research, as creating a tension that allows new ways to understand life and the potential generativity of imagination. He sees it as a form of expression that defies conventional classification as science or art. In "Artificial Life under Tension—A Lesson in Epistemological Fabulation," Bec characterizes artificial-life research:

[A-life] is the result of marked tension between the living and the technologically created near-living. . . . This tension describes a distinctive trajectory in the overall relationship between the arts and the sciences. Thus it opens up entirely new fields of exploration and plays a part in the current reconfiguration of knowledge and forms of expression. This trajectory traverses the scientific, artistic, and technological domains in all their diversity, evolutions, and mutations. . . .

The artistic and scientific convergences and divergences artificial life testifies to are based on a primeval tremor. An imperceptible tremor of the living, a vibration going back to time immemorial. By giving rise to a logical proliferation wave, it compels recognition of the "pro-creation" of techno-biodiversity as a fundamental mode of human expression.[8]

A-Life Sculpture and Autonomous Agents

Bec sees modeling, the creation of artificial creatures, as a critical operation:

As the bearer of variable, manipulable parameters, this median object allows for the processing of emotional, imaginative information in an artistic setting in the same way as for logical, rational information in a scientific one.[9]

He laments that many researchers are caught in the grips of positivist epistemology, leaning toward bio-mimesis rather than freer imagination. He is hopeful that the strategy of creating evolutionary systems and setting them in open expanding networked environments in which they can evolve will give room for new life-forms. He sees this research as a new kind of "technozoosemiotics," which can create new understandings of life as it is and as it might be.

Yves Klein

Yves Klein has been working for several years on what he calls "living sculpture." These sculptures integrate organic forms with complex interactions guided by neural network programming. Klein feels that the behavior of sculpture is a critical element:

The process of creating a "living sculpture" is challenging due to the complexity involved in having that sculpture integrate multiple changing inputs and react in real time. From finding appropriate materials, to developing technologies for gesture, locomotion, sensory input, and behavior, countless technological and physical obstacles have to be overcome to achieve a unified sculpture. I consider my work to be an attempt to find a symbiotic balance between classical artistic expression and contemporary technologies. . . .

As a design progresses from the inner basic systems to the complex outer systems, surprising and elaborate aesthetic and behavioral effects may emerge that exceed our expectations. A living sculpture should be interesting and exciting from any perspective.[10]

Klein has created a series of organic form-inspired works, including the *Scorpiobot, Lady Bug,* and *Octofungi. Octofungi* is an eight-sided polyurethane sculpture that uses a neural network to integrate current events sensed via multiple sensors and shape-metal alloy for silent, nonlinear motion:

Yves Klein: ⟨http://www.netzone.com/~yklein/index.html⟩

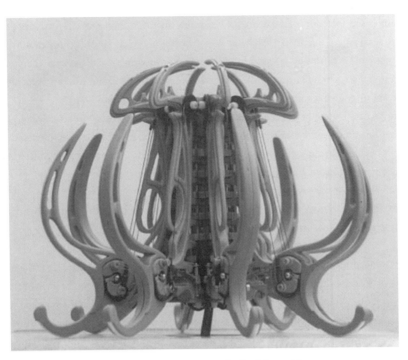

Fig. 4.3.5. Yves Amu Klein, *Octofungi,* a living sculpture.

Octofungi learns its environment by adjusting the strengths of its neural connections towards the long-term average of what it sees around it. Usually, thirty seconds or so of a nonchanging environment is enough to relax *Octofungi*. Then, if something changes in the environment, *Octofungi* will recognize the change, and will evaluate the magnitude and impetuousness of the change. The change will trigger either fear or curiosity, and the brain will instruct the legs to move accordingly.[11]

The future development of *Octofungi*'s software aims toward the creation of "brain breeding" software that will simulate the interplay of genetics, harmonal flow, and fitness testing by the environment:

The neural structure of the brains is genetically determined, as are the brains' hormonal flow patterns. In addition, the software emulates the electrical "noise" associated with biological neural structures. The combination of neural networks and time-dependent hormonal flow can potentially create brains that are much more flexible and "fuzzy" than conventional neural networks.

For example, the injection of a particular hormone into the brain can simulate an emotional change, thus changing the brain's behavior. As the hormone slowly spreads throughout the brain, it may trigger other hormone releases, modifying or exacerbating the behavioral change. As this is happening, electrical noise may allow unrelated sections of the brain to communicate weakly with one another, giving the partitions a sort of gestalt of the brain's state.

As the program runs, brains are built on the fly according to their genetic model. They are then tested for interesting behavior, and are allowed to breed with one another depending upon their performance. These processes mimic the biological notion of survival of the fittest and tend to produce populations of brains with a desired type of behavior.

Ultimately, Klein hopes to create colonies of these artificial entities capable of exploring their environment, learning, evolution, and self replication:

My ultimate goal is to create sculptures that can replicate, and consequently, have the ability to change their form and behavior as the generations pass. One idea I have been investigating is creating a cellular sculpture. The basic concept involves a multitude of self-contained cells that can interconnect themselves much like a self-propelled Erector Set. The parent sculptures would need to be sophisticated enough to assemble a copy of themselves and imbue the child with a new genetic code.

Like many of the artists described in this book, Klein recognizes that he is working in an area that is both art and science. He feels that the conventional categories will need to change to accommodate what artists-scientists are doing: "As art continues to evolve, classifications between art and science will continue to become more fuzzy and the world will continue to become more interesting."

Other Artists and Projects

In *Scavengers*, **Louis-Philippe Demers and Bill Vorn** author an artificial ecology made of shadowy robots, building on biological concepts such "chain reactions, propagation and aggregation behavior, herds and swarms, etc." to create a "hybrid world between nature and the artificial" (see chapter 5.4). **Survival Research Labs** is also seen as unleashing artificial-life forms (see chapter 5.4). **Gerard Boyer** created *Machine Palmipède,* an abstract robot composed mostly of gears flailing with its own evolutionary behaviors:

Gerard Boyer: ⟨http://www.telefonica.es/fat/alife/aboyer.html⟩

"Violent mechanical stubbornness, left to their own devices . . . blocked, camisoled in their primitive energetic needs." **Simon Penny and Jamieson Schulte** built a sound installation called *Sympathetic Sentience Three,* in which a multitude of simple sound generators create sound through "emergent complexity" responses to each other's infrared communications and visitors' movement through the darkened space. **Erwin Driessens and Maria Verstappen** constructed the "Tickle" robot to navigate with tickling motions over viewer bodies, using artificial-life algorithms to make decisions about its movement. **Nell Tenhaaf**'s *You Could Be Me* installation presents visitors with a position-sensitive interactive artificial-life form that extracts information from them to evolve artificial empathy and integrates video and artificial-life-influenced graphics and LED (light emitting diodes) displays that expose its processes.

Genetic, Evolutionary, and Organic Art

Evolution is the major focus of many A-Life researchers. They concentrate on creating digital evolutionary systems in which "offspring" are created and some kind of selection processes pick the most "fit," which then go on to create the next generation. The computer graphic "Game of Life," described earlier, illustrates a simple approach to this research. In what is called "weak" A-Life, researchers claim that they are simulating life and evolutionary processes. In "strong" A-Life, researchers claim that they are actually creating silicon life.

Several artists and researchers have become involved in creating artworks that probe these ideas, for example, by creating computer graphic environments in which synthetic creatures can interact and evolve. Often viewers are invited to interact with the environment and thus become part of the evolutionary pressures. As with the other art discussed in this chapter, the creation of the behavioral and evolutionary rules are a prime part of the artwork.

In addition to the usual "what should the creatures look like?" artists must ask a series of other questions: What variety of forms should there be? What should be the behavior repertoire of each form? How do the creatures address the usual biological issues of locomotion, exploration, eating, reproduction, biological rhythms of sleep and wakefulness, and so forth? How should they interact with each other and viewers? What kind

---

Simon Penny and Jamieson Schulte: ⟨http://www.telefonica.es/fat/alife/apenny.html⟩
Erwin Driessens and Maria Verstappen: ⟨http://www.xs4all.nl/~notnot/⟩

of ecological niches should exist? Should the creatures compete for resources? Should they cooperate? What evolutionary pressures should select for continued life and reproduction? What range of change should be designed into the system? How should any implicit or explicit cultural commentary be manifested in these decisions? What position should the work manifest in regard to philosophical issues implicit in the concepts?

It is not much of a stretch to think about this work as a continuation of the historic artistic tradition of landscape. For hundreds of years, artists have been able to play mini-gods and create artificial places populated with whatever images of organic life and inorganic forms they wished. But they only had to freeze one moment in time. Evolutionary art requires that the artist create the rules and underlying tendencies of the landscape as well as the visual and performative aspects.

How is an artist working on the creation of a digital organic environment any different than a computer scientist or other researcher creating a similar environment? As with so many of the fields in this book, that question is not easily answered. In fact, many of the most renowned creators of A-Life environments have professional identities outside of art. Researchers and artists have collaborated many times. For example, Thomas Ray, a major researcher in the field, has collaborated extensively with Laurent Mignonneau and Christa Sommerer in some of their award-winning A-Life-based artworks.

Thomas Ray

Thomas Ray, a well-respected researcher in A-Life, has published numerous academic papers and been a major participant in A-Life conferences. He has also collaborated with the art group Knowbotics Research. He is famous for creating the *Tierra* environment, in which computer programs self-modify their own code in order to create new capabilities. Evolutionary pressures select which new code entities will prosper. He is a proponent of strong A-Life, believing it makes sense to think of this stuff as silicon life. He describes his approach in *Tierra:*

Life on Earth is the product of evolution by natural selection operating in the medium of carbon chemistry. However, in theory, the process of evolution is neither limited to occurring on the Earth, nor in carbon chemistry. Just as it may occur on other planets, it may also operate in other media, such as the medium of digital computation. And just as evolution on other planets is not a model of life on Earth, nor is natural evolution in the digital medium.[12]

Thomas Ray: ⟨http://www.hip.atr.co.jp/~ray/⟩

Fig. 4.3.6. Thomas Ray, *Skull3*. A representation of a genetic program with geometric objects symbolizing computer programs and the skull representing death, which eliminates old or defective programs. Photo credit: Anti-Gravity Workshop.

Ray offers an eloquent explication of his ideas about evolution as art in his paper "Evolution as Art." He notes that evolution is an amazing force that has organized biological matter on the earth into remarkably beautiful and diverse forms. He believes that in the future the artist's role may be to launch systems that allow for evolution to work its magic:

Evolution is a creative process, which acting independently, has produced living forms of great beauty and complexity. Today, artists and engineers are beginning to work together with evolution. In the future, it may be possible for artists to work in collaboration with evolution to produce works of art whose beauty and complexity approach that of organic life. [13]

Ray asserts that some of the beauty is that which can be seen. Other parts, however, seem to be more about the elegance of the systems that have evolved:

As we move outside of the range of what we can normally visualize, we encounter forms in living systems with a similar quality of richness, subtlety, and complexity, but which require an unconventional aesthetic to appreciate. For example, the forms of ecosystems and metabolic pathways are based on the flows of matter and energy through these systems.

If evolution is a force with its own ways, how can an artist be involved? Ray notes that there is a continuum of control stretching from just establishing the system to an ongoing active intervention to exert selective interventions. For example, an artist could create a synthetic environment and endow it with rules of selection and genetic algorithms and let it run. Or an artist (or audience) could look at each generation's offspring and decide which ones are allowed to proceed—applying a kind of "artificial selection." Having seen what evolution has accomplished in the biosphere, Ray leans in the direction of minimal control:

In order to maximally exploit the creative potential of evolution, it is necessary for the human collaborator to give up most of their control over the process. The human only sets up the environment for evolution to operate in, provides it with raw materials, and then watches as evolution expresses its creativity. This means that the human does not provide any guidance to evolution, and thus can not necessarily expect evolution to produce a useful product. But it is under these conditions that evolution has the maximal freedom to express its own creativity.

We do not know yet, if we can ever expect evolution in the digital medium to express a level of creativity comparable to what we have seen in the organic medium.

What a radical shift in aesthetics and artistic practice! If one accepts the A-life approach to creating art, then the artist's role is that of systems creator. The art would be evaluated on criteria such as the openness of the system to evolution, the richness of the repertoire of forms and behaviors, the cleverness of the genetic algorithms, and the linkage of underlying forces to externally observed phenomena such as the appearance, motion, and interactions of digital entities.

Biota.org

Ray noted that evolution has accomplished much more than breeders within species—creating entirely new species and complexity. In part, this derives from the diversity of populations that interact and the freedom for mutation and selection to act. He urged the creation of a "Digital Biodiversity Preserve." Biota.org is attempting to accomplish just that by creating a worldwide networked space for artificial-life entities to interact and evolve. Its tasks include: (1) to define an open-standard grammar for design, and

---

Biota.org: ⟨http://www.biota.org⟩

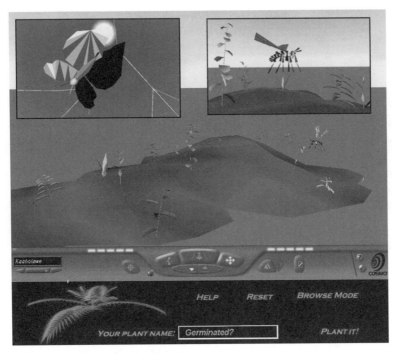

Fig. 4.3.7. Biota.org (Bruce Damer, Karen Marcelo, Frank Revi), *Nerve Garden*. A VRML-networked environment in which artificial life forms can develop and interact ⟨http://www.biota.org⟩.

(2) to establish a public networked repository for these creations that supports their evolution.

The organizational statements include strong A-Life perspectives on the possibilities of silicon life and the power of networks to allow these life-forms to evolve and prosper:

Freed from the constraints of chemistry and able to travel literally at the speed of light, bitomic life could transit the solar system in hours, but would only take hold if some storage and expressor mechanism was out there to receive it. The digital realm may itself be a kind of surrogate, for if molecular nano-fabricators one day permit the recrossing of the bitomic-atomic barrier, like the ancestors of whales returning to the ocean, digital biota will evolve and emerge well adapted to the vast ocean of space. Thus, life finds a way through the keyhole of human technology escaping both the bounds of Earth and the mortal coils of the double helix.[14]

The organization created two *Nerve Gardens* into which researchers and artists could "plant" A-life forms created in the VRML (virtual reality modeling language) format.

Fig. 4.3.8. Christa Sommerer and Laurent Mignonneau, *A-Volve* (copyright 1994). Fishlike artificial life forms evolve through interactions with viewers. (Supported by ICC-NTT Japan and NCSA, Urbana, Illinois.)

The ultimate goal is to create a distributed Internet environment in which the A-Life forms could live and be watched by observers anywhere.

Christa Sommerer and Laurent Mignonneau

Christa Sommerer and Laurent Mignonneau have created a well-known series of artworks that explore A-Life concepts. The installations typically also investigate other areas of emerging technologies. In *Interactive Plant Growing* and *Transplant* (see chapter 7.4) visitor body movements impact on the evolving A-Life forms. In *Lifespacies* (see chapter 6.3), remote participants can jointly influence A-Life entities via telecommunications and Web links. *Interactive Plant Growing* (see chapter 2.5) allowed visitors to shape artificial plant forms by touching real plants growing in the exhibition space. *A-volve, Phototropy,* and *GENMA* are discussed here.

A-Volve was created as part of an artist in residency at the ICC in Tokyo. Thomas Ray collaborated in its development. In this work, computer-synthesized 3-D projected creatures appear to be living in a pool of real water. Audience actions are crucial in

Christa Sommerer and Laurent Mignonneau: ⟨http://ms84.mic.atr.co.jp/~christa/WORKS⟩

creating creatures and shaping forces that affect the artificial ecology. For example, the fish will alternate between avoiding and seeking contact with fingers placed in the water. Here is the artists' description:

A special editor provides that the visitors can create any kind of form with their finger and change and modify them in real time. The three-dimensional forms will be immediately "alive" and they will move and swim in the water of the pool. These virtual creatures are products of the rules of evolution; they are influenced by human creation and decision.

The movements and behaviour of the organisms will depend mainly on their forms, on how the viewer was designing them on the drawing screen. Each creature moves, reacts, and evolves according to its form, creating unpredictable and always new lifelike behaviour. Since the organisms will capture all slightest movements of the viewer's hand in the water, the form and behaviour of these organisms will change constantly.[15]

In *Phototropy,* the artists explore the boundary between physical reality and A-Life. *Phototropy* is a biological expression describing the tendency that keeps organisms or organs, for example bacteria or plants, following the light in order to get nutrition, and hence to survive. The installation has several features:

1. Computer-generated insects follow and fight for light "nutrition" provided by visitors moving flashlights.
2. When they have sufficient light intensity, the creatures reproduce and create offspring carrying genetic code from their parents.
3. Insects that stay in the hot part of the beam too long die and float to the floor. They describe the visitor's impact on the A-Life forms:

The visitor has to be careful with his lamp. Though it is very easy to use (a normal torch lamp functions as interface), it requires the viewer's responsibility and care for the creatures. If he moves too fast, the insects will hardly follow, and thus will have no time and occasion for reproduction. If he moves the lamp too slowly, the insects will reproduce rapidly, but reach the center of the beam too quickly: hence they will burn and die as fast as they were born. . . . To really appreciate the creation and the development of new populations and individuals, the viewer becomes responsible for their creation, their evolution, and their survival.[16]

GENMA: Genetic Manipulator invites visitors to shape virtual microscopic creatures by using a touch screen to select genetic characteristics. Sommerer and Mignonneau see

GENMA as a useful tool for provoking visitors to consider the new powers to manipulate organisms that are being developed in research labs:

GENMA is a machine that enables us to manipulate "Nature." Nature exemplary is represented as artificial nature of a micro scale: abstract amoeboid artificial three-dimensional forms and shapes. Principles of artificial life and genetic programming are implemented in those forms or "creatures," allowing the visitor to manipulate their virtual genes in real time. . . .

Selecting and merging different parts of the genetic string, recombining them and modifying their genetics, he can engaging [sic] in more intense experiments and learn how to create complex forms out of seemingly simple structures at the very beginning. . . .

GENMA is a kind of dream machine that allows us to "play" scientist and as we watch ourselves in doing so; it also mirrors the absurdity of this action and interaction. By using science, namely principles of artificial life, as source for creation, *GENMA* also wants to address the question of what it means to manipulate and what impact it will have on us in the future. Nevertheless, *GENMA* doesn't take position in the common "good-bad" classification or in the sense of "political correctness," but wants to reflect our fascination for the unknown and unexplored.[17]

Karl Sims

Karl Sims, who had been a computer animator with Whitney/Demos Production in California and a researcher with Thinking Machines, created a work called *Panspermia* based on genetic art ideas. His environment generated a rich array of computer graphics and then allowed the viewers to select which images they liked the best, subsequently using techniques of artificial evolution to promote features of those images in the next generation. He hopes his installation will instill an appreciation of biological life:

In this work, I attempt to bring together several concepts: chaos, complexity, evolution, self-propagating entities, and the nature of life itself. This unusual botanical form of life, reproducing itself from planet to planet throughout space, is in many ways analogous to other self-replicating systems, including humans, entire species, and even ideas. A window into this panspermic system will hopefully expand awareness of self-propagating systems in general, as well as inspire thoughts about our entire planet of life as a single entity.[18]

Karl Sims: ⟨http://www.ntticc.or.jp/permanent/karl/⟩

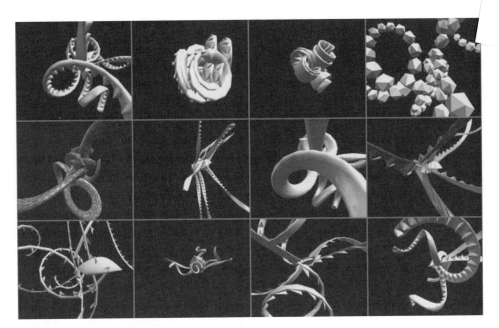

Fig. 4.3.9. Karl Sims, *Galapagos*. Users apply selection pressure to evolving artificial life forms by selecting the best of a generation from a series of monitors.

Sims used fractal techniques to generate organic-looking 3-D forms, including a "set of twenty 'genetic' parameters describing fractal limits, branching factors, scaling, stochastic contributions, phototropism, etc." The audience chooses which entities are most interesting and then the next generation begins. He explains the audience's impact:

"Artificial evolution" techniques were used to find sets of values for these genetic parameters that specified plants of a wide variety of shapes and sizes. The computer generates random mutations of sets of parameters, and the user interactively selects those resulting tree structures that survive and reproduce to create each new generation. This process of "survival of the prettiest" repeats until interesting shapes emerge. Plants can also be mated with one another to mix traits from separate individuals.

Sims created another version of this work, called *Galapagos,* for the ICC in Tokyo. Again, viewers had an important role in guiding the evolution through their "artistic selection" by stepping on foot pads in front of monitors displaying the graphic forms they wanted to promote. Sims notes, "Perhaps someday the value of simulated examples of evolution such as the one presented in this exhibit will be comparable to the value

Genetic, Evolutionary, and Organic Art

that Darwin found in the mystical creatures of the Galapagos Islands." He describes to audiences their role as breeders:

The process in this exhibit is a collaboration between human and machine. The visitors provide the aesthetic information by selecting which animated forms are most interesting, and the computers provide the ability to simulate the genetics, growth, and behavior of the virtual organisms. But the results can potentially surpass what either human or machine could produce alone. Although the aesthetics of the participants determine the results, the participants do not design in the traditional sense. They are rather using selective breeding to explore the hyperspace of possible organisms in this simulated genetic system. Since the genetic codes and complexity of the results are managed by the computer, the results are not constrained by the limits of human design ability or understanding.[19]

Life 2.0 and 3.0 Competitions and Other Shows

Several artists and theorists organized the Life 2.0 and 3.0 competitions to showcase work by artists and researchers to create interesting forms of artificial life. Organizers included Nell Tenhaaf, Rafael Lozano-Hemmer, and Susie Ramsay, Several artists described in this and other chapters won awards. The Life 2.0 Web site describes the rationale:

"Life 2.0" is the first international competition seeking to reward excellence in artistic creation that has embedded in it the practices of Artificial Life (A-Life). We are looking for artworks that are premised on the strategies of A-Life research, its conceptual approaches as well as its methods of digital synthesis.

At the end of this century we are facing redefined boundaries between humans, animals, and inorganic life. Some of the markers which already signal our "posthuman condition" are genetic interventions, simulations of evolutionary systems and emergent behaviours, nano-technologies, surgical implants of machinic parts, the implementation of automated systems of data capture and control. We're interested in art that reflects upon the panorama of potential interaction between synthetic "life" and organic life, for example:

- Autonomous agents that shape and perhaps interpret the data-saturated environment we have in common

---

Life 2.0: ⟨http://www.telefonica.es/fat/life2_0.html⟩
Life 3.0: ⟨http://www.telefonica.es/fat/vida3/eprensa.html⟩

- Portraits of intersubjectivity or empathy, shared between artificial entities and us
- Intelligent anthropomorphising of the data sphere and its inhabitants
- User-defined exploration and interaction that is designed to mitigate fear and enhance curiosity in the face of emergent phenomena, which are by definition beyond our control.

The international jury will grant awards to the most outstanding electronic art projects employing techniques such as digital genetics, autonomous robotics, recursive chaotic algorithms, knowbots, computer viruses, avatars and virtual ecosystems.

Several previous art exhibitions have featured A-Life and genetic art, including "Like Life" in England, and one at UCLA in Los Angeles.

Other Artists and Projects

Influenced by Participants   *Technosphere* offers another example of artificial ecologies. **Jane Prophet** created a Web-based environment in which digital creatures could grow and prosper. Web visitors were invited to assemble creatures by selecting body parts and behavior characteristics. The person's creature would then be put in the ecosystem to thrive best as it could. The creator was automatically notified by e-mail as their creature evolved. **Doris Vila**'s *Flock of Words* uses an A-Life algorithm to project words moving in and out of legibility in an installation space influenced by visitor motion. **Tsuneya Kurihara** presented viewers with an interactive VR world populated with artificial-life dolphins. **Scott S. Fisher, Michael Girard, and Susan Amkraut** created *Menagerie* (see chapter 7.3), which offers a VR world populated by A-Life creatures that can interact with visitors. **Rebecca Allen**'s *Bush Soul* and *Emergence* projects let visitors enter multi-participant networked worlds populated with artificial life. Each visitor has an avatar that integrates its own A-life tendencies with user actions. Visitors can use voice and gestures, and the artificial life forms can respond with tactile feedback. **Richard Brown and Igor Aleksander**'s *Biotica* lets viewers influence 3-D animated cellular A-Life forms.

"Like Life": ⟨http://www.cogs.susx.ac.uk/ecal97/like_life.html⟩
UCLA A-Life art show ⟨http://www.sscnet.ucla.edu/anthro/gessler/artshow/⟩
Jane Prophet: ⟨http://www.technosphere.org.uk⟩
Doris Vila: ⟨http://www.telebonn.gmd.de/cai/bec/th-masch.htm⟩
Tsuneya Kurihara: ⟨http://www.siggraph.org/conferences/siggraph96/core/conference/bayou/1.html⟩
Fisher, Girard, and Amkraut: ⟨http://www.cda.ucla.edu/Pages/fisher.html⟩
Rebecca Allen: ⟨http://emergence.design.ucla.edu/home.htm⟩
Richard Brown and Igor Aleksander: ⟨http://webserver1.wellcome.ac.uk/en/old/sciart98/08Biotica.html⟩

In *Las Meninas,* **Michael Tolson** allowed audiences to "feed" the organisms by manipulating a sensor. This installation also incorporated viewing arrangements that forced the audience to explore concepts of voyeurism. Some artists have also created software that allows viewers to "breed" images on their own systems. **Troy Innocent**'s *Shaolin Wooden Men* is billed as the first simulated recording group. *Psyvision* is a fifty-minute music video built from artificial worlds. Rejecting the imaginative failure of much computer graphics, Innocent considers the rules and grammars that generate the artificial worlds as important aesthetic focuses. In *Iconica,* he created an artificial world in which visitors can interact with artificial-life forms, based on a multitude of different models, which respond to questions with iconic language. **Sharon Daniel** creates installations and networks that apply artificial life and dynamic systems concepts, procedures, and imagery to intellectual systems.

Self-Evolving   Jon McCormick's *Turbulence* presents viewers with a video-disk world populated by a "menagerie" of artificial-life forms. McCormick explains *Turbulence* and his ambivalent attitude toward the possibility that physical wild spaces may some day need to be supplemented by virtual life-forms: "I have seen many unimaginable things, just as fascinating as any natural scene or any great painting. . . . In a strange and tragic way, this work is yet another turning point that marks the expedited loss of nature and the wild to ourselves." **Ulrike Gabriel** constructed a VR world called *Perceptual Arena* in which visitors confronted evolving artificial-life forms (see chapter 7.3). **Peter Broadwell, Rob Myers, and Rebecca Fuson** developed *Plasm: Yer Mug* (see chapter 7.4), which offers visitors a strange electronic mirror which "reflects" their face transformed by the actions of A-life graphic forms living in the computer's memory. **Mogens Jacobsen**'s *Entropy Machine* presented a box surrounded by bacteria growing in petri dishes, with projected images drawn from biology that changed with A-life algorithms influenced by room temperature. **Rodney Berry** created *Feeping Creatures,* which offers the user an ecology of A-Life sound creatures that evolve by sending messages to each other. **Willy LeMaitre and Eric Rosenveig**'s *Appearance Machine* offered a self-evolving

---

Michael Tolson: ⟨http://193.170.192.5/prix/1995/E95az3I-las.html⟩
Troy Innocent: ⟨http://www.iconica.org/⟩
Sharon Daniel: ⟨http://arts.ucsc.edu/sdaniel⟩
Jon McCormick: ⟨http://www.cs.monash.edu.au/~jonmc/art.html⟩
Ulrike Gabriel: ⟨http://www.t0.or.at/arena/arena_img.htm⟩
Broadwell, Myers, and Fuson: ⟨http://www.plasm.com/peter/public_html/YerMug.html⟩
Mogens Jacobsen: ⟨http://www.artnode.dk/contri/jacobsen/em/em_index.html⟩
Rodney Berry: ⟨http://www.cofa.unsw.edu.au/research/rodney/feepvid_dl.html⟩
Willy LeMaitre and Eric Rosenveig: ⟨http://www.appearancemachine.com/⟩

Web video system that changed images, sounds, and camera positions based on analysis of wind-blown refuse. *Paul Brown*'s *Sandlines* built a self-modifying graphic based on cellular automata principles.

Michael Grey created an artificial-life environment called *Jelly Life,* in which jelly-fishlike creatures evolved. Grey is interested in the interconnectedness of the emergence of life and theories of being. *Diane Ludin, Ricardo Dominguez, and Fakeshop* developed the *Genetic Response System,* which is a "viroid" that shapes itself "via the outcome of the knowledge it gains in its seeking out specific information about genetics on-line," such as the price of biotechnology stocks or the state of the human genome project.

Genetic Art, Music, and Software

These ideas of genetic art and A-Life have captured the imagination of many artists around the world. Artists and musicians have created software based on genetic algorithms and Web events in which a genetic algorithm generates images or music, and Web visitor votes affect which lines of development are pursued in the next generation. For example, *William Latham,* a researcher at IBM U.K., created the MUTATOR program in collaboration with *Stephen Todd.* The organic art offers over "four hundred million combinations" for the viewer to explore. Latham and Todd also wrote a book explaining their ideas, called *Evolutionary Art and Computers. Scott Draves* wrote programs called "Bomb" and "Atelier Ho," which use genetic algorithms to function like a "visual parasite" evolving images and sounds, reacting to keyboard entry and other sounds, and growing versions and offshoots. Artist *Nell Tenhaaf* integrated "Bomb" into her *Neonudism* event. Also, many artists have created Web genetic art galleries and events, some of whose links are listed below. Many conferences have addressed this research—for example, a workshop titled "Genetic Algorithms in Visual Art and Music" at the 2000 Genetic and Evolutionary Computation Conference.

Paul Brown: ⟨http://www.paul-brown.com/GALLERY/TIMEBASE/SANDLINE/INDEX.HTM⟩
Michael Grey: ⟨http://www.aec.at/prix/einstieg/List_kats.html#Grafik⟩
Diane Ludin: ⟨http://www2.sva.edu/~dianel/genrep/intro.html⟩
William Latham: ⟨http://www.artworks.co.uk/⟩
Scott Draves: ⟨http://draves.org⟩
Genetic art galleries: ⟨http://gracco.irmkant.rm.cnr.it/luigi/alg_art.htm⟩
    ⟨http://www.cgrg.ohio-state.edu/~mlewis/Class/mutation.html⟩
    ⟨http://www.t0.or.at/msguide/ai/genart.htm⟩
    ⟨http://www.seas.gwu.edu/faculty/musgrave/genetic.html⟩
Genetic Algorithms Workshop: ⟨http://galileo.dc.fi.udc.es/workshop/gecco2000/⟩

Musicians have similarly explored genetic art ideas. For example, **Bruce L. Jacob** has written algorithms to generate music for acoustic instruments. Noting that not all of it is good, he has also written filters to listen to it and grade it. *Jeremy Leach* also applies genetic algorithms to music seeking to relate changes in nature to changes in sound.

Theoretical Perspectives on Genetic and A-Life Art

A-Life installations such as Sommerer and Mignonneau's *A-Volve* create quite engaging environments. The interface draws people in, and the interactivity, "evolution," and behavior of the creatures is fascinating. However, the work also highlights a critical dilemma posed by evolutionary art and indeed by much of the art discussed in this book that works the borders of art/science and technology: How can this art best be analyzed, evaluated, and discussed? Art historical and critical traditions do not provide much guidance, and alternative traditions are still in the primitive stages of development.

Focusing on more traditional aspects, one could discuss genetic art in terms of the visual power of the creatures. Borrowing from traditions of film animation, one could discuss the aesthetics of motion. Building on the still young field of interactive multimedia, one could evaluate the interactive experience of users. While useful, all of these skim the surface of critical elements of the art. At the beginning of A-Life's status as a research discipline, it may now be enough that artists identify the field as worthy of artistic attention and find some engaging way to explore some of its core ideas. As the field matures, however, the art must be evaluated with more depth. The fact that an artwork is "genetic" will not be sufficient.

How do genetic art environments differ from each other? What makes some more engaging, provocative, or perceptive? What scientific or research issues do the artists identify as interesting? How does the environment they create succeed in exploring those questions? How do the visual, aural, and interactive elements they construct highlight these questions, for example, those listed previously about the evolutionary potentials, appearance, and behavior repertoire of the synthetic creatures? What philosophical questions or issues in cultural analysis does the work address? *A-Volve* and other similar works are rarely discussed in these terms. The "Life 2.0" competition previously described begins to focus artistic attention on these conceptual features.

As stressed so much in this book, literacy is a problem. Artists, audiences, art critics, and historians will need to learn about the research areas that the art explores. A-Life (and

---

Bruce L. Jacob: ⟨http://www.ee.umd.edu/~blj/⟩
Jeremy Leach: ⟨http://www.bath.ac.uk/~mapjll/algo-comp.html⟩

other research areas) cannot be considered just a minor technical niche. Its meaning to the culture is larger than that, and the discourse must include that from outside the research field. Artists have begun that work; audiences and interpreters must continue it.

Notes

1. K. Rinaldo, "*The Flock*," ⟨http://mitpress.mit.edu/e-journals/LEA/PROFILES/FLOCK/flock.html⟩.

2. J. Slayton, "Project Description," ⟨http://surveil.sjsu.edu/⟩.

3. S. Wilson, "Aesthetics and Practice of Designing Interactive Computer Events," ⟨http://userwww.sfsu.edu/~swilson/papers/interactive2.html⟩.

4. K. Rinaldo, "Technology Recapitulates Phylogeny," ⟨http://mitpress.mit.edu/e-journals/LEA/ARTICLES/alife1.html⟩.

5. S. Penny, "*Petit Mal* Description," ⟨http://www-art.cfa.cmu.edu/www-penny⟩.

6. L. Bec, "Prologema," ⟨http://www.aec.at/fest/fest93e/bec.html⟩.

7. Artec96, "Description of Bec Research Agenda," ⟨http://www.babel.fr/arles-festival/us/hypozoologie.htm⟩.

8. L. Bec, "Artificial Life under Tension—A Lesson in Epistemological Fabulation," in C. Sommerer and L. Mignonneau, eds., *Science@Art* (New York: Springer-Verlag 1998), p. 92.

9. Ibid., p. 93.

10. Y. Klein, "Living Sculpture," ⟨http://www.netzone.com/~yklein/doc/Leonardo.html⟩.

11. Y. Klein, "*Octofungi* Description," ⟨http://www.netzone.com/~yklein/index.html⟩.

12. T. Ray, "What is Tierra," ⟨http://www.hip.atr.co.jp/~ray/tierra/whatis.htm⟩.

13. T. Ray, "Evolution as Art," ⟨http://www.hip.atr.co.jp/~ray/pubs/art/art.html⟩.

14. Biota, "Statement of Purpose," ⟨http://www.biota.org⟩.

15. C. Sommerer and L. Mignonneau, "*A-Volve* Explanation," ⟨http://www.ntticc.or.jp/preactivities/gallery/a-volve/explan_v_e.html⟩.

16. C. Sommerer and L. Mignonneau, "Phototropy Explanation," ⟨http://www.ntticc.or.jp/preactivities/gallery/⟩.

17. C. Sommerer and L. Mignonneau, "*GEMNA* Explanation," ⟨http://www.ntticc.or.jp/preactivities/gallery⟩.

18. K. Sims, "*Panspermia* Description," ⟨http://www.aec.at/prix/1991/E91gnA-panspermia.htm⟩.

19. K. Sims, "*Galapagos* Description," ⟨http://www.ntticc.or.jp/permanent/karl/karl_e.html⟩.

Kinetics, Sound Installations, and Robots

Robotics and Kinetics

Introduction: Robots—Creatures of Art and Science

Robots are quintessential creatures of our time. Intelligent machines, once a oxymoron, are now becoming commonplace. Robots are as much at home in art, cinema, and literature as they are in science and technology. They are of increasing importance in mundane everyday worlds such as manufacturing and entertainment. Even more profoundly, robots raise intriguing cultural questions that seem to engage philosophers, artists, scientists, and technologists:

What is the limit of human abilities to create autonomous machines?
Where does the desire to create robots come from?
What is the nature of embodied intelligence?
Must robots be made to look anthropomorphic or zoomorphic?
How do our views of robots reflect on our views of humanity?
What practical and moral questions are raised by robots?
What are the dangers of creating autonomous machines?

This section reviews research undertaken by scientists, technologists, and artists to explore kinetics and robotics. Contemporary developments have begun to approach the successful creation of autonomous machines. Note that other chapters include other related scientific and artistic research, namely, artificial intelligence (7.6), artificial life (4.3), bionics (2.5), and telepresence (6.3).

Brief History and Definitions

The literature and history of the pre-industrial age is full of attempts in many cultures to create or describe moving and autonomous objects and lifelike automata. It is an age-old, worldwide dream. Here is a brief list:

- Egyptian priests created talking and moving statues that amazed worshippers.
- King-shu Tse of China (c. 500 B.C.) created magpies and horses that worked by internal springs.
- Hero of Alexandria, Ctesibius, and other engineers created elaborate automata, often activated by water. Hero wrote the *Treatise on Pneumatica*.
- Al-Jazari at Amid (c. 1206), a famous Muslim scientist, created the lifelike Peacock Fountain and wrote the *Book of the Knowledge of Mechanical Contrivances*.
- Albertus Magnus (1204–1272) created a life-sized automaton servant.
- Elijah of Chelm and Rabbi Low of Prague (1550–1580) wrote Golem stories.

Fig. 5.1.1. European automata. A: Crowing cock atop Strasbourg cathedral; B: striking jacks atop Venice St. Mark's cathedral; C: Professor Faber's Euphonia talking head; D: Jaquet Droz's "Scribe," which duplicated human writing.

- Jacques de Vaucanson (1709–1782) created a duck that eats, drinks, splashes, and digests food and the Flute Player.
- Baron Wolfgang von Kempelen (1734–1804) created a talking machine and a sham chess player (with hidden midget).
- Pierre Jaquet-Droz (1721–1790) and Henri-Louis Jaquet-Droz (1752–1791) invented the Scribe, the Draughtsman, and the Lady Musician.

What is a robot? The term was originally coined by the Czech author Karel Capek in the 1917 short story "Opilec." It comes from the Czech *robota,* which means obligatory work or servitude. It also appears in Capek's play *R.U.R.* (Rossum's Universal Robots). In this play the robot is defined as an artificial humanoid machine created in great numbers for a source of cheap labor.

The term now has assumed a range of meanings. Some emphasize the humanoid (or animal) appearance, while others hold that robots do not need to resemble lifelike forms. For example, Webster's definition is: "an automatic device that performs functions normally ascribed to humans or a machine in the form of a human." Others stress the ability of the machine to do repetitive tasks or its sophistication in behavior, for example, dexterity or balance. Some hold autonomy or intelligent adaptability as critical components. For others, just the appearance, or humanlike or animallike movement is sufficient, for example, movie robots. Exploring the continua of appearance, function, and intelligence are some of the issues that intrigue scientific and artistic researchers.

The Robot Institute of America defines robots as "programmable, multifunctional manipulators designed to move material parts, tools, or specialized devices through variable programmed motions or for the performance of a variety of tasks." With this kind of definition, it becomes obvious that the distinction between the sophisticated machine and the robot is possibly fuzzy. The core elements seem to include: mechanisms that act on the physical world with something more than simple repetition. But even this core definition is eroding as researchers apply the term to software-only intelligences, for example, knowledge "robots" that act as information agents on the Internet.

Is It Art or Science?

Technological and artistic research probes both the "minds" and bodies of robots. What kind of sophisticated motion and manipulation can be created? How can robots deal "intelligently" with the physical and human world? What are the implications of the spread of robots into increasingly wide niches of culture? What can robots teach us about what it is to be human or animal? Some of the work produced by research labs could easily be considered as extensions of historical art traditions of kinetic sculpture or theater, and some of the works created by artists could function as research.

In the rush to focus on robots, it is easy to forget that our culture is dominated by many machines that are not robots. These, too, can use state-of-the-art technology and be culturally provocative. Some artists choose to explore the possibilities and implications of these nonrobot machines. Thus, this section includes chapters describing artists' work with kinetics and experimental sound.

An Overview of Scientific and Technological Research Agendas

Several trends have accelerated robot research and development in recent years. Intellectually, the fields of computer science, cognitive science, biology, and artificial life have seen the field of robotics—with its entities acting in the physical world—as a

ENTER THE WORLD OF ROBOTICS.
The Robotics Discovery Set™ enables users ages nine and up to easily enter the world of robotics.

This new set provides everything needed to bring your smart LEGO® creations to life. Using the Scout's hands–on Command Center, create over 3000 different behaviors – all at the touch of a button.

Coming in the fall of 2000 you will be able to unleash the full power of your Scout

Fig. 5.1.2. LEGO Mindstorms. Web advertisement for the Robot Discovery Set. LEGO and Mindstorms are trademarks of the LEGO Group. Copyright 1999 by the LEGO Group, used with permission, ⟨http://www.legomindstorms.com/⟩.

fertile environment to investigate concepts such as intelligence, agency, artificial evolution, communication, and the like. Technologically, the development of miniaturized, inexpensive microcomputers and electronic sensors and actuators have spurred research.

Practical applied worlds such as manufacturing, inventory management, space exploration, dangerous materials handling, medicine, disability assistance, and entertainment have sought to extend the microelectronic revolution into robot devices, seeking lower costs and extended capabilities. Applications are being developed in fields such as agriculture, building construction, domestic robotics, medical robotics, space robotics, traffic and highways and underwater robotics.

Around the world, many research institutes and labs have been formed. Researchers are actively pursuing a wide variety of inquiries. This section briefly reviews some of

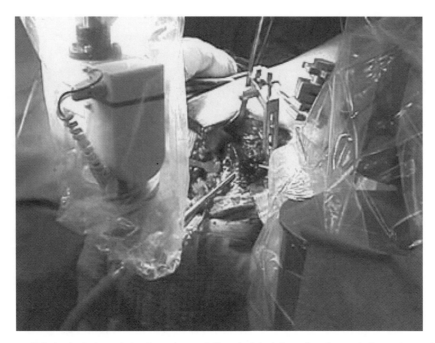

Fig. 5.1.3. *Robodoc.* A robotic surgical assistant that can drill precise holes in bones ⟨http://www.robodoc.com/surgery.htm⟩.

the research agendas, with examples of some of the research pursued. Robot institutes in Japan, the United States, and Europe are pursuing a wide variety of theoretical and practical research topics. To provide an overview of research agendas, this list presents a composite of topics compiled from several advanced research centers and some samples of specific work:

Theoretical topics: adaptation and learning in biological and artificial systems, artificial life, artificial muscles, autonomous systems, biology, biorobotics, cognition, control, cooperation, evolution, graphic interfaces, humanoids, hybrid systems, intelligent decision systems, man-machine interfaces, manipulation, mechatronics, micro-robotics, mobile robots, nanorobotics, neural networks, object recognition, olfactory sensing, tele-operation, touch and vision guided manipulation, three-dimensional localization and planning, virtual reality, and vision.

Japan Robot Institute: ⟨http://www.gmd.de/People/Uwe.Zimmer/Lists/Robotics.in.Japan.html⟩

Fig. 5.1.4. No Hands across America Project (Carnegie Mellon University Robotics Institute). Road traffic visual analysis system is used to robotically steer a car ⟨http://www.ri.cmu.edu/projects/project178.html⟩.

The "scientific" agendas for robotic research claim scientific neutrality. Some artists, however, such as the **Critical Art Ensemble** take a perspective of "Contestational Robotics." They note the unstated biases in robotics research, which predominantly emphasize increasing the surveillance and police powers of established authorities (for example, robot police vehicles that can disperse protestors), and they seek to develop robots to serve as resistant forces, such as robot pamphleteers that can safely counter the police robots.

Examples of Conceptual Challenges and Approaches

Vision

Robot vision is another classical problem. Human and animal vision systems are tremendously sophisticated, filling in details that literally cannot be seen. For example, when humans look at a tabletop they think they see all of the legs even though the image of some

---

Critical Art Ensemble and Robert Pell: ⟨http://www.critical-art.net/lectures/robot.htm⟩

legs are missing from the actual visual field. Carnegie Mellon researchers have a number of vision projects under way, such as: 3-D vision for navigation, SAPIENT (Situational awareness for driving in traffic); RACCOON (car following at night); NLIPS (lip-reading gesture—speech integration); and ARTISAN (object recognition for tele-robotic manipulation). MIT research includes: Wheelesley (a wheelchair navigation system) and Pebbles (a single-camera system for obstacle avoidance in rough, unstructured environments).

Sophisticated Motion

Researchers seek to understand the subtleties of human and animal motion. How do insects, birds, fish, animals, and humans move through and manipulate the world? How do they choreograph the ways that the senses and muscles work together? Early robot research helped accentuate the sophistication of everyday human and animal motion. For example, some early robot arms would obliterate eggs that they tried to pick up because the designers failed to understand the way feedback and control worked to help humans delicately adjust the way that force is applied. Researchers now are getting close to creating robots that can fly like insects, swim like fish, and conduct sophisticated human activities such as surgery:

Tim Smithers and Miles Pebody working at the Artificial Intelligence Laboratory at the Vrije Universiteit Brussel created *The Fish,* which was shaped like a fish and could move through water. Illustrating the crossovers between research and art, the Web description of this project notes that the research has terminated and that "now it is only exhibited as a piece of art."[1]

The Institute for Flexible Automation at the Technical University of Vienna has developed a robot that is capable of following objects with its camera "eyes." After being shown an object, it will follow it visually as the object is moved.

University of Southern California robotics research includes: AFV (autonomous flying vehicle) and cerebellar control of walking robots.

Humanlike walking is a major challenge. John Bares, a Carnegie Mellon researcher, notes that it is not easy to duplicate the sophisticated balance, sensor, and actuator accomplishments of our bodies:

Carnegie Mellon University: ⟨http://www.ri.cmu.edu/ri-home/research.html⟩
University of Southern California: ⟨http://www-robotics.usc.edu/nav.html#projects⟩
University of California at Berkeley: ⟨http://robotics.eecs.berkeley.edu/⟩

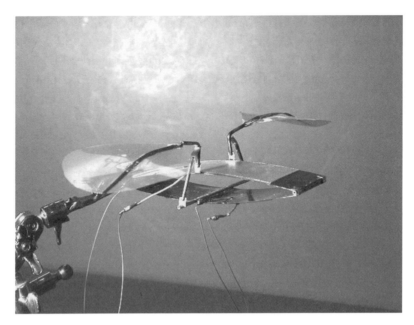

Fig. 5.1.5. Daniel Monopoli, DMI prototype. Research to create mesoscale flying robotic insects. Center for Intelligent Mechatronics at Vanderbilt University ⟨http://www.vuse.vanderbilt.edu/~meinfo/labs/cim/projects/flying.htm⟩.

It's very difficult to duplicate the human tendon-muscle-bone structure and power, . . . Humans walk dynamically, a process in which we must quickly plan foot repositioning—and then realign the body's center of gravity as we change feet and move forward. That's tough for robots, engineers have found . . . Control of legs and body to maintain an upright position is a worldwide challenge.[2]

But the difficulty just acts as a challenge to researchers. MIT's Leg Lab has developed a series of running and hopping robots. Isao Shimoyama at the University of Tokyo, and Atsui Takanishi and others at Tokyo's Waseda University, have also developed walking robots, including some that walk on stilts.

The University of Tokyo Robot Research Lab sponsors a project called the "remote-brained robots," which investigates the idea that sophisticated movement in the world will require massively parallel computing and that the brain can be separated from the physical robot:

A remote-brained robot does not bring its own brain within the body. It leaves the brain in the mother environment, by which we mean the environment in which the brain's software is developed, and talks with it by wireless links. . . . The brain software is developed in the mother environment, which is inherited over generations. It can benefit directly from the mother's "evolution," meaning that the software gains power easily when the mother is upgraded to a more powerful computer. . . . The remote-brained approach allows us to tie AI directly to the world, enabling the verification of high-level AI techniques which could previously only be used in simulation. . . .

There has been a missing link in research, between "AI, which couldn't survive if embodied in the real world" and "robots with feeble intelligence." Our approach, through building remote-brained robots, aims to open the way for engineering advances which will bridge the gap.[3]

Autonomy

Researchers have identified a series of conceptual challenges confronting the developers of practical robots. Autonomy is an issue underlying much contemporary research. How does one endow the robot with the ability to solve its own problems? Carnegie Mellon offers examples of research projects to address these issues, such as: an autonomous helicopter, cross-country navigation, an autonomous harvester, and neural net navigation.

Top-Down vs. Bottom-Up Subsumption Architectures

The last years have seen a debate grow about the most productive ways to conceptualize robot intelligence and deal with autonomy. Historically, robot intelligence projects stressed systems with massive integrated computing resources. In this view, sophisticated robots would need computing power for perception, planning, interpretation, and activation. Critics call this approach "Good Old-Fashioned Artificial Intelligence" (GO-FAI). While the approach achieved some successes, it did not achieve its larger research goals.

Many contemporary researchers question the approach, believing instead that mastery of complex environments can be achieved by endowing robots with many small intelligences, for example, the ability of a robot leg to have a withdrawal reflex when it encounters an obstacle. Creating a framework in which these intelligences can learn from experience and intercommunicate can result in the emergence of sophisticated behaviors that look quite intelligent. Rodney Brooks at MIT is one of the most well-known proponents of this approach. His "insects" are famous for demonstrating quite adaptive behavior even though they start with a simple set of skills, sometimes called "subsumption." Brooks reasoned that GOFAI's attempt to create compehensive "maps" might work in labs, but what about the real world, with its vast spaces and unpredictable obstacles?

Fig. 5.1.6. Rodney Brooks. Ants project at MIT. Small robots demonstrate subsumption architecture by evolving complex behaviors from a simple repertoire (http://www.ai.mit.edu/projects/ants/).

This kind of approach is related to research on Artificial Life. The project description for the *Ants* micro-robotics project at MIT explains:

The software on each robot is made up of many little programs, or behaviors. Each behavior monitors a few of the robot's sensors and outputs a motor command based on those sensor's readings. These commands are then sent to the motors based on a hierarchy; the outputs of more important behaviors override, or subsume, the outputs of less important ones.[4]

Brooks and Anita Flynn wrote a 1989 paper called "Fast, Cheap, and Out of Control: A Robot Invasion of the Solar System," in which they proposed using a swarm of microbots for space exploration. They claimed that the project would be shorter in development, less expensive, and more flexible than the traditional NASA approach of building highly sophisticated single robots. Other labs have similar projects under way in subsumption-oriented evolution of motion:

- USC robot researchers are working on the Rodney project, which explores the idea of "genetic walking," in which a robot develops walking skills through learning and evolving abilities.

- The University of Sussex, Centre for Computational Neuroscience and Robotics, created an eight-legged insect called Maggie. Equipped with whiskers, bumpers, and infrared sensors, it uses a bottom-up strategy to evolve walking abilities. The evolutionary process of its learning to walk took 3,500 generations of simulation on a Sun workstation.

Social Communication between Robots

The Interaction Lab at the University of Southern California, which is attempting to learn how to build robots that can interact successfully with humans and other synthetic creatures, studies interaction in a variety of scales and settings, including herds of physical robots, animal populations, and economies. They seek to develop methods for the "principled synthesis of group behavior and learning by imitation." Their multifaceted research program includes:

Multi-agent robotics: including dynamic task division, specialization, formations, learning behaviors and social rules, and distributed spatial representations. . . .

Fig. 5.1.7. Barry Brian Werger and USC Interaction Lab, study of social learning and behavior among communicating robots, ⟨http://www-robotics.usc.edu/~barry/SociallyMobile.html⟩.

Multi-modal representations: learning by imitation (perception, representation, motor control, and sensory-motor mapping). . . .

Multi-agent systems: methods for synthesizing and analyzing complex group behavior, including multi-agent/robot learning, dealing with nonstationary conditions, uncertainty, partial observability, and credit assignment; also cooperation and competition, and dominance hierarchies.[5]

Exploring bottom-up concepts of intelligence, the Artificial Intelligence Laboratory at the Vrije Universiteit Brussel is developing "language game robots," which communicate with each other by radio links. Researchers want to understand how complex intelligence can be built up and how language and cooperation might evolve. In one experimental "ecosystem" setup, robots competed for food (charging station access) and were allowed to negotiate and communicate in building up workable intelligence.

MIT's artificial intelligence lab sponsors a series of projects looking at social learning in microbots. "The Ants" project consists of a series of subprojects. In "Clustering Around Food," the robot ants send messages when they find food or when they see other robots with food. Building on simple behaviors, complex social foraging behaviors can emerge. The project description notes that the behaviors might be useful in projects like hazardous waste removal or planetary exploration. In "Tag," the robots enact a game like the children's game of tag, with one robot as "it," which tries to tag other robots and pass the "it" status. In "Manhunt," a complex game of team tag is played.

The Department of Cybernetics at the University of Reading supports a research project called "7 Dwarf Robots." These robots move about an enclosed space sensing the world via high-frequency sound waves and communicating with infrared signals. Endowed with intelligence primitives, they evolve complex behaviors, each exhibiting individual characteristics. Demonstrating the possibilities of social communication, the project description notes that "last year one robot (in Reading) programmed another robot (in upstate New York) with what it had itself learned, without human intervention." This lab has also produced insect walkers and the world's "first half-marathon robot."[6]

Humanoid Robots

Early robot research focused on animal and human models of intelligence and motion. Later, researchers abandoned these models, instead proposing that robot form ought to be free to assume whatever shape it needed to optimize its goals. Recently, however, researchers are again assessing whether there might be important reasons to build on organic models. This speculation arises from several considerations: appreciation of the "engineering" accomplishments of animal and human sensory-motor structures; analyses of intelligence that emphasized the importance of embodiment in shaping intelligence; and realization that many robot goals assume interactions with humans and physical environments, which are facilitated by human and animal forms for robots.

Rodney Brooks heads up the "Cog" project lab at MIT, which has undertaken many projects to explore humanoid robots. The project description explains its rationale:

If one takes seriously the arguments of Johnson and Lakoff, then the form of our bodies is critical to the representations that we develop and use for both our internal thought (whatever that might mean . . .) and our language. If we are to build a robot with humanlike intelligence, then it must have a humanlike body in order to be able to develop similar sorts of representations. . . . Since we can only build a very crude approximation to a human body, there is a danger that the essential aspects of the human body will be totally missed. . . .

A second reason for building a humanoid-form robot stands on firmer ground. An important aspect of being human is interaction with other humans. For a human-level intelligent robot to gain experience in interacting with humans it needs a large number of interactions. If the robot has humanoid form then it will be both easy and natural for humans to interact with it in a humanlike way.[7]

Fig. 5.1.8. Rodney Brooks. Cog project studies the importance of humanoid form in developing robotic intelligence, ⟨http://www.ai.mit.edu/people/brooks/⟩.

The description also responds to the question: Why not simulate the robot rather than build it? It claims that important understandings come from constructing robots to deal with the physical world that might be impossible to learn by simulation: "To do a worthwhile simulation you have to understand all the issues relevant to the simulation beforehand; but as far as human-level intelligence is concerned, that is exactly what we are trying to find out—the relevant issues."

The project will ultimately include many features of human life—motion, balance, vision, hearing, touch, vocalization, and the like. Brooks is famous for his robotic insects with their simple-behaviors subsumption architecture. In a *Popular Mechanics* interview, he indicated that Cog is based on similar principles: robots can build up complex behaviors based on simple ones (hearing, seeing, and moving for a humanoid robot), and experience in the world is useful for refining a robot's behavior. Cog can lean and turn its body and head but has no skin, arms, or fingers. It cannot walk or move. He explains the attempt to let it learn through experience:

Cog must learn to relate what it sees in the camera to its own head motion, to know what motion is in the world and what is due to its own head . . . We're trying to find ways for Cog to learn about the world by itself—let it get its calibration from the world, just as humans do.[8]

Robots and Popular Culture

Robots have long been a feature of popular culture in literature and cinema, but for the public they have stayed in the realm of fantasy. In recent years, however, the increasing sophistication and decreasing cost of digital electronics has combined with the development of inexpensive sensors and actuators to bring robot construction within the grasp of nonengineers. Hobbyist robots have become available, and many more people are beginning to experiment. Public interest and literacy about robots has begun to accelerate. Sony's sophisticated $3000 dog robot sold out on its first day.

The difference in the public spread of computer knowledge and robot knowledge is instructive. In the 1960s, computers and robots were probably equally esoteric for nonspecialists. However, the advent of home microcomputers induced large numbers of people to educate themselves and experiment. Artists and teenagers became experts. This shadow world of nonofficial researchers fueled much of the innovation.

Curiously, robot experimentation did not experience the same surge in popularity. Although a full analysis of this divergence is outside the scope of this book, some reasons can be suggested. The practical business application of the desktop computer provided an economic motivation to push its development. The technical infrastructure for a plain computer—that is, CRTs such as a video monitor and keyboard—was already part of many people's experience. The problems of real-world action, for example, sensing and moving, were much more difficult than those for a device sitting in the limited domain of the desktop.

Now, the tide seems to be turning. Robots are becoming popular, like early microcomputers were. Children's toys such as Capsella and Robot Legos allow for easy sophisticated experimentation and construction. Popular events such as robot races and robot wars attract nonprofessional constructors and wide audiences. Hybrid art show/technical meetings such as Robotronika mix up traditional categories of who should be interested in this research. Artists and researchers fill the ranks of both presenters and audience. The Web makes robot news easily available; for example, NASA's "Cool Robot of the Week" offers a changing display of hobbyist and research robots.

Several events give examples of the expanding interest. The Robot Wars event is described in chapter 5.3. In the mobile robot competitions sponsored by the American Association of Artificial Intelligence (AAAI), competing robots are confronted with challenges they must solve. Both professionals and nonprofessionals are invited to participate. There were also challenges for the Fifth Annual Mobile Robot Competition. In the Call a Meeting challenge, a robot must schedule a meeting for two ficitional professors by shuffling between rooms to determine if they are occupied and then optimize the time while informing participants. In the Clean Up the Tennis Court challenge, a robot must collect a number of tennis balls and one powered Squiggle ball into a pen.[9]

Summary: Robot Hopes, Fears, and Realities

Wired magazine often presents a feature called "Reality Check," in which they ask a panel of experts when (if ever) various technological goals will be achieved. They convened a panel on the future of robotics to assess progress toward these goals: a self-driving taxi, a housecleaning robot, a self-replicating robot, and C-3PO (George Lucas's robot from *Star Wars*).[10]

Although not in total agreement, the panel suggested that achievement of most of these goals was still distant. The robot taxi would require sophisticated vision skills to differentiate the elements of urban clutter far beyond current capabilities, but limited autonomous vehicles might be available in areas of a city with specialized infrastructure.

NASA's Cool Robot: ⟨http://ranier.hq.nasa.gov/telerobotics_page/coolrobots.html⟩

Although work is proceeding on humanoid robots, the ease with which C-3PO moved in the world and interacted with humans is also still distant. Self-replicating robots could be achieved relatively soon if financial resources could be made available. Housecleaning robots were seen as similarly achievable, although they might take the form of swarms of small robots rather than the more common image of the robot maid.

Robots offer an interesting instance of research of importance simultaneously to science and art. They have mixed parentage: electrical and mechanical engineering, cybernetics, and artificial intelligence on one side and sculpture, cinema, automatons, and portraiture on the other. The urge to usurp God's ability to make something like a human is an ancient challenge and a taboo. The further development of the technology absolutely requires a multitude of perspectives from technical and artistic sources.

In a story written decades ago, the science fiction writer Isaac Asimov wrote his famous laws of robotics to govern the ethics of robot research, for example: (1) don't injure humanity; (2) obey human orders; and (3) protect itself. They seemed quite fanciful when first written, although now they seem highly practical and relevant. They are an interesting example of the arts anticipating research.[11]

Robots have not been the focus of much cultural theory except obliquely. Critical attention focuses on the poles of virtual disembodied existence and on organic bodies and the interrelationship between the two. Robots represent an interesting middle state— "embodied" in matter, not organic but controlled by an artificial mind. As research proceeds to investigate the links between intelligence, robotic embodiment, and the possibilities of alternatives to organic embodiment, robots may demand more theoretical analysis.

Notes

1. T. Smithers and M. Pebody, "Robotrinka Fish Description," ⟨http://robot.t0.or.at/exhib/⟩.

2. J. Bares, "Robot Research," ⟨http://popularmechanics.com/popmech/sci/9507STROAM.html⟩.

3. University of Tokyo Robot Research Lab, "Remote Brain Project," ⟨http://www.jsk.t.u-tokyo.ac.jp/⟩.

4. R. Brooks et al., "*Ants* Description," ⟨http://www.ai.mit.edu/people/brooks/⟩.

5. University of Southern California Interaction Lab, "Robot Communication," ⟨http://www-robotics.usc.edu/~agents/⟩.

6. Department of Cybernetics, the University of Reading, "7 Dwarf Robots," ⟨http://robot.t0.or.at/exhib/⟩.

7. R. Brooks et al., "Cog Project Description," ⟨http://www.ai.mit.edu/projects/cog/⟩.

8.   R. Brooks, "Interview with Rodney Brooks," ⟨http://popularmechanics.com/popmech/sci/9507STROAM.html⟩.

9.   American Association of Artificial Intelligence, "Robot Challenge," ⟨http://tommy.jsc.nasa.gov/~korten/competition96.html⟩.

10.   *Wired*, "Reality Check—Robots," ⟨http://www.wired.com/wired/4.03/reality.check.html⟩.

11.   I. Asimov, "Robot Rules," ⟨http://www.cc.gatech.edu/aimosaic/robot-lab/MRLHome.html⟩.

5.2

Conceptual Kinetics and Electronics

Artistic Research

In the last decades, artists have been active exploring robotics and machine-activated motion. Some of the art research shares similar agendas with the scientific/technological research, for example, exploring the limits of "intelligent" machines, the dexterity of machine motion, the relevance of artificial life and emergent behavior concepts, and the implications of telepresence and telerobotics.

Other artists, however, pursue divergent interests. For example, some artists create robotic/kinetic devices that reflect on the military/industrial origins of much contemporary research. Others create devices that demonically comment on issues of control and the relationship of machines to human activity. Still others abandon the utilitarian emphases of scientific research to explore the qualities of the devices' motion or appearance, such as beauty, mystery, intrigue, danger, or foreboding. Still, for others, the robots and machine environments are primarily used as dramatic settings to explore a panoply of personal or formal issues similar to those pursued by artists working in conventional media.

This artistic activity illustrates an important difference in the ways artists and researchers approach research. Even though scientists and technologists may give some heed to the context of their funding or research agendas, artists are much more likely to deeply explore the cultural context underlying the research activity. Similarly, robotics researchers usually emphasize the functional qualities of robot appearance or quality of motion or ignore them; artists can make these the focus of their work. Questioning what is taken for granted in other disciplines is often the heart of the artistic enterprise.

A more radical position would hold that robotic research is intrinsically art. In *Beyond Modern Sculpture* the art theorist Jack Burnham suggested that self-replication was at the core of art and that robotics was an inevitable continuation of that quest. These ideas are intriguing to some artists who work in the field. Building on Burnham's ideas, Bruce Cannon, whose works are described in a following section, wrote an essay called "Art in the Age of the Microcontroller," which considers the inherent aesthetics of electronics and robotics:

[T]he automata of the last few centuries and the electronic robots emerging in the 1960s both represented for him the logical extension of this striving. He suggested that robots themselves were the ultimate extension of sculpture, and should be judged as such without any other esthetic criteria. That their striving [of artists toward self-replication] made them inherently art, regardless of their physical form.

Despite the fact that he later recanted all this, it was and remains an amazing conceptual leap, one that I respect and admire. As an artist using computers, interested in artificial intelligence and robotics, I strive toward the purity of this vision, but fail. I long to be able to strip away the superficial trappings in which I feel I must dress technological work in order for it to fit into the dialectic of the art world. I crave the unary pursuit of sentience and autonomy over the rote schematicization of the prevailing cultural fad.[1]

Kinetic Art Precursors

Contemporary artists working with robotics can trace their lineage to kinetic art. Kinetic art is art that moves, motivated by human touch, natural forces such as wind, or by motor. In the early part of the twentieth century, kinetic artists were crucial pioneers seeking to expand the arts to address contemporary culture. In that era, when the norms of the art world were firmly dominated by historical media such as painting and sculpture, making art that moved was radical. Also, in this era artists willing to work with electricity, motors, metal fabrication, and new materials were as much technological researchers as digital artists are today.

As will be explained in my book *Great Moments in Art and Science,* kinetic artists worked from a variety of perspectives. Some, like the Bauhaus artists, Futurists, and Constructivists, and artists such as László Moholy-Nagy, sought to create art that reflected on the new opportunities offered by industrial/technological "progress." Others such as the Dadaists, Surrealists, and artists such as Marcel Duchamp were more dubious about progress. Others such as Alexander Calder saw motion, change, and time as just more formal elements to be explored in composition.

In the 1960s and 1970s, artists continued the exploration of kinetics, refining old themes and expanding its concerns. Examples include Frank Malina, Nicolas Shoeffer, Otto Peine, Takis, Jean Tinguey, EAT (Experiments in Art and Technology), Lygia Clark, Helio Oiticica, Jesus Rafael Soto, Alejandro Otero, Pablo Neruda, Agam, Alexander Calder, and David Medalla. Interested readers should consult the histories of technological art listed in the bibliography.

Eduardo Kac's article "Foundations and Development of Robotic Art" identifies several artists from the 1960s and 1970s as especially significant precursors of contemporary robotic work. Nam June Paik and Shya Abe created *Robot K-456* in 1964. They rolled this "robot," which had a vaguely anthropomorphic/electronic look without very sophisticated behaviors, around the streets in attempts to create public events. In 1966, James Seawright created *Watcher* and *Searcher,* which were interactive kinetic sculptures. In 1970, Edward Ihnatowicz created *Senster,* which was a roboticlooking arm that sensed the presence of humans. In the 1970s, Norman White created the *Helpless Robot,* which asked humans to interact with it in order to make it function.

Contemporary kinetic artists update this work by incorporating more sophisticated technology and using the technology to explore cultural commentary or conceptual investigations. Also, note that the boundary between robotic and kinetic art is not clear because of the wide range of meanings the word *robot* has assumed. When does a sophisticated machine cross over to "robothood"? Conversely, many things called robots are not very sophisticated in their behavior. Humanoid appearance is not a requirement. Most likely, many of the artists who now think of their work as experiments in robotics would have considered it kinetic a few years back.

Kinetics and Light Sculpture

Milton Komisar

Milton Komisar's career spans decades. He was one of the first kinetic light artists to apply computers to control. His light sculptures are famous for their elegant movement of light and their progression in time created via computer program. Komisar describes his approach:

Developing a system to work with Light in this particular way has led me to the idea of COMPOSITION IN TIME. This is traditionally a musical concept. There is no sound in my work. I do not want to create a multimedia art form. I believe it is possible to "mold" Light through time in such a way that a coherent composition is experienced by the viewer. The physical structure and the electronics are simply necessary tools to this end. I have been working with this goal in mind for the last twenty-three years.[2]

Gregory Barsamian

Gregory Barsamian creates kinetic sculptures that use stroboscopic technology to freeze and manipulate motion. The events combine the reality of constructed objects with the dreamlike quality of mediated vision. Barsamian has shown his work in several locales, including the ICC. He explains animation as a doorway to the unconscious:

My technique adds the fourth dimension of time and allows the viewer to share the same physical space and time with an animated sequence. Animation is ideally suited to the realization of

Milton Komisar: ⟨http://www.xlnt.com/neonart/mkomisar/mkomisar.html⟩
Gregory Barsamian: ⟨http://www.concentric.net/~Venial/⟩

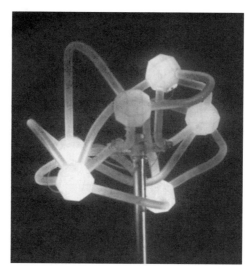

Fig. 5.2.1. Milton Komisar, *Sign of the Fish*. Modular units used in a computer-controlled light sculpture.

subconscious images and alternate realities. My passions lie in bringing these images to life in this most vivid form.[3]

Barsamian questions science's claim that it presents absolute vision. He questions the givens of perception and emphasizes the importance of the unconscious, noting that no one angle of view captures totality and that science is distrustful of the imagery of dreams:

In creating alternate realities, I confront the viewer physically in the language of the subconscious with a skepticism of our perceptions. The power of sharing the same space with these surreal three-dimensional images lies partially in witnessing your own act of interpretation. . . . It is the nature of that order that defines us as human beings. Order, however, is not what I offer you. Instead, I offer a three-dimensional window into an ontological bazaar where self-deception is an oxymoron.

In one installation called *Putti,* tiny cherubim fly around in circles, change direction, and transform back and forth with helicopters. The curator Janet L. Farber describes the event:

Putti is perhaps the clearest illustration of Barsamian's intentions. Hovering overhead, spinning figures of cherubs (putti) turn into helicopters and back again into winged babes. The nature of this transformation is purposefully ambiguous: Do the cupids become helicopters first or do the

Fig. 5.2.2. Gregory Barsamian, *Putti*. A kinetic sculpture is illuminated by a stroboscopic light.

whirlybirds turn into ministering angels? . . . Yet, what does it say about human nature that the interpretation most frequently given of this transformation is negative? It conjures up the loss of innocence, the encroachment of police states, the buzz of Valkyrian war machines.[4]

Other Artists and Projects

James Seawright, director of visual arts at Princeton, has a long history of developing interactive kinetic sculptures, such as *Watcher,* which modify their behavioral and sound patterns based on changing light patterns produced by other sculptures or viewers. Recent work incorporates more sophisticated technologies, such as *Mirror I,* which focuses the sun's light in a complex way, arranging 225 mirrored blocks to precisely focus rays on an *X* on the sidewalk twelve feet away. ***Eric and Deborah Staller*** create kinetic and light public art, such as *Bubbleheads,* in which multiperson bicycles are driven by

---

James Seawright: ⟨http://www.tezcat.com/~divozenk/plaza/mirrori.html⟩

people, each wearing light sculptures. A yearly kinetic sculpture race called Da Vinci Days challenges artists, engineers, and others to design human-powered moving sculptures that must negotiate city streets, mud, the river, and a sand trap. ***Paul Friedlander*** creates stroboscopic kinetic light installations. Coordinating the movement of sound and light, ***Guy Marsden***'s sculptures confound traditional categories. Promoting the humanistic study of light, *Seth Riskin* creates dance and body movement events that incorporate light phenomena—for example, by including projectors or reflectors on his body. ***Jennifer Steinkamp*** sets up room-sized installations in which shadow, projection, and light movement "de-center" and "dematerialize" space and respond to visitor motion and proximity.

Conceptual Kinetics

What conceptual kinetic artists do is quite remarkable. They convert the mundanities of motors, gears and levers into philosophical and artistic discourse.

Bryan Rogers

Bryan Rogers was in the forefront of conceptual kinetics in the 1970s and 1980s. He appropriated state-of-the-art mechanical and electronic technology to create families of devices focused on particular concepts or cultural niches, for example, his "Timepieces," "Umbrella," and "Coffin" series. Illustrating this approach, his "Coffin" series featured multiple variations on the theme, for example, rotating coffins, self-propelled coffins, and coffins whose lids automatically opened to welcome the viewer. His constructions typically played with disjunction—finely crafted advanced engineering applied to the creation of unlikely objects, puns, and conceptual explorations.

Roger's approach is also evident in other works. His spearfishing piece is famous for the uproar it created. An aquarium was fitted with a rapid-action hydraulic harpoon that would periodically jut in and out. The event was picked up by the tabloid *National Enquirer* and caught the attention of the Society for the Prevention of Cruelty to Animals, even though the probability of harm to the fish was remote. In his multipart *Odyssetron* project, he undertook to create a seaworthy robot that could navigate itself

---

Kinetic sculpture races: ⟨http://www.rdrop.com/users/batie/davinci97/kinetic.html⟩
Seth Riskin: ⟨http://web.mit.edu/mit-cavs/www/Seth.html⟩
Jennifer Steinkamp: ⟨http://jsteinkamp.com/html/art_statement.htm⟩
Bryan Rogers: ⟨http://www-art.cfa.cmu.edu/www-rogers/⟩

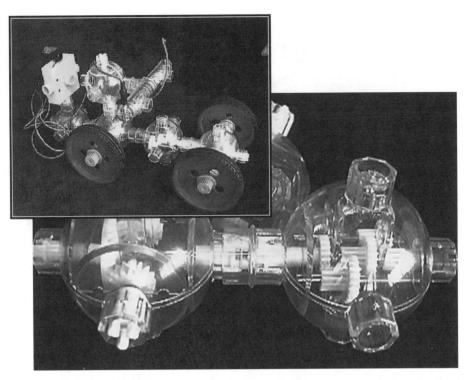

Fig. 5.2.3. Capsella mechanical experimenter kit, composed of modules such as motors, gears, and pulleys, to explore the principles of energy conversion and transfer. Photo: Stephen Wilson.

around the globe. Rogers holds degrees in both art and engineering and founded the Conceptual Design program at San Francisco State University. He developed the Studio for Creative Inquiry at Carnegie Mellon University.

Perry Hoberman

Perry Hoberman creates installations that expose the cultural underpinnings of technology. Typically, his works are simultaneously humorous and troubling. *Faraday's Garden* presented the viewer with a hodgepodge of consumer appliances such as radios and power tools and image-projecting machines. Ironically commenting on issues of control, the appliances automatically sprung into action, tripped by security foot pads, as viewers

Perry Hoberman: ⟨http://www.hoberman.com/perry/⟩

Conceptual Kinetics

Fig. 5.2.4. Perry Hoberman, *Faraday's Garden*. Household devices and image machines are activated by visitor's motion.

moved about the space (see also chapters 7.3 and 7.4). Here is the description of the installation from Hoberman's Web site:

The machines wait silently, ready to be activated at any moment by the footfalls of the public. When stepped upon, the switch matting triggers the various machines and appliances, creating a kind of force field of noise and activity around each viewer. As the number of participants increases, the general level of cacophony rises, creating a wildly complex symphony of machines, sounds, and projections. The machines and accessories (such as tapes, films, slides, and records) are collected from thrift stores, flea markets, and garage sales. Since they span the entire twentieth century, movement around the room also functions as a kind of time travel. All wires and switches are left exposed, creating an intense environment of electrical current.[5]

Alan Rath

Alan Rath creates sculptures built out of the paraphernalia of the electronic age. His sculptures incorporate electronics, video screens, speakers, microprocessors, voice chips, and robotic elements. He was one of the first artists to create tapeless digital video in which image sequences were drawn directly from chip memory. He has an electrical engineering degree from MIT and has worked as an artist since the 1980s.

Chapter 5.2: Conceptual Kinetics and Electronics

His works have been described as playful, humorous, ironic, and beautiful. He typically makes the electronics and other constructive structures of his work visible to the viewer, and indeed works with the electronic infrastructure of connectors and components as aesthetic elements. Rath's view that the selection and construction of components is an important element of the art was expressed in an interview he gave on the "San Francisco Gate" Web site:

Often there's not a single optimal solution, so things like the selection of an electronic part are open to interpretation. I'm picking components based on what they look like. To me, transformers can be attractive or ugly. The pieces are made in a certain meditative state. A lot of emotion goes into the building, and I hope they somehow contain that. You know, the Mars lander is a beautiful piece of sculpture. The people who built it identified with it, so it has a lot of soul. I want power from art at that level of commitment and mastery.[6]

Typically, the electronics are embedded in other artifacts of everyday culture as part of the cultural commentary. Dana Friis-Hansen, senior curator at the Houston Contemporary Arts Museum, wrote this introduction to Rath's "Bio-Mechanics" show:

Alan Rath's "live machines" are eerily engaging—we are immediately drawn in by their uncanny, humanlike actions. On video screens, eyes move, mouths open, faces wince, tongues lick, hands gesture to spell out messages. Simple speaker cones seem to whisper, breathe, or pulse like a heartbeat. We cannot help but project human emotions—fear, curiosity, desire, pain, excitement, or the will to communicate—onto these otherwise confusing configurations of circuitry.[7]

Examples of the video and kinetic work include *Info Glut 3*, in which video screens display sign language as the sculpture speaks; *Message in a Bottle*, in which a small video screen enclosed in a bottle shows the image of a hand signing phrases by sign language; *Arecibo*, which uses hand signals to spell out the digital encoding of DNA being sent out in a radio beam's search for extraterrestrial life; *Ultra Wallflower*, in which several speakers hung on the wall in an arrangement suggesting plant life vibrate in isolated sociality; *Ouch*, in which the video image of the artist's face responds to being held in a vice; *Likker*, which has a tongue extended on a long metal rod; *ScannerII*, in which eyes glance to and fro as under surveillance. Rath has created a wealth of pieces that include enlarged video eyes and mouths that are simultaneously funny and ominous.

Later work includes robotics, such as *Robot Dance*, in which two kinetic structures vaguely representing hands play hand jive games together; *One Track Minds*, in which two small carts on a single track play approach-and-avoidance games with each other, and

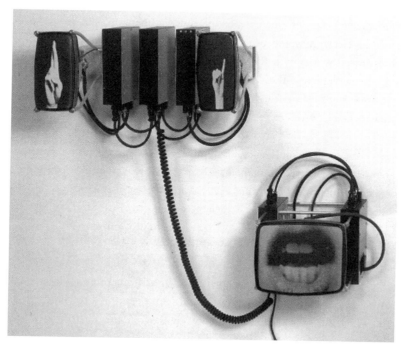

Fig. 5.2.5. Alan Rath, *Info Glut3*. A digital video sculpture speaks with sign language.

Five on the Wall, in which five constructions hanging on the wall engage in synchronized performance. He suggests that machine autonomy is a fascinating concept that he intends to pursue in further work: "Interesting is halfway between nothing and random."[8]

Rath indicates several motivations in his work. He is upset by our culture's ambivalent and shallow attitude toward technology. He notes that many of those whose decry technology fail to see its relationship to the everyday life they take for granted, such as shoes or glasses, and make unwarranted distinctions:

I don't know why people are so alienated from machinery. The next Freud will figure out why we perceive that stuff as external and different. I am amused by the idea that people might draw the line at machinery in a gallery. Somehow the technology and chemistry of paint is OK but other technologies are out of place. It's only that way because of the time we grew up in. Future generations won't see anything strange about it.

Similarly, many people—even those intimately involved with technology in their work—cut themselves off from appreciating the beauty of the technology. Rath believes that machines are an important part of the human exploratory and play potentials. He would like people to get more connected to that aspect of technology:

Our problem is not getting their experiential or toy possibilities [that is, the potential of the technology]. Play is one of the most significant human activities and machines help us play. Probably because our tax dollars are used to put robots on Mars, people don't like that to be called play. But it's obviously play. And then it becomes beautiful.

Bruce Cannon

Bruce Cannon creates conceptual kinetics and robotics that explore a variety of topics, such as life, death, time, social convention, personality, responsiveness, and relationships. His installations will often integrate surplus, ignored, and neglected cultural items with the latest in electronic and robotic technology. Several of the works incorporate electronic speech and proximity detection. Often the minimalist works are ironic and/or meditative. Here are excerpts from his artist's statement:

[Engineering aesthetics] function for me as grounding devices, reality checks on the often arrogant projects of both art and technology. They also invoke a reductive coding which I find interesting, and in fact have adopted as one of my principal techniques. While some of the pieces manage to exhibit lifelike behaviors despite their technical limitations, my general approach is to construct objects whose behaviors or characteristics are in some ways lifelike yet which embody little of the richness of being.

These machines' failure to transcend their artificiality is their most significant aspect. The pieces are not so much lifelike as referential to being, and what is missing is what resonates for me. I have come to think of this negative space as the place where the work happens, at its best a sort of electro-mechanical Haiku in which randomness and absence generate issues of sentience and presence which I would be unable to evoke directly.[9]

Doublespeak attempts to deconstruct social conventions of polite speech. It uses the subtleties of electronic sensors to accentuate a person's role in choosing what to hear, with a person's shadows triggering one facing sculpture to say what is expected of it while the other says what it's really thinking. *Contact II* comments on human relationships by reacting to a viewer's interposing in its space. It speaks phrases of love and admiration

---

Bruce Cannon: ⟨http://www.siliconcrucible.com⟩

Fig. 5.2.6. Bruce Cannon, *Time Tree*. A robotic tree that moves at slow pace similar to that of plants. Courtesy of the artist and Gallery Paule Anglim.

until one gets too close, when it says, "I wish you were dead." In what appears to be a framed mirror, *Reflection* captures one digital portrait a day and scrolls through the time archive, showing an image of an accumulating life.

Tree Time (in collaboration with Paul Stout) is an installation that reanimates a lightning-struck tree transported from a forest. Robotic elements have been added so that the branches move at an almost imperceptible organiclike speed, its motion perceptible only over many days:

This machine is I think equal parts meditation on slowness and bastardization of nature. The obvious reference to Mary Shelley's Frankenstein in the lightning-struck tree, the garish reassembly, the electrification, the technological "improvement" upon the original organism, is intentional. Paul calls it eco-porn, which I think is nice . . . I associate both words with *Tree Time,* because of the pleasure and the pain, the beauty and the obscenity of the endeavor. In that sense, *Tree Time* is a morality story about limits.[10]

Cannon's paper "Anti-Speed" reflects on our culture's preoccupation with speed as a fear of mortality. *Tree Time* attempts to reflect that concern with time and its underlying meanings:

We're preoccupied with speed because we fear death. In other words, immortality is our goal, and since we can't have that we settle for the compression of time. The faster we can do things, the more things we can get done in the amount of time we do have. And this speeding up of the pace of life, the squeezing of more and more events into a given period of time, is the culture's desperate attempt to live longer.[11]

Paul DeMarinis

Paul DeMarinis's works straddle the world of art, music, invention, technological archaeology, and social commentary. DeMarinis carefully studies the history of technology and invention to discover its underexplored underbelly. He simultaneously celebrates its innovation and analyzes what has been sacrificed in its unfolding. He then develops elegant, unprecedented installations that communicate his investigations, mixed with personal and social references. DeMarinis's magic is that the accomplishment of these installations often require him to invent and extend technologies. Thus, his creative process is intimately intertwined with the very processes of invention, on which he comments:

My pieces deal, in part, with the way technologies mediate the relationship of people to their memories and to question the situation of technology in our lives, the mythos of technology. The fact that I use technology itself to delineate these themes means that I must develop alternate or sometimes "impossible" technologies. Without overly stressing the apparent impossibility of making a hologram of a record play the music in the record's groove, or making a clay pot recording of a voice, or making a bathtub make music, I must admit that many of the technologies in my pieces did not exist when I set out to make them. I have had to invent them. It is an important requisite of my art that the pieces actually work. I wouldn't be comfortable with a piece that created an illusion by conventional means. For me the real illusions are the ones that still mystify even when the technology is revealed and explained. Nor would I be satisfied if the works stopped there. There are many other cultural and personal themes woven into them.[12]

DeMarinis's brief descriptions of several of his installations illustrate this mixture of invention, humor, and analysis:

Fig. 5.2.7. Paul DeMarinis, *Messenger* (1998). Skeletons shake their letters to spell out a sent message.

- *RainDance/Musica Acuatica:* Twenty falling streams of water, modulated with audio signals, create music and sound when intercepted by visitors' umbrellas.
- *Living with Electricity:* Three domestic settings, each containing a throw rug, a lamp, a transduced rocking chair, and a sound-making device fitted with actuators. The three areas are interconnected via local area network so that rocking in one chair produces movement and sound in a different one.
- *Gray Matter:* Interactive electrified objects that produce sound and sensation when stroked with the hand.
- *Edison Effect:* Ancient phonograph records, wax cylinders, and holograms are scanned with lasers to produce music at once familiar and distant, like some faintly remembered melody running through the head.
- *Fireflies Alight on the Abacus of Al-Farabi:* A sixty-foot-long music wire with little dancing loops of monofilament is stretched in a dark room and illumined by an emerald laser beam. The loops dance on the harmonic nodes of the wire, producing flickering points of light and aeolian harplike sounds.

Each of his installations has a depth of historical reflection that is easily missed by the casual observer. *The Messenger,* presented in the Galerie Metronom in Barcelona, demonstrates this depth. Here is DeMarinis's short description:

The Messenger: E-mail messages received over the Internet are displayed letter by letter on three alphabetic telegraph receivers: a large array of 26 talking washbasins, each intoning a letter of the alphabet in Spanish; a chorus line of 26 dancing skeletons, and a series of 26 electrolytic jars with metal electrodes in the form of the letters *A* to *Z* that oscillate and bubble when electricity is passed through them.[13]

The installation seems a fascinating event to observe. Its historical referents make it even richer. It is based on a 1753 landmark telecommunications event in which an unnamed researcher with the initials C.M. sent messages via twenty-six charge-carrying lines that caused movement in distant static-electricity-detecting Leyden jars, thus indicating letters of the alphabet. DeMarinis sees this era's interest in electricity intimately tied to cultural developments in democracy. In the catalogue for *The Messenger* installation, DeMarinis shares some of his analysis of Francisco Salvá, a Catalan scientist's, experiments:

Electricity, though observed since ancient times, only became a subject of intense interest in certain enlightened circles during the first half of the 18th century. . . . In electrical demonstrations during the ancien regime, little distinction was made among the message being transmitted, the path of conduction, and the recipient. On one occassion in a demonstration before the king, organized by the Abbé Nolle, 180 guards were said to have been made to jump simultaneously [by shocks]. . . .

[In Salvá's experiments] 26 wires each carried a voltage corresponding to a letter. Salvá specifies a number of people, one for each wire. Upon receiving a sensible shock, each of these people, presumably servants, was to call out the name of the letter of the alphabet to which he corresponded. . . . Toward the final years of the 18th century, after Galvani's discovery of animal electricity, Salvá formulated a revised proposal for the telegraph using freshly severed frogs' legs as the indicators. Each leg, when stimulated by the spark, would dance and, in so doing, jerk a slip of paper on which the corresponding letter of the alphabet had been written. In the first decade of the new century, after Volta's invention of the electrochemical battery, Salvá proposed a scheme that proves politically correct to this day: electrical current flowing through the wires causes electrolytic decomposition of water, the resulting bubbles of hydrogen serving to indicate the letter selected.

Background materials for other installations provide other fascinating discourses on culture, technological history, and personal reflection. His description of his work *The Edison Effect* (reflecting on Thomas Edison's inventions and cultural context) shows a similar mining of technological history for its deeper meanings and its conversion of these ideas into visual poetry:

[I]t invokes a metaphorical allusion to the physical phenomenon known as the "Edison effect," wherein atoms from a glowing filament are deposited on the inner surface of light bulbs, causing them to darken. . . . the metaphorical image of the darkening of the light is an ancient one, recurring in the I-Ching, in Mazdaism, and in Shakespeare's oxymoronic "when night's candles have burnt out." Enantiodromic reversal at the atomic level can be used to symbolize opposing primal forces and may serve to mythicize otherwise commonplace occurrences. . . .

Eminent authorities, including French scientist Sainte Claire de Ville, upon reading announcements of the talking machine, pronounced it a fraud and a hoax perpetrated by a concealed ventriloquist. . . . Perhaps the very notion of compressing the vitality of human utterances, of squeezing the flights-of-fancy of musical invention into the unidimensional coffin of machine reproduction, was abhorrent on some primal level. Or, perhaps, there persisted the stubborn notion that sounds are inherently transitory and must always be synthesized or intoned anew . . .

A dream of early phonographers was to read with their eyes the wiggly line inscribed by the needle as a lasting trace upon the wax . . . Until very recently—the 1980s,—the memorative act of audition still consisted of dragging a diamond stylus, fingernail-like, across a vinyl blackboard. . . . [With the CD] the laser touches but fleetingly upon the groove, the impact of its photons abrading no material whatsoever. The rupture is complete. The emancipation of memory from touch has been fulfilled.[14]

DeMarinis claims that neither extreme of antitechnological Luddism nor unbridled technophilia are sufficient to reflect on the role of technology in culture. Art that attempts to make ideas physical is a fertile locus for considering the myths of technology:

Art is a response to belief and acts as a consolidating force within culture. It gives place, time, image, and sound to myths. But the myths of science are not content to be represented by picture, poems, and symphonies. The scientific revolution threw away the idea that things were connected by appearances and replaced it with the idea that things are connected by how they work. Thus the artist's role is to animate with the imagination the way things work.

I think of technology as having a dual-being. It is simultaneously a dream, or product of our dreams, and the medium in which our dreams are exchanged and elaborated. . . . To disentangle these two functions of technology is difficult. One could, of course, stand aside and take an anti-technological approach. I have chosen what is perhaps a more difficult path—to use technology itself to express and investigate this dilemma. I try to do this by standing technology on its head. Exploring alternative technologies, using physical principles that have not found any place in the dominant technology, re-connecting the dream and the mechanism. . . .

The promise of technology enabling us to be conscious masters of our experience, overlords of the material world, is long past. We have more the impression of being swallowed by our own doing. . . . There is no way out, but we are hopefully capable of an occasional lucid moment within our dream where we can savor and marvel at the whole process even as we are swept away by it, that being the nature of our experience.[15]

Other Artists and Projects

Comment on Popular Culture *Sheldon Brown*'s outdoor installation *Video Wind Chimes* presented a video projector that looked like a streetlight but projects video from television stations that changes channels as it blows in the wind. ***Tammy Knipp***'s *Case Study* presented viewers with a simulation of dyslexic perception by asking them to vulnerably lay down under big video monitors that kinetically move toward them. ***Marque Cornblatt*** created kinetic and quasi-robotic devices that expressed the ambivalence of the cyber era. Using cyberpunk techniques of assemblage, he created installations that confuse and bemuse, such as self-propelled TV stands doing whirlies, and Icarus figures with windable wings. One critic described the works as being assembled out of the three *T*s—toys, trash, and technology. ***Joseph DeLappe*** created works such as *The Mouse Series* which customized computer mice as forms of social commentary, and *Masturbatory Interactant,* in which a bar-code scanner on a mechanical arm randomly selects self-erotic videos to be projected onto a plastic inflatable party doll. ***Jim Pallas***'s interactive scultpures usually comment on social processes, such as *The Senate Piece,* commissioned by U.S. Senator Carl Levin, in which kinetic objects respond to Senate processes—an inflatable senator comes to life during quorum calls and a dollar bill drops during Senate activities. ***Neil Grimmer*** created electronically controlled kinetic sculptures out of commercial items such as vibrators. The ***Art and Robotics Group*** of Canada created *SenseBus,* which allows everyday home objects to sense and communicate with each other without any central brain or traditional digital interfaces, and *SpaceProbe,* which was a collection of electro-art that incorporated unusual sensors and telemetry. ***Steve Gompf***'s *Televisors and Early Motion Picture Technologies* kinetically activates old wooden boxes and other found objects in an attempt to comment on the early days of television. ***Arthur Ganson***'s whimsical contraptions activate diverse materials and found objects to manifest "qualities least associated with machines."

Sheldon Brown: ⟨http://www.cra.ucsd.edu/~sheldon/⟩
Tammy Knipp: ⟨http://siggraph.org/artdesign/gallery/S97/art_knipp.html⟩
Marque Cornblatt: ⟨http//www.falsegods.com.transhuman.html⟩
Joseph DeLappe: ⟨http://digitalart.artsci.unr.edu/delappe.html⟩
Jim Pallas: ⟨http://www.ylem.org/artists/jpallas/JPALLAS1.HTM⟩
Art and Robotics Group: ⟨http://www.interaccess.org/arg/index.html⟩
Arthur Ganson: ⟨http://web.mit.edu/museum/exhibits/ganson.html⟩

Comment on Interaction ***Peter Dittmer*** created kinetic devices that explored language and communication. His systems typically engaged the viewer in a dialogue via onscreen text and spoken words. Things sometimes go wrong. The *Wet Nurse* kinetically acted on a glass of milk as part of its interactions. ***Mark Madel***'s interactive electronic sculptures, such as *Timesharing,* demand user interaction with the piece and each other, calling forth "blurred distinctions between interaction and relationship and between coercion and invitation." ***Laura Kikauka*** created kinetic works that question technophilia. A *Leonardo* article describes her *Hairbrain2000* harness, which creates a wearable viewing chamber in which sparks and relay clicks are activated by viewer motion. The artist group ***Sine::apsis*** experiments creates kinetic and robotic art events that attempt to introduce concepts from biology and artificial life into their installations in order to investigate complex behaviors and interactive structures. The ***Soda*** artist group's kinetic installations challenge the inertia of architectural spaces and the nature of autonomy—for example, with autonomously flexing panels and electrochromic mirrors with dissolving messages and *The Priest and the Dying Man* sculptural robots, which speak together about free will.

Extensions of Puppets ***Heri Dero*** creates shadow puppetlike kinetic assemblages by which he tries to capture the "machine as mystery, play, magic, and metaphor." In *Childhood/Hot and Cold Wars,* ***Ken Feingold*** built an installation reflecting on personal memories of the cultural history of the 1950s and 1960s by offering a grandfather clock full of rear-projected video images shaped by a globe of the earth that viewers could turn. In *Interior,* viewers touch a medical torso to control puppets speaking, seen outside of a window.

Linkage of Motion and Virtual Space In *Room for Walking,* ***Daniel Jolliffe*** offered a wagon with a video projection in its bed that reveals aspects of a virtual object as the viewers pulled it around. ***Doug Back***'s *Small Artist Pushing Technology* consisted of a monitor on wheels with an image of a small person pushing coordinates the same way that the viewer moves the monitor. ***Sigi Möslinger***'s *Sweetcart* revealed a digital landscape as a monitor is rolled about.

Peter Dittmer: ⟨http://www.foro-artistico.de/english/program/literat.htm⟩
Mark Madel: ⟨http://www.mggrd.com/mm⟩
Laura Kikauka: ⟨http://mitpress.mit.edu/e-journals/Leonardo/gallery/gallery291/kikauka.html⟩
Sine::apsis experiments: ⟨http://www.sine.org⟩
Soda: ⟨http://www.soda.co.uk/index.htm⟩
Heri Dero: ⟨http://www.ntticc.or.jp/permanent/heridono/introduction_e.html⟩
Ken Feingold: ⟨http://www.kenfeingold.com/⟩
Daniel Jolliffe: ⟨http://www.interaccess.org/touch/jolliffe.htm⟩
Doug Back: ⟨http://www.interlog.com/~steev/exhibition/networks/gallery/back.html⟩

Summary: More Than Robotics

Although not high technology, machines are a critical infrastructure for contemporary society. They are the brawn that ultimately translates intelligence to the world of stuff. Research attention is mostly focused on robots as the ultimate machines that incorporate intelligence and flexibility. But the lowly machine without robot aspirations is an important cultural icon. Indeed, some feel that the electricity that makes the technology go deserves more cultural analysis.

Since the beginning of this century, isolated artists have experimented with machine motion and electrical light. Usually they approached these technologies within the aesthetic traditions of seeking beauty of motion and illumination. Within the last decades, however, artists have explored the cultural implications of machines and light. The artists described in this chapter have been free to create mechanical installations that use the latest technologies but pursue cultural agendas unaddressed by mainstream industrial applications. In some ways they are the logical extension of the ancient art form of sculpture.

Notes

1. B. Cannon, "Art in the Age of the Microcontroller," ⟨http://www.jps.net/bcannon/⟩.

2. M. Komisar, "Statement," ⟨http://www.xlnt.com/neonart/mkomisar/mkomisar.html⟩.

3. G. Barsamian, "Description of Work," ⟨http://www.concentric.net/~Venial/⟩.

4. J. Farber, "Commentary on *Putti*," ⟨http://www.concentric.net/~Venial/⟩.

5. P. Hoberman, "*Faraday's Garden* Description," ⟨http://www.portola.com/PEOPLE/PERRY/perry.html⟩.

6. "San Francisco Gate" Web site, "Interview with Alan Rath," ⟨http://sfgate.com/eguide/profile/⟩.

7. D. Friis-Hansen, "Curatorial Statement 'Bio-Mechanics' Show," ⟨http://www.camh.org/cam_exhandprograms/cam_onlineexh/cam_rath/rath-index.html⟩ [expired link].

8. "San Francisco Gate" Web site, "Interview with Alan Rath," ⟨http://sfgate.com/eguide/profile/⟩.

9. B. Cannon, "Artist Statement," ⟨http://www.jps.net/bcannon/⟩.

10. B. Cannon, "*Tree Time* Description," ⟨http://www.jps.net/bcannon/⟩.

11. B. Cannon, "Anti-Speed," ⟨http://www.jps.net/bcannon/⟩.

12. *ICC Journal*, "Interview with Paul DeMarinis," ⟨http://www.well.com/user/demarini/shiba.html⟩.

13. P. DeMarinis, "Description of *The Messenger*," ⟨http://www.well.com/user/demarini/messenger.html⟩.

14. P. DeMarinis, "Description of *Edison Effect*," ⟨http://www.well.com/user/demarini/edison.html⟩.

15. *ICC Journal*, "Interview with Paul DeMarinis," ⟨http://www.well.com/user/demarini/shiba.html⟩.

5.3

Kinetic Instruments, Sound Sculpture, and Industrial Music

Music has a long tradition of technological experimentation with devices to produce sound. Those that became accepted into the repertoire came to be called the standard instruments, such as pianos, trumpets, violins, and the like. Contemporary sound artists are attracted to the possibilities of using the tools and materials of industrial technological culture in sound making. They are interested in sound outside of music and speech. Sound artists were also among the first to explore electricity, recording technology, synthesis, sound editing, radio, and electronics. Contemporary artists have created a multitude of forms that integrate the visual and sonic arts.

They have worked in many ways. Some have pursued sensual novelty, trying to create sounds never heard before. Some develop alternative instruments. Some try to explore the sounds of the natural- and the human-built world. Others create kinetic sculptures and interactive installations that investigate critical themes in cultural analysis within the context of sound. Other chapters have described other science or technologically inspired investigations, such as the search for artificial intelligence and genetic music generating algorithms (4.3 and 7.6); acoustic ecology (2.4); auralization of information (7.7); and speech synthesis and computer manipulations of 3-D sound space (7.5).

Technological and electronic sound experimentation is a major chapter in the relationships among art, science, and technology. Sound artists and musicians have demonstrated many of the cross-disciplinary themes described in this book. Musicians and sound artists entered into the scientific discourse about sound and hearing early. They pioneered technological investigations of electronic sound and were among the earliest to work with personal computers.

Unfortunately, the consideration of sound-artist experimentation with technology is an enormous topic beyond the scope of this book. There are books, journals, and organizations specifically devoted to these inquiries. For example, some of those resources include: *Leonardo Electronic Music Journal, Electronic Music Journal,* and *SoundCulture.* This section attempts to provide only an brief sampling of the ways to consider the relationship of science and technology to sound art. It concentrates primarily on artists who work within the traditions of kinetic sculpture and installation.

A Brief Theoretical Overture

Some theorists consider sound grossly underexplored in analysis of the arts. Douglas Kahn and Gregory Whitehead edited a landmark book, *Wireless Imagination: Sound, Radio, and the Avant-Garde,* which collects theoretical analysis and historical documents. Some of the historical eras they consider include the age of the phonograph's invention, the Futurists' art of noise, Duchamp's gap music, the Dadaists' experiments with sound art, the Russian Constructivists, the Surrealists, and John Cage's experiments. Technology was simultaneously a provocation and a tool for this work, although Kahn and Whitehead shy away from technological determinism, searching instead for broader cultural themes.

They give voice to the view of many about the strange neglect of sound, given its unique kind of access to people's consciousness. They find the paucity of systematic theoretical attention a major lack:

The human ear offers not just another hole in the body, but a hole in the head. Moreover, the absence of obstructive anatomical features such as earlids would seem to assure a direct and unmediated pathway for acoustic phenomena, with sonic vibrations heading straight into the central nervous system.[1]

One would expect to find amid the accumulation of studies of modernism, postmodernism, the avant-garde, and postwar experimentalism a more faithful attendance to the cultural preoccupations of hearing—one of the two major senses, the "public" ones, as John Cage described them, for their ability to make contact from a distance.[2]

In part, they attribute the lack to the dominance of the "regime of the visual," a tendency in modern Western culture to focus on sight, a subject of much contemporary analysis. They claim that sounds cannot be objectified as easily as sights:

Yet another problem exists in merely thinking about sound within a culture that so readily and pervasively privileges the eye over the ear. Visuality is so embedded that attempts at redress seem doomed to tautology. Many contemporary theories and philosophies, in fact, invoke aural, sonic, musical, and preguttural metaphors at the points where they are unable to speak, at the limits of language. How can we then rely on the same theories and philosophies to query the very sounds heard during such moments of matriculation? How, for instance, can listening be explained when the subject in recent theory has been situated, no matter how askew, in the web of the gaze, mirroring, reflection, the spectacle, and other ocular tropes? Visually disposed language, furthermore, favors thinking about sound as an object, but sound functions poorly in this regard: it dissipates, modulates, infiltrates other sounds, becomes absorbed and deflected by actual objects, and fills a space surrounding them.[3]

Other parts of this book will suggest that expanding the scientific, technological, and artistic attention given to other senses (touch, taste, and smell) will also open up new cultural issues.

The authors also note the privileging of "music" in Western thought about sound. Tendencies such as modernist autoreferentiality, the quest for nonobjectivity, the dynamics of musical rules, and music's mystical associations with the sublime are all mitigated against a culturally expansive approach to sound. Even the avant-garde traditions the book described—except possibly Surrealism—often framed their work as the expansion of music. As a result of its august tradition, Western music did not deal well with mass media, new technologies, and interactions with folk and non-Western traditions.

Only recently has a significant sound-arts movement of sound installation, audio, and radio experimental artists emerged. These artists have undertaken projects such as "the radicalization of sound/image relationships" and of "acoustics in architectural, environmental, or virtual space." Technological innovations and sound-art experimentation have always gone hand in hand. Kahn and Whitehead explain this relationship in part by teasing out the profound cultural implications of the new sound technologies.

For example, sound recording had a major effect on thought. It raised the spectre of technologizing the body and suggested access to previously unexplored regions, such as the afterlife and the unconscious. Radio similarly inspired the artistic imagination. It raised ideas of disembodiment—a sound could exist in two distant spaces at once. It referred to "the expanses of the ocean, to crowds, to other lands, and to the otherness of the unexplored globe."[4] Even more, it could shake up categories of thought, suggesting that the inflow of ideas in a nonlinear way from many sources at once. Innovations inspire artists and then they take it forward, both exploring its cultural meanings missed by technical practitioners and themselves inventing new technical extensions. It is a pattern often repeated.

Experiments in Sound Installation

Barry Schwartz

Barry Schwartz is engaged by the technological paraphernalia of industrial culture. He creates large sound installations with demonic machines, high voltage, mangled video monitors, chemical baths, and the like. The works convey a mixture of mesmerization and danger. Generally, he employs transducers to make sound and "probes the nexus and relationships among the electrical, mechanical, aural, visual, and theatrical systems." The "Robotronika" show description of Schwartz in the publicity for *Arcus Interruptus* provides some insight into his aesthetics and way of working:

Schwartz's work incorporates metal, mechanics, computer-controlled hardware, chemically reactive agents, high-voltage electricity, and live video feeds. Reclaimed technological refuse, electrified transit wires, telephone poles, an electrical tower removed from the landscape, and use of pre-existing structures are employed in the fabrication of his artwork. Past installations have featured electrically charged piano strings attached to high-voltage utility tower structures. During the installation and performance, Schwartz plays the strings, coaxing arcs of electricity to dance between his fingers. Creating an auto-electronic environment, Schwartz stands in fountains and waterfalls of nonconductive fluid, manipulating various mechanical devices.[5]

In describing one of Schwartz's "turntable" installations, one Web site calls Schwartz's work a "theater of shocks":

Equal parts death wish, weird science, and hot-wired art, Barry's theater of shocks and jolts gives brand new meaning to the word *turntable*—his version is a larger-than-life physical apparatus

Fig. 5.3.1. Barry Schwartz, *Turntable* installation. Industrial machinery is used in live performance.

which consists of a huge stainless steel disk mounted on a waist-high gimbal and amplified through STC via contact mics. Spinning this oversize metal plate like a large record, he utilizes an extended constructed "tone arm" that houses the stylus, a chunk of dry ice, which elicits low sobbing moans, squeals, and inhuman howls from the steel surface.[6]

In *Beam Gantry,* an installation performed as part of the European Media Arts Festival (EMAF), Schwartz constructed an "instrument" based on the structures used to suspend high-voltage wires all over the world. The work exploited both the iconic and functional qualities of the technologies. High-voltage piano wires drape down the sides with fluid pouring over them into a pool below. Video monitors, closed-circuit cameras, and concave dishes, suspended on cables, reflect sound and image from the event. Disembodied television sets sit sparking in the fluid. A performer sits on the structure generating sound, video, and mechanical interactions.[7]

In *I-Beam Music,* created in the 1995 Kampnagel (Internationale Kulturfabrik) in Hamburg, Germany, Schwartz collaborated with Nicolas Anatol Baginsky to create a large industrial-technology sound event with sophisticated sound analysis computer intelligence guiding its actions. The program describes the installation, which includes machines, computers, surveillance cameras, video projectors, water, high voltage, fire, and chemicals. A computer adds to the sound by recursively processing the sounds created by the mechanisms:

During the performance, the six-string instrument undertakes an automatic journey through an environment twenty-five meters in length. Similar to the functional principle of a car wash, the string instrument travels through different situations and is being played there in very different ways: mechanical fingers pick the strings, chemicals create tones, extreme heat and cold tune the instrument.[8]

Gordon Monahan

Gordon Monahan experiments with a variety of technologies for producing new kinds of instruments. His *Large Hot Pipe Organ* makes use of large pipes and propane systems, tuned by experimentation with pipe lengths, shapes, and ignition systems.

Several performances play with the Doppler effect which changes the perception of a sound source's frequency depending on its movement. His catapult performance offered

---

Gordon Monahan: ⟨http://www.tacheles.de/raven/artists/bastiaan/history.html⟩

Fig. 5.3.2. Gordon Monahan, *Speaker Swinging* performance. The Doppler effect changes the frequencies of the sound.

strangely modified sound as speakers were thrust out of catapults. His *Speaker Swinging* performances typically involve performers swinging speakers above their heads with the effect sometimes augmented by electronic lighting. Monahan also creates Aeolean harps in which sound is generated by moving air or water.

Ed Osborn

Ed Osborn creates kinetic sculptures and public art installations that explore a variety of sound technologies. Usually the works function metaphorically, using the sound to stimulate viewers to think about concepts of interest.

Swarm used the electronic control of fans to create a semblance of biological flocking. The fans seemed to engage in leading and following behaviors, just as though they were many animals: "Alluding to themes of instability and physical force, technological acuity

---

Ed Osborn: ⟨http://roving.net⟩

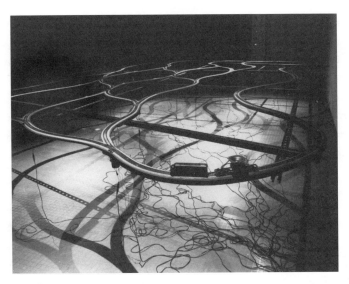

Fig. 5.3.3. Ed Osborn, *Parabolica*. An electric train with speaker attached creates a variety of moving sound events.

and hazard, and collective will, Swarm is an image of a living entity awakened and intoxicated by a sound of its own making."[9]

Skeletons investigates the coupling of audio and visual shadows and the physics of subaudible sound: "[T]hese works bring into hearing range the normally inaudible artifacts of physical movement, thus allowing a sonic depiction of a common but silent taxonomy of real motion in real space." In *Parabolica,* Osborn illustrates the tendency of artists to extend technologies in unexpected ways by extending the technology of model trains. He creates an installation in which the movement of the trains interacts with the sound sources they are "broadcasting," all relating to concepts of random distributions and chance:

Each time the train goes around the track, the switches that determine what route the train will take get reset randomly so that it takes a different one through this matrix of paths. . . . So over time the train is describing with its motion the form of the bell curve. . . . As the train does this it broadcasts sounds of people talking about making decisions and plans, describing confidence and certainty, and sounds of highly stressed mechanical systems.[10]

Osborn also creates sound-based public art installations and conceptual works. The *Bus Shelter* project, at the University of Washington, creates sound representations of the flow of information systems in this university campus environment by "putting an ear to the ground"—reading the flow of fluids in surrounding plant life and the flow of Internet data: "Together, these two areas illustrate twin systems of sustenance and provide listeners with an aural image of very different kinds of living forces. The contrast also serves to highlight the relationships between and the demands placed upon these two support systems."[11] *Audio Recordings of Great Works of Art* is a ten year project in which he recorded the ambient museum sounds that could be heard while standing in front of specific famous works of art.

Nigel Helyer

Nigel Helyer is a sound sculptor who often works with public spaces. He uses sound to reawaken ideas that get submerged in everyday living and in the advance of technology, such as disembodiment. He believes that sound sculpture is potentially very powerful because sound "flows between everything," and tries to use technology to reclaim public space. The digital world is isolating people even further from important levels of experience. He describes his method of working:

[M]y aim is to identify and scrutinize methods of defining, entering, and occupying public territories within urban space. Urban space in this context refuses the conventional representation of that homogenous continuum of spatial, economic, and ethical structures in which corporate and civic systems are ideologically interchangeable. Rather, urban space (be it the entire city or an individual built structure) is to be regarded as a vascular body rendered coherent by those flows of transactions, eddies of relationships, and the digestions of transmissions. . . .

Electronic communications are to a great extent divorced from our quotidian "metabolic" life. The digital world is shielded from chaos; the talking clock will naturally speak while the city is burning down. We have attempted to establish a technological immune system, buffered from external influence, scanned for infection, backed-up, passworded and air-conditioned. It is a vascular system which cannot and does not share the chaos and flux which we "enjoy" on the street. This lack of biological frailty forms the basis of the sinister and dehumanising aspect of the digital process, but by some token we are fascinated by a network which lacks density but which operates at infinite speed. Our speech now exists beyond metabolism.[12]

Nigel Helyer: ⟨http://www.arts.monash.edu.au/visarts/globe/issue7/nhtxt.html⟩

Fig. 5.3.4. Nigel Helyer, *Silent Forest*. A sound installation exploring the sounds of technology and war juxtaposed with sounds of Vietnam's forests.

One of his most known works, *Silent Forest,* explores the soundscape of the grand opera house in Hanoi as a locus where war, nature, and culture interact. He uses air raid sirens and digitally processed sounds of the Southeast Asian forest to create the event. The forest sounds of animals near extinction are rendered by old 78 rpm field recordings, themselves a reminder of technologies on the verge of extinction.

Physically the work consists of two multisource sound delivery systems—one modelled on early warning sirens, which act as an incongruous image to ironically link the technologies and soundscapes of Warfare with the technologies of Music culture. . . . This siren configuration acts as a physical and sonic leitmotiv, delivering the audio works which interrogate the politics of French colonialism (and subsequent North American neo-colonialism). The multiple, overlapping audiotexts juxtapose reconstructed fragments of Western art music which, by mimesis or concept, represent the "forest," with a series of "siren performances" to be finally overlaid by reconstructed fragments of traditional Vietnamese music. . . .

The arrow of time is resolute—the loss, silence, or absence evoked by such an inversion is manifold—in reality many of the animals and habitats preserved on these recordings no longer exist.[13]

Helyer explores the contradictions in broadcast and electronic sound reproduction. He uses the metaphors of silence and collaged sound to explore silences in larger cultural events. Technology serves multiple roles, supporting negative trends and allowing listeners to understand the trends:

The enigma of describing the tangible (phenomenal) through the organs of transmission is a central feature of the paradoxical logic that we accept on a daily basis (the telephone, radio, television, and telematics). My interest as an artist is to explore our habitual suspension of disbelief which these technologies demand, with research which parallels and inverts this enigma—to pursue the definition, locus, and function of silence and silencing is an exhumation of structure beneath the sedimentary layers of surface noise. . . .

At this juncture we have the opportunity to re-discover the original utopic possibilities of electricity, by re-immersing ourselves in the lost histories, the forgotten seminal points of an unformed discourse of transmission.

Trimpin

Trimpin creates sound installations that often use advanced technology to make sounds with conventional acoustical musical instruments. For example, he has devised methods for the machinelike activation of trombones, pianos, cymbals, and the like. He also does not restrict himself to conventional musical instruments; he created an installation to play composer Conlon Nancarrow's piano-roll composition via mallets striking one hundred hanging wooden shoes. Another installation called *Circumference* placed the audience in the middle of an array of plastic pitchers, galvanized pails, garbage can lids, circular-saw blades, dryer exhaust vents, along with more conventional percussion instruments. He has studied both music and mechanical engineering and was awarded a MacArthur "genius" award for his work, which was described by one newspaper account as "magical vision and technical ingenuity join forces in his work in ways that haunt, delight, and confound."

Trimpin: ⟨http://mitpress.mit.edu/e-journals/Leonardo/gallery/gallery291/trimpin.html⟩

Fig. 5.3.5. Trimpin, *Phffft*. An air-pulsated kinetic sound environment.

Trimpin also creates interactive works. A piece called *Phffft* activates two hundred acoustic tuned devices (reeds, flutes, pitched pipes, whistles, etc.) through viewer action. User motion and air currents activate a computer that activates the instruments. Trimpin is also interested in exploring the power of natural forces to activate sound. Some of his pieces use dripping water to activate drum heads. As part of SoundCulture96, he created an interactive fire organ for an Exploratorium exhibit on AIDS: "*FireOrgan* is a contemplative sanctuary of fire and sound, an extraordinary new instrument that uses fire, glass tubes, resonance, and thermodynamics to create an environment of deep, low tones activated by the presence of visitors and sometimes even fueled by their voices."[14]

Fig. 5.3.6. Bill Fontana, *Acoustical Visions of Venice.* Live sounds from twelve spots in the city are collected into one spot.

Bill Fontana

A pioneer of sound sculpture, Bill Fontana has created sound installations throughout the world that highlight aspects of the historical, cultural, architectural, and physical qualities of locations. He first started working with the simple technology of the field tape recorder and speakers. He then developed a series of innovative strategies for connecting the field situation with the viewer's space. The series of installations constitute a remarkable history of the use of technology to open new artistic possibilities. His paper "Resoundings" explains his approach and some of its history. He explains how recording became an intense compositional act:

I began my artistic career as a composer. What really began to interest me was not so much the music that I could write, but the states of mind I would experience when I felt musical enough

---

Bill Fontana: ⟨http://www.resoundings.org/⟩

to compose. In those moments, when I became musical, all the sounds around me also became musical.[15]

He describes experiences that were seminal in shaping his approach. In 1976, while he was recording a total eclipse of the sun in an Australian rain forest, he noted the radical changes in the forest as the eclipse proceeded:

During the minutes just before the moment of totality (having a duration of two minutes), the acoustic protocol between birds, determining who sang at the different times of day, became mixed up. All available species were singing at the same time during the minutes immediately proceeding totality, as the normal temporal clues given by light were obliterated by a rain forest suddenly filled with sparkling shadows. When totality suddenly brought total darkness, there was a deep silence.

This recording was seminal for my work because a total eclipse is always conceived of as being a visual experience, and such a compelling sonic result was indicative of how ignored the acoustic sensibility is in our normal experience of the world. From this moment on, my artistic mission consciously became the transformation and deconstruction of the visual with the aural.

Fontana's experience illustrates qualities often observed with high-tech artists. Scientistlike carefulness of observation and depth of analysis are integrated with a willingness to be amazed and a desire to communicate the experience. Fontana proceeds to explain how he came up with his trademark approach of linking the space of sound origins with the space of listening, overcoming the ephemeral qualities of sound:

One of the most useful methods has been to create installations that connect two separate physical environments through the medium of permanent listening. Microphones installed in one location transmit their resulting sound continuums to another location, where they can be permanently heard as a transparent overlay to visual space.

As these acoustic overlays create the illusion of permanence, they start to interact with the temporal aspects of the visual space. This will suspend the known identity of the site by animating it with evocations of past identities playing on the acoustic memory of the site, or by deconstructing the visual identity of the site by infusing it with a totally new acoustic identity that is strong enough to compete with its visual identity.

One project called *Perpetual Motion* illustrates the linking of technology, acoustical analysis, and cultural history that underlie the work. He proposes to create a sound sculpture of bells in Cologne's center that will be modified to pick up the resonant frequencies of the sounds that surround them—thus conceptually turning the bell inside out. Building on the natural tendencies of materials such as metal to resonate, he is attempting to fine tune the bells so they can "capture" sound instead of emanating it.

Other Artists and Projects

Unorthodox Instuments Previously associated with Survival Research Labs, ***Matt Heckert*** creates machines out of industrial materials, "big machinery" that create sound. Often he presents them as the "Automated Sound Orchestra." Samples include: *Rotary Violin,* a set of twelve piano strings on two revolving hoops, and *Big Boxer,* a motor-driven dome weighing about 240 pounds. *Weather Station* used 310 piezo speakers to create sounds based on outdoor light, temperature, and humidity. ***Peter Bosch and Simone Simons***'s *Krachtgever* (*Invigorator*) presents an installation built of many wooden packing crates that generate sounds via electonically controlled vibrators and strikers located inside. ***Ron Kuivila*** creates his own instruments and installations that use electronics to disassemble sound. For example, *Fast Feet, Slow Smoke* works with slow-motion sound, and *Comparing Habits* uses ultrasonics to convert even the subtlest of motions, such as the passage of air currents through a room into sound. *Radial Arcs* coordinated ninety-six stun guns, and *Spark Armonica* converted extreme spark sounds, which cannot be recorded or reproduced, into audible signals. ***Ken Butler***'s sculptural instruments represent bricolage of found objects, consumer culture, and kinetic activation. Converting sub-audible sounds into perceptible events, ***Nicolai Carsten***'s *Frozen Water* uses a bass speaker to activate ripples in a container of water facing it and overlays other frequencies to still the ripples with wave impedance. ***Shawn Decker***'s sound installations activate everyday objects such as *The Night Sounds,* in which motors strike piano strings attached to variously moving buckets, and *Wire Fields,* in which a gallery was strung with 32 piano wires activated by small motors that were affected by visitor motion, other strings' vibration, and sounds from a nearby garden. ***Nick Collins***'s installations often read user actions to activate events—for example, *Table de Seance,* in which movements of the Ouija board are converted to ethereal MIDI sounds from the surroundings,

Matt Heckert: ⟨http://www.mattheckert.com/⟩
Peter Bosch and Simone Simons: ⟨http://prixars.orf.at/press/english/musikwin.htm⟩
Ron Kuivila: ⟨http://www.lovely.com/bios/kuivilla.html⟩
Spark Armonica: ⟨http://nttad.com/asci/archive/arttech/rkwork.html⟩
Nicolai Carsten: ⟨http://www.ntticc.or.jp/special/sound_art/carsten_e.html⟩
Shawn Decker: ⟨http://webspaces.artic.edu/~sdecke/⟩
Nick Collins: ⟨http://members.xoom.com/_XMCM/Nicollins/installations.html⟩

and *The Primrose Path,* in which walking on five paths activates five different sound effects used by the movie industry to represent walking.

Site-Specific and Architectural Installations ***Dan Senn*** builds sound-history-referencing installations that use technologies such as peizo transducers to activate spaces. His pendulum-based pieces range in size from eighteen inches to hundreds of square feet. One project called *Catacomb Memories,* suggesting a burial space, explored evocations of subaudio sound via unusual transducers. ***Robin Minard*** creates sound installations that integrate ambient sounds and attempt to heighten audience awareness of their surroundings. He often uses emerging technologies in their construction. *Neptun* completely surrounded a space with loudspeaker ribbon. *Still/Life* mounted piezo speakers on the floors and walls to create "plantlike" installations. ***Bill and Mary Buchen*** create "sonic architecture," in which architectural components are sonically activated. ***Minoru Sato***'s *Finding of a State of Light: Distribution of Luminous Intensity and Its Fluctuation* focuses on the lighting conditions in the gallery by making audible sounds produced by energy equivalent to them.

Other Technology-Exploring Sound Installation Artists Not Presented Elsewhere in the Book Gregory Whitehead, Kathy Kennedy, Christof Migone, Hank Bull, Tetsuo Kogawa, Dan Lander, Julia Loklev, Helen Thorington, Jean François Denis, Paul Panhysen, Terry Fox, Mineko Grimmer, Maryanne Amacher, Heri Dono, Ellen Fullman, Alvin Lucier, and *Martine Riches.*

Experimental Instruments — Bart Hopkins

Although not quite kinetic sound installations, the creation of experimental instruments offers a related example of technological experimentation. Many of the installations previously described could be conceived of as instrument construction. Some of this activity concentrates on extending state-of-the-art electronics and computers to generate new kinds of sounds. Some has little interest in this "high" technology; rather, the sound artists build instruments out of natural and man-made materials that have not been previously used for these purposes.

It would be a mistake, however, to dismiss this work as unrelated to science and technology just because it doesn't use electronics. Some of it exemplifies the key qualities

---

Dan Senn: ⟨http://www.nwrain.net/~newsense/⟩
Robin Minard: ⟨http://www.mhsg.ac.at/iem/bem/minard1.htm⟩
Bill and Mary Buchen: ⟨http://www.users.interport.net/~sonarc/maintext.html⟩
Minoru Sato: ⟨http://www.ntticc.or.jp/special/sound_art/ms_e.html⟩
Bart Hopkins: ⟨http://windworld.com/emi⟩

of observation and experimentation that inform high-tech artists. Similar to scientists, these instrument-making artists are paying attention to their world—noting materials, physical principles, and relationships missed by others. Like high-technology developers, they are willing to take risks and experiment to achieve their visions. The instrument makers provide a vigorous example of ways artists can be connected to the processes of research.

Bart Hopkins, an instrument maker who works with flame organs among other forms, has been editor of the journal *Experimental Musical Instruments,* which serves a worldwide forum for this activity. Excerpts of an article he wrote on "pyrophones" illustrates the sciencelike attitude of many in this community. He is amazed, enthused, and curious about his topic. He is eager to share his discoveries and to learn more from other researchers. He is interested in natural phenomena but also motivated by aesthetic aims. The article explains the processes of this art/science—reviewing practical and theoretical research, analyzing of the physics, speculating on experimental actions, and attempting manipulations. For example, the article presents a detailed analysis of how flame tones come about.[16]

Hopkins edited a book called *Gravikords, Whirlies, and Pyrophones: Experimental Musical Instruments,* which compiled some of this experimental work. A sample of the contents give a feel for the range of activity—the nature of experimentation and the variety of technologies investigated:

Michel Moglia—Fire Organ
Richard Waters—The Waterphone
Godfried-Willem Raes—Pneumaphones
Don Buchla—Thunder and Lightning
Leon Theremin—The Theremin
Oliver DiCicco—Möbius Instruments
Sarah Hopkins—Whirly Instruments
Wendy Mae Chambers—The Car Horn Organ
Brian Ransom—Ceramic Instruments

Summary: Research as Art

Kinetic sound installation has been an active area of integration of research and art. Many of the sound artists have had to undertake their own research to realize their goals. In doing so they demonstrate the poetry that can come from investigation of new technologies.

Notes

1.   D. Kahn and G. Whitehead, *Wireless Imagination* (Cambridge, Mass.: MIT Press, 1992), p. ii.

2.   Ibid., p. 1.

3.   Ibid., p. 4.

4.   Ibid., p. 21.

5.   "Robotronika" show, "Description of *Arcus Interruptus*," ⟨http://robot.t0.or.at/exhib/schwartz.htm⟩.

6.   "Recombinant" show, "Description of Turntable Installation," ⟨http://shell4.ba.best.com/~asphodel/recombinant/artists/schwartz.html⟩.

7.   European Media Arts Festival, "Description of *Beam Gantry*," ⟨http://www.emaf.de/1994/hochsp_e.html⟩.

8.   1995 Kampnagel Web site, "Description of *I-Beam Music*," ⟨http://www.foro-artistico.de/english/program/I-BM/INDEX.HTM⟩.

9.   E. Osborn, "Project Descriptions," ⟨http://www.sirius.com/~edosborn/artworks.html⟩.

10.   *Switch,* "Interview with Ed Osborn," ⟨http://switch.sjsu.edu/switch/sound/articles/interview_ed/Ed.html⟩.

11.   E. Osborn, "*Bus Shelter* Description," ⟨http://www.sirius.com/~edosborn/artworks.html⟩.

12.   N. Helyer, "Artist Statement," ⟨http://www-personal.usyd.edu.au/~nhelyer/⟩.

13.   N. Helyer, "*Silent Forest* Description," ⟨http://www-personal.usyd.edu.au/~nhelyer/⟩.

14.   Exploratorium, "Trimpin AIDS project," ⟨http://www.exploratorium.edu/AIDS/trimpin.state.html⟩.

15.   B. Fontana, "Resoundings," ⟨http://www.resoundings.org/Pages/Resoundings⟩.

16.   B. Hopkins, "Pyrophone," ⟨http://windworld.com/emi/pyrophone.htm⟩.

5.4

Robots

Artists have explored robotics from a variety of perspectives, including: theater and dance; autonomy; extreme performance, destruction and mayhem; social metaphor; extending robot motion and interfaces; and robot architecture. (Artistic experimentation with tele-operated robots is considered in chapter 6.3.) The art world's growing interest is manifested by show/conferences such as "Robotronika," in Vienna, and Japan's ICC show "Evolving with Robots," which explores the idea that robots are no longer passive slaves but rather intelligent sensing communicating entitites.

Robotic Theater and Robotic Dance

Barry Brian Werger—Ullanta Robot Theatre

The Ullanta performance group creates scripted dramatic presentations that include robots as performers. The robots incorporate autonomous behavior so that each performance is slightly different, being altered by the robots' view of each other and their environment. At the "Robotronika" show, Ullanta presented a play called *The Self-Made Man and the Moon.*

Margo K. Apostolos

Margo Apostolos is a dancer and professor of dance who has been interested for a long time in the aesthetics of robot motion. She has choreographed dance works, worked with industrial designers, and collaborated with researchers on a variety of projects. In her paper "Robot Choreography: Moving in a New Direction," she describes this interest:

Just as in dance the human body moves through space efficiently and artistically, just as a dancer performs in seemingly effortless movement, so may a robot. The graceful movement of the human form can provide a standard for the study of an aesthetic dimension of robotic movement.[1]

Her article analyzes historic definitions of dance and the philosophical questions in applying the term to robotic motions. She notes that choreographers such as Alvin Nikolais have been interested in "dehumanized" dance, which offers interesting conceptual crossovers. While robots are not as graceful as humans, there are many opportunities

---

"Robotronika": ⟨http://robot.t0.or.at/exhib/⟩
ICC, "Evolving with Robots": ⟨http://www.ntticc.or.jp/special/robot/index_e.html⟩
Ullanta Robot Theatre: ⟨http://www-robotics.usc.edu/%7Ebarry/ullanta/⟩
Margo K. Apostolos: ⟨http://www2.hmc.edu/~alves/auscult.html⟩

Fig. 5.4.1. Barry Brian Werger, *Ullanta Performance Robotics*. Robot theater in which robot actors enact plays.

for programming smooth motion that are typically not exploited. She also considers the unique motion capabilities of robots, such as 360-degree wrist motion, that can be explored. She shows how compositions can build on the rhythms of robotic motion and sound. In one later performance, a composer created a work called *Auscultations*, which amplified and modified the sounds of robotic joints and motions as the sounds for the dance. Apostolos's later research work involves working with NASA on space telerobotics and with the Annenberg Center for Communications on facial expressions and human-computer interactions.

Finally, she suggests that robotic dance is a unique area of inquiry in which scientists and artists can learn from each other. The description of her talk, entitled "Sensual Science: From Dance Performance to Space Telerobotics," explains this view:

Sensual Science is presented as a way of looking at the world which employs creative thinking and artistic expression. Scientific discovery and artistic creation progress in various ways, and integration of the two processes may result in exciting new discoveries. Robot choreography will serve as an exemplary case of a blend in an artistic—scientific integration.[2]

Other Artists and Projects

In *Robot Puppet,* presented at the "Robotronika" show, **Wolfgang Hilbert** attempted to update the historical form of puppetry. Pulling on cords allowed the viewer to control a computer-activated virtual robot as a metaphor for the hope humans have of controlling their technology. Periodically, however, the robot acted self-sufficiently and exited from the interaction. Swedish choreographer **Asa Unander-Scharin** collaborated with mathematician and systems developer **Magnus Lundin** to create *Orfeus Kagan,* a dance for an industrial robot that attempted to mimic human motion. The U.K. performance group *VOID* produced *2 Minutes of Bliss,* in which performers shared a labyrinthine space with moving robots that carried video cameras and projectors so that the images were mergers of humans and robots. **Paul Granjon and Zlab** build robots that they use in performance, including *Toutou,* a singing and swinging dog; *Robot Tamagotchi,* a fluffy robot that needs care; and *Robot Head,* a radio-controlled speaking robot head controlled by the user's eye motions.

Autonomy

One of the major cultural issues of robotics focuses on autonomy because it is often identified as a distinguishing characteristic of being human. Is it possible to create autonomous machines? How does robotic autonomy reflect on human autonomy? Artists working with artificial-life concepts, such as Ken Rinaldo and Yves Klein, also explore the autonomy of robots (see chapter 4.3).

Simon Penny

Simon Penny creates installations and robotics that explore a variety of issues in technoculture. As described in the chapter on artificial life (4.3) and considered below, his *Petit Mal* installation is a landmark attempt to create robots with genuine autonomy.

Prior to this work, Penny created a variety of kinetic installations that highlighted aspects of cultural life, such as the culture of surveillance and the ramifications of the "datasphere" (the flow of information through culture in electromagnetic and other forms, see chapter 7.7).

His paper "Embodied Cultural Agents: At the intersection of Art, Robotics, and Cognitive Science," presented at the AAAI Socially Intelligent Agents Symposium at MIT in 1997, explains part of his agenda with *Petit Mal* and his thoughts about some of the unique features of art research:

Wolfgang Hilbert, "Robotronika": ⟨http://robot.t0.or.at/exhib/⟩
Asa Unander-Scharin: ⟨http://www.speech.kth.se/unander.scharin⟩
VOID: ⟨http://www.voidp.demon.co.uk/exeter.htm⟩
Zlab: ⟨http://www.zprod.org/zLabPrez.html⟩
Simon Penny: ⟨http://www-art.cfa.cmu.edu/www-penny/index.html⟩

Fig. 5.4.2. Simon Penny, *Pride of Our Young Nation.* A robot cannon locates visitors and thrusts its barrel at them.

Central concerns are an holistic approach to the hardware/software duality, the construction of a seemingly sentient and social machine from minimal components, the generation of an agent interface utilising purely kinesthetic or somatosensory modes which "speak the language of the body" and bypasses textual, verbal, or iconic signs. General goals are exploration of the "aesthetics of behavior," of the cultural dimensions of autonomous agents and of emergent sociality amongst agents, virtual and embodied. The research emerges from artistic practice and is therefore concerned with subtle and evocative modes of communication rather than pragmatic goal-based functions. A notion of an ongoing conversation between system and user is desired over a (Pavlovian) stimulus and response model.[3]

Penny suggests that artists bring unique perspectives to technological research that can advance inquiry. For example, with *Petit Mal,* the minimal funding of the project, his interest as an artist in its performative aspects, his sensitivity to cultural representation and audience projection, and his idiosyncratic goals, which differed from conventional research agendas, all resulted in robots that opened up new areas of inquiry for researchers:

My goal has been to focus on the social and cultural aspects of the question "how much can be left out" by concentrating on the dynamics of projection and representation (I mean this latter in a visual and critical theory sense). The tool for this exploration was *Petit Mal,* an autonomous Robotic Artwork. *Petit Mal* constitutes an Embodied Cultural Agent. . . .

I would like to emphasize here that I am an artist and *Petit Mal* was conceived as an artwork constructed according to an artistic methodology. That means: I approached the project holistically, I made most of it: from metal fabrication to circuit board prototyping to pseudo code with my own hands. My formal training is in art, I am an amateur in fields of robotic engineering, artificial intelligence, and cognitive science. My knowledge is unsystematic, it has been acquired on the basis of need and interest. However, my outsider status has allowed me an external and interdisciplinary perspective on research in these fields. . . . The formulation "autonomous robotic artwork" marks out a territory quite novel with respect to traditional artistic endeavors as there is no canon of autonomous interactive esthetics. . . . I am particularly interested in interaction which takes place in the space of the body, in which kinesthetic intelligences, rather than "literary-imagistic" intelligences, play a major part.

Artistic goals and technological research interplayed in the development of the project. Penny believes that systematic unpredictability and "nonoptimization" are artistically fertile and scientifically provocative:

I wanted to avoid anthropomorphism, zoomorphism, or biomorphism. . . . I wanted to present the viewer with a phenomenon which was clearly sentient, while also being itself, a machine, not masquerading as a dog or a president.

The heart of the mechanical structure of the robot is a double pendulum, an inherently unpredictable mechanism. Emblematically, this mechanism stands for the generative principal that the machine, as a whole, is unpredictable, and a little "out of control." This is the logic behind the choice of name for the robot. In neurological terminology, a *Petit Mal* is an epileptic condition, a short lapse of consciousness. The humour of this notion originates in the way in which it is contrary to the conventional idea of "control" in robotics. . . .

My approach has been that the limitations and quirks of the mechanical structure and the sensors are not problems to be overcome, but generators of variety, possibly even of "personality." I believe that a significant amount of the "information" of which the behavior of the robot is constructed is inherent in the hardware, not in the code. . . . In this sense then, my device is "anti-optimised" in order to induce the maximum of personality. Nor is it a simple task to build a machine which malfunctions reliably, which teeters on the threshold between functioning and nonfunctioning. This is as exacting an engineering task as building a machine whose efficiency is maximised.

Penny proposes that the art world's concern with cultural narratives can be a useful set of ideas for robot and autonomous-agent research:

People immediately ascribe vastly complex motivations and understandings to the *Petit Mal*. . . . This observation emphasises the culturally situated nature of the interaction. The vast amount of what is construed to be the "knowledge of the robot" is in fact located in the cultural environment, is projected upon the robot by the viewer, and is in no way contained in the robot.

Such observations, I believe, have deep ramifications for the building of agents. I believe it is a fallacy to assume that the characteristics of an agent are in the code and are limited to what is explicitly described in the code. In fact, the opposite is much closer to the truth. Agents, like any other cultural product, inhere meaning only to the extent that they are understood by, or represent to, the viewer or user.

The project following *Petit Mal, Caucus,* attempts to explore robot social behavior. Penny investigates the use of linguistic communication to build up an artificial society of machines.

Nicolas Anatol Baginsky

Nicolas Baginsky makes robots that use chaos theory and sophisticated intelligence to interact with viewers. Building on techniques from neural net research, Baginsky creates robots that create complex image or sound events.

One installation, called the *Elizabeth Gardner Project,* used analysis of images taken by its video eyes to control its reactions. The *Narcissism Enterprise* installation did more complex analysis of its visual field. Using neural nets, it analyzed the current viewers' image and built composites with other previous viewers' images through patterns it extracted on the fly:

The system decides which images it will return to the visitors as their desired optical and acoustic mirror images. It reflects selected aspects of the optical and acoustic appearance back on the visitor. It does this in a sensuous and playful way. . . .

This subsystem finds, extracts, and recombines facial features (eyes, noses, mouths) from the incoming video images. . . . Thus, composite images are created that unite in a new face the

---

Nicolas Anatol Baginsky: 〈http://www.provi.de/~nab/n_roblst.htm〉

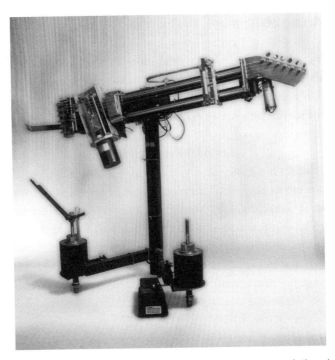

Fig. 5.4.3. Nicolas Anatol Baginsky, *Aglaopheme*. Robots learn by listening to their own and other robots' performances.

significant features of all those, for instance, with dark hair and a moustache. The classification of such "significant features" is not predetermined but are created by the system itself based on the perceived data.[4]

At the "Robotronika" show, Baginsky presented his "Aglaopheme" robot from the "Muses of the Other World" series. These robots use neural nets to learn to perform music. They listen to their own performances and that of other devices near them and try to extract rules. The system can invent "melodies" and improvise. It also develops pure "inhumane music." Human teachers can also interact with it to influence its learning:

The three sirens, "the Muses of the Other World," are partly robots, partly music instrument, and teach themselves to make music. In the assigned programs no knowledge of harmony and oscillation teachings is implemented. Instead, neural networks—coincidentally initialized—learn organizing and unsupervised melody improvisation and Instrumental Virtuosity.[5]

Extreme Performance, Destruction, Mayhem, and Control

For some artists, the military and industrial roots of robotics and machinery research are not neutral. They create artworks that use these technologies to reflect on these antecedents and history. For others, the machines offer opportunities for intense interactions that can offset the passivity of contemporary mediated culture.

Survival Research Labs

Survival Research Labs has a long history of creating events in which machines, teleoperated devices, and robots engage in extravaganzas of destruction and other kinds of extreme behaviors. It was founded by Mark Pauline in 1978, and since then has engaged hundreds of artists, engineers, and others in producing shows seen around the world. Elements have included everything from rocket motors and flame throwers to dead animal carcasses, the latest miniaturized electronics, and Internet control. The performances span the full range of what is called robotics. Many of the devices are radio-controlled by a human crew, while others are programmed for autonomous activities. Often there is an aura of danger for the SRL artists and the audience.

Typically, performances feature weird machines moving toward and away from each other, clobbering or shooting projectiles at each other, spewing flames, and making strange noises. One important goal is converting the technology from its original military or scientific purposes:

Since its inception, SRL has operated as an organization of creative technicians dedicated to redirecting the techniques, tools, and tenets of industry, science, and the military away from their typical manifestations in practicality, product, or warfare.

We build machines of a fairly large size—they are very extreme. . . . [T]hey are constructed by a basic plan which is the basic cry of physicists everywhere: you want to release the most energy in the shortest period of time.[6]

SRL's performances try to create a nonverbal narrative, a series of connected events that create settings for the machines and their activities. It is an art of spectacle that provides a "counter-narrative" to military research:

Survival Research Labs: ⟨www.srl.org⟩

Fig. 5.4.4. Mark Pauline and Survival Research Labs, *The Unexpected Destruction of Elaborately Engineered Artifacts.* Robots made from industrial waste engage in extreme behavior. SRL Video.

At SRL the lines [between art and technology] are very blurred. The kind of skills and ideas that go into the machines at SRL and the way that technology is portrayed is similar to the way that technology is portrayed in the schemes of the military. The similarities between SRL and the military's use of technology is that we're both trying to extract the most extreme performances out of the devices that we are dealing with, and trying to make a deep impression on people. In our case, we are trying to get an audience to sit still for an hour while trying to present a narrative production with machines as the actors. . . .

The reason you want to have a narrative sense to it is because the ultimate goal of the show is to make it look like it is a real world, like a habitat for these devices—that they belong here and this is what the machines do here. The aesthetics pretty much revolve around that.

SRL demonstrates the uneasy balance of fascination with technology and caution or aversion about its demonic potentialities. In most interviews, SRL tries to avoid specificity about its goals. Rather, they confront audiences with the disequilibrating experience of machines pushed in strange ways and leave the details to the audience and interpreters. Some see the shows as maniacal boys' infatuation with weapons; others see it as a preview of new kinds of mechanical beauty and ritual that may fill a world dominated by synthetic devices:

I think that the relationship that most people have with technology is very formal. In fact, most people have no relationship with technology except through their work: to make money at their jobs. At SRL, we try to take these kinds of things, and use those kinds of clichés and the way they are usually analyzed. We take them, pick them apart, and re-combine them into the images and ideas that we present at the shows. I think for some people it calls into question—reminds or even haunts them—of the things that connect with their day-to-day relationship with technology. . . . I'm accused of having all sorts of political stripes, from neo-Nazi to far left.

The shows are often banned by regulatory agencies. SRL considers it a political act to try to mount shows on its scale, which don't kowtow to safe, acceptable standards and take risks:

I mean, there's so much lame performance art that rich people are into. If the artist wants to get out of the ghetto, they have to be more traditional. My approach is more the opposite—I try to be more out of control. . . .

Even though it's difficult for us to get shows, we always eventually do them. The fact that people seek to interfere with us is only a measure of how threatening it is—which is a measure of how important it is. That's just the way it goes: it comes with the territory. I could obviously organize myself so that I didn't pose a threat. I would be able to get shows left and right and probably be rich and living in a nice house. But to me, that's not my role.

Pauline believes that artists must become much more proactive in working with technology. They need to avoid becoming infatuated with it for its own sake but still try to push it in ways beyond its normal uses. Only in this way can an artist provoke an audience to new understandings:

The fact of the matter is that if artists don't become conversant with technology then they will just be left out of the culture more than they are now. . . . I think that a lot of people who start getting into technology just to get into it for its own sake. You have to be very careful of that. But on the other hand, you can do stuff with technology that you can't do in any other way— and that's the only reason to use it. It's the whole thing that this society respects.

SRL sees technology as a primal force in human culture. Its performances attempt to put audiences in touch with those forces:

There is this book where Nietzsche basically expounded the idea that technology was the will to power—where we basically will ourselves to be our own gods. We remake ourselves as god, and that's part of technology. You can use it to create forces on a level that can't be explained within the historical realm of the power of individuals like atomic weapons and rockets—things that could not have been imagined as being the domain of humans. Basically, its about the harnessing of natural forces and re-doing them in a more useful image. I think that's what we do at SRL: that's part of the extreme angle that the machines are developed with.

Some see SRL's activities as illustrative of possible human-machine co-evolution. Manuel De Landa's book *War in the Age of Intelligent Machines* posits the idea, based on research on nonlinear systems, that military machines can be seen as a symbiosis between humans and machines. *Wired* magazine organized a trialouge between De Landa, Pauline, and Mark Dery as an interesting way to explore this idea. De Landa offers his view of SRL work:

What Mark does is push things far from equilibrium, to that point of unpredictability. From the videos I've seen of his performances, I gather that a lot of the experience has to do with the fact that you don't know when these machines are going to attack the audience; the question in everybody's mind is, "Hey, are these guys really in control?"[7]

Pauline describes the way his normal sense of his body fades into the background when he is working on one of the teleoperated machines. De Landa suggests that this connection is an example of the kind of machine-human symbiosis about which he writes. Pauline suggests that it is essential that humans entertain this kind of connection for future survival:

[T]hat, to me, is the mark of a true machine consciousness—when a mechanical system gets to a point where there's a disjunction between you and what's going on because what's going on is just too complicated or too intense. Systems are getting so complicated that they're out of control in a rational sense.

The role model for the future of human interaction with machines, if we want to avoid our own destruction and regain control, is to start thinking of our interaction with technology in terms of the intuitive, the irrational.

Fig. 5.4.5. *Robot Wars.* Two contestants joust for surpremacy.

"Robot Wars"—Mark Thorpe

Mark Thorpe, whose career included performance art and model building for Industrial Light and Magic (ILM), Lucasfilm's special effects house, organized a new kind of media event called Robot Wars. In these competitions, engineers and artists create robotic devices that do gladiator battle against each other. Elegance, strength, and cleverness are combined to create very strange entities, called sculptures by some and mechanical monsters by others. The "Robot Wars" Web site describes the events:

Robot Wars is a mechanical sporting event that features radio-controlled robots (and autonomous robots) in a contest of destruction and survival. As a unique blend of sport, theater, art, and engineering, this event draws contestants from all over the U.S. . . . some from overseas. Robots are limited by weight, not size. Heavyweights weigh up to 170 lbs. . . . 300 lbs. for legged robots. A specially designed arena with mechanical hazards, custom lighting, and techno music all add to the drama and excitement of this new sport.[8]

Thorpe traces his development of Robot Wars to his interest in "dangerous toys" and his other experiences in special effects, sculpture, and performance. He attempts to cultivate its wide appeal across particular social groups. Some observers criticize the emphasis on

Robot Wars: ⟨http://www.robotwars.com⟩

violence. They see Robot Wars as a glorification of destruction in a society already obsessed with violence. Thorpe believes it is a positive antidote to normal violence because of its emphasis on clean competition and its linkage of battle to creativity and innovation.

RW is so popular because of it's unique mix of art, technology, sport, and theater in a way that explores and celebrates basic life issues of survival and destruction without compromising human values—a rare combo in this age of dehumanization and political correctness. . . . RW is necessarily and wonderfully violent. But it is violent in a healthy way. . . . The main thing is that in Robot Wars no one is hurting anyone or being hurt, including animals.[9]

Eric Hobijn

Eric Hobijn produces computer-controlled events that often feature flamethrowers and other body-threatening arrangements. *Delusions of Self-Immolation* was a twelve-meter-long machine that audience members could subject themselves to. A European Media Arts Festival documentation describes *The Dante Organ* flamethrower performance:

Hobijn plays with the observer's secret wishes and hidden desires for spectacular images, his desire for kicks. Different than other media, however, where violence and danger is concealed from the viewer behind the television screen, a form of technology called Hot Hardware has been used here which, via an immediate intensity, removes any sense of distance.[10]

Seemen

Seemen is a loosely organized group of "postindustrial folk artists," engineers, and others who explore the potential of contemporary technology to create events that generate intense interaction for the audience. They see that many media and art forms create a passivity and remoteness from experience that can be counteracted. Dangerous machines and violence are often part of their agenda, but so is intense love. The actions of the robots "poetically symbolize man's struggles and triumphs."

An interview with Kal Spelletich explored some of the ideas underlying Seemen activities. The interviewer asked whether Spelletich sees it as part of the industrial dangerous art movement, such as SRL. He sees technology-mediated intensity as a possible antidote to cultural sloth:

Erik Hobijn: ⟨http://www.v2.nl/Organisatie/V2Text/PresentsE.html#Hobijn⟩
Seemen: ⟨http://www.seemen.org⟩

Fig. 5.4.6. Seemen. Robots built from industrial waste engage in acts of violence and love. Photography © Nicole Rosenthal.

The more people stay at home sitting in front of computers, playing video games and watching movies on video (staying passive audience members) instead of going out and getting real experiences (escapist vs. reality), the more they will be drawn to events that allow them to witness and experience their own mortality and humanity. We will always want to experience live things, To experience something live, not Memorex, Where something may not work as expected, where the outcome is different each time. People want real-life adventures. The more deprived you get the sicker you become. Thrill sports are booming because of this. . . .

I attempt both to give people a cathartic release and to challenge their preconceptions of what art and art content can be. The key is to keep the aesthetic image connected to a narrative or concept. Feasting, celebrating, bingeing, drug abuse, alcoholism, violence, sex, the human condition breaking down and pulling itself back up are all real-life things that can shape and change lives, like Carnival.[11]

The interviewer asked whether using military technology enables an audience to understand war better. It then asked if there wasn't a danger that using violent technology—for example, flamethrowers—just makes audiences more insensitive:

Fire needs context though, just like any other medium, be it paint, plaster, or film. No medium is the be-all end-all that instantly transforms into art. I have used fire as a landscape cleanser, metaphor for male ejaculate, as a river of fire to follow or cross, a waterfall of fire, as a salute to the day, as bad breath from Rush Limbaugh, trees armed with flamethrowers, underwater sea monsters spitting fire from watery depths, flaming robots fucking, fighting, and lovemaking, a bride carrying her own torch. . . . There is nothing wrong with being enraptured by fire. This was the original TV. You can stare at it for hours. It is as beautiful as a landscape with deer, a waterfall, flowers, and trees. But it is as retinal a trap as thick lush gobs of paint.

Seemen is active in the *Burning Man* "techno-pagan" festivals realized in the Nevada desert. They create technological installations that combine cyberpunk, industrial, and techno-experimental elements. Artists working with extreme technology span a wide range, from those focusing on violence to other kinds of intensity. Seemen straddles an interesting area where violence becomes life affirming. The publicity for an event called *Violent Machines Perform Acts of Love* challenges the audience to use machines as tools for engaging a variety of experiences:

Operate machines and robots performing sex acts, caring gestures, violent outbursts, dysfunctional behavior, superhuman effort, effusions of valor, loving strokes, visceral performances, fits of neurosis, fey gesticulations, nearfelt conduct, poetic eruptions, enigmatic follies, glutinous greed, resolving negative feelings, confrontation and resolution, outbursts of sincerity, proclamations of death, emanations of survival and feats of agility, all reckoning your most intimate interactions.

Other Artists and Projects

Christian Ristow, Chip Flynne, and Kimric Smythe (The People Hater Group) create carnival midway robots such as the *Drunken Master,* which has meathook teeth that "can mangle just about anything," and the *Randy Weaver* robot, a lifelike, gun-toting replica of the survivalist. The "orgies of mechanical destruction" are a "reaction to the wired society where we experience more and more of the world through TV screens and computer monitors. . . . Disneyland meets Faces of Death." Other groups exploring the potentials of kinetically activated military and industrial debris include the European groups ***Mutoid Waste Company*** and ***Spiral Tribe.***

The People Hater Group: ⟨http://www.discovery.com/area/technology/robotics/rob1.1.1.html⟩
Mutoid Waste Company: ⟨http://www.third-eye.org.uk/dc11/mutoid.htm⟩
Spiral Tribe: ⟨http://www.nwnet.co.uk/hulmecc/n-23/spiral.htm⟩

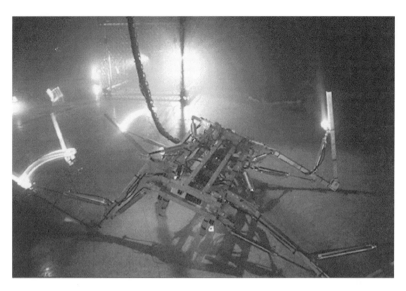

Fig. 5.4.7. Louis-Philippe Demers and Bill Vorn, *Court of Miracles*. A machine that moves by limping metaphorically suggests one part of the human condition.

Social Metaphors

Some artists attempt to exploit robots' dual status as artificial machine and vaguely anthropomorphic stand-in to comment on human society.

Louis-Philippe Demers and Bill Vorn

Louis-Philippe Demers and Bill Vorn create rich dramatic installations inhabited by robotic devices that can interact with humans. Their *Scavengers* installation attempts to enable visitors to directly experience a kind of machine life, society, and ecosystem. Here are extracts of the description from the Ars Electronica Web pages:

The installations are deployed in dark hazy spaces, in unusual architectural sites where the viewer is invited to consider an invented habitat created solely for the robot-organisms. . . . The robotic genders are designed on the basis of their behaviors in the habitat, and there are metaphors of natural societies: parasites, scavengers, overpopulation, flocks, etc. The machines are reduced to

Louis-Philippe Demers and Bill Vorn: ⟨http://www.billvorn.com⟩

their most nominal expression to implement their intended behaviors. A simple hammer machine becomes a rhythmic element at the same time and a parasite when installed judiciously on another robot-organism. . . .

Rituals, hierarchy, chaos, aggregation, the collective versus the individual are among the potential behaviors addressed by the installations. . . . The installations also convey a displacement of sensations, perceptions and expectations: duality, ambiguity, and contradiction are part of the sculptures. The aesthetics of the societies are continuously in conflict when they are not animated. . . . As they move and react, the initial perception is destroyed. What was first seen as the external inert perception, the known experience of the objects, is continuously transformed.[12]

Their *Court of Miracles,* presented at ISEA 97, illustrates their approach of presenting machine worlds that are not "clean," like conventional industrial robots. Here are excerpts from their Web page:

A Cour des Miracles (The Court of Miracles) is an interactive robotic installation . . . spaces that are Surrealistic immersive sites where viewers become both explorers and intruders.

By creating this universe of faked realities loaded with "pain" and "groan," the aim of this work is to induce empathy of the viewer towards these "characters," which are solely articulated metallic structures. . . . Six of these characters have been created to populate the installation: the Begging Machine, the Convulsive Machine, the Crawling Machine, the Harassing Machine, the Heretic Machine, the Limping Machine.[13]

A commentary on the installation by Norie Neumark for ABC National Radio gives some of the flavor of the work, which vibrates between a sixteenth-century tableau and an advanced cyber future scenario:

Although I'm not really a robot lover, there was something about these metallic skeletal pieces, caught in cages, chained to walls, freakishly dismembered, screaming and writhing their agony that engaged me despite my prejudice. The work played at an edge of human-machine that, thanks especially to the sounds—whispering, howling, groanlike—and to the pained distortions of the movements, evoked a disturbing border state that much of the cyborg-mania misses. These miraculous/horrific, simple and strange machine freaks expressed and evoked an alienation from the smooth high-tech control-desire of the computer world as well as suggesting the impossibility of escape.[14]

Other installations include *The Frenchman Lake, Escape Velocity,* and *The Trial.* In a paper entitled "Real Artificial Life as an Immersive Media," Demers and Vorn suggest

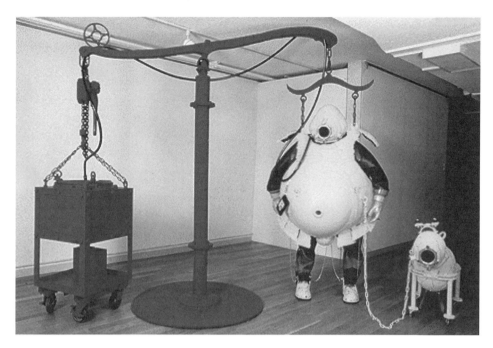

Fig. 5.4.8. Kenji Yanobe, *Yellow Suit.* Survival gear for unknown catastrophes.

that much interactive art is extremely limited in what it does—concentrating attention in a crippled way. Many installations are focused on direct reactive control waiting for commands rather than organically unfolding and evolving in genuine interactivity.[15]

Kenji Yanobe

Kenji Yanobe builds very elaborate robotic devices on the scale of small cars. He also builds robotic "suits" that resemble divers' suits. Many of the devices are large enough for a human to sit inside and operate. These robots play with the ambiguousness of humans' relationships to technology, referencing comic books, animation, and science fiction. Are the humans trapped or augmented? Often, works such as *Survival System Train,* containing cars for food production and water and air purification, seem to suggest preparation for some disaster. Marque Cornblatt's review of a 1997 Yanobe show presents some perspective on the work:

Yanobe's work, made from industrial salvage, is designed to serve as the artist's personal survival gear as well as tools for conquest and protection in the event of nuclear accident or other disaster.

Looking at this assortment of personal survival suits and protective vehicles made me keenly aware of the isolation caused by our technologically aggressive culture. This armor of radiation-proof lead and glass seems to suggest that in spite of the wonders of science and technology, we are still essentially alone in the world, perhaps even more so, except maybe for the dog.

Yanobe claims that attitudes toward technology in Japan are much more positive than in the West, noting, for example, that Manga comic-book robots are usually portrayed as cute and friendly:

Yanobe . . . believes that in the West, historically we were often guinea pigs for technology and industrial development. The inevitable disasters, nuclear accidents, and environmental damage have fueled a distrust of technology and the big business behind it. In Japan, Yanobe states, technology is often seen as a benevolent force helping the nation develop and expand. . . . The cultural view is a level of intimacy and comfort with technology.[16]

He believes that the Japanese engagement with robotics and futuristic scenarios is a unique cultural strength and wants to build on that in his art. But Yanobe's optimism is mixed with wariness. Technology can run amok. He was concerned with nuclear accidents and the possibilities of widespread disaster:

My generation doesn't know about the war, . . . We were just comfortable in life. I don't know real crime, real war. That's why I want to know what is reality. Even in Japan, I can't understand reality, what is atomic disaster. One day, I plan to go to Chernobyl. I would like to try to see what happens in the world.[17]

He also worries about *otaku* (a Japanese word meaning "obsession"). He sees a shadow side to the fascination with technology and science fiction—an inability to distinguish fantasy and reality. He hopes to create works that straddle both worlds: "My generation is one that went into fantasy and fell in love with characters and stuff like that and animation. I'm from the *otaku* generation, I can see the danger and confusion between fantasy and reality."

Norman White

Norman White was one of the pioneers in electronics- and telecommunications-based art. His robotic installations typically offer social metaphors for relationships between humans. His aesthetics emphasize the importance of behavior in addition to appearance,

Fig. 5.4.9. Norman White, *Helpless Robot*. A robot tries to get help from passersby.

and his exploration of non-art settings. In *Helpless Robot,* a robot tried to engage passing humans to help it get more comfortable. It asked viewers to think about beggars, hustlers, and others who ask passersby for help. As people helped, it became more demanding. Another work reflects on sexual relationships:

Them Fuckin' Robots (1988). Fellow artist Laura Kikauka and I each built an electro-mechanical sex machine (hers, female; mine, male). . . . We then brought these two machines together for a public performance. The male machine, the first and last anthropomorphic robot I've ever built, responds to the magnetic fields generated by the female organ, thereby increasing its rate of breathing and moving its limbs, simultaneously charging a capacitor to strobing "orgasm." The

Norman White: ⟨http://www.bmts.com/~normill/artpage.html⟩

female machine, on the other hand, is a diverse assemblage including a boiling kettle, a squirting oil pump, a twitching sewing machine treadle, and huge solenoid on a fur-covered board—all hanging from an old bedspring and energized by an electronic power sequencer.

Other Artists and Projects

Jim Whiting's *Unnatural Bodies* explores the metaphoric world of struggling humanity played out with robots. *Ars Electronica*'s description of *Unnatural Bodies* portrays maimed robot-machines, "cruel-aggressive, electronically controlled figure-ensembles, sensible robots, and desolate machines, who are moving at a horrible scenario. It is populated with legless creatures in cages who can't run away, degenerated twitching, snapping rumps. At the end the danse macabre compresses, heightens, spastically, poundingly, crookedly creeping or hanging, garishly into that cry."[18] *Mathew Sanderson* creates robotlike sculptures that reflect on beauty and struggles. He describes his sculptures as "fossils" of a process of transformation from the human state: "The crawling female figure is supposed to look vulnerable and possibly wounded, but yet still full of strength and courage, fighting her way forward towards shelter." *Ken Feingold*'s *OU* presents a robot half body similar to arcade fortune-telling robots that solicit money and offer to "answer" questions about the future. *Frank Garvey*'s *Omnicircus* provides an array of robots—beggars, hookers, junkies, thieves, and musicians—surrealistically commenting on the social world. *The Institute for Applied Autonomy* is a group of artists who develop "Contestational Robotics" and "Robotic Objectors," which attempt to "invert the traditional relationship between robots and authoritarian power structure" by developing low-cost robots that can be used by "culturally resistant forces" such as *Little Brother,* which gives out pamphlets, and *GraffitiWriter,* which can mark public spaces. *Adrianne Wortzel* creates "robotic pageants," such as *Sayonara Diorama* and *Nomad Is an Island,* which combine elements of installation art, theater, and Internet art. In *Globe Theatre* robots moved through the audience taking notes. The events are often broadcast over the Internet via CUSeeMe videoconferencing systems. She received an NSF grant to work in collaboration with engineering professors to explore the control of robotics over the Internet.

Jim Whiting: ⟨http://www.aec.at/fest/fest91e/whitin.html⟩
Mathew Sanderson: ⟨http://www.newwave.co.uk/lou2/description.html⟩
Ken Feingold: ⟨http://www.kenfeingold.com/⟩
Frank Garvey: ⟨http://www.omnicircus.com/⟩
The Institute for Applied Autonomy: ⟨http://www.appliedautonomy.com⟩
Adrianne Wortzel: ⟨http://artnetweb.com/wortzel/⟩

Fig. 5.4.10. Chico MacMurtie, *Amorphic Robot Works.* Panorama of a robot-musician performance installation. Photo: Brian Kane ⟨www.briankane.net⟩.

Extending Robot Motion and Interfaces

Chico MacMurtie

Chico MacMurtie and the Amorphous Robotics Works creates award-winning robotic acrobats and musicians. His installations include operas, dances, and musical forms, often orchestrated by computer MIDI control. His robots perform a great variety of sophisticated motion, pushing the boundaries of how robots can move and methods of linking controllers to the robots. Appreciation of movement and sound is a critical organizing principle for the work. MacMurtie, the director, and tour director Mark Ruch, describe their motivations:

The work is an ongoing endeavor to uncover the primacy of movement and sound. Each machine is inspired or influenced, both, by modern society, and what I physically experience and sense. The whole of this input informs my ideas and work. [MacMurtie]

. . . [A]s there is a beauty and elegance in movement itself, there is equally potent an experience in watching a machine (human or organic in form), struggling to stand, attempting to throw a rock, or playing a drum. These primal activities, when executed by machines, evoke a deep and sometimes emotional reaction. It is the universality of emotional experience which intrigues us, and it is the contrapuntal use of machinery as artistic medium and organic movement as form which, perhaps ironically, combine to provoke these emotional reactions most readily. [Ruch][19]

Work such as *Tumbling Man* and *Walking Trees* concentrated on experiments with motion and alternative forms of control. *Tumbling Man* was a person-shaped robot

---

Amorphous Robotics Works: ⟨http://cronos.net/~bk/amorphic/info.html⟩

controlled by sensors that were placed on two people's bodies and caused the robot to try to mimic their motions:

Different parts of each participant are, however, connected to the limbs of the robot. Therefore, for instance, it is possible for the arms of one participant to control the arms of the robot, while the legs of the robot are controlled by the legs of the other participant. In order to make the robot carry out the intended movements, the two participants must collaborate. As the work proceeds, the connections can be changed so that the participants must discover the new roles of the interaction. This work is an example of kinetic robotic sculpture which involves interactivity between both participants and the robot.[20]

In the 1998 show "Robotronika," in Vienna, Amorphous Robotics Works and MacMurtie presented *Telescoping Totem-pole,* which explored the merging of robotics with organic forms shaped from malleable rubber inflatables. The "totem pole" attempted to model internal body processes and "represent the human condition":

In various stages, images of viscera unfold from the fabric, heads and premature forms emerge, rising to a promontory position over the elevating form. The symbolic representation of the viscera, heads, and forms being birthed from an organic mass is a reflection of our own internal cycle. Like all things that grow, a recollection of the impulse to arise, and subsequently subside, is generated by an internal clock.[21]

Austin Robot Group

For many years the Austin (Texas) Robot Group, composed of engineers, artists, and hobbyists, have pursued a wide variety of robotic experiments, what they call "cultural robotics." Projects have included robot musicians, robot blimps, cross-country navigators, and humanoid robots: "[T]he Robot Group is busy creating a rattletrap hybrid of weird science and Outsider Art. . . . the thirty-member organization synergizes the artists' desire to make art that reflects the Information Society around them, and the engineers' urge to give vent to their creative impulses."[22]

Austin Robot Group associates David Santos and James Perez create robot blimps. These small devices can be tele-operated or move under autonomous control. Santos's and Perez's projects include experiments such as blimp acrobatics and blimp eye-view

---

Austin Robot Group: ⟨http://www.polycosmos.org/robotgrp/rghome.htm⟩

Fig. 5.4.11. Ulrike Gabriel and Otherspace, *Terrain_02*. Brain waves of two persons control the movement of solar-powered robots. Work in the collection of the NTT InterCommunication Center, Tokyo.

projects via wireless cameras. One of their projects was selected as the public art to be installed in the Austin, Texas, airport.[23]

Ulrike Gabriel

Ulrike Gabriel created a series of installations called "Terrain," in which viewers' brain waves affected the motion of a colony of small solar-powered robots. In *Terrain_01*, a single viewer controlled the environment. The installation consists of small mobile robots that have the rudimentary intelligence to pursue light and avoid each other. The viewer's brain wave affects the lights, illuminating the vehicles. These simple behaviors and brain-wave influence result in complex events. A 1993 Ars Electronica description explains:

In *Terrain_01*, a colony of autonomous beings moves within a circular world. On account of system-immanent nonlinear interactions, a complex structural formation develops with the continuous adaptation process of the individuals to the external environmental influences. The "living" movement pattern of the permanently active living beings lures the observer away to fantastic psychological interpretations of their behaviour.[24]

Gabriel worked with Bob O'Kane in creation of "Terrain." O'Kane was a technical specialist who had helped many of the artists at the Institute for New Media realize their works. An interview with O'Kane, connected to the display of "Terrain" in Japan, reveals cultural differences in the ways Western and Japanese viewers responded to the work:

In the West, people sit down and want to be in control of the installation. With the Asian ideas of meditation and relaxation, . . . the willingness to be removed from a situation seems much stronger here. It is generally easier for users here in Tokyo to make the system work, rather than in Frankfurt. There, people were really bothered by the idea that they are responsible for making it move. The viewers stand around the user, looking at him, expecting him to "hurry up and relax!" At the Ars Electronica, someone would sit down and it just wouldn't work. They would close their eyes, so they remove themselves from the situation. They relax, the robots start moving, they open their eyes to watch them, and then the lights would go off again, because they started to think about it, and the brain sensors picked up that activity. This would happen several times, because they wanted to be in control. The ultimate goal is to be able to watch the robots while they move, to "bear it with open eyes," as Gabriel describes it, to be rewarded, to observe a function of what you're doing.[25]

The next version, *Terrain_02,* used the comparison of two viewers' brain waves to orchestrate the control.[26]

Martin Spanjaard

Martin Spanjaard creates robotic talking balls named *Adelbrecht.* These unassuming balls periodically start rolling themselves about, knocking into things, and occasionally talking about life. They sense position, bumps, ambient sound level, touch, and low batteries. The ball also computes two emotional states—mood and lust. Various actions affect these states, for example, petting heightens lust but getting stuck decreases lust and increases anger. *Adelbrecht* was shown in the European Media Arts Festival and SIGGRAPH art shows. Spanjaard describes his interest in the human-machine boundary:

To describe this ball named *Adelbrecht* in the shortest possible terms: it's an anthropomorphised protozoa robot in the form of a ball forty cm in diameter. It, or let's say "he," talks about his life: rolling, bumping, people touching him, and so on, about the things happening in the life

Fig. 5.4.12. Martin Spanjaard, *Adelbrecht*. The inside of a robot ball that rolls about, sensing position, touch, and sound, and talks about life.

of a ball. He confronts us with the boundaries between Being and Machine, with the transition from It to Him. Because I have written his behaviour, I function as example and source of inspiration. He is therefore also a self-portrait. Finally, he is an actor, trying to interest us enough to follow him for a spell. . . . An outer loop program uses all this (plus a diversity of sensorial information) to generate speech and behaviour: understandable, meaningful, but not predictable.[27]

In future versions, Spannjaard proposes to explore more complex computing based on fuzzy logic and a kind of "dreaming" in which the ball would contemplate its record of interactions stored from the time it was "awake" and moving.

Stephen Wilson

I created an installation called *Demon Seed,* which explored human tendencies to project evil onto technologies not fully understood. Four small moving robot arms organized themselves into a "robot dance troupe." Audience members could use a squeeze interface to influence the troupe. The Web site documentation explains:

Four computer-controlled robots "danced" in front of digitized images of demons from various world cultures. The computer choreographed the movements of the robot dance troupe. The choreography created synchronized motions that were simultaneously graceful and frightening.

Stephen Wilson: ⟨http://userwww.sfsu.edu/~swilson/art/demon/demon.html⟩

Fig. 5.4.13. Stephen Wilson, *Demon Seed*. A robot dance troupe is influenced by viewers' actions with a squeeze rod.

Each robot was costumed with objects appropriate to the culture of the images it danced in front of. One of the robots periodically could be controlled by audience members via two velvet-covered squeeze rods. Sometimes the squeezer would be invited to speak into a microphone. The robot they were controlling would then move and speak with computer-transformed versions of their voice. The installation probed the human tendency to project our fears and demons into objects we don't understand and can't control. It also explored touch and body motion as new visceral channels with which we will be able to communicate with computers and robots.

Robot Architecture

Many decades ago Nicholas Negroponte developed the idea of flexible robotic architecture. He asked why our architectural structures should have fixed forms when evolving computer and robotic technology created the opportunity to make buildings that could be dynamically adapted to different functions, persons, and contexts. He founded the Architecture Machine Group at MIT, which eventually evolved into the Media Lab. His early work called *Seek* created a computer-mediated environment for gerbils in

Fig. 5.4.14. Ted Krueger, *Responsive Rods.* Studies for robotic architecture.

which a robotic arm dynamically tried to arrange wooden blocks in a pattern based on the natural runs of the gerbils. Contemporary researchers continue this kind of inquiry.

Ted Krueger

Ted Krueger updates the ideas of flexible architecture to include ideas about new materials, artificial life, and sophisticated computer-mediated interactive intelligence. His paper "Like a Second Skin," presented at ISEA95, suggests that our present architectural models are unnecessarily timid:

Is this only a metaphoric operation? We make architecture out of the most inert and durable materials available: stone, glass, steel, reinforced concrete, or wood. We're unsure about plastics. To speak of a body, especially in the vicinity of epidermis, is to recall almost the opposite quality on every count—infirm, perishable, mutable, and frequently anti-inert. . . .

In this paper, I argue for the possibility of an intelligent and interactive architecture conceived of as a metadermis referencing recent work in the fields of mobile robotics, intelligent structures and skins, and interactive materials. These developments can serve as both a source of technical information and as a methodology by which architecture may develop qualities which are currently considered to be available only within the organic realm.[28]

He notes that much can be learned and adapted from other technologies. For example, fiber optics allow for the sensing of almost all aspects of life, such as position, orientation, rotation, pressure, strain, velocity, acceleration, and vibration. The shrinking of computers means that everything can be intelligently interactive—the idea of inertness may disappear. He suggests that current thought about "intelligent buildings" is primitive in its pursuit of optimization and efficiency. Organic life uses other methods to achieve robustness: "There is no necessary link between optimization and intelligence. It is possible that they are antagonistic. By inspection, one finds that many systems occurring in nature are not optimized but are rather a collection of apparently redundant and residual processes."

Later papers, "Autonomous Architecture," presented at the CAIIA conference Consciousness Reframed: Art and Consciousness in the Post-Biological Era, and "Symbiotic Architecture," presented at ISEA97, suggest that autonomy is a critical component for intelligent architecture. Enrichment of the variety of sensor inputs is useful for developing these kinds of autonomous agents. He extensively builds on the ideas of Rodney Brooks:

The availability of a wide range of sensing technologies suggests that the kind of awareness developed by architectural entities may be foreign to human experience. . . . The deepest levels of integration will arise from situations in which the representational systems are derived from the experiential.

The robot may, in fact, communicate with itself via the environment. The results of an activity are given by changes to the robot's context and may be directly perceived and made use of by other sensors. This is more efficient than passing the projected consequences of the action to a centralized comprehensive model and then verifying the model relative to the actual context.[29]

"Meterotic Architecture," presented at Consciousness Reframed II, argues that new "smart biomaterials" will enable artists and architects to embed distributed intelligence in the same materials that are used for construction.

Summary: Kinetics and Robots—Hybrids of Art and Science

Mechanics and electronics have allowed artists to create sculptures and installations that play in time. One could think of them as machine theater. Kinetic and sound artists augmented technology to create fantastic objects and spaces that were sensually interesting, comments on the scientific/technological nature of the world, and often explored other personal or cultural themes. In later years, much of the kinetics energy is focused on robotics.

Fascinating robotics research is progressing in scientific research institutes and artists' studios. The robot heritage stretches back to both sculpture and machines. Researchers seek to develop both the "bodies" and "minds" of these entities, and attempt to enhance the sophistication of the robots' abilities to move, sense, interact with humans, understand, plan, and act. Some work on models derived from animals and humans; others seek to create totally new kinds of behavior. Both seek to push the limits of human ingenuity and wonder about the implications of what they create. The artists' robots sometimes explore concepts of characterization and social metaphor.

Eduardo Kac and Marcel.li Antunez Roca, both artists who work with robotics and technological extensions of the body, drafted a statement after their participation in a robot art show called "Metamachines." They attempted to clarify what they saw as the unique qualities of robot art (autonomy and sophisticated intelligence) and to assert the need for changing art and cultural perspectives:

Expanding the narrow definition of robots in science, engineering, and industry, art robots make room for social criticism, personal concerns, and the free play of imagination and fantasy. Robots are objects that work in time and space. Their open and diverse spatio-temporal structures are capable of specific responses to differing stimuli. Some of the visual forms that robotic art can take include autonomous real-space agents, biomorphic automata, electronic prosthetics integrated with living organisms, and telerobots (including webots).

Robots are not only objects to be perceived by the public—as is the case with all other art forms—but are themselves capable of perceiving the public, responding according to the possibilities of their sensors. Robots display behavior. Robotic behavior can be mimetic, synthetic, or a combination of both. Simulating physical and temporal aspects of our existence, robots are capable of inventing new behaviors.

One of the crucial concerns of robotic art is the nature of a robot's behavior: Is it autonomous, semi-autonomous, responsive, interactive, adaptive, organic, adaptable, telepresential, or otherwise? The behavior of other agents with which robots may interact is also key to robotic art. The interplay that occurs between all involved in a given piece (robots, humans, etc.) defines the specific qualities of that piece. . . . Robots belong to a new category of objects and situations disruptive to the traditional taxonomy of art.[30]

Notes

1. M. Apostolos, "Robot Choreography," *Leonardo* 23(1990): 1.

2. M. Apostolos, "Sensual Science," ⟨http://www-leland.stanford.edu/class/ee380⟩.

3. S. Penny, "Embodied Cultural Agents," ⟨http://www-art.cfa.cmu.edu/www-penny/index.html⟩.

4. N. Baginsky, "Description of the *Narcissism* Project," ⟨http://www.provi.de/~nab/nab_int3.htm⟩.

5. N. Baginsky, "Description of Aglaopheme," ⟨http://robot.t0.or.at/exhib/baginsky.htm⟩.

6. M. Pauline, "Interview," ⟨http://www.conceptlab.com/interviews/pauline.html⟩.

7. M. Dery, "Out of Control: Trialogue," ⟨http://www.srl.org/interviews/out.of.control.html⟩.

8. M. Thorpe, "Description of Robot Wars," ⟨http://www.robotwars.com/rwi/⟩.

9. Ibid.

10. European Media Arts Festival, "Description of Hobijn Performance," ⟨http://www.emaf.de/english/perform.htm⟩.

11. K. Spelletich, "Interview for a Book Called *Transitions*," ⟨http://www.laughingsquid.com/seemen/⟩.

12. Ars Electronica, "*Scavengers* Description," ⟨http://193.170.192.5/prix/1996/E96auszI-scavengers.html⟩.

13. B. Vorn, "*Court of Miracles* Description," ⟨http://www.comm.uqam.ca/~vorn/chaos.html⟩.

14. N. Neumark, "Commentary on *Court of Miracles*," ⟨http://www.comm.uqam.ca/~vorn/CdM/CdMcritique1F.html⟩.

15. Paper presented at the Convergence: Fifth Biennial Symposium for Arts and Technology, in 1995. (Also available at ⟨http://www.comm.uqam.ca/~vorn/chaos.html⟩.)

16. M. Cornblatt, "Review of Yanobe Show," ⟨http://www.sfbg.com/AandE/31/25/Art/index.html⟩.

17. Yanobe, "Interview," ⟨http://www.giantrobot.com/issue8/kenji/index.html⟩.

18. Ars Electronica, "Description of *Unnatural Bodies*," ⟨http://www.aec.at/fest/fest91e/whitin.html⟩.

19. C. MacMurtie and M. Ruch, "Artist Statements," ⟨http://cronos.net/~bk/amorphic/info.html⟩.

20. C. MacMurtie, "*Tumbling Man* Description," ⟨http://cronos.net/~bk/amorphic/⟩.

21. C. MacMurtie, "*Telescoping Totem-pole* Description," ⟨http://cronos.net/~bk/amorphic/⟩.

22. Austin Robot Group, "Description of Organization," ⟨http://www.polycosmos.org/robotgrp/rghome.htm⟩.

23. Austin Robot Group, "Airport Proposal," ⟨http://www.polycosmos.org/alta/sculptur/flying.htm⟩.

24. Ars Electronica, "Description of 'Terrain'," ⟨http://www.aec.at/fest/fest93e/gabr.html⟩.

25. B. O'Kane, "Interview," ⟨http://race-server.race.u-tokyo.ac.jp/RACE/TGM/Texts/okane.html⟩.

26. Virtualistes Gallery, "Description of *Terrain_02*," ⟨http://www.babelweb.org/virtualistes/galerie/terrainf.html⟩.

27. M. Spanjaard, "Adelbrecht Description," ⟨http://www.xs4all.nl/~marspan/⟩.

28. T. Krueger, "Like a Second Skin," ⟨http://comp.uark.edu/~tkrueger/metadermis/meta.html⟩.

29. T. Krueger, "Autonomous Architecture," ⟨http://comp.uark.edu/~tkrueger/autonarch.html⟩.

30. E. Kac and M. Roca, "Metamachines," ⟨http://www.ekac.org/articles.html⟩.

6

Telecommunications

Telecommunications Research Agendas and Theoretical Reflections

Introduction: Overcoming Distance

Humans have long dreamed of travel, of visiting faraway places and peoples. They have been both curious and wary about "the Other." They have sought domination and cooperation. They have been explorers and nomads. As people dispersed, connections became important for work, empire, and affiliation. But travel and communication have also been arduous and/or time-consuming.

The last century and a half's technology have dramatically changed travel and communication. One set of technologies—transportation—has decreased the time necessary to physically move from place to place. Trains, cars, powerboats, mass transit, and aircraft have all emerged. The hope of almost instant teletransportation remains a long-term dream.

Another set of technologies—communication—has focused on the movement of messages rather than persons. Written messages have been exchanged. Distance communication via techniques such as drums, fire, and smoke have emerged. As the physics of electricity became clearer, changing electrical fields have been drafted as messengers, first via wires and then via electromagnetic waves through the air. The modalities of what could be transmitted has extended, moving from text to sound to still images to moving sound images. A variety of social structures of telecommunication has evolved, including one to many (e.g., mass media), one to one (e.g., telephone, E-mail), and many to many (e.g., conference calls). Telecommunications have been extended from not only connecting people to people but also to allowing people access to distant information sources and databases (e.g., libraries, bank accounts, and the World Wide Web). People may be eventually able to access all books, images, and other material in archives located anywhere in the world.

Researchers and entrepreneurs are feverishly working to extend telecommunications in almost every way imaginable. They seek to increase its speed, expand the modalities of what can be sent, free it from wires, and decrease its cost. They also are inventing new social forms built on telecommunications, for example, answering machines, call forwarding, caller ID, permanent personal telephone numbers, telemedicine, telecommuting, dispersed work groups, distance learning, video on demand, virtual communities, and phone sex lines.

The advent of telecommunications has affected much more than the practical details of communications. For example, philosophers note that it raises epistemological questions: What can one know about the existence of the entity at the other end of a telecommunication link? Psychologists and anthropologists study its impact on issues such as identity, competency, relationships, social interchange, national boundaries, privacy, and social control. For example, computer-mediated communications such as E-mail and virtual communities allow for unprecedented experimentation in gender and multiple identities.

Up until the last two decades, artists have been strangely inactive in explorations of telecommunication. In spite of the profound philosophical and practical impact of technologies such as the telephone, radio, and E-mail, the art world mostly ignored these developments. Starting in the 1970s, however, artists began to pay attention. With the advent of the World Wide Web, artistic interest has increased significantly. Artists have approached telecommunication in a variety of ways: exploring the opportunities and limits of the technologies and extending and deconstructing the social forms it creates. After briefly outlining the future trends in technology research and theoretical analysis of cyber culture, this section reviews contemporary artistic experimentation with telecommunication, including such areas as telephone-based art, wireless communication, telepresence, and Internet and World Wide Web investigations.

Other chapters review developments closely related to telecommunications. Since the computer has become integrated into communication, anything that enhances the computer's ability to sense a person's action (for example, touch or gesture, position, or movement detection) can become something to be sent. Similarly, any enhancements in the computer's ability to represent information (for example, 3-D graphics, 3-D sound, or force feedback) allows more complex representations to be received. Thus, virtual reality technology (discussed in chapter 7.3), which integrates sophisticated sensor and rendering capabilities to create an experience of being in some virtual place, can be a tool in telecommunications if it is set up to represent the telecommunicators' environments instead of fictional places. Similarly, since much telecommunication focuses on accessing information resources, information visualization and manipulation (see chapter 7.7) are relevant to this chapter.

Telecommunications Research and Development

This section briefly reviews current and future trends in telecommunications based technological research and development. It also identifies researchers' activities in creating new social forms based on these developments. These promise to have great cultural impact in the future and thus are candidates for artistic attention.

Bandwidth

Researchers are working to increase the speed with which computer-based information can be sent at all levels, from the international telecommunication backbones to the connections to an individual's computer. Using sophisticated new fiber optics and multiplexing software technologies, researchers have demonstrated multiple gigabit systems that can send data around the world as quickly as a local computer's CPU can now

Explore Tele-Immersion

Image courtesy of Samroeng Thongrong, Jason Leigh, Electronic Visualization Laboratory, University of Illinois at Chicago, 1999

Supercomputing 1998

Electronic Visualization Laboratory, University of Illinois at Chicago
www.evl.uic.edu/cavern

Fig. 6.1.1. Banner graphic for a Supercomputing 98 demo connecting international locations through a variety of CAVE-compatible applications developed with the CAVERNsoft Tele-Immersion library. Image courtesy of Samroeng Thongrong and Jason Leigh, Electronic Visualization Laboratory, University of Illinois at Chicago, 1998, ⟨http://www.ncsa.uiuc.edu: 80/EVL/docs/Welcome.html⟩.

communicate with its hard drive. Researchers are also attempting to break the modem 56K bottleneck using a variety of techniques, including cable modems (which send computer data over cable TV lines), xDSL ("digital subscriber line" technologies, which can send digital information over regular phone lines at speeds of up to 15mb), and direct satellite (which can download data at fast speeds). Bandwidth is essential infrastructure in enabling many new kinds of developments.

Wireless Communication, Mobile Computing, Location-Sensitive Communication, and Ubiquitous Computing

Many researchers envision a world where there is instant, inexpensive, wireless telecommunication access everywhere on the globe. Using low-cost, low-power cells to transfer calls automatically around urban areas, cellular technologies revolutionized the ability

Fig. 6.1.2. Motorola's Iridium. Graphic representation of a project that proposed to use low-orbit satellites to give wireless data and voice access anywhere on the globe. Copyright 1999 by Motorola, Inc. All rights reserved. Iridium is a registered trademark and servicemark of Iridium IP. LLC, ⟨http://motorola.com/GSS/SSTG/projects/iridium/overview.html⟩.

of many developing countries that could not afford to build a wire-based infrastructure, to institute phone service. Similarly, low-altitude satellite systems such as Iridium (ominously declared bankrupt as this book was in final production) and Teledesic, which require only simple, small antennas, promise worldwide communication everywhere, even outside of urban areas. PDAs and miniaturized portable computers promise to make full computer-mediated communication such as image-based communication available everywhere. Using global positioning system technologies, researchers propose location-sensitive communication that change depending on where the communicators are located. Ubiquitous computing can make every aspect of remote spaces alive with communication possibilities; for example, a French product called Le Flashing sends coded emanations that indicate one's sexual preferences and flashes when similar people are nearby. Possible social implications include more erosion of the meaning of national boundaries, the entry of third world cultures into international discourse, and changing expectations about continual access and availability to persons and information.

Fig. 6.1.3. Thomas Zimmerman, *IBM's Personal Area Network*. A system that uses body voltages to transmit and communicate data by touching. The system can also power itself by capacitance generated by walking.

The Electromagnetic Spectrum

The electromagnetic spectrum extends from very low frequencies up through radio frequencies (AM, FM, and short wave), television frequencies (VHF and UHF), light (infrared, visible, and ultraviolet), to microwaves and beyond. Each frequency/wavelength has its unique qualities of propagation, human perceptibility, penetration, and the like. Researchers are investigating the telecommunications potentials throughout the spectrum.

Desktop Video, Video on Demand, Voice over Internet, and Convergence

As bandwidth increases, video and voice will become part of normal computer communications. Desktop video conferencing will make the long term fantasy about picture phones a reality. But the extension of normal telecommunications to include

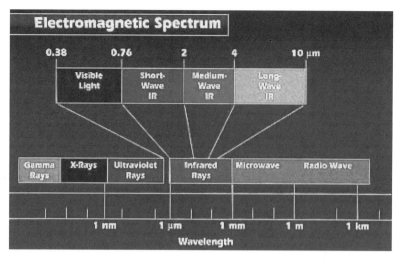

Fig. 6.1.4. Diagram of the electromagnetic spectrum (from the Blasdel Web site ⟨http://www.gpbx.net/greensburg/blasdel/charta.html⟩).

video in addition to voice will raise cultural questions about presentation and mores. Video on demand may have major impact on social structures such as broadcasting and video rental stores. Primarily text-based information archives may be increasingly expected to include video information. The convergence of telephone, television, and the computer will disrupt historical distinctions between computing, communication, and mass media.

Computer Telephone Integration (CTI), Enhanced Telephone Services, Unified Messaging, Call Centers, and Speech Recognition

Considering voice as just another data type, researchers seek to apply the full intelligence capabilities of computing to the manipulation of telephone communication. Already, services such as answering centers, call forwarding, and caller ID are available. Future services include intelligent answering and forwarding services (which will offer different messages and forwarding options to different callers), permanent person-identified telephone numbers (which will stay with a person throughout their life and attempt to track down the person in order to complete a call), unified messaging (which will interconvert fax, E-mail, and voice), enhanced caller ID (which pops up all information the system has on a caller), and intelligent call centers and help systems (which will use voice recognition and artificial intelligence to attempt to answer questions and, when it can't respond, will route calls to human assistants the system judges appropriate to the question).

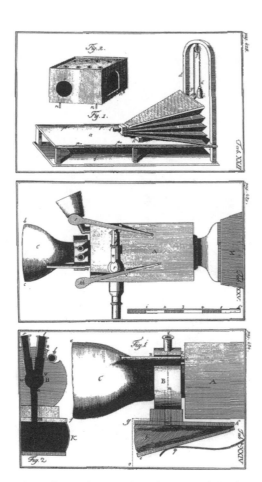

Fig. 6.1.5. Wolfgang von Kempelen's talking machine. A mechanical contrivance designed to reproduce speech sounds. Illustration from his book *Mechanismus der menschlichen Sprache nebst Beschreibung einer sprechenden Maschine* (1791). Illustrates the long history of speech research (from Hartmut Traunmüller's Web site, ⟨http://130.237.171.100/STAFF/hartmut/kemplne.htm⟩).

New social forms have emerged, such as anonymous chat lines, 976 pay-call systems, and international call centers (which forward calls based on international time differences). Researchers are investigating new CTI possibilities that will integrate video, sound, and text into a data unity.

Computer-Supported Group Work, Virtual Communities, and Collaboration

New communication technologies will enable people to work and meet together in unprecedented ways. Voice and video conferences can link people scattered all over the

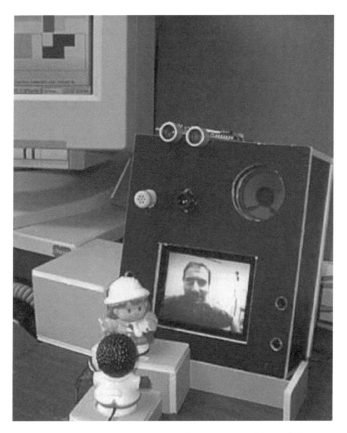

Fig. 6.1.6. Human-Computer Interaction and Computer-Supported Cooperative Work Research Group, University of Calgary (Canada) and University of Tsukuba (Japan). Mediating awareness, communication and privacy through digital but physical surrogates, ⟨http://www.cpsc.ucalgary.ca/grouplab/⟩.

globe into single, shared communication spaces. Linked whiteboards can transmit writing on the board to similar devices in other places and create live, shared work spaces. Background information and video records of past meetings can be instantly available as part of the current communication. Researchers are working to enhance computer-based virtual communities by techniques such as avatar representation systems, 3-D virtual graphic worlds, and new social forms governing cyberspace inhabitation. The MIT Media Lab's Sociable Media Group investigates issues concerning identity and society in the networked world, addressing such questions as: How do we perceive other people on-line? What does a virtual crowd look like? How do social conventions develop in the networked world?

Fig. 6.1.7. Privacy.org. Big Brother Award, ironically given each year for the most intrusive use of technology for surveillance or data consolidation, ⟨http://www.bigbrotherawards.org/⟩.

Privacy, Centralized Data Systems, Surveillance, Security, and Verification

Routine data telecommunications raises the danger of too easy accumulation of information on individuals. Easy video communication allows the possibility of remote video surveillance, and already some police departments are developing remote monitoring systems for public areas. Responding to the ease with which advanced technologies can snoop, researchers are developing encryption systems. The reliance on computer communication raises issues of verification of persons and messages. Digital certificate systems has been one response. Other approaches attempt to verify telecommunication messages by required ancillary physical verification systems such as smart cards and biosensors (e.g., fingerprint, voice, or retinal vein patterns).

Telepresence and Enhanced Telesensing

Researchers are attempting to enable individuals to functionally control devices at a distance, for example, telesurgery, tele-operation of robots, and the remote exploration and manipulation of dangerous or inaccessible environments such as other planets, the depths of the sea, crime sites, and toxic areas. To accomplish this telepresence, a controller needs enhanced ability to sense the remote location (perhaps beyond normal video access) and subtlety of control of devices. Researchers are developing systems with techniques such as force feedback (which allows remote sense of touch), 3-D video with

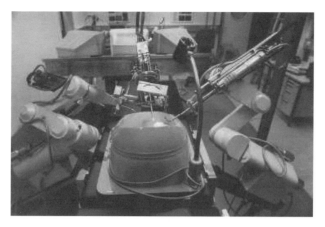

Fig. 6.1.8. ARTEMIS (Advanced Robot and Telemanipulator System for Minimal Invasive Surgery) experimental telesurgery system. Photo courtesy of Forschungszentrum Karlsruhe, (http://www.iai.fzk.de/~artemis/arbeitssys/arbeitssysengl.html).

multiple views, and VR-like control systems. Others invent new forms based on telepresence, such as virtual travel (which will allow a person to "rent" a robot for a day to tour exotic locations, such as Tahiti).

World Wide Web Experimentation

The World Wide Web has created unprecedented environments for communication and access to distant information. Researchers are working hard to probe the possibilities of this kind of connectedness. We can only offer a few examples of this research here. Addressing the overwhelming vastness of information resources available, researchers have developed artificially intelligent agents to help individuals navigate and act. They have developed new forms of search aids and information visualization (discussed in chapter 7.7). They have created radically new forms of social organization, such as online stores, auctions, chats, galleries, and concert halls. They have created Web cameras that provide real-time images from thousands of locations. They have connected thousands of devices to the Internet, letting remote viewers control telescopes, microscopes, electric trains, vending machines, and the like. Researchers have created inexpensive embedded Internet chips so that all vehicles and appliances could become Internet nodes. They have invented new kinds of nonphysical money, such as E-cash. They have accumulated and amalgamated diverse information sources, such as allowing simultaneous images of the entire earth's weather. Building on the Web's infrastructure, they are creating new models of entertainment, government, advertising, research, education, commerce, medicine, and most other social institutions.

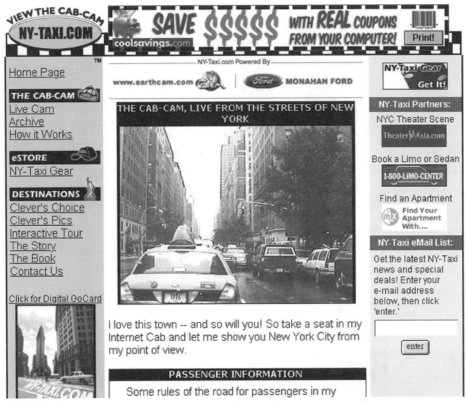

Fig. 6.1.9. The *N.Y. Taxi Cab-Cam*. A live Internet Web cam broadcasts from a New York City Internet lab's cab as it cruises the streets, ⟨http://www.ny-taxi.com/⟩.

Research Trends in Telepresence

Researchers are working to extend telepresence into many niches of life. This section briefly reviews some that may have significant impact and call for artistic response.

Ubiquitous Telepresence
The Ubiquitous Telepresence project, located at the University of Colorado, seeks to develop low-cost, easy-to-maintain telerobotic devices that can be put into operation

---

Ubiquitous Telepresence: ⟨http://www.cs.colorado.edu/~zorn/ut/index.html⟩

throughout the world, becoming "as ubiquitous as television sets." Once there are enough of them, viewers using the Internet as the access framework will be able to experience worldwide telepresence.

Remote Museums and Observatories

Museums are becoming expensive to operate and cannot reach many who would like to access the collections and shows. Similarly, scientific settings such as observatories are inaccessible. Analysts are beginning to explore the possibilities of telepresence for

Fig. 6.1.10. John Baruch, director. The University of Bradford's robotic telescope. Scientists and others can direct the telescope's direction of view from anywhere in the world via the World Wide Web, ⟨http://www.telescope.org/rti/intro.html⟩.

enlarging audience "access." For example, in "To Be There or Not be to Be There: Presence, Telepresence, and the Future of Museums," William J. Mitchell and Oliver B. R. Strimpel acknowledge that telepresence is less intense than physical presence but may be a good substitute:

[I]n our daily lives, we continually have to choose among different grades of presence with different properties and different associated costs. In other words, there is an emerging economy of presence. Within it we make choices among available alternatives and allocate resources to meet demands that are made on us and to achieve what we want.[1]

Unfortunately, it is not difficult to imagine a new class system developing in which access to physical presence will be highly related to economic position.

Remote Military Medicine

Developers are trying to create remote integrated telepresence medical systems that will constantly monitor the vital signs of soldiers in the field with instant alerts of woundings, and remote telesurgery if possible. R. M. Satava wrote "Virtual Reality and Telepresence for Military Medicine," which describes some of the future scenarios:

Through the use of remote telepresence surgery and tele consultation, virtual reality (VR) technology, global position sensing (GPS) technology, and identification of friend or foe (IFF), a framework for a digital medical battlefield can be constructed. Within this framework, the "digital physician" will bring his or her expertise to wounded soldiers using a computer workstation.

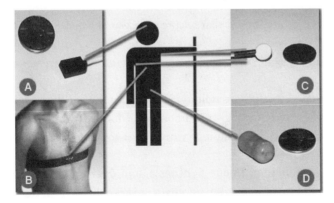

Fig. 6.1.11. Biopack (developed by MIT researchers). A medical telesensor, developed for an Everest mountain climbing project, that remotely measures pulse, blood oxygen, and internal body temperature via a pill sensor, ⟨http://www.everest.org/Everest/bio.html⟩.

Further, the digital physician will train the way he fights. "Virtual cadavers" will be used to practice surgery before operating on real patients.[2]

It is easy to see trends in this vision: our doctors can be always present with us—remotely knowing the precise states of our bodies, remotely able to apply tests and administer drugs, and ready to operate from the comfort of their remote offices if necessary.

Social Ecology and Telepresence

The Ontario Telepresence Project was a well-funded, multidisciplinary research project designed to understand the ways that telepresence fits into normal work situations. It brought together researchers from areas such as computer science, communication studies, interface design, psychology, and anthropology. They had access to high-end video-conferencing hardware and software that could be deployed in a variety of ways. They defined telepresence to be high bandwidth, almost ubiquitously present remote video (as opposed to physical sensors and actuators). They built innovative environments such as video mail, video answering machines, and unified messaging that included video.

Researcher Ron Riesenbach summarized the work in a report called "Lessismore." Although it was focused on work groups, the research is provocative for those in the arts trying to understand the possibilities and limitations of telepresence versus physical arrangements.

He is skeptical about claims that videoconferencing will revolutionize the work space. He notes that there are "more interesting things to do with video than to project real-time talking heads," and he warns that many people hold "the mistaken belief that quantity, size, improved technical performance, and speed, by themselves, make something better." The project found that telepresence was enhanced somewhat by superior technology, but the "social ecology" of the workplace was even more important. Even in highly mediated geographical dispersion, the long-term effectiveness of work groups often suffered.[3]

The project found that the serendipity, ubiquitousness, and the peripheral awareness of physical settings were missing in technology-mediated environments. It also found that intrusion on privacy in ubiquitous video presence systems also generated concern.

3-D Model Building, Image Synthesis, and Overlay for Telepresence

Researchers have been investigating automatic-image-enhancement systems that could enhance telepresence. Some integrate virtual worlds and live video; others take in a video stream and automatically generate 3-D models of the video environment. Once the

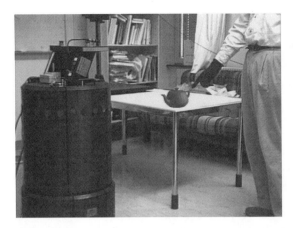

Fig. 6.1.12. Kristian T. Simsarian, Swedish Institute of Computer Science (SICS). Use of a shared virtual world and gestures such as pointing as communication between a robot and local and remote humans, ⟨http://www.sics.se/~kristian/⟩.

models are built a user can inspect the distance environment in new ways, for example, looking from points of view not represented by the original camera positions.[4]

Autonomous Flying Vehicles

Military research labs around the world are devoting signficant resources to the development of miniaturized airplanes and helicopters for purposes of surveillance and combat. Some developers have adopted them to entertainment applications, such as low-cost aerial cameras.

The Meaning of the Telecommunications "Revolution"

Almost all agree that telecommunications are radically changing our culture. Not all agree about the implications or the underlying meaning of the changes. Optimists note that telephone, radio, and TV communications have created a global village in which the diverse cultures of the world can see each other as resources in the arts, commerce, science, and most other endeavors. Even more, cyber futurists proclaim that computer-mediated communications open the doorway to an unprecedented future with increased opportunity, productivity, freedom, and access. Telecommunications are seen as increasing access to information and persons in accordance with need and interest rather than geographical accident. In *Being Digital,* Nicolas Negroponte notes that moving bits rather than atoms accelerates social and intellectual progress. Some have called the World Wide Web, with its almost universal access to publishing, the greatest invention since

Fig. 6.1.13. Dr. Jim Montgomery, University of Southern California Robotics Research Laboratory. Autonomous Vehicle Aerial Tracking and Retrieval (AVATAR) project. A flying robot uses sensors and loosely coupled computing modules to intelligently control its own flight, ⟨http://www-robotics.usc.edu/~avatar/⟩.

the printing press. For example, George Gilder, in *Microcosm,* expresses this optimism; furthermore, he asserts that technology will ease telecommunications to the point that bandwidth will essentially be free:

The central event of the twentieth century is the overthrow of matter. In technology, economics, and the politics of nations, wealth in the form of physical resources is steadily declining in value and significance. The powers of mind are everywhere ascendant over the brute force of things. This change marks a great historic divide. Dominating previous human history was the movement and manipulation of massive objects against friction and gravity. . . . Wealth and power came mainly to the possessor of material things or to the ruler of military forces capable of conquering the physical means of production: land, labor, and capital. Today, the ascendant nations and corporations are masters not of land and material resources but of ideas and technologies. Japan and other barren Asian islands have become the world's fastest-growing economies. Electronics is the world's fastest growing major industry. Computer software, a pure product of mind, is the chief source of added value in world commerce. The global network of telecommunications carries more valuable goods than all the world's super tankers. Today, wealth comes not to the rulers of slave labor but to the liberators of human creativity, not to the conquerors of land but to the emancipators of mind. Impelled by an accelerating surge of innovation, this trend will transform man's relations with nature in the twenty-first century. . . . Finally, the overthrow of matter will stultify all materialist philosophy and open new vistas of human imagination and moral revival.[5]

Ted Nelson's idea of every author's work universally available in hypertext and his coordinate vision of a way for authors to be automatically paid for their work (as expressed in his book *Computer Lib/Dream Machines*) illustrates an idea stressed repeatedly in this book. Nelson's idea was far ahead of his era's science and technology; he proposed it more as a creative vision than a scientific principle. Yet it inspired technologists and scientists who followed and it helped to shape their research. The developers of the World Wide Web noted that Nelson's work was critical in their formulation of the Web. Arthur C. Clarke's ideas about communication satellites, expressed in science fiction writing, had a similar effect on the later research and development of real satellite systems. These kinds of contributions of the arts to science and technology could be cultivated much more than they are.

In his book *Computer Lib/Dream Machines,* Ted (Theodor) Nelson, the man who coined the term *hypertext,* described this vision long before most had any experience with computer-mediated communication: "Now that we have all these wonderful devices, it should be the goal of society to put them in the service of truth and learning. . . . Obviously, putting man's entire heritage into a hypertext is going to take a while. But it can and should be done."[6]

Cyber culture is seen as offering new opportunities for increased democracy, expanded community, global access to intellectual and cultural resources, amplified consciousness, and freedom from the body and gender roles (as expressed by cyber feminists). Telecommunication allows for decentered, multi-vocal authorship more in tune with postmodern realities than conventional institutions. (See additional anlaysis of cyber culture in chapter 7.1.)

Critiques of Cyber Utopianism

Cyber pessimists and cyber realists are not so sure about the results of telecommunication developments. They wonder about what is given up in technologically mediated communication versus face-to-face communication. Are essential elements—for example, body language, smells, shifting perspectives, and aura—missing? They wonder about what happens to the community and public life when people meet in cyberspace instead of physical space. They wonder about the ease with which telecommunications can be monitored and controlled. And they fear the cultural imperialism and suppression of diversity that may hide behind a boundaryless cyberspace.

Even those who generally promote the opportunities of computer-mediated telecommunications warn of the dangers. For example, in *Virtual Communities,* Howard Rheingold identifies several possible dangers that constrain the democratizing possibilities of computer-mediated communications (CMC), including commodification, panoptical surveillance, and Baudrillardian confusion about reality:

[E]lectronic communications media already have preempted public discussions by turning more and more of the content of the media into advertisements for various commodities—a process these critics call commodification. Even the political process, according to this school of critics, has been turned into a commodity.

When people use the convenience of electronic communication or transaction, we leave invisible digital trails; now that technologies for tracking those trails are maturing, there is cause to worry. The spreading use of computer matching to piece together the digital trails we all leave in cyberspace is one indication of privacy problems to come. . . .

What these technologies support, in fact, is the same dissemination of power and control, but freed from the architectural constraints of Bentham's stone and brick prototype. On the basis of the "information revolution," not just the prison or factory, but the social totality, comes to function as the hierarchical and disciplinary panoptic machine.[7]

He also warns that communication technologies can promote a superficial concern with style, simulation, and surface. Some critical theorists might take issue with Rheingold's implicit faith in the ability to penetrate to the underlying authentic reality, but they would agree with him in questioning the unexamined hype of the telecommunications-ushered golden age:

These critics believe that information technologies have already changed what used to pass for reality into a slicked-up electronic simulation. . . . The television programs, movie stars, and theme parks work together to create global industry devoted to maintaining a web of illusion that grows more lifelike as more people buy into it and as technologies grow more powerful.

One good reason for paying attention to the claims of the hyperrealists is that the society they predicted decades ago bears a disturbingly closer resemblance to real life than do the forecasts of the rosier-visioned technological utopians. . . . "[T]he society of the spectacle" . . . offered a far less rosy and, as events have proved, more realistic portrayal of the way information technologies have changed social customs.

Artists working with telecommunications position themselves throughout this spectrum of thought. Some lean toward belief in the possibility of radical cultural change;

others assert that much of telecommunications hype is a smoke screen for old imperialistic tendencies. Others explore the new telecommunications environments while reserving judgment about their larger cultural meanings. Several art and cultural theorists give voice to this range of views.

For many postmodern critics, cyber culture needs to be problematicized and deconstructed. They resist the utopian rhetoric and instead try to understand the social practices underlying the new environments. They warn of what lurks behind the propaganda and call for resistance and analysis. These views are especially strong among theoreticians outside the United States. Essays such as "The Californian Ideology" decry the naiveté of Americans who are seen as more taken in by the hype, perhaps because much of the research and entrepreneurial energy behind cyber technology is of American origin and because of the heritage of New Age thinking.[8]

In an influential essay, "Utopian Promises—Net Realities," the Critical Art Ensemble (CAE) attempts to demystify some of these ideas. While acknowledging some useful and potentially humanitarian possibilities of the new technologies, such as increased access to vital information and increased opportunities for cross-cultural artistic collaboration, they suggest that the same misguided euphoria that greeted radio (such as Brecht's optimism) and portable video now attends the new technologies, while the "most significant use of the electronic apparatus is to keep order, to replicate dominant pancapitalist ideology, and to develop new markets."

They trace the origins of the Internet to the search for robust military command and control, its extension to academics as a way of fostering corporate and military research, and its evolution into a marketing, entertainment, and surveillance infrastructure:

Thus was born the most successful repressive apparatus of all time; and yet it was (and still is) successfully represented under the sign of liberation. What is even more frightening is that the corporation's best allies in maintaining the gleaming utopian surface of cyberspace are some of the very populations who should know better. Techno-utopianists have accepted the corporate hype, and are now disseminating it as the reality of the net. This regrettable alliance between the elite virtual class and new age cybernauts is structured around five key virtual promises. These are the promised social changes that seem as if they will occur at any moment, but never actually come into being.[9]

The five promises are: new body, convenience, community, democracy, and new consciousness. Since all the promises underlie much contemporary technology-oriented artistic practice, the critiques are worth understanding, and a sampling is presented here:

New Body   Cyberspace promises the freedom to reimagine the body and liberate it from organic constraints, such as biological sex. The CAE suggests that the shadow side of this abstraction of the body is the "data body," in which people become data accumulations useful for surveillance and marketing:

The virtual body is a body of great potential. On this body we can reinscribe ourselves using whatever coding system we desire. We can try on new body configurations. We can experiment with immortality by going places and doing things that would be impossible in the physical world. For the virtual body, nothing is fixed and everything is possible. . . .

What did this allegedly liberated body cost? Payment was taken in the form of a loss of individual sovereignty, . . . The data body is the total collection of files connected to an individual. . . . The second function of the data body is to give marketeers more accurate demographic information to design and create target populations.

Convenience   Corporations need to maximize the work they get out of workers. The CAE sees the promise of convenience as intricately tied up with this need:

The question then becomes how to make humans more like robots, or to update the discourse, more like cyborgs . . . People must be seduced into wanting to wear them [wearable computers], at least until the technology evolves that can be permanently fixed to their bodies. . . . The means of seduction? Convenience. Life will be so much easier if we only connect to the machine. . . . The seduction continues, persuading us that we should desire to carry our electronic extensions with us all the time . . . Now the sweatshop can go anywhere you do!

Community   The Net is promoted as place to create new communities. The CAE suggests that this use of the term does not respect the true sociological and psychological forces that generate real human communities, such as shared geography, history, or values. They see the mediation of community via networks as yet another instance of control:

That someone would want to stay in his or her home or office and reject human contact in favor of a textually mediated communication experience can only be a symptom of rising alienation, not a cure for it. . . . Why the marketing apparatus would desire such a situation is equally clear: the lonelier people get, the more they will have no choice but to turn to work and to consumption as a means of seeking pleasure.

Democracy The CAE asserts that while a platform for diverse voices is useful, real democracy requires effective autonomous action. It proposes several problems in the net vision of democracy:

1. Information is only as effective as the geographic localization of the body allows.
2. Institutional oppression is still unaffected by the net, which usually functions outside validated frameworks.
3. The transparency of net actions makes them available to authorities and more likely to control.
4. Net access is still unevenly distributed, with a great bulk of the world's people outside its purview.
5. The Net could promote increased passivity as well as action.

New Consciousness Analysts such as Roy Ascott (discussed in a following section) believe that the Net could be a tool for global spiritual awakening. The CAE sees this as ethnocentrism and class myopia, with large segments of the world's people uninvolved in the digital revolution. It sees the danger of a new kind of information imperialism:

There is a belief promoted by cyber gurus (Timothy Leary, Jaron Lanier, Roy Ascott, Richard Kriesche, Mark Pesci) [sic] that the net is the apparatus of a benign collective consciousness. It is the brain of the planet which transcends into mind through the activities of its users. It can function as a third eye or sixth sense for those who commune with this global coming together. . . . Even if we accept the good intentions and optimistic hopes of the new age cybernauts, how could anyone conclude that an apparatus emerging out military aggression and corporate predation could possibly function as a new form of terrestrial spiritual development?

Even accepting these kinds of analyses, some believe that meaningful action is possible. For example, in his essay "Net.Art, Machines, and Parasites," Andreas Broeckmann expressed his belief that artists can function as "parasites," disrupting the normal practices of the telecommunications and information networks:

The relationship between network art and parasitism was earlier suggested by Erik Hobijn, who introduced a concept for Techno-Parasites: "Parasites live and feed on other plants and animals. Techno-Parasites use whatever technical systems or apparatuses they can find as hosts, drawing on their output, their energy supplies, and cycles to procreate and grow. . . . Techno-Parasites suck other machines empty, disrupt their circuits, effect power cuts, disable them, destroy them." . . .

The parasite is a strategist and an ecologist; it knows its environment and, like a nomad, it is good at "passing through" and at conquering through movement, rather than at occupying, settling, and conquering by force. . . . The hypothesis put forward here is that the parasitological aesthetics described by Michel Serres is, at least in part, applicable to the net.artistic practice.[10]

Ars Electronica's 1998 "InfoWar" event (see chapter 7.7) expresses a related view. InfoWar's organizers propose that the Net is dominated by military and commercial forces using information networks as tools of domination. Resistance and analysis are critical activities for Net artists. The information society is not a utopian dream.[11]

Some postmodern critics might question the effectiveness of such subversion activities because they see trends that are deeper than political and economic domination.[12] For example, in their essay "Mapping the Multimedia Terrain of Postmodern Society," Jonathan S. Epstein and Patrick Lichty focus on Baudrillardian notions of the imploding meaning of the hyperreal. The Internet and Web just accelerate the already dizzying circulation of images and information into a space where authenticity and meaning almost disappear:

When formulating a methodology for speaking of a mediated information society of floating signifiers, shifting genders, and cultural panic sites in which the individual can no longer find social or cultural mooring, one would automatically begin with McLuhan's notion that the form of media dictates its content. . . .

The French sociologist Jean Baudrillard (1993) wrote of the implosion of reality into the hyperreal of the digital domain of simulations and simulacra. This, in some extent, has borne itself out in the creation of the Internet, with its varied social, discursive, and simulated physical (VRML) spaces, and in the emergent communications and multimedia technologies . . . the cultural spheres explode beyond their limits, infecting the rest of the culture until all of its aspects are now interrelated; the politics of sport, the aesthetics of sexuality, ad infinitum. Baudrillard writes: "Thus every category is subject to contamination, substitution is possible between any sphere and any other; there is a total confusion of types.

Throughout his writings, Baudrillard repeatedly describes the cultural simulacrum placed before us, a mise en scene of event-scenes with no apparent cause, a free-floating mediascape in which the endless reproduction of cultural forms speeds by us on computer and TV screens. . . . The problem becomes the multiplicity of these images, which are presented at such a rate that they can not be adequately deciphered individually.[13]

Paul Virilio focuses on the obsession of our culture with speed and notes that the acceleration of images rendered through telecommunications contributes to the hyperreality:

In postmodernity, motion, speed, and flux are deciding features as we implode our preconceptions of time and space with ever-increasing advances in digital communications technologies. Following from Virilio, we become telepresent bodies with no sense of near or far, with monitors for eyes that perceive the imagery of the mediascape approaching us with increasing intensity and speed.

"The techniques of rationality have ceaselessly distanced us from what we've taken as the advent of an objective world: the rapid tour, the accelerated transport of people, signs, or things, reproduce—by aggravating them—the effects of pyknolepsy, since they provoke a perpetually repeated hijacking of the subject from any spatial-temporal context". . . . that is, a continuum in which the media/discursive loci operate in intricate webs of interaction at high speeds of movement which collapse the space to the point into one where little depth in content can be introduced, sustained, or perceived when content is actually present at all. The user is presented with so much information, any depth of comprehension in its entirety is impossible. This is the realm of multimedia, the World Wide Web, and of postmodern culture in general.

The Exploration of New Possibilities

Freedom of the Digital Body

As described in the chapters on biology and the body (chapters 2.2 and 2.5), many artists are intrigued by the way cyberspace and telematic space introduce new freedoms of representation. In text-based cyberspace, anyone can present themselves in a variety of ways, choosing elements of identity such as gender or personal history. The traditional cues (and constraints) of appearance are missing so that fiction and reality can be melded in flexible ways. For example, one could choose to represent oneself as a gender different than that suggested by the physical body. One might wonder: Will the advent of video-based telecommunications reduce this freedom by reintroducing visual cues?

Telecommunications is an ideal environment for constructed identities to interact, for example in E-mail, chat, MUDs, and MOOs. As an introduction to these ideas, which guide much artistic work in telecommunications, this book can offer only a brief summary of the theoretical foment in this area. Many theorists, such as Sherry Turkle in *Life on the Screen,* and Sandy Stone note that the cyberspace reality is not as far from everyday physical reality as it may seem. Critical theory asserts that the idea of a coherent, unified, continuous self is an artifact of modernism. One's identity is constantly being constructed by the metanarratives that underlie one's life's activities. Especially, in the postmodern world, so dominated by mass media, advertising, and circulating images, identity is fluid. In the most radical version, every life is a fiction being written by each person and all those they come in contact with. As suggested by Roland Barthes, the authentic "author" is dead. In a paper entitled "Violation and Virtuality," Stone explains the value of CMC

(Computer Mediated Communication) environments, which she sees as useful to analysis because the narrow-bandwidth interactions are both "real and schematized":

If we consider the physical map of the body and our experience of inhabiting it as socially mediated, then it should not be difficult to imagine the next step in a progression toward the social—that is, to imagine the location of the self that inhabits the body as also socially mediated—not in the usual ways we think of subject construction in terms of position within a social field or of capacity to experience, but of the physical location of the subject, independent of the body within which theories of the body are accustomed to ground it, within a system of symbolic exchange, i.e., information technology.[14]

Telematic cyberspace can take on a compelling "reality" that engages participants with an intensity that matches physical reality. For example, for those involved in the famous "rape in cyberspace," in which one character in a MOO virtually raped another, created an experience that many claimed was similar to what would have been felt in a physical rape.[15] Humans have always lived both in the physical and the virtual world (e.g., storytelling and myth). Telecommunications allows persons to interact with each other while in the virtual state. (See chapter 7.1 for more analysis.)

Roy Ascott and New Consciousness

Roy Ascott, a well-known art theorist and artist, was one of the early pioneers in telecommunications-based art. Even before the general culture was paying attention, Ascott recognized the import of telecommunications and created installations such as the *La Plissure du Texte,* realized at the "Electra" show, organized in Paris by Frank Popper in 1983. In 1991, along with Carl Loeffler, he co-edited the special issue of *Leonardo,* "Connectivity," which focused on telecommunications. Ascott's editorial in that issue, "Connectivity: Art and Interactive Telecommunications," articulates some of his faith in the profundity of the changes he saw being wrought by telecommunications and his belief that artists are ideally suited to explore the new possibilities:

Interactive telecommunications . . . speaks a language of cooperation, creativity, and transformation. It is the technology not of monologue but of conversation. It feeds fecund open-endedness rather than an aesthetics of closure and completion. . . .

The new telematic systems of computerized communications are giving rise to a new, felt qualities of human presence, a fascination with presence, an eroticism of presence. Simply put, this is a quality of being both here, at this place, and also there, in many other places, at one and the same time. . . . This is a strange experience, new in the repertoire of human capabilities.

To meet others in dataspace, mind to mind, virtually, face to face, at no matter what geograpahical location, or in multiple, dispersed locations, in real time or in computer-mediated asynchronous time, is exhilarating. It is also demanding. . . .

Art is no longer seen as a linear affair, dealing in harmony, completion, resolution, closures composed and ordered finality. Instead it is open-ended, even fugitive, fleeting, tentative, virtual. Forming rather than formed, it celebrates process, embodies system, embraces chaos. The technology of these transformative systems fulfills a profound human desire: to transcend the limitations of body, time, and space; to escape language, to defeat metaphors of self and identity that alienate and isolate, that imprison mind in solipsistic systems. . . .

The fear that new technology will lead to a homogenized, uniform, lobotonized culture has been found to be entirely groundless.[16]

Since that time he has focused on the impact of digital and telecommunications networks on consciousness. CAIIA (Center for Advanced Inquiry into the Interactive Arts) which he directs, has organized several Consciousness Reframed conferences, which attempt to integrate the perspectives of artists, scientists, psychologists, anthropologists, philosophers, and non-Western sources such as shamans in investigations of the "noosphere," the shared global awareness.

For Ascott, the interlinked digital telecommunications networks have a meaning far beyond the quick sending of messages to distant places. They are the infrastructure that is promoting the evolution of consciousness. Artists are essential in exploring these changes. Art historian Edward Shanken analyzed Ascott's ideas that neworks could help build global spritual interchange, as expressed in his article "Is There Love in the Telematic Embrace?"

[T]he transformation he proposes is akin to the ideas of global consciousness expounded by French paleontologist and theologian Teilhard de Chardin, who theorized the "noosphere," and by futurologist Peter Russell, who theorized the "global brain." Ideas like these were invaluable to Ascott in his quest to imagine a parallel development through the visual languages of art.[17]

Shanken suggests that although Teilhard de Chardin's ideas of the "noosphere" and Russel's global mind meet with widespread skepticism in the scientific and the postmodern critical communities, they offer a provocative base on which to build art theory:

Despite their problems, the theories of the "noosphere" and the global brain provided Ascott with provocative models on which to build his own artistic theorization of telematic consciousness and the future. Criticisms of Teilhard and Russell apply only partially to Ascott's work, unconstrained as art is by the rational conventions of biology, neuroscience, and philosophy. As acultural system established simultaneously adjacent to and apart from other disciplinary conventions, art often makes use of systems of thought unacceptable anywhere else and at odds with convention. Art, for Ascott, functions as both theory and practice, and simultaneous as neither, as an entity unto itself drawing on the theory and practice of other disciplines for its own ends.

Ascott has been criticized for his utopianism. Shanken notes, for example, that it is not clear that participants in international, dispersed authorship events really experience anything like a global consciousness. Similarly, he suggests that the self-reflection fostered by the computer screen interrupts the unification of intuition and technology. Shanken explains that Ascott attempts to maintain a dynamically unstable unification of diverse theories:

His aim was to address technophobia in general, and in particular to answer critics of electronic art who feared that technology would overwhelm and dehumanize the arts, a last bastion of humanist values. If it could be shown that telematic art had the potential to embody love, then art could be electronic and serve humanist principles simultaneously. In constructing his argument, Ascott strategically opposed seemingly incompatible ontologies. . . . Indeed, the artist's resistance to conform to either an Enlightenment or a postmodern ontology is a particularly provocative aspect of his work. His insistence on maintaining paradox, on permitting and encouraging the simultaneous coexistence of logically incompatible systems is, I believe, one of his important achievements.

Ascott is aware of the critiques of technological optimism, such as the worldwide disparities of access, the dangers of surveillance, the commodification of thought, and the shallowness of circulating media. Nonetheless, he believes that art can help move against these:

Ascott resists the inevitability of such sober prophecies. He summons the social force of art against technology. Unencumbered by the destructive history of technology and the demands of rational epistemology, perhaps art—as the cultural convention charged with the embodiment and maintenance of the loftiest of human ideals (which includes the rigorous questioning of them)—can circumvent the scenario described above.

Summary: Telecommunications—The Grand Cyber Debate

As described in many places in this book, technological artists inevitably enter into the agora of these ideas—some prophesying new possibilities, others proclaiming decay, and others the dissolution of categories. Telematic artists cannot help but confront this debate, since telecommunications and CMC are seen to be at the heart of all scenarios: Is place and the physical body still important? Is cyberspace a new place or just a suburb of physical space?

Notes

1. W. Mitchell and O. Strimpel, "To Be There or Not Be to Be There," ⟨http://www.net.org/html/tele.html⟩.

2. R. Sataava, "Virtual Reality and Telepresence for Military Medicine," ⟨http://www.ncbi.nlm.nih.gov/htbin-post/Entrez/query?uid=9140589&form=6&db=m&Dopt=b⟩.

3. R. Risenbach, "Lessismore," ⟨http://www.videoconference.com/lessis.htm⟩.

4. KTH (Swedish Technical Insitute), "3-D Models for Telecommunication," ⟨http://media.it.kth.se/SONAH/ANALYSYS/race/pl4/rac_acts/htm/ac109-t.htm⟩.

5. G. Gilder, *Microcosm* (New York: Simond Schuster, 1989), p. 17.

6. T. Nelson, *Computer Lib/Dream Machines* (Redmond, Wash.: Microsoft Press, 1987), introduction.

7. H. Rheingold, *The Virtual Community,* ⟨http://www.rheingold.com/vc/book/10.html⟩.

8. R. Barbrook and A. Cameron, "The Californian Ideology," ⟨http://alamut.com/subj/idiologies/pessimism/califIdeo_I.html⟩.

9. Critical Art Ensemble, "Utopian Promises—Net Realities," ⟨http://mailer.fsu.edu/~sbarnes/lectures/ars.htm⟩.

10. A. Broeckmann, "Net.Art, Machines, and Parasites," ⟨http://miki.wroclaw.top.pl/wro97/anghtml/brekmtxt.htm⟩.

11. Ars Electronica, "'InfoWar' Description," ⟨http://web.aec.at/infowar/eng.html⟩.

12. For more, see S. Wilson, "Dark and Light Visions," ⟨http://userwww.sfsu.edu/~swilson⟩.

13. J. Epstein and P. Lichty, "Mapping the Multimedia Terrain of Postmodern Society," ⟨http://www.solisys.com/users/yasa/cybertheory.html⟩.

14. S. Stone, "Violation and Virtuality," ⟨http://actlab.rtf.utexas.edu/art_and_tech/stone_papers/violation-and-virtuality⟩.

15. J. Dibbell, "Rape in Cyberspace," ⟨http://www.dc.peachnet.edu/~mnunes/pres_95.html⟩.

16. R. Ascott, "Connectivity: Art and Interactive Telecommunications," in R. Ascott and C. Loeffler, eds., *Leonardo:* (special issue, "Connectivity") 24(1992):2, pp. 115–17.

17. E. Shanken, "Is There Love in the Telematic Embrace?" ⟨http://www-mitpress.mit.edu/e-journals/Leonardo/isast/articles/shanken.html⟩.

6.2

Telephone, Radio, and Net.Radio

A Brief History of Telematic Art

Visionary artists have been exploring telecommunications since the first primitive satellite communications systems were put in the sky and E-mail had just appeared on the scene. Although a full history of this activity is impossible here, it is useful to contextualize contemporary work by noting these precursors. This work demonstrates the artists' role in claiming technological territory early in its development. Also, some of it was so far ahead of its time that its implications have still not been fully realized. For more details on this work, interested readers should consult the special 1991 issue of *Leonardo*, "Connectivity," edited by Roy Ascott and Carl Loeffler,[1] and Frank Popper's book, *Art of the Electronic Age*.[2] Examples include:

- Kit Galloway's and Sherrie Rabinowitz's *Hole in Space* and *Electronic Cafe* (which links remote strangers via satellite)
- Douglas Davis's *Live Telecast* (a Documenta 6 linking of thirty cities)
- Liza Bear, Willoughby Sharp, Sharon Grace, and Carl Loeffler (ArtCom and Send/Receive Center fifteen-hour hookup)
- Tim Klinkowstein's *Levittown* (an international connection between a U.S. suburban and European McDonald's)
- Eric Gidney's *Telesky* (connecting Australia and MIT)
- Natan Karcymar's *Contact* (twenty-four telephone artworks)
- Fred Forest's *Kunstland* (an interactive video and telephone network installation)
- Roy Ascott's "La Plissure du texte" (an E-mail collaborative story in eleven cities)
- Norman White's *Hearsay* (a story passed and translated around the world)
- Robert Adrian's *Telephone Music* (linkage between musicians in various countries)
- Morton Subotnik (interactive net music)
- Don Foresta (collaborative multi-city painting)
- Stephen Barron's and Sylvia Hansmann's *Lines* (faxes sent to eight cities from a trek across the Greenwich meridian)
- Bruce Breland's *DAX (a digital art exchange)* (satellite exchange between multiple cities)
- Stephen Wilson's "Parade of Shame" (an interactive videotext event)
- Mit Mitropoulos's "Face to Face" (point-to-point cable TV)
- Judy Malloy's "Uncle Roger" (an on-line hypermedia story)
- Karen O'Rourke's *City Portraits* (a multi-city collaboration to create cities never seen)
- Chip Farmer's and Randy Morningstar's *Habitat* (a pioneering on-line graphic community)

Theoretical Perspectives on Telephone Art

The telephone is one of the oldest of the electricity-based telecommunications technologies. It has spread throughout the world, presents an extremely friendly user interface, is robust and reliable, and has profoundly influenced all aspects of life, from personal relationships to high finance. It has spawned an array of fascinating offspring technologies, such as answering machines and pay phones, and its underlying network infrastructure seems to invite investigation. As described earlier, artists have seemed mostly uninterested until recently.

Historical Reflections on the Cultural Meaning of the Telephone

This lack of attention is remarkable given the philosophical questions stimulated by the telephone. In "Aspects of the Aesthetics of Telecommunications," Eduardo Kac outlines some of these:

In some ways it [the telephone] was the cleanest way to reach the regime of any number of metaphysical certitudes. It destabilizes the identity of self and other, subject and thing; it abolishes the originariness of site; it undermines the authority of the Book and constantly menaces the existence of literature. It is itself unsure of its identity as object, thing, piece of equipment, perlocutionary intensity, or artwork . . .

The beginnings of telephony argued for the artistic merits of the telephone based on its capacity of transmitting sound over long distances, i.e., based on its resemblance to what we know today as radio. It would be possible, Bell and other pioneers hoped, to listen to operas, news, concerts, and plays over the phone.[3]

But many disliked the invasion of privacy introduced by incoming telephone calls. They reacted to the intrusion of the "destinal alarm," the unsettling demand for speech launched toward others. H. G. Wells longed for one-way telephones. Kac notes the phonocentrism of our culture:

[T]he telephone emphasizes the linearity of signs by splitting sound off from all other senses, by isolating the vocal element of communication from its natural congruity with the facial and the gestural. By cutting the audile out of its interrelation with the visual and the tactile, and by separating interlocutors from the speech community, the telephone abstracts communication processes and reinforces Western phonocentrism, now translated into an outreaching telephonocentrism. It is to destabilize this phonocentrism, and subsequently to contribute in undoing hierarchies and centralization of meaning, knowledge and experience, that theorists like Ronell and telecommunications artists invest their calls.

In her article "Circuits of the Voice: From Cosmology to Telephony," Frances Dyson similarly explores some of the metaphysical and anthropological implications of the telephone. She connects the underlying meanings of the telephone with older Dogon and Christian traditions that sought to understand manifestations of God's voice and the ritual and myth that surround its appearance and circulation. She traces that interest to other aspects of culture, including spiritualism (a medium's access to faraway voices), psychoanalysis (access to the voices of the unconscious), the gendering of those who have access to faraway voices (witches, mediums, and operators), and the interest of early technological researchers (e.g., Thomas Watson and Guglielmo Marconi) in using their technologies to access voices not of the normal physical plane. Avatal Ronel's book *The Telephone Book* is another source for this kind of analysis. The following excerpts offer a sample of Dyson's analysis, which draws parallels between the underlying meanings of telephone calls and older ideas about the circuit of communication of the holy spirit, which is the translation from speech to action and flesh:

It is to the telephone that the metaphor of the circuit is most readily applied, prompting certain connections with the 'transmissive' mappings of the voice and speech already described. The telephone shares with the 'elemental/theological' circuits of the Dogon and Christian cultures similar ideas and mechanisms of transference. . . . The medium in this case is electricity.

The medium is the life-force or 'nama'. Common to all three (telephone, Christian, and Dogon) systems is the belief that vocal transmission is primordially generative: creating dialogue which is not just restricted to speech, but which causes transformations within bodies, between bodies, between radically different spheres such as the heavens and the earth, and radically different forms of being—human and celestial. And these exchanges are themselves embedded within a symbolic system which endows each factor in the flow—voice, air, breath, movement, ear, and mouth—with multiple metaphoric values and relational possibilities. . . .

Yet the telephone's ambit is not purely communicational—by bringing the outside into the home and daytime into nighttime, by transmitting invisible voices from the electronic ether (from the heavens) at great speed, by delivering a "call," the telephone penetrates and transforms spatio-temporal, conceptual, and cultural barriers. It transmits the voice of the "other". . .

The eradication of distance between voices is also the raison d'être of the telephone; it is the attempt to install an anechoic vacuum, a space of no distance, an absolute space which bodies, being voluminous things, cannot occupy, but through which disembodied voices can travel.[4]

Artistic Lack of Interest in the Telephone

Given the symbolic and cultural richness of the telephone, it is strange that artists have not done more with the technology. Kac traces some isolated examples of artists' involvement with telephones, such as László Moholy-Nagy's experiments in using the telephone to transmit directions for fabricating enamel tile paintings. In Chicago's Museum of Contemporary Art's experimental 1969 "Art by Telephone" conceptual art exhibit, thirty-six artists were invited to communicate with museum staff by telephone to indicate their contribution to the show. Most artists just used the telephone to instruct staff to create objects or installations rather than explore the medium itself. Only Robert Huot developed a conceptual proposal that uniquely exploited the telephone context. Kac recounts that the telephone numbers of twenty-six men named Arthur were systematically chosen from the phone books of twenty-six cities (each last name starting with sequential letters of the alphabet), and visitors to the museum were invited to call the numbers and ask for "Art."

Kac suggests that artists did not create telephone art (and indeed all kinds of telecommunications art) until recently because of fear of abandoning historical attention to the visual and the reluctance to surrender the artist's control to a dynamic, participative process. The advent of conceptual and performance art helped prepare artists for this kind of work, although even in the open process of telecommunications events, many artists still clung to their privileged positions. The following section provides some brief samples of artists who investigate the telephone and its associated cultural context.

Examples of Telephone-Based Art

Disembodied Art Gallery

The Disembodied Art Gallery is a British group exploring conceptual and telecommunications-based art. Its members realized several telephone-based works. They emphasize the importance of broad participation and the use of everyday technology. Their Web site lists several of the principles underlying "Disembodied Art":

[We] employ communications technology to allow any individual to contribute to the exhibit or event, if they wish to do so. . . . [T]he public are encouraged to contribute to each event, from wherever they are in the world. The method of contribution may be by telephone, fax, or another means of communication—depending upon the nature of the particular event. . . . [We] develop an "appropriate technology" approach to electronic arts. We are interested in exploring the relationship between people and the everyday technology that they use to communicate with each other . . . in a manner which has much in common with the ideas of cultural jamming.[5]

Disembodied Art Gallery: ⟨http://www.disembody.demon.co.uk/home.html⟩

Fig. 6.2.1. Disembodied Art Gallery, *Temporary Line*. A telephone sculpture. Photo: Mark Monk-Terry.

The gallery has mounted several telephone and fax-based art events. Typically, the events conceptually focus on the cultural and psychological meanings of the technology used:

- *Disembodied Voices:* A telephone-accessible multi-room virtual space containing voices left by previous visitors.
- *The Answering Machine Solution CD:* A large collection of thirty-second tracks that may be used as answering machine messages (collected from sixty world artists).
- *Babble:* A telematic art installation that received over seventy voice contributions from the United States, Australia, Japan, and Europe. Callers telephoned a U.K. number and recorded their poetry, stories, and thoughts on an answering machine tape.

The Disembodied Art Gallery tries to explore the range of complexity associated with the technologies used. It tries to make the interfaces clear and stimulating for the general public. At the same time, the gallery installations attempt to subvert the technologies by reworking them from their standard configurations. Installations simultaneously explore "confrontational and unsettling qualities that new technology often possesses" and

TELEPHONE Nos.

0171 278 2207	....	0171 387 1736
0171 278 2208	....	0171 387 1756
0171 837 6028	....	0171 387 1823
0171 837 5193	....	0171 278 2179
0171 837 6417	....	0171 278 2163
0171 278 4290	....	0171 278 2083
0171 837 1034	....	0171 387 1362
0171 837 7959	....	0171 278 2017
0171 837 1644	....	0171 387 1569
0171 837 7234	....	0171 387 1526
0171 837 1481	....	0171 387 1587
0171 837 0867	....	0171 837 0298
0171 278 7259	....	0171 837 0399
0171 278 2502	....	0171 837 1768
0171 278 2501	....	0171 387 1398
0171 278 2275	....	0171 837 3758
0171 278 2217	....	0171 837 0933
0171 278 2260	....	0171 837 0499

Please do any combination of the following:

(1) call no./nos. and let the phone ring a short while and then hang up
(2) call these nos. in some kind of pattern
(the nos. are listed as a floor plan of the booth)
(3) call and have a chat with an expectant or unexpectant person
(4) go to Kings X station watch public reaction/answer the phones and chat
(5) do something different

Fig. 6.2.2. Heath Bunting, *Kings Cross Phone-In*. Text inviting callers to call banks of phones at a train station at synchronized times.

simultaneously the positive qualities—"i.e., their scope for bringing distant communities or individuals together."

Heath Bunting

Heath Bunting is an artist, associated with the Disembodied Art Gallery, who also explores telephone-based work. He is known for his global telephone directory of public phone booths placed on the Internet to stimulate telephone-based art. One of his most well-known events was the *Kings Cross Phone-in*. He gave his friends the numbers of public telephones at Kings Cross railway station, and they called these numbers at a fixed time and involved accidental passersby in lively on-line discussions right in the station hall. One commentator noted that "the incident was a resounding success; at six o'clock one August afternoon, the station was transformed into a massive techno-crowd dancing to the sound of ringing telephones."

Heath Bunting: ⟨http://www.backspace.org/hayvend/box.rand/86.html⟩

For the ICC in Tokyo, he created an event that explored unified messaging technology that could interlink E-mail, fax, and the Internet. The publicity describes the event: "You can send a message anywhere with this service. Your messages to individuals/groups via a variety of mediums are routed to the appropriate network bridge (telephone/fax/postal/road/water/railway/signboards, etc.). . . . The service is free. The results/repercussions of the message delivery attempt will be added to the Web site daily. . . . Anonymous service will be offered." The project is an attempt to "demystify electronic networks and mystify conventional networks."[6]

Culture jamming is one element of Bunting's aesthetic focus. Seeking to disrupt the smooth functioning of data systems, he hopes to raise awareness. An interviewer describes his approach:

Heath is dreaming up ways to sabotage other technologies like CCTV and marketing databases. But he is not going to go around smashing cameras, that's not his style: "By smashing cameras you only reinforce the system. You need to get people to begin to doubt the system. That's what I do—I create disbelief. The idea is to introduce bad data into such systems using techniques of illusion so that they cease to become trustworthy—optical illusions for cameras, inconsistencies and false identities for the databases."[7]

Ian Pollock and Janet Silk

Ian Pollock and Janet Silk have created telephone-based works for many years. Their work probes technologies such as voice mail and pay phones in the attempt to empower people and to make visible the history and culture of particular locations. Their Web-site statement explains their interests and documents several of their installations:

Focusing on the individual's experience, survival, and testimony, we are interested in their and our relationship to an environment and how time and history determine a course of action. Our work has been described as being reflective of the fragility of existence, preoccupied with memory, impermanence, and the desire to connect in spite of political, physical, or historic limitations.[8]

The Museum of the Future used voice mail to accumulate participants' texts about the future. A phone tree allowed callers to leave messages and to explore this unusual voice database. The documentation notes, "*The Museum of the Future* serves as a time machine

---

Ian Pollock and Janet Silk: ⟨http://www.sirius.com/~ps313/⟩

Fig. 6.2.3. Ian Pollock and Janet Silk, *Local 411*. View of urban area from which visitors could call special information numbers from pay phones to hear about former residents who had been exiled by urban renewal.

to search the past, present, and the future for visions of a future that inspire us with loathing or reverence." Listeners were invited to call the special number where "they can explore concepts of the future by accessing narratives, interviews, theories, information, and sound bites related to the shifting perceptions of the future." The telephone-constructed environment "offers suggestions as to why this and other cultures have strived to learn what the future holds. *The Museum of the Future* will show predictions of the past, hopes of the present, and why we can never know what tomorrow may be."

Area Code was a public art event that used pay phones to introduce levels of meaning into particular geographical places. "Participants pick up maps indicating locations of specific phone booths, and then call from these booths to hear stories, in the form of fictional letters." By linking the narratives with physical places, the artists tried to make "the history (and in some cases the future) of the site visible."

Local 411 used pay phones to reintroduce ideas about persons displaced in an urban renewal project that became home to museums, concert halls, and upscale hotels and restaurants. Visitors could call special numbers to hear stories of the displaced. The documentation explains the interest in displacement:

The phone is physical; it creates an intimate space that unites people over distance, even across time. The casualness of its power, its transparent movement across vast geographical locations, its ease at accessing people of different classes and cultures, make it an exciting media for the subject of public address. . . .

Here, four thousand former residents of residential hotels have been displaced to make room for what has been called the jewel in the crown that is San Francisco. . . . Gone are the Rock Hotel, the Rex, and many others. All traces of the former use of the area have been erased. . . . *Local 411* was a temporary public monument that challenged the erasure of memory from the site and questioned the position of the arts in the process of gentrification.[9]

Stephen Wilson

In *Is Anyone There?* I used the pay phone network both as subject and as tool to investigate the life of the city. This event also explored the cultural meaning of other aspects of techno-culture: artificial intelligence, database information networks, and the safety of art spectatorship. The documentation Web site explains the features of the installation and its conceptual investigations:

Overview   For one week a computer telemarketing device makes hourly calls to selected pay telephones, engages whoever answers in conversations about life in the city, and digitally stores the conversations. The installation later allows viewers to interactively explore the city via a database of these recorded calls and digital video of life near the phones. It appropriates the often intrusive computer-based telemarketing technology and uses it in a new way, involving people who don't traditionally participate in the art world in an event that probes the diversity of life in the city and the relation of truth to fiction.

Interactivity, Art Audiences, and the Safety of Art Spectatorship   This event challenges two common features of art viewing: the typically elite nature of high-culture consumption and the passivity of much art appreciation. All those on the street who answer the ringing pay phones—many who would be unlikely to attend any conventional art institutions—become participants in this art event. The drama of their dialogue with the computer system is an essential aesthetic focus. In addition, the event systematically questions the safety of passive art viewing by requiring viewers to generate strategies to search the images and sounds of the stored calls. More radically, the event periodically shifts the viewer in the gallery from the safety of spectator to the challenging position of full participant. It places live calls to the phone that the viewer had been vicariously experiencing, and demands that the viewer engage in a real conversation with a live stranger.[10]

_____

Stephen Wilson: ⟨http://userwww.sfsu.edu/~swilson/isany/isany.des.html⟩

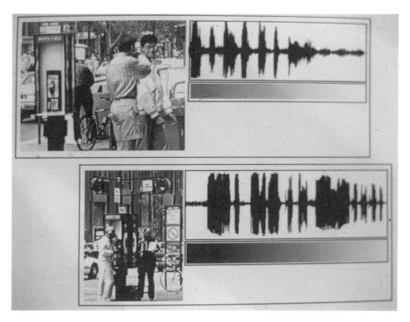

Fig. 6.2.4. Stephen Wilson, *Is Anyone There?* A computer automatically calls pay phones twenty-four hours a day for a week and tries to talk to people.

Other Artists and Projects

Jim Pallas started exploring answering-machine-like technology in the 1970s. He created *Phoney-Vents*, in which he played specially prepared sound tapes to whomever he chose to call. Later he developed *Dialevents*, which let people call in to hear tapes. In part he was investigating the intrusive quality of the phone call, but later chose to abandon that element. *Fred Forrest*, a French telecommunications artist, created a well-known telephone piece called *Watch-Towers of Peace*, which investigated the concepts of borders and tried to create worldwide interest in the difficult situation in the former Yugoslavia. He created an installation in which international telephone calls shaped sounds broadcast into what was Yugoslavia from towers constructed on the Austrian border—a modulating whistle changing with the number of calls and the distance of their origin.

Famous for his work with the Audio Ballerinas (see chapter 3.4), ***Benoit Maubrey*** works in a variety of acoustic ecology settings. He has also completed a series of telephone

Jim Pallas: ⟨http://www.ylem.org/artists/jpallas/phone/phoneyvent.html⟩

pieces that allow outsiders to participate in distant events via the telephone. In *Cellular Buddies,* outsiders could project their voice into art exhibits via amplified cellular phones carried by guards. A long-standing series called "Electronic Speakers Corner" creates public sculptures through which distant people can talk by telephone. ***Nick Wray*** created *The Living Garden,* which functioned as a "memory bank" in which physical locations in the garden were portals to recorded memories that could be deposited physically or by telecommunications.

Brian Springer created impromptu conference calls by linking people, via multi-line conference call, whom he conceptually collects from the telephone book. For example, he would call people who have the same name and then place them together in an unexpected call to figure out what happened and why the call was placed. In another example, he would place different religious groups, for example, Buddhists and Methodists, together in one of the chance conference calls. ***The Center for Metahuman Exploration***'s *Boundary Link* explored the telephone's ability to bridge social categories by offering visitors to an art gallery an installation of four telephone booths that were connected to youths in a maximum security detention facility. ***Igor Stromajer and the Intima Arts Base*** creates a variety of mobile phone and Net art events including their *How + Where to Vibrate a Human Being?* project, which orchestrated the use of judiciously placed cellular phones to create a telemasturbatory event. Called a "telephone terrorist," ***Robin Rimbaud*** (*Scanner*), is a sound artist who builds works out of scavenged electronic communications such as intercepted cell-phone conversations.

Radio, Television, and Wireless

Radio and television are also telecommunications. Their history and their relationship to the arts are revealing. Like cinema, they grew so extensive in their purview and their affiliation with mass media models that they created their own cultural niches and distinctive aesthetics such that they were no longer considered part of the arts. Still, many in the arts were reluctant to surrender them totally to this fate. Alternative traditions of art radio and pirate media attempt more free-form experiments with the technologies outside of the dominant mass

---

Benoit Maubrey: ⟨http://www.snafu.de/~maubrey⟩
Nick Wray: ⟨http://www.livinggarden.net/⟩
Center for Metahuman Exploration: ⟨http://www.metahuman.org⟩
Intima: ⟨http://www.intima.org⟩; vibration event: ⟨http://kid.kibla.org/~intima/gsmart/vibraf.html⟩
Scanner: ⟨http://www.dfuse.com/scanner/⟩

media models. Although a full discussion of these traditions is impossible in the limited space of this book, a brief consideration identifies some of the significant themes.

The overwhelming role of mass broadcasting media in shaping life makes artistic exploration of alternatives and stances of resistance more critical. Postmodern interests in limitations of national boundaries makes boundary-ignoring technologies such as radio more intriguing. Concerns with omnipresent surveillance invite artistic countermeasures. The ability of wireless technologies to go wherever a person goes, spread into less urbanized areas, free themselves from infrastructure, and adjust to locations of sender and receiver (discussed more in chapter 3.5 on GPS—global positioning system technology) all invite artistic attention. The magic of being free of wires is compelling.

Wireless technologies will become even more important artistic focuses in the future for several technological and cultural reasons:

- Radio and television broadcasting electronics is becoming less expensive.
- Cellular phone and cellular data systems are becoming widely and inexpensively available.
- Digital technology makes it easier to send wireless video information than ever before.
- Several low-orbit satellite systems, such as Iridium and Teledesic, will make worldwide, boundary-independent wireless communication possible.

A Brief History of Radio Art

In his SIGGRAPH paper "History of Telecommunications Art," Eduardo Kac identifies some of the unique qualities of radio—its disembodiment of remote voice and its potentials as both private and mass media:

In the late 1920s, commercialization of airwaves was in its infancy. Radio was a new medium that captured the imagination of the listeners with an auditory space capable of evoking mental images with no spatio-temporal limits. A remote and undetected source of sound dissociated from optical images, radio opened listeners to their own mindscapes, enveloping them in an acoustic space that could provide both socialization and private experiences. Radio was the first true mass medium, capable of remotely addressing millions at once, as opposed to cinema, for example, which was only available to a local audience.[11]

In his essay "Out of the Dark: Notes on the Nobodies of Radio Art," Gregory Whitehead describes both the conceptual opportunity that radio offered to connect with other minds emerging from the ether and the failure of the arts to come to terms with the import of the medium:

For most of the wireless age, artists have found themselves vacated (or have vacated themselves) from radiophonic space. . . . Radio's gradual drift into such a flatly pedestrian state of mind contrasts sharply with the high flying and exuberant aspirations first triggered by Marconi's twitching finger: promises of communication with alien beings, the establishment of a universal language, instantaneous travel through collapsing space, and the achievement of a lasting global peace. [Whitehead notes intriguing oppositions—seeing] the radio signal as intimate but untouchable, sensually charged but technically remote, reaching deep inside but from way out there, seductive in its invitation but possibly lethal in its effects. Shaping the play of these frictions, the radio artist must then enact a kind of sacrificial auto-electrocution, performed in order to go straight out of one mind and (who's there?) then diffuse, in search of a place to settle.[12]

Kac and Whitehead each offer a few milestones in radio-art history to illustrate some of the possibilities and fantasies. In 1921, Velimir Khlebnikov, a Russian Futurist, created a proposal for radio as "the spiritual sun of the country." Radio would have the "power and means to mesmerize the minds of the entire nation, both healing the sick via long-distance hypnotic suggestion and increasing labor productivity." In 1928, the German experimental film maker Walter Ruttmann was invited to create works for radio. He created *Weekend,* which was "a movie without images, a discontinuous narrative based on the mental images projected by the sounds alone." According to Kac, Ruttmann's innovative work for radio hoped to "open the airwaves to the aesthetic of the avant-garde, challenging the standardization of programming imposed by commercial imperatives." In 1933, the Italian Futurists F. T. Marinetti and Pino Masnata penned the "Manifesto Della Radio":

In the manifesto, they proposed that radio be freed from artistic and literary tradition and that the art of radio begins where theater and movies stop. . . . Marinetti and Masnata proposed the reception, amplification and transfiguration of vibrations emitted by living beings and matter.[13]

In 1938, Orson Welles produced *War of the Worlds* for the Mercury Theatre of the Air. The portrayal of an invasion from Mars was so convincing that thousands of people fled in fear, abandoning their work and homes. Kac noted that

Welles's simulated Martian invasion revealed, for the first time, the true power of radio. It exhibited the unique ability of radio to play with the breath of speech and the plastic sonority of its special effects to excite the imagination of the listener. It showed how the technical reliability of the medium built its credibility, giving veracity to news transmitted through it. . . . It was "hyperreal" in Baudrillard's sense of the word.

In spite of some experimentation, radio has become clearly dominated by its commercial and official contexts. Whitehead describes the situation:

A revitalized practice of radio art languishes in cultural limbo because today's wireless imagination applies itself exclusively—fervently!—to questions of intensified commodity circulation and precision weapons systems. . . . If the idea of radiophony as the autonomous, electrified play of bodies unknown to each other (the unabashed aspiration of radio art) sounds at times like it has been irretrievably lost, it is most likely because the air has already become too thick with the buzz of commerce and war.[14]

Into this mire, technologically oriented artists have entered, trying to construct alternative radio-art traditions.

Free ("Pirate") Radio and TV

Radio technology allows the same functions as the telephone—transmissions of sound—but does away with the wires. In the early days of radio, many thought it would be used much like the telephone—communication of many to many. Quickly it fell into the mass media model of one-to-many communications, except for niches like CB radio. In fact, governments quickly moved to regulate and control the airwaves—both radio and television. In many countries, governments asserted that the airwaves were national resources and should be totally controlled by governments for the sake of insuring high standards and educating the public (for example, Soviet Russia and the British Broadcasting Corporation). In countries like the United States, the airwaves were not nationalized but they were—in the interest of the "orderly" use of the commons—highly controlled and regulated, with entities given the right to broadcast only after they passed bureaucratic inspection and purchased licenses. In current times, many nationalized systems have been abandoned for mixed public and private models.

Artists, alternative musicians, and others interested in political disenfranchisement mounted movements of free radio (and TV). They pointed out that these systems of regulation systematically restricted the aesthetic and political range of what was available on broadcast media. Licenses and legal services could end up costing $100,000 to get a station legitimated. The airwaves offered such wide reach and had become so integrated

---

Free radio network: ⟨http://www.frn.net/⟩
Pirate radio links: ⟨http://www.sasquatch.com/~zane/radio.html⟩
Free radio Web ring: ⟨http://home.swbell.net/bmjohns/radio.htm⟩

A
B

Fig. 6.2.5. Covers from books providing resources for the international free radio movement. A: Andrea Borgnino, *Radio Pirata* (http://www.alpcom.it/hamradio/freewaves/); B: Ron Sakolsky and Stephen Dunifer (editors), *Seizing the Airwaves* (AK Press), an anthology of articles on the liberation radio movement (http://www.freeradio.org/).

into the fabric of everyday life that wider access was critical. The Federal Communications Commission regularly fights these micro stations for the sake of protecting the clarity of frequency channels.

Interviews with the founder of Berkeley Free Radio express this access-promoting view, noting that it shouldn't be called pirate radio:

I prefer to call it free radio or rebel radio, . . . The FCC uses "pirate" in a pejorative sense. "Pirate" implies stealing something that belongs to someone else. We feel the airwaves belong to the people of this country.

The question is, should the airwaves be used as a primary means of fostering a democratic, pluralistic, and vibrantly diverse society through a free and open exchange of ideas, news, information, art, and culture, or should they be a concession stand for narrow, anti-democratic corporate interests working hand in hand with a government whose main goal is domestic pacification, control, and the maximization of private profit?[15]

Alternative music artists offer an important part of the impetus toward free radio. An interview with the music group Pearl Jam expresses these views:

"My philosophy is that the media is too controlled by large corporations. . . . What you hear on radio is very limited. The public is too much a consumer of media and not a maker of media. This is a way to give you some creative input into it." [Eddie] Vedder, who regularly gives out his home phone number to fans, will broadcast a cellular phone number during the radio shows so people can talk to him on the air.[16]

A worldwide free-radio movement has developed that links artists, musicians, hobbyists, hackers, and political activists around the world in the establishment of alternative broadcast stations. Web sites are full of references to access frequencies, philosophical rationales, legal advice, samples of broadcasts, and technical details on low cost, miniature transmitters, and antennas. Alternative art and music have been a critical part of the impetus, and artists have had to become conversant with the physics of electromagnetic radiation and masters of arcane technologies such as how to hide antennas or to convert consumer devices into transmitters. They see even public radio as being constrained by its financial burdens and its needs to please appropriate target audiences.

Japanese theorist Joshiya Ueno's article "Who is a Media Activist?" attempts to contextualize free radio in the larger context of media activism. He considers relationships with Hakkin Bey's idea of TAZ (tactical art zones), and notes that some media activists dislike the links to political activism and see their work differently, drawing parallels with the squatter movement:

Once Walter Benjamin, a great ancestor of the thinking of media activism, said, "The destructive character knows only one watchword: make room; only one activity: clearing away." This is also a crucial slogan for us. Because for activism, whether it is squatting a building or broadcasting pirate media or weaving and hacking on a computer, to invent space and occupy virtual or real space is the most radical and important gesture. A media activist concerned with free radio possesses an on-air frequency, a squatting activist occupies buildings and real spaces, and a free-Net-media activist makes room in cyberspace. In fact, these different gestures come to look like one action. Insofar as they have a critical sense about clearing space for the "in-between" relation, media activists are very conscious of the scarcity of lands or things or bandwidths or net space . . . the media activist is opposed to, and is skeptical about, all professions. The media activist is a mediator among many activities as far as (s)he can go beyond and outside each profession and role.[17]

Radio-Based Art and Theater Installations

Ian Pollock and Janet Silk

Ian Pollock and Janet Silk, known for telephone art, also explored the media archaeology of radio in "Dead Air, Radio from beyond the Grave." In its early days, some hoped that radio would provide access to the dead. Watson, Edison, and Marconi all hoped the technology could connect with spirits, and the artists created an installation that reawakened that idea:

The installation offers a humorous dialogue with the dead in order that we may overcome our dubious perception of life beyond the grave. . . . There are several radio receivers at various stages of (dis)-assembly. All are tuned to unoccupied frequencies and have headphones attached to them. The visitor is invited to listen for messages from the beyond.

Pollock and Silk reflect on the attitudes of the early inventors: "They all saw radio waves as the threads of the cosmic fabric, connecting every vibrating molecule with every other vibrating molecule; as a way for those who have something to say to reach those who want to hear, as a means for the dead to explain their situations and predicaments in the afterlife to those not yet there."

Neil Wiernik — Radio-Art Installation

Neil Wiernik investigates metaphoric meanings of radio. In his work *How to Become an Amateur Radio,* he created an installation made of two homemade FM transmitters and four receivers, and created sound works to be broadcast and received as people walked around the gallery. He notes that "each audio work is an exploration of 'human-made electromagnetic activity,' which uses a kind of storytelling technique combining text and sound. . . . What if we communicated with each other using not only words, signs, and symbols but also via self-generated electronic signals?"

Other Artists and Projects

Jessie Drew, another artist interested in free radio, creates installations that use small transmitters. He regularly gives out diagrams and instructions on how to create

---

Neil Wiernik: ⟨http://www.interaccess.org/aurora/index.html⟩

inexpensive transmitters as part of his art. *Antenna Theatre* presents theatrical events in which the audience walks around and encounters dialogue via the wireless transmitters used in places like museums. **Nina Czegledy** organized an "Aurora" show in Canada focused on the electromagnetic spectrum as a context for art. **Patrick Ready** created a *Radio and Beans* installation in which beans were planted among various radio-wave electronic devices. Commenting on the "datasphere," **Simon Penny**'s *Lo Yo Yo* is a kinetic sculpture that tunes diverse radio stations as it swings.

Art Radio

The Past and Future of Art Radio

Recognizing artistic involvement in radio as culturally valuable, several countries established art-radio networks or programming, for example, DeutschlandRadio, Kunstradio-Radiokunst, Radio Lada, YLE Finnish State Radio, and cultural programming on the Australian Broadcasting Corporation. In the United States, New American Radio provided experimental radio to public broadcasting stations for many years.

Art radio allowed for more challenging material than would normally appear even on the Public Broadcasting Service or the nationalized networks. Much of the work offered material that was acoustically unusual or gathered within nontraditional documentary traditions but stayed within the recognizable frameworks of radio. Some of it, however, challenged the cultural framework of radio or experimented with technological innovations. Some brought in visual and performance artists to experiment with the technology. Several radio projects are briefly described.

Kunstradio

Horizontal Radio was a twenty-four-hour experimental radio/Internet event that involved over twenty radio stations, over two hundred artists, and writers in various cities stretching from Moscow and Helsinki, to Denver and Sydney. It was coordinated by EBU (the European Broadcasting Union), Transit, the ORF's Kunstradio, and Ars Electronica 95:

Horizontal Radio functioned as an experimental field of tension generated by the highly differing characteristics of transmission and communcation of the classical isosynchronous properties of radio and the asynchronous context- and download-related properties of digital data networks (on demand, random access, caching). *Horizontal Radio,* instead of perpetuating the vertical hierarchy between clearly defined transmittors and receivers, turned into a platform for the exchange of transmissions. . . . Such a network environment implies an artistic conception that places less emphasis on primary production and more on dialogic distribution and administration.[18]

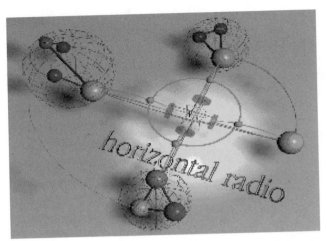

Fig. 6.2.6. ORF Kunstradio, *Horizontal Radio*. A twenty-four-hour experimental radio/Net event that involved over twenty radio stations and over two hundred artists. Conceptualized and coordinated by Gerfried Stocker and Heidi Grundmann.

ORF's Kunstradio also created *Familie Auer,* a commentary on radio sitcom, as part of Ars Electronica97. This project encouraged dispersed artists to collaborate in creating a fictional family drama.[19]

Another project, *Static between the Stations,* juxtaposed radio sounds with the electronic noise of transmissions and tried to create events not "tyrannized by the clock." *Rivers and Bridges* was an international radio and Internet event created as part of Ars Electronica96. It used these metaphors to expand the ways that artists dealt with radio: "Flowing and bridging are often used metaphorically in describing the movement of information in the networks and the globe-spanning transmission systems that comprise them." The organizers sought to explore ideas such as "darkness, education of the ear, thinking loud, different data in different senses, the cryptographic, translation, and dismantling borders. They hoped to enhance the activity of listening: getting the same consciousness toward the audio-acoustic environment as to the visual one."[20]

New American Radio

In the book *Breaking the Broadcast Barrier: Radio Art 1980–1994,* Jacki Apple describes the approaches taken by American experimental radio artists. The book considers work by many artists, from Terry Allen to Gregory Whitehead. She describes the artists' approaches:

Some approach radio as an architectural space to be constructed sonically and linguistically, or as the site of an event—an arena, a stage, a promenade, a public square, a cafe, a telephone booth, an intimate interior. Some use it as a gathering place, or a conduit, a means to create community. Some artists employ the media landscape itself as the narrative, while others look into the body as the site and the source; the voicebox, the larynx, become medium and metaphor. Still others gather the sounds of the world as evidence and construct maps of imaginary geographies.[21]

Another Canadian book, called *Radio Rethink,* by Daina Augaitis and Dan Lander, reviews similar experiments in radio art.

The Future of Radio Art

In the United States, radio art flourished in the 1970s and 1980s. Analysts disagree about its ultimate impact and its prospects. In "The Beat and the Box," David Moss questions the efficacy of art radio. In his view, commercial radio has built such a strong approach to flattening information that these gestures are doomed:

In the 1960s the artists moved in, and the fringes of sound art began to infiltrate, inhabit, and subvert the radio waves. It was a crucial moment with a logical rationale: if radio was a symbol of normalcy, and if music/art from the edges was normal, then it was part of everyday life. Getting on the radio was a way around the "art" dilemma. Artistic work made for radio could reach people as part of their everyday lives. Perhaps they would be shocked, amused, mesmerized, or informed by these other events and unexpected sounds. . . . After seventy successful years in the wallpaper business, however, radio has mainly the power to flatten, smoothe out, disembody, and trivialize the information it conveys.[22]

In "Art Rangers in Radioland," an introductory essay to *Breaking the Broadcast Barrier: Radio Art 1980–1994,* Jacki Apple summarizes the accomplishments and prospects of art radio. She notes how much an integrated part of life radio has been for this century and explains how artists who grew up with it tried to reclaim it. She suggests that the advent of college radio, listener supported radio, and the development of accessible recording and broadcasting technologies encouraged experimentation outside of the dominant commercial pattern. American artists drew from different sources than the European avant-garde:

Thus, in the 1980s radio and audio artworks—sound art, experimental narratives, sonic geographies, pseudo documentaries, radio cinema, conceptual and multimedia performances, a whole panoply of broadcast interventions that confronted the politics of culture, subverted mass media news and entertainment, and challenged aural perceptions, infiltrated the broadcast landscape, and acquired an audience. . . .

Although these works encompass a diversity of esthetics and styles, the artists share a sensibility radically different from that of their predecessors, whose roots are in a European avant-garde tradition. It is a distinctly postmodern American sensibility of blurred boundaries between realities—a convergence of art concepts and forms and media culture, of history, memory, fantasy, and fiction, of public and private space.[23]

Apple sees radio art as continuous with the alternative spaces that developed during the 1970s, which nurtured performance and conceptual art. She notes that radio art had special access to the inner ear in a way that cinema and video could not. She concludes, however, that decreasing support for public radio, the increasing dominance of audience demographic analysis, and increasing opportunities in cyberspace were making radio a less likely destination for artists.

While film and video remain always outside the body, a facsimile on a screen, and words remain bound to the page of the book, aural media both surround and penetrate the body. Radio in its most creative manifestations is the original holographic virtual space. Projected onto the visual field of the inner eye, resonating along aural pathways in the boom box of the brain, words and sounds become living presences. Think of radio as words with wings, . . .

In the next century radio as we have known it may disappear, swallowed up by multimedia cyberspace. Or, as an obsolete technology relegated to the subculture fringes, it might exist only in pirate form, a weapon of the world's underclasses, a tool of artists, revolutionaries, shamans, and other questioning voices in our brave new tech world.

The Migration to Net.Radio

Many of those involved in free radio and art radio have begun to switch their interest to the Web—what is often called net.radio or Web radio. Like free radio, it allows anyone to be a publisher, and furthermore, it extends its reach globally. There are, however, important ways in which it is more limited. It relies on an infrastructure controlled by telecommunications companies and requires listeners to have computer expertise and relatively expensive receiving equipment (a computer versus a transistor radio). Groups often integrate radio and Web technologies. As the migration continues, however, the distinction between radio artists and Web artists begins to disappear and the reference to radio becomes dimmer.

VanGogh TV—Ponton European Media Art

The European VanGogh TV artist group, famous for their "pirate" broadcast initiatives, created "Media Bus" events for Ars Electronica, Documenta, and European Media Arts Festivals. In more recent years it has turned its interest to Internet and VR technologies. Its Web statement "Subjective Speed" gives a flavor of their approach: "VGTV is not monitoring the usual media strategy, selling paradise, or fear, for money. . . . VGTV is developing systems for so-called senseless, irrational, poetic or artistic use, and for so-called senseful, rational, working applications."[24]

Their *Piazza Virtuale* (Virtual Square), presented in 1993 at Documenta IX, in Kassel, Germany, created an unprecedented media space that linked participants via satellite, live TV, a computer network, ISDN phones, videophone, and fax. Eduardo Kac describes the installation and the desire of the artists to create a kind of television that is analogous to the Internet in its structure, even before the Internet was widely understood.

With no pre-set rules or moderators, up to twenty viewers called, logged on, or dialed-up simultaneously, and started to interact with one another in the public space of television, occasionally controlling remote video cameras on a track in the studio's ceiling. . . . This kind of work is deeply rooted in the idea that art has a social responsibility, and that new media create new forms of social relations. Ponton artists act on it directly, in the domain of mediascape and reality. Among other implications, this project takes away the monologic voice of television to convert it into another form of public space for interaction, analogous to the Internet.[25]

Radio Lada, Aura Radio, Cellular Pirate Radio, TNC Network, Free Planet Radio, and E-lab

Groups around the world are offering experimental Web radio. Occasionally they will include "ether" radio as part of their work. Examples include Aura Radio (Finland), Cellular Pirate Radio (Canada), Radio Lada (Italy), TNC Network (France), Free Planet Radio (the Netherlands), and E-lab (Latvia). There are many more of these groups around the world, their actions facilitated by the ease of Web access.

Radio Internationale Stadt (RIS) won an Ars Electronica 98 special mention in the Net category for their Web.radio site. It provides an archive for all kinds of experimental music and an opportunity to reach a world audience (as opposed to the limited audience of traditional ether broadcasting):

VanGogh TV: ⟨http://www.vgtv.com/about.html⟩

RIS represents a new generation of radio. Unlike the conventional radio systems that only work with transmitter/receiver broadcasting, RIS is an open-audio archive system, based on the Internet. Recipients can choose their specific audio content according to their personal interests, can listen to it anytime, and can add content to the system.[26]

E-lab also won an Ars Electronica 98 special mention in the Net category for their *X-change* project. Their Web site provides an entry for experimental artists around the world (especially those working in net.radio) and a collaborative "meeting place for creative minds."[27]

Radio Lada similarly sees its artistic reference in "nonstop user-definable art broadcast" and emphasizes the range of practice they try to incorporate—directors, actors, visual artists, philosophers, media artists, composers/sound artists, writers, and radio makers. Free Planet Radio offers a similar explanation of what they do, emphasizing their broadcasting of live events in addition to sound and their focus on "subculture/audio art/media experiments, and other creative initiatives not being part of the mainstream and commercialism." Gerfried Stocker's *X-space* was another experiment investigating new kinds of media collaborations.[28] TNC connected local events at clubs and museums on three continents.

Listening to the Mediterranean

Artists seek to combine radio, net.radio, and public-space installations in innovative ways. For example, the Lada98 sound-art festival "L'Arte dell'Ascolto," under the direction of Roberto Paci Dal, used the collaboration of UNESCO and several Italian and Austrian radio networks to create an event manifested both in local physical spaces and on networks. It sought to explore possible interactions between the contexts and multiethnic diversity:

One of the main purposes of the Festival is the development of artistical projects based on the utilization of telecommunication technologies in order to create communities of people operating simultaneously in electronic space and in the real world and to relate traditional cultures to modernity. Making the most of a culture that naturally and everywhere is becoming multiethnical is to us fundamental, and so is the need to aggregate those who are interested around events able to conjugate entertainment with research.

An adventure that combines art with scientific research. A project aiming to define the chances of wireless communication with special regard to those areas of the Mediterrenean which are disadvantaged under the point of view of the systems of communications; and for which, therefore having access to Internet through the radio could be of great importance.[29]

Analysis of Net.Radio

In her article "Net.radio and the Public Space," free-radio activist Josephine Bosma describes her interest in the Internet and her desire to extend its reach to a larger public:

In 1991 I started using pirate radio as a tool, coming from an artist background. . . . Of all media serving as social 'battlefields,' the Internet seemed to be the place where it all came together and where interventions could still take place. . . .

In the context of tactical media use, connecting the net to physical, preferably public, spaces is the most important step that has to be taken next. As a net.journalist, which I have mostly come to be, I feel the lack of knowledge about the net amongst a majority of people is becoming a larger obstacle in my work. . . . Creating extensions of the net outside of it could help solve this problem. [30]

She also notes that the Net is expanding its reach to include people outside of high-art culture, with high-energy infusions from "party culture" (for example, raves). She laments the lack of documentation for these often ephemeral events and describes an E-mail she received describing an experiment of XRL (eXtended Live Radio) to link Web, radio, and physical space in a nexus between Berlin and Ljubljana, Slovenia: "We transmitted EVERYTHING on the ether, no censorship. At the same time, we always tried to make the process transparent, to tell something about the measures that had to be taken to make the transmissions possible, and about the possibilities in general."

She describes the activities of another group, Convex TV, which also maintains its interest in the broadcast media. Through various actions, they helped the audience to build low-power FM transmitters, organized people to listen to transmissions on car stereos, and did live shows in the midst of symposia. For Bosma, the ideal is the "pluriform" use of multiple media:

It is important to give more support to initiatives which connect the net to physical/public spaces or to get directly involved in these connections yourself. It prevents the net from becoming a technically and socially inbred, and thus paralyzed, entity. It offers us the challenge of finding new languages, in any sense of the word, to express and extend net.cultural specific moods, techniques, and young (unstable?) traditions outside of the net. The public and physical space is naturally most interestingly, entered via live events which utilise a combination of several media and/or 'technologies.'. . .

She suggests that the Web is superior to radio in meeting many artistic agendas:

To make a good judgment of what radio is in the age of digital media, the traditional concept of radio has to be overthrown completely. Where in the beginning of this century the communicative possibilities of radio were diminished in favor of control over the airwaves for reasons of censorship and security, today the system of cables, airwaves, and satellites that shape the internet needs first of all no strong security measures for the sake of basic services like ambulance and airplane traffic control, and secondly, censorship seems a lot less easy task to perform. Radio no longer needs to be a single stream of sound that is transmitted from a central point to its listeners.[31]

Bosma notes, however, that net.radio raises the "question of the screens." Part of radio's appeal has been its ambience; net.radio's arrival via computer screens creates both opportunities and limitations:

Besides from all this, a very sensitive question arises with radio on the net, which is: What to do with those screens? I have talked to many media artists, radio and television people about this, trying to get a grip on what future radio would "look" like. The special quality of radio or audio in general is of course its "omnipresence," compared to TV or video, which is locked in a box in the corner. Now with radio on the net, it has a shiny prison as well. . . . Robert Adrian, media artist from Vienna, thinks the screen will add new qualities to radio, like every new extension does. . . . Gerfried Stocker made the suggestion to just leave the screen black, or not use it at all. . . .

What will happen when moving images are added, though? Will radio become television on the Internet? . . . The new possibilities offered by the internet for radio are various and they are enormous, and this cannot be said often enough. Radio finally might regain some of its freedom.

Emerging technologies may obsolesce conventional radio. Great development effort is focused on inventing efficient sound compression schemes such as MP3 and wireless radio Internet capabilities so that people will be able inexpensively to access the Internet (including sound sites) anywhere at any time without wires. At that point distinctions between net.radio and broadcast radio will become moot.

Summary: Dangers and Opportunities in Convergence

Telephone, radio, and television are the most widespread communication media. Radio reaches people in both the developed and developing countries. It is accessible via inexpensive receivers, and because it is sound based, it can be engaged as people go about other business, such as work or commuting. Telephones are similarly extensive in their spread. This chapter has shown that artists have developed some interesting interventions, but they must be considered minimal in the face of the cultural reach of these mass communication media.

Analysts predict that we sit on the eve of convergence. In this view, the disparate technologies of telephone, radio, television, computers, and data communications will all converge into a single system. The rhetoric of progress paints a glorious picture of intelligent communications and instant access of everything and everyone. The World Wide Web is seen as the most likely infrastructure for convergence, with entertainment video and telephone racing to adopt themselves to that context.

Like so many technocultural developments, the future holds both possibilities and dangers. Is convergence a development to be welcomed? Is there any wisdom in having several separate parallel communication media? Is there something to be gained by separating sound from image? Should artists really abandon any attempt to make inroads in the old media of radio and telephones? What new possibilities inhere in converged media? These are the questions that face the telecommunication artists of the next millennium.

Notes

1.   R. Ascott and C. Loeffler, eds., *Leonardo* (special issue, "Connectivity") vol. 24(1991): 2.

2.   F. Popper, *Art of the Electronic Age* (New York: Henry N. Abrams, 1993).

3.   E. Kac, "Aspects of the Aesthetics of Telecommunications," ⟨http://www.ekac.org/Telecom.Paper.Siggrap.html⟩.

4.   F. Dyson, "Circuits of the Voice: From Cosmology to Telephony," ⟨http://sysx.apana.org.au/soundsite/csa/essays_in_sound/meaning_of_voice.html⟩.

5.   Disembodied Art Gallery Web site, "Principles," ⟨http://www.dismbody.demon.co.uk/home.html⟩.

6.   InterCommunication Center, "Description of Bunting Event," ⟨http://red.ntticc.or.jp/preactivities/ic95/profile/heath-e.html⟩.

7.   H. Bunting, "Interview," ⟨http://www.backspace.org/hayvent/box.rand/86.html⟩.

8.   I. Pollock and J. Silk, "Artistic Statement," ⟨http://www.sirius.com/~ps313/⟩.

9.   I. Pollock and J. Silk, "*Local 411* Description," ⟨http://www.sirius.com/~ps313/⟩.

10.   S. Wilson, "*Is Anyone There?* Description," ⟨http://userwww.sfsu.edu/~swilson/isany/isany.des.html⟩.

11.   E. Kac, "History of Telecommuncations Art," ⟨http://www.ekac.org/Telecom.Paper.Siggrap.html⟩.

12.   G. Whitehead, "Out of the Dark: Notes on the Nobodies of Radio Art," ⟨http://somewhere.org/NAR/Writings/Critical/whitehead/Main.htm⟩.

13. E. Kac, op. cit.

14. G. Whitehead, op. cit.

15. Berkeley Free Radio, "Interview," ⟨http://www.sasquatch.com/~zane/frbwash.txt⟩.

16. Kava, Brad, "Interview with Pearl Jam," ⟨http://www.sasquatch.com/~zane/jam.txt⟩.

17. J. Ueno, "Who is a Media Activist?" ⟨http://193.2.132.70/nettime/zkp4/13.htm⟩.

18. Horizontal Radio, "Description," ⟨http://gewi.kfunigraz.ac.at/x-space/horrad/⟩.

19. Orfkunstradio, "*Familie Auer* Description," ⟨http://thing.at/orfkunstradio/AUER/⟩.

20. Orfkunstradio, "Rivers and Bridges," ⟨http://www.thing.or.at/thing/orfkunstradio/PROJECTS/projects1.html⟩.

21. J. Apple, "Description of Breaking the Broadcast Barrier," ⟨http://somewhere.org/NAR/Writings/Critical/thorington-apple/intro_broad.htm⟩.

22. D. Moss, "The Beat and the Box," ⟨http://somewhere.org/NAR/Writings/Critical/moss/Main.htm⟩.

23. J. Apple, "Art Rangers in Radioland," ⟨http://somewhere.org/NAR/Writings/Critical/thorington-apple/intro_broad.htm⟩.

24. VanGogh TV, "Subjective Speed," ⟨http://www.vgtv.com/about.html⟩.

25. E. Kac, "Description of *Piazza Virtuale*," ⟨http://www.ekac.org/InteractiveArtontheNet.html⟩.

26. Radio Internationale Stadt, "Statement," ⟨http://orang.orang.org/⟩.

27. E-lab, "*X-change* project," ⟨http://xchange.re-lab.net/⟩.

28. G. Stocker, "*X-space* project," ⟨http://gewi.kfunigraz.ac.at/~gerfried/kurzportrait.html⟩.

29. Lada98 festival, "Listening to the Mediterranean," ⟨http://giardini.sm/lada98⟩.

30. J. Bosma, "Net.Radio and the Public Space," ⟨http://thing.at/orfkunstradio/FUTURE/RTF/SYMPOSIUM/LECTURES/BOSMA/bosma.html⟩.

31. J. Bosma, "Net.radio," ⟨http://www.giardini.sm./radio/net-radio.htm⟩.

6.3

Teleconferencing, Videoconferencing, Satellites, the Internet, and Telepresence

This chapter reviews artistic work with technologies that evolved after the telephone, radio, and television. It presents artists who work with sophisticated communication systems such as satellites, teleconferencing and videoconferencing, and computer-mediated communications. It explores the theorization and artistic investigations of telepresence. As with so many of these technologies, some of the artists seem primarily interested in elaborating new possibilities while others concentrate on subversive questioning. Generally, this chapter focuses on work that did not rely on the World Wide Web, although much telecommunications (and many of this chapter's artists) is now moving to use the Web infrastructure and most current work assumes the use of the Web. The next chapter focuses explicitly on Web-based artistic work.

Teleconferencing, Videoconferencing, Satellites, and Internet Collaboration

Sherrie Rabinowitz and Kit Galloway—Electronic Cafe

Sherrie Rabinowitz and Kit Galloway have led experimentation in telecommunications-based art and media over the last twenty years. Their *Hole in Space* is legendary in the ground it broke in geographically dispersed collaborative art. They created the first Electronic Cafe, which explored telecommunications context as art, media, and entertainment. Since that time over forty affiliates have been established around the world. As they state, what "started as art experimentation has evolved into an institutionalized context that integrates technological innovation with multicultural sensitivities." They continue to pioneer new approaches, for example, creating events such as ambient music created by DJs in several cities at once, with multiple city crowds dancing together in composite image space, and performance in which an artist's image is projected life-size in a remote place while the artists remotely control instruments, lights, and props. Their Web site describes their underlying principles, which emphasize multicultural collaboration and media experimentation with emerging technologies:

Not Just Another Cybercafe   ECI is the mother of all cybercafes. ECI is, first and foremost, a networked cultural research lab: a unique international network of multimedia telecommunications venues with over forty affiliates around the globe. For over a decade, ECI has functioned not only as a pioneer but as a leading multicultural community conducting groundbreaking aesthetic research in the exploration of real-time networked collaborative multimedia environments.

Electronic Cafe: ⟨http://main.ecafe.com/index2.html⟩

Fig. 6.3.1. Sherrie Rabinowitz and Kit Galloway, Electronic Cafe. *Bliss:* A motion-controlled VRML avatar performance. *Concept:* Kit Galloway & Sherrie Rabinowitz

Blitcom: Mark Pesce, Jan Mallis, and Margo; *Bliss:* Motion Controlled VRML Avatar; *Bliss Performer:* Mary Ann Daniel; *Bliss figure:* Bay Raitt; *Bliss's room/world:* Jim Ludtke; *Bliss's theme:* Paul Godwin; *Bliss anima. Software:* Protozoa, ALIVE, Eric Gregory & Brad DeGraf; *Bliss's Creative Dir.:* Celia Pearce, Momentum Media Group

For Electronic Cafe International—Santa Monica: John Sokol, IDB Systems; John & Matt Graham, Webcasting (GTS); Mona Jean Cedars, Performer & Video Avatar at ECI; Noah Bogan, The Nohow Group; Bob Rice, Audio; Kit Galloway, Composite Imaging

Sponsor: Cosmo Software (an SGI Company), Bill McCloskey; *Equipment sponsors:* Cosmo Software, a Silicon Graphics Company (primary sponsor); Ascend Communications; Graham Technology Solutions; John Sokol, International Digital Broadcasting Systems; Metawire; EIS (Video projector)

Cash sponsors: Cosmo Software, a Silicon Graphics Company (primary sponsor); HollyWorlds; LA VRML Users Group *INKIND support:* CyberTown; LA Bridge; Performance Animation Society

Not Just Another Cyber Entertainment Network   ECI has been using technology to explore co-creation and collaboration in real-time networked environments. The prerequisites for this are: (1) employing a multitude of disciplines; (2) using the performing arts as modes of investigating these new ways of being in the world; and (3) creating a new context so that new forms and content can emerge. . . . They participate in and model the essentials for an information and telecommunications environment that is both functional and inclusive.[1]

Their events investigate the artistic possibilities of the newest technologies, integrating technologists and research scientists along with artists. Here is a sampling of a recent year's activities from their Web site:

- Telepoetry: The Last Poets: linking the Kitchen in New York City and the Cybercafe in Santa Monica.
- Performance Animation Society: The Performance Animation Society is a member/sponsor-supported special interest group dedicated to the emerging field of Performance Animation, virtual theater, digital puppeteering, and motion capture.
- Los Angeles: The VRML Barn Raising of ECI will bring together most of southern California's top VRML companies, designers, and authoring tools for a unique, two-day weekend of VRML 2.0 production, experimentation, and collaborative creation.
- Cyber Hum: The Kitchen in New York City and Cybercafe in Santa Monica. Connecting the experimental media communities of the East and West coasts in an evening of ambient bliss.

Richard Kriesche

Richard Kriesche is an Austrian artist who has been in the forefront of explorations of satellite and Internet-based art. His *Artsat* project (see chapter 3.4) interactively communicated with cosmonauts on an orbiting spacecraft. He has also created a series of telematic sculpture installations that link physical environments with worldwide data flows, and in 1994 edited a book entitled *Teleskulptur.*

In the 1995 Venice Biennale, he created *Telematic Sculpture 4,* which linked the movement of kinetic sculpture to the ratio of art news-group Internet traffic to general news-group traffic. The documentation explained that the machine threatened to break through a wall if the ratios were wrong: "Internet users were invited to help, preventing the machine from reaching the wall by accessing these pages, sending E-mail to *T.S.4,* or discuss it in one of the above mentioned newsgroups." Kriesche believes that the Internet is a new kind of extra-human machine: "[I]t is becoming obvious that the Internet in no longer in the tradition of conventional machines, that the Internet is the first out-of-human machine, a kind of meta-machine, and that we are on the way to transgressing ourselves."[2]

Kriesche is interested in several aspects of telecommunications technologies: its possibilities for new kinds of interconnections and its tendencies to make an atomized amorphous social sphere. He believes art-science can help realize some of these positive potentials and fight against the tendency to segment everything into particles, which results in a "socially isolated, segregated, ego-centered individual." In his participation

---

Richard Kriesche: ⟨http://iis.joanneum.ac.at/kriesche/⟩

Fig. 6.3.2. Richard Kriesche, *Telematic Sculpture 4*. A project that linked the movement of kinetic sculpture to the ratio of art news-group Internet traffic to general news-group traffic.

in the Conference on a New Space for Culture and Society, he explained some of his views on research as a kind of art:

This technology is still the result of a research and scientific approach in which hardly any emphasis is given on the cultural implications. Worse, what can be seen right now is that the cultural implications are only seen in their outcome toward communications media, but have no impact for the development and the innovative aspect of technology and the media themselves. This is where art and creativity comes into play. Scientific and technological research, because of its interconnectedness with the whole of society and its construction, must be embedded in the multitudes of cultural awareness of all its members. The reason for this is to finally understand how and by what means the global society can and must be built. Then science and research will become, as it has already been highlighted by some of the researchers and scientists, not only the common ground of our culture, but furthermore a particular kind of art.

This science-art connectedness has been basic in our evolution. According to the potentials of our information technology, this connectedness can and must now be regarded as basic again for our daily life processes. Then art, in its most advanced format, both participatorial, communicative, interactive, and sublime, will become the general awareness of going beyond the technological, political, sociological, etc., constraints to imagine the not yet given culture.[3]

Fig. 6.3.3. Paul Sermon, *Telematic Dreaming.* Two people in remote places share beds.

Paul Sermon

Paul Sermon is an artist who has been actively exploring telematic art since the 1980s. His *Telematic Dreaming* series won many prizes and has been shown throughout the world. It is acknowledged as one of the most significant telematic works. It creates an environment of telematically linked beds or couches located in geographically dispersed areas. Participants are enabled to be together with each appearing in the other's bed—another telepresent person seems to be next to the viewer via projection. The linking of the cool, analytical telecommunications technology with the hot meaning infused situation of beds and couches highlights the possibilities and limitations of distance communication. The ICC's Web page offers Sermon's description:

Telematic Dreaming deliberately plays with the ambiguous connotations of a bed as a telepresent projection surface. The psychological complexity of the object dissolves the geographical distance and technology involved in the complete ISDN installation. The ability to exist outside of the user's own space and time is created by an alarmingly real sense of touch that is enhanced by

Paul Sermon: ⟨http://www.hgb-leipzig.de/~sermon/index.html⟩

the context of the bed and caused by an acute shift of senses in this telematic space. The user's consciousness is within the telepresent body controlled by a voyeurism of itself. The cause-and-effect interactions of the body determine its own space and time. By extending this through the ISDN fiber-optic network, the body can travel at the speed of light and locate itself wherever it is interacting. In *Telematic Dreaming*, the two users exchange their tactile senses and touch each other by replacing their hands with their eyes. . . .

As an artist I provide the context, I design the dynamics of the system around an object of psychological complexity, such as a bed or a sofa. For this reason, the work is extremely intense, and audiences are sometimes reluctant to take up the role of the performer. This is usually because they are initially concerned about performing in front of an audience. However, once the viewer takes on the role of the performer they lose contact with the audience and discover that the actual performance is taking place within the telematic space, and not on the bed or sofa. . . . Bringing your self back to your actual body is as hard as getting your self onto the bed or sofa in the first place, and being able to communicate in the actual space and the telematic space simultaneously is almost impossible.[4]

Sermon provides an environment for the investigation of several cultural themes raised by telecommunication, such as the meaning of the physical body in a world dominated by virtual representations and the nature of communication with anonymous partners. Machiko Kusahara, a media curator at the ICC, describes the paradoxical nature of telemediated virtual intimacy:

Telematic Dreaming surely has the most powerful impact because of the dissimilating effect of the bed, a sign shared by everyone. . . . [D]espite the fact that the body is the only means of communication therein, the body of the other party is ghostlike, without substance. This contradictory situation not only confounds the audience, but also, after first releasing them from the logic and restrictions of daily life and dismantling the various elements of signatory identity and the biological environment of the body, it enables experimentation with and enjoyment of the role the body plays in communication. The virtuality of the space enables it to maintain both theatricality and the context of daily life at the same time.[5]

Telematic Dreaming allows not necessarily an escape from the body but the opportunity to observe oneself from a new perspective. It also allows the viewer to explore the relationship of touching and looking: "Just because language dictates that we touch with our hands and see with our eyes doesn't mean that's all. Its absolutely conceivable that we can also see with our hands and touch with our eyes in just the same way; it's a mater of manipulation and definition."[6] Sermon is often amazed by the new responses he sees, especially when the installation is mounted in new cultural settings, such as Japan. He has accumulated tapes from the years of public installations. He has created several other related installations, including: *Tables Turned,* which projects people at each other's tables; *Disappearing Act,* which mutually interprojected stage sets into two remote locations; and *Heaven,* which made images of heaven available via a CUSeeMe reflector (in collaboration with Joachim Blank).

Keith Roberson

Keith Roberson creates events focused on videoconferencing. Some are planned, aesthetically configured events; others are subversive actions that work with found situations. Roberson creates events in which he uses the strategy of disruption to stimulate awareness. For example, he created a cyber terrorist performance in which he took over a business videoconferencing center by generating so much noise that business as usual was impossible. He sees a danger in the emerging teleconferencing conventions that adopt the bland broadcast standards and marginalize the potentially innovative and democratizing imagery of cyber experimenters.

In another variation, artists would build events using audiences who were individuals who happened to be on Internet CUSeeMe channels. CUSeeMe conferences often make use of "reflector" sites, which allow anyone to sign on to see who happens to be connected. The artists explicitly crafted the interventions based on the social expectations of CUSeeMe environments, for example, by integrating the process of engaging anonymous videoconferees into the core of the event and by the introduction of culturally loaded props. He notes that "victims' responses ranged wildly, from highly amused to blasé, from truly shocked to ignoring us completely." The props were often significant: "The use of props—imbued with potent popular cultural meaning, like a Michael Jackson mask, a gun, G.I. Joe and Yoda dolls—helped the participants feel comfortable interacting and establish a primitive basis for structure to build."

In another study, Roberson and his class created an "exquisite corpse" using Internet videoconferencing software that allows people in dispersed locations to appear in multiple small computer windows, which could be repositioned at will. He appreciates and builds upon the unpredictable and democratizing aspects of videoconferencing but notes that the aesthetics are still in flux:

Keith Roberson: ⟨http://garnet.acns.fsu.edu/~kroberson/⟩

Judging by the aesthetic criteria of other media, artworks based on videoconference might appear disjointed and digitally distorted, especially in comparison to the mediums of video and television. However, videoconference allows for the audience and artist to be on potentially equal ground. Unlike video, where artifacts limit the medium to source-to-receiver communication, everyone who participates in my artworks communicates simultaneously. . . . [A]n artist relying on spontaneous interaction risks total failure if the intended interaction does not occur. . . . Unless the overall concept overrides the action, there is no assurance of success.[7]

The realities of videoconferencing, with its delays and compressed images, creates opportunities for reconsidering one's own body and its relationship with others:

Through a layering of images, two-dimensional bodies may intersect and recombine; where we once saw ourselves as separate, enclosed, and intact, we now become dissipated, dispersed, and vulnerable. The physics interrelating time, space, and motion metaphorically diverge in the videoconference medium. . . . Through its distancing from the body, my new self gains certain freedoms, but becomes diffused with other participants. My image's instability reflects the instability of my own existence. . . . Current image-transmission technology is based on movement or "significant" change within the frame in order to use the available transmission bandwidth. The areas of the image which move the most are transmitted the most frequently. While my still torso is transmitted only every ten seconds, my gesturing hand is transmitted at ten times per second. In the bordering areas between my hand and torso, fragments of both accumulate over time.

Nell Tenhaaf

Tenhaaf focuses on the mythologies and self-representations of technologies and scientific inquires such as biology. One set of works called *Neonudism* explores ideas about self-revelation and discourse around network-mediated sex. It uses the low-tech CUSeeMe videoconferencing environment to investigate Netopian hopes for new communication technologies. *Neonudism* sets up communication sessions in which excitements of voyeurism are aroused and deconstructed. It is purposely based on "low tech" as a strategy to reveal structures that underlie even advanced forms, such as virtual reality:

Nell Tenhaaf: ⟨http://www.yorku.ca/faculty/academic/tenhaaf/⟩

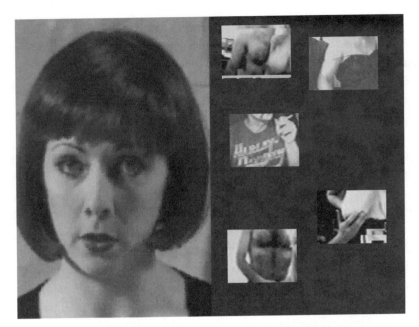

Fig. 6.3.4. Nell Tenhaaf, *Neonudism*. CUSeeMe-based event that explores voyeurism.

This approach to the study of subjectivity enriched enormously the conceptualization of inter-subjectivity, the everyday and intimate exchanges that continuously shape the human psyche. At the same time that it examines subjectivity, *Neonudism* is about the banality that the Net can inflict on intersubjectivity, . . . Here I'm referring to the exaggerated promise of electronic erotic connection in a medium such as CUSeeMe, with no intention to deny the new forms of communication that the Net permits. . . . Voyeurism remains voyeurism, in whatever medium. Cyberspace is portrayed as a space of perceptual revolution, especially when it is linked to technological hype such as the Virtual Reality fad. *Neonudism* seeks to strip bare that apparatus. . . .

The premise of *Neonudism* is to divest ourselves of the clothing of elaborate tools with a view toward disclosing ourselves as fully as possible. The contradiction is that we require technology to do this, we require teledelegation to link ourselves to the revealing potential of Net sexuality. Our solution is simple technology that exposes itself as a prosthesis of the imagination: the desiring machine itself stripped bare. . . .

The intention is to suggest that the CUSeeMe encounter is more with self than with others: you don't really contact anyone else in this medium; rather you experience yourself in a new way, in simulated erotic exchange with others.[8]

Other Artists and Projects

Cultural Commentary **Brian Springer** creates artworks within the culture-jamming tradition of learning how to intercept telecommunications such as cellular phone calls. His well-known videotape called *Spin* was constructed from the raw (before editing) satellite transmissions sent by news organizations during U.S. political campaigns. Even before the Web had become so pervasive, **Margaret Crane and Jon Winet**'s *General Hospital* attempted to stimulate discussions of alternatives to mental hospitals by stimulating communication in an Internet news group called alt.society.mental-health.

Extended Transmisions Questioning technological infatuation, **Gregory Garvey** presented *Smart Stall* as part of the 1996 Siggraph art show. In this event, remote toilet stalls were linked by high-bandwidth lines that transmitted graffiti on connected whiteboards. Anything drawn in one stall appeared in corresponding whiteboard space in the remote stall. Infrared motion detectors activated an abusive semi-intelligent agent, who gave instructions and tried to simulate action. **Stahl Stenslie**'s work with projects such as *CyberSM* experiment with communication of tactile information (see chapter 2.5). In *Between the Worlds,* **Agnes Hegedüs** sets up an unusual communication system that includes composites for faces and processed real-time images of the hands of the communicators.

Augmented Collaboration **Christa Sommerer and Laurent Mignonneau**'s *Life Spacies* invites remote visitors to interact with each other via artificial life-forms in a virtual world, which also can add their own agendas. *Kazuhiko Hachiya*'s *Inter Discommunication Machine* exchanges the sight and hearing experiences of two people in a space so that each experiences what the other would see and hear. In one version of **Bill Seaman**'s *The World Generator,* remote viewers can collaborate in creating a poetic world of image and text (see chapter 7.2). **Akitsugu Maebayashi**'s *Disclavier* allows several people to jointly produce music by actions of their avatars in a shared virtual space. In *Point of Vue, Point of You,* **Elsa Mazeau** invites remote videoconferencees to dynamically shape the shared projection by arranging information by pointing at elements. **Shibuya Suzuki**'s *Three Men and Three Legs* asks three networked collaborators

---

Brian Springer: ⟨http://www.mrf.hu/spring.html⟩
Gregory Garvey: ⟨http://142.232.132.45/dedocs/ggarvey/GGsmartPage.html⟩
Stahl Stenslie: ⟨http://televr.fou.telenor.no/stahl/⟩
Agnes Hegedüs: ⟨http://www.aec.at/fest/fest95/mythos/catalog/hegedus.html⟩
Christa Sommerer and Laurent Mignonneau: ⟨http://ms84.mic.atr.co.jp/~christa/WORKS/LifeLinks.html⟩
Kazuhiko Hachiya: ⟨http://www.p3.org/p3-light/P3exhibitions.html⟩
Akitsugu Maebayashi: ⟨http://www.din.or.jp/~mae884⟩
Elsa Mazeau: ⟨http://www.siggraph.org/s97/conference/garden/vue.html⟩

to jointly control a virtual 3-D puppet by each manipulating one leg. In *Rengei-Za,* **Rieko Nakamura** created a networked collaborative visual-art-creation environment loosely based on the traditional Japanese *renga* literary form, in which authors sequentially build a poem. **Karen O'Rourke and Art Reiseaux** created the *Paris Réseau* event, in which multi-city participants took a virtual trip through Paris. **Judy Malloy and Cathy Marshall** created *Forward Anywhere* to explore the potential of the MUD and MOO forms for hypertext art. **Art+Com**'s *Ping* asked Internet viewers to map items to a metaphorically geographic data space through which camera agents could navigate. **Maurice Benayoun**'s *Tunnel under the Atlantic* and *Tunnel around the World* creates a televirtual meeting place in which viewers in remote locations must "dig" through virtual strata in coming to a full visual-sound meeting place. The German organization **Virtual Company** created the *Talking Head* project, which projects video images of all participants in a videoconference onto 3-D physical busts of heads. Under **Linda Stone**'s leadership, Microsoft's Virtual Worlds research group fosters collaborations between artists and technologists in developing new kinds of telecommunication-mediated artificial worlds. **Time's Up,** in association with BIOMACHINES and radioqualia, presented *Closing the Loop 2000,* which encouraged sound collaborations across the net.

CUSeeMe Events   As part of the PORT show at MIT in 1997, **Adrianne Wortzel** presented an event called *Starboard.* Using CUSeeMe, real audio, and the Web, the event created a performance/improvisation that was open to CUSeeMe visitors from anywhere in the world; they could become part of the events. Wortzel offered provocative themes as an inducement to join in, such as the "Ambiance of Recursive Nesting" and "Formulas for Clandestine Cross-Platform Meetings." **Ebon Fisher**'s *Wailing in the Alula Dimension* uses CUSeeMe to allow visitors to enter into a strange ongoing performance. Instead of the usual talking heads so common in this medium, distant viewers

---

Shibuya Suzuki: ⟨http://www.iamas.ac.jp/~zuckey/⟩
Rieko Nakamura: ⟨http://www.renga.com/rengeiza_e/index.htm⟩
Karen O'Rourke: ⟨http://panoramix.univ-paris1.fr/CERAPLA/Q1E.html⟩
Judy Malloy: ⟨http://www.artswire.org/jmalloy/ccac.html⟩
Art+Com: ⟨http://www.artcom.de⟩
Ping: ⟨http://www.artcom.de/projects/ping/⟩
Maurice Benayoun: ⟨http://panoramix.univ-paris1.fr/UFR04/benayoun/Tuntitrg.htm⟩
Virtual Company: ⟨http://www.siggraph.org/s97/conference/garden/talking.html⟩
Microsoft, Virtual Worlds: ⟨http://www.research.microsoft.com/scripts/main/research.asp⟩
Time's Up: ⟨http://www.timesup.org/⟩
Adrianne Wortzel: ⟨http://artnetweb.com/wortzel/⟩
Ebon Fisher: ⟨http://www.interport.net/~outpost/ebon.html⟩

encounter strangely costumed actors climbing poles and ropes. Fisher is known for devising rituals that incorporate new technologies.

Telecommunicated Performance   **Kurt Heinz** helped set up the Telepoetics networks in which poets and media artists link their performances via live videoconference technology. **La Fura dels Baus** is famous for its intense experimental theater, sometimes called "ferocious" and "explosive." Its *Work in Progress* event linked several cities via videoconferencing. In one version, each city sequentially built on the previous interactions. **Void Productions** presented plays where actors and audience interacted with live performers via Internet chat. **Helen Thorington**'s *Adrift* coordinated sound, text, and VRML interactions between remote and physcial performers. **Laura Knott** coordinated the *World Wide Simultaneous Dance,* with over one hundred dancers in many cities dancing together with videoconference projection. Exploring the potentials of MOO text-based online environments for theater, **The Plaintext Players** have performed many events, including *Gutter City,* set both in the present and the Civil War, and *Roman Forum,* which reflects on U.S. presidential politics. **Adriene Jenik and Lisa Brenneis** organized a series of events called *Desktop Theater* in which they made interventions in on-going chat and game spaces.

Telepresence Definitions

Telepresence represents a major goal of telecommunications in both research and art. In some ways, every kind of telecommunications is telepresence—a technology for a person to be present *in some form* in a distant place. Thus, the telephone, videoconference, and even E-mail could be thought of as telepresence. Any synchronous, near real-time system for exchange could qualify. Some early observers of the telephone were so unnerved by the unnatural presence afforded by the disembodied voice that they ran frightened from the room.

In the experience of presence, there is a continuum that roughly correlates with the variety and richness of the sensual cues presented. Thus an interactive videoconference

---

Kurt Heinz: ⟨http://www.tezcat.com/~malachit/deposit/⟩
La Fura dels Baus: ⟨http://lafura.upc.es/wip/whatis.htm⟩
Void Productions: ⟨http://www.voidp.demon.co.uk/void.htm⟩
Helen Thorington, *Adrift:* ⟨http://www.turbulence.org/Adrift/index.html⟩
Laura Knott, *World Wide Simultaneous Dance:* ⟨http://www.dowhile.org/physical/people/knott.html⟩
The Plaintext Players: ⟨http://yin.arts.uci.edu/~players/⟩
Adriene Jenik and Lisa Brenneis: ⟨http://www.desktoptheater.org/⟩

presents more of that quality than a telephone call, which presents more than an interactive computer chat text. Contemporary researchers are working hard to develop ways to transmit additional aspects of real physical experience, such as haptic and kinesthetic information.

For some, telepresence is not complete until one has the ability to act on the perceived environment from a distance, not only perceive it. This is the sense in which many researchers use the term: presence has no chance of being achieved unless one can act on the environment, for example, via a surrogate robotic arm. Telepresence researchers are interested in systems such as tele-exploration of Mars and telesurgery. They are exploring sophisticated VR-like arrangements to orchestrate a person's perception of a distance space (e.g., head-movement-controlled displays), and the execution of acts in a distant place by moving their local body (e.g., local hand-movements being translated into equivalent motions of the distant robotic hand).

Researchers and artists have ventured a variety of definitions, with many different nuances. Lars Rosenberg offers this definition in the journal *Telepresence:*

The fundamental purpose of a telepresence system is to extend an operator's sensory-motor facilities and problem-solving abilities to a remote environment. Telepresence has been defined by Sheridan (1992) as a human/machine system in which the human operator receives "sufficient information about the tele-operator and the task environment, displayed in a sufficiently natural way, that the operator feels physically present at the remote site." Very similar to virtual reality, in which we strive to achieve the illusion of presence within a computer simulation, telepresence strives to achieve the illusion of presence at a remote location. The end results of both telepresence and virtual reality are essentially the same, a human-computer interface which allows a user to take advantage of natural human abilities when interacting with an environment other than the direct surroundings. The ultimate goal of these efforts is to produce a transparent link from human to machine, a user interface through which information is passed so naturally between operator and environment that the user achieves a complete sense of presence within the remote site.[9]

Scott Fisher and Brenda Laurel offer a succinct definition—a technology that "enables people to feel as if they are actually present in a different place or time."[10]

Philosophically, telepresence raises some other perplexing questions. As discussed in following sections, artist and researcher Ken Goldberg notes that the epistemological world we interact with in real physical space can be problematicized. How do we verify its reality? Is the distance across a room or at the gap of one's fingers fundamentally different than the longer distances of telecommunication space? Critical theory would suggest that the physical world is as much a construction of mind (for example, the role of gender in perception) as the world constructed from telecommunication cues. Finally, the history of the arts would suggest that communication is not the cut-and-dry sending of sensual cues as engineering would like to imagine; for example, one televideo image of a scene is not necessarily interchangeable with all others. Using the same technologies, an artist is likely to be able to send a richer sense of presence than someone not attending to the communicative act.

Artists are increasingly interested in the issue of telepresence. What kinds of environments can they create to allow audiences to confront the questions and meanings of communication over a distance? These questions become more critical as the richness of information being sent increases and the ability to act at a distance is augmented. In an introduction to the *Ylem Newsletter* on telepresence art that he edited, Eduardo Kac noted that telepresence art grew out of the telecommunications art described earlier:

Telepresence art branches out of this context, to merge the virtual quality of telematic space with the hardscape of physical environments—that is the basic premise of telepresence art. Whereas telepresence art was originally developed as an investigation of the unique aesthetic possibilities of telerobotics, today artists from different backgrounds contribute their own ideas and perceptions about the problem of remote presence in art.[11]

Artists Exploring Telepresence

Ken Goldberg

Ken Goldberg moves simultaneously in the research and art worlds. He is a professor of robotics at the University of California at Berkeley and an active participant in the international art community. His installations are shown all over the globe and he has won numerous awards. Many of his artworks explore what he calls "tele-epistemology." He describes his interest in an article in the *Ylem Newsletter:*

I'm interested in the distance between the viewer and what is being viewed. How does technology alter our perceptions of distance, scale, and structure? Technologies for viewing continue to evolve, from the camera obscura to the telescope to the atomic force microscope; each new technology raises questions about what is real versus what is an artifact of the viewing process. Currently I'm interested in issues that arise in electronic art that involves "telepresence." . . .

Ken Goldberg: ⟨http://www.ieor.berkeley.edu/~goldberg/⟩

How does technology alter our perceptions of distance, scale, and structure? The epistemological question is: "How do I know this is real?" The visitor acts and perceives this "reality" through an instrument with no objective scale. How does the framed vision of an instrument such as the microscope differ from the framing induced by the World Wide Web?[12]

Illustrating the cross influences of art, technology, and philosophy, Goldberg edited a book called *The Robot in the Garden: Telerobotics and Telepistemology on the Internet*, which pursued this theme in depth with contributions from several disciplines. In considering classic questions of epistemology, Goldberg's outline of the book considers philosophers' search for appropriate criteria of knowledge and doubt. He suggests that the Internet is especially subject to skepticism:

But there is a crucial difference between telerobotic installations on the Internet and such familiar cases as the evening news and NASA's Mars mission. When specialists at NASA interact with a telerobotic system, they are acting within a system of authority that gives them confidence about the veridicality of the images they are viewing. The same is true of television news broadcasts: the authority of the network stands behind the veridicality of the evening news. On the Internet, however, there is no corresponding authority. Any determined teenager can set up a live Web camera or telerobotic site. This lack of centralized authority is one of the Internet's primary virtues. But it also suggests a fundamental epistemological difference between television and the Internet.

He also looks at the art world's concern with authenticity and representation. Many artists are beginning to work with telerobotics as an fruitful arena to probe these ideas. He refers to Lev Manovich's analysis of art's history in regard to deception:

As the idea of the real comes to seem uncertain, so too does the concept of the artwork itself. Walter Benjamin's claim that the technological reproduction of works of art brings about a change in the very character of the artwork is a point we noted briefly above. Benjamin saw the artwork as in some sense diminished by the employment of new reproductive techniques inasmuch as such techniques diminished the distance that the artwork establishes in its occupation of a unique space. Throughout human history, representational technologies have served two functions: to deceive the viewer and to enable action, i.e., to allow the viewer to manipulate reality through representations. Fashion and makeup, paintings, dioramas, decoys, and virtual reality fall into the first category. Maps, architectural drawings, X-rays, and telerobotics fall into the second. To deceive the viewer or to enable action: these are the two axes which structure the history of visual representations.[13]

Fig. 6.3.5. Ken Goldberg, *Telegarden.* Web visitors can plant and tend seeds via a Web-controlled robot arm.

Telegarden, one of his most well-known works, presented a garden tended by a robot arm to the World Wide Web. Anyone could observe it, but visitors willing to register could plant a seed and then repeatedly visit it to water it and observe its growth. The garden allowed the telegardeners to share messages with other gardeners. Reportedly, gardeners became very attached to their seeds and took steps to ensure their health, for instance, by arranging for surrogate caregivers if they were going to be out of Web contact. Ars Electronica's Web site describes the project: "It's connected across the Web, one side in cyberspace, the other in a very physical environment. There are problems with material reality: things break, there are insects in the garden, certain plants die from lack of water."[14] Reviews posted on the site give some flavor of people's reactions:

For the experienced gardener, the *Telegarden* offers a search for the soul of gardening. Sowing a single, unseen, and untouched seed thousands of miles away might seem mechanical, but it engen-

ders a Zen-like appreciation for the fundamental act of growing. Though drained of sensory cues, planting that distant seed still stirs anticipation, protectiveness, and nurturing. The unmistakable vibration of the garden pulses and pulls, even through a modem.
—Warren Schultz, *Garden Design*, December/January 1996.[15]

An earlier work, called *The Mercury Project,* allowed Web visitors to help excavate artifacts in order to understand a hypothetical culture. The site offers an elaborate fictional text describing the discovery of the site and the scientific problems it poses. Web visitors could cause a robotic device to move sand around in the search for artifacts. Web archaeologists shared their insights via the Web in their search for knowledge.

Shadow Server is another in Goldberg's series of epistemological inquiries into ways of knowing. Web visitors are confronted with a light-tight box with mysterious objects inside. Web controls allow visitors to turn various combinations of lights on in order to cast shadows that might help reveal what the objects are. Annick Bureaud analyzes the site in a review for *Leonardo Digital Reviews:*

This work strongly questions the motto of these past few years in electronic art, which was "make visible the invisible." *Shadow Server* makes the normally visible invisible and gives you only its shadow. The "thing" is hidden, the objects, the "real" are no longer available. Moreover, what is interesting and relevant are the shadows themselves, in other words, a certain kind of information about the "thing." You have no way to get any information about the objects themselves, the shadows have become the "thing." . . . We cannot but think of Plato and the Cave. Ken Goldberg, in a kind of reverse situation, demonstrates here that the shadows are as important, if not more than, the objects.[16]

Memento Mori (chapter 3.3) offers Web visitors access to a real-time display of a seismograph as a reminder of mortality:

Memento mori, Latin for "Reminder of Death," is an art historical motif. *Memento Mori* paintings often include skulls or skeletons alongside lush interiors, young maidens, or fresh fruit. Normally they are interpreted as a warning about vanity and overindulgence: "All flesh is grass." On the other hand, they can be interpreted as gentle reminders: "Don't forget to smell the roses." Both may be appropriate for the digital age.[17]

Legal Tender (in collaboration with Eric Paulos) also explored the question of truth at a distance. Web viewers were able to devise remote tests to check whether a

$100 bill was authentic. One of Goldberg's major accomplishments is the creation of events that straddle science, art, and technology. They use state-of-the-art technology to probe the meanings of that technology. They trouble and perplex at the same time that they engage and amaze. In *Ouija 2000* Goldberg presented an online telerobotic Ouija board that responded to questions about the new millennium. Web visitors could control the movement of the planchette and observe the results on a web camera. Physical visitors to the installation thus observed a self-moving Ouija board acting very much the way historical boards did—moved apparently by spirits from afar.

Eduardo Kac

Kac believes that the development of technologies of virtual reality and telepresence represent significant cultural events. Artists have an important role to play in elaborating the meaning of these technologies. He sees the involvement of artists with telepresence as a natural outgrowth of prior artistic experimentation with telcommunications and interactivity. Over the years he has created a remarkable series of installations that investigate telepresence within an artistic context. Artists have interests that move beyond the practical goals of other researchers:

Telepresence is being pursued by scientists as a pragmatic and operational medium that aims at equating robotic and human experience. The goal is to reach a point in which the anthropomorphic features of the robot matches the nuances of human gestures. In this search for an "operational double," to use Baudrillard's term, humans wearing flexible armatures will, scientists believe, have a quantifiable feeling of "being there." While it is clear that actions will be performed by telepresence on a more routine basis in the future, I do not think that the ability to execute specific tasks which captivates scientists is what will interest artists working with telepresence. It is certainly not what stimulates me.

The idea of telepresence as an art medium is not about the technological feat, the amazing sensation of "being there," or any practical application, the success of which is measured by accomplishing goals. I see telepresence art as a means for questioning the unidirectional communication structures that mark both high art (painting, sculpture) and mass media (television, radio). I see telepresence art as a way to produce an open and engaging experience that manifests the cultural changes brought about by remote control, remote vision, telekinesis, and real-time ex-

Eduardo Kac: ⟨http://www.ekac.org/⟩

Fig. 6.3.6. Eduardo Kac, *Rara Avis* (1996). A networked telepresence installation in which local and remote participants experienced a large aviary from the point of view of a telerobotic macaw.

change of audiovisual information. I see telepresence art as challenging the teleological nature of technology. To me, telepresence art creates a unique context in which participants are invited to experience invented remote worlds from perspectives and scales different than human, as perceived through the sensorial apparatus of telerobots. The rhythms created by this new art will be accented by intuitive interfaces, linking and networking concepts, telerobot design, and remote environment construction. . . .[18]

His paper "*Ornitorrinco* and *Rara Avis:* Telepresence Art on the Internet" describes two of his installations and presents some theoretical background. He starts the paper by distinguishing virtual reality and telepresence while noting that even though they seem opposite, the future may see increasing integration since they both rely on the same electronic space. Telepresence is

remote control of a nonautonomous robot in a distant physical space. I understand virtual reality as related to the creation and experience of purely digital worlds. . . . [V]irtual reality relies on the power of illusion to give the observer a sense of actually being in a synthetic world. . . . By contrast, telepresence transports an individual from one physical space to another. . . . Telepresence virtualizes what in actuality has physical, tangible existence. . . . Digital or synthetic worlds may become "equivalent" to tangible realities, since both telepresence and virtual reality technologies can project human action beyond its ordinary, immediate grasp.[19]

Kac creates art that explores telepresence, robotics, holography, language, wearable computing, and many other topics that are considered in other chapters of this book. This section focuses on the works exploring telepresence. *Ornitorrinco in Eden* is part of a series of related events. It connected people in Washington and Kentucky to simultaneously control a wireless mobile robot named Ornitorrinco (Platypus) in Chicago via simple telephone touch pads activated in a conference call. Those controlling the robots and others on the Internet were able to see the world from Ornitorrinco's eyes. Ornitorrinco's environment was populated with obsolete media such as old records, tapes, and circuit boards because of Kac's desire to "comment on the disposable environment we live in, made of products that become obsolete faster than users manage to understand and master their functionality."

Even more, Kac wanted to explore the phenomenon of isolation in cyberspace. He believes that the unprecedented telecommunication technologies provide opportunities for new kinds of "intercourse and negotiation of meaning" in public space. He noted that most information technologies atomize interactions, and explains his strategy in *Ornitorrinco* as creating a space that required collaboration, moved against hierarchical tendencies in media, and led audiences to consider emancipatory possibilities:

In the new interactive and participatory context generated by this networked telepresence installation over the Internet, communicative encounters took place not through verbal or oral exchange but through the rhythms that resulted from the participants' engagement in a shared mediated experience. Viewers and participants were invited to experience together, in the same body, an invented remote space from a perspective other than their own, temporarily lifting the ground of identity, geographic location, and physical presence. . . . With *Ornitorrinco*, we transform electronic space from a representation medium into an actuation medium. . . .

The expansion of communications and telepresence technologies will prompt new forms of interface between humans, plants, animals, and robots. The *Ornitorrinco* project has pursued this strategy while at the same time insisting on undermining current trends toward stabilization of standards and other regulatory practices. The aesthetic of hybridization explored by Ornitorrinco calls for alternatives to the hegemonic configuration of the mediascape.

In *Ornitorrinco on the Moon,* a link was established between Chicago and Graz, Austria. Austrian visitors could control a mobile robot via a telephone touch pad and see sequences of still images from the robot's visual perspective. By navigating, each person could construct their own idea of the space. In an interview, Kac summarized a critical element of the *Ornitorrinco* installations involving shifting points of view:

What the telepresence installation with the Ornitorrinco telerobot is all about is to metaphorically ask the viewer to look at the world from someone else's point of view. It's a nonmetaphysical out-of-body experience, if you will. You are asked to remove yourself from your direct experience of the space that surrounds you and transport yourself, in space and time, to another body, to another situation, to another identity. You're asked to put yourself in somebody else's shoes. . . .

We live in a world where our mental images of places, cultures, and people are no longer acquired through direct observation. . . . We think of places and we have developed concepts about cultures that we have never seen, never experienced, based only on clichés that are circulated by the media, Hollywood, television, magazines, and so forth. Ours is a very unstable world in which everything seems to fluctuate and be inconsistent, therefore the inconsistency between what the name means and what the place means. Well, it doesn't mean anything until you're there moving around and making your choices. Again, nothing exists until you make it your own, until you claim it, until you create your own narrative, until you construct it. . . . The artist is no longer someone that creates a closed structure to be pondered on, or gazed at. . . . I think that artists must have a sense of being uncomfortable, of investigating, of asking questions, of experimenting, of taking risks.[20]

Rara Avis confronts the viewer with a large aviary populated by thirty flying finches and one large, strange-looking, immobile, tropical macawlike bird called a macowl. The macowl was a telerobot with CCD video cameras for eyes. In front of the cage was a VR headset that allowed the visitor to see the installation from the perspective of the large bird. The VR headset also caused the bird to turn its head in correspondence with the participant's movement. Remote Internet viewers could make sounds out of the macowl's mouth via microphones. Thus, parallel to Ornitorrinco's state, several persons shared the body of the telerobot. Internet viewers could hear the sounds of the space in addition to seeing out of the eyes of the large bird. *Rara Avis* presented the opportunity to explore several agendas. Kac hoped it would stimulate participants to think about telecommunication's assault on traditional boundaries, the place of self in relation to "the other," and new interaction possibilities in virtual space:

By enabling the local participant to be both vicariously inside and physically outside the cage, this installation created a metaphor that revealed how new communications technology enables the effacement of boundaries at the same time that it reaffirms them. . . . This image of "the different," "the other," embodied by the telerobotic Macowl, was dramatized by the fact that the participant temporarily adopted the point of view of the rare bird. . . . As the piece combined physical and non-physical entities, it merged immediate perceptual phenomena with a heightened awareness of what affects us but is visually absent, physically remote . . . [W]e must not embrace the technofilic or the technophobic extremes. Both positions are dangerous in that they fail to address the deeper, more complex social implications of new technologies—either by blindingly embracing them or by fearfully refusing to accept their impact in our lives.

In *Teleporting to an Unknown State,* presented at the 1996 Siggraph Art Show, Kac created an installation that allowed viewers throughout the world to transmit light to a seed in one location, thus collaborating in promotion of its growth. In Kac's *Uirapuru* visitors communicate with a telerobotic lighter-than-air floating animal symbolizing the mythical bird that gave life to the Amazonian forest, singing "when it hosts the spirits of those who are far away." Through a Web and local interface and sensors placed inside an artificial forest, visitors animate its activity through the level of Internet activity and can track the motion via a live VRML avatar that tracks the floating animal.

Like many artists working at the frontiers of techno-culture, Kac monitors the worlds of science and technology as part of his artistic practice. For example, he sees the arrival of the Mars Pathfinder on Mars as a critical event for our culture. In an editorial, Kac comments on the momentousness of the event and suggests some of the aesthetic issues that call for artistic attention:

This historic event rekindled the drama of distance and the cultural meaning of telepresence on the imagination of the general public, reverting the numbing and soothing effect of habitual televised entertainment and newscasting. . . . While the aesthetic dimension of this experience will go unnoticed by most directly involved in the project and telespectators alike, it is precisely this aspect of the media event I witnessed today that I find particularly significant. Some of the aesthetic features unique to this telepresence event are the relativity of space and time (seven months to get there, ten minutes to transmit a picture); the nature of the human-machine interface (combination of tele-operation and autonomy); remote space negotiation and navigation (unpredictability of the terrain, feeling of remote presence); tele-operation (at-a-distance control of a robot); capture, transmission, reception, processing and unveiling of the images; the instantness of the pictures; the realization of all this live on television (integration between the one-to-one experience of remote control with the public space of television); and the impact of this telepresence event on the collective consciousness.[21]

Fig. 6.3.7. NASA's Mars Pathfinder. A robot that combined autonomy and telepresence in exploring Mars (NASA/JPL/ Caltech/NSSDC).

Marking the cultural import of the development of telepresence, Kac entertains both positive and negative ramifications. His paper "Telepresence Art" considers several theoretical implications of telepresence: its creation of new mutivocal pathways of communication; its potentially disruptive dislocations of self and place; and its investigation of senses outside the usually privileged ones of vision and hearing. He explains the changes brought on by telepresence:

Telepresence is an individualized bidirectional experience, and as such it differs both from the dialogic experience of telephony and the unidirectional reception of television messages. . . . I'm interested in understanding to what extent these changes will reinforce social and cultural codes already in place and to what extent they will create new ones, generating unpredictable contexts where different art forms will emerge.

"As we enter the age of telepresence," writes [Abraham] Moles, "we seek to establish an equivalence between "actual presence" and "vicarial presence." This vicarial presence is destroying the organizing principle upon which our society has, until now, been constructed. We have called this principle the law of proximity: what is close is more important, true, or concrete than what is far away, smaller, and more difficult to access (all other factors being equal). We are aspiring, henceforth, to a way of life in which the distance between us and objects is becoming irrelevant to our realm of consciousness. In this respect, telepresence also signifies a feeling of equidistance of everyone from everyone else, and from each of us to any world event. . . . The subordination of three-dimensional bodily space to real time is a process of abstraction that continuously blurs the distinction between images and reality. It brings the most tragic news from Bosnia or Somalia to the same sphere of entertainment as sitcoms and talk shows.[22]

Eric Paulos

Eric Paulos focuses on telepresence. Many of his projects attempt to set up situations in which individuals can control remote viewing apparatuses—for example, *PRoP: Personal Roving Presence, Mechanical Gaze,* and *Ubiquitous Tele-embodiment via Blimps.* He hopes that the personal roving presences can give people the power to truly enter and act in remote environments, using the Internet as a highly accessible interface:

In the rush into cyberspace we leave our physical presence and our real-world environment behind. The Internet, undoubtedly a remarkable modern communications tool, still does not empower us to enter the real world of the person at the other end of the connection. We cannot look out their window, admire their furniture, talk to their office mates, tour their laboratory, or walk outside. We lack the equivalent of a body or Personal Roving Presence (PRoP) at the other end with which we can move around in, communicate through, and observe with. However, by combining elements of computer graphics, the Internet, and telerobotics it is possible to transparently immerse users into navigable real remote worlds filled with rich spatial sensorium and to make such systems accessible from any networked computer in the world, in essence: Globally Accessible Tele-embodiment.[23]

In *Mechanical Gaze,* Paulos and his collaborator John Candy offered Web visitors a remote controllable robot arm (with six degrees of freedom and a camera attached) that could inspect museum exhibit objects from multiple points of view and zoom levels. They sought to create a device that would be useful in many contexts, for example, artistic, scientific, educational, and commercial. In *Space Browsers,* they offered mobile blimp robots. In a paper called "Space Browsers: A Tool for Ubiquitous Tele-embodiment," they explain their approach to telepresence. They conclude that videoconferencing lacks crucial qualities by which people get a sense of place: mobility, autonomy, and the ability to change points of view. Paulos and Candy furthermore assert that a free-roaming telerobotic blimp can more naturally fit into the social ecology of remote locations and achieve even superior access in some situations (for example, taking aerial overviews):

Eric Paulos: ⟨http://www.cs.berkeley.edu/~paulos/⟩

Fig. 6.3.8. Eric Paulos, *Space Browser* (1997). One of several small Internet-controlled telerobotic blimps with video and two-way audio that allows people to explore and interact within an installation space.

The blimp offers the possibility of a wide range of spontaneous, group interactions. Telephones and teleconferencing are intrusive media. The recipient must interrupt whatever they are doing to answer the call. The interaction is either one-on-one, or within a prearranged group. A tele-mobot cruising by a group can overhear the conversation, recognize the group members, and decide if it is appropriate to enter the conversation.

Compared to a true visit, the mobot televisit lacks very little. We do not have enough ex-perience yet to say whether something essential is lost. Clearly there is an asymmetry in a mobot-human encounter. The mobot does not manifest body language or facial gestures. This may cause unease or defensiveness in one or both parties.[24]

Paulos speculates that these remote devices might have significant cultural impact, expanding the ability to travel, enhancing the reach of the disabled, and introducing new possibilities such as remote viewers meeting together in a third place via their proxy blimps. Paulos straddles both the robotics research and art worlds, presenting demos in both art and technical settings, and organizing seminars with titles such as "Displaced Perceptions: Intriguing Questions on the Desires of Uninhibited Technology," which bring together artists and researchers. Paulos suggests that security and safety is an emerging issue in telerobotics. As these devices spread, it is possible that they could inflict physical damage. In an abstract to his Siggraph panel "Interfacing Reality: Exploring Emerging Trends between Humans and Machine," Paulos asks:

What are the issues of responsibility in terms of property damage and human injury with such systems? Does the responsibility rest in the hands of the creator of such a system or in the remote individual controlling the system? In the case of the latter it is entirely possible, due to the ease of anonymity on the Internet, that the identity of the remote user may never be know.[25]

Rafael Lozano-Hemmer

Rafael Lozano-Hemmer is an artist who creates events that integrate performance art, telepresence, and virtual reality. He heads a company called Transition State Theory, which has created events for many international art festivals. One well-known event called *The Trace* enabled people in remote installations to telepresently occupy each other's space. Each location includes a large projection screen, robotic lamps that can move their focus, and ultrasonic trackers that monitor people's movements. The telepresence is created via four methods:

(a) computer graphics—geometric figures projected on the ceiling "float" exactly above the location of the local and the remote participants; (b) robotic lights—two white narrow light beams follow automatically and intersect at the exact 3-D location of the remote tracker, while two blue beams follow and intersect at the position of the local participant; (c) positional sound—audio samples indicating the relative distance between participants are panned around to match the

---

Rafael Lozano-Hemmer: ⟨http://www.telefonica.es/fat/elozano.html⟩

Fig. 6.3.9. Rafael Lozano-Hemmer and Will Bauer, *The Trace*. Participants can trace the movement of persons in a distant space via light beams, graphics, and 3-D sound that re-create their relative placement in the local space. Work in collection of Telefonica Foundation, Madrid.

movements of the remote participant; the sounds seem to originate from his or her relative position, giving a very clear "feeling" of where he or she is; (d) statistics screen—each station has a giant monitor that shows statistics, messages, and graphics designed to give the participants quantitative information about movement in both stations.[26]

The Trace moves beyond traditional conceptualizations of telepresence. Lozano-Hemmer is not interested in maximizing the reality of remote presence but rather in investigating interesting ways of abstracting awareness of remote individuals. He is also interested in the new possibilities of multiple people occupying the same telepresence space:

Participants know nothing about each other except for their relative 3-D movements and positions. *The Trace* is a telepresence piece in the sense that it constructs a deterritorial transmission of presence, but unlike most other telepresence technologies it does not seek to "amplify" the senses of the participants but to construct three-dimensional shadows that may occupy and encompass the real space of their bodies.

"Telembodyment" happens when the two participants share the same telematic coordinates by entering the other's 3-D representation. Telembodyment can be seen as a metaphor for those moments in which humans are inside other humans: physically, as in pregnancy, sex, or surgery; or virtually, as in Mikhail Bakhtin's "intersubjectivity" or the holy communion's "the body of Christ."

Idle Hands investigates nonconventional collaboration made possible by computer networking and tracking technology. A giant projected hand follows the movement of a person in an installation space (via an ultrasonic tracking wand). Wherever the person goes, the ominous hand follows. The installation produces sounds that are a composite based on the person's movements (interpreted without direct causal linkage) and the movements of another participant on a percussion controller. *Vectorial Elevation, Relational Architecture #4* invited Web viewers to control the position of several real search lights over Mexico City. In *Repositioning Fear,* Internet participants could help control the words projected on the facade of a building (see chapter 7.4).[27]

Nina Sobell and Emily Hartzell

VirtuAlice is an event consisting of an electric wheelchair controllable by local participants and an adjustable camera telerobotically activated by Web visitors. Nina Sobell and Emily Hartzell created the event to explore the intersection of different spaces. The wheelchair could be occupied by gallery visitors to explore the gallery. The camera was pan-tilt-controlled by Web visitors, who could only see the art they wanted via cooperation of the local driver. A later work called *Starfish Noises* allowed physical and remote viewers to explore Manhattan. They explain the interdependence created in *VirtuAlice:*

Participants from each dimension share control over *VirtuAlice,* and it is together that they create the dynamic theater that is the artwork. . . . The rider acts as the Web participant's chauffeur as they ride through the gallery. Neither has complete control over the experience, and it is only through their interaction that *VirtuAlice* is brought to life. Each eventually becomes aware that he or she is putting on a performance, for each other and for the gallery audience—the Web visitor through the gesture of camera movement, the chauffeur through body language.[28]

Nina Sobell and Emily Hartzell: ⟨http://www.cat.nyu.edu/parkbench/⟩

Fig. 6.3.10. Nina Sobell and Emily Hartzell, *VirtuAlice*. Visitors to the gallery collaborate with remotely-connected viewers to explore the gallery space. The gallery visitor positions the electronic vehicle while the Web voyeur controls the camera direction.

Lynn Hershman

In *Difference Engine 3* (in collaboration with CONSTRUCT), physical and remote viewers can encounter each other in physical and virtual tours of the ZKM media museum. Three EBUs (bidirectional browsing units) are available to inspect the museums via live digital cameras. Physical viewers can aim the EBU to see different projected views of the virtual museum, and remote Internet visitors can position the EBUs to see different views of the physical museum. Physical viewer images are digitized and attached to avatars roaming the virtual museum, which can interact with other physical viewers or remote viewer avatars. Hershman designed the installation to reflect on "concepts such

Lynn Hershman, *Difference Engine 3:* ⟨http://www.construct.net/who/fog/zkm/site/html/install.html⟩

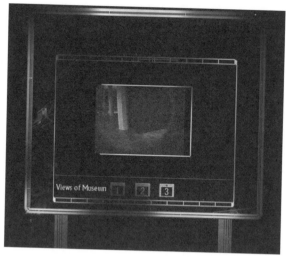

Fig. 6.3.11. Lynn Hershman, *Difference Engine 3*. Remote and physical viewers can virtually visit the ZKM museum via Web cameras and avatars.

as surveillance, voyeurism, and digital absorption." An earlier work called *Tillie the Tele-robotic Doll* had allowed viewers to control the directional gaze of two dolls in a setting where views were confounded by the presence of mirrors.

Ken Feingold

Ken Feingold has been critical of much computer art's claims to "interactivity," and claims of telematic art to radically facilitate new kinds of communications. He uses the technologies to inquire into their meanings but claims they do not interest him as much as the intention and experience created by a work. *Where I Can See My House from Here So We Are* uses telerobotically controlled ventriloquist's dummies to create a very disorienting "hell" space for inquiries into communication:

Three small robot-puppets, each with a video camera-eye and microphone ears, are together in a mirrored space. In this version of the work, the mirrored space is in an exhibition site; its walls are high enough so that the robot-puppets can't see directly over them, but low enough so that viewers to the exhibition may watch them and speak with them. Another version has the mirror-

Ken Feingold: ⟨http://www.kenfeingold.com/⟩

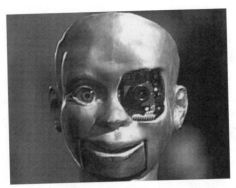

Fig. 6.3.12. Ken Feingold, *Where I Can See My House from Here So We Are.* Telerobotically controlled ventriloquist's dummies with camera eyes are used to create a disorienting "hell" space for inquiries into communication.

space in a closed room—which one could only enter through cyberspace. Each robot-puppet is connected, via the Internet, to another space, elsewhere, in which their sight and hearing is seen and heard by a distant viewer-ventriloquist as projected video and amplified sound. In each remote space there is a control device, consisting visibly of an opened attaché-case-like object, revealing a joystick and a microphone. . . . [T]he viewer-ventriloquist may drive around the robot-puppet to which they are connected in the mirrored space, and when the remote viewer-ventriloquist speaks, their voice is "projected" through the robot-puppet, amplified within it, and moves the robot-puppet's mouth. In this way, three viewer-ventriloquists may meet in the fourth (mirrored) room and exchange with each other as they please, or participate in directed theatrical performances. The space has physical limits for each robot-puppet, like national borders which they cannot cross, and their wires are only so long . . . vain tropes, perhaps, but reminders.

The work creates an immediate sense of dislocation, a making-ambiguous of the sense of "here" and "there." . . . So in this state of mind, conversations tend to expressions of "Where am I?" "Is that me or you?" "Which way am I going?" etc. And I have to admit that, to a great extent, I was aiming a critical jab at this notion of purely open-interaction, at on-line sentimentality, and at the kind of conversation that I was finding on-line up to those days when I began working on it.[29]

Other Artists and Projects

Telepresent Visuality **Martin Reiter**'s project called *Neteye* combines sensors and cameras to create mobile Internet accessible cameras, built out of recycled industrial waste.

Martin Reiter: ⟨http://www.tacheles.de/schrec/nurschre/neteye/nete.htm#info⟩

Eventually this "game" will allow users to access these cameras in several cities, using the robots as avatars. Reiter is also known for his work with the German artist group ***Tacheles,*** which offered low-cost Internet access to artists and street people. Questioning demarcations between the real and the virtual, ***Joel Slayton***'s *Telepresent Surveillance* project, built out of moving robot "eyes," is described in chapter 5.3. ***Cassandra Lehman***'s *Wango: Compost* enabled remote viewers to control vehicles collecting video and sound information. ***Rosa Trujillo Bonalo***'s project called *Amazonas Interactive* integrates concerns of "humankind, ecology, nature, environment, evolution, and communication." In part, the project will attempt to use telepresence technology to transport viewers into the Amazon region in hopes of stimulating actions to help protect this unique environment.

Devices and Environments ***Netband,*** which consists of a group of European artists, initiated a project called *Egg of the Internet* as part of a Dutch Electronic Arts Festival. Internet visitors were able to help hatch an egg by telerobotically applying heat for its incubation and then caring for it. As part of the "Robotronika" show, exploring the shadow side of telepresence, ***Survival Research Labs*** (see chapter 5.4) also created events that built on telerobotic control of dangerous devices. *Further Explorations in Lethal Experimentation,* presented at ZKM and in Austin, Texas, allowed Internet visitors to control a potentially lethal air launcher located in San Francisco by targeting objects and launching projectiles. "Increasing the Latent Period in a System of Remote Destructibility" was another similar event presented at the ICC in Tokyo. ***Masaki Fujihata***'s *Light on the Net* project allowed people all over the world to control specific lightbulbs in a 7-×-7 grid. ***June Houston***'s "Ghost Watcher" site uses the marvels of telerobotics to help her sleep. Her site describes a worry that ghosts might inhabit her house. She has set up an elaborate Web-controlled system that lets visitors turn lights on an off, activate a variety of cameras, and help to try to "catch" the ghosts in action via Web video. The Australian organization ***ANAT*** sponsored "Fusion99," which allowed artists in Germany and Australia to undertake a series of remote control projects. ***Michael Rodemer*** created a telepresent event called "About Now in Chicago: About Here, Now,"

Cassandra Lehman: ⟨http://art.net/TheGallery/Holmes/AEA/AEA.html⟩
Rosa Trujillo Bonalo: ⟨http://www.mediartech.com/en/amazonas.htm⟩
Netband: ⟨http://www.v2.nl/DEAF/persona/netband.html⟩
Survival Research Labs: ⟨http://www.srl.org⟩
Masaki Fujihata: ⟨http://www.flab.mag.keio.ac.jp/⟩
June Houston: ⟨http://www.flyvision.org/sitelite/Houston/GhostWatcher/⟩
"Fusion99": ⟨http://www.uni-weimar.de/~fusion/⟩
Michael Rodemer: ⟨http://www.ylem.org/newsletters/sep_oct1997⟩

which tried to give gallery visitors in West Virginia part of the experience of being on the lakefront in Chicago. They saw projections of Lake Michigan and felt gusts of wind controlled by a computer-controlled damper linked to wind velocity data from Chicago. *Ana Giron*'s *Artificial Time* allowed Web visitors to control a halogen light, which casted artificial shadowy directions on a sundial-like installation.

Remote Robots **A Fam.Brandt** created *Kybermax,* a pneumatically actuated, humanoid robot that could be simultaneously operated by three Internet viewers, highlighting the opportunities and discord possible when several remote users try to control a single device. The **Monochrome** artist/research group built a telerobotically controllable "animal" as part of "Robotronika." They called it Exot (alien). Several remote visitors could contend to manipulate the Lego-constructed robot. **Garnet Hertz** has created telepresence robot projects called *Interface* and *Doppelganger.* The goal was to create inexpensive, easy-to-access robots that distant users could control via low-bandwidth connections.

Physical/Virtual Relationships The **F.A.B.R.I.CATORS**'s *Robot's Avatar Dreaming with Virtual Illusions* confounds physical and virtual space. In a physical and virtual arena there are ten ambiences among which physical robots and virtual avatars can wander. The robots and avatars can be influenced by each other's behaviors in the corresponding worlds and by physical viewers and Net viewers. **Kouichirou Eto and ArtLab** developed the *Sound Creatures* installation, in which Internet-linked users could collaboratively produce sound by interchanges in virtual space and in physical space via interacting robots. My *Crimezyland* (see chapter 7.7) invited physical and remote visitors to control and observe an outdoor robotic installation on crime but gave control preference to those brave enough to physically venture out. In *Netskin,* **Christian Möller and Joachim Sauter** created an installation that allowed physical and Net visitors to control projections on the facade of the Ars Electronica center. **Shu Lea Cheang** created a two-month installation, *Buy One Get One,* at the ICC in which two digital suitcases modeled after bentoboxes were linked. Commenting on the nature of national identities in a cyberworld, Cheang traveled the physical world providing real-time updates from her traveling case to the one in the installation. Exploring the boundaries between the virtual and the physical, *The*

A Fam.Brandt: ⟨http://max.t0.or.at⟩
Garnet Hertz: ⟨http://142.232.132.45/dedocs/ggarvey/GGsmartPage.html⟩
F.A.B.R.I.CATORS: ⟨http://www.fabricat.com⟩
ArtLab: ⟨http://www.canon.co.jp/cast/artlab/⟩
Stephen Wilson: ⟨http://userwww.sfsu.edu/~swilson⟩
Christian Möller and Joachim Sauter: ⟨http://www.canon.co.jp/cast/artlab/pros2/works⟩
Shu Lea Cheang, *Buy One Get One:* ⟨http://www.ntticc.or.jp/HoME⟩

TeleZone was a telerobotic art installation in which visitors designed architectural structures using a CAD interface which got translated into a robot arm's placement of physical structural elements. *TeleSculpture 99*, sponsored by Arizona State University, allowed sculptors in several world cities to send 3-D models via the Internet that were translated into physical models via rapid prototyping technologies (see chapter 3.3).

New Information Spaces   In *Scanner++, Dump Your Trash*, **Joachim Blank and Karl Heinz Jeron** created a joint physical/Internet installation that links the sterility of cyberspace with recycling in the physical realm. In one section, visitors walk on a catwalk that scans and projects the soles of their feet; another motion creates Web sites and searches. **Marcos Novak** is developing ideas of "Transarchitecture," which examine the use of physical spaces as portals to virtual space. **Susan Collins**'s *In Conversation* arranged for unexpected Internet teleconferences by projecting animated mouths onto the street pavement through which remote viewers could communicate with surprised passersby who could communcate back through hidden microphones and cameras. **The Centre for Metahuman Exploration**'s *Project Paradise* fostered vicarious communication between remote viewers via the control of body motions of physical human "avatars" in two telelinked booths. Commenting on the questionable efficacy of voting, their *Absentee Ballot* invited an audience to control the ballot marking of a robotically augmented voter seen on television via telephone.

Linked Visual Reality Environments   In *Neither Here Nor There*, **Dan Sandin** explores the potential of the CAVE VR environments to support immersive telecollaborations (see chapter 7.3). **Simon Penny**'s *Traces* allows participants in networked CAVEs to communicate body motion via abstracted traces, which eventually take on lives of their own (see chapter 4.3).

Devices Connected to the Web

One of the most interesting subcultures of the Web is the enormous effort to connect Web cameras, strange devices, and sensors to the Web. The Web camera phenomenon

---

The TeleZone: ⟨http://telezone.aec.at⟩
TeleSculpture 99: ⟨http://is.asu.edu/events/prism⟩
Joachim Blank and Karl Heinz Jeron, *Scanner++*: ⟨http://www.sero.org/sero/doku/nbk/nbk_english.html⟩
Marcos Novak, Transarchitecture: ⟨http://www.aud.ucla.edu/~marcos/⟩
Susan Collins: ⟨http://www.ucl.ac.uk/slade/sac/SACbio.html⟩
Centre for Metahuman Exploration: ⟨http://www.metahuman.org/⟩
Thingys on the Net: ⟨http://www.oink.com/thingys⟩
Mark's list of devices: ⟨http://www.awe.com/mark/fave-inter.html⟩

is discussed in the next chapter on Web art (6.4). One can instantaneously view thousands of locations all over the globe.

Many Web experimenters, however, go beyond simple viewing by enabling Web visitors to sense a variety of phenomena and to control devices. For example, Web viewers can sense weather conditions (wind direction and speed, temperature, etc.) at a variety of locations, identify what radio or television stations people are watching, the state of doors in homes and offices, hot-tub conditions, the toilet flushing status, and so forth.

Casual Web visitors can also tele-operate a variety of devices at will. Samples include the following: robots, telescopes, stereos, moving message signs, speech synthesizers, printers, paper shredders, music generators, electric trains, Christmas tree lights, and pagers. These telepresence experiments raise several interesting questions: Are they art, science, or pranks? What is this cultural phenomenon that leads so many people to invite distant people to sense or control something in their local environment? The experiments are examples of how technological investigations and cultural explorations can merge, and links to them can be found in the urls provided.

Visualizing Net Activity

Some analysts think of the Net as a quasi-organic entity, almost like an interlinked nerve network allowing for the interchange of messages or for organizing the activities of the earth and its peoples. For some it begins to assume a life of its own. Several artists have tried to find ways to visualize the activity on the Net. See also chapter 7.7 on new information systems.

Stelarc

In the last decade, Stelarc has created an extraordinary set of performances and installations that explore the implications of technology for concepts and practicalities of the body (see chapter 2.5). Many of his events reflect on the ways that our bodies are not under our own control. In several events he wires his body up to the Internet, either to reflect general activity levels or to allow specific individuals to tele-operate his body. In *Parasite,* he uses images acquired via the Internet to map particular muscle stimulation. As Stelarc describes it, the "cyborged body enters a symbiotic/parasitic connection with information":

Yahoo list, devices: ⟨http://dir.yahoo.com/Computers_and_Internet/Internet/Interesting_Devices_Connected_to_the_Net/⟩
Stelarc: ⟨http://www.stelarc.va.com.au/⟩

Fig. 6.3.13. *Icepick* offers a hyperwired house that provides details to Internet visitors on the times of doorbell rings, garbage (UPC scans of packaging), the toilet (flush frequency and duration), temperature, and various Webscans. The image shows Webcam images of recent doorbell ringers and statistics of the last day's rings by the hour.

The body has been augmented, invaded, and now becomes a host—not only for technology, but also for remote agents. Just as the Internet provides extensive and interactive ways of displaying, linking, and retrieving information and images, it may now allow unexpected ways of accessing, interfacing, and uploading the body itself. And instead of seeing the Internet as a means of fulfilling outmoded metaphysical desires of disembodiment, it offers, on the contrary, powerful individual and collective strategies for projecting body presence and extruding body awareness.[30]

In one *PingBody* event, Stelarc (located in Luxembourg) had his muscles stimulated by Web viewers in remote cities (Helsinki, Paris, and Amsterdam). In another *PingBody* event, he linked his body's muscles directly to Net activity levels (with the length of time for a message to be sent to a location and return used as an indicator of the Internet's traffic in particular locations).

Stelarc is a passionate spokesperson for a vision of humanity that reaches beyond the present limits of bodies. Defying categorization into either cyber optimism or cyber

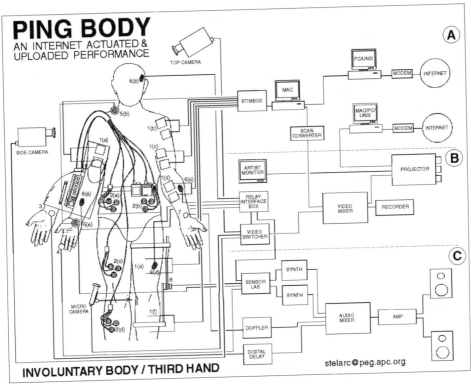

Fig. 6.3.14. Stelarc, *PingBody*. Stelarc's muscle groups are stimulated by viewers in various cities.

pessimism, Stelarc has created provocative events that sit uneasily in both worlds. His works reflect on current theoretical speculations on cultural control of the body but do not fit neatly into its categories. They go beyond the place that most serious researchers have yet ventured. Telecommunications is part of what fascinates Stelarc: How can we create composite, remotely linked bodies that question traditional notions of "ownership" and locating the self? Telepresently linked bodies can create

a shifting, sliding awareness that is neither "all-here" in this body nor "all-there" in those bodies. This is not about a fragmented body but a multiplicity of bodies and parts of bodies prompting and remotely guiding each other. This is not about master-slave control mechanisms but feedback-loops of alternate awareness, agency, and of split physiologies . . . the body not as a site of inscription but as a medium for the manifestation of remote agents.

In our Platonic, Cartesian, and Freudian pasts these might have been considered pathological, and in our Foucaultian present we focus on inscription and control of the body. . . . Bodies must now perform in techno-terrains and data structures beyond the human scale, where intention and action collapse into accelerated responses. . . .

Imagine a body remapped and reconfigured—not in genetic memory but rather in electronic circuitry. What of a body that is intimately interfaced to the WWW—and that is stirred and is startled by distant whispers and remote promptings of other bodies in other places? A body that is informed by spiders, knowbots, and phantoms? Consider the Internet structured so that it would scan, select, and switch—automatically interfacing clusters of on-line bodies in real time.[31]

Project Taos

Project Taos, a Japanese group of artists, anthropologists, musicians, and network engineers, has won competitions throughout the world for their set of projects called *Sensorium,* which visualize and use the Internet in unprecedented ways. This section describes two projects that attempt to convert the abstractions of Net activity into perceivable forms (see chapter 3.3 for descriptions of projects such as *Breathing Earth* and *BeWare*):

Sensorium's key thematic concept is "sense." *Sensorium* seeks to create a Web site that allows one to sense the world in a variety of ways. As a tool via the Net, *Sensorium* is designed to expand the meaning of the term *sense* from various viewpoints. The true capabilities of the Net do not just lie in mundane scenarios, such as virtual shopping at home, transmitting information, or as a medium for expression, but rather in helping us sense such aspects of human activity as how and what we eat, the air we breathe, what are the relationships that we share with our family, other people and creatures. We hope that the Net allows the audience to see the more important things in our lives.[32]

Web Hopper visualizes one's own trace of Internet pathways in the context of other's activities. It translates packet information about the IP address's "just connected moments" into latitude and longitude and visualizes it on maps. *Net Sound* translates the types of Internet protocols being requested into different MIDI sound events. Its developers hoped it might also support the management of networks.

Other Artists and Projects

Commenting on the historic connection between stock market prices and hemlines, **Nancy Paterson**'s *Stock Market Skirt* uses Net-collected information about stock market

---

Sensorium: ⟨http://www.sensorium.org/⟩
Nancy Paterson: ⟨http://www.bccc.com/nancy/skirt.html⟩

Fig. 6.3.15. Sensorium, *Web Hopper*. The system translates Internet packet-transport information into geographical data to show the traces of Net surfing in real time on a world atlas.

activity to control the height of a hemline of a robotic skirt. ***Joel Slayton and the C5*** (see chapter 7.7) have attempted to "map" the entire IP domain-number topology, showing ownership and activity levels in number blocks. ***Art+ Com***'s *Ride the Byte* visualized the movement of bytes on the network. *Global String* by ***Atau Tanaka and Kasper Toeplitz*** offered a musical string that vibrated in accordance with the flow of digital data. ***Randall Packer and Steve Bradley***'s *Telemusic* spatially positioned sound based on the relative network location of its source. As part of Lucent's New Experiments in Art and Technology program, sound artist ***Ben Rubin*** created a sound event that responded to levels of Internet activity.

C5: ⟨http://c5.sjsu.edu/index.html⟩
Art+Com: ⟨http://www.vision-ruhr.de/artists/art/?lang=en⟩
Atau Tanaka and Kasper Toeplitz: ⟨http://www.sensorband.com/atau/globalstring/main.html⟩
Randall Packer and Steve Bradley: ⟨http:/www.zakros.com/projects/projects.html⟩

| CURRENT IMAGE | 5 MINUTES AGO | 10 MINUTES AGO |
| SIE 105.75 | SIE 104.50 | SIE 105.00 |

Fig. 6.3.16. Nancy Paterson, *Stock Market Skirt.* Utilizing a PERL script running under Linux operating system, the skirt hemline dynamically tracks the rise and fall of individual stock values on-line ⟨www.bccc.com/nancy⟩.

Parapsychological Communication

Although considered a fringe science by some, scattered artists and industrial, government, and academic labs around the world continue to search for concrete understandings of parapsychological communication, considered by some as the ultimate telecommunication.

Kathleen Rogers

Kathleen Rogers is investigating the relationship of telepresence and parapsychology. She is exploring the possibility of telepresence technology to communicate inner states:

Recent experiments in parapsychology are proving that nonexplicable phenomena such as telepathy are replicable within an electronic void. . . . My current work is a synthesis of nonexplicable

---

Kathleen Rogers: ⟨http://web.ukonline.co.uk/Members/kathleen.rogers/psinetweb/psimain.html⟩

psi-phenomena, elements extracted from the parascientific lab, virtual technology, and art. . . . My fundamental interest is the remote replication of human perception and cognition, and I use the metaphor of the void to elaborate on this theme, applying the following hypothesis:

- The void is the absolute inner space possessed by each human being.
- Remote electronic space is a projective echo of this absolute inner space.
- Interactive telematic systems make the void a distributable phenomena.

In studying the parasciences, the aim is to identify ways in which telepresent technologies might allow us to distribute a persistent manifestation of identifiable personal energy.[33]

ICC Show—"Portable Sacred Grounds: Telepresence World"

In 1998, the ICC organized a show called "Portable Sacred Grounds: Telepresence World," which featured artists who investigated the application of telepresence concepts to extra-normal states of consciousness. Could, for example, unconscious or spiritual states be communicated? The publicity explains the scope of the show:

["Portable Sacred Grounds: Telepresence World"] reviews telepresence technology and its concept from the viewpoints focused on the nature of "collaboration," "memory and unconsciousness," "spiritual exchange," "primitivism," "multilayered reality," and "sacredness" to discover a new vision of the world. It proceeds through information space and presents new forms of sacred ground, such as "remote coexistence," where participant accesses to a virtual space foster a new world, and "telepast," where past and present resonate together. Participants will experience the fascinating world of images where boundaries—as for West/East, past/present, and interior/exterior—do not exist.

The curator attempted to expand the concepts of telepresence. None of the works included conventional telepresence arrangements with identifiable specific remote locations linked. In *Terra Present/TerraPast*, Art+Com connected the viewer with live images of the entire earth. In *The World Generator/The Engine of Desire*, Bill Seeman integrates VR and telepresence concepts to let remote viewers construct and navigate a fictitious world together. In *Neo Shamanism*, Tjebbe Van Tijen and Fred Gales create a giant drum/projection screen that serves as a portal to shamanistic spaces.[34] In *Garden of Memory*, Minato Chihiro and Moriwaki Hiroyuki create an imaginary world that viewers can traverse. They provide props to facilitate the travel: "Its devices—an orb hanging down from the ceiling, a trapeze watching the world and a stroboscope in sync with a fountain rendering a sphere—enable the viewer to search the images (and texts) which represent the outline of our world."[35]

Summary: Being There

One of the grandest appeals of telecommunications is the ability to transcend the ancient barrier of time and space. Starting with the telegraph, we have now found ways to know what is going on in a distant space and to be virtually present. The trajectory of research moves toward increasing both the reach of our distant perception and the power of our remote ability to act. Theorists have explored questions such as: What aspects of our perception can we trust? How does virtual presence reflect on our physical bodies? What cultural forces drive telecommunications research? Ultimately we must ask: Why is it so important to be virtually present in distant places?

Notes

1. K. Galloway and S. Rabinowitz, "E-Cafe Description," ⟨http://main.ecafe.com/index2.html⟩.

2. R. Kriesche, "Biennale Statement," ⟨http://pconf.terminal.cz/participants/kriesche.html⟩.

3. R. Kriesche, "New Ideas in Science and Art," ⟨http://pconf.terminal.cz/participants/kriesche.html⟩.

4. P. Sermon, "*Telematic Dreaming* Description," ⟨http://www.ntticc.or.jp/preevent/ic95/revival/paul/index-e.htm⟩.

5. M. Kushara, "Commentary on *Telematic Dreaming*," ⟨http://www.ntticc.or.jp/preevent/ic95/revival/paul/index-e.htm⟩.

6. Ars Electronica, "Interview with Paul Sermon," ⟨http://telematic.aec.at/inter.html⟩.

7. K. Roberson, "Description of Work," ⟨http://www.nomadnet.org/massage/nascence/index.html⟩.

8. N. Tenhaaf, "*Neonudism* Description," ⟨http://www.36MC-IDEA.org.uk/speakers/nt/neonudism.htm.⟩.

9. L. Rosenberg, "Definitions of Telepresence," ⟨http://cdr.stanford.edu:80/html/telepresence/definition.html⟩.

10. Quoted in L. Rosenberg, op. cit.

11. E. Kac, "Introduction," *YLEM Newsletter* on "Telepresence," available at ⟨www.ekac.org⟩.

12. K. Goldberg, Article in *YLEM Newsletter*, ⟨http://www.ylem.org/newsletters/sep_oct1997/article1.html⟩.

13. L. Manovich, quoted in K. Goldberg, "Tele-epistemology," ⟨http://www.ieor.berkeley.edu/~goldberg/⟩.

14. Ars Electronica, "Description of *Telegarden*," ⟨http://telegarden.aec.at⟩.

15. Ars Electronica, "Reviews of *Telegarden*," ⟨http://www.ieor.berkeley.edu/~goldberg/⟩.

16. A. Bureaud, "Review of *Shadow Server*," ⟨http://mitpress.mit.edu/e-journals/Leonardo/reviews/websites/burshad.html⟩.

17. K. Goldberg, "Description of *Memento Mori*," ⟨http://www.ieor.berkeley.edu/~goldberg/⟩.

18. E. Kac, "Telepresence," ⟨http://www.ekac.org/Telepresence.art._94.html⟩.

19. E. Kac, "Ornitorrinco and Rara Avis," ⟨http://www.ekac.org/ornitrara.html⟩.

20. E. Kac, "Interview about Ornitorrinco," ⟨http://www.ekac.org/intervcomp94.html⟩.

21. E. Kac, "Reflections on Mars Pathfinder," ⟨http://www.ekac.org/mars.html⟩.

22. E. Kac, "Telepresence Art," ⟨http://www.ekac.org/Telepresence.art._94.html⟩.

23. E. Paulos, "Description of PRoP," ⟨http://www.cs.berkeley.edu/~paulos/⟩.

24. E. Paulos and J. Candy, *"Space Browsers,"* ⟨http://www.cs.berkeley.edu/~paulos/⟩.

25. E. Paulos, "Interfacing Reality," ⟨http://www.cs.berkeley.edu/~paulos/⟩.

26. R. Lozano-Hemmer, *"The Trace,"* ⟨http://www.ylem.org/newsletters/sep_oct1997/article7.html⟩.

27. R. Lozano-Hemmer, "*Idle Hands* Description," ⟨http://www.media.hks.se/media/teach/gibson/idletext.html⟩.

28. N. Sobell and E. Hartzell, "VirtuAlice," ⟨http://www.ylem.org/newsletters/sep_oct1997/article3.html⟩.

29. K. Feingold, "*Where I Can See My House from Here So We Are* Description," ⟨http://www.kenfeingold.com/⟩. Also ⟨http://www.tech90s.net/kf/transcript/kf_11.html⟩.

30. Stelarc, "*Parasite* Description," ⟨http://www.stelarc.va.com.au/parasite/index.htm⟩.

31. Ibid.

32. Project Taos, *Sensorium,* "Artistic Statement," ⟨http://www.sensorium.org/⟩.

33. K. Rogers, "Telepresence and Parapsychology," ⟨http://www.uiah.fi/bookshop/isea_proc/spacescapes/j/11.html⟩.

34. T. Van Tijen and F. Gales, "Neo Shamanism," ⟨http://red.ntticc.or.jp/special/telepre/tjebbe_e.html⟩.

35. M. Chihiro and M. Hiroyuki, *"Garden of Memory,"* ⟨http://red.ntticc.or.jp/special/telepre/minato_e.html⟩.

Web Art

The World Wide Web has exploded with artistic activity in the last few years. Some even identify a unique category of Web art, with associated festivals, competitions, and debates about its aesthetics. The Web, however, has integrated itself into so many areas of human activity—for example, art, entertainment, education, science, commerce, government, and information archives—and assumed so many forms that it is a mistake to think of it as one activity. Like books, its form and intent can vary enormously.

Nonetheless, it is a critical context for many kinds of artistic activity in the current era. Some consider it the most significant techno-cultural development since the printing press. This chapter attempts to clarify the concept of Web art and surveys the work of those artists who specifically emphasize the telecommunications features of the Web. Other sections of this book describe artists who use the Web in other ways, for example, artists working with artificial life who use it as the "breeding ground" for synthetic entities (chapter 4.3), artists interested in geological or ecological concepts of the earth as a system and who use the Web as a place to integrate information (chapter 3.3), telepresence artists who use it to explore the limits of projecting presence (chapter 6.3), and artists looking at the structure and visualization of information (chapter 7.7). The Web has become the infrastructure for many kinds of art.

Many artists are using the Web primarily as a distribution system. Artists who create images, sound, video, animations, or 3-D digital worlds can make their art available on the Web with minimal economic or technical barriers. For example, a young experimental band in Singapore can be heard by audiences in the United States who would have never heard it before. Although the unprecedented ease of finding international audiences is a cultural development of major import that calls out for analysis, these artists are not focused on explorations of the significance of the new technological capabilities per se. The use of the Web only as a transport system is outside the focus of this book.

The work of hypermedia artists is a slightly more complex issue. Many are using the Web to create exquisite, nonlinear multimedia events that can be navigated differently by each visitor. Many sites won awards because of the innovative ways the artists pushed the visual and media limitations of the Web. Technological developments such as shockwave, dynamic html, streaming video, and Web programming capabilities such as javascript enhance the Web's interactive media capabilities such that they approach everything possible in CD-ROMs. Complex hypermedia is an artistic form that has been uniquely enabled by developments in information technology. Indeed, many consider it one of the most radical developments in the arts because of its promotion of interactivity, distributed authorship, decentered presentation, and the interleaving of image, sound, and text. Artists who create navigable 3-D worlds similarly take advantage of the easy linking structure to support cyber-geographical exploration.

For much of the work in hypermedia or 3-D worlds being produced, however, there is nothing indispensable about the Web and its telecommunications infrastructure. The Web is an ideal base for this kind of work because the hyperlink is a core concept of the Web and has been built in from day one, but much of the work appearing on the Web could just as well have been distributed via CD-ROM or DVD. If all of the resources for the hypermedia work have been preproduced by the author, then the Web is unnecessary. The Web is essential, however, for art that builds on connections between remote persons or makes use of information located in distant locations. This chapter focuses on Web works that uniquely investigate the Web's telecom scenarios. Chapter 7.2's focus on computer media addresses hypermedia and interactive media (both Web and non-Web) as a form of technology-inspired art.

What then are the unique telecommunications-enabled features of the Web that have inspired artistic research?

Connectivity between Persons The Web continues and enhances traditional Internet environments for person-to-person distance communication (for example, chat or news groups). Web artists are exploring manifold ways of enhancing this activity, for example, by building artificial worlds to support chat, developing avatar systems, exploring the possibilities of integrating sound and video communication, and constructing unusual conceptual structures to orchestrate communication.

Collaboration and Group Work In the non-art world, networked "groupware," which enhances group creativity and productivity on work tasks, is a major focus of innovation. Artists have developed a variety of Web environments for real-time synchronous activity (for example, use of the Net for coordinated musical performance by remotely separated individuals) or asynchronous work (for example, "exquisite corpse" kinds of arrangements by which remote artists can all work together to create artworks).

The Creation of Distributed Archives Telecommunications does not only allow access to remote persons. It also allows access to remote information sources and indeed can facilitate wide participation in the generation, accumulation, and distribution of information (and also the control, surveillance, and perversion of information). Artists have, for example, created Web sites focused on unusual topics and collect ideas from widely disparate sources.

Internationalism The Web enables unprecedented international communication across political boundaries. Some artists have built works that exploit this unique capability.

Comment on the Web Context Since the Web has quickly become a central feature of social life, some artists have created Web sites that reflexively use the Web and its technologies to deconstruct and recontextualize developments in areas such as representations of knowledge, commercialization, advertising, privacy, censorship, surveillance, intellectual property, and cultural imperialism. They have also created Web sites that play with the interface and linking conventions of the Web.

Because of the enormous amount of artistic activity on the Web, this chapter can only briefly survey the work to give an idea of the range of approaches. Web art deserves a book of its own. The Web is in such flux that difficult analytical problems confront those trying to understand and evaluate Web art. Should those using the Web mostly as a distribution system or hypermedia-authoring environment be judged by the same criteria as those pushing the technological boundaries of the Web's connectivity?

From the perspective of this book's agenda, there is even a more challenging problem. The Web has attracted a wide range of cultural experimenters drawn from widely ranging disciplines and practices who seek to explode its boundaries and invent its new capabilities and contexts. They include artists, media workers, computer scientists, entrepreneurs, psychologists, anthropologists, architects, information scientists, digital folk artists, and so on. Especially in this time of wild innovation and postmodern questioning of the claims of the arts' uniqueness, the boundaries of the arts are indistinct and permeable. Is the creation of a new kind of visual-oriented Web browser that elegantly and provocatively represents the information structure of a Web site being visited a work of art, industrial design, or information science? I and others who have served on Web-art juries have had to confront this kind of question repeatedly.[1]

Critical Perspectives on Web Art

Ars Electronica has pioneered attention to Web art by including the category in its international electronic art competitions since 1995. The statement of its Net-art jury is instructive for understanding what it saw as the essential core of the Web:

There is a perpetual debate about art: about its definitions, boundaries, function, aims, and means; about accessibility, politics, status; about high culture versus popular culture; about inclusion versus exclusion; about funding and money; about art and anti-art. . . . Net art does not equal taking whatever is on your gallery's walls, converting it into something the computer can digest, and making that accessible via the Net. Net art deals with the consequences of what it means to be on the Net and the implications of the choice of making that particular technology one's medium. For instance, film as a medium has its own rules and restrictions. . . .

The novelty of the Net and its prime importance is that it links. Connectivity is its structure and its means. The Net "connects computers, people, sensors, vehicles, telephones, and just about anything together in a global network which is fast and cheap," as Joichi Ito put it; this interconnectedness is the context. "The fun and substance of the [Net] is that it is meant to connect living minds at work in all manners of complex and purposeful configurations," Derrick de Kerckhove wrote in the 1995 "Prix Ars Electronica" catalogue. Using this connectivity, showing an awareness of this connectivity, and making sense of all the archived and real-time information that abounds in this context is what net art should be dealing with. Anything "stable," as in: done, finished, not growing, unchanging, is not Net art. What we are looking for is places on the net that reflect, stimulate, enhance this connectivity, that draw from it and thrive on it and that need the net in order to be able to exist.[2]

The jury identified six criteria that it used to guide its deliberations:

1. Grammar
2. Structure (nonlinear access)
3. Public service and Net-awareness (acknowledging its role in the global network)
4. Cooperation (stimulating cooperation and contribution)
5. Community and identity (promoting new senses of community)
6. Openness (availability for modification and augmentation).

In "Net.Art, Machines, and Parasites" (see also chapter 6.1), Andreas Broeckmann, who has been associated with the Dutch V2 group, analyzes Net and Web art with a special eye toward deconstructing the cultural and economic context of the Net. He notes the importance of developing creative responses to the forces of commercialism and censorship, and provides examples of sites (such as those described below, focused on commentary on the Web) that infiltrate the Web's structure in order to deconstruct the functioning of its underlying premises. He suggests that the aesthetic of this kind of work entails moving away from humanist notions of artistic agency into more "machinic" concepts of working with the social engines of the culture:

The aesthetics of such projects is dependent not so much on the intention of a single or collective author, but on the process initiated by and within the complex machine of people, the network infrastructure, desires, technical hardware, design tools, interfaces, behaviours. Machines in the sense in which I am using the word here are not only technical apparatuses, they are assemblages of heterogeneous parts, aggregations which transform forces, articulate, and propel their elements, and force them into a continuous state of transformation and becoming. Machinic assemblages are made up of singularities which dynamically transform the environment by which they are being transformed and recomposed. And the machinic assemblage as a whole has an aesthetic effect. The artistic explorations of the machinic are attempts at formulating an understanding of production, of transformation, and of becoming that is no longer dependent on a humanist notion of intentional agency. Its place is taken by an ethics and an aesthetics of becoming machine.[3]

Eduardo Kac's essay "Interactive Art on the Internet" expresses more faith in the multivocal possibilities of the Web. While aware of the dangers, he believes that the Web truly does create a new situation in the arts and seeks to map "the unique aesthetic investigations made possible by this seamlessly integrated worldwide computer network." He sees it as a combination of the telephone, television, and totally new forms:

Hybrids also allow artists to go beyond the creation of on-line pieces that conform to the emerging design and conceptual standards of the Internet, therefore evading what very often could seem repetitive solutions to design exercises.

Undoubtedly, the Internet represents a new challenge for art. It foregrounds the immaterial and underscores cultural propositions, placing the aesthetic debate at the core of social transformations. Unique to the postmodernity, it also offers a practical model of decentralized knowledge and power structures, challenging contemporary paradigms of behavior and discourse. The wonderful cultural elements it brings will continue to change our lives, beyond the unidirectional structures that currently give shape to the mediascape. However, as participants in a new phase of social change, facing international conflicts and domestic disputes, we must not lose sight of the dual stand of the Internet. If dominated by corporate agendas, it could become another form of delivery of information parallel to television and radio . . . The Internet also exhibits the risk of making all cultural artifacts look very similar, with virtual surfaces, standard interfaces, and regulated forms of communication.[4]

Judy Malloy, a member of the SIGGRAPH98 Web-art jury, expressed an interest in hypermedia and in the artist's care to create textual and visual impact: "I looked for an understanding of the Internet as a public art space, an understanding of the Web's hypertextual capabilities, content, depth, innovation, and impact, both visual and verbal."[5]

Many Web art initiatives were developed too late in this book's production cycle for detailed description. For example, ZKM, under direction of **Peter Weibel,** presented a major show called *Net Condition,* which was "not about 'net-art for net-art's sake'; rather, it's about the artist's look at the way society and technology interact with each other, are each other's 'condition.'" The Walker Art Center Web sites under direction of *Steve Dietz* initiated a major effort to support and make available experimental Web art, including *Shock of the View* and another set of projects called the **Art Entertainment Network,** which explored the "intersections between art, entertainment, the network, commerce, and life." **Natalie Bookchin**'s *NetNetNet* series presented works and presentations by internationally recognized artists who attempted to deconstruct the Web. The Whitney Museum of American Art's Biennial included Internet art for the first time, and San Francisco's Museum of Modern Art announced a $50,000 prize for Web art.

Here is a brief review of some notable Web art, arranged in accordance with the categories indicated above. Note that many of the sites cross over many categories.

Archive and Information Sites

Artists create sites that use the Web creatively to collect, update, and recontextualize information about specific topics. Since ancient times humans have worked to make themselves "experts" on particular realms of experience; they have collected objects, ideas, and references. In the past this expertise usually has been only known to friends or the local community. The Web makes this knowledge knowable to an international audience and enables the collector to gather information from worldwide sources and to constantly update it.

Artists typically differ from other information workers such as librarians in several intriguing ways. They can be more idiosyncratic in their definition of an area of interest or in the way they analyze a topic. They are more likely to use conceptual strategies such as mapping, decomposition, and unlikely linkages to represent a body of information. Also, they use image and media creatively as a way of presenting and navigating their Web offerings. This section describes a few examples of artist "information" sites.

ZKM: ⟨http://on1.zkm.de/netCondition.root/netcondition/start/language/default_e/⟩
Walker Art Center Projects: ⟨http://www.walkerart.org/⟩
Walker Art Center: ⟨http://aen.walkerart.org⟩
Natalie Bookchin, *NetNetNet:* ⟨http://calarts.edu/~ntntnt⟩
Whitney Museum of American Art, Biennial Web Art: ⟨http://www.echonyc.com/~whitney/exhibition/2kb/internet.html⟩

Antonio Muntadas— *The File Room* and *On Translation*

Antonio Muntadas devised *The File Room* as an archive on cultural censorship. The site, which won many awards, used its international reach to solicit Web visitors to contribute instances of censorship. The site then made these anecdotes available via a variety of interactive map interfaces that organized the information in different ways, for example, location, date, media of cultural production, and grounds for censorship. The Web archive was also accompanied by a physical installation in a Chicago gallery.

Muntadas produced another site called *On Translation,* which builds on the children's game called telephone. In the project, the Internet was used to send phrases through a

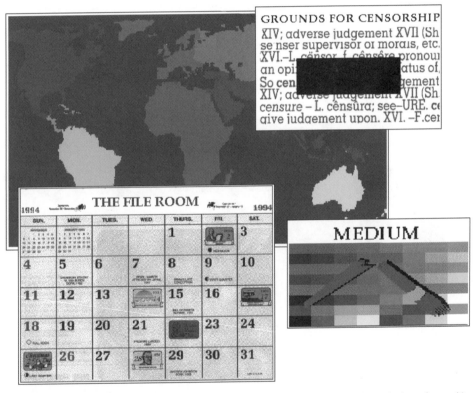

Fig. 6.4.1. Antonio Muntadas, *The File Room*. Web visitors can navigate a collaborative, cumulating database of censorship incidents by location, date, grounds for censorship, and medium, ⟨http://www.thefileroom.org/⟩.

Antonio Muntadas: ⟨http://www.thefileroom.org⟩

chain of twenty-three translations, moving from country to country. One phrase used was "Communication systems provide the possibility of developing a better understanding between people: in which language?" He sought to focus on "translation, interpretation, and cultural differences." The Internet was used to post various versions of the message as it moved and to solicit contributions of texts and links related to translation and technology.

Pipsqueak Productions— *The Company Therapist*

Pipsqueak Productions devised the hyperfictional environment of a therapist's practice. Web visitors were invited to become part of the project by generating texts coming from particular patients. A description in the National Information Infrastructure (NII) award documentation describes how the project coordinated the often chaotic nature of distributed hyperfiction:

Enter psychiatrist Charles Balis's world, who's primarily treating employees of a large San Francisco computer company. Updated daily, there's over a year's worth of well-organized patient

Fig. 6.4.2. Pipsqueak Productions, *The Company Therapist*. Web visitors generate texts coming from fictional patients.

The Company Therapist: ⟨http://www.TheTherapist.com⟩

transcripts, doctor's notes, correspondence, and other materials ensnaring readers in the doctor's fictional world. Written by its audience, *The Company Therapist* is both compelling entertainment and an educational venue to help writers improve their craft. We've solved a vexing problem in collaborative fiction by allowing each writer's voice to represent individual characters. Each author's work is distinct and yet moves the entire narrative forward. This interactivity makes *The Company Therapist* unique among Web-based dramas.

The "Disinformation" Web Site

Soliciting contributions from the international Web audience, the "Disinformation" Web site endeavors to find the news missed by mainstream news sources (e.g., public affairs, science, and culture) and to offer a wider range of perspectives than one would get elsewhere. It seeks to provide access to "hidden information that seldom seems to slip through the cracks of the corporate-owned media conglomerates."

Fig. 6.4.3. The "Disinformation" Web site provides access to hidden news stories missed or ignored by mainstream media. Courtesy of the Disinformation Company ⟨www.disinfo.com⟩.

"Disinformation" Web site: ⟨http://www.disinfo.com⟩

La Finca
The Homestead
Una colonia en el espacio ciber
A colony in cyberspace

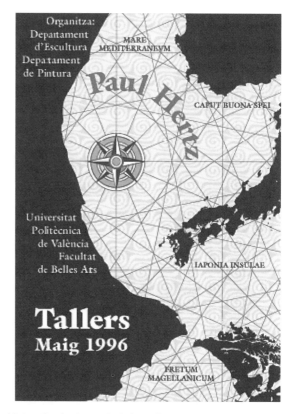

Fig. 6.4.4. Paul Hertz, *The Homestead (La Finca)*. International artists are invited to reflect on concepts of colonization and home.

Paul Hertz—*The Homestead (La Finca)*

Paul Hertz, working in conjunction with the ARTEC center at the University of Valencia, Spain, created a Web exhibition project called *The Homestead (La Finca)*. This event invited an array of international artists to reflect on the concepts of colonization and homesteading in new territories:

The Homestead aims to be "a colony in cyberspace," with work by artists and critics on the theme of colonization, particularly:

- The effects of historical colonization on the technological present
- The colonizing effects of technology as a means of cultural dominance
- Media as extensions of the nervous system
- The colonization of the body by media

Karen O'Rourke—*Paris Réseau*

O'Rourke and her collaborators have been working on a multi-year CD-ROM and Web project to create a unique archive of people's reminiscences about Paris. Images and

Fig. 6.4.5. Karen O'Rourke, *Paris Réseau.* Visitors can take various itineraries through archived reflections about Paris.

Paul Hertz, *The Homestead (La Finca):* ⟨http://www.rtvf.nwu.edu/Homestead/⟩

texts are collected and made available for different "itineraries," for example, searching for information by geography, image, or a text fragment (for example, calling out every item with a 3 in it).

Other Artists and Projects

Experimental Focus   The *Pocket Protector* site, set up by **Rose Stasuk,** comments on the merging of geek and artist culture. Artists were invited to submit their own renderings on the theme of pocket protectors, the pen holders often associated with geeky engineers. *Portrait in Space,* by **Akke Wagenaar,** creates word portraits of a man and a woman, focusing on the importance of communication and transportation methods in the definitions of these persons' lives. Many words link to Web resources that enhance the meaning. **Igor Stromajer** created *Intima,* a Web-based fictional biography of an astronaut to be created cooperatively by Web visitors. The biography will accumulate until the astronaut dies, when a CD-ROM will be made. **Douglas Davis,** linking Web experimentation with his earlier pioneering work to encourage artistic experimentation with new media such as video, constructed a Web site whose aim is to create the "World's Longest Collaborative Sentence," which invited Web viewers to examine the sentence up to the point of their visit and then add something. The *Art Crimes* site attempts to document the activities of graffiti artists worldwide in order to preserve this work, which rapidly dissappears. *Last Will and Testament* looks like a legal will with phrases crossed out. Each line is a bequest, with certain words being links to sites that elaborate on the item, many including experiments with Web interface forms, such as E-poems with dynamically transforming words.

Personal Topics   **Anne Wilson** created the *Inquiry about Hair* site to share her own art research on attitudes about hair and to solicit visitors' contributions. Part of the site's appeal is the conceptual range it introduces to the central ideas, for example, disease, fetishes, beauty, romance, body art, and weaving. The *Mother Millennia* site gathered ideas about the concept of mother accumulated from multiple cultural perspectives.

Karen O'Rourke: ⟨http://panoramix.univ-paris1.fr/CERAPLA/preseau.html⟩
Rose Stasuk: ⟨http://web.nwe.ufl.edu/~rstasuk/PocketPro/pocprotr.html⟩
Akke Wagenaar: ⟨http://www.sva.edu/salon/salon96/net-works/akke/Portrait/index_small/description.html⟩
Igor Stromajer: ⟨http://www2.arnes.si/~ljintima2/in/about.html⟩
Douglas Davis: ⟨http://math240.lehman.cuny.edu/art/index.html⟩
Art Crimes: ⟨http://www.graffiti.org/index/story.html⟩
Last Will and Testament: ⟨will.teleportacia.org⟩
Anne Wilson, *Inquiry about Hair:* ⟨http://www.anu.edu.au/ITA/CSA/textiles/hairinquiry/hair_inquiry.html⟩
Mother Millennia: ⟨http://mothermillennia.org/⟩

Private Loves/Public Opera, by **Barbara Lee and Beverly Reiser,** accumulated love stories from international Web collaborators.

Social Issues   **Akke Wagenaar** also created the *Hiroshima Project* Web site, a 3-D cumulating database of information about the atomic bomb dropped on Hiroshima, its effects on its victims, and the debate generated by the event. Intelligent agents assist in the searches. She sees the Internet as an especially appropriate context for this topic because of its international bridging of victim and perpetrator. *ArtAIDS,* organized by **Peter Ride and Ming Wei Lee,** collected a "time capsule" snapshot of the world's experience with AIDS at one particular moment in 1997, soliciting contributions from around the world in order to allow future generations to understand the impact of AIDS. Linking progressive cultural development with political structure and investigative journalism, **Paul Garrin**'s Media Filter is a "tactical media" organization devoted to collecting and providing alternative news via the World Wide Web, with features such as "GloboCopWatch."

Regions   *Lesko's Codebox,* created by **Wendy Vissar,** focuses on investigating life in a region of Albania, with a special emphasis on the rules and mores that govern social interchange. But it also extends out to encourage Web visitors to contibute reflections on the codes that govern their own lives both on- and off-line. Using the Web and IRC chat, **Melentie Pandilovski** created *Site of the Empire* to focus artistic attention on Macedonia, asking, "What would the world have been like if Alexander the Great (from this region) had not died so young?" It accumulates documentation of the region in multiple media by a variety of visiting and local artists. *Tumblong* focuses on the connection between Australia and Britain in the past, present, and future by presenting culturally loaded artifacts and providing a forum for artists and the public to comment and elaborate on those objects. With the assistance of the Soros Center for Contemporary Art, **J. Bradley Adams and Steven West** curated *Multiple/Homes,* which linked two specific places—Moldova and New York City. Artists in each place were invited to reflect on the concept of home and produce work that was posted on the Web and as physical posters in each location. In the *HUMBOT* installation and Web project, **Daniel Burckhardt** and others retraced

Barbara Lee and Beverly Reiser: ⟨http://www.ylem.org/private_loves/public_opera.html⟩
Akke Wagenaar: ⟨http://www.aec.at/fest/fest95/mythos/wagenaar.html⟩
Peter Ride and Ming Wei Lee: ⟨http://www.artaids.org.uk⟩
Paul Garrin: ⟨http://Mediafilter.org/MFF/mfhome/⟩
Wendy Vissar: *Lesko's Codebox* ⟨http://www.codebox.com/⟩
Site of the Empire: ⟨http://www.soros.org.mk/scca/empire/aboutthe.htm⟩
Multiple/Homes: ⟨http://www.artlink.org/visual_arts/feature/moldova/text.html⟩
Daniel Burckhardt et al.: ⟨http://www.humbot.org⟩

the journey of explorer Alexander von Humboldt, two centuries ago through Central and South America, by combining their own Internet reports with the historic reports.

Projects to Accumulate Web-Viewer Opinions

Vitaly Komar and Alex Melamid—*The Most Wanted Paintings on the Web*

Vitaly Komar and Alex Melamid invented the conceptual art event called *The Most Wanted Paintings on the Web,* which commented on market research. They hired a public

Most Wanted and *Least Wanted* Paintings

in chronological order

USA's *Most Wanted* :
(dishwasher size)

USA's *Least Wanted* :
(paperback size)

France's *Most Wanted* :
(television size)

France's *Least Wanted* :
(wall size)

Fig. 6.4.6. Vitaly Komar and Alex Melamid, Screen capture from *The Most Wanted Paintings on the Web,* a project for the Web at Dia Center for the Arts. The painting is generated from Web visitors' votes about most and least desired characteristics of paintings, ⟨http://www.diacenter.org/km⟩.

Vitaly Komar and Alex Melamid, *The Most Wanted Paintings on the Web:* ⟨http://www.diacenter.org/km/intro.html⟩

survey firm to find out what people wanted most and least in paintings and then produced paintings based on these findings, which were exhibited in a show called "Peoples Choice." Under the sponsorship of the Dia Center, they expanded the project to cover twelve countries and used the Web to solicit votes and to display the results. The description of the project notes:

In an age where opinion polls and market research invade almost every aspect of our "democratic/consumer" society (with the notable exception of art), Komar and Melamid's project poses relevant questions that an art-interested public, and society in general, often fail to ask: What would art look like if it were to please the greatest number of people? Or, conversely: What kind of culture is produced by a society that lives and governs itself by opinion polls?

Other Artists and Projects

In *Truism 33,* **Jenny Holzer** created a Web event that used truisms as a way to understand contemporary culture. Confronted by blinking, changing "truisms," Web visitors select a conceptual category—for example, belief and change—and then get to vote via checkboxes. A cumulating total shows the fate of the various truisms. Here is a sample of the sayings on "belief": "a little knowledge can go a long way; a lot of professionals are crackpots; absolute submission can be a form of freedom." As part of her explorations of surveillance, **Julia Scher**'s Web site includes a "security-land" section, which solicits visitors' commentaries on surveillance (see also chapter 7.7).

Genetic Art Using Web-Visitor Voting

Artists have developed sites with "genetic" components. That is, the form of display at any time is dependent on the "votes" of previous visitors. Several of the works described in the chapter on artificial life and genetic algorithms (4.3) incorporated similar structures, for example, *Technosphere* and *Biota.* This section describes several others that let Web visitors make judgments that shape what the next visitors will see.

Jenny Holzer, *Truism 33:* ⟨http://www.adaweb.com/project/holzer/cgi/pcb.cgi⟩
Julia Scher: ⟨http://adaweb.com/project/secure/⟩

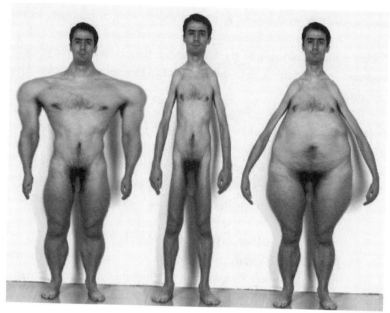

Fig. 6.4.7. John Tonkin, *Elastic Masculinities*. Web visitors can vote on body characteristics, ⟨http://www.johnt.org/meniscus/index.html/⟩.

John Tonkin—*Meniscus*

John Tonkin's *Meniscus* Web events, including *Elective Physiognomies, Elastic Masculinities, and Instant Eugenics,* explore ideas relating to "the body, identity, and subjectivity." Typically the viewer is confronted by an image, for example, of a face—and asked to make value judgments about the form. His Java applets then dynamically transform the face. The judgments sometimes affect what subsequent viewers will see.

On a deeper level, Tonkin seeks to comment on culturally defined norms about appearance and behavior and their ultimate manifestation in eugenics and genetic engineering. The linking of this cultural analysis with the technological form of genetic algorithms is an ideal combination:

I am interested in how we classify others according to their physicality, both in personal structures and in more systematic scientific structures such as physiognomy and genetics. . . . Physiognomy

John Tonkin: ⟨http://www.johnt.org⟩

demonstrates a desire for exactness in the knowledge of these inexact things. Its reductionist methodologies often involved positioning the subject on a scale of binary opposites (e.g., spiritual versus sensual), or as deviations from idealised norms. . . . Perhaps the major conceit of both physiognomists and eugenicists was that the ideal which they promoted was basically themselves: white, male and privileged.

Paul Vanouse, Eric Nyberg, and Lisa Hutton—"Persistent Data Confidante"

The *Persistent Data Confidante* site collects people's confessions. It solicits: "Have you ever felt the need to share with others a part of you which was very private?" Once a person confesses, the site shares someone else's confession in anonymous form. The reader then rates the quality of the confession; the cumulating references affect the future likelihood of that confession being shared. Periodically, the site will mate two highly rated confessions to generate an offspring.

☐ Have you ever had a personal conflict which you needed to share or confess?

☐ Have you ever felt the need to share with others a part of you which was very private?

☐ Do you often long to live vicariously through the most exciting confessions of people you've never met, possessing values and lifestyles you may not completely understand?

☐ Have you ever sought out an anonymous contact or forum with persons who share such cravings?

If you have answered "yes" to one or more of these questions you should seriously consider trading your confessions or secrets with the Persistent Data Confidante.

Fig. 6.4.8. Paul Vanouse, Eric Nyberg, and Lisa Hutton, *Persistent Data Confidante*. Web visitors are encouraged to share confessions and see those of other visitors.

Paul Vanouse, Eric Nyberg, and Lisa Hutton: ⟨http://www-crca.ucsd.edu/~pdc/⟩

From across the Internet, elementary particles of **information** are accelerated to nearly the speed of light. They travel in opposite directions around the **giant fiber ring** (above left), meeting in the **collision chamber**, (above right). The collisions release small but massive **information particles**, scattered in all directions. Some of them are collected by the information detector, and stored for later viewing. **Pressing 'reload' will get you a new random collision.** Be sure to try it a few times, because on any given run you might get something really boring. Note: collisions tend to be image-rich. Images come from whatever web site they were originally found on, rather than from www.eecs.harvard.edu.

click HERE to see a collision

Fig. 6.4.9. Trevor Blackwell. The *SuperCollider* Web site collides two randomly selected Web sites to create new composites.

Recomposing Web Resources

Trevor Blackwell—*Super Collider*

Using the metaphor of an atom smasher, Trevor Blackwell's *Super Collider* randomly accesses Web sites and smashes the sites together to return a new composite to the Web visitor. The creators were interested in new sets of ideas and connections that could be generated by the random pairing of information from the storehouse of the Web:

From across the Internet, elementary particles of information are accelerated to nearly the speed of light. They travel in opposite directions around the giant fiber ring. . . , meeting in the colli-

---

Super Collider: ⟨http://www.eecs.harvard.edu/collider.html⟩

sion chamber, . . . The collisions release small but massive information particles, scattered in all directions. Some of them are collected by the information detector and stored for later viewing.

Stephen Wilson—*50 Points of Light*

I created a hybrid CD-ROM event that explored the Web's ability to provide a visual portal into internationally dispersed locations. What would it be like to be able to see fifty places on the earth at one moment? The event invites explorations of panoramas, street scenes, and inside spaces from each city and permits recapitulations in time. World weather, the flow of night and day, and snippets of cultural materials such as art and music provide an "information visualization extravaganza." It explores cultural diversity and the cultural forces working to weaken that diversity. In the periodic "globalization" event, local cultural information is superseded by multinational corporate logos and jingles.

Fig. 6.4.10. Stephen Wilson, *50 Points of Light.* A composite image made from live Webcam images from fifty places on the globe.

Stephen Wilson: ⟨http://userwww.sfsu.edu/~swilson/art/50points/50points.html⟩

http://194.
216.252.65

Fig. 6.4.11. Elizabeth Diller and Ricardo Scofidio, *Refresh*. An array of images, some containing live Webcams, and others fictional reenactments.

Elizabeth Diller and Ricardo Scofidio—*Refresh*

Architects Elizabeth Diller and Ricardo Scofidio have created a Web event, *Refresh,* sponsored by the Dia Center, which investigates "liveness" and authenticity. Their Web site collects twelve Webcams they selected from the thousands available on the Web. For each of these Webcam settings they created fictional narrative text and images, created by hiring actors to enact scenarios that digitally composite them into the original setting. A Web visitor coming to the site encounters a grid of images, one of which is the live updating Web camera and others that are the fictional versions. The site explores the role of mediation in our perceptions of "truth":

For technophobes who blame technology for the collapse of the public sphere, liveness may be a last vestige of authenticity—seeing and/or hearing the event at the precise moment of its occurrence. The unmediated is the immediate. For technophiles, liveness defines technology's aspiration to simulate the real in real time. This skepticism [about truth] aids, to a degree, these artists'

Elizabeth Diller and Ricardo Scofidio: ⟨http://www.diacenter.org/dillerscofidio/index.html⟩

desire to tease the distinctions: to undermine the authority of "live" overmediated experience and to collapse the two into an indeterminate unity.

Other Artists and Projects

In *The Multi-Cultural Recycler*, **Amy Alexander** has created a Web environment that "recycles" Web camera images, implementing an appropriately postmodern aesthetic by dynamically creating new composite images from Webcams. Reflecting on standard browser views as only one possible manifestation, **Mark Napier**'s *The Shredder* takes a Web page defined by the user and creates a new image by reshaping text and images to make "a chaotic, irrational, raucous collage," converting "information into art." In *re-m@il* **Blank & Jeron** comment on our use of E-mail by creating a public art E-mail answering event in which the public can anonymously respond to personally received E-mail messages that people have chosen to forward to the system. **Shane Cooper**'s installation *Remote Control* presents a familiar set of couch and TV to comment on news by dynamically creating a news program from Internet sources with the truth of each sentence reversed. *Bits and Pieces* by **Peter M. Traub** automatically scours the Web for sounds in order to compose sound collages. **Natalie Bookchin and Alexei Shulgin**'s *The Universal Page* is a Web site that attempts to create a Web page composed from the average derived from merging every Web page it finds via its special script.

Collaborative Environments and Person-to-Person Communication

Bonnie Mitchell—*As Worlds Collide*

Bonnie Mitchell has created a series of Web-based collaborative events in which remote artists add elements to evolving works. *As Worlds Collide* uses QuicktimeVR technology to create synthetic landscapes (with sections contributed by many people) that viewers can explore in 360-degree panoramas as though they were real geographical spaces.

Amy Alexander: ⟨http://shoko.calarts.edu/~alex/recycler.html⟩
Mark Napier: ⟨http://www.potatoland.org/shredder/⟩
Blank & Jeron: ⟨http://www.sero.org/re-mail⟩
Shane Cooper: ⟨http://www.shanecooper.com⟩
Peter M. Traub: ⟨http://music.dartmouth.edu/~peter/bits/⟩
Natalie Bookchin and Alexei Shulgin: ⟨http://universalpage.org⟩
Bonnie Mitchell: ⟨http://creativity.syr.edu/~worlds/x⟩

Fig. 6.4.12. Ed Stastny, *Infinite Grid.* Web visitors contribute images to a dynamically growing array.

Ed Stastny—"SITO," *HyGrid,* and *Infinite Grid*

Ed Stastny's "SITO," originally called OTIS, was one of the pioneering collaborative art sites on the Web. Several "synergy" events promoted the intention: "'SITO,' at its most basic interpretation and intention, is a place for image-makers and image-lovers to exchange ideas, collaborate, and, in a loose sense of the word, meet." Another project, *Hy-*

Ed Stastny: ⟨http://www.sito.org/sito/synergy/infgrid/index.html⟩

Grid, provided a spatial hyperdimensional system into which Web visitors could contribute images. The grid was constantly evolving and allowed experimental navigation, such as "tri-linear gridding," "wormholes," and "weirdlinks." *Infinite Grid* allowed additions forever.

Knowbotic Research—*IO_DENCIES*

Knowbotic Research is an interdisciplinary German/Austrian artist group known for its investigations of knowledge systems and information visualization (see chapter 7.7). Its

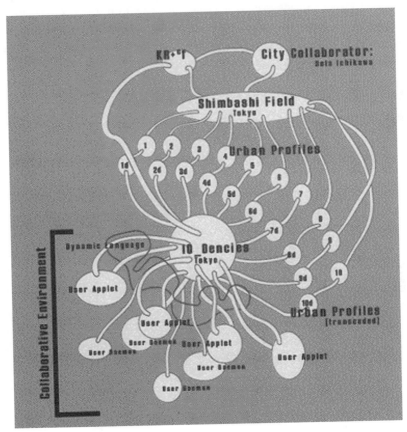

Fig. 6.4.13. Knowbotic Research, *IO_DENCIES*. An information system for analyzing conditions and orchestrating action in the cities of Tokyo and Sao Paolo.

Knowbotic Research: ⟨http://www.khm.de/people/krcf/IO/⟩

IO_DENCIES environment, which focused on diverse international urban environments, in this instance the megacities of Tokyo and Sao Paolo, facilitated collaboration and the emergence of insights about urban processes. Its system represented the forces of the city visually via live Internet data and let remote participants intervene directly on the representation:

By using special tools ("attractors"), Internet users may intervene in these force fields. Every modification in these zones changes the flows, and these interventions are displayed immediately. However, users see not only the changes and consequences of their own interventions, but also the interventions and desired changes effected by other users "working" in the same hypothetical urban area. In this way, users may gain a deeper insight into the complexity and inner connections of networked, connective agency.

Other Artists and Projects

Art Collaborations Remote C/ontrol investigates possibilities of communications technology for supporting artistic work by creating a temporary World Wide Web collaborative space. It was organized for Ars Electronica97 by **Andreas Broeckmann and Diana McCarty.** Each geographically remote artist posted work and cross commentary in the shared server space and attempted to extend the on-line material into other contexts via telephone, radio, and CD-ROM. **Nina Sobell and Emily Hartzell**'s *Park Bench* created a network of Web kiosks in New York City that allowed real-time collaboration and video communication. As part of Ars Electronica97, a networked environment called *OpenX,* was created as a "walk-in network," process open to change and membership as different connections came into being. *Honoria in Ciberspazio* is an on-line interactive "opera," set up by **Madelyn Starbuck,** that explores life in on-line communities. Web visitors are encouraged to contribute events by writing poetry in reaction to a posted story line. Inviting "participants to construct their own narratives—at the intersection of chance and interpretation," **Sharon Daniel**'s *Narrative Contingencies* Web site creates an evolving database in which viewer contributions of text, image, and interpretative actions cause shiftings, accretions, and deletions of the structure. The Society for Old

---

Andreas Broeckmann and Diana McCarty: ⟨http://remote.aec.at⟩
Nina Sobell and Emily Hartzell: ⟨http://www.cat.nyu.edu/parkbench⟩
OpenX: ⟨http://www.aec.at/fleshfactor/openx.html⟩
Madelyn Starbuck: ⟨www.cyberopera.org⟩
Sharon Daniel: ⟨http://metaphor.ucsc.edu/~sdaniel/⟩

and New Media, in association with the Guggenheim Museum, presented an *Online Virtual Court* in its *Brandon* project, in which Web visitors could agree to act as jurors in trials of cases of sexual assault involving cross-gender victims. ***Andy Deck and Mark Napier****'s Graphic Jam* allows several Web visitors to simultaneously draw on a shared digital canvas and to activate animations to show recent changes.

3-D and VRML Worlds **Helen Thorington** created the "Turbulence" Web space for artists specifically focused on the investigation of unique features of the Web. Participants attempted to extend features of VRML, multimedia, and cumulating archives. It was part of the *OpenX* initiative. ***Moove*** created a 3-D environment, *Roomancer,* to facilitate a range of work with computers, including on-line communication. It provides a meta-phor of a 3-D version of a building in which people can rearrange their own resources, such as files and images, for example, by placing images on a wall in a virtual gallery. On-line communication can take place in "rooms" occupied by avatars representing each of the communicating parties. ***Andy Best and Merja Puustinen*** (part of the Ampcon group) developed a multiuser VRML/chat environment called *Conversations with Angels.* Commenting on the pedestrian nature of many VRML worlds, they populated their VRML world with avatar robots and a variety of other characters modeled on those that make real life interesting, such as a serial killer, a middle-aged single mother, a redneck survivalist, and a lesbian princess. ***Phase(x)3*** is a highly visual collaborative dynamic VRML environment in which professionals such as architects can collectively and cumu-latively author 3-D models. Each author places models into the space, and they are then evaluated and reinterpreted by subsequent authors. The system allows for multiple perspectives in looking at the body of work, such as the visual representation of genealogy or spatial visualization of the criteria to be used in evaluation.

Virtual Worlds In *Homeport,* **Lawrence Weiner** worked within the "Palace" virtual world system to create an interactive navigation experience. Each person's adventure was affected by others on-line at the same time, for example, some of the passageways in the virtual world could be occupied by only one person at a time. The group ***Sherwood***

Guggenheim: ⟨http://brandon.guggenheim.org⟩
Andy Deck and Mark Napier: ⟨http://bbs2.thing.net/jam/⟩
Helen Thorington: ⟨http://www.turbulence.org⟩
Moove: ⟨http://www.moove.de⟩
Andy Best and Merja Puustinen: ⟨http://ampcom.kaapeli.fi/⟩
Phase(x)3: ⟨http://space.arch.ethz.ch/ss99/index.phtml⟩
Lawrence Weiner: ⟨Homeport http://www.adaweb.com/project/homeport/⟩

Forest uses the Alphaworld Web context, which provides a 3-D world with avatar-populated "chat" capabilities. Using a forest metaphor, "the purpose of Sherwood is to try to design a very natural, attractive setting with woodlands, flowers, and water and then build a village community."

Music ***ResRocket*** created a collaborative Net environment that allowed real-time collaboration among remote musicians working on MIDI works. Taking advantage of Internet interconnectivity, ***Jerome Joy***'s projects attempt to create collaborative Web environments for music composition. Using an interface of an intriguing icon grid pointing to different kinds of music resources each time it loads, ***James Stevens*** created the "Backspace" site, which allows artists and musicians all over the world to contribute work to a cumulating archive. Conceptually, the site is linked to "the ebb and flow of the Thames." ***Koji Ito*** created *Net Rezonator* to facilitate real-time musical and visual communication and collaboration via features such as the visual representation of sounds.

Oral History ***Carol Flax***'s *Ex/Changing Families* explores the world of emotions attached to the process of adoption. It is a traveling collaborative work that includes installation, artists' books, and a cumulating World Wide Web environment. *Bubbe's Back Porch* extends ***Abbe Don***'s prior multimedia installation work, such as *We Make Memories,* to create a Web environment that encourages family storytelling and oral history. ***Eva Wohlgemuth and Kathy Rae Huffman*** created a collaborative Web environment called *Face Settings,* especially targeted at facilitating women's involvement in the Net. Conceptually, it builds on women's involvement with food and intimate conversation, such as at the dinner table.

Miscellaneous ***SETIathome*** enables tens of thousands of Networked individuals to set up their computers so that spare computer time is automatically used to help analyze data from the scientific search for extraterrestrial intelligence. *E-motions* ironically focuses on aloneness and connectivity in a connected world. Using images of the Russian cosmonauts, the various texts lament and plead for connection, for example: "Is there anybody

---

Sherwood Forest: ⟨http://www.ccon.org/events/sherwood.html⟩
ResRocket: ⟨http://www.resrocket.com/⟩
Jerome Joy: ⟨http://www.imaginet.fr/manca/joy⟩
James Stevens: ⟨http://bak.spc.org/⟩
Koji Ito: ⟨http://www.hilab.mag.keio.ac.jp/~nr/⟩
Carol Flax: ⟨http://www.cmp.ucr.edu/Ex/ex_changing_right.html⟩
Abbe Don: ⟨http://www.bubbe.com/⟩
Eva Wohlgemuth and Kathy Rae Huffman, *Face Settings:* ⟨http://thing.at/face⟩
SETIathome: ⟨http://setiathome.ssl.berkeley.edu/⟩

out there—Help me." Actual live E-mail submission spaces on the page allow the visitor to send E-mail to the fictional persons and receive messages back. In **Seiko Mikami**'s *Molecular Clinic,* each Web visitor was able to download a single molecule of a 3-D spider living on a Web site. They could manipulate the visual and behavioral qualities of the molecule and then reinsert it back into the "spider," thus cumulatively contributing to its evolution.

The Development of New Capabilities

Kazuhiko Hachiya—Post Pet

Kazuhiko Hachiya created *Post Pet* as a comment on E-mail systems and as an alternative context for using this technology. The sending and receiving of E-mail messages are visualized as objects carried by cartoon creatures, for example, a cat or teddy bear:

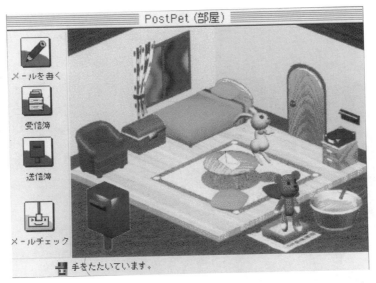

Fig. 6.4.14. Kazuhiko Hachiya, *PostPet.* An E-mail system that visualizes message movement via cartoon characters. Copyright, 1996–99 by Sony Communication Network Corp. All rights reserved. PostPet is a trademark of Sony Communication Network Corp.

Seiko Mikami: ⟨http://www.cast.canon.co.jp/cast/al5/special-e.html⟩
Kazuhiko Hachiya: ⟨http://www.postpet.de/⟩

The *Post Pet* appears on the recipient's screen and delivers the message "personally." Now the recipient is not only obligated to treat the "Post Pet" well, to feed it, care for it, play with it, but also to respond to the message so the *Post Pet* can find its way back to its owner. There are already homes on the Internet for lost *Post Pets* and "Post Pet Hotels," to which owners may entrust their pets.

The *Post Pet* is a kind of "E-mail tamagotchi." Unlike the solitary tamagotchi, however, it is an aid to communication, and it turns the purely technical, hard to understand process of data transfer via Internet into an emotional and connective experience.

I/O/D — *Web Stalker*

The British group I/O/D claimed that conventional Web browsers do a poor job of visually representing the information structures of sites being visited. It created a highly visual browser called *Web Stalker,* which deconstructs existing sites. Designed to cut out advertising and other nonessential "noise," it uses circles and lines to portray the linkage

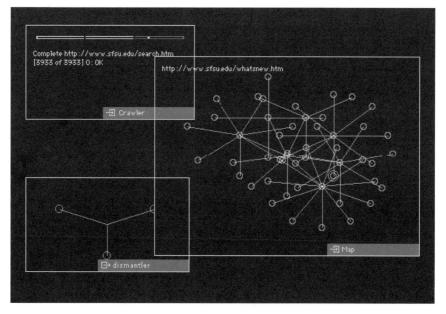

Fig. 6.4.15. Backspace/I/O/D. *Web Stalker.* An alternative system for browsing and visualizing the Web.

I/O/D: ⟨http://www.backspace.org/iod/⟩

structures of the Web and allows the user to operate on the representation. Matthew Fuller, one of its creators, explains the search for alternative "maps" of Web space:

A lot of the working capabilities within the [standard commercial] browser have been determined by the needs of advertisers, corporations, and so on, rather than experimentation with the format of the Web. . . . *Web Stalker* is based on the belief that the user should be able to define the different functions they want to apply to a Web document, rather being than launched through a finished Web site.

GINGA

GINGA (Global Information Network as Genomorphic Architecture) is a customizable VRML system for organizing and searching "information-scapes." The spaces can either

Fig. 6.4.16. Fumio Matsumoto, *GINGA* (Global Information Network, as Genomorphic Architecture). A VRML information visualization.

GINGA: ⟨http://www.plannet-arch.com/ginga.htm⟩

be preexisting Web or locally published resources. Users can choose from a variety of representational schemes (Nebula, Ring, Network, Forest, Strata, Text, Image, Polyphony, and Cemetery) and can design the appearance and behaviors of avatars to act on the worlds. Other users' avatars can be invited in to explore the world and interact. The user can shape the worlds with informational "cyberspacial codes" such as distribution, depth, relationships, and dates, which change the functional and visual qualities of the display. A later Web VRML project called *Infotube* attempted to create a visualization of a Japanese shopping mall based primarily on information about the stores rather than on physical spatial arrangements.

Other Artists and Projects

Tools   Probing the creation of tools as a cultural act, Ars Electronica in various years gave Net prizes to ***Tim Berners-Lee*** for his invention of the World Wide Web and to ***Linus Torvalds*** for his development of Linux. Linux is considered extraordinary because of its status as a free and open source, its technical excellence, and its mobilization of the world community of altruistic developers to compete with dominant commercial values. ***SmartMoney****'s Marketmap* converts stock-market activity into a pulsing field of color that could dynamically be reconfigured to inspect concepts such as historical context and personal holdings. ***Inxight*** developed an innovative visualization of Web information called "hyberbolic trees," which resemble mandalas and dynamically adjust the detail of the display with the user's changing focus. In *Name Space,* ***Paul Garrin and Andreas Trager*** developed a comprehensive system for people to assign and manage their domain names on the Internet. ***Toshiya Naka and Yoshiyuki Mochizuki*** developed the multiuser *Wonderspace,* a VRML world in which visitors could communicate via voice and in which sound and motion could be synchronized. ***Konsum Art.Server****'s LinX3D Console* presents a 3-D environment in which Net text is translated into textures and "data avatars." ***Maciej Wisniewski****'s Netomat*™ is an alternative browser that "dialogues with the net to retrieve information as unmediated and independent in form,"

Linus Torvalds, Linux: ⟨http://www.linux.org⟩
SmartMoney: ⟨http://www.smartmoney.com/marketmap/⟩
Inxight: ⟨http://www.inxight.com/⟩
Paul Garrin and Andreas Trager: ⟨http://name.space.xs2.net/⟩
Wonderspace: ⟨http://www.siggraph.org/s97/conference/garden/wonder.html⟩
Konsum Art.Server: ⟨http://www.konsum.net/linx3d⟩
Maciej Wisniewski: ⟨http://www.netomat.net/⟩

moving beyond the conventions such as Web pages and links, allowing a dynamic inter-action with resources that flow to the screen.

Conceptual Extensions   Henning Timcke and Andreas Trottmann created the "Color of E-Mail" site, which enables visitors to define VIRT (very important receiving targets) and provides a flexible automatic system for converting messages among electronic options (E-mail, fax, pagers, and phones); for selecting the precise time of delivery; for adding headers such as "Very Important"; and for considering likely targets, for example, "your boss." Working as part of the PAIR artist-in-residence program with Xerox PARC researchers ***Jock MacKinlay and Polle Zelleweger,*** I developed *Search as Portrait,* which visually represented the path of a Web search as a portrait of the person who conducted the search, and *Shadow Server,* which offered a browser that constantly supplemented current Web pages. Acting as a metaphor for life, it offered visual information in the background derived from pages not chosen at the last-choice point and from pages that potentially could be accessed in future choices.[6]

Reflections about the Net

Guillermo Gómez-Peña, Roberto Sifuentes, and James Luna—CyberVato and "Temple of Confessions"

Guillermo Gómez-Peña, along with associates Roberto Sifuentes and James Luna, believe the Internet cannot be separated from the larger socio-political context in which it sits, for example, the examination of who has Internet access. Believing nonetheless that the new communication technologies can be a tool for encouraging analysis of the narratives that underlie contemporary culture, they created a series of innovative performances that engage audiences in reflection. In the 1994 *El Naftaztec:Cyber TV for 2000 A.D.,* Gómez-Peña created a spoof in which it seemed Indian "pirates" had seized the airwaves to talk about Mayan innovations in cyber culture. The event was also broadcast on the Internet Mbone, and viewers were encouraged to send messages back.

Often the performances link physical and virtual spaces. For example, *CyberVato* pre-sented on-line "bandito" characters. The description of the event at the Rice University Gallery demonstrates this use of on-line environments.

Henning Timcke and Andreas Trottmann: ⟨http://virt.uals.com/⟩
Shadow Server: ⟨http://userwww.sfsu.edu/~swilson/⟩
"Temple of Confessions": ⟨http://www.echonyc.com/~confess/⟩

THE SHAME-MAN AND EL MEXICAN'T MEET THE CYBER-VATO

at the Ethno-CyberPunk Trading Post & Curio Shop on the Electronic Frontier

Fig. 6.4.17. Guillermo Gómez-Peña, Roberto Sifuentes, and James Luna, *CyberVato*. A Web site solicits the anonymous fears and stereotypes of Web visitors.

Sifuentes, as the character CyberVato, the "Information Superhighway Bandito," encased in a Plexiglas box, will extend the installation through Internet access. Viewers at all sites may observe the CyberVato's worldwide communications on topics ranging from the Chiapas activist movement and affirmative action to NAFTA and academic discourse. *The Shame Man* and *El Mexican't Meet the CyberVato* explore issues of cultural identity and stereotypes, while contrasting the intimacy of private ritual with the clamor of the public realm. The work also provides a subtle commentary on the skewed demographics of Internet participation.[7]

The artists often use the global reach and anonymity of the Internet to solicit visitors fears and desires about racial stereotypes, for example, "lazy wetbacks." Seeking to deconstruct these ideas, they then turn these stereotypes into performance personas that they enact in physical and on-line performance. The article "Gómez-Peña as Technoprophet/ Artist" describes the approach:

The information Gómez-Peña collects also serves another purpose. In what he calls "reverse anthropology," the typical ethnocentric worldview is inverted to treat the dominant American culture as exotic and unfamiliar. His "Temple of Confessions" site is based on the idea of "cultural confessionals" in which users confess their fears and desires. He finds that the anonymity of the Internet encourages people to respond with no inhibitions or fear of repercussion, which makes their input more revealing than traditional ethnographic fieldwork, where answers are generally monitored and filtered. As Sifuentes puts it, "What they reveal about their innermost fantasies and fears—toward Latinos, the Spanish language, immigration, and urban violence—and their search for spiritual fulfillment is beginning to take shape as a barometer measuring the climate of racial intolerance in this country."[8]

The artists suggest that cyberspace is quickly following the hegemonic patterns of the physical world. They endeavor to find interventions that question and retard this tendency and incorporate outreach into their art. Gómez-Peña notes:

I resent the fact that I am constantly told that as a "Latino" I am supposedly "culturally handicapped" or somehow unfit to handle high technology; yet once I have the apparatus right in front of me, I am tempted and uncontrollably propelled to work against it; to question it, expose it, subvert it, and imbue it with humor, radical politics and linguas polutas such as Spanglish and Frangl. I venture into the terra ignota of cyberlandia without documents, a map, or an invitation at hand. . . .

In a manifesto for remapping cyberspace, Gómez-Peña declares:

What "we" desire is to remap the hegemonic cartography of cyberspace; to "politicize" the debate; to develop a multicentric theoretical understanding of the (cultural, political, and aesthetic) possibilities of new technologies; to exchange a different sort of information (mythopoetical, activist, performative, imagistic); and to hopefully do all this with humor, inventiveness, and intelligence. Chicano artists in particular wish to "brownify" virtual space; to "spanglishize the Net," and to "infect" the lingua franca. . . . To attain all this, the many virtual communities must get used to a new cultural presence—the Webback (el virus virtual); a new sensibility; and many new languages spoken on the Net.[9]

Jodi.org

The *Jodi* site is famous for its visual pyrotechnics and its commentary on the "clean" information design of the Web. The site (created by artists who use the collective name

---

Jodi: ⟨http://www.jodi.org⟩

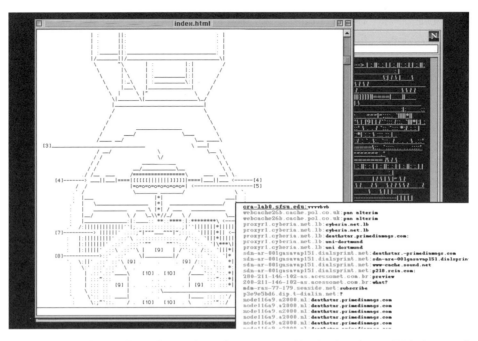

Fig. 6.4.18. The *"Jodi.org"* Web site takes over the interface to create graphic animations of text and Web elements such as the domain names of visitors.

Jodi) is a complexly interlinking set of visually challenging pages that contain Web interface and mark-up elements. Many of the animations move objects around in ways not often seen, for example, bouncing text. The site is full of conceptual jokes about how the Web works. For example, if you enter a page address that doesn't exist, the site takes you to an elaborate visual sequence built around graphic renditions of the text "404," the code for a page not found.

Other Artists and Projects

Comment on Interface Conventions ***Form Contest*** creates Web pages entirely out of interface elements such as buttons, text fields, and dialogue boxes. It creates interesting visual displays based on the way these elements are laid out and invites interaction, but the interface elements lead nowhere. The site invites reflection on our interface conven-

---

Form Contest: ⟨http://www.c3.hu/hyper3/form⟩

tions and the shadows of meaning they suggest. In *The Past, the Present, and the Future,* **Nicos Souleles, Dawn Riley, and Patricia Vogler** presented a Web event that plays with Web publishing conventions such as frames and linking. It presents the semblance of an abstract game that builds on the metaphor of time, and visually intrigues viewers while suggesting that the Web does not have to look as it does. *Ablink,* set up by **Vuk Cosic,** experiments with Web interface capabilities to create fascinating visual displays such as moving lines and innovative formats for links. Its links point toward other sites that similarly experiment with Web displays. **Heath Bunting** created a work called *ReadMe,* in which every word was linked to a site that used the word in its name. **Melinda Rackham** has created a Web site, called *subtle.net,* that contains many events that challenge traditional interface elements, including *resistant media* that don't quite obey; *line,* which is a visual pun on living on-line; *carrier,* which "investigates viral symbiosis"; and *tunnel,* which "explores the slimey areas of cybersex." **Igor Stromajer**'s *INTIMA* Web site moves against the impersonality of most Web sites creating an experience that explores the "intimate, ascetic, and interactive aesthetics. The key words for all activities are seclusion and ascetics."

Discarded Resources **Nick Philip**'s *Nowhere.com* installation took advantage of the fake return domain name nowhere.com used by spam E-mailers. It intercepted reams of angry responses returned from people who had been spammed, converting them into faxes that were then automatically fed in real time to many garbage cans. ***Digital Landfill*** comments on the ease with which digital materials can be created and abandoned. It solicits Web visitors to contribute unwanted materials and offers them up as sources for others.

Reflections on Infrastructure Such as Chat or E-mail Exploring the churning of intellectual property ideas in a connected world, **Kristin Lucas** created *Between a Rock and a Hard Drive* as a comment on chat and avatar-enabled Web worlds. In an unnerving and humorous encounter, every object on the Web pages is active, with voice bubbles made of text like what would be found on-line so that one can't tell which represent

Nicos Souleles, Dawn Riley and Patricia Vogler: ⟨http://www.voyeurmagic.com.au/rules.htm⟩
Vuk Cosic: ⟨http://www.ljudmila.org/~vuk/⟩
Heath Bunting: ⟨http://www.irational.org/heath/_readme.html⟩
Melinda Rackham: ⟨www.subtle.net⟩
Igor Stromajer: ⟨http://www.intima.org⟩
Nick Philip: ⟨http://www.best.com/~nphilip/instal.htm⟩
Nick Philip: ⟨http://www.potatoland.org/landfill/⟩
Kristin Lucas: ⟨http://www.diacenter.org/lucas/index.html⟩

real people. ***Jean-Louis Boissier***'s *Round Trip Ticket* jumps off from an event in Jean Jacques Rousseau's life in which he distributed copied train tickets as a reaction to the public's indifference to his work. He set an agent loose on the Web to find names related to the word *pleasant,* sending them the round-trip ticket that Rousseau sent. Flaming is the phenomenon observed on the Internet and other on-line environments in which people use language that is quite extreme, rude, or far beyond what would be expected in face-to-face interchanges. ***Judy Malloy***'s site "Flame Wars" comments on flaming by soliciting and archiving flames from the Web audience, which is encouraged to flame "the government, the boss, Bill Gates, or a Bozo of my choice." ***C5***'s *IP Project* uses a database of who occupies IP number space, to create five interfaces for navigating the Web and to generate a new topography of the Web. ***Noah Wardrip-Fruin***'s *Impermanence Agent* is a browser modification that uses a person's Web browser to reshape a starting story; it "tracks each user's Web browsing and uses this information to customize its story. It customizes until none of its own story is left." ***Janet Cohen, Keith Frank, and Jon Ippolito***'s *The Unreliable Archivist* allows multiple dynamic reorganizations of the resources available on the AdaWeb site.

Social Cultural Context of the Web   ***Etoy*** (see chapter 7.7) is a pseudo corporation created by an artist group with all the trappings of a major corporate Web presence. ***Digi-crime*** offers the semblance of a Web-based company that can be engaged for a variety of digital crimes. The site exudes both humor and commentary on Web commerce and digital culture. All the elements are there: corporate descriptions, FAQs, and hiring information. ***Fred Forest,*** who has a long history with telecommunications art, created *Time-out* as a virtual artwork. It exists only on the Web, and purchasers are given an exclusive access code that allows them to see the work. Forest followed appropriate legal and cultural procedures to confirm its existence. It was authenticated by an official appraiser and legally defined as an artwork subject to the rules of taxes and intellectual property. ***Christiane Robbins***'s *Amidst the White Noise. . . . My Fifth Amend-*

Jean-Louis Boissier: ⟨http://labart.univ-paris8.fr/BILLET/JJRbillet.html⟩
Judy Malloy: ⟨http://www2.links2go.com/go/www.artswire.org/~jmalloy/flamewar.html⟩
C5: ⟨http://www.c5corp.com/⟩
Noah Wardrip-Fruin: ⟨http://www.cat.nyu.edu/agent/⟩
Janet Cohen, Keith Frank, Jon Ippolito: ⟨http://www.walkerart.org/gallery9/three/⟩
Etoy: ⟨http://www.etoy.com⟩
Digi-crime: ⟨http://www.digicrime.com/dc.html⟩
Fred Forest: ⟨http://www.gpdoc.com/htm/filnet05.htm⟩
Christiane Robbins: ⟨http://www.banffcentre.ab.ca/mva/deep_web/dprojects.html⟩

ment Privilege uses the Web to deconstruct its unacknowledged reconstitution of the racism present in other cultural institutions. ***Vuk Cosic*** committed "Net theft" by entirely copying the "Documenta X" Web site just before it closed down and making it available on his site. As part of the Dia Center–sponsored Web works, ***Hermann Hack*** is part of a group of German artists who have undertaken projects to deconstruct the social context of the Net and to take actions to move against the dominant paradigms. *Homepage of the Forgotten People* created Internet domains for those supported by Amnesty International, and *The Virtual Roof* provided chat rooms and other Web resources for homeless people. In "MMF/Make Money Fast," ***Rolf Schmidt*** wanted to develop a way within the Web context to countervene against the get-rich scams being propagated through the Internet. His "MMF" site provides a space for information about the scams and sources for countermeasures, such as publicizing the MMFers and their addresses and ways of ridiculing and embarrassing them. ® ***TMark*** undertakes insurgency projects that question the commercial assumptions and infrastructure of the Internet. ***Daniel Garcia Andujar***'s *Technologies to the People* project is a virtual company dedicated to bringing technologies and network access to disenfranchised people, including the third world as well as "homeless, orphaned, expatriated or unemployed, fringe groups, runaways, immigrants, alcoholics, drug addicts, people suffering from mental dysfunctions and any other categories of 'undesirables.' " ***Mongrel***'s *Natural Selection* provides a search engine that returns seemingly normal results that have been critically modified to deconstruct race politics, which pervade Net life. ***The Redundant Technology Initiative*** collects discarded computers from businesses to build art installations and to offer Net access to disenfranchised people. ***Fabian Wagmister***'s *Cultural Specificity* project deconstructs software and Web structures in terms of "consonance-dissonance to Latin American cultural conditions and needs."

New Cultural Niches   The dynamic foment of the Web challenges artists to respond to a constant flow of new cultural, social, and economic structures that never existed

---

Vuk Cosic: 〈http://www.ljudmila.org/~vuk/dx/〉
Hermann Hack: 〈http://www.hack-roof.de〉
Rolf Schmidt: 〈http://ga.to/mmf/〉
®TMark: 〈http://rtmark.com/〉
Mongrel: 〈http://www.mongrel.org〉
Daniel Garcia Andujar: 〈http://www.irational.org/tttp/primera.html〉
Redundant Technology Initiative: 〈http://lowtech.org〉
Fabian Wagmister: 〈http://fabian.resnet.ucla.edu/〉

before. For example, entrepreneurs created eBay, which became a tremendously popular Web structure that allows anyone anywhere in the world to put just about anything up for sale to Web-connected bidders. Seeing online auctions as an opportunity for commentary on commercial life, artists and culture jammers have pushed the boundaries of what is offered for sale. For example, they have offered themselves, body parts, their virginity, erotic pictures of ex-girlfriends, an artist's demographics, exhibition space in prestigious shows, and the right to broadcast into a display at an art show.

Arrangements That Use Readings of the Physical World to Affect the Web

Lingua Elettrica

The Nervous Objects Group created an integrated Web/physical installation called *Lingua Elettrica* at Artspace in Sydney, Australia, and at ISEA97. In this reactive work, configurations of physical objects and projections and Web pages were shaped by the movements of people in the gallery. A later show, called *Terra Nova,* incorporated other works that bridged physical boundaries, including a wind activated Web work. CUSeeMe low-bandwith Internet video allowed *Lingua Elettrica* visitors in the two locations to converse:

Lingua Elettrica integrates virtual processing systems with physical space. People visiting the installation at Artspace are able to trigger the content of the *Lingua Elettrica* Web site through their physical relation to sensors and objects suspended within the gallery. The sensors trigger scripts that reconfigure random compositions from the stored information. Movement and vicinity to the objects result in people altering the visual and aural conditions within the gallery as well as the content of the pages on the *Lingua* Web site.

Shu Lea Cheang—Bowling Alley

The *Bowling Alley* event hooked up Minneapolis bowling alleys to the Web. Sensors in the bowling alleys used the action there to control the Web site and projections in the alley. Bowling activities scrambled videodisk projections, text from E-mail conversations,

---

Nervous Objects Group: ⟨http://no.va.com.au/lingua/information.html⟩
Shu Lea Cheang, *Bowling Alley:* ⟨http://bowlingalley.walkerart.org/⟩

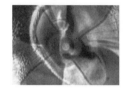

Fig. 6.4.19. Nervous Objects Group, *Lingua Elettrica*. Exhibited at Artspace Australia as a part of ISEA98. Web page from a sensor-driven artwork that linked Web and physical spaces.

and also Web visitor's navigation paths through the Web site much like the pins are scrambled when they are hit.

Konrad Becker—*Remote Viewing*

Remote Viewing, created by Konrad Becker, provided a unique art collaboration and display system that linked Web and physical space. Groups of invited artists were given access to a special Web space into which they could load their work and commentary. The work is physically projected into the gallery space. Other Web viewers can request

---

Konrad Becker, *Remote Viewing:* ⟨http://www.t0.or.at/~konrad⟩

JOIN THE EVER-FORMULATING SELF-SCRAMBLING HYPERTEXTS ON
P O W E R   A C C E S S   &   D E S I R E

OVERVIEW
of Bowling Alley
3-site installation

SPARE
traverses Minneapolis
Ten's texts

NEXT
continues through a
single message

WRITE-IN
write-into the site,
adding to the body of
alien texts

STRIKE
remixes a scramble of
Minneapolis Ten &
alien texts.

This site requires Netscape 1.1 or higher to view

Bowling Alley is a cybernetic installation linking 3 public spaces through ISDN lines and digital sensor data: Walker Art Center's Gallery 7, Minneapolis; Bryant Lake Bowl, Minneapolis; and the Bowling Alley Website. Bowling at Bryant Lake Bowl triggers changes in the chain of ISDN connections; scrambling the gallery's laserdisc projection and interfering with viewers' paths through the website.

Bowling Alley installation commissioned by the Walker Art Center 1995-96 and conceived by Shu Lea Cheang. Bowling Alley Website artists: Sawad Brooks, Christa Erickson, and Beth Stryker. [credits]

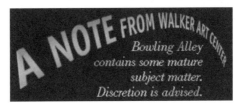

Fig. 6.4.20. Shu Lea Cheang, *Bowling Alley*. Activity at the bowling alley affects Web-visitor images and navigation paths. A collaboration with Web artists Sawad Brooks, Christa Erickson, and Beth Stryker; Netlink by Tim Desley.

access to submit work and commentary to be included in the show. Becker is "a hyper-reality researcher/developer and interdisciplinary event designer" and chairman of the Institute for New Culture-Technologies, in Austria. The group explores a variety of innovative network-oriented work spaces and art venues, such as "e-scape."

Other Artists and Projects

Fred Forest's *Internet-Graffiti* allowed physical visitors to the Short-Circuit 2000 show and Web visitors to jointly modify a composite poem. Commenting on the loss of the physical, ***Jenny Marketou***'s *SmellBytes*™ creates a digital action environment that in-cludes installations with live projections from the Web, an artificial agent whose passion

---

Fred Forest: ⟨http://www.fredforest.net⟩
Jenny Marketou: ⟨http://smellbytes.banff.org⟩

is odors, and a community of "smellbuddies" with digital odor IDs. In addition to creating Web works that use rapidly flashing text to explore language, **Valery Grancher** makes installations that push the physical/virtual boundary such as *Feelings,* in which Web surfers can send words into an installation where they get spoken into the space. **Jordan Crandall**'s *Drive, Track #3* places visitors into a disorienting installation of surrounding video projection in which part of the projections come from the Internet determined by the rhythm and speed of visitor movement. In the installation *The Rules Are No Game,* **Markus Huemer** uses the movements of visitors on a representation of Jackson Pollock's *No. 32* to extract and project text gathered from the Internet. In *Seance Box No. 1* **Ken Feingold** challenges Web surfers to manipulate a telepuppet that tries to communicate with an artificial actor. **Steven Greenwood**'s *Woven Presents* activates a computer controlled sewing machine to embroider on cloth text that has been extracted from search engine results on the word *war.* **Jeffrey Shaw**'s *Distributed Legible City* enables geographically dispersed people riding home on bikes to navigate simulated urban landscapes constructed of giant text (see chapter 7.4). **Michael Samyn and Group Z** offer Web events such as *Home* and *I Confess* that defy the usual conventions of interface.

See chapters 6.3 and 7.4 for other artists who investigate the borders between virtual and physical objects and spaces.

Cool Sites

Creating a kind of Art Brut, many individuals besides artists have used the Web for cultural exploration. Creating what have come to be called cool sites, they have invented Web sites that win international recognition for their imagination, cultural commentary, intensity, media exploration, orchestration of worldwide collaboration, idiosyncratic topics, unusual configurations of information, or sometimes just their weirdness. The appeal and cultural innovation challenges traditional boundaries of art and popular culture.

Valery Grancher: ⟨http://www.imaginet.fr/nomemory/⟩
Jordan Crandall: ⟨http://blast.org/crandall/⟩
Markus Huemer: ⟨http://www.khm.de/~huemer⟩
Ken Feingold: ⟨http://www.kenfeingold.com⟩
Steven Greenwood: ⟨http://members.tripod.com/sgwood/⟩
Jeffrey Shaw: ⟨http://escape.lancs.ac.uk⟩
Yahoo Cool Links Directory: ⟨http://dir.yahoo.com/Entertainment/Cool_Links/⟩

A few examples of this experimentation follow:

Attacks on childhood icons, such as the *Distorted Barbie;* urban mythology, such as *Exploding Whale;* gender wars, such as *All Men Must Die,* which collects stories of rotten boyfriends and husbands; Consumer and employment sites, such as *ComplainUSA,* which encourages employees to complain about their companies' actions and policies; cryptozoology, such as *Bigfoot;* Webcams, such as the *Nesting sites,* which provide live feeds from nests of endangered birds (see also chapter 6.3 for more on Webcams); guerrilla art, such as *Adbusters,* which visually deconstructs corporate advertising; pets, such as *Burner.kitty,* which offers free Web sites for pets; political commentary, such as *"Conspiracy On-line,"* which collects conspiracy theories and related information; anthropological information, such as *InsectsasFood,* which collects recipes and practices of insect eating, and *Contortion,* which collects photos of contortion exploits; history, such as *History of Torture* and *History of the Lobotomy;* travel, such as *Roadside Attractions;* and found items, such as *Photos People Left,* which posts discarded photos found by contributors.

Summary: The Web as an Art Arena

The onrush of Web developments validates the visionary intuitions of early telecommunications artists who pointed to these esoteric technologies as a critical cultural arena. As argued repeatedly in these chapters, artists can often lead the way in understanding the implications of research. Most analysts are amazed with the speed and scope of Web-related initiatives to enter into every walk of life, extending from commerce and government to entertainment and art. Some wonder where it will end. As described in this chapter, artists have been among the leaders in exploring the technological and cultural possibilities of the Web. They have also been among those most willing to question the euphoria.

Notes

1. For more information, see my book *World Wide Web Design Guide* (Indianapolis, IN: Hayden, 1995).

2. Ars Electronica, "1997 Web Jury Statement," ⟨http://193.170.192.5/prix/jury/E97www.html⟩.

3. A. Broeckmann, "Net.Art, Machines, and Parasites," ⟨http://miki.wroclaw.top.pl/wro97/anghtml/brekmtxt.html⟩.

4. E. Kac, "Interactive Art on the Internet," ⟨http://www.ekac.org/Telepresence.art._94.html⟩.

5. J. Malloy, "Siggraph Jury Statement," ⟨http://www.siggraph.org/s98/conference/art/artsite.html⟩.

6.   See S. Wilson, "The Web as an Art Context," in C. Harris, ed., *Art and Innovation* (Cambridge, MA: MIT Press, 1999).

7.   Rice University Gallery, "Description of CyberVato," ⟨http://www.rice.edu/CyberVato⟩.

8.   M. Musgsrove, "Gómez-Peña as Technoprophet/Artist," ⟨http://www4.ncsu.edu/eos/users/m/mcmusgro/public/ggptechnop.htm⟩.

9.   G. Gómez-Peña, "Virtual Barrio @ the Other Frontier," in L. Hershman Leeson, *Clicking In* (Seattle: Bay Press, 1996), p. 173.

Digital Information Systems/Computers

7.1

Research Agendas and Theoretical Overview

Introduction: The Computer Revolution

The computer has been the defining technology of the current era. Together with supporting technologies such as miniaturized digital electronics and communications, it has radically transformed most areas of life—media, commerce, research, industry, education, entertainment, health care, the military, and day-to-day living. It has also revolutionized thought about fundamental concepts about what it is to be human, such as the nature of intelligence, the fabric of information, the structure of society, the legacies of culture, and the nature of work. Prognosticators suggest that even more profound changes lie ahead. This section reviews research and artistic activity focused on the past, present, and future of computers.

For many, the computer is an appliance sitting on desks. But computers actually have much wider reach, and researchers are feverishly working to extend their application into every corner of life. Invisible computers lurk everywhere—in the toaster, the toy, the automobile, the television, the stove. Computers underlie many systems that people rely on—telephones, manufacturing, transportation, health care, finance, and government. Even more profoundly, digital information systems have changed cultural patterns, for example, in the ways people are abstracted and represented, the ways social decisions are made, the ways we attribute value to information, and the ways that images are used to shape meaning.

The Diverse Histories of Technological Imagination

Much early artistic activity with computers focused on their capabilities as manipulators of images and sound, probably because these were closest to historical forms of the arts. Current audiences marvel at the special effects and transformations of reality enabled by computers. Interactive media and immersive virtual reality engage contemporary artists working within the traditions of image and media. Ironically, the review of the theories presented later in this chapter will consider some that are skeptical about the "revolution"—placing computer imaging in a direct line stretching from Renaissance perspective through painting, photography, cinema, and television, to the present moment. In this analysis, the computer is not a radical breakthrough but rather a continuation of narratives emphasizing the unitary point of view, the controlled frame, and the manipulation of desire through image.

It is a mistake, however, to focus on only one area of activity in trying to understand the underlying meaning of computers, their impact, and their relationship to culture and art. Computers and their associated technologies represent the culmination of many techno-cultural streams. The recognition of this diversity of historical lines of technological imagination can be useful in understanding future trends in the expansion of research and artistic activity. My book *Using Computers to Create Art* presents a chapter called "An Unorthodox History of Computers," which identifies several of these origins:

1. Cybernetics/automatic control: how machines and animals can use information to control behavior, including examples such as Egyptian pneumatic sculptures, Tivoli Garden fountains, and steam engine and weapon control systems

2. Automata and robots: mechanical simulations of human and animal life, including examples such as Greek deus ex machina contrivances, the clock automaton, Vaucanson's duck automaton, and movie robots.

3. Calculation and statistics: machines that could calculate and organize information, including slide rules, Babbage's analytical engine, IBM's business machines, and World War II artillary computers.

4. Image and sound machines: machines for recording, manipulating, and presenting, such as Egyptian water-activated instruments, Eastern and Western puppets, Renaissance camera obscura, Athanasius Kirshner's slide projector, and nineteenth-century devices for telecommunications and image and sound presentation.

Current theoretical discourse often identifies some aspect of these histories as the key to understanding the cultural meaning of computers and related technologies, for example, military weapons development, corporate and governmental desires for surveillance and control, or image manipulation and mass media as the reinforcer of cultural narratives. But there is also a danger in reducing the totality of techno-cultural development to these themes. The analysis validly deconstructs techno–self-delusion, but it runs the danger of discounting the role of true invention and inquiry. Contemporary cyber culture did not develop only out of military and corporate control contexts, but also out of religion, entertainment, science, and art. Its motivations include power, domination, economic acquisitiveness, fascination with surface, and also play, wonder, spiritual quests, and free-floating curiosity. One cannot understand contemporary research or technological art without acknowledging these diverse frameworks.

This section previews frontier research agendas in digital technology fields such as new interfaces, artificial intelligence, and information visualization. It also analyzes theoretical speculation on cultural and aesthetic implications and reviews artistic experimentation in a variety of computer-related areas: digital video installation, interactive multimedia, virtual reality; installations that sense motion, gaze, facial expression, and touch; artificial intelligence, speech, surveillance, and information visualization. The focus on computers in these chapters is somewhat artificial. Digital technology has become essential to so many fields that it forms part of the background in almost all the chapters of this book.

Research Agendas

The computer is a strange machine—brain amplifier, number cruncher, image manipulator. Researchers and artists seem fascinated by challenging its limits; they are intent on seeing how far they can extend its capabilities. Sometimes this quest means trying to enable computers to manifest skills that are quite unremarkable for humans but extraordinary for a machine, for example, understanding spoken words or extracting the meaning of a children's story. Or it might mean developing skills that are beyond human capabilities, for example, being able to instantly analyze a database composed of millions of records in thirty different ways. The research agendas to extend capabilities and reach are critical elements of this era's cultural history. This activity in think tanks and worldwide labs is the flow that must be a source for current and future artistic activity.

Inputs—Systems Recognizing Speech, Gestures, Faces, Objects, Motion, Touch, Emotions, and Biological Signals

Speech Recognition

How can a computer understand the words of human speech? Commercial speaker-independent-recognition products are already available. The task of understanding the meaning of speech is much more difficult and still challenges researchers. Extensions include the development of "auditory consciousness" and "auditory scene analysis," which will allow systems to track multiple human speakers in complex sound environments and identify their relative physical locations. Other research seeks to track speaker changes, topic changes, and changes in emphasis. "Meeting capture" and "speaker segmentation" will enable the scan and analysis of a record of a complex sound event, such as a meeting, in order to systematically summarize the event and reconstruct the flow of the conversation by speaker and the thematic thread, and allow automatic browsing and gisting. Reportedly, the CIA has a system that can monitor thousands of phone calls simultaneously, listening for specific key phrases. A project called *Net Sound* is attempting to extract the underlying acoustic structure of sounds in hopes of finding algorithmic representations, like a vector or postscript representations of images, which would allow more efficient network transmissions. "Emotional computing" projects seek to enable computers to analyze the emotional content of speech via attributes such as

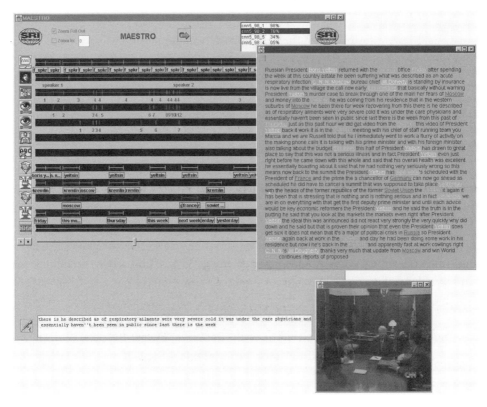

Fig. 7.1.1. SRI International, MAESTRO (Multimedia Annotation and Enhancement via a Synergy of Automatic Technologies and Reviewing Operators). DARPA-funded research system to identify speakers and semantic content in speech 〈http:// www.speech.sri.com〉.

pace, intonation, inflection, rhythm, and context, and also to produce more emotionally expressive synthetic speech in response.[1]

Music Recognition and Synthesis

Research is attempting to develop music recognition systems that can extract information about instrument identification and spatial location, and notes and other musical features, in hopes of supporting automatic annotation and browsing. Another project attempts to automate the analysis of singing voices, extracting sufficient information to be able to resynthesize the voice. Many researchers around the world are working on automatic composition systems to write music judged interesting and of high quality, or to mimic the styles of particular composers. The *Opera of the Future* project attempts

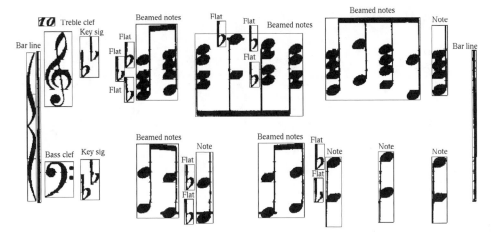

Fig. 7.1.2. David Bainbridge, optical musical score recognition system ⟨http://www.cs.waikato.ac.nz/~davidb/publications/acsc96⟩.

to create musical performance augmentation systems that can work in both professional and casual contexts. For example, *Hyperinstruments* seeks to understand a performer's intention by gesture and feeling, and self-modify to enhance the performance.

Object Tracking and Identification

How can a computer recognize the presence and identity of objects in its environment? Many diverse technologies are being experimented with—some requiring special markings or transponders and some that do not. Bar codes and RFID (radio frequency ID) are perhaps the oldest and most developed. Bar codes are the familiar Universal Product Code zebra patterns of lines found on many items, such as cans. These codes can flexibly be read in any orientation by bouncing a scanning laser beam off them. Researchers are currently experimenting with two-dimensional markings that will allow the placement of much more information in denser displays and methods of unobtrusively embedding the information in other printed material on the products. RFID requires the placement of a small electronic device on each object that can then be read at a distance without the requirement of visual access. Examples are the security tags placed on store merchandise or library books that activate alarms when someone passes through a portal. Passive tags, which can be produced very inexpensively, use resonant electronics to reply when activated by the inquiring signals. More expensive active tags require battery power, can provide extensive information when queried, and can be dynamically updated.

Fig. 7.1.3. Examples of 2-D bar-code symbologies that communicate much more information than standard bar codes. *Right,* 417 code; *left,* Data Matrix code. From the article "Two Dimensional Codes" in *Automatic ID News,* October 1995 ⟨http:// www.autoidnews.com⟩.

Researchers are working to decrease the cost and increase the range and amount of information contained on the tags. For example, a Micron product used for more complex inventory schemes such as tracking railroad cars and trucks can be read at distances of up to three hundred meters. They even have a system that combines cellular phone and GPS technology and will eventually allow items such as cargo containers to be tracked anywhere in the world. Related technology has been applied to tracking prisoners on house arrest, children at amusement parks (to prevent them from getting lost), taxicabs (to insure that drivers don't take too lengthy breaks), luxury automobiles (to track them if they are stolen), and military assets (in order to deploy them in battle).

Other researchers are attempting to increase the intelligence of video-image processing systems so they can identify objects in natural scenes without special markers or transponders. MIT's Physics and Media group is studying the possibility of avoiding circuitry and video altogether by using natural physical features of objects and persons for tracking and identification. For example, the Spin Resonance project is seeking to determine whether the atomic spin patterns of molecules could be sensed and used for identification purposes.

Recognizing the Presence and Movement of Persons

How can a computer system recognize the presence of persons? As with object identification, some approaches use special transponders while others require no special devices. A variety of systems (and even some commercial products) employ technologies such as ultrasonic range finding, capacitance field changes, or video analysis to determine the presence and relative position and motion of persons. For example, after scanning a

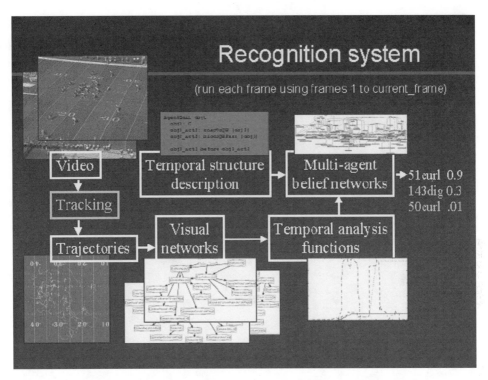

Fig. 7.1.4. Stephen Intille and Aaron Bobick, MIT Media Lab's Vision and Modeling group. A system that recognizes American football plays from trajectories of players from video sequences 〈www.media.mit.edu/vismod/demos/football〉.

video image for patterns that match prestored templates of persons, these systems can make good guesses about numbers of persons and their positions in a scene. Other systems that require persons to wear small devices such as radio signal emitters can achieve highly accurate spatial placement by triangulating the signals. MIT's Smart Fish program (so named because of some fishes' ability to scan their environment via weak electrical fields) can make complex position determinations of nearby persons and objects. Adaptive Optics Associates created a game, *Cypress Adventure*, that tracked human motion by use of special markers placed on the hands. Pedersen and Sokoler's AROMA system experiments with the abstract representation of all people in a space.

Recognizing the Identity of Persons

How can a system identify people in its visual field? This task can require considerably more intelligence than merely recognizing whether or not a person is present. The task is simplified if people wear transponders. For example, the Xerox PARC

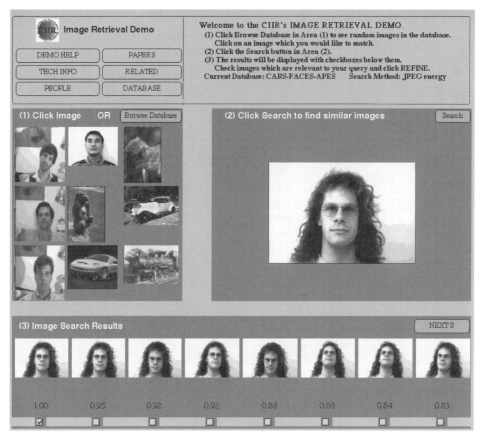

Fig. 7.1.5. Center for Intelligent Information Retrieval, University of Massachusetts. A demo Web page from a multimedia indexing and retrieval (MIR) system for identifying the content of photos, including facial identity, ⟨http://cowarie.cs.umass.edu/~demo/Demo.html⟩.

ubiquitous-computing project used small, worn tabs to communicate presence to a central computer. Anyone else in the system could then locate coworkers at all times. The MIT Media Lab's Personal Identification Network similarly used tags for people and objects that could then be used to create smart objects and spaces. For example, *Mort the Penguin* could greet people as they arrived in its space and customize its behavior based on who was present.

Nontransponder systems confront a more difficult image-recognition challenge. For example, face-recognition systems must analyze video images to identify what particular persons are present. The Media Lab's Face Recognition project detects and codes faces in order to enable automatic face recognition and database queries. The SmartCam project enables a video camera to automatically track particular actors in complex scenes. A variety of commercial systems oriented toward law enforcement use these technologies to identify likely suspects in mug shot databases. International conferences such as the Automatic Face Recognition Conference present a wide variety of approaches. Researchers are working to enhance the systems so they can find people by scanning real-time video.[2]

Gesture Recognition and Synthesis

How can a computer interpret body actions made by humans, such as pointing? Gesture-recognition systems fall into two categories: those which require special instrumentation, such as gloves or body marking, and those that assess human free-motion in space via techniques such as video analysis. Several commercial glove-based systems are available that assess the positions of the fingers. The data glove is a common feature of immersive virtual reality systems and has even been made available in commercial game systems. Experimental projects attempt more subtle reading, such as a variety of projects that translate sign language into speech. Whole-body suits allow the determination of movements of more than the hand. Position sensors placed on the head, another feature of some VR systems, allow the determination of rotation and tilt of the head and orient the scene presented to a helmet-mounted video to match the orientation of the head. Other systems track eye movement by reading the position of the pupil through video analysis or a light beam bounced off special contact lenses. Other motion-tracking systems require special reflectors or marks to be placed at critical joints in order to make the video analysis simpler, since it can search for known indicators. Commercial versions of this technology are used to improve the performance of athletes and dancers, or to extract motion data to be used to animate synthetic actors.

Recognizing the specific meaningful movements of bodies in free space is more difficult. How can a computer look at a video image to understand what a person is doing, for example, pointing, lifting an item, or facing in a particular direction? Researchers have developed systems that can extract motions, match them against templates, and use the context to interpret the motions. For example, the Media Lab's Seeing Action and Smart Room projects attempt to create spaces where gestures can be interpreted. The *Virtual Personal Aerobics Instructor* watches user motion in order to customize its instruction, responding to the physical activities of the user. A research project called Synthetic Characters uses stuffed animals as interfaces to indicate gestures, after which autonomous characters take over. Several projects attempt to automate the recognition of sign language or handwriting.

Fig. 7.1.6. MIT Media Lab Vision and Modeling group's *KidsRoom*. An automated, interactive play space for children that recognizes gestures and helps to create a narrative fantasy, ⟨http://vismod.www.media.mit.edu/vismod/demos/kidsroom/kidsroom.html⟩.

Akira Utsumi at ATR Media Integration and Communications Research Labs is working on a real-time, multi-camera system that uses its ability to interpret hand gestures to allow viewers to manipulate virtual objects in computer-generated 3-D space. Jakub Segen at Bell Labs created a similar system that can interpret hand positions in free space. Jun Rekimoto at Sony's computer science lab developed the *Holo Wall*, which senses user actions and generates sounds or movements of simulated insects in response to these actors. Hirotada Ueda at Sony's research lab developed the *Ultra Magic Paper Interface*, which makes a sheet of normal paper into the interface by using stereo video cameras to interpret gestures such as touching or turning a page. John Sibert at George Washington University set up a system that can read a pointing finger by using a small infrared emitter. In a system called *Stretchable Music*, Pete Rice and Joshua Strickon at MIT used laser tracking to read the movement of hands, which could then shape musical sounds by manipulating virtual objects. Working at the ATR Media Integration and Communication Research Lab, Kazuyuki Ebihara created an installation called *Shall We Dance?* that could read human movement and facial expression to control a virtual puppet. His *Virtual Kabuki* was a facial animation system based on the explicit analysis of facial muscles.[3]

Facial Expression Recognition and Synthesis

Facial expressions are one extremely important set of human gestures that are difficult for machines to recognize. How can computers interpret human facial expressions such as smiling and frowning? How can they generate synthetic characters with expressive characteristics? Researchers have been studying the problem for the last two decades using approaches such as neural networks and hidden Markov models to extract facial features. Some use databases of facial expressions to compare against current expressions. Others model the facial muscles. Several commercial products translate the facial expressions of actors (via special sensors placed on critical facial muscles) to build repertoires of expressive capabilities for computer-animated characters. A variety of research projects link recognition in more complex mediated systems. One project attempting to create "objects with psychosocial capability" seeks to develop animated avatars that can represent a user's nonverbal intentions and generate appropriate facial expressions in gestures in conversations. FaceView creates a subtle animation system based on the observation of faces. Several systems attempt to use lip reading as a way to enhance the accuracy of speech recognition.[4]

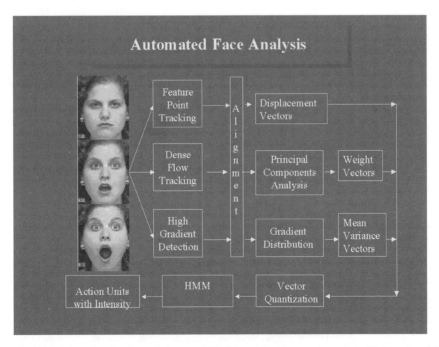

Fig. 7.1.7. Jeffrey Cohn, automated face analysis research identifies emotions by computer analysis of facial motion, ⟨http://www.cs.cmu.edu/~face/⟩.

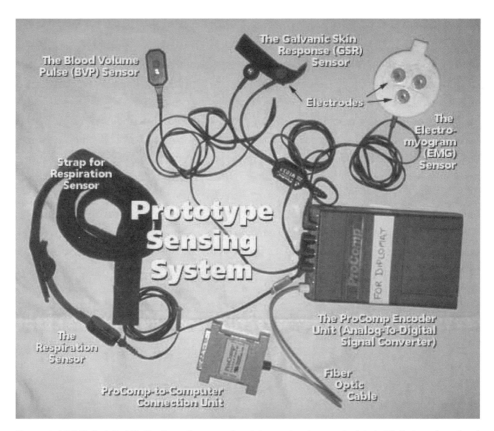

Fig. 7.1.8. MIT Media Lab, Affective Computing group. A prototype system to sense physiological indicators of emotional states, ⟨http://www.media.mit.edu/projects/affect/ACresearch/sensing.html⟩.

Recognizing and Synthesizing Emotional States; Affective Computing; and Biosensors

How can a computer recognize people's emotional states? How can a computer create synthetic productions that seem appropriately emotional? Researchers are attempting to enable computers to use a variety of biological signals to understand human emotional states. Some of this biosensor research focused on detecting odors and body chemicals is considered in the biology chapter (2.1).

Researchers are also investigating other body signals, such as body posture and non-semantic elements of voice, such as intonation and phrasing. The Media Lab's Affective Computing is a full research project that includes basic research on the psychology of human emotions and the creation of computer systems for the sensing, understanding, modeling, synthesizing, and recognition of emotions. Demo projects include emotionally expressive animated avatars, clothing that senses emotion, and computers that can respond to frustration. Some commercial products, such as voice stress analyzers, attempt to use emotional sensing in order to determine whether a conversational phone partner is lying or if they are ready for a sales closing. Carnegie Mellon's Oz project creates semi-autonomous characters that understand emotional relationships.

Integrated Recognition Systems

How can computers use a variety of sensing sytems to increase their capability to understand human situations? Humans regularly use context and multiple senses to understand their environments. Researchers are investigating similar approaches for computers. As part of the MIT Perceptual Computing project, Ali Azarbayejani and others created the Smart Spaces project, which presents an activated space that uses the combination of 3-D motion tracking, facial-expression tracking, gesture recognition, and speech recognition to allow computers to interpret human actions. Demos of the capabilities (some shown at the 1996 SIGGRAPH "Digital Bayou" show) included *Waldorf* (in which animated characters mimic the gestures and facial expressions of the observer), *Seagul* (in which an observer controls the flight of an animated bird over simulated landscapes), and *Netspace* (in which an observer navigates WEB space via speech and body movement).

Haptics and Force Feedback

How can computers exploit the fact that humans regularly use their body (through touch, motion, and pressure) to understand and manipulate the world? How can computers integrate humanity's long history with physical objects into digital information systems? Bill Buxton, an interface researcher at the University of Toronto, is famous for his observation that conventional mouse interfaces are analogous to asking someone to act with one hand and both legs tied and their mouth gagged. Moving far beyond simple touch screens, researchers are developing innovative systems that use motion, vibration, texture, and pressure as the medium of exchange. Systems enable people to communicate with the computer via gestures such as touch, squeeze, push, pull, stroke, kinesthetic motion, and the like. Many new intelligent toys have been developed that react intelligently to children's touch.

Illustrating a more advanced form of this technology, Bruce M. Blumberg at MIT Media Lab's Synthetic Characters group developed the Swamped! project, that uses multiuser-, touch-, and kinesthetically sensitive stuffed animals to control synthetic computer-generated characters. The characters' behaviors were also influenced by perceptions of the environment and motivational and emotional states.

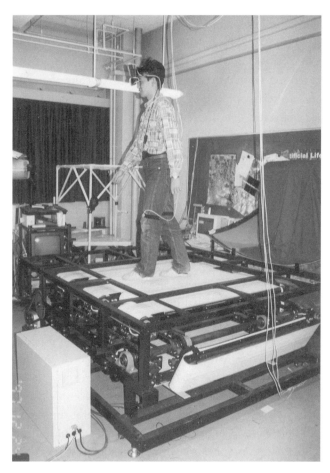

Fig. 7.1.9. Hiroo Iwata, Virtual Reality Laboratory, University of Tsukuba. *Infinite Plane*—a computer-controllable walking plane that can read and respond to a walker's action, ⟨http://intron.kz.tsukuba.ac.jp/index.html⟩.

Systems also respond back. Some texture-simulation systems use subtle mechanical systems to give computer users the feeling of touching different kinds of palpable surfaces. Using the pressure cues of normal physical life, many force feedback systems simulate the sensation of touching physical objects, for example, a 3-D mouse that responds with variable resistance as a person touches virtual objects or tries to move in virtual environments. Some commercial game systems—for example, *I-Force*—already include simple versions of this, such as force-feedback control sticks for flight simulations. To create the effects, researchers use a variety of technologies, such as motors, magnetic surfaces, pneumatics, fans, exoskeletons, and heat.

Several labs illustrate the scope of research. MIT's Touch Lab undertakes a unified research agenda stretching from the physics, physiology, and psychology of touch to the technological contrivances necessary to sense and simulate touch. For example, Scott Brave's *In Touch* installation allows two people in different locations to have the experience of tactile communication by jointly manipulating two linked rotatable cylinders. The University of North Carolina uses force feedback to enhance virtual reality research environments, such as their nano-manipulator, which lets viewers manipulate hypothetical nano-scale structures. Makoto Sato's lab developed a variety of force feedback installations, including *Virtual Basketball,* which used force feedback to give users the feel for the shape and weight of the ball in conjunction with appropriate audio and visual cues. Yuichiro Kume's *Fantastic Phantom Slipper* tracked the motion of feet and provided tactile stimulation to the soles of the special slippers that simulated the experience of walking.

Hiroo Iwata's lab explores a variety of haptics studies, including the *Haptic Screen,* which is a force-feedback device that changes its physical shape to give the experience of shapes of virtual objects; *Haptic Master,* in which users can feel the rigidity or weight of virtual objects; and *Infinite Plane—Torus Treadmill,* which creates the sensation of real physical walking in virtual space by force-transmitting wires that synchronize physical walking pressures with the virtual display and ultrasonic sensing of user action.[5]

Output Systems—Synthesized Speech, 3D Sound, Virtual Reality, and Motion Systems

Synthesized Speech

Speech synthesis has a long history stretching back to the development of speaking automatons of the nineteenth century. Electronic speech synthesizers have been around for decades, but they usually produced mechanical inhuman sound. Current research focuses on several issues, for example, low-level synthesis techniques for simulating speech sounds, acoustic linguistic analysis to understand the fundamental sound components of various languages, the enhancement of text to speech algorithms so they can use meaning to improve the quality of speech, and the use of the knowledge of human physiology to improve the linkage of synthesized speech to the motion of animated artificial characters. Many major research laboratories, such as Microsoft's, the MIT Media Lab, Bell Labs, and the Swiss National Labs are developing enhanced speech-synthesis capabilities as part of efforts to create believable artificial characters. Some analysts believe that speech recognition and synthesis will be the major interface for future computer systems.[6]

3-D Sound

Researchers have analyzed human hearing physiology sufficiently to create systems that synthetically place sounds anywhere in 3-D space. Projects seek to extend the technology in ways such as clarifying teleconferences by placing each speaker in an apparent physical location, and enhancing the believability of virtual reality environments. MIT's Audio Spotlight project creates a sound space in which the sound appears to physically come from anywhere a person shines a flashlight.

Immersive Virtual Reality

How can computer systems create integrated sensing and representation systems that give the user the illusion of being in another location? This technology has moved quickly from the laboratory to real-world applications. Typical instrumented versions link position-sensing helmets that read the direction and tilt of a person's view with a data glove that reads hand gestures. Three-dimensional sound and eyepiece video displays render synthetic world views that match the movement of the observer such that they feel they are present. Other approaches, such as the Electronic Visualization Center's CAVE installations, try to eliminate some of the body gear by creating the illusion of presence by projecting images on all the walls and the ceiling of a space without using the helmet. Researchers are attempting to heighten the illusion in several ways: creating synthetic, computer-generated worlds that are more realistic; speeding the process up so that it tracks motion better; and building more sensitive motion trackers. They also are extending the areas into which immersive VR can be applied, such as journeys inside of bodies, abstract informational visualization displays, and amusement parks.

Motion Simulation

How can a digitally controlled motion system enable users to feel they are undergoing particular physical experiences? Working beyond simple force feedback, researchers are attempting to build complex simulators that reproduce experiences such as flying an airplane, flying a spacecraft, diving underwater, and driving a racing car. The simulators physically move small environments through the control of position changes, acceleration, and vibration. The motion is tightly linked to immersive visual and audio presentations to enhance the illusion. Military research has a long history of perfecting this technology through flight and battleground training simulators, feeling that it is better to crash simulated planes and destroy simulated tanks than real ones. Entertainment researchers, such as those at Disney, have similarly pushed the illusion technology to create trips into imaginary environments. Contemporary researchers seek to increase the

Fig. 7.1.10. *CyberAir Base*. A computer-controlled motion platform used in the design of arcade and theater installations. Virtogo, Inc. ⟨http://www.virtogo.com/⟩.

subtlety of motion control, the linkage to other senses, and the interactive response to user actions. For example, Disney researchers have created a roller-coaster simulator in which a person can design their own roller coaster and then physically experience their creation.

Interpreting and Manipulating Information

Interpreting and Representing Video and Image Information

How can computers extract the meaning structures of video and other image information? How can computers create enhanced displays that represent the underlying meanings and facilitate their storage, retrieval, and analysis? Contemporary society stores much significant information in image forms, such as photographs, video, and cinema. Until recently, the primary ways to search and analyze this information required human abstracters or catalogers; computers could not automatically determine what was portrayed in images. Researchers are now challenging many aspects of the problem. Several automatic image-search systems have been developed that can search large databases of images

Fig. 7.1.11. Nuria Oliver, Barbara Rosario, and Alex Pentland, *Learning and Understanding Action in Video Imagery.* MIT Media Lab's Vision and Modeling group. A system to identify and track people in video sequences, ⟨http://vismod.www.media.mit.edu/darpa-vsam/⟩.

to find specific kinds, such as pictures of dogs or sunsets. For example, the Getty Research Center, in collaboration with NEC, developed *Armore,* which identifies Web images based on size, color, structural relationships, and the semantic information surrounding the image.

Researchers are also working to develop automatic video-logging systems that can determine camera and scene changes. For example, companies such as Virage, Islip, and Excalibur are working on features such as automatic face-tracking and the identification of zooms and pans. MIT's Media Stream project represents the flow of images with a perspective box composed of time-slice stills that support storyteller, retrieval, and editing systems. The Salient Stills research attacks the problem of representing a time medium like video in stills. It automatically constructs panoramic stills that amalgamate background context with high-resolution close-ups of important elements in one still image. The Object-Oriented Video research effort attempts to capture the semantics and physical meanings of scenes by constructing discrete representations of their composite objects. Starting with several conventional video streams from different points of view, it uses sophisticated algorithms to create texture-mapped 3-D models of a setting and its parts. Once the model is constructed, one can inspect the scene from any arbitrary point of view even though there was never a camera in those positions. Objects can be inspected in their entire circumference, and camera view can be customized for each observer. Video production decisions, such as camera position, lighting, composition, and focal length, can be decided in postproduction. DARPA (Defense Advanced Research Projects Agency) supports several major research efforts to automatically extract video information, including VSAM (video surveillance and monitoring). New commercial services such as Space-Imaging provide military one-meter-resolution satellite images with image analysis of just about any place on the earth, for a price.

Ambient Sound, Location-Specific Sound, Earcons, Feature Extraction, and Sonification

How can systems be created in which sound conveys new kinds of information? For example, the *Ambient Sound* project explores several uses of ambient sound. One variation assigns specific sounds to each person in a work group, all of whom wear ID responders that let the central system know where they go physically. As one walks around the space, ambient sounds offer cues about who is nearby, and volume levels indicate the length of time since they were present. Another project maps one's personal E-mail flow via peripheral sounds, for example, sound level indicating new E-mail messages. Several Earcon projects are exploring the power of composed sounds to function as interface indicators of the states of computerized information systems. A project called *Nomadic Radio* offers a wearable computer system that customizes the sound one hears by the physical location and previous responses to the sound. *Garble Phone* tries to extract the gist of conversational speech while masking the details. One application would allow

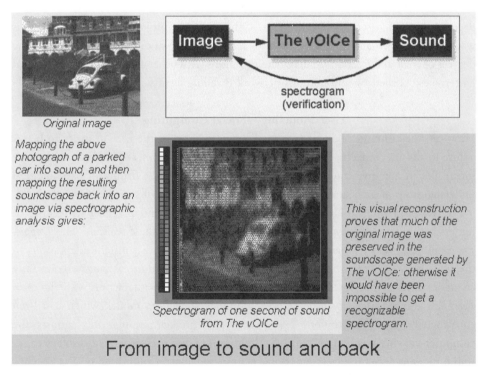

Fig. 7.1.12. Peter Meijer, vOICe system. A sonification system for translating visual information into an array of sounds ⟨http://ourworld.compuserve.com/homepages/PeterMeijer/voice.html⟩.

trusted telephone callers to hear the processed version of a call in progress in order to decide whether it was appropriate to break in. Sonification projects attempt to convert visual information to sound, for example, to allow blind people to know what objects are around.

Autonomous Software Agents

Can software agents be developed that can learn a user's needs or desires in order to represent the user's interests? MIT's Autonomous Agent group defines agents as long-lived, semi-autonomous, proactive, and adaptive. It is developing agents able to help with information filtering, navigation, remembering, recommending, matchmaking, and negotiation. Example demos include *Letszia,* which finds Web pages reachable from a currently viewed page that might be useful; *Expert Finder,* which finds people whose skills could help the user; and *Remembrance,* which brings up information from past events relevant to the current situation. The News of the Future project seeks to create agents that select news based on content filters that know the user's likes and dislikes, and social filters that use information from the entire user community to identify information deemed relevant to others like the user. Other research projects study topics such as ways to teach agents or to enable them to learn by observation, methods of specifying degrees of trust and freedom to negotiate in actions with other agents, and the ecological patterns established by autonomous network-dwelling agents.

User Modeling—Social Computing

How can the computer build an internal representation of the user in order to better adjust to the user's needs, styles, and intentions? Interface researchers believe that the future lies in adaptive interfaces and computer systems built firmly on psychological understandings of users and anthropological understandings of social settings. Both Xerox PARC and Interval support studies of work settings and practices in seeking to understand how information systems should be adapted to these realities. The *Doppelganger* project seeks to maintain and update information with constant adaptations to the who, why, where, when, and how of the user's situation. *Storyagent* constantly changes stories it tells based on the model it builds through interaction with the user. *PeerGlass* allowed users to inspect and manipulate their own model, which had been constructed by the computer. *The Shutter Bug* project automatically adjusts the computer-viewing position and artificial lighting of computer-generated displays to respond to its understandings of emotional states. The Social Computing project at Stanford found that people extended interpersonal mores to their relationships with computers, for example, showing sensitivity to the computer's "feelings."

Fig. 7.1.13. Bruce Blumberg, MIT Media Lab's Synthetic Characters group. A surprised raccoon from the Swamped project, which develops animated characters with social intelligence, ⟨http://characters.www.media.mit.edu/groups.characters/⟩.

Machine Understanding, Recognizing the Meanings of Texts, and Language Translation

How can a computer read, extract the meanings, and summarize the gist of a text? This automatic abstractor function has been a long-term research goal. Early artificial-intelligence research discovered, however, that even children's stories posed a major challenge for computers. Human text is highly elliptical and requires a great deal of background information and sophisticated knowledge representation structures in order to understand it. Research continues, however, toward that goal. For example, MIT's News of the Future group supports several projects using a variety of techniques to automatically understand news stories. Interval researchers are investigating the importance of demographics to the way information can be understood, for example, that related to gender, life stages, style, geographic, and spiritual orientation. Another project attempts to create personalized information assistants that filter information based on personal experiences. Also, many researchers around the world are working on the specialized problem of automatic translation among languages.

Information Foraging, Information Visualization, and Augmented Reality Systems

How can computers enhance the ability of people to work with large flows of textual information and provide tools for discovery, searching, analysis, clustering, and manipulation? Contemporary culture produces enormous volumes of information in areas such as research results, correspondence, legal proceedings, and the like. Analysts warn of information overload and anxiety. Sophisticated information-processing tools facilitate

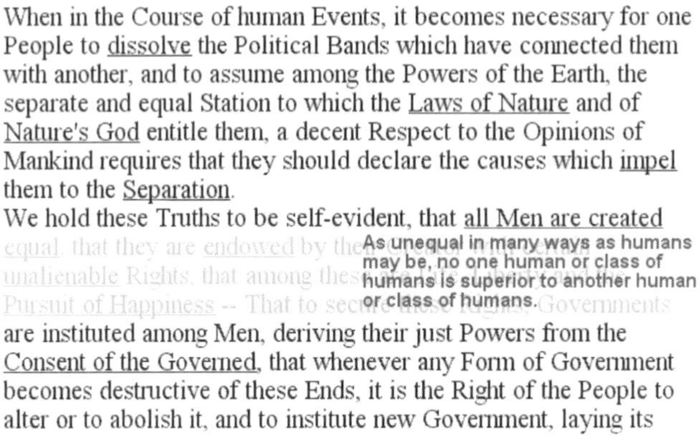

When in the Course of human Events, it becomes necessary for one People to <u>dissolve</u> the Political Bands which have connected them with another, and to assume among the Powers of the Earth, the separate and equal Station to which the <u>Laws of Nature</u> and of <u>Nature's God</u> entitle them, a decent Respect to the Opinions of Mankind requires that they should declare the causes which <u>impel</u> them to the <u>Separation</u>.

We hold these Truths to be self-evident, that <u>all Men are created</u> <u>equal</u>, that they are <u>endowed</u> by the As unequal in many ways as humans may be, no one human or class of <u>unalienable</u> Rights, that among thes humans is superior to another human Pursuit of Happiness -- That to sec or class of humans. Governments are instituted among Men, deriving their just Powers from the <u>Consent of the Governed</u>, that whenever any Form of Government becomes destructive of these Ends, it is the Right of the People to alter or to abolish it, and to institute new Government, laying its

Fig. 7.1.14. Bay-Wei Chang, Jock Mackinlay, and Polle Zellweger. Xerox PARC's *Fluid Documents* project. An experimental interface that presents annotations in context using techniques such as marginalia, interlinear, and overlay, ⟨http://www. parc.xerox.com/istl.projects/fluid/⟩.

the location of information and the analysis of its meanings. The breadth of the World Wide Web especially has exacerbated the need for information analysis and visualization tools. The familiar metaphor of the iconic desktop computer and mouse-based manipulation is a well-known result of this kind of investigation.

Contemporary researchers advance that kind of inquiry by creating new kinds of experimental 2-D, 3-D, tree, animation, time-based, and metaphoric visualization methods to represent the structure of information and facilitate overviewing, filtering, and selective focusing. They integrate cognitive science, linguistics, and visual and media design. For example, Xerox PARC's *Citation Visualizer* creates a visual network of interlocking shapes and connectors to represent the usually invisible structure of all the journal articles that quote a particular article and all the articles that it quotes. Another project called the *Magic Lens* provides a dynamic interface "magnifying glass" that allows a user to view a vast display of informationally enhancing elements that fit some specified criteria. The *Hyperbolic Tree* interface represents both the breadth and depth of searches, simultaneously and dynamically adding visual details to particular areas of search results as a user moves a mouse over displays. Another project explores the usefulness of the biological and anthropological metaphors of "foraging" to design search techniques, for example, exploiting the "scent" of a good lead and allowing "competition" between search paths. MIT's Information Landscape project maps information to a geographical metaphor. *Liquid Pages* provides supplementary material within the context of existing pages by opening up marginalia or superimposition.[7]

Artificial Intelligence

How can digital information systems be enhanced so they can duplicate the sophisticated intelligence of humans? Artificial Intelligence has a long history of research, with the researchers pushing the boundaries of machine intelligence in areas such as making sense of an environment, learning, adapting, knowledge representation, planning, and creative problem-solving.

Early achievements with feats such as playing chess led enthusiasts to proclaim that after a few more years of research computers would surpass human abilities in most realms. After researchers encountered major difficulties in tasks simple for humans, such as image and language understanding, and major theoretical critiques such as Hubert Dreyfus's *What Computers Can't Do,* the glamour faded. Nonetheless, research has continued to advance and it underlies many of the research agendas described in this book, for example, speech, image, gesture, face, and video recognition; information visualization; software agents, and robotics. Current research projects focus on topics such as knowbots and infobots; autonomous vehicles (such as the Mars Explorer); and the continuation of language understanding.[8] In one project called *Cyc,* Douglas Lenat proposes to build up the commonsense database of an artificial intelligence through long-term education similar to a child's experience.

Physical Information Systems

Ubiquitous Computing, Tangible Bits, and Smart Spaces

How can physical work, play, and living spaces be enhanced through digital information systems? How can the long biological experience of humans in manipulating physical objects be exploited as an interface to information systems? Researchers propose that contemporary models that focus on the computer as a separate appliance will seem like an anachronism in the digitally enhanced future. Sensing, computing, and communication functions will become invisible and integrated into the manufacture of many objects and the architectural arrangements of spaces. Persons, objects, and surfaces will be tightly linked into integrated digital systems. Major industry efforts such as *Bluetooth* are working to create the necessary infrastructure.

This cluster of research activities goes by a variety of names, including Ubiquitous/ Pervasive/Invisible/Embedded Computing, Tangible Bits, Augmented Reality, Smart Spaces, and Smart Objects.

Xerox PARC's *Ubiquitous Computing* project attempted to activate all meeting spaces so that even persons who were not present could virtually drop in at will. MIT's Tangible Bits group attempts to make objects into information collectors and indicators of information states. Projects include the creation of smart manipulatable objects as interfaces

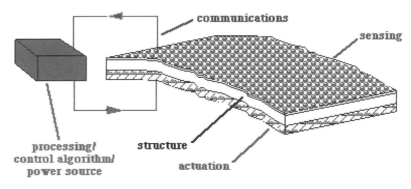

Fig. 7.1.15. SPIE Smart Structures Working Group. A diagram showing the basic components of smart materials that include sensors and actuators as part of their structure, 〈http://www.spie.org/web/workinggroups/smartstructures/smartstructuresdesc.html〉.

and the control of ambient phenomena such as air, water, sound, and light to convey information. In the *Luminous Room* and *IO Bulb* projects, lightbulbs are also digital cameras and information indicators. The *Touch Counter* keeps a record of everyone who has touched an object. *Ping Pong Plus* creates a reactive table in which a projected image on it creates simulated ripples of water everywhere the ball hits. The Things That Think group works on creating smart objects, such as *Ventus,* an automatic medicine dispenser that is aware of past action. Smart Structure and Materials research groups seek to create materials with instrinsic sensing, computing, and actuating capabilities. Several research groups at MIT's Media Lab are working with the Museum of Modern Art in New York City to develop new kinds of smart public spaces that will embed information display into intrinsic features of the museum.

Other projects emphasize manipulation as an interface. *Media Blocks* lets users edit and compose media by connecting physical blocks. Another MIT group called Tools to Think With explores the use of digitally linked objects for idea stimulation and educational purposes. Building on constructivist notions that learning comes through active manipulation, they have created systems such as digitally linked Lego blocks and beads that intrinsically provide feedback about the structure of information systems through their reactions. A spin-off from Interval Research called Zowie Toys is creating "smart toys" that link physically manipulated objects with digital worlds on nearby computers.

Wearable Computing

What are the possibilities of digitally activating everyday objects such as clothes? Examples of wearable computing research includes sensory amplifiers that allow a user to see and hear usually unavailable sources, personal heads-up display eyeglasses that provide information relevant to the context in which users finds themselves, and elements that augment a user's performance, such as remembrance agents or shoes that know dance steps. Wearable camera systems provide unobtrusive ways to record information about specific locations. The *GroupWear* project used transponder tabs to ascertain who near the wearer might share common interests or other relevant "memes," and indicated that by changing a wearable display.

Research addresses several challenges: expanding the kinds of inputs, such as tactile; developing new kinds of outputs, such as heads-up displays on normal glasses; increasing the context awareness of wearable systems; signaling emotional states; and extending applications in areas such as business, medicine, and group work. The chapter on biology research (2.1) described several military research projects that build sophisticated biosensors and information displays into combat uniforms. MIT's Parasitic Power project is developing methods of powering digital electronics through the energy generated in

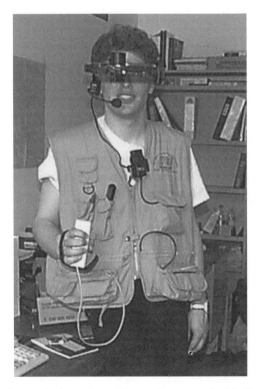

Fig. 7.1.16. Gerd Kortuem. An example of current wearable computing components, ⟨http://www.cs.uoregon.edu/research/wearables⟩.

natural processes such as walking. Augmented reality systems provide information on a transparent visor tuned to the physical location and gaze direction of a viewer.

Smart Houses and Intelligent Transportation Systems

How can digital information systems enhance the intelligence of homes and the automobile transportation system? Researchers are rushing to make buildings and homes more intelligent. They are trying to increase the sophistication of sensors, the capabilities of actuators and controls, the reach of and interconnectedness of architectural and movable objects, and integrate these systems into the lives of the inhabitants. Architects have long sought the ability to make smart buildings and cities, as exemplified by Negroponte's early work with programmable cities and the Architecture Machine group. Similarly, designers seek to increase the intelligence of the automobile transportation system with innovations such as smart highways, self-driving cars, automatic navigation systems, and intelligent traffic-pattern coordination systems.[9]

Theoretical Reflections on Digital Culture and Art

Introduction: Areas of Analysis

Theoretical analysis of the cultural implications of digital technology has flourished in the last decade. Theorists explain that computers can no longer be conceptualized merely as isolated office appliances. Rather, digital technology and its underlying conceptual frameworks and associated socio-cultural infrastructure profoundly affect life and thought. This section reviews the theoretical analysis of their implications in several areas: basic conceptions of reality and our epistemological ability to know it; the meaning of bodies and physical space in a world increasingly dominated by virtuality; the nature of identity and gender; the interrelationships between digital technologies and larger socio-economic-cultural forces; the historical place of digital technologies and media; interactivity; and the special challenges confronting artists who work with digital technology. This short review can only skim the surface of what has become an enormous corpus of scholarship. Readers who are unfamiliar with basic perspectives in critical theory should consult my paper "Dark and Light Visions: The Relationships."[10]

What Is Reality and How Can We Come to Know It?

Throughout the history of philosophy, epistemologists have debated our ability to know the essence of reality. For example, in the "Dialogue of the Cave," Plato suggests that nature consists of a world of essential forms that can only be known imperfectly, like shadows on a cave wall. In the Renaissance, scientific thought asserted a renewed faith in observation and reason as techniques to know reality. For a few hundred years this epistemological faith was verified by significant practical and theoretical accomplishments.

Starting at the beginning of the twentieth century, physics and biology increasingly moved into worlds not accessible by direct observation. The increasing reliance on sophisticated instrumentation and chains of reasoning renewed epistemological doubts—the atom and the gene could not be seen. Telecommunications similarly introduced situations in which faith in remote realities had to rely on electromagnetic reproduction and representation. Contemporary theorists note that epistemological challenges are exacerbated even further by the spread of digital technologies. By its very nature, digital representation requires the breaking apart of phenomena and their representation by symbolic bits. Increasingly we rely on representations in all corners of life.

Sandy Stone describes the paradoxical situation of the apparent increase in our knowledge while at the same time relying on epistemologically shaky distant representations. She suggests that the realistic accomplishments of the illusion factories of digital Hollywood help to undermine our faith in the images coming from scientific instrumentation:

We find ourselves in the paradoxical situation that the more we call "that which becomes known" by the name "reality," the further we distance ourselves from it. Because with time and increasingly sophisticated tools, reality seems more and more intelligible—as images on screens, images of events that we will never ourselves experience, subatomic collisions, the DNA helix, movement of ions within synapses. What's actually happening is that our understanding of the world and of "nature" increasingly becomes secondhand, like a story.[11]

In his essay "Digital Apparitions," Willem Flusser suggests that the density of data points has been an underlying criterion for distinguishing "real" and synthetic worlds. He notes, however, that research promises to continue to enhance that density to the point where we will not be able to distinguish it. The epistemological question then becomes prime. Distrust derives also from the fact that artificial worlds are human made.[12]

Flusser recounts the history of attempts to represent experience with images. Eventually, science used calculation and geometrical analysis to represent growing understandings. Doubts arose, however, that the equations were imposed on nature, and the calculatory flexibility of the computer augments doubts. Since we can construct convincing worlds to represent theories, the realizations lead to a radical doubt: "whether everything, including ourselves, may have to be understood as a digital apparition."[13] McKenzie Wark reflects on the implications of telecommunications on notions of reality. He notes that the speed of information flows relative to the movement of persons and things changes our relation to them: they become less palpable and less important.[14]

In deconstructing the history of image and representation, other theorists such as Timothy Druckrey show how the fall of faith in the power of the unitary point of view to ascertain truth leads to more profound questioning of the possibility of knowing reality. Moreover, consciousness and perception, the core tools of knowing reality, are themselves becoming defined by technology:

If there is a common denominator within the divergent discourses of postmodernity, it is the concept that a system of scientific visualization and any totalizing model of either the "real" world or its representations cannot be put into place. . . . The unrepresentable "real" collides with the unreflected "virtual."[15]

Peter Weibel similarly notes that digital technologies offer new kinds of flexible control over perception and representation that further undermine epistemological faith. Phenomena cannot be separated from observers and their observational interface. Weibel notes that there is a price to pay: information is seen as floating and resistant to fixation: "The doubts cast by endophysics (subsequent to the theory of relativity, quantum and chaos theories) over the classical, objective nature of the world and its concomitant terms and programs amount to a description of our media and computer worlds.[16]

What, then, is the role of the arts in this epistemologically shaky world? Weibel suggests that the arts and media can provide a powerful setting for understanding the artificiality of even what was formerly called the real world and for experimenting with the observer effect.

Flusser suggests, much like the science historian Paul Feyerabend, that science can be considered much like art. Scientists rely on digitized data that is subject to radical epistemological doubt, making them similar to the products of art. No one can claim direct access to reality:

What we call "the world," what our senses, by not entirely clear methods, have computed into perceptions, into emotions, desires, insights, even the senses themselves, are reified processes of computation. Science calculates the world as it has already been conceived. It deals with facts, with things made, not with data. The scientists are computer artists *avant la lettre,* and the results of science are not some "objective insights," but models for handling the computed. Understanding that science is a form of art does not debase it. Quite the contrary: science has become a paradigm for all other arts.[17]

Current research moves into areas that will further challenge faith in our perceptions even more. For example, interface research into synthetic tactile, kinesthetic, and olfactory experience will destabilize our faith in these previously trusted, phenonomenologically compelling senses, and the precision stimulation of the brain to generate artificial cognition will render all perception problematic.

Although these analyses are popular in the arts, many scientists still accept the ontological status of a real world. The canons of scientific verification provide methods to increase faith in external realities even if they can't guarantee their existence. Similarly, the phenomenological experience of the physicality of the real world and the bodily sense of sickness, hunger, sex, and death speak strongly for a world outside of perception. The arts would best be wary of prematurely taking sides in this digitally inspired debate about reality, which promises to continue into the next decades.

Are Bodies and Physical Space Relevant?

Radical analysis suggests that the physical body and physical space are becoming increasingly irrelevant. Digital technologies such as virtual reality allow people to inhabit simulated worlds of their own arbitrary design using synthetic bodies. These virtual creations can model physical reality or freely improvise from the imagination. Digital communications and telepresence allow people to perceive and act at a distance, disregarding the old constraints of physical space. As people spend more time in digitally produced environments, these worlds become more realistic, and as their significance in people's lives increases, the body and physical space may decrease in importance.

Historical Precedents

Speculation about the relationship of virtuality and physicality is not new. Some of the Greeks wondered if they should ban theater because audiences became so involved in the artificial realities that they seemed to become oblivious to everyday physical reality. Over time, similar concerns were expressed by the power of novels, cinema, and television to kidnap participants outside of everyday life.

The tendency to deprecate the physical body has other precedents in Western traditions of mind-body dualism and distrust of the flesh. For example, Plato felt that the body was imperfect and distant from the essential forms, which were the most important aspect of the universe. Christian traditions saw the flesh as a distraction from attending to more important spiritual concerns and saw the sojourn on the earth in body and physical space as an unfortunate detour on the way to a more ethereal heaven. In the Enlightenment, the mind was seen as much more important than the body. It is easy to see the contemporary ascendancy of virtual bodies and places as a continuation of these themes.

Perfectionism does not need to be expressed in religious terms. Florian Rotzer reflects on the techno-utopian dreams of downloading consciousness, with the body seen as an imperfect vessel.[18] Ultimately, some hope to bypass the body by intervening directly with the brain. Frances Dyson deconstructs the great variety of narratives underlying the interest in virtuality and its expression of the desire for new kinds of sociality and metaphysical experience:

The cultural imaginary propelling virtuality inhabits a curiously sophisticated rhetorical landscape, shaped by the endless permutations of futuristic, libertarian, psychedelic, cybernetic, anarchistic, subcultural, utopian, mystical, and science-fiction mythologies permeating late-twentieth-century culture.[19]

The Nature of the Virtual Experience

Marcos Novak's "Transarchitecture" attempts to prepare for the ascendancy of virtual space as our principal architecture. He notes that the "alien" used to be experiences that were outside of a community; the ability of virtual reality to spatially manifest our ideas means that "we will discover the alien that is so near as to be outlandish. We will become citizens in the spaces of our varied consciousnesses."[20]

N. Katherine Hayles describes the phenomenon as the ascendancy of pattern over presence. She describes the posthuman as a "coupling so intense and multifaceted that it is no longer possible to distinguish meaningfully between the biological organism and the information circuits in which it is enmeshed."[21] Frances Dyson explores aurality and the phenomenology of sound as a tool for understanding virtual reality's interest in immersion. Like sound, VR enables the consideration of nondualistic, non-Cartesian modes of experience:

Like virtuality, the phenomenal invisibility, intangibility, multiplicity, and existential flux of sound challenges an understanding of the real based upon the visible, material, and enduring object. Sound cannot be held for close examination, nor can it be separated from the aural continuum and given a singular identity. In a constant state of becoming, sound comes into and goes out of existence in a manner that confounds ontological representation. Similarly, being both heard outside and felt within, sound blurs the distinction between the interior and exterior of the body, annihilating the distance between subject and object, self and other.[22]

Doubts about Liberation from the Body and Space

Debate rages about the possibilities of the new technologies to experientially explore postmodern concepts of body and space. Some are hopeful; some are doubtful. Virtual reality provides some illusion of body awareness. Typically, VR travelers are imaged by bodily representations in the virtual worlds. Also, body suits full of sensors translate body actions into those in the virtual world. Simon Penny notes, however, that the real body is abandoned and that VR reaffirms traditional notions of space and dualism in spite of its rhetoric:

One leaves it at the door while the mind goes wandering, unhindered by a physical body, inhabiting an ethereal virtual body in pristine virtual space, itself a "Pure" Platonic space, free of farts, dirt, and untidy bodily fluids. . . . [I]t is a clear continuation of the rationalist dream of disembodied mind, part of the long Western tradition of denial of the body. This reaffirms the Cartesian duality, reifying it in code and hardware.[23]

Arriving at a similar conclusion, Frances Dyson observes forces within techno-culture that try to restrain VR's exploratory tendencies and draw it back into traditional conceptualizations of Cartesian space populated with rigidly bounded entities:

Rather than entering a "free-space," subjectivity is recontextualized within the programmatic grid of technology, and embedded in this grid are all those elements that drive the fixed and rigid reality, the prescribed subjectivity one might, through VR, be trying to escape. Causality, linearity, hierarchy, the discrete unit, the id one, the individual—all are situated within a teleology geared towards increasing control over systems of representation.[24]

The Political Underlife of the Interest in Virtual Bodies and Virtual Places

The emphasis on disembodiment and cyberspace also can be seen as the manifestation of political narratives. In "Code Warriors," Arthur and Marilouise Kroker describe the process of withdrawal induced by the cyberworld. They see the emergence of a "virtual class" that uses the fascination with cyberworlds to discredit independent sensual experience and ease the path to domination. They describe the "bunker self," which dumbs down its participation: "Privileging information while exterminating meaning, surfing without engagement, digital reality provides a new virtual playing field for tuning out and turning off."[25] Physical place also becomes a shell. Florian Rotzer describes cities as holding places for data communication.

In "The Information War," Hakim Bey draws parallels between the antimaterialist bias of both religion and science. He sees the growing etherialization as helping support the evolution of the control and image-manipulation potentials of the modern state: "[T]he state now consists of no more than the management of images. It is no longer a force but a disembodied patterning of information. . . . [T]he media serves a religious or priestly role, appearing to offer us a way out of the body by redefining spirit as information."[26]

Bey notes that money and media have already consumed the first world in abstractions. He warns that this view has lost touch with the physical base of life and its reliance on others to do physical work:

Americans and other "first world" types seem particularly susceptible to the rhetoric of a "metaphysical economy" because we can no longer see (or feel or smell) around us very much evidence of a physical world. Our architecture has become symbolic. . . . [W]e spend our leisure largely engrossed in media rather than in direct experience of material reality. The material world for us has come to symbolize catastrophe. . . . And yet, this "first world" economy is not self-sufficient. It depends for its position (top of the pyramid) on a vast substructure of old-fashioned material production.[27]

Artist Guillermo Gómez-Peña questions the typical presentation of cyberspace as a politically neutral, raceless, genderless, and classless "territo" open to all comers.[28] He suggests that the sanitization of cyberspace from ethnic and other physical references serves to distract attention from important physical-world phenomena. Kevin Robbins wonders about the implications of the dream of virtual reality on the "real" world, and worries about the severance of social relationships.[29]

War, historically a very physical event, becomes a testing ground for the attempted escape from physicality. In analyzing the Gulf War, Frances Dyson notes that digital smart weapons obscure the consequences of actions in virtuality:

[H]igh-tech weaponry eliminates the "Vietnam syndrome" by sanitizing the body through technology. The body of the victor sparkles in its metal jacket, while the body of the victim disappears without a trace. With virtuality, the circuit is completed: floating above the carnage, the pilot initiates actions, the consequences of which are seen only via the snow of signal termination.[30]

According to some analysts, even those few remnants of life where the flesh in unexpendable will be brought under the discipline of digital infomatics. Bill Nichols describes the impulse to control reproduction:

As one expert in the engineering of human prototypes put it, reproduction in the laboratory is willed, chosen, purposed, and controlled, and is, therefore, more human than coitus with all its vagaries and elements of chance. . . . These opportunities shift reproduction from family life, private space, and domestic relations to the realm of production itself by means of the medical expert, clinical space, and commodity relations.[31]

Optimism about Liberation from the Body and Space

Others praise the liberatory possibilities: gravity and time can be defied; one can explore fantasies that are impossible in the biological and material world, break free from the Cartesian box, encounter other people in unprecedented ways, and transform one's identity in countless ways.

Florian Rotzer sees the technology as offering a new kind of imagination experience if it is developed in appropriate ways that avoid deterministic tendencies.[32] Some artists and theorists coming from a cyber-feminist perspective propose that the technologies might provide an intriguing counterpoint to traditional modernist ideas of the mind/body split and a unitary self. For example, Catherine Richards suggests that virtual reality may move against traditional ego boundaries:

I saw in such technologies as VR a site to try out and try on the projects reoccupying postmodern debate: the project of inventing new images of the body where it could be seen as a threshold, a field of intensities rather than half of the mind/body dualism; and the feminist project of redesigning female subjectivity.[33]

Referring to Elaine Scarry's research on pain, Diane Gromala proposes that pain provides a test case for virtual reality by offering an intense body-based experience that is difficult to share with others. Gromala proposed to undertake a project to explore the limits of virtual reality to deal with subjectivity and experiences such as pain:

The disembodied experience, combined with qualities of VR that seemingly do not replicate "reality," serves to upset notions in our relationship to the symbolic realm, as well as binary mind/body, subject/object, and material/immaterial distinctions.[34]

Digital Technology and Identity

Postmodern thought challenges traditional concepts of identity. From the Enlightenment on, Western culture fought to establish the individual as a unitary, volitional entity with powers of perception and action. Its literature glorified the individual's metamorphosis and acts of self-assertion and identity. Critical theory suggests a less romantic, more complex view, in which an individual's identity is fluid, shaped by circulating narratives of gender, class, nation, history, media, and situation. Digital technology accelerates the process and provides a laboratory for experiments in identity. The digitalization of information provides great flexibility in representation, while telecommunications and on-line environments sever the connection between physical persons and their communications. Theorists have sought to elaborate on these new views of identity and to analyze the impact of digital technologies.

Bill Nichols sees the self as a potentially outdated concept. The old unitary self may have lost its relevance in a world dominated by digital representation and interdependency:

Liberation from any literal referent beyond the simulation, like liberation from a cultural tradition bound to aura and ritual, brings the actual process of constructing meaning, and social reality, into sharper focus. This liberation also undercuts the Renaissance concept of the individual. "Clear and distinct" people may be a prerequisite for an industrial economy based on the sale of labor power, but mutually dependent cyborgs may be a higher priority for a postindustrial postmodern economy.[35]

In the essay "Digital Apparition," Flusser draws an analogy between modern physics and identity in the digital world. He notes that personal identity can be viewed as confluent densities of information, just as physical reality can be viewed as the density of matter points.[36] Frances Dyson suggests that the virtual body acting in virtual space transgresses traditional notions of physical-body boundaries and location. In this fluidity it more radically challenges the basic Western notions of dualistic demarcations, which underlie some concepts of identity.[37]

Virtual on-line communities invite experimentation with identity. These worlds are often constructed on the fly by participants and allow people to present themselves in any way they want. They are freed from the physical body cues of gender, age, and appearance to enact various personas. Anonymity allows for people to try out idealized or negative identities, to cross genders, or to manifest as multiple identities. Commentators draw parallels between on-line and physical life. Sandy Stone, well-known for her writing and creative work related to identity experimentation, describes the experimental possibilities of on-line representation: "They learn how to manipulate those personalities—take them out of the box, dust them, run them, put them back in the box, put them away, take out another one."[38]

In her books *The Second Self* and *Life on the Screen*, Sherry Turkle investigates the anthropology and implications of on-line activities. She sees on-line communities functioning as places to experiment with identity, much like psychotherapy. She draws a connection between the on-line experiments with multiplicity and contemporary notions of the fluid, postmodern self:

Virtual personas are objects-to-think-with. When people adopt an on-line persona, they cross a boundary into highly charged territory. Some feel an uncomfortable sense of fragmentation, some a sense of relief. Some sense the possibilities for self-discovery, even self-transformation. . . . [M]any manifestations of multiplicity in our culture, including the adoption of on-line personae, are contributing to a general reconsideration of traditional, unitary notions of identity. Contemporary psychology is being challenged to conceptualize healthy selves which are not unitary but which have flexible aspects to their many aspects.[39]

Sigfried Zielinksi proposes that multiplicity is built into the structure of digital communications; it is not just a feature of special on-line communities. The speed and rapid reconfigurations of networks makes it difficult to maintain old identities: "The Net is thus an impossible place. . . . [I]t is not a suitable place for intentionally acting subjects to stay, not even temporarily."[40]

Cultural Narratives at the Heart of Techno-culture

Digital technology is often presented within a narrative of progress and revolutionary change. Digital media is represented as a radical break with its precedents. Analysts note that this is a mistake. Technologies and media can be better understood as part of larger cultural trends. Media history that tries to ignore larger cultural forces is doomed to be misleading and incomplete.

Erkki Huhtamo, a practitioner of "media archeology," explains the need for an "anonymous history" of digital developments that does not accept the self-definitions used by researchers and practitioners within the field:

Such an "anonymous history" should include not only the industrial developments, but also the social history of the computer user, the history of the computer as an object of design and as a source of style and fashion, the histories of the computer in counter- and subcultural contexts, the history of the computer's encounter and gradual merger with media culture. . . . and, indeed, the "mental history" of the computer—the computer as a "dream machine," an immaterial object of desires, fantasies, fears, and utopias.[41]

George Legrady similarly notes that media technologies cannot be separated from larger cultural forces. The relationship of technology and human consciousness can serve a multitude of functions—"as an extension of the human body, as a mirror of the self, as a mediation between nature and culture, as a potential discursive medium or a tool of alienation and control. . . . All technologies distort. By expanding our abilities to perceive, they simultaneously diminish us."[42]

Simon Penny focuses on the utopian rhetoric of the computer world. Drawing on previous media history, he notes that the real impact may well be the inverse of the rhetoric:

It becomes clear that the realities of new technologies as they are actually implemented is generally in direct opposition to the rhetorics that heralded them into the market. Artists and inventors imbued with a sincere utopianism often become part of the mechanism by which such technologies become products. . . . One of the classic techno-utopian myths of computers is that access to information will be a liberation, and its results will be, by definition, democratizing. The reality of this technology is an effective centralizing of power. This democratizing myth is strongly reminiscent of some of that surrounded the introduction of television.[43]

Acting as a media archaeologist, Huhtamo explains that new technologies often cyclically recapitulate previous cultural themes. The technologies are simultaneously new and old. Their function in culture can best be understood by becoming aware of the prior historical "layers." In several articles he reveals the power of this approach by unearthing historical trends underlying contemporary technologies. For example, in "From Kaleidoscomaniac to Cybernerd: Notes Toward an Archeology of Media," he traces the history of media spectacle through the nineteenth and twentieth centuries, and analyzes the similarity of audience amazement and "panic" in the face of projected images stretching from eighteenth-century Parisian Fantasmagorie shows, through early cinema, to Disneyland captain EO 3-D, laser extravaganzas. He extracts "topoi," "cultural building blocks" that are molds for experience that continually get reactivated, and asks what psychological and cultural purposes are being served by these phenomena in different eras.[44]

In another analysis, he traces the history of the telectroscope, a precursor of television. This nineteenth-century-proposed device would have allowed individuals to view each other from a distance, as in a picture phone. He explores the way television got diverted from this one-to-one communication model to the one-to-many model of broadcast TV, and he shows how rhetoric surrounding virtual reality as a person-to-person communication device recapitulates some of the same themes.[45]

In "Encapsulated Bodies in Motion: Simulators and the Quest for Total Immersion," Huhtamo analyzes the quest for "immersive" technology. He notes parallels among diverse cultural phenomena such as computer games, simulators, theme park rides, drugs, Eastern meditative experiences, and cinema virtual-zoom travel scenes. He documents the historical advertising rhetoric for immersion in remote realities, stretching from 1850s descriptions of stereographic photography, through early promotions for buying TVs and 1950s Cinerama. He suggests that a historical analysis of the Victorian fascination with stereography offers insights into the contemporary interest in technologies such as immersive virtual reality. Stereography provided an escape to physically inaccessible places, by allowing the objectifying male gaze to penetrate virgin lands. Huhtamo quotes Charles Baudelaire: "A thousand hungry eyes were bending over the peep-holes of the stereoscope, as if they were the attic-windows of the infinite."[46]

Huhtamo ties in another historical thread by linking contemporary simulators with amusement parks, streetcars, and railroads. Amusement park rides attempt to provide visitors with specially truncated and amplified experiences of bodies in motion. The popularity of physical rides, however, cannot be explained only as physical thrill; rather, they grow out of a metapsychological need for people to deal with increasing mechanization and control of the body.[47]

Doug Kahn similarly analyzes sound technologies such as phonography, telephones, and sound film by looking at them in terms of deeper cultural and social practices. For example, he identifies cultural themes implicit in phonography, such as the severance of voice from the body, movement of sound into the realm of representation, and the search for the voice of the soul.[48]

The Dominance of Vision

Tracing the cultural significance of perspective and the dominance of vision has been a major focus of many analysts. In this view the development of perspective in the Renaissance was not just a technical innovation. It instantiated cultural themes such as the importance of sight, the privileging of particular points of view, the disregard of other senses, and a faith in the ability to organize and dominate space. The power of contemporary media and representation derive from this dominance. Contemporary technological developments can be understood as parts of this cultural trend, sometimes recapitulating them and at other times breaking new ground. Timothy Druckrey explains the importance of vision in broad cultural issues of power and the role of technology in the process:

A politics of seeing, recording, and accumulation emerged. Experience was circumscribed by a series of stages in which the displacement of vision by representational systems was both scientifically legitimated and culturally necessary. Photography, cinema, and scientific visualization coalesced with systems of illusion, recording, spectacle, information, and the public sphere. In a panoptic culture, the management of visuality identifies consumption as passive and production as empowering—essentials in the system of capital.[49]

Peter Weibel notes the importance of technology in making vision the dominant sense of the modern era:

The primacy of the eye, . . . as the dominant sense organ of the twentieth century is the consequence of a technical revolution that put an enormous apparatus to the service of vision. The
rise of the eye is rooted in the fact that all of its aspects (creation, transmission, reception) were
supported by analog and digital machines. The triumph of the visual in the twentieth century
is the triumph of a techno-vision.[50]

Lev Manovich wrote a series of articles tracing the interrelationships of vision technologies and cultural practices. In "Labor of Vision," he investigated how technologies of
vision changed in the transition from modernity to postmodernity. Drawing on Walter
Benjamin, he noted that broad psycho-cultural trends affect both leisure and work. For
example, the factory worker and the filmgoer confronted similar perceptual tasks "of
keeping pace with the rhythm of production." In the contemporary world, the perceptual
task has changed in both leisure and work to monitoring data displays, ready for events.[51]

Manovich traces the history of the human-machine interface from Taylorism to cognitive science. In the early part of the twentieth century, researchers studied the psychology and physiology of workers with the aim of increasing their productivity with
machines. In the late twentieth century, the focus has shifted to efficiency in mental
work. Radar operators, fighter pilots, computer-game players, and VR navigators are all
working on similar perceptual tasks.[52]

Sophisticated computer-image recognition systems can correct distortions, focus, and
blur. They can use understandings about contexts to extrapolate from visual information.
Manovich suggests that perspective is losing its ascendancy, becoming just one space-
mapping and visualization technique among many.[53]

Analysts such as the Critical Art Ensemble assert that vision serves important functions
of domination and control. Vision technologies are used in military intelligence, surveillance, population control, geographic management, and also in maintaining the symbolic
order through the representation and manipulation of spectacle. The war machine and
the sight machine are closely linked.

The analysis notes that the "flesh machine" will eventually complete the triumpherate
by bringing the body and biosphere into disciplines of control. Vision again plays a
crucial role by subjecting the "target" to surveillance and symbolic positioning. It uses
the military metaphors of "visual intelligence" to explain the role of visual surveillance
in domination of the biosphere:

The significant principle here—the one being replicated in the development of the flesh machine—is that vision equals control. Therefore the flesh machine, like its counterparts, is becoming increasingly photocentric. . . . From the macro to the micro, no stone can remain unturned.

Every aspect of the body must be open to the vision of medical and scientific authority. Once the body is thoroughly mapped and its mechanistic splendor revealed, any body invader (organic or otherwise) can be eliminated, and the future of that body can be accurately predicted.[54]

The Relationship of Digital Technologies to Gender, Class, and Socioeconomic Forces

Class and Ethnic Identity

Arthur and Marilouise Kroker offer a critique of what they call "virtual manifest destiny." Many atrocities to the human spirit are being committed in the name of digitally mediated "improvements." They warn about the unacknowledged consequences of a mindless simplistic march to the digital utopia spearheaded by capitalists and digital visionaries, "a virtual war strategy where knowledge is reduced to data storage dumps, friendship is dissolved into floating cyber interactions, and communication means the end of meaning. Virtualization in the cyber hands of the new technological class is all about our being dumbed down."[55]

Guillermo Gómez-Peña challenges the assertion that the digital world is classless, genderless, and free of ethnic discriminations. He analyzes circulating stereotypes that characterize third world peoples as being incompetent and uninterested in digital developments: "We continue to be manual beings—*homo fabers* par excellence, imaginative artisans (not technicians)—and our understanding of the world is strictly political, poetical, or metaphysical at best, but certainly not scientific." He attributes relative nonparticipation to lack of access instead of supposed cultural traits and describes the spread of digital culture into the third world.[56]

Some analysts seek to problematize the neutrality of digital media. Drawing on Walter Benjamin, Bill Nichols draws parallels with the cultural functions of film. Benjamin claimed that mechanical reproduction technologies had the potential to revolutionize society. The film industry served to "contain" this potential. He sees a similar danger as the "explosive" potentials of digital culture, such as the elimination of drudgery and the promotion of collectivity, get "defused" and channeled into controllable expressions. Furthermore, historically internal processes such as intelligence and perception become incorporated as commodities. "[T]he automated intelligence of chips reveals the power of postindustrial capitalism to simulate and replace the world around us, rendering not only its exterior realm but also its interior ones of consciousness, intelligence, thought, and intersubjectivity as commodity experience."[57]

Simon Penny similarly notes that consumer culture is embedded in the hardware and software systems. He is skeptical about its revolutionary potentials:

At the computer, as in the supermarket, one submits to the interactive scenario and the limited freedoms it offers: total freedom among a set of fixed options. A postmortem capitalist paradise. In postmodern times, we build a personal identity from novel combinations of manufactured commodities. "I shop, therefore I am.". . . . Computer technology, hardware architecture, and software design reify value systems.[58]

In an essay called "The Californian Ideology," Richard Barbrook and Andy Cameron debunk the rhetorics of digital utopianism by analyzing its underlying libertarian assumptions and blind spots. They trace the origins of digital nirvana in the cultural and political upheavals of the 1960s and follow the evolution of the ideas to their present state, which posits a utopian world that could be ushered in by giving free reign to unfettered development of digital technologies, communication, media, and associated dispersed social structures and industries. They call it Californian because many of the trends find their most extreme expression in places like Silicon Valley:

This new faith has emerged from a bizarre fusion of the cultural bohemianism of San Francisco with the hi-tech industries of Silicon Valley. . . . [T]he Californian Ideology promiscuously combines the free-wheeling spirit of the hippies and the entrepreneurial zeal of the yuppies. This amalgamation of opposites has been achieved through a profound faith in the emancipatory potential of the new information technologies. In the digital utopia, everybody will be both hip and rich.[59]

Barbrook and Cameron raise several questions about this view. They note that this digital future depends on the unacknowledged labor of digital underclasses. It is supported by a drive of capitalist expansionism and attempts to free itself from obligations to laborers. The shadow side of the digital class's freedom and individuality is a lack of connection to other workers and an unrealized acceptance of work as the main life value. The digital class has accepted the ideology of the free market and the withering away of government without careful analysis of the consequences.

In the more extreme visions, the digital elite narcissistically seek to become hyper-evolved "Extropians," freeing themselves from the mundane details of everyday life and community interdependence and responsibility. They believe in the possibility of roboticizing all labor, failing to acknowledge the ultimate reliance on human labor.

Cyberfeminist Critique

Investigating the way digital technologies continue classic Western approaches to visualization, cyberfeminists comment on the often unrecognized gendering of vision. Digital environments such as VR often recapitulate the problematics of the male gaze, with its assumptions of authority, privilege, and penetration. Simon Penny describes the instantiation of the male gaze in VR: "What the eye wants, the eye gets."[60]

Nell Tenhaaf notes that the modernist philosophies underlying technological development are essentially male and marginalize other female approaches to working with technology:

The philosophy of technology, however, has been articulated entirely from a masculinist perspective in terms that metaphorize and marginalize the feminine. . . . The modernist philosophical framework for technology is the discourse of the will, specifically the will to power postulated by Friedrich Nietzsche in the late nineteenth century. Expanded upon by subsequent philosophers, in particular Martin Heidegger, this discourse views technology as the manifestation of an essentially masculine will that is the driving force of the whole modern era. In its language and imagery, the will to power is interwoven with a deeply entrenched and mythic concept of duality that describes commanding (and the power of the machine) as a masculine attribute, while submission (and the rule of feeling) is described as a feminine one.[61]

Sandy Stone's studies of people's interactions with computers revealed this same focus on domination. She refers back to Francis Bacon's ideas that "nature was a woman who had to be seized and wooed, and her secrets were to be wrested from her by the controlling man."[62] Nancy Paterson traces male attitudes about women and the power of knowledge back to before the electronic era, to examples in literature like Eve and Pandora. The update links the danger to digital technology: "The power which these women wield is evil, technological, and, of course, seductive."[63]

Tenhaaf, Paterson, and some of the other feminist theorists discussed in the previous sections on the body and identity (chapter 2.5) believe that there are ways to work with technology outside of the narratives of domination. For example, in "The Future Looms: Weaving Women and Cybernetics," Sadie Plant draws analogies between software and weaving. She considers new developments such as the Net and artificial life, which defy direct control, and sees them as manifestations of the unpredictable "Other," to be engaged rather than dominated. Women must master the technologies and introduce new models that strike out beyond the patriarchal emphases of much technology, such as the military origins of the Internet or the violent goals of computer games.[64] Paterson asks what models might work:

Cyberfeminism as a philosophy has the potential to create a poetic, passionate, political identity and unity without relying on a logic and language of exclusion or appropriation. . . . New electronic technologies are currently utilized to manipulate and define our experiences. Cyberfeminism does not accept as inevitable current applications of new technologies which impose and maintain specific cultural, political, and sexual stereotypes. Empowerment of women in the field of new electronic media can only result from the demystification of technology, and the appropriation of access to these tools. Cyberfeminism is essentially subversive.[65]

Like artists such as Char Davies and Brenda Laurel, discussed in chapter 7.3, she sees the liberatory possibilities of VR and other digital technologies as a fertile area to work. Cyber culture offers the opportunity to experiment with new identities for the self, and to force others to interact without gender bias:

[They ground] themselves in personal physical experience. This skill will serve well as we venture into other dimensions and back home again. However skilled we become at navigating these spaces and temporarily leaving our bodies behind, it is doubtful that we will ever achieve immortality. Virtuality is patriarchy's blind spot. . . . Transgressing order and linear organization of information, cyber feminists recognize the opportunity to redefine "reality" on our terms and in our interest, and realize that the electronic communications infrastructure or "matrix" may be the ideal instrument for a new breed of feminists to pick up and play.

Other cyberfeminist writers and artists believe that interface conventions are not as neutral as typically presented. They carry messages of domination and control often associated with male perspectives. Interface models involving more aspects of the body allow for more convivial exchanges between humans and digital technology. For example, in paper called "Posthuman or Para-ego? Interactive Models of Human-Technology Relations," Zo Sofoulis explores an agenda in which physicality is emphasized. She notes the growing interest in topics such as cyber sex and teledildanics, and the reliance on verbal mediation. She is also interested, however, in other relationships between the body and digital technology, such as polymorphous corporeality. Experimental art installations offer an opportunity for artists to engage audience bodies in unprecedented ways. Bruno Latour's actor-network theory provides useful concepts for thinking about the dynamic meaning that arises from exchanges between bodies and technology:

What it looks like and what it can do (its performative competency) is not determined in advance along some foregone technological trajectory, but emerges only through contestation, contingency, and a delicate "dance of agency" negotiated within a heterogeneous network or assemblage of human and nonhuman agents.[66]

Sofoulis suggests that female artists have special perspectives to contribute to this analysis of the core of interface because of historical values of corporeality and relationship:

[O]ur capacities to make intimate connections with and through high-tech equipment, capacities much celebrated in cyber culture . . . have little to do with the higher reaches of reason enabling transcendence of the flesh via technology into a posthuman state; and perhaps have more to do with the agencies and competencies of our bodily beings, which are essential to our own humanity, a humanity understood as extending its sociality within a life world shared with many other kinds of physical and virtual entities.

In "Cyberfeminism with a Difference," Rosi Braidotti analyzes the possibilities for a feminist agenda in cyberspace. She explores the possibility of re-embodiment and parody but warns about the danger of reproducing patriarchal patterns. She sees virtual reality as a possible space for feminist experimentation, although with inherent dangers. Old feminist models of female identity may not address the realities of the cyberworld. Women need to find new dynamic modes that use the technical possibilities in innovative ways:

Yes, the girls are getting mad; we want our cyber dreams, we want our own shared hallucinations. You may keep your blood and gore, what's at stake for us is how to grab cyberspace so as to exit the old, decayed, seduced, abducted, and abandoned corpse of phallo-logocentric patriarchy; the death squads of the phallus, the geriatric, money-minded, silicon-inflated body of militant phallocracy and its annexed and indexed feminine "Other." The riot girls know that they can do better than this. . . .

I would like to argue therefore that the central point to keep in mind in the context of a discussion on cyberspace is that the last thing we need at this point in Western history is a renewal of the old myth of transcendence as flight from the body.[67]

Information Structures and the Fabric of Life

Developments of digital technology reduce everything to information and change processes of daily life. There is expansion and acceleration but also condensation and loss. For example, in "Between Nodes and Data Packets," Florian Rotzer notes that telecommunications already started a process of spatial condensation. As a consequence, human relationships experienced acceleration and intensification. Communication and decision-making became linked in an unprecedented way.[68]

Several theorists analyze the loss of information inherent in digital systems. Friederich Kittler comments on the basic conundrum that real numbers and the analog values of nature can only be approximated in the digitizing process.[69]

Siegfried Zielinski draws on Georges Bataille, Gilles Deleuze, and Félix Guattari to describe features of life that can easily be neglected by the Net and other digital environments. He urges engagement of the "Other." Digital systems exclude the contingent and the inconsistent. He sees the value of adopting an agenda as similar to Bataille—seeking to represent "everything that is excluded from the system and everyday routine; everything that resists being understood by science: the repellent, the vile, the violent, the instinctiveness of the death drive."[70]

In "From the Analytical Engine to Lady Ada's Art," Regina Cornwall similarly warns of the new dangers of rationalism and consumerism. She deconstructs the kinds of perspectives promoted in digital systems to the exclusion of others, and warns that the computer is not a neutral mind amplifier. Digital systems privilege explicit versus implicit, ambiguous, and metaphoric knowing; the objective over the interpreted; and data and information over knowledge and wisdom.[71]

In books such as *The Dynamic Ideography* and *Trees of Knowledge,* Pierre Lévy suggests that the digital era has brought forth radical new developments of thought, which transcend old forms of text and image and in which digital mediated linkages allow unprecedented forms of interpersonal idea exchange and symbiosis.

In his book *Technosis,* Erik Davis proposes a different kind of analysis, showing how the spiritual and millennialist dreams and fears of the last decades of the twentieth century influenced digital and information culture and how the language and ideas of the information society have shaped contemporary spirituality. The book suggests that a new "network path" offers pluralistic perspectives that are capable of "grappling with some of the forces that are currently tearing us apart: spirit and science, modernity and nihilism, technology and the human."[72]

The Role of the Arts in Digital Culture

The contemporary world confronts the arts with significant challenges. How can artists address the profound social and cultural changes implicit in the advance of digital culture, such as the impact on concepts of truth, identity, the body, and economic structure, identified in previous sections? How does the fact that the tools of digital art are the same that underlie commerce, government, and military activities limit and empower digital artists? Although this entire book addresses these questions, this section reviews theorists' reflections on them.

Limitations of Artistic Intervention

In "Consumer Culture and the Technological Imperative: The Artist in Dataspace," Simon Penny explains that aesthetic distantiation is no longer tenable when artists are engaging the same systems used in general communication and commerce. Penny explains that artists cannot afford to ignore the ideological and consumer rhetoric surrounding the digital world, audiences will interpret the work in this context. Even more, artists themselves are becoming part of the flow of this world, for example, making aesthetic judgments based on innovation and underlying tropes of progress: "An artist cannot engage technology without engaging consumer commodity economics."[73]

Penny notes that artists' involvement with the new technology creates dilemmas. Artists often end up functioning as beta testers for industrial research, generating ideas that result in products for which they often do not receive acknowledgment or financial rewards. If artists allow themselves to get caught in the fascination with research, they then run the danger of needing to constantly keep up. Artistic work that can be quite innovative and challenging when first invented can have its position as cultural provocation undermined by the arrival of consumer products that popularize and commodify the "innovation." Because the artists' technologies are known in many other contexts, such as entertainment and communication, audiences use many paradigms besides art to interpret what they encounter.

Richard Wright notes a similar development in the spread of computer imaging programs that embed aestheticized techniques that used to be available to only an art elite. For example, the commercial program *Painter* automatically transforms images in the style of various painters. Wright proposes that Walter Benjamin's ideas about the loss of aura surrounding art objects can be extended to the means of production. He warns that technologized arts develop a dependence on industry that can compromise artistic agendas, and sees some similarities to photography.[74]

Artists struggle to develop aesthetics that move against the dominant, modernist, engineering models underlying digital tools. Artists working with digital systems often do not care about issues such as efficiency and clarity, which interest the technical community. Penny notes: "Contrary to the clear and direct presentation of the technical community, these artists exploit innuendo, connotation, allusion, and sometimes self-contradiction."[75] He warns that artists finding an independence from the technological imperative is critical but will not be easy.

Peter Weibel similarly warns of artists' uncritically accepting the industrial context and yielding to the seduction of techno-fascination:

The standardised weave of norms of technical apparatus, from frequencies to software, is accepted without criticism and provides standardised artistic packages. Instead of experimentally investigating artistic practice in laboratories, evolving discursive collages beyond and against the industrial empire, instead of investigating the conditions of production and consumption of art in a cultural laboratory, creating a new framework for an existence in the data world, most media artists become voluntary victims within the mighty text of technology. They celebrate their own fascination with fetish technology instead of developing a distance to this fascination.[76]

Some artists and theorists believe that one way artists can work against the domination of the commercial infrastructure is by rejecting the off-the-shelf software and inventing new tools. There is significant debate about the necessity of this approach. For several years, Ars Electronica paid special attention to artists who took this step. The 1994 Ars Electronica jury issued this statement:

This year the jury chose to recognize with honourable mentions a number of works which represent the development of new software tools with potential for rich, artistic development. These new developments are very important, since most software systems have been created for commercial, scientific, or mass entertainment purposes and are often not well suited for artistic work. . . . The jury debated at some length about where the boundary between innovative software tools and artistic work might be, but finally decided that these kinds of works were of such artistic interest that their categorisation was secondary.[77]

Other artists and theorists disagreed that this was essential and asserted that significant art could be generated within systems designed for commercial, scientific, or other purposes. For example, the artist Henry See wrote an open letter to Ars Electronica questioning this approach. Like Penny, See noted that artists are subject to new developments in hardware and software, constantly changing demands for mastering new skills, and of being in endless danger of appearing "out of date." He warns, however, that the focus on the development of new tools runs the danger of accepting these cycles and values of the industry—a too facile branding of artists as "out of date." He believes that artists can develop artistic work with off-the-shelf tools without being dominated by the underlying assumptions of the tools.[78]

The Potential of Artistic Intervention

In spite of these dangers, many theorists believe that artists can significantly contribute to the discourse about technology and culture. In "Media Art to the Rescue," Derrick de Kerckhove writes that science presents a very limited view of man and misses the profound changes brought by technology. Technological artists can enable us to see, hear, and feel more and to ensure that technology becomes a tool for enhancement rather than a chain of limitation: "The role of the artist today, as always, is to recover for the general public the larger context that has been lost by science's exclusive investigations of text."[79]

In "Art Making as Forging Evidence," Luc Courchesne suggests that artists can mitigate against the loss of faith in the ability to know the world. Like many writers described in this chapter, Courchesne notes the dominance of vision in human knowing and the decay of faith in the senses brought on by scientific inquiry into the unseeable and the advent of chaos theory and critiques of science. When reality is up for grabs, trust becomes extremely important. "Beauty," in the sense of something standing out as special, "can help build trust." Because artists can create works that command interest and attention, they can become trusted sources: "Artists, designers, and other breeds of form givers are usually aware that the experience of beauty can transform forged evidence into facts and reality."[80]

McKenzie Wark believes that part of the potential power of the arts comes from their ability to enter into the settings where the new technologies are playing out: "Electronic artists negotiate between the dead hand of traditional, institutionalized aesthetic discourses and the organic, emergent forms of social communications."[81] Soke Dinkla draws parallels with the art and life movements of earlier decades: artists sought ways to integrate art more with everyday life. Now, as life becomes increasingly technologized, technological artists may find a way to realize that old dream.[82] Huhtamo sees digital culture challenging the high-art/low-art distinctions of earlier generations. Young artists go to work in digital industries without angst about its status as art, and digital industries begin to develop standards of quality that resemble those of the art world.[83]

Many theorists believe that digital work continues what many call "postphotographic" practice. The interest has shifted from the precious object to a process of engagement. Artists no longer believe in freezing particular moments in time from particular viewpoints, but rather seek to explore multiplicity. Kevin Robbins describes the power of digital techniques to explore postmodern concepts of fluid meanings and reality.[84] Roy Ascott suggests that digital artists can help lead the way to understanding the evanescence of reality and the importance of levels below the surface.[85]

Peter Weibel similarly identifies the arts as an appropriate place to explore these new ideas about reality. He calls it the "endo" approach, which acknowledges the observer's role in shaping what comes to be called reality. Electronic arts are precisely at that nexus where the features of the postmodern world can be made clearer and worked with:

Electronic art moves art from an object-centered stage to a context- and observer-oriented one. In this way, it becomes a motor of change, from modernity to postmodernity, i.e., the transition from closed, decision-defined, and complete systems to open, nondefined, and incomplete ones, from the world of necessity to a world of observer-driven variables, from mono-perspective to multiple perspective, from hegemony to pluralism, from text to context, from locality to non-locality, from totality to particularity, from objectivity to observer relativity, from autonomy to covariance, from the dictatorship of subjectivity to the immanent world of the machine.[86]

In developments such as quantum and relativity theory and psychoanalysis, our culture experienced a series of shocks to conventional notions of reality. The electronic arts, with their focus on virtuality, provide a laboratory for understanding endophysics and experimenting with the meanings of these changes and helping a culture negotiate change even if it involves loss:

These trends have always found their way into the arts, where they were simultaneously promoted, lamented, delayed, aesthetically idealized, brought to attention, or ignored. An attendant sense of loss, be it aesthetic or epistemological, has been inevitable. It is the price each alteration of reality and any new era has to pay.[87]

Artists can use the digital tools to investigate those messy features of life likely to be ignored in digital environments such as the Net. Manuel De Landa, who wrote *War in the Age of Intelligent Machines,* focuses on organizational structures in society and digital technologies. He is especially interested in artificial intelligence and autonomous agents. He sees a growing together of biological and machinic tendencies, represented especially in military technology. He uses the term *strata* to describe self-organizing tendencies in society and technology. Artists provide an important possibility in moving against the ossification of these strata: "The artist is that agent (human or not) that takes stratified matter-energy or sedimented cultural materials and makes them follow a line of flight, or a line of song, or of color."[88]

Siegfried Zielinski describes the ways artists can stand against and outside of the main flow of our culture: "I deduce that it is our aesthetic duty to take that which is versus that which is turned over, that which is turned inside out, seriously, and to combine it with diversity and incalculability."[89] One important way to do this is to problematicize the interface and other assumptions of the digital world—to make its apparatus visible and to work on alternatives such as haptic, gestural, or spoken mediation. In an article called "Perverting Technological Correctness," Rafael Lozano-Hemmer proposes nine approaches that artists can take to question common views of the digital world, including strategies such as misusing a technology's function, presenting in a way that does not attempt to maximize pleasure, and stretching roles of who is supposed to use technology.[90]

In his essay "Points of Departure," created for V2's Next Five Minutes Conference, Andreas Broeckmann explores the possibility of Tactical Art Media. Building on Félix Guattari, he sees artists as potentially strong in making clear the domination of new media on culture and in developing countermeasures. Artists must analyze the media ecology and find places to intervene in those structures. They must nurture disruption and hetereogenesis (undoing the mass, unquestioning conformity). He quotes Druckrey's warning that many are getting caught in the search for technological progress rather than engaging deeper questions about machine culture:

Rather than an encounter with technology as the crucial mechanism in the culture of the late twentieth century, the discourse is shifting into the implementation of software solutions that veil the staggering impact of machine culture. Instead of radical questions concerning the sundering of ethics and the refiguration of communication, we are hypnotized by innovations in imaging and processing that unhinge so many of our assumptions about the fallacies of progress that yet hold our imagination in the balance. . . . For so much work utilizing electronic media, the characteristics (often seen as limitations) of the delivery system represent a hurdle to be overcome rather than a form to be interrogated.[91]

But neo-Luddism does not provide a viable answer, according to Druckrey. Rather, artists must make themselves sufficiently knowledgeable that they understand the system well enough to countervail. Artistic flexibility and the willingness to pursue nonstandard paths are resources in this work,

yet this interrogation of our tools should not lead into a new form of Luddism. Seeing the symbolic and political implications of certain technologies is an important prerequisite of identifying the cracks in the system, for identifying the breaks where usages can be moulded into new and productive forms and strategies. . . . The potential of media to be machines of difference, to be machines of heterogeneity, must be exploited by media tacticians in ways that find creative solutions for specific situations. In this, subjectification can function as a useful guideline for the choice of tools and strategies.

The process is not easy. The cultural forces driving digital development can easily assimilate many artistic gestures. Artists must constantly reevaluate the effectiveness of their media interventions.

Interactivity and User-Interface Conventions

Interactivity is often considered the distinguishing feature of computer-based media. The audience is invited to take action to influence the flow of events or to navigate through the data hyperspace. In the early days this relationship between the audience and work was considered quite radical. Artist and audience were seen as cocreators and the likelihood of intellectual, spiritual, and aesthetic engagement was seen as heightened. As the field has matured, artists and theorists have sought to deconstruct interactivity. As discussed in my paper "The Aesthetics and Practice of Interactive Events,"[92] the mere act of making choices does not necessarily result in significant artistic interchange. Also, the interactive paraphernalia of computing (e.g., menus and mouse navigation) cannot be separated from the history and conventions and social niches of computer use in the mainstream. Conventional interactivity comes out of the disciplines of computer-human interface design and engineering, whose agendas focus on efficiency and productivity rather than on more artistic goals such as provocation, discovery, nuance, and exploration. Theorists have sought to investigate the constraints and elucidate the artistic opportunities and challenges.

In his essay "It Is Interactive—but Is It Art?" written for the 1993 SIGGRAPH art show "Machine Culture," Erkki Huhtamo describes the rhetorical dangers of interactivity. Interactive systems are not automatically revolutionary and can be quite the opposite, disguising strategies for "marketing, surveillance, and exercise of authority."[93]

Artists can push interactivity in many ways. Huhtamo is interested in art that provides metacommentaries on interactivity by using the technology itself to mythicize and de-automate the discourse. He notes that many artists succumb to unquestioning celebrations of the technology and its underlying conventions. It is not easy for artists to work with experimental technologies without being socialized into the researcher's frameworks.

Many artists believe that they can simultaneously explore the frontiers of interactivity and not succumb to unwarranted assumptions. For example, artists described in chapter 4.5 on artificial life, such as Simon Penny and Ken Rinaldo, attempt to create unpredictable, evolving digital systems that generate a very different kind of interaction than the usual menu-based systems. Other artists, such as those described in chapter 7.4, create kinematic systems that read a variety of human actions to help determine the flow of events and challenge the solitary conventions of computer use.

David Rokeby, an artist who has won many awards for his interactive installations that react to the movements of visitors, has written extensively on interactive aesthetics. His essay "Transforming Mirrors" reviews the history and psychology of interactive art and explores the difference between conventional and interactive works. He notes that interactive art derives its power in part from the history of "inert" art and only becomes interesting when it moves outside the paradigms of control to explore "encounter."

Rokeby describes several kinds of interactive art, including "Navigable Structures, the Invention of Media, Transforming Mirrors, and Automata." His own work typically offers transforming mirrors, in which the user's actions influence but do not control the work. Through the works, viewers typically learn something about themselves and others: "By providing us with mirrors, artificial media, points of view, and automata, interactive artworks offer us the tools for constructing identities, our sense of ourselves in relation to the artwork, and by implication, in relation to the world."[94]

Rokeby sees interactive installations as microcosms in which viewers can assume responsibility for their actions and reflect on their status in larger social systems. He warns that utopian rhetoric can oversell and oversimplify interactive technologies. Artists are in a privileged position of understanding the tools but must avoid becoming merely public relations cheerleaders. Their role is "to explore, but at the same time, question, challenge, and transform the technologies that they utilize."

In chapter 7.2, Jim Campbell describes his interest in interactivity that is not control. He is interested in systems that are responsive but unpredictable in nonrandom ways. Many computer systems focus on maximizing control; Campbell is interested in systems that respond and are influenced by feelings and intuitions.[95]

Monika Fleischmann, an artist known for interactive VR research, works on developing new kinds of interfaces that enable visitors to act on their imaginations. She experiments with new kinds of body sensors and rich spaces that allow visitors' imaginations room to shape events. She hopes to "tempt the viewer out of the role of consumer."[96]

Known for his interactive kinetic installations, Perry Hoberman attempts to move against the commercial conventions of interactive systems. He finds the most interest in open-ended systems, for example, intersubjective systems that facilitate people interacting with each other, as on the Internet. He also questions the typical mouse, keyboard, and graphic user-interface conventions. In a talk offered at the "Seen and Heard Conference," he suggests involving more aspects of the body and moving away from multiple-choice arrangements. He defines part of the problem as how to get people to stop interacting to regard what they are looking at.[97]

In the essay "An Invitation to Interactive Art," curator Itsuo Sakane introduced the "Interactions 97" show by reflecting on interactive art. He asks, "Where, then, in the long history of humanity did interactive art originate? And where is it going? What meaning does it have in the culture produced by human creative behavior?" He notes that through history all art asked for audience interaction. In this century he sees influences from Marcel Duchamp, happenings, new-art audience participation in the 1950s and 1960s, and MacLuhanism.

These trends culminated in a faith in the power of audience participation. Drawing on Regina Cornwall's analysis of interactive computer art, Sakane traces the similarities and differences between games, military software, and interactive art. He notes that self-discovery is a key distinguishing characteristic. Sakane sees interactive computer art as a step in the transformation of culture. He sees play as a major cultural force that can be integrated into art:

[As] people had begun to recognize the new power of the media, new hope emerged that through participation we might rediscover the world for ourselves through our own senses. . . . [E]ven though they share common ancestors, the objectives and values embodied in interactive art, which were born of a liberated consciousness, are clearly different from those of the current video games, whose values are based on the marketplace. . . . Because it is connected with the broader human reality and the joy of discovery, it should be seen in the context of [Roger] Caillois and [Johan] Huizinga's ideas of play as a basis of culture.[98]

In "Interactivity Means Interpassivity," Mona Sakis suggests that interactive art may indicate a certain desire for passivity. She links it with other human desires to avoid responsibility for one's actions: "The mania to deliver oneself up to 'technologically produced' intoxications is a tricky way to reduce one's responsibility."[99]

Ken Feingold also traces the history of interactive art. He notes that artists need not be dominated by the military and video game origins of their tools. He explains important precedents with the surrealists and Duchamp, and emphasizes the importance of touch in human learning about the world and relationships. He sees interactive art building on this biological basis, "to touch, to acquire, to investigate, to examine the results of one's production . . . to affirm one's own existence in the world—the earliest and most durable forms of agency."[100]

Don Ritter, an artist who creates movement and touch-sensitive installations (see chapter 7.5), also questions the conventions of interactive systems. He notes that the normal situation of a solitary person manipulating a mouse and keyboard in front of a monitor is extremely limiting and in part a reflection of the economic constraints of the computer industry rather than a necessary part of aesthetic experiences. He developed "physical aesthetics" as a way to analyze the physical situation of interactive artworks. In his paper "My Finger's Getting Tired: Unencumbered Interactive Installations for the Entire Body," he describes the need to attend to bodily experience. He also notes that most experiences do not require people to act in solitary fashion and that many aesthetic experiences involve groups. Again, he traces the solitary pattern to the efficiency, control, and economic needs of the computer industry. He sees great opportunity in artists exploring interactive events that use the whole body and engage groups of people with each other.[101]

User Interface

Several theorists seek to deconstruct interface conventions. The system of windows, menus, icons, and mouse manipulation are only one of many sets of cultural constructs. Paul Brown, an artist and organizer of the *Fine Arts Forum,* is known for his iconoclastic views about the "user-friendly interface." He warns that these systems come with many hidden assumptions that often lull artists and audience from a more radical critique:

User-friendly tools work by adopting existing paradigmatic metaphors. In essence they tell the user . . . "there is nothing new to learn, your existing knowledge and skill can be applied to these new systems." It's not surprising therefore that they cauterise creative development and could possibly delay (and may even prevent) the evolution of new methodologies and critical dialogues.[102]

In "Cinema as a Cultural Interface," Lev Manovich suggests that the computer is becoming an interface to culture rather just a limited data manipulator as it becomes the main access point to ever-widening forms of information: "All culture, past and present, is beginning to be filtered through a computer, with its particular human-computer interface."[103]

He analyzes the three conventions of printed word, cinema, and general-purpose computer-human interface design that are increasingly being conflated, into the computer experience, extracting the assumptions that underlie each, for example, HCI's emphasis on the manipulation of objects and cinema's immersion in an imaginary world. He explains the assumptions of each and suggests that

hybrid cultural interfaces attempt to mediate between these two fundamentally different and ultimately noncompatible approaches. . . . Cultural interfaces try to accommodate both the demand for consistency and the demand for originality. . . . It is a strange, often awkward mix between the conventions of traditional artistic forms and the conventions of HCI—between an immersive environment and a set of controls; between standardization and originality.

Manovich sees this combination as creating tensions. He notes that current manifestations are not the last word and that many other possibilities are open to artistic and media experimentation.

In "A Postscript on the Emerging Aesthetics of Interactive Art," Simon Penny also warns that the lack of audience experience with artistic interactivity offers challenges because users lack experience and a language, which results in a "crisis of meaning." He suggests that new models will need to be developed drawing on the metaphors of artificial life rather than the hypertextual model of a navigable datasphere.[104]

Summary: Debate in the Art Community—Possibilities of an Enhanced Future

Are digital systems the beginning of a grand age or are they the culmination of dark forces of dehumanization and domination? Or are they both? The digital artists described in the next chapters fall all along that continuum, although they tend to be a bit more optimistic than the theorists. The theoretical analysis presented in this chapter is a critical inoculation to runaway techno-euphoria.

Some theorists, however, do believe that the possibilities outweigh the limitations. They suggest that the critique of the rhetoric of progress may be missing some genuine new possibilities. For example, Jaron Lanier, one of the pioneers of virtual reality, believes that digital technologies can create unprecedented ways for humans to know themselves and to communicate with others. He calls it "postsymbolic communication," a kind of "conscious shared dreaming."[105]

Mark Pesce, one of the pioneers of 3-D VRML Web technology, believes that digital and telecommunications technologies are making possible a new kind of global consciousness and that artists have an important role in elaborating the possibilities. In a paper called "Proximal and Distal Unity," he speculates that we are in the midst of creating a "noosphere":

The fact of the ubiquity and simultaneity of the advent of the Web—which in any reasonable historical sense has occurred in an instant—contains within it the most significant indicator of the presence of the "noosphere.". . . How then should we act when confronted with the quite-likely-but-in-the-end-unprovable existence of a cybernetic superbeing? We must begin to develop ways to communicate with it.[106]

Roy Ascott asserts a view that digital systems are radically transforming human possibilities in expansive ways. He uses the term *cyberception* to describe a consciousness enhanced in its depth and scope. Artists have an important role in elaborating and communicating these possibilities. In a paper called "Turning on Technology," Ascott suggests that the digital age has helped art complete its movement out from its visual base to address core questions about consciousness. That movement is facilitated when artists move beyond seduction with technology to total spiritual engagement:

In this reconfiguration of ourselves and our culture, the process of transformation lies between what I call cyberception, technologically extended cognition and perception, and the technoetic aesthetic, art allied to the technology of consciousness.

Engaging constructively with the technological environment, [art] sets creativity in motion, within the frame of indeterminacy, building new ideas, new forms, and new experience from the bottom up, with the artist relinquishing total control while fully immersed in the evolutive process. . . . And it is a noetic enticement, an invitation to share in the consciousness of a new millennium, the triumphant seduction of technology by art, not the seduction of the artist by technology.[107]

Analyses such as this forcefully demonstrate that there are many readings of the research and art described in this book. The utopian dreams of the researchers and frontier artists must somehow be synthesized with less benign, alternative readings.

Notes

1.   These research agenda sections compile a review of projects too extensive to list individually. A few links are listed, but interested readers should consult the think tank sites listed in chapter 8.1 or my "Emerging Technologies" Web site: ⟨http://userwww.sfsu.edu/~swilson/emerging/wilson.newtech.html⟩.

2.   Automatic Face Recognition Conference, ⟨http://www.mic.atr.co.jp/events/fg98/adprogram.html⟩.

3.   Jun Rekimoto *Holo wall:* ⟨http://www.csl.sony.co.jp/person/rekimoto/holowall/⟩; ultramagic paper: ⟨http://www.csl.sony.co.jp/person/masui/APCHI/posters.html#Usuda:UltraMagi⟩; VISLab Gesture resource page: ⟨http://vislab.eecs.uic.edu/FaceGest/Gesture.html⟩; machine gesture and sign language recognition; ⟨http://www.cse.unsw.EDU.AU/~waleed/gsl-rec/⟩.

4.   Peter Kruizinga's "Face Recognition" home page: ⟨http://www.cs.rug.nl/~peterkr/FACE/face.html⟩; "Facial Analysis" page: ⟨http://mambo.ucsc.edu/psl/fanl.html⟩.

5.   MIT's Synthetic Characters Group: ⟨http://www.media.mit.edu/groups/characters/⟩; *Haptic Master:* ⟨http://intron.kz.tsukuba.ac.jp/HM/txt.html⟩; *Torus Treadmill:* ⟨http://www.psicologia.net/pages/survey. htm⟩; University of North Carolina nano-manipulator: ⟨http://www.cs.unc.edu/Research/nano⟩.

6.   "Talking Heads" site: ⟨http://www.haskins.yale.edu/haskins/HEADS/contents.html⟩.

7.   Ben Shneiderman's list of resources: ⟨http://www.otal.umd.edu/Olive/⟩; Stanford information visualization resources: ⟨http://www-graphics.stanford.edu/courses/cs348c-96-fall/resources.html⟩.

8.   Yahoo artificial intelligence resources: ⟨http://www.yahoo.com/Science/Computer_Science/Artificial_ Intelligence/⟩; Teaching resources for artificial intelligence: ⟨http://yoda.cis.temple.edu:8080/IIIA/ai.html⟩.

9.   MIT's Intelligent Environment research: ⟨http://www.ai.mit.edu/projects/hal/⟩; intelligent transportation system web tour: ⟨http://www.itsonline.com/traftech2.html⟩.

10. S. Wilson, "Dark and Light Visions": ⟨http://userwww.sfsu.edu/~swilson/papers/postmodern. pap.html⟩.

11. R. Stone, "Preface," in T. Druckrey, ed., *Electronic Culture* (New York: Aperture, 1996), p. 6

12. W. Flusser, "Digital Apparitions," in T. Druckrey, ed., *Electronic Culture,* p. 241.

13. Ibid., p. 243.

14. M. Wark, "Suck on This, Planet of Noise," in S. Penny, ed., *Critical Issues in Electronic Media* (Albany: SUNY Press, 1995), p. 16.

15. T. Druckrey, "Introduction," in T. Druckrey, ed., *Electronic Culture,* p. 20.

16. P. Weibel, "The World as Interface," in T. Druckrey, ed., *Electronic Culture,* p. 346.

17. W. Flusser, op. cit., p. 245.

18. F. Rotzer, "Attack on the Brain," in L. Hershman Leeson, ed., *Clicking In,* (Seattle: Bay Press, 1996) p. 202.

19. F. Dyson, "In/quest of Presence," in S. Penny, ed., *Critical Issues in Electronic Media,* p. 27.

20. M. Novak, "TransArchitecture," available at ⟨http://www.archi.org⟩.

21. N. K. Hayles, "Virtual Bodies and Flickering Signifiers," in T. Druckrey, ed., *Electronic Culture,* p. 266.

22. F. Dyson, op. cit., p. 29.

23. S. Penny, "Consumer Culture and the Technological Imperative," in S. Penny, ed., *Critical Issues in Electronic Media,* p. 69.

24. F. Dyson, op. cit., p. 32.

25. A. Kroker and M. Kroker, "Code Warriors," in L. Hershman Leeson, ed., *Clicking In,* p. 249.

26. H. Bey, "The Information War," in T. Druckrey, ed., *Electronic Culture,* p. 370.

27. Ibid., p. 372.

28. G. Gómez-Peña, "Virtual Barrio," in L. Hershman Leeson, ed., *Clicking In,* p. 175.

29. K. Robbins, in T. Druckrey, ed., *Electronic Culture,* p. 163.

30. F. Dyson, op. cit., p. 69.

31. B. Nichols, "The Work of Culture in the Age of Cybernetic Systems," T. Druckrey, ed., *Electronic Culture,* p. 141.

32. F. Rotzer, "Virtual Words: Fascination and Reactions," in S. Penny, ed., *Critical Issues in Electronic Media,* p. 120.

33. C. Richards, "Fungal Intimacy," in L. Hershman Leeson, ed., *Clicking In,* p. 261.

34. D. Gromala, "Pain and Subjectivity in Virtual Reality," in L. Hershman Leeson, ed., *Clicking In,* p. 237.

35. B. Nichols, op cit., p. 141.

36. W. Flusser, op. cit., p. 244.

37. F. Dyson, op. cit., p. 30.

38. S. Stone, "Interview," in L. Hershman Leeson, ed., *Clicking In,* p. 113.

39. S. Turkle, "Rethinking Identity through Virtual Community," in L. Hershman Leeson, ed., *Clicking In,* p. 121.

40. S. Zielinski, "Rethinking the Border and the Boundary," in T. Druckrey, ed., *Electronic Culture,* p. 281.

41. E. Huhtamo, "From Kaleidoscomaniac to Cybernerd: Notes toward an Archeology of Media," in T. Druckrey, ed., *Electronic Culture,* p. 298.

42. G. Legrady, "Image, Language, and Belief in Synthesis," in S. Penny, ed., *Critical Issues in Electronic Media,* p. 187.

43. S. Penny, op. cit., p. 63.

44. E. Huhtamo, op cit., p. 302.

45. Ibid., p. 303.

46. E. Huhtamo, "Encapsulated Bodies in Motion: Simulators and the Quest for Total Immersion," in S. Penny, ed., *Critical Issues on Electronic Media,* p. 161.

47. Ibid., p. 167.

48. D. Kahn, "Track Organology," in S. Penny, ed., *Critical Issues in Electronic Media,* p. 209.

49. T. Druckrey, op. cit.

50. P. Weibel, op. cit.

51. L. Manovich, "Labor of Perception," in L. Hershman Leeson, ed., *Clicking In,* p. 183.

52. Ibid., p. 188.

53. L. Manovich, "The Automation of Sight," in T. Druckrey, ed., *Electronic Culture,* p. 237.

54. Critical Art Ensemble, "The Coming of Age of the Flesh Machine," T. Druckrey, ed., *Electronic Culture,* p. 396.

55. A. Kroker, and M. Kroker, "Code Warriors," in L. Hershman Leeson, ed., *Clicking In,* p. 254.

56. G. Gómez-Peña, op. cit., p. 176.

57. B. Nichols, op. cit., p. 131.

58. S. Penny, op. cit., p. 56.

59. R. Barbrook, and A. Cameron, "Californian Ideology," ⟨http://alamut.com/subj/ideologies/pessimism/califIdeo_I.html⟩

60. S. Penny, op. cit., p. 60.

61. N. Tenhaaf, "On Monitors and Men and Other Unsolved Feminine Mysteries," in S. Penny, ed., *Critical Issues in Electronic Media,* p. 219.

62. S. Stone, op. cit., p. 108.

63. N. Paterson, "Cyberfeminism," ⟨http://internetfr.auen.w4w.net/archiv/cyberfem.html⟩.

64. S. Plant, "The Future Looms," in L. Hershman-Lesson, ed., *Clicking In,* p. 123.

65. N. Paterson, op. cit.

66. Z. Sofoulis, "Posthuman or Para-ego? Interactive Models of Human-Technology Relations" ⟨http://www.imago.com.au/WOV/papers/posthum.htm⟩.

67. R. Braidotti, "Cyberfeminism with a Difference," ⟨http://www.let.uu.nl/womens_studies/rosi/cyberfem.htm⟩.

68. F. Rotzer, "Between Nodes and Data Packets," in T. Druckrey, ed., *Electronic Culture,* p. 249.

69. F. Kittler, "There Is No Software," in T. Druckrey, ed., *Electronic Culture,* p. 335.

70. S. Zielinski, "Thinking the Border and the Boundary," in T. Druckrey, ed., *Electronic Culture,* p. 282.

71. R. Cornwall, "From the Analytical Engine to Lady Ada's Art," in T. Druckrey, ed., *Iterations* (Cambridge: MIT Press, 1993), p. 41.

72. E. Davis, "Description of *Technosis,*" ⟨http://www.levity.com/figment/index.html⟩.

73. S. Penny, op. cit., p. 47.

74. R. Wright, "Technology as the People's Friend," in S. Penny, ed., *Critical Issues in Electronic Media,* p. 89.

75. S. Penny, op. cit., p. 58.

76. P. Weibel, "Nettime Interview," ⟨http://www.factory.org/nettime/archive-1996/0378.html⟩.

77. Ars Electronica, "Interactive-Art Jury Statement 1994," ⟨http://www.aec.at⟩.

78. H. See, "Open Letter to Ars Electronica Jury," ⟨http://www.merzcom.com/people/henry/papers/AEJuryOpenLetter.html⟩.

79. D. de Kerckhove, "Media Art to the Rescue," ⟨http://www.ctraces.com/Circuit_Traces/CT1_4/kerckhov.html⟩.

80. L. Courchesne, "Art Making as Forging Evidence," ⟨http://www.din.umontreal.ca/courchesne/conferences.html⟩.

81. M. Wark, op. cit., p. 9.

82. S. Dinkla, "Form Participation to Interaction," in L. Hershman Leeson, ed., *Clicking In,* p. 289.

83. E. Huhtamo, "Digitalian Treasures," in L. Hershman Leeson, ed., *Clicking In,* p. 310.

84. K. Robbins, op. cit., p. 156.

85. R. Ascott, "Photography at the Interface," in T. Druckrey, ed., *Electronic Culture,* p. 165.

86. P. Weibel, "The World as Interface," in T. Druckrey, ed., *Electronic Culture,* p. 342.

87. P. Weibel, op. cit., p. 346.

88. G. Wittig, *Switch* magazine, "Interview with Manuel De Landa," ⟨http://switch.sjsu.edu/web/v3n3/DeLanda/delanda.html⟩.

89. S. Zielinski, op. cit.

90. L. Lozano-Hemmer, "Perverting Technological Correctness," ⟨http://www.conceptlab.com/interface/theories/technology/index.html⟩.

91. A. Broeckmann, "Points of Departure," ⟨http://www.v2.nl/n5m/text/abroeck.html⟩.

92. S. Wilson, "The Aesthetics and Practice of Interactive Events," ⟨http://userwww.sfsu.edu/~swilson⟩.

93. E. Huhtamo, "It Is Interactive—but Is It Art?" ACM/SIGGRAPH, Visual Proceedings, 1993, p. 133.

94. D. Rokeby, "Transforming Mirrors," ⟨http://www.interlog.com/~drokeby/⟩.

95. J. Campbell, "Delusions of Dialogue," ⟨http://www.adaweb.com/context/events/moma/bbs4/⟩.

96. M. Fleischmann, "Artistic Statement," ⟨http://viswiz.gmd.de/VMSD/PAGES.en/mia/vci/⟩.

97. P. Hoberman, "Interactivity," ⟨http://math.lehman.cuny.edu/tb/issue3/scene/hoberman.html⟩.

98. I. Sakane, "Introduction to Interaction 97 Show," ⟨http://www.iamas.ac.jp/interaction/97/chief_Sakane.html⟩.

99. M. Sakis, "Interactivity Means Interpassivity," ⟨http://www.gu.edu.au/gwis/akccmp/MIA/abstracts/sarkis69.html⟩.

100. K. Feingold, "History of Interactive Art," ⟨http://www.tech90s.net/kf/transcript/kf_01.html⟩.

101. D. Ritter, "My Finger's Getting Tired," ⟨http://www.users.interport.net/~ritter/articles-myfinger.html⟩.

102. P. Brown, "User Interfaces," ⟨http://www.home.gil.com.au/~pbrown/WORDS/NETART.HTM⟩.

103. L. Manovich, "Cinema as a Cultural Interface," ⟨http://jupiter.ucsd.edu/~manovich/text/cinema-cultural.html⟩.

104. S. Penny, "A Postscript on the Emerging Aesthetics of Interactive Art," ⟨http://www.adaweb.com/context/events/moma/bbs5/transcript/penney16.html⟩.

105. L. Hershman Leeson, "Interview with Jaron Lanier," in L. Hershman Leeson, ed., *Clicking In,* p. 48.

106. M. Pesce, "Proximal and Distal Unity," ⟨http://www.hyperreal.org/~mpesce/pdu.html⟩.

107. R. Ascott, "Turning on Technology," ⟨http://www.cooper.edu/art/techno/essays/ascott.html⟩.

7.2

Computer Media

Introduction: Extensions of Photography, Cinema, Video, and Literature

The most prevalent forms of computer-mediated art update existing traditions from photography, cinema, video, and literature. (Section 5 on kinetics and robotics describes work that continues the traditions of sculpture.) These computer-media works add critical new dimensions of interactivity; reflection on computer function, interface, and historical context; and commentary on contemporary techno-mediated culture. They also create media events that are recognizable relatives of previous forms. These are the types of art most often thought of when someone mentions computer art.

Artists work in fields called computer graphics, computer animation, digital video, interactive cinema, interactive multimedia, and hypermedia. They create technological 2-D works, CD-ROMs, works for the World Wide Web, and installations. Conceptually, the work explores genres already familiar from their ancestors, such as poetic, expressionistic, surrealist, pop culture, documentary, and deconstructivist approaches. The most interesting work reflexively comments on the impact of technology on culture or uniquely exploits the new expressive or analytic capabilities of the computer, such as interactivity, speed, memory, the concretization of algorithms, and the ability to control and manipulate multiple video and sound streams, synthesize visual perspective and texture, and mix synthetic and photographic imagery. The World Wide Web has overcome distribution barriers by establishing instant publishing and distribution opportunities for artists.

These forms of art have even entered the mainstream. What were once exotic artistic experiments have now become commonplace features of industries such as software, entertainment, and advertising. Even interactive media have begun to widely disperse into areas such as games, education, and commercial applications, and into the Web and interactive TV.

Artistic experimentation with computer-influenced media is an important part of the contemporary context of art and technology's mutual influence, and thus a relevant focus for this book. This chapter, however, can offer only a brief summary for several reasons. The relatively long history of these forms means that there is now an enormous body of diverse work calling for its own specialized analysis. The assimilation of the experimental approaches of computer graphics and digital video into mainstream industries complicates their consideration as art, raising questions such as distinctions between art, design, and media, which are outside the scope of this book. The rapid pace of development in these fields means that they have passed into a more mature stage that is no longer the kind of experimental work that is the focus of this book. Much of this work is now being incorporated as part of their "parental" fields, such as photography, cinema, and literature, and thus is becoming a part of the discourse in those fields.

This chapter surveys a small sample of some award-winning artists who work with the unique interactive and hypermedia aspects of computer media. It does not consider the art of computer graphics, computer animation, 3-D synthesized worlds, image processing, or digital video, all of which are also illustrations of technology-influenced art that deserve their own analysis. This chapter considers artists using the technology in a variety of ways, such as the following: deconstructive and feminist critique of cultural trends; extending poetic and expressive capabilities; commentary on game form; multi-person events; video installation; interactive layered documentary; authoring rich hypermedia literary structures; analysis of the role of media in culture; and reflexive analysis of the computer interface. These categories are only an organizational convenience; works could easily have been placed in multiple categories. More details about interactive media work can be found in sources such as the documentation from Mike Leggett's "Burning the Interface" show, which looked at artists' CD-ROMs.

Deconstructive and Feminist Critiques of Cultural Trends

Jill Scott

Jill Scott has worked with electronic media since the days of video art. She uses the flexibility of interactive media to explore a variety of themes—the nature of memory, the state of the body in an electronic era, shifting attitudes toward technology, and feminist critiques of history and its presentation. Usually she creates installations with many display screens and interface surfaces through which visitors can move. She explicitly addresses the possibilities of interactivity and the "relationship between design and desire in the realm of new machine/human interfaces."

Scott is concerned with the narratives of body representation and its technological transformation. The role of gendered history/herstory underlie many of her installations. She notes that art has historically been a major arena for explicating the discourses of the body. In a paper called "The Digital Body and the Bionic World View," she traces contemporary thought about bionics and raises questions about the body's ultimate transformations and representation.

Mike Leggett: ⟨http://www.thehub.com.au/biff/other.html⟩
Jill Scott: ⟨http://beham.zkm.de/~scott/information.html⟩

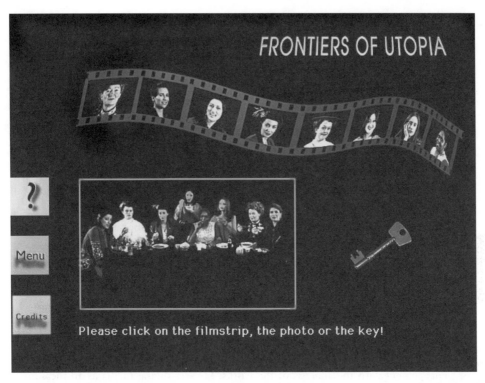

Fig. 7.2.1. Jill Scott, *Frontiers of Utopia*. Viewers can traverse various twentieth-century historical eras by accessing the lives of eight female utopian characters.

Scott created a series of interactive installations that asked visitors to explore our culture's attitudes toward technology, with a special emphasis on the way women were represented in these narratives. In *Paradise Tossed,* she looked at the relationship of desire and design. The interactive structure and use of 3-D computer modeling are not just embellishments but intrinsically related to the core analysis underlying the work itself, suggesting modernist idealism. Visitors can enter dream homes from the history of design (Art Nouveau, Art Deco, Op Art, and Space Age) and explore the subjects of technological utopia, design cliché, and the manipulation of the woman's workplace.

Frontiers of Utopia continues the exploration of attitudes toward technology. It creates a kind of Brechtian theater in which visitors can span historical eras by conversing with eight virtual video female characters who illustrate various "moods, criticisms, and attitudes toward Utopia" and reflect on class, gender, and history. Scott sees the associationistic structure of interactive multimedia as simulating the web of human memory, evoking associations and the richness of human experience as people tap "into the open concept of a multimedia labyrinth."

Lynn Hershman

Lynn Hershman is considered one of the most productive new-technology artists. Her wide-ranging works are discussed in the telecommunications (6.4) and alternative interface chapters (7.4). Hershman was a pioneer in interactive work, with her introduction of the *Lorna* videodisk. Using a remote control, viewers tried to liberate Lorna from her fear-dominated life and access aspects of her thought via objects in the scene.

Deep Contact, another videodisk project (completed in collaboration with Sara Roberts), required viewers to voyeuristically explore a garden populated by mysterious persons by touching the image of their guide's body on the screen.

George Legrady

George Legrady uses interactive media to deconstruct social narratives of power and progress and to understand the role of mediated experience in contemporary life. He creates unprecedented interfaces that lead visitors to take actions that facilitate the visitor's engagement with the analysis of the works. He sees digital media as an ideal locus for investigating fragmented identity: "The nonlinear capabilities of digital interactive media promise to be an ideal environment for the (re)presentation of fragmented histories and the construction of narratives based on hybrid cultural identities."[1]

Legrady's paper "Interface Metaphors and New Narratives in Interactive Media" analyzes the aesthetic and conceptual importance of the interface:

Digital interactive media require metaphor-based, organizational models by which to conceptually situate the viewer and to provide a way of accessing and understanding data. . . . By knowing "the story" or metaphor, the viewer can successfully navigate inside the interactive program. As a result, these metaphor environments promise to be the key site for innovative developments of a linguistic, symbolic, aesthetic, sensory, and conceptual nature, redefining the interactive viewer's experience within the digital environment.[2]

Lynn Hershman: ⟨http://arakis.ucdavis.edu/hershman/intro.html⟩
George Legrady: ⟨http://userwww.sfsu.edu/~legrady/Art/projects.top.html⟩

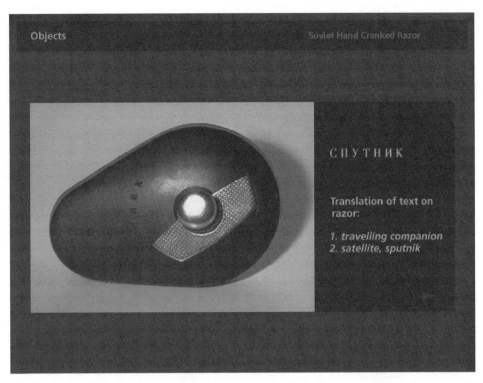

СПУТНИК

Translation of text on razor:

1. *travelling companion*
2. *satellite, sputnik*

Fig. 7.2.2. George Legrady, *An Anecdoted Archive from the Cold War.* An interctive event using the metaphor of the Budapest Workers' Movement Museum as the interface to explore personal and official texts from the Stalinist era.

Legrady's family left Hungary as part of the 1950s uprisings. In *An Anecdoted Archive from the Cold War,* Legrady introduces the metaphor of the Budapest Workers' Movement Museum as the interface to explore personal and official texts from the Stalinist era. The visitor navigates the interactive event by moving around the virtual museum. Legrady integrates old black-and-white family movies, everyday objects, and official propaganda as a means of exploring multiple references and codings and to explode the presentations of official history.

The Clearing deconstructs various news accounts of the events in Bosnia by allowing the user to explore the surface of an old hunting photograph. Color-coded texts describe diverse points of view about the conflict from different countries' media. Legrady used a gamelike metaphor to induce exploration and to refer to the nature of military interfaces; for example, in the first stage the mouse is in the shape of a targeting frame.[3]

Slippery Traces invites the viewer to navigate a database of 240 postcards that Legrady collected. Each choice brings up many possible links to other cards. The links are not of one kind. For example, they combine literal, metaphoric, semiotic, and historical frameworks. The associations that viewers make in setting up their navigation is a critical element of the work. Legrady's motion-sensing *Tracing* event is described in the alternative interface chapter (7.4).

Other Artists and Projects

Peter Weibel, former director of Ars Electronica and the author of texts analyzing media culture, created *The Curtain of Lascaux,* which looks at the ancient question of how we know what we know. Commenting on the difficulties of old models of objective reality, he created a situation in which the image of the viewer seemed to be caught inside a brick wall, thus forcing the observer to be also be the observed. **Christine Tamblyn,** an artist and theorist who explored gender issues surrounding technology, created *She Loves It, She Loves It Not: Women and Technology* in collaboration with **Marjorie Franklin**, which introduced an atypical interface to highlight its challenge to traditional design assumptions and included advertisements, science fiction movie excerpts, and personal anecdotes. In *Mistaken Identities,* Tamblyn created an interactive work that allowed viewers to explore the ways in which new media can deconstruct assumptions about the stability of identity by looking at the lives of ten famous women. **Joan Truckenbrod**'s *Everyday Family* installation seeks to encourage support for alternative family structures via a homey living-room situation, with an activated scrapbook that calls up relevant images and sometimes collages the current viewer's hidden-camera captured image in the changing display. In *Think About the People Now,* **Paul Sermon** used hypermedia to recreate the moment at Whitehall in London when a protestor lit himself afire.

Sonya Rapoport's Arbor Erecta presents an interactive media event published on the Web that resembles an anthropological site investigating New Guinea initation rites but actually focuses on a transgender change. **Marita Liulia**'s *Ambitious Bitch* CD-ROM

Peter Weibel: ⟨http://thing.at/ejournal/ArtSite/weibel/weibel.html⟩
Christine Tamblyn: ⟨http://melissa.simplenet.com/intersections/reflections/christine.html⟩
Joan Truckenbrod: ⟨http://www.inform.umd.edu/EdRes/Colleges/ARHU/Depts/ArtGal/.WWW/digvil/efamily/truckenb.htm⟩
Paul Sermon: ⟨http://www.aec.at/prix/1991/E91gnI-think.html⟩
Marita Liulia: ⟨http://www.av-arkki.fi/taiteilijat/marita_liulia.html⟩
Ambitious Bitch: ⟨http://www.medeia.com/⟩

brings together references from art and media, and high and low culture to reflect on feminist issues in a style called "randy," " lush," "self-ironic," and "postfeminist." ***Gregory Garvey***'s *Gender Bender* is an interactive event published on the Web that questions concepts of gender by allowing the visitor to determine their gender as interpreted by a psychological instrument called the Bern Sex Role Inventory, with instant feedback via techniques such as the Morph-o-meter, which teeters between male and and female visual representations. *Brandon,* a Web work produced by ***Shu Lea Cheang*** and several collaborators, provides experimental interfaces and animations in its exploration of cross-gender experiences.

   Perry Hoberman, also discussed in the kinetics (5.2) and robotics (5.4) chapters, created a computer-media work called *The Subdivision of the Electric Light,* in which electric light is used as a metaphor to understand the development of technology over the last century, what some people call media archaeology. ***Marjorie Franklin***'s *Digital Blood* (created in collaboration with Paul Thompkins) explores the conceptual framework of two female software developer friends who inadvertently engender an artificial life. *The Consensual Fantasy Engine,* by ***Paul Vanouse and Peter Weyhrauch,*** comments on news portrayals of events by allowing audiences to dynamically shape a digital video of the O.J. Simpson trial events by levels of applause at choice points. ***Claude Closky***'s *Do You Want Love or Lust?,* presented as part of the Dia Center's Web art, deconstructs pop psychology as presented in mass-culture magazines, building the endless questions of pop psychology into an interactive event.

Extending Poetic and Expressive Capabilities

Bill Seaman

Bill Seaman has been the champion of what he calls "emergent and recombinant poetics." Building on the computer's flexibility in orchestrating image, video, sound, and text events, Seaman attempts to create expressively rich environments in which the visitor

---

Brandon Project: ⟨http://brandon.guggenheim.org/title/JP2.html⟩
Perry Hoberman, *The Subdivision of the Electric Light:* ⟨http://www.hoberman.com/perry/⟩
Marjorie Franklin, *Digital Blood:* ⟨http://www-mitpress.mit.edu/e-journals/Leonardo/isast/wow/wow303/digital.html⟩
Claude Closky: ⟨http://www.diacenter.org/closky/intro.html⟩
Bill Seaman: ⟨http://research.umbc.edu/~seaman/⟩

Fig. 7.2.3. Bill Seaman, *Passage Sets/One Pulls Pivots at the Tip of the Tongue* (1995). Detail. An interactive system that enables visitors to dynamically recombine text, sound, and image.

and the author can collaborate in producing a large set of possible manifestations. The interface itself often becomes part of the poetry.

Seaman's extraordinary installations offer new perspectives on what poetry can be in the contemporary era. He creates structures in which language in many of its forms, such as text, spoken text, sound, still image, and moving image, can flexibly interact and be influenced by the visitor. In *The Exquisite Mechanism of Shivers,* participants chose words from scrolling lists. These choices immediately modify the ongoing image, sound, and text event. Seaman notes the importance of recombination: "The work explores pluralistic meaning through the presentation of material in continuously changing, alternate contexts. Humor, visual puns, word/image/sound play, modular musical composition, "canned chance," as well as sense/nonsense relations are all explored.[4]

The Watch Detail interactive videodisk event focuses on time. Viewers can control a shifting image, text, and sound event that reflects on time frames extending from seconds to geological time. Seaman uses the interactive process itself to engage viewers in experiences of time, for example, by changing options based on the duration of interaction. *Passage Sets/One Pulls Pivots at the Tip of the Tongue* confronts the visitor with a stage facing three projection screens. They are invited to interactivity compose poems

Fig. 7.2.4. Tamas Waliczky, *The Way*. Viewers can encounter scenes in ways that are perceptually impossible in normal life.

made of moving text, images, and sounds, many derived from urban architecture. The visitors have some control over two of the screens, and the third, an "autonomous poem generator," makes its own in response. The arrangement comments on "sensuality/sexuality, identity in cyberspace, and future communication/sensual feedback systems." *World Generator/The Engine of Desire* extends emergent poetics into telecommunications spaces. Visitors can apply multiple transformations of media objects and even unleash objects with their own genetic behaviors. Remote visitors can interact with each other in poetic-world creation.

Tamas Waliczky

Tamas Waliczky constructs poetic works in which viewers encounter extraordinary forms of time and space. Challenging Western traditions of photography and cinema, Waliczky creates his own rules for his photographiclike processed worlds, denying laws

Tamas Waliczky, "Butterfly Effect": ⟨http://ns.c3.hu/butterfly/Waliczky/waliczky.html⟩

of perspective that we take for granted to create a dynamic dreamlike space. The "Butterfly Effect" show describes his *Landscape* installation:

Landscape . . . was originally designed for an interactive opera. In the opera's story, the moment was to be visualized, in which God stopped time. Waliczky describes this moment using the visualization of raindrops. The first picture shows a little village in the rain. Suddenly the rain stops, the drops freeze, and the camera moves around and through this world standing still.[5]

In the *Focus* installation, Waliczky created an artificial street scene out of hundreds of photographs composited into many layers. The scene looks somewhat like a regular photograph, but the viewer has absolute interactive control over what layers are in focus. In the *Forest* installation, viewers are lost in a processed black-and-white foggy forest with no orientation points of sky, ground, or horizon, and seemingly endless extension in x, y, and z axes—the "visual expression of an impasse." Art theorist Lev Manovich describes Waliczky's deconstruction of perceptual commonplaces in his paper "The Camera and the World":

Waliczky systematically maps out an important part of the new postcomputer aesthetic space. It is the part where new ways to structure the world and new ways to see it meet. The interactions between the virtual camera and the virtual world—this is the main subject of Waliczky's aesthetic research. . . . Waliczky thus is neither a virtual filmmaker who works only with images nor an virtual architect who works only with space. Rather, he can be described as a maker of virtual documentaries. In every one of his works, he creates a world structured in a unique way, and then he documents it for us. . . . Each of his worlds establishes a cosmology of its own, a unique logical system which governs all of the world's elements.[6]

Other Artists and Projects

Erwin Redl creates "digital minimal opera" interactive works that slow down the usual pace of computer media. In *You Me and,* abstract silhouette characters enact a variety of mysterious actions. In a Web work called *Truth Is a Moving Target,* a sentence spread out in a grid begins to morph into other words. In *Bio-Morph Encyclopedia,* ***Shibayama Nobuhiro*** presents a work in which the interface itself is part of the poetry, building

Erwin Redl: ⟨http://www.hotwired.com/rgb/redl/index_bio.html⟩

on Muybridge–like imagery to consider the philosophical and poetical implications of our culture's attempt to unravel the mysteries of time. *Joseph Squires*'s Web-published *Urban Diary* offers a rich interactive event based on lush images of a diary and other artifacts. Users navigate by clicking on the images and reading diary text. In *Pedestrian: Walking as Meditation and the Lure of Everyday Objects*, **Annette Weintraub** offers the metaphor of walking in the city as the navigational entry in meditation about life in the city. Building on detailed studies of Jean-Jacques Rousseau, *Jean-Louis Boissier*'s *Second Promenade* uses the poetic power of interactivity to build a kind of reverie that integrates the user's attitude with Rousseau's experience of "consciousness of being, with a relationship with things, beings, and events constantly the object of a dual observation of oneself." *Globus Oculi* conveys the childhood state of mind when learning to read.

As part of the Dia Center–sponsored series of Web art, **Constance DeJong, Tony Oursler, and Stephen Vitiello** created *Fantastic Prayers,* in which one could interactively visit an "urban landscape, which is inscribed with memories of lives lived, objects possessed or discarded, and places inhabited," an Arcadia with its "residents slumbering deeply, rocking, rocking in the cradle of dreams." In *The 10 Loveliest Things I Know,* **Chris Hales** offers an unusual interactive video focused on the experiences of childhood, employing the techniques of slapstick and pathos to contextualize childhood events. **Juliet Martin** created the Web event *Please Stay on the Line,* which uses the metaphor of the telephone to create an interactive media poem. Aspects of telephoning, such as the touch-tone pad, become part of expressive presentation. **Mike Mosher**'s *Posada Space* offers a VRMI environment in which the user can naviagate a space filled with animated skeletons based on José Guadelupe cartoons and works by Diego Rivera and Frida Kahlo.

The "Doors of Perception" Web site offers an intriguing animated media experience as its Web access into the archives of information from its conferences. Exotic text-sound poems fill **Erik Loyer**'s Web work called *Lair of the Marrow Monkey.* Starting with the simple rectangle, *REaCT* presents an animated abstract extravaganza. **Beverly**

Joseph Squires: ⟨http://www.art.uiuc.edu/ludgate/the/place/urban_diary/intro.html⟩
Annette Weintraub: ⟨http://www.turbulence.org/Works/pedestrian/intro.html⟩
Jean-Louis Boissier: ⟨http://www.ntticc.or.jp/preactivities/gallery/jlb/boiss_e.html⟩
Constance DeJong: ⟨http://www.diacenter.org/rooftop/webproj/fprayer/fprayer2.html⟩
Juliet Martin: ⟨http://www.rsub.com/thenvelope/pstol⟩
Mike Mosher: ⟨http://catalog.com/ylem/artists/mmosher/Opening.html⟩
"Doors of Perception": ⟨http://www.doorsofperception.com/doors/⟩
Erik Loyer: *Lair of the Marrow Monkey:* ⟨http://www.marrowmonkey.com/⟩

Reiser creates interactive works that explore magical realism. For her, one important use of technology is the exploration of new human possibilities. *Life on a Slice* allowed viewers to navigate by virtually touching virtual objects that appeared in a projection of themselves. *Voice Garden: Labyrinth of Love, Labyrinth of Desire* explored love and desire in a mystical setting. **Mark Amerika**'s *PHON:E:ME* Web event offers an alternative interface focused on exploring the sonority of voice sound works. In *Redsmoke Lew Baldwin* ostensively presents a Web page of a fictitious rock band, although he reports it "quickly evolved into a depository for the subconscious."

Art Games

Stephan Eichhorn, Tjark Ihmels, KP Ludwig John, Michael Touma

This collaboration of artists creates interactive events that sit ambiguously in space, combining cinema and computer games. They describe *Die Veteranen,* one of their most well-known works, as a "game of shape, color, sound, and association. It is a game without rules, a fantasy game that is inventing its own rule":

The starting picture opens an electronic storybook of unlimited opportunities. A visit at the movies with a unique program, an entire catalogue of all words of this world (that consist of four letters), associative interviews with representatives of five continents—all this can be the beginning of a voyage of discovery of a very special kind and more. . . . With the help of the observer's voice, a free interaction between computer and observer becomes possible. . . . There is neither a goal/aim/purpose in the game *Die Veteranen,* nor a plot. . . . The focus is on the observer. His wish to discover the unknown presses him on and serves his relaxation. He can be assured that he will find something new or that he sees things in a different light every time he comes to the dream world.[7]

Another "game" called *Ottos Mopps* confronts users with a countdown clock that starts with the entering of their names and offers only one chance for users to save themselves by finding missing people.

Beverly Reiser: ⟨http://www.idiom.com/~beverly/⟩
Mark Amerika, *PHON:E:ME:* ⟨http://phoneme.walkerart.org⟩
Lew Baldwin, *Redsmoke:* ⟨http://www.redsmoke.com⟩
Die Veteranen: ⟨http://www.uni-leipzig.de/veteranen/english.html⟩
Ottos Mopps: ⟨http://www.mediopolis.de/transmedia/english/cdrom.htm⟩

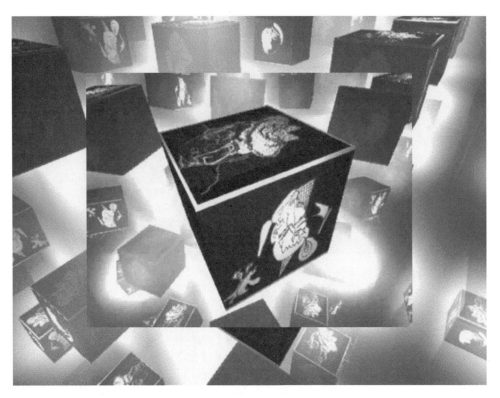

Fig. 7.2.5. Stephan Eichhorn, Tjark Ihmels, KP Ludwig John, and Michael Touma, *Die Veteranen*. A computer game without rules.

Webster Lewin, Bill Barminski, and Jerry Hesketh

Webster Lewin, Bill Barminski, and Jerry Hesketh created a spoof on CD-ROMs called *Encyclopedia of Clamps*, which opens with a deadpan treatment of information about clamps but then moves on to a unprecedented digital carnival. In one section a person roams halls filled with events: "Click on one poster and you become a hapless guest on the *Blimp Rambo Show*, where you're subjected to a cruel "pink-o-meter" test and beaten into unconsciousness with *Blimp*'s "stick of justice." Click on another poster and you might find yourself "jamming on a musical fish."

Encyclopedia of Clamps: ⟨http://www.deluxoland.com/what.html⟩

Fig. 7.2.6. Webster Lewin, Bill Barminski, and Jerry Hesketh, *Encyclopedia of Clamps.* An interactive CD-ROM take-off on educational and information systems.

Other Artists and Projects

Jutta Kirchgeorg creates interactive Internet and CD-ROM works that reflect on digital culture by making the experience of working with the interface itself as a commentary on the relationship between people and machines. For example, in *Information Age,* the machine seems to be breaking down as the user frantically seeks to stop the events. **Youn H. Lee** created *Land of Time,* a 3-D digital animated puzzle of a clock. As the pieces are snapped together they start functioning. Other images that reflect on time emerge as progress is made. *Jacques Servin* Web published a "game" called *Beast* that experientially led participants into a parody of technology lust. Every page warned users that their systems would probably be insufficient to view the event. In the Web-based *Blotto,* **Paula Levine** comments on the construction of reality by letting users dynamically react to obscure Rorschach-like images. At the ***open source media art project***'s *Esc to Begin* event, visitors advance through thirteen online games in which "visitors can explore, save,

Jutta Kirchgeorg: ⟨http://www.artswire.org/circuits/artatconf/cybercafe_cdrom.html⟩
Youn H. Lee: ⟨http://www.siggraph.org/s97/conference/garden/time.html⟩
Jacques Servin: ⟨http://www.quake.net/~jacq/Beast⟩
Paula Levine: ⟨http://cet.sfsu.edu/FMPro4/Web/Blotto/index.htm⟩
Open source media art project: ⟨http://www.zkm.de/~bernd/etb-os/⟩

and conquer the various levels of gaming of the interface design, artificial intelligence, or data visualisation." **Timothee Ingen-Housz**'s *A-aktion.com* presents a game-like event in which the author engages in auto-aggression and self-mutilation. **Ben Benjamin** created *Superbad* as a nonlinear funhouse environment commenting on popular culture. **John F. Simon**'s *Every Icon Is a Java Applet* systematically generates every black-and-white combination of the 32-pixel grid that composes digital screen icons.

Multi-Person Events

Toshio Iwai

Toshio Iwai's well-known work explores several areas of emerging technologies. For example, his work on reading gesture is discussed in the sensor chapter (7.4). *Resonance of 4* allows four people to collaborate in creating sounds by manipulating an abstract graphic grid:

Resonance of 4 is an interactive audio-visual installation which allows four people to create one musical composition in cooperation with each other. In this installation, four players are given different tones with which they can compose their own melodies. Each person uses a mouse to place dots on four grid images projected onto the floor. My hope is that each player listens to the melodies which are being created by the other players, and then tries to change their own melody to make better harmony. In this way, the installation will not only generate a resonance of sounds, but will create a resonance of minds between the four players.[8]

One commentator described the unprecedented kind of communication experienced in working with *Resonance of 4* as a kind of "sublime crossword," as the visitor tries to process feedback from one's own actions and the actions of others.

Other Artists and Projects

Peter Broadwell and Rob Myers's *Plasm: Not a Crime* event comments on cryptology and communications systems by offering users an environment in which the authorship

---

Timothee Ingen-Housz: ⟨http://salon-digital.zkm.de/a-aktion/⟩
Ben Benjamin: ⟨http://www.superbad.com⟩
John F. Simon: ⟨http://www.numeral.com/everyicon.html⟩
Toshio Iwai: ⟨http://www.iamas.ac.jp/~iwai/artworks/piano.html⟩
Peter Broadwell and Rob Myers: ⟨http://www.plasm.com/~peter/⟩

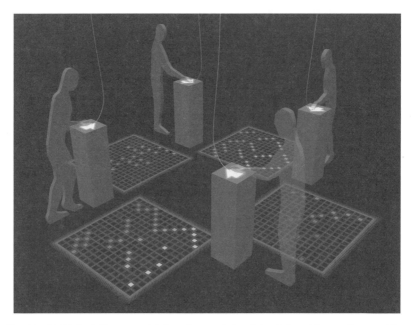

Fig. 7.2.7. Toshio Iwai, *Resonance of 4* allows four people to collaborate in real-time music composition.

of sentences is disassociated to the extent that it is impossible to identify sources. **Scott Sona Snibbe**'s installation *Motion Phone* allows networked users to control an evolving abstract image generator by their drawing gestures. Ars Electronica's Web description notes: "People are surrounded by the beauty of their natural motion when they see it sampled over time, rather than the instantaneous slice of reality we are used to."

Video Installation

Video artists were among the first to investigate the aesthetic, technical, and cultural implications of video freed from major studio infrastructures. Many felt that the development of devices such as Sony's Porta-Pak promised to be a major art development. Artists created conceptual work, community-based video, video sculpture, and personal video. Some experimented with video image processors. Examples are Nam June Paik, Vito

Scott Sona Snibbe: ⟨http://193.170.192.5/prix/1996/E96azI-motion.html⟩

Acconci, John Baldessari, Douglas Davis, Woody and Steina Vasulka, Bill Viola, Doug Hall, Ant Farm, Video Free America, Dan Sandin, Diana Thater, and Tony Oursler.

Several artists experimented with interactive installations that invoked viewer actions in creation of the works. These could be thought of as the first interactive electronic art. Video art has a long, multifaceted history that is impossible to capsulate here. Many of the pioneers have now begun to move into experimentation with contemporary technology. This section focuses on only a sample of those specifically working with contemporary digital extensions of the technology.

Elizabeth Diller and Ricardo Scofidio

Elizabeth Diller and Ricardo Scofidio collaborate in the production of works that span architecture, dance, and electronic media. *Indigestion,* supported by the Banff Centre, generates a dinner conversation between two virtual characters. The visitor gets the conversation going by choices of gender and class. The "Serious Games" show documentation describes the event as

an interactive video installation that converges old and new genres (film noir, video games, video installation art, Exquisite Corpse structures) into an ironic mix. Two characters meet across a dinner table and only their animated hands appear on screen—reaching, spearing, gesturing—their food and witty dialogue revealing a mysterious tale. The viewer/guest chooses the characters from a "menu" of gender and "class." Indigestion uses technologies of "choice" to question the rhetoric of choice surrounding interactive technologies, politics, and class distinctions.[9]

Jim Campbell

Jim Campbell combines video-image processing with computer-mediated effects to produce poetically transformed events featuring the viewer. The viewer moves in front of the camera but the image seen is not ordinary reality. Campbell believes that the computer is a profound cultural invention and that computer art will advance through understanding the computer's underlying capabilities and limitations and their relationship to human functioning. His paper "Delusions of Dialogue: Control and Choice in Interactive Art," originally presented in a series at the Museum of Modern Art, explains the importance of investigating memory, interface, programming, and control; working with the com-

---

Elizabeth Diller and Ricardo Scofidio: ⟨http://www.ace.co.uk/seriousgames/diller.htm⟩
Jim Campbell: ⟨http://www.kemperart.org/dwessy.htm⟩

puter's strengths; and the challenge of "transforming emotions or an intution into a logical representation":

I believe that some understanding of the way that a computer functions might help us to critically analyze the state of the art and examine why the art has so clearly not reached the level of transcending the technology. . . . A different approach is to start with an idea from technology and let the work flow from the set of technological possibilities. This avoids the problem of finding a mathematical equivalent by starting with one, but certainly a problem with this approach is that it is difficult to take the work beyond self referentiality. Often these works are only about the technology that they use and their processes and effects.[10]

Interactivity is of special interest. In a statement for the "Mortal Coil" show, Campbell explains that he is interested in interactive works that do not rely on the control metaphor of traditional interfaces and move more toward what he calls "responsive"—moving away from "a viewer dominating a work, and more about viewers participating in the developing personality of a work":

Attempting to create systems that respond and progress in recognizably nonrandom, but at the same time unpredictable ways, I have tried to create works that have destinies of their own. Having always been fascinated with the philosophical analogies of certain scientific disciplines, my work has been very influenced by science, in particular some of the ideas relating to chaos and quantum mechanics. Using technological tools and scientific models as metaphors for memory and illusion, my work seeks to interpret, represent, and mirror psychological states and processes, and their breakdown. Time and memory, individual and collective, electronic and real are the elements of my work.[11]

"Delusions of Dialogue: Control and Choice in Interactive Art" explains more about control:

The computer industry's goal of making computers and programs smarter is simply to make computers more efficient at being controlled by the user to get a job done. Why should they do anything else? It's generally what we want computers for. We want them to be passive slaves. One can see this in the software, hardware, and interfaces that are currently being used. This model is fine until it collides with art. For example, look at the concept of icons as an interface device. They are designed to be precise and accurate and discrete, on or off. They are designed to present a closed set of possibilities. They are not capable of subtlety, ambiguity, or question. An interface of choice and control makes sense for a word processor or an information retrieval system or a game, but not as a metaphor for interactivity or dialogue.[12]

Fig. 7.2.8. Jim Campbell, *Hallucination*. A video installation with the current viewer's live image appearing on fire.

Campbell's installations are exquisite illustrations of this careful attention to interactivity and other underlying structures, such as memory. In *Digital Watch*, the viewer's real-time image is superimposed with a watch and pulsed images that the computer has stored from recent times. But the images come in discrete delayed segments so that the viewer experiences loss of control of their body. In her description of the installation at the Krannert Art Gallery, a commentator suggested that the work "disturbs the linear concept of time and the power we believe we exert not only over our bodies, but over images of our bodies."[13] In *Hallucination*, viewers see their images on the projected screen consumed in fire. The fire follows their movement like a demon shadow. The buring sound gets louder the closer they get to the "mirror" screen. Also, a ghost person appears in the apparently live image, but in physical reality there is no one there. In *Memory Recollection Transformation*, current visitors' video are composited with their own actions from the recent past. It is a provocation to think about memory and reality.

Hallucination: ⟨http://gertrude.art.uiuc.edu/@art/signal/campbell.html⟩
Memory Recollection Transformation: ⟨http://www.artscenecal.com/ArticlesFile/Archive/Articles1997/Articles0697/JCampbell.html⟩

Video Installation

Fig. 7.2.9. Ed Tannenbaum, *Recollections*. A video installation offering processed versions of users' live action. Copyright 1999 by Ed Tannenbaum.

Ed Tannenbaum

Demonstrating a prime example of the artist-researcher integration described in this book, Ed Tannenbaum has invented interactive video installations that intrigue visitors in over forty-five institutions around the world. His artist statement on his Web site explains:

Being dissatisfied with the existing tools available to the video artist, I began designing and building my own in 1975. I soon discovered that using the tools was sometimes more interesting than simply watching the results that they produced. During my residencies at the Exploratorium, I developed user interfaces that enable a casual participant to almost immediately understand and use my machines, often putting them into a very strange videospace. . . . Most of my work provides an impossible "mirror," one that distorts or reveals in surprising ways.

Recollections is one of Tannenbaum's most famous works. People move in front of it and it projects processed images that enhance and reshape those movements. *Sym-*

Ed Tannenbaum: ⟨http://www.et-arts.com/index.html⟩

ulations allows visitors to create compositions out of their own faces. It invites visitors to play with various kinds of symmetry, for example, creating a full version of their face made out of a mirrored image of half of their face. *Do-Undo* stores short segments of visitors' motions so that they can view them in various speeds and directions. *Elastic Surgery* allows visitors to distort their faces in impossible ways.

Other Artists and Projects

Working with the ATR Media Integration and Communication Research Laboratories, **Sidney Fels** developed an interactive video kaleidoscope called the *Iamascope*. An image-tracking algorithm generated MIDI music in accordance with people's gestures. **Tamiko Thiel** works with a variety of computer media, including the development of the Starbright Foundation's 3-D virtual worlds, which are used as therapy for hospital-confined children. Her *Totem of Heavenly Wisdom* installation employed digital video to use body motions as a doorway into philosophical associations. **Gary Hill** creates "subperceptual" dark installations in which viewers encounter room-sized faint images of text that come and go at the thresholds of perception. In installations such as *Tall Ships* and *Learning Curve,* projections play with the shape and brightness of video as tools for inspiring reflection. **Studio Azzurro**'s works, such as *Frammenti di una Battaglia,* project digitally controlled video onto sand, leaves, and bamboo. **Suh Do-Ho**'s *Sight-Seeing* installation projects dual videos made with a special camera that recorded both scenes being looked at and the artist's facial reactions.

Interactive Documentary

Graham Harwood

In *Rehearsal of Memory,* Graham Harwood exploits the conventions of interactive media to engage viewers with residents of a mental hospital in a virtual dialogue. Interactive media brings the stories to life and forces the user to engage the material by interactively moving through it. Commentators claim that the view into life in the hospital generates an intensity rarely matched:

Tamiko Thiel: ⟨http://art-tech.org/html/virtual/thiel.html⟩
Gary Hill: ⟨http://www.sva.edu/MFJ/JournaPages/MFJ29/BillHornGaryHill43096.html⟩
Studio Azzurro: ⟨http://www.imprese.com/video/net/st.htm⟩
Suh Do-Ho: ⟨http://www.ntticc.or.jp/event/sight_seeing/kaisetsu_e.html⟩
Graham Harwood: ⟨http://www.ace.co.uk/seriousgames/harwood.htm⟩

Fig. 7..2.10. Graham Harwood, *Rehearsal of Memory*. An interactive documentary examining concepts of sanity via encounters with mental hospital inmates.

A simple single-screen projection features the bodies of residents—eerie self-portraits created by pressing their bodies against scanners. Touching their scars triggers their stories of self-damage and pain—an embodied history. Viewers can navigate across many bodies, painfully aware of their responsibility in which wounds they choose to probe.[14]

Harwood describes some of his motivations in undertaking the work. He questions the ease with which modern society categorizes people and the role of technology in perpetuating these assessments:

This artwork is about the recording of the life experiences of the client group that are a mirror to ourselves ("normal society") and our amnesia when confronted with the excesses of our society. . . . This work is about people everywhere who are trying to remember the faces of the extras in the cinema of history. This artwork is a rehearsal of memories not quite forgotten. . . . How did I come to know that the acts of some people are sick, while the acts of others I accept as normal? . . . Insanity, it seems to me, never exists except in relation to strong fictions of sanity.

It is no accident that social control reproduces itself into technological forms. The reduction of information to binary representation leads to a levelling process of data . . . the modern machine is currently perceived as a neutral decision-making space. This image of anonymity creates a sufficient distance from events to create a situation in which we are ritually free to give up our ability to feel the consequences of our actions.

Commentator Annick Bureaud notes that Harwood intimately tied the interface to the subject. Clicking on the bodies on the screen feels intrusive, and exiting the program is made difficult.[15] In the history of cinema, documentary was often used for progressive purposes. Harwood completed some of the work on *Rehearsal* while he worked at an organization called ARTEC, which tried to train people that society had forsaken. Upon leaving, he recorded that he had hoped that new technology could be a tool of liberation, functioning "as a window of opportunity for the activist in me and so [I] did not mind being employed to prove its power and universality when dealing with social problems." As he left, however, he saw old structures moving in to "strangle the technologies' new-born promise of social opportunity."

Other Artists and Projects

In *Good Daughter, Bad Mother, Good Mother, Bad Daughter: Catharsis,* **Susan E. Metros** presents an interactive reconstruction of childrearing seen from a daughter's perspective. The narrative evolves through accumulating diary fragments, text fragments, spoken voices, and images. **Lisa Prah** creates interactive events that straddle the worlds of documentary and fiction. Her *Bernadette* allows viewers to explore different points of view around events in the life of St. Bernadette. **Peter d'Agostino** was one of the early pioneers in videodisk and other video experimentation. He is co-editor with David Tafler of the book *Transmission: Theory and Practice for a New Television Aesthetics.* For many years he has been working on a project called *Traces,* which attempts to document the atomic age and its effects. *Traces* has several versions, including a multichannel video installation, Web site, and videotape. D'Agostino has collected an extraordinarily rich body of information about the dropping of the atomic bombs on Japan and the long-term results. **Jennifer and Kevin McCoy**'s *Small Appliances* is an interactive video projection installation that explores the role of technology in women's lives, allowing viewers to interactively switch between "subjective" and "objective" accounts of ten women's stories. In *Sampling Broadway* **Annette Weintraub** evokes the complexity of urban life in her layering of text, image, and sound in documenting downtown areas along Broadway. **Lev Manovich**'s *Freud-Lissitzky Navigator* takes visitors through "architectural simulations of Freud's theories" using various visualization attempts stretching from Sergei Eisenstein to computer games.

Lisa Prah: ⟨http://web.aec.at/infowar/PROJEKTE/bernadette.html⟩
Peter d'Agostino: ⟨http://www.temple.edu/newtechlab/TRACES/⟩
Kevin and Jennifer McCoy: ⟨http://www.lightfactory.org/smallappliances/⟩
Annette Weintraub: ⟨http://www.turbulence.org/Works/broadway/index.html⟩
Lev Manovich: ⟨http://visarts.ucsd.edu/~manovich/FLN⟩

Hyperfiction

Judy Malloy

Judy Malloy was one of the early pioneers in hyperfiction even before the advent of CD-ROMs or the World Wide Web. Her *Uncle Roger* provided an early example of nonlinear fiction in which each viewer constructed the story by patterns of choices. Another work called *Forward Anywhere,* done as part of Xerox PARC's artist-in-residency program with researcher Cathy Marshall, explores the interweaving of two lives. Now Malloy, like most hyperfiction authors, has turned to the World Wide Web as a more easily distributed framework for this work. In a documentation of her work, Malloy explains her approach:

In my work, small increments of information—images or words (sometimes fictional, sometimes nonfictional)—are used as molecular units to form a whole narrative. For the most part, the narrators of these works are fictional women from different walks of life. I am interested in putting the reader inside the minds of these women.[16]

Mark Amerika

Mark Amerika is interested in the new possibilities for literature and narrative. *Grammatron* is one of the most ambitious projects yet exploring this new model of writing and media composition. The links and information structure themselves become part of the work. The *Grammatron* Web site describes the event and its new kind of "cyborg-narrator":

A story about cyberspace, Cabala mysticism, digicash paracurrencies and the evolution of virtual sex in a society afraid to go outside and get in touch with its own nature, *Grammatron* depicts a near-future world where stories are no longer conceived for book production but are instead created for a more immersive networked-narrative environment that, taking place on the Net, calls into question how a narrative is composed, published, and distributed in the age of digital dissemination. . . . The cyborg-narrator, whose language investigations will create fluid narrative worlds for other cyborg-narrators to immerse themselves in, no longer has to feel bound by the self-contained artifact of book media.[17]

Judy Malloy: ⟨http://www.well.com/user/jmalloy/⟩
Mark Amerika: ⟨http://www.grammatron.com/about.html⟩

Fig. 7.2.11. Mark Amerika, *Grammatron*. Web welcome page, for hypermedia fiction.

Other Artists and Projects

Darcey Steinke's *Blindspot* explores the form of hypermedia by presenting a short story via subdivisions of the web page into many "frames." ***Giselle Beiguelman*** attempts to explore the future of reading and writing in *The Book after the Book,* a work focused on nonlinear narrative. Additional links to hyperfiction projects and theory can be found at ***Tom Goldpaugh***'s *Postmodern Theory, Culture Studies, and Hypertext* Web site.

Summary: Computer Media—The Next Stages of Cinema and Television?

The artists described in this chapter have pushed photography, literature, cinema, and television in ways that could not have been anticipated a few years ago. They explore new capabilities such as interactivity, multiple streams of media, and hypermedia to

---

Darcey Steinke: ⟨http://adaweb.walkerart.org/project/blindspot⟩
Giselle Beiguelman: ⟨http://www.desvirtual.com/giselle/⟩
Tom Goldpaugh: ⟨http://www.marist.edu/humanities/english/eculture.html⟩

pursue goals such as deconstruction, personal expression, and documentary. They do it for personal vision, and some proclaim the revolutionary implications of the new capabilities. Mainstream media developers push their industries in similar directions, such as interactive television and Web-based media entertainment. The way artistic vision and media expansion interplay will be a significant feature of the next decade.

Notes

1. G. Legrady, "Introduction," ⟨http://www.artun.ee/center/i1/legrady.html⟩.

2. G. Legrady, "Interface Metaphors and New Narratives in Interactive Media," ⟨http://www.cda.ucla.edu/faculty/legrady/InterfaceMetaphors.html⟩.

3. G. Lovink, "Interview with George Legrady," ⟨http://www.mediamatic.nl/magazine/8*2/Lovink Legrady.html⟩.

4. B. Seaman, "Documentation of *Exquisite Mechanism*," ⟨http://research.umbc.edu/~seaman/⟩.

5. C3 Organization, "Description of Waliczky Installation," ⟨http://ns.c3.hu/butterfly/Waliczky/waliczky.html⟩.

6. L. Manovich, "The Camera and the World," ⟨http://jupiter.ucsd.edu/~manovich/text/waliczky.html⟩.

7. Stephan Eichhorn, Tjark Ihmels, KP Ludwig John, Michael Touma, "*Die Veteranen* Description," ⟨http://www.uni-leipzig.de/veteranen/english.html⟩.

8. T. Iwai, "Documentation of *Resonance of 4*," ⟨http://209.133.8.181/bc/bcsbritn.html⟩.

9. "Serious Games," "Description of *Indigestion*," ⟨http://www.ace.co.uk/seriousgames/diller.htm⟩.

10. J. Campbell, "Delusions of Dialogue: Control and Choice in Interactive Art," ⟨http://www.adaweb.com/context/events/moma/bbs4/⟩.

11. J. Campbell, "'Mortal Coil' Show Statement," ⟨http://arts.ucsc.edu/mortal/Campbell/campbell.html⟩.

12. J. Campbell, "Delusions of Dialogue: Control and Choice in Interactive Art," op. cit.

13. Krannert Art Gallery, "Curatorial Statement," ⟨http://www.kemperart.org/dwessy.htm⟩.

14. G. Harwood, "Documentation of *Rehearsal of Memory*," ⟨http://shaman.dds.nl/~n5m/texts/graham.html⟩.

15. A. Bureaud, "Review of *Rehearsal of Memory*," ⟨http://mitpress.mit.edu/e-journals/Leonardo/reviews/bureaudisea.html⟩.

16. J. Malloy, "Artist Statement," ⟨http://www.well.com/user/jmalloy/⟩.

17. M. Amerika, "Description of *Grammatron*," ⟨http://www.grammatron.com/about.html⟩.

7.3

Virtual Reality

Introduction: Artists as Architects of Virtual Reality

Many artists consider their work experiments in virtual reality. As explained in other chapters, *virtual reality* is a term much subject to confusion. All synthetic worlds generated by computers could be considered virtual worlds; indeed, fictional worlds of all kinds—for example, those created by literature, theater, cinema, or art—can be considered virtual realities. In contemporary discourse, however, the term *virtual reality* is most often applied to experiences generated by computers that simulate sensual cues of physical reality more closely than that generated only on a screen and by regular speakers.

Typically, virtual reality presentations aspire toward immersive visual and audio experience through innovations in perceptual and interface technologies. Perceptually, they augment usual computer displays via technologies such as stereoscopic 3-D eye displays, surround projection on all surfaces, and/or 3-D spatially localized sound. Building on physiological responses such as interpreting differences in what the two eyes see as depth, or translating the millisecond delays between signals at the ears as spatial location, these systems fool users into constructing what they are seeing and hearing as more realistic than normal "reality."

These systems also attempt to enhance the computer's response to user actions by expanding the interface beyond mouse and keyboard. For example, they use head-mounted position sensors to track the direction and tilt of the head in order to realistically present spatially correlated imagery and sound. Some use data gloves to read actions such as pointing. The systems are highly navigable, using the metaphor and cues of 3-D space to allow users to move through them. Some enable users to manipulate virtual objects, for example, by using data gloves or data suits to read gestural attempts to move objects.

Researchers and commercial developers have used these displays for applications in a wide range of fields, such as education, technical training, military simulation, games, virtual travel, scientific research, and architecture. Researchers continue to search for ways to expand the perceptual richness and responsiveness of these systems. They are developing less cumbersome arrangements than the typical ones that require head-mounted displays and data gloves; working on position and gesture recognition that will allow movement in free space; developing immersive 3-D projection techniques that use lightweight glasses instead of heavily instrumented goggles; and investigating the use of force feedback to add tactility and kinesthetics to the sensory cues that the computer can use to simulate real physical experience.

Artists have been intrigued since VR technology first emerged. For example, in Canada, the Banff Centre supported major Bioapparatus and Art and Virtual Environments initiatives, which included residences, publications, and conferences to encourage cultural analysis and experimentation with virtual reality. This chapter reviews the work of artists experimenting in this field, including some supported and inspired by those efforts. Many researchers in other fields are drawn to virtual reality as a kind of superrealistic cinema, and seek to create simulations with enhanced illusions of reality. Note, however, that this commonsense view is problematic; many theorists question the cultural assumptions underlying the coding of VR types of displays as "realistic." In his book *Artificial Reality,* Myron Krueger asserts that the shaping of artificial worlds is a moral responsibility: "The design of such intimate technology is an aesthetic issue as much as an engineering one. We must recognize this if we are to understand and choose what we become as a result of what we have made."

Artists generally have been less interested in simulating identifiable realities. They use VR's conventions and techniques of representing reality as a tool upon which to expand, and they confront users with a paradox. Virtual reality codes itself as offering realistic exploration through 3-D spatial conventions and perceptual cues, but the worlds they can explore are not typical. These transformations are often provocative, conceptually rich, and suggest future directions for VR research: the exploration of unorthodox spaces and worlds; the VR world as metaphor; alternative objects and creatures; relationships between the physical and the virtual; information visualization; and virtual reality in theater.[1] The University of Washington's virtual reality lab offers a useful resource on art and virtual reality on its Web site.

Note that artists working with VR concepts and systems are discussed in several other chapters. Chapter 7.4, on alternative interfaces such as touch, gesture, and gazed, describes many artists trying to expand the sensory capabilities of virtual reality. Similarly, section 2, analyzing biological and medical technologies, describes several artists who use VR techniques and explore its underlying characteristics.

Unorthodox Spaces and Characters

Brenda Laurel and Rachel Strickland

Brenda Laurel has a long history as an artist, researcher, and theorist. Her groundbreaking Ph.D. dissertation in theater (later published as *Computers as Theater*) analyzed the

---

Brenda Laurel: ⟨http://www.tauzero.com/Brenda_Laurel/⟩

experience of interactive media long before systems were capable of realizing them. Over the years she consulted with Atari, Apple Computer, Interval Research, and other companies, helping to shape the emerging field. She worked with Apple's Interface Design Group, championing the idea that the perspectives of the arts could help inform research. Her *Guides* project, which explored the idea of agents in narrative, is described in the chapter on artificial intelligence (7.6). The book she edited, *The Art of Computer-Human Interface Design,* exemplifies the range of her approach. In more recent times she conducted research on gender and computers and went on to found the company Purple Moon, which specialized in developing and marketing computer games that would appeal to girls. Laurel is often invited to speak at technology, art, and media conferences, and some of these talks and essays are gathered together in the book *Severed Heads.*

Rachel Strickland is an architect, videographer, and interaction designer at Interval Research. She was an experimental filmmaker working with Ricky Leacock at MIT during the Architecture Machine Group days. Later she worked at Atari's research center and helped shape Apple's Vivarium research on computers and education. Research they did together on Apple's *Playground* Vivarium project, exploring the role of narrative and non-Western concepts of space in children's play, suggested some of the directions pursued in later VR experiments.

Working with sponsorship from Interval Research, Laurel and Strickland collaborated in a VR experiment called *Placeholder,* executed at the Banff Centre for the Arts in their experimental spaces program. *Placeholder* broke new ground in several directions, such as the role of narrative in VR simulations, the sense of place, the integration of captured imagery and synthetic worlds, the utilization of representation schemes outside of Western perspective, the importance of letting people "mark" space in virtual worlds; the role of artificial characters, multi-person collaborations, real-time position, and gesture tracking; and opportunities for creating alternative geographies:

Placeholder was an installation which explored narrative action in virtual environments. The geography of *Placeholder* took inspiration from three actual locations in the Banff National Park—a cave, a waterfall, and a formation of weathered earthen spires overlooking a river. Three-dimensional videographic scene elements, spatialized sounds and voices, and embodiment as petroglyphic spirit animals were employed to construct a composite landscape that could be visited concurrently by two physically remote participants wearing head-mounted displays, who were guided by a disembodied "Voice of the Goddess" as they walked about, conversed, used both hands to touch and move virtual objects, and recorded fragments of their own narratives in the three worlds.[2]

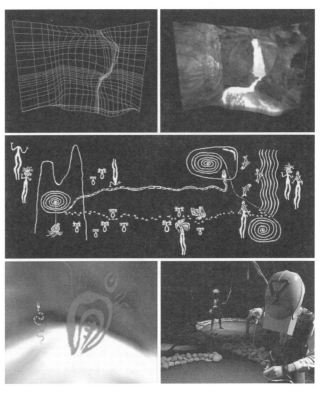

Fig. 7.3.1. Brenda Laurel and Rachel Strickland, *Placeholder: Landscape and Narrative in a Virtual Environment*. Produced through a collaboration between Interval Research Corp. and the Banff Centre, 1993. Images reproduced here include a conceptual map of the terrain and video views of the immersive waterfall.

Laurel and Strickland wanted to base their VR work on complex understandings of how humans organize space. They built on anthropological and psychological research that went beyond the mappable physical qualities, studied the naturalist concept of *umwelt* (the physical world as lived and perceived by different animals), and explored concepts of space that included spatial organization and "character" influenced by qualities such as atmosphere and the passage of time. Their research investigated issues such as multiple simultaneous points of view, 3-D sound's role in creating sense of space, and the leaving of "markers" as a way to define space. Their strategy included

experimenting with approximations of the sensory-motor experiences of nonhuman creatures, experimenting with narrative strategies for supporting transformation, locomotion, and spatial and temporal discontinuities, and exploring concepts of space and time that underlie representation systems which have been developed by cultures whose views of the world differ dramatically from our own.[3]

Building on ideas about the importance of body sense and anthropological insights about the power of animal symbols to represent the quality of places, they created animal "critters" (Crow, Spider, Fish, and Snake) to inhabit the virtual world. Visitors could embody themselves by moving toward these critters. Their virtual representations would then take on "smart costumes" that gave them certain perceptual and movement characteristics. They noted that body sense is typically devalued in postmodern discourse, resulting in a "profound disembodiment of cognition and feeling," and needs to be recognized as an important element in new experimental media. Objecting to the limitations of the customary VR-hand interface, *Placeholder* invites visitors to engage in complex body actions as part of the exploration.

Laurel and Strickland believe that the ability to leave marks is a basic human tendency that helps to organize both place and identity. They implemented a voice-holder feature in *Placeholder* that allowed visitors to leave voice records in places they visited in the virtual world.

Laurel is well-known as a spokesperson for the potential of new technolology to incorporate "imagination, ethics, and humanism," and for the power of research to sometimes transcend more provincial commercial or political agendas. She claims that artists and media innovators can make effective interventions in popular culture. The popular forms of computer graphics and games, and new forms, such as virtual reality, are not value free. Realism is itself a style decision. In "Ethos of Computing," prepared for ACM 97, she analyzes the importance of technologies, such as the computer, as a reflector and stimulator of change:

The computer serves as a projection surface for our own hopes and fears about what it means to be human in these times. . . . Culture comprehends technology through the means of narratives or myths, and those narratives influence the future shape and purposes of technology. The culture-technology circuit is at the heart of cultural evolution. . . . I think that what is needed is an intervention at the level of popular culture. Many of you are making such conscious interventions—mindfully creating technologies that cause us to produce new myths, and mindfully making art that influences the shape of technology.[4]

Virtual reality is seen as an especially potent form of media. It goes far beyond the computer as appliance. It allows us to turn things inside out. Laurel urges developers to insure that it does not take on a mangled, narrow vision. Unlike normal computers, virtual reality "uses our bodies as its instrument" and "gives us a first-person, body-centric view" instead of framed pictures or pull-down menus.

Laurel notes two kinds of interventions that she sees as essential in VR work such as *Placeholder:* (1) using it as a tool to "honor and celebrate the natural world and the ways that it articulates with our imaginations" rather than searching for stereotyped abstract out-of-body experiences, and (2) finding a way to incorporate deep personal storytelling as it has been practiced throughout human history—"purposeful action that is intended to communicate, to teach, to heal." She notes that some participants in Char Davies's breath-activated *Osmose* VR work (described below) claimed to encounter spiritual experience, and she campaigns against the tendency of some in the art world to abandon hope of new possibilities: "How tragic that artists have come to hold the self-marginalizing belief that cynicism is superior to hope. To my mind, there is nothing healthier than rolling up one's sleeves and trying to give the world fresh visions of joy, fresh uses of technology that indeed 'exteriorize the soul.'"

Michael Naimark

Michael Naimark is known for being a pioneer in alternate media and for his integration of art and research. As a member of the Architecture Machine Group at MIT, the precursor of the Media Lab, he helped develop the *Aspen Movie Map,* which was one of the first interactive movie events. He also consulted with Apple computer's multimedia and interface groups and other companies on the design of interactive technologies During these years he also worked as an independent artist exploring the potentials and limitations of visualization technologies. He also produced several papers on the challenges of real-space imaging.

His white-room event was one of the first virtual reality installations completed without computers. He created a living room installation in which all surfaces and objects were painted white. Before painting the space, he made a movie of people moving within it. A specially modified motorized tripod rotated a movie projector such that the movie images sequentially and precisely fill and empty sections of the white space with life; for example, the empty white couch came alive with upholstery and people talking. This uncanny illusion intimated virtual reality before most of us had any experience of it. Other installations included the *San Francisco Bay Flyover,* completed for the Exploratorium, in which visitors could navigate on a grid in the sky to view the city from above,

---

Michael Naimark: ⟨http://web.interval.com/projects/be_now_here/⟩

Fig. 7.3.3. Char Davies, *Osmose* (1995). A breath-based interface allows viewers to navigate an organic immersive VR environment. Real-time-frame capture from the immersive virtual environment. Copyright 1995–2000 by Char Davies and Softimage.

ing with transparency, luminosity, spatial ambiguity, implicit rather than explicit meaning, and temporality as well as a body-centered user interface of breath and balance with the intent of affirming the role of the subjectively felt physical body in virtual space.[6]

In a paper called "Changing Space: VR as an Arena of Being," Davies reflects on the experience of the five thousand people who had experienced *Osmose*. She was surprised by the intensity of experience for many of the immersants. She draws on psychologist Arthur Deikman's theory of deautomatization, which suggests that destabilizing psychic structures can result in increased attention and "perceptual expansion." *Osmose* put people in a safe but unfamiliar environment in which they could experience their bodies and perceptions in new ways. She refers to the literature on meditation, which reports similar phenomena. Many visitors to *Osmose* reported experiences such as "being in another place," "losing track of time," "unable to speak rationally afterwards," and euphoria.

Davies believes that a conventional VR environment would be unlikely to stimulate these kinds of effects. The aesthetic decisions she made about the luminous nature of the virtual world and the noncontrol-oriented breath interface set the scene:

[I]n *Osmose* we used transparency and luminous particles to "desolidify" things and dissolve spatial distinctions. . . . [R]ather than relying on conventional hand-based VR interface methods such as joystick, wand, trackball, or glove—which tend to support a disembodied, distanced, and controlling stance toward the world—we used an interface based on breath and balance to allow participants to simply "float" by breathing—in to rise, out to fall, and leaning to change direction. . . . [T]he hands-off interface approach freed them from the urge to "handle" things and from habitual gravity-bound modes of interaction and navigation.[7]

In *Éphémère,* visitors navigate an organic world that explores the temporary nature of existence and the linkage between the earth and the body. Davies's Web site describes the installation:

Éphémère is an exploration of the ephemerality of being and the symbolic equivalence of body and earth. The work's iconography, evolved through Davies's long-standing practice as a painter, is grounded in nature as metaphor: archetypal elements of root, rock, and stream, etc., recur throughout. Spatially, the work is structured into three parts: Landscape, subterranean Earth, and interior Body. The body, of flesh and bone, functions as the substratum beneath the fecund earth and the bloomings and witherings of the land.[8]

Davies explains that the structure of the event itself calls forth reflections on fleeing life. Processes of change surround the immersant—day and night, seasons, and decay and rebirth. The fate of some of the objects, such as seeds, rely on the proximity, movement, and gaze of the visitor. Davies believes that the VR environment, with is simulated realness and ghostlike presentation, is ideal for exploring ideas and feelings about this flow of the universe:

This river of life and time, the inexorable force that pours through all things, is what concerns me. . . . According to Heidegger, the Greeks called this flow *physis*. In truth, *physis* means outside of all specific connotations of mountains, sea, or animals, the pure blooming in the power of which all that appears and thus "is."

The very immateriality, temporality and apparent three-dimensionality of immersive virtual space is well suited for manifesting such a concept. In *Éphémère,* besides the various comings-into-being, lingerings, and passings-away, and the transformations of illumination and spatial contexts, there are "flows" of rivers, root flows, and body fluids streaming through the work.

The artist in the era of cyberspace cannot escape responsibility. Davies hopes her works awaken awareness of the physical world and its future. She notes the domination of logical thinking in cyberspace construction and warns that it may "distract from earthly responsibilities and the very wonder of being embodied among all this":

The function of the artist in correcting the unconscious bias of a given culture can be betrayed if he merely repeats the bias of a culture without readjusting it. In this sense the role of art is to create the means of perception by creating counterenvironments that open the door of perception to people otherwise numbed in a nonperceivable situation. . . .

While our habitual perceptions may lead to the forgetting of being, the paradoxical qualities of immersion in a virtual environment, if constructed so as not to reinforce conventional assumptions and behaviour, can be used to open doors of perception. In this context, *Éphémère* is an attempt to reaffirm our poetic and mythic need for Nature, returning attention to our fragile and fleeting existences as mortal beings embedded in a vast, multichanneled flow of life through time.

Other Artists and Projects

Lawrence Paul Yuxweluptun, a Canadian Native American artist, primarily works as a painter creating surrealistic compositions that incorporate imagery from Western and Native American traditions. Yuxweluptun also did some innovative work at the Banff Centre exploring VR capabilities to represent Native American concepts of space and confronting biases built into VR's conventions. His *Inherent Rights, Vision Rights* offers the visitor an opportunity to experience some aspects of a sacred longhouse ceremony in which viewers find themselves in featureless planes surrounded by strange animal sounds and flies buzzing at the head. He purposely rejected some VR capabilities, such as defying gravity or walking through walls, because these were counter to Native American traditions that saw the ground and objects of Nature as sacred.

Lawrence Paul Yuxweluptun: ⟨http://mitpress.mit.edu/e-journals/Leonardo/gallery/gallery291/yuxweluptun.html⟩

Richard Brown believes that artists can use new technologies to give people experiences of alternative space-time frameworks, physical experiences impossible in the real world, "free of the Cartesian Prison." Using VR technology, his *Alembic* installation allows people's movements to affect the continual metamorphosis of forms from "geometric crystalline structures to amorphous liquid shapes" and other indicators of fourth-dimensional "hyperprocess." Brown hopes to cultivate a "sense of mystery and magic, a sense of beyond, a spirtual depth that is often absent in the cool rationality of much electronic art." Brown's Web site contains extensive links to philosophical, mathematical, and artisitic speculation on 4-D space.

Michael Scroggins and Stewart Dickson created a VR installation called *The Topological Slide,* built on mathematical concepts from studies of topologies. Visitors are able to explore this strange space that does not exist in physical space. Working with Canon's ArtLab, *Ulrike Gabriel and Bob O'Kane* created a VR world called *Perceptual Arena,* which viewers with head-mounted displays could fill with objects. Using a data glove, they could sculpt the objects to their desire, much like clay. In *Arena Life,* the artists later augmented the installation by adding artificial-life techniques to guide the development of the objects along self-organizing tendencies as well as interactive manipulation. *Mario Canali and Marcello Campione*'s *Satori* offered helmeted visitors an interlinked space to explore, with no goal except self-knowledge. *Margaret Watson*'s CAVE-based *Liquid Meditation* let visitors encounter an architecture of liquid. Using the CAVE, *Teresa Wennberg* created *The Parallel Dimension* environment, which allows visitors to explore loss of identity and the importance of touch. *Margaret Dolinsky* has created a series of CAVE worlds that create unorthodox spaces in which to explore metaphor and different kinds of consciousness; in *Dream Grrrls* one follows a labyrinth full of unusual and whimsical events, "where you can ride a see-saw in the clouds—be shaken when you look into the face of an unfamiliar psyche. Follow the voices in the air or let experience be the guide in a place where one can only expect the unexpected." The *F.A.B.R.I.CATORS* (discussed later in this chapter, in a section on VR information

Richard Brown, *Alembic:* ⟨http://www-crd.rca.ac.uk/~richardb/⟩
Michael Scroggins and Stewart Dickson, *The Topological Slide:* ⟨http://www.wri.com/~mathart/portfolio/topo_slide/topo_slide_top.html⟩
Ulrike Gabriel and Bob O'kane, *Perceptual Arena:* ⟨http://www.v2.nl/DEAF/94/parts.html⟩
Mario Canali and Marcello Campione, *Satori:* ⟨http://www.telefonica.es/fat/ecanali.html⟩
Margaret Watson: ⟨http://www.evl.uic.edu/anstey/AECevents/mca.html⟩
Teresa Wennberg: ⟨http://www.nada.kth.se/~teresa/⟩
Margaret Dolinsky: ⟨http://www.avl.iu.edu/dolinsky/⟩

visualization) developed *Kali the Goddess of the Millennium* which they see as creating a cinematic mythic space, "a compendium of intrinsic illusions which goes from the most extreme of that of the natural, the apparent, and the real . . . to the most refined of myths." ***Diane Gromala,*** who has written extensively on the metameanings of virtual reality and the importance of the subjective and corporeal, collaborated with ***Yacov Sharir*** in *Dancing with the Virtual Dervish,* which used virtual reality to help participants connect with inner-body experiences.

The Virtual Reality World as Metaphor

Maurice Benayoun

Maurice Benayoun is interested in the poetry and metaphysics of new technologies. He has created a series of award-winning works that probe the expressive power of new technologies to ask questions about the world. His *Quarxx* computer graphics world (see chapter 3.2), explores variant worlds where alternative physical laws pertain. His *Tunnels* projects (see chapter 6.3) open new perspectives on communication at a distance. This section focuses on two works, *World Skin* and *Is the World Flat,* which use virtual reality as a poetic tool.

In *World Skin,* created in collaboration with Jean-Baptiste Barrire, visitors enter the immersive landscape of a war-ravaged area in which they can move through wasteland scenes of destruction. Benayoun uses CAVE technology in which polarized projections fill walls, the ceiling, and floor to surround the visitor in a 3-D image-scape viewable through shuttered polarized glasses. Three-dimensional sound enhances the devastating reality. One visitor is given a position-tracking wand that controls the direction of movement and gaze for the group.

But this landscape is not realistic in the usual sense. Something is wrong about this world: there is a horrible abstraction. Even though depth is everywhere, many of the persons and items in the landscape are rendered almost flat, like stage props. These flats sit in 3-D space in relation to each other. The color is polytone only within restricted parts of the spectrum, for example, shades of brown. Also, several of the visitors are given special digital cameras that they may point at the persons and objects they see.

F.A.B.R.I.CATORS: ⟨http://www.fabricat.com⟩
Diane Gromala: ⟨http://carmen.artsci.washington.edu/gromala/effect/gromala.html⟩
Maurice Benayoun: ⟨http://www.benayoun.com⟩

Fig. 7.3.4. Maurice Benayoun, *World Skin*. Visitors tour a war scene via CAVE virtual reality, and the objects they photograph disappear from virtual reality to become printed shots. (Production support by Ars Electronica Center, Z-A.NET, and SGI.)

They click the camera and the object they focused on disappears, only to be replaced by a silhouette.

World Skin asks viewers to think about their relationship to events happening in the world and questions the role of image capturing and representational technologies, such as photography, in presenting that world and mediating our thoughts and actions. Benayoun sees virtual reality, with its ability to sit astride realism and virtuality, as an artistic tool for forcefully bringing those questions to mind. Benayoun's Web-site documentation of the event presents his reflections on the work:

Armed with cameras, we are making our way through a three-dimensional space. The landscape before our eyes is scarred by war-demolished buildings, armed men, tanks and artillery, piles of rubble, the wounded and the maimed. This arrangement of photographs and news pictures from different zones and theaters of war depicts a universe filled with mute violence. The audio reproduces the sound of a world in which to breathe is to suffer. Special effects? Hardly. We, the

Alternative virtual reality technologies—The CAVE and *Garnet Vision.* The CAVE is a virtual reality technology developed by computer art pioneers Tom DiFanti and Dan Sandin of the Electronic Visualization Center at the University of Illinois, in Chicago. (DiFanti and Sandin have a long history of inventing new technologies as art exploration.) Seeking to escape the encumbering restrictions of head-mounted displays, they invented the CAVE, which projects polarized 3-D images on all the surfaces of a cube surrounding the visitors and immerses them in 3-D sound. Viewers only need to wear lightweight polarized sunglasses to experience the effect. A position-sensing polhemus device and interactive wand allows the lead visitor to control direction and movement through the virtual world. Many artists have begun to explore the expressive capabilities of the system. The Ars Electronica center offered an artist in residency for research with the CAVE.

Garnet Vision is another immersive projective system developed by Hiroo Iwata. A room-sized dodecahedron dome surrounds the viewer, creating 3-D spatial and sound perception.[9]

visitors, feel as though our presence could disturb this chaotic equilibrium, but it is precisely our intervention that stirs up the pain. We are taking pictures, and here, photography is a weapon of erasure. The land of war has no borders. Like so many tourists, we are visiting it with camera in hand. Each of us can take pictures, capture a moment of this world that is wrestling with death. The image thus recorded exists no longer. Each photographed fragment disappears from the screen and is replaced by a black silhouette. With each click of the shutter, a part of the world is extinguished. Each exposure is then printed out. As soon as an image is printed to paper, it is no longer visible on the projection screen. All that remains is its eerie shadow, cast according to the viewer's perspective and concealing fragments of future photographs. The farther we penetrate into this universe, the more strongly aware we become of its infinite nature. And the chaotic elements renew themselves, so that as soon as we recognize them, they recompose themselves once again in a tragedy without end.

We take pictures. First by our aggression, then feeling the pleasure of sharing, we rip the skin off the body of the world. This skin becomes a trophy, and our fame grows with the disappearance of the world. . . . War is a dangerous, interactive, community undertaking. Interactive creation plays with this chaos, in which placing the body at stake suggests a relative vulnerability. The world falls victim to the viewer's glance, and everyone is involved in its disappearance. The collective unveiling becomes a personal pleasure, the object of fetishistic satisfaction. We keep to ourselves what we have seen (or rather, the traces of what we have seen).[10]

An earlier work called *Is God Flat?* used technology to ask metaphysical questions about the nature of the world and how we discover it. Visitors were presented with a world full of virtual brick walls that they could dig through to create tunnels. Then they could try to escape or at least discover its structure. Even though the space offered some of the cues of reality, one continually got caught in labyrinthian traps. Benayoun works from a fundamental question about the knowability of the world. Virtual reality presents obvious fictions, but its pretense toward realistic representation lets it be a tool for thinking about the knowability of the nonmediated world. In *Crossing Talks* Benayoun confronted visitors with a CAVE-based space in which projections caused the floor to appear to tilt as people variously moved toward projected images on the walls. Visitors could stabilize the scene only by communicating with the other strangers in the space.

Sheldon Brown

Sheldon Brown has worked extensively in conceptual kinetics and multimedia installation. In one project called *Mi Casa Es Tu Casa,* Brown created a networked virtual reality playhouse in which children from the United States and Mexico could jointly play. *Apparitions* (in collaboration with Vital Signs artist group) uses virtual reality to consider issues associated with medicine, identity, and the body, and to reflect on VR's representational possibilities and constraints as a child of military research, considering the way it privileges and denies various ways of seeing. The Web site describes the usefulness of virtual reality to question social structures:

Gallery visitors engage in activities which call into question notions of "real" and "virtual" by interacting with physical objects, exploring ideas on the World Wide Web, and moving through a virtual environment via large-scale video projections.

Social institutions, as sites of knowledge and authority, draw lines between public and private, deciding for us what parts of our lives and bodies we control. In medicine, imaging technologies are used both to inform these choices, and to legitimate medical authority. . . . Our exhibition is designed to examine the revolutionary and restrictive possibilities of these technologies, as well as their interplay with the human body and identity. . . . The virtual environment itself is a three-dimensional, photo-realistic representation of a medical clinic composed of several distinct rooms or spaces, connected by a system of doors and hallways. Some of the spaces closely resemble what one might expect to find in a contemporary medical clinic. Others take advantage of the

Sheldon Brown: ⟨http://www-apparitions.ucsd.edu/project.html⟩

Fig. 7.3.5. Sheldon Brown, Vital Signs, *Apparitions*. An installation in which viewers engage physical objects, the World Wide Web, and video projections in an exploration of the institutions of medicine.

fact that they are virtual spaces by becoming fantastic allegorical spaces which deal conceptually with virtual technologies, information spaces, and technologically mediated zones of popular culture.

Apparitions looks at these technologies in both their liberating potential and their history of constraint. . . . The way control is exerted in a system depends on how things appear to be knowable. The objective gaze of science is not disinterested; it both produces and is grounded in a whole range of ideological assumptions. At the same time, the technologies of visualization produce images that motivate public opinion and infiltrate the aesthetic texture of daily life.[11]

Alternative Objects and Creatures

Jeffrey Shaw

Jeffrey Shaw has constructed numerous installations that challenge conventional concepts of space. He also directed the ZKM media academy in Karlsruhe, Germany, which

---

Jeffrey Shaw: ⟨http://www.iamas.ac.jp/interaction/i97/artist_Shaw.html⟩

Fig. 7.3.6. Jeffrey Shaw, *Place, A User's Manual*. An immersive 360-degree environment of many places in the world, seamlessly made with a motorized observation deck.

supported many other artists in their explorations of concepts of virtual reality. His *Legible City* installation (see chapter 7.4) allowed visitors to bicycle around a virtual city constructed of words. *Place—A User's Manual* comments on Western concepts of place and location by offering a 360-degree scene seamlessly composited from images of many different places in the world. A motorized observation deck in the middle allows visitors to orient their observation window. Projected texts exploring concepts of place and language are determined by user actions such as speech. The ground on which these panoramas are positioned is marked by a diagram of the Sephirotic Tree of the later Qabbalists.

In an essay presented at the Doors of Perception conference, Shaw explains his idea that virtual reality represents a new architecture in which the public can shape their relationships to space and each other. He traces the historic quest for transparency and dissolution through experiments in temporary inflatables and cinema projection:

Architecture is both an existential and ideological enclosure. In the latter half of the twentieth century we have witnessed the progressive dematerialisation of architecture motivated by a search for singular lightness and mutability. This lightness and mutability expresses the desire of every person to become architect of his/her own surroundings. The scale of such a fulfillment is best identified as the home—the inner sanctum of that urban theater where we enact our lives. A paradoxical oscillation in this development comes from the desire to also find conjunctions between the personal and the social enclosure—between privacy and engagement, between detachment and immersion. . . .

Now in the 1990s, the virtual landscape becomes a networked cosmography which mirrors the real world into a televirtual imaginative and social space. Tele-virtual-reality is the appropriate domain of our architectonic desires today. Space, time, and interaction become the design parameters of this buoyant ambiance, where we deconstruct the literal, circumvent the constraints of identity, and evoke a fluid poetics of space, person, and intimate experience.[12]

In *Golden Calf,* Shaw creates a virtual pagan idol that users can inspect by moving a handheld LCD interface monitor. They can move closer and further, look underneath and even penetrate inside. The image on the screen synchronizes with the relative position of the monitor to the virtual idol, for example, getting bigger as the viewer comes closer and showing the appropriate perspective as the viewer rotates around the idol.

Shaw uses the metaphoric qualities of the virtual object that link contemporary technology with ancient religious searches for meaning. The visitor can try to enter the virtual calf, but like many idols, it is not what it first seemed. Furthermore, the inspection movements about the calf seem quasi religious, unwittingly inducting the viewer into a sort of ritual dance. Art historian Edward Shanken, commenting on the tension between the feeling of freedom and control, observes that "the viewer can laugh at the playfulness of the work, the irony and absurdity of praying to an archaic idol resurrected in silicon and software. Alternatively, the viewer can contemplate the more serious ramifications of human seduction by technology into a state of preprogrammed obedience."[13]

ICC critic Toshiharu Itoh suggests that Shaw is interested in virtual reality not for its transport to artificial worlds, but rather for its ability to bring us to that frontier between the real and the virtual:

Shaw has suggested that the most fascinating aspects of life in our era do not lie in things limited to the real world or some fixation with imaginary worlds, but are found instead in the creation of imaginary spaces of dialogue with the bodily experiences of history and its real spaces. In other words, Shaw emphasizes the creativity of the border region where one foot rests in the real world, and the other in the world of fantasy. The same can be said for Shaw's attitude towards new technologies. While one cannot deny the compelling and fascinating new areas that have been born from the development of new, ever-evolving technologies, still one has to recognize their frequent, relative poverty in comparison to the rich complexities that real experiences impart to humans as living beings. Put otherwise, we are obliged to constantly traverse this divide. And within these conditions, we are called upon regularly to give new meanings to technology.[14]

Other Artists and Projects

Robert McFadden used Sense8's desktop VR apparatus to create a VR installation called *Picture Yourself in Fiction.* Visitors could navigate the space near a virtual floating box that could emanate poetry. *Scott Fisher* is an artist and researcher long concerned with creation of immersive first-person worlds in contexts such as MIT's Architecture Machine group and NASA's Virtual Environment Workstation Project, which conducted innovative research on key VR technologies such as head-coupled displays, data gloves, and 3-D audio technology. He then went on to establish Telepresence Research to continue research on first-person media. Fisher has created a variety of 3-D immersive environments. *Menagerie* confronted users with a world full of synthetic creatures that manifested animal behaviors such as flocking and curiosity, and different kinds of responsiveness toward the visitor's actions. The animals enter and leave through portholes that materialize and dematerialize. *Chemerium,* a VRML world created in collaboration with *Perry Hoberman,* similarly confronted viewers with synthetic animals. It also allowed them to create their own strange combination chimera. Ampcom, a collaboration between *Andy Best and Merja Puustinen,* created *Conversations with Angels* as a VRML environment in which visitors could explore and meet with artificial characters and other people's avatars. All kinds of encounters are possible, for example Marg, the bored housewife, Carl, the serial killer, and Fat Bob, the redneck. *Hiroo Iwata's Anomalocaris* confronts participants with a virtual creature represented both visually and tactilely via force feedback.

Relationships between the Physical and the Virtual

Masaki Fujihata

Masaki Fujihata creates complex worlds that disrupt conventional expectations of spatial relationships. Metaphysical confusion is one of his main agendas. He wants to "show us ways to make new holes in our solid brains." The installations create geographies of

---

Robert McFadden: ⟨http://www.hitl.washington.edu/scivw/EVE/II.J.Art.html⟩
Scott Fisher: ⟨http://www.portola.com/PEOPLE/sfisher/index.html⟩
Andy Best and Merja Puustinen: ⟨http://ampcom.kaapeli.fi/index.html⟩
Masaki Fujihata: ⟨http://www.flab.mag.keio.ac.jp/index.html⟩

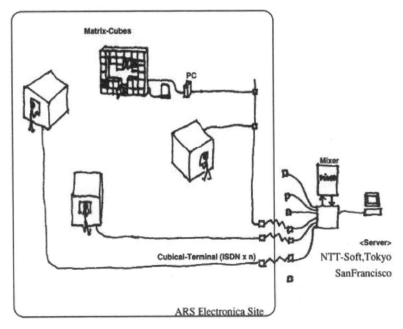

Fig. 7.3.7. Masaki Fujihata, *Global Interior Project.* Visitors explore complex relationships between the physical and the virtual by navigating in a matrix of physical box modules that are linked by telecommunications.

physical, virtual space, and representations of virtual space that all get confounded. His *Beyond Pages* installation (see chapter 7.4), blurs the boundaries between a projected book and the objects it represents. The Ars Electronica interactive jury statement summarizes the *Global Interior Project:*

Global Interior Project is a networked multiuser virtual environment where people can meet, talk, and discover the metaphysics of reality. It is an example of communication media design which demonstrates new possibilities of connecting people in a new way. The aim of this project is to visualize the metamechanism of electronically networked space and communication, while moving back and forth between "real" and virtual space. Hence, the word *real* means your living space, where you can manipulate a virtual world with a device such as a trackball and see the calculated image on your screen. A remarkable feature of this project is the possibility of setting up a model as an actual object to represent the architecture/construction of virtual space. It is constructed with computer-controlled boxes that have an actual door. The status of this door signals the existence of someone in virtual space. As a result, it is possible to make an extraordinary link between the virtual and the real.[15]

The installation consists of both a physical setup and a virtual world. Each physical and virtual room has a symbolic object to help orient travelers. A sculpture of matrix cubes with movable doors maps the presence of people in virtual space. When more than one person is in the same virtual room they can communicate via sound and video, and visitors can be in more than one space at a time. A special room called the Self allows instant transport. The documentation describes the setup.

The Cubical-Terminal is a large white box with a window. The user stands outside of this box, looking in through the window to both watch the image of virtual space and navigate through it by manipulating a trackball device. It is also equipped with a speaker, microphone, and video camera. Matrix-Cubes is a sculptural map of the virtual space constructed of eighteen boxes and placed nearby the Cubical-Terminal. The status of the door of each box shows whether that room is occupied in the virtual space.[16]

All the interrelationships offer a rich field of action and reflection. Each person's identity is distributed in several places. The Ars Electronica description explains:

The Matrix-Cubes thus function as a miniature/metaphor/map of the virtual space. In this system, it is possible for a participant to have a threefold existence: Real Me, Virtual Me, and Virtual Me in the Actual World. For example, an action conducted with the image of Real Me through virtual space causes a reaction on the part of Virtual Me.[17]

Fujihata believes that virtual reality is best not used for simulating physical reality. Inventing artificial spaces seems much more appropriate:

VR also deals in the manipulation of 3-D data. This data, however, is not used for generating photo-realistic images, but to create a space to interact, and describe relations between user and object. In VR space, the only way to make the world complex is making links between object, vertex, polygon, or character. The process for describing the world in VR is same as writing a story using text. To design its objects is to define its concept. To combine these objects is to write the world. In this manner, it is nonsense to put the data of a photo-realistic cup, coffee, chair, and door into VR space, because these are deeply related with our real body's functions, eating, and tactility.

Fig. 7.3.8. Agnes Hegedüs, *Handsight*. User manipulations of a physical eyeball sculpture affect the positions of virtual objects.

Agnes Hegedüs

Agnes Hegedüs has created a variety of works that explore virtual reality and telecommunications. One well-known work called *Handsight* uses the metaphor of the eyeball as its interface. The visitors' manipulation of a real object affects the life of virtual objects:

This work is constituted by a circular projection screen, a handheld interface which has the form of an eyeball, and a transparent sphere with a hole, into which the viewer can insert this interface outside. The transparent sphere is represented as a virtual eye on the projection screen. When entering the sphere, the viewer enters the virtual eye through its iris and sees a computer-generated 3-D tableau which is located within its virtual interior. This tableau is iconographically related to a Hungarian folk art tradition that created miniature religious scenes in glass.[18]

Agnes Hegedüs: ⟨http://www.c3.hu/butterfly/Hegedus/cv.html⟩

Other Artists and Projects

A collaboration of **Eben Gay, Amatul Hannan, Brenden Maher, Colin Owens, Eric Pierce, and Ann Powers** created *Faery Garden,* which blurred the real and virtual. Visitors entered into a setting that included real objects likes bushes and drapes and a VR projection screen. Events happen in both spaces using theatrical techniques to create an illusion, for example, sounds in the bushes and a tiny virtual waterfall that splashes. **Steina Vasulka**'s *Theatre of Hybrid Automata* allows a camera to explore a physical space while another space visualizes a virtual space explored in correspondence.

Ellen Sandor and art(n) Laboratory developed a physical 3-D photographic process for representing 3-D objects and worlds. The laboratory includes many collaborators, and they hold patents on this special technology, which has been used to represent CAVE spaces. In *Clouds of, Robin Petterd* uses VR technology to reflect on the possibility of technology distancing people from experience. Entering the gallery, one sees a video projection, but putting on the VR glasses reveals a more complex reality that underlies the projection. Reflecting on the changing status of the physical body, **Catherine Richards**'s *Virtual Body* confronted the viewer with a nineteenth-century stereoscope viewing column with a hole into which the hand could be inserted, with resulting views of distortion and metamorphosis. **Marcos Novak**'s *TransArchitecture* attempts to analyze the new relationships between the physical and the virtual in which they each act as portals to the other. **Jean-Louis Boissier** creates interactive installations such as *Moments de Jean-Jacques Rousseau,* in which a computer watches a viewer's reading of a physical book to bring up related virtual pages. **Jay Lees and Bill Keays**'s *Suspended Window* surrounded viewers with two large windows, one a "real" window looking at a scene and the other a virtual window showing the same scene. **Fakeshop** is a combined physical and online art group whose intention is "to create and promote works that simultaneously straddle the physical and 'online' realm." *Multiple Dwelling*'s physical installation refers to the room full of involuntary donors of body parts from the movie *Coma;* online residents can "inhabit" this space via avatars.

Faery Garden: ⟨http://www.siggraph.org/s97/conference/garden/faery.html⟩
Ellen Sandor: ⟨http://www.artn.nwu.edu/⟩
Robin Petterd: ⟨http://www.otheredge.com.au/robin/⟩
Catherine Richards: ⟨http://www.telefonica.es/fat/erichards.html⟩
Marcos Novak: ⟨http://www.aud.ucla.edu/~marcos/⟩
Jean-Louis Boissier: ⟨http://contactzones.cit.cornell.edu/artists/boissier_flora.html⟩
Jay Lees and Bill Keays: ⟨http://www.media.mit.edu/~keays/research/suspwin.html⟩
Fakeshop: ⟨http://www.cyberia.co.jp/tokyo/04/galerie/1998/fakeshop-e.htm⟩

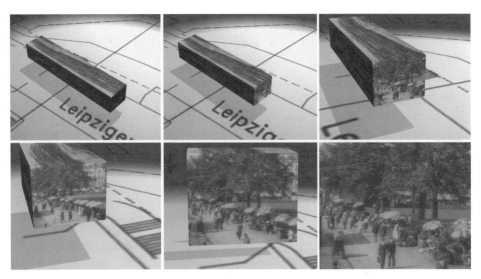

Fig. 7.3.9. Art+Com (Joachim Sauter, Dirk Luesebrink, and Iris Schoell), *The Invisible Shape of Things Past.* A system allows visitors to "draw" film objects in the virtual space with the aid of a video camera.

Information Visualization

Art+Com, Dirk Luesebrink, and Joachim Sauter

Art+Com is an innovative German organization that brings together researchers, technologists, information scientists, and artists to explore opportunities created by the new technologies. Projects explore a variety of issues in virtual reality, interface, visualization, and communication. Their projects—for example, *Terrapast/Terrapresent, Home of the Brain,* and *Technopolis*—are described in several places in this book.

The Invisible Shape of Things Past, devised by Dirk Luesebrink and Joachim Sauter, linked real documentary film and video of Berlin with a virtual environment that gave great flexibility in how information could be conceptualized and displayed. Their system made 3-D objects out of film clips that then could be spatially located in the artificial space. Individual frames were lined up sequentially in space in accordance with camera movements such as tracks, zooms, and pans. The film objects were also placed on a space-time continuum of their history and geography. Users could navigate to see events

Art+Com: ⟨http://www.artcom.de/projects⟩

that happened at the same time or in the same place. In awarding an Ars Electronica prize, the jury warned that the openness of this system might be quickly usurped by corporate agendas:

These 3-D objects can be created from historic footage or from contemporary film images. In this case, they were from historic footage of parts of Berlin (streets, squares, buildings). The film objects can be placed within a representation of the city, a virtual Berlin. This virtual city can be accessed via a time line (Berlin 1910, 1930, and so on). The images then occur in the virtual city on the actual location where they were shot. It is also possible to move the images to the same location in a different time period, resulting in rich connections between designing, organizing, and thinking about the city in its many aspects.[19]

Steffan Meschkat, one of the Art+Com collaborators, expresses the view that the production of information systems as opposed to the creation of material objects has become the main cultural and artistic activity. He also notes that as information systems gain the capability of self-evolution, the creative task will be in defining the structure beforehand.

Benjamin Britton

Benjamin Britton has created an intense VR recreation of the caves at Lascaux, the location of the world's most famous Paleolithic wall paintings. The caves were being wrecked by the natural traffic of visitors. Britton worked intensively with archaeologists to create a VR world that allows the public to explore the caves without causing damage.

The closing of the caves and reliance on representations such as that by Britton raises fascinating issues about objectivity and representation. Britton suggests that objectivity is impossible. He suggests that a more important task is for artists to understand the intellectual and spiritual core of what they are working with and to craft their representations to connect audiences with that core while at the same time not dominating the audience with that vision:

We who seek to interpret the past and to promulgate our perceptions to the public must ask ourselves, what do we want the audience to learn? The accuracy of all archaeological data is often

---

Benjamin Britton: ⟨http://www.daap.uc.edu/soa/benb/⟩

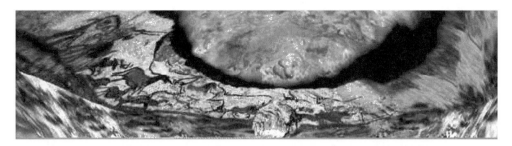

Fig. 7.3.10. Benjamin Britton, *LASCAUX*. A VR re-creation of the animal paintings in the caves at Lascaux.

drastically flawed and it is never absolutely complete. . . . This work is a forensic and interpretive process which is inexorably bound to the culture of the creative producer. . . .

The reconstructionist must ask: What does this site mean? With the lightest of touch, the reconstructionist must acknowledge that the fingerprints of subjective interpretation will be left on the model; and the scientist—now an artist—should make the fingerprints sing. Instead of focusing your thoughts on photo montage, 3-D models, and new technology, look within, look up, and recognize that your personal purpose is no different than that of your audience. . . . It will never be accuracy, clarity, sobriety, and intention which imbue an object with meaning. Give your reconstruction passion, opinion, hearsay, and myth—but give it lightly so your site can speak for itself, too. Remember to lightly veil your mystery so your viewers can sense the truth of its actual existence.[20]

Other Artists and Projects

The Italian VR group *F.A.B.R.I.CATORS* (including collaborators Yesi Maharaj Singh and Franz Fischnaller, among others) uses the CAVE to create navigable 3-D environments in which visitors can explore the historical eras of the Renaissance and the cyber age. They seek to create a new integration of research and art aiming at the "elaboration of bizarre and efficient inventions." Cultural accomplishments such as architecture, paintings, and inventions are given a 3-D life that could be explored via the metaphor of spatial movement. In the installation *Multi Mega Book*, the "ambiance of the fifteenth century features Italian buildings of the epoch integrated in one unique space. Sculptures extracted from their original context appear in imaginary squares. Paintings are displayed

---

F.A.B.R.I.CATORS: 〈http://hpux.dsi.unimi.it/imaging/LAST_SUPPER/lastsupper.html〉

as gigantic scenery. Users explore the revival of the classics, the heliocentric theory, movable type, the printing press, and the printed book."

In *L'Ultima Cena,* visitors can virtually tour Leonardo da Vinci's masterpiece. But virtual space offers some improvements on physical space. Visitors are able to see graphic representations of ideas underlying the work, animations of characters in the paintings, and access to juxtapositions and points of view that are impossible in physical reality, such as "navigating" the new innovations in perspective and walking inside the room where Christ and the Apostles are having dinner, taking points of view not available to viewers of the painting.

The Swedish artists *Bino, Cybeard, and Cool* used a six-sided CAVE to create *Giza Virtual Nights,* which integrated real information about the Pyramids at Giza (for example, requiring bending to go through restricted spaces in the real pyramids) with encounters with cyber-fictional Egyptian characters. In *SMDK Simulation Space Mosaic of Mobile Datasounds,* **Knowbotic Research** enabled helmeted visitors to navigate a virtual 3-D visual sound-scape of abstract forms representing sounds contributed from the Internet that had synthetically been grouped by intelligent-analysis software. In *Microworlds, Sirens, and Argonauts* **Agueda Simó** provides a "living narrative landscape" in which viewers can navigate "multiscale microscopic worlds that grow and transform as the users interact with them" with some entities serving as special "attactors" (hence the term *sirens*).

Researchers *Flavia Sparacino, Alex Pentland, Glorianna Davenport, Michal Hlavac, and Mary Obelnicki* from MIT's Media Lab used CAVE technology to create the dynamic *City of News.* Information was mapped onto a representation of the urban landscape through which visitors could virtually move. Web text and images form the skyscrapers and alleys of the city. *Monika Fleischmann* explores the artistic potential of new VR technologies. Several of her works are reviewed in next chapter (7.4) on alternative interfaces. In *Home of the Brain,* Fleischmann creates a virtual reality representation of ideas, a new kind of public space. A set of locations is created into which visitors can navigate to explore the ideas of a set of theorists critical to contemporary thought about cyberspace. Physical space is shaped by "interpretation of their ideas with

---

Giza Virtual Nights: ⟨www.nada.kth.se/giza-vr; http://www.sics.se/~bino⟩
Knowbotic Research, *SMDK:* ⟨http://netbase.t0.or.at/~krcf/smdk/smdk1.html⟩
Agueda Simó: ⟨http://anim.usc.edu/simo/⟩
MIT Media Lab, *City of News:* ⟨http://www-white.media.mit.edu/~flavia/⟩
Monika Fleischmann, *Home of the Brain:* ⟨http://viswiz.gmd.de/VMSD/PAGES.en/mia/homepage.html⟩

the concepts of hope—adventure—utopia and catastrophe which correspond primarily to symbolic colours." The text of the analysts appears as Möbius strips: "Chains of thoughts twist around the houses or the objects like thin paper snakes." ***Rita Addison, Marcus Thiebaux, and David Zeltzer*** developed *Detour: Brain Deconstruction Ahead,* a VR immersive event that explored the brain space of Rita Addison as she tried to regain brain function after traumatic injury. The artists wondered in the Web documentation of the event if the spatial metaphors of virtual reality would be effective in conveying a sense of this inordinate reality: "Can virtual reality allow us to really share what goes inside another person's mind?"

Distributed Virtual Reality

Carl Loeffler

Carl Loeffler was an early integrator of video, performance, and conceptual art and founded ArtCom, which was for a long time a major resource center for information about experimental arts and access to technologies such as telecommunication. More recently he has been project director of Telecommunications and Virtual Reality at the STUDIO for Creative Inquiry at Carnegie Mellon University.

Illustrating the ability of artists to innovate in the research world, Loeffler works internationally with corporations and institutions around the world, exploring the potentials of distributed virtual reality and regularly making presentations at international conferences. This technology allows geographically separated individuals to enter and work together in simulated 3-D worlds from their dispersed locations. One set of projects plans to create world-spanning virtual museum exhibits. He produced prototypes, including the *Virtual Ancient Egypt: The Temple of Horus,* which was presented at the Guggenheim Museum, and *Virtual Pompeii.* In *Virtual Reality Casebook,* which he authored with Tim Anderson, he describes the potential of distributed virtual reality: "There can be little doubt that networked immersion environments, cyberspace, artificial or virtual reality . . . will evolve into one of the greatest ventures ever to come forward." Other networked VR environments are described in chapter 6.3.

Rita Addison, Marcus Thiebaux, and David Zeltzer: ⟨http://www.evl.uic.edu/EVL/VROOM/HTML/ PROJECTS/45Addison.html⟩
Carl Loeffler: ⟨http://www.cyberstage.org/archive/cstage12/carl12.htm⟩

Virtual Reality, Music, and Theater

Jaron Lanier

Jaron Lanier coined the term *virtual reality* and was an early evangelist for its application in research, commerce, education, and art. He headed a company called VPL, which provided technology for many of the first projects working on virtual reality. He was invited as a major visionary to shows and conferences to articulate the practical and theoretical implications of virtual reality. He wrote many articles and a forthcoming book, *Information Is Alienated Experience.* He is a lead scientist with the National Tele-immersion Initiative and a visiting artist at places such as the Tisch School of Telecommunications at New York University.

Lanier is also a musician and has created many innovative musical applications of virtual reality. He exemplifies the hybrid artists/scientists who are helping to shape emerging technologies. Lanier heads a VR-oriented music group called Chromatophoria. He "plays virtual musical instruments that couldn't exist in reality, and uses physical instruments as sophisticated interfaces to the virtual world."

In an address entitled "Frontiers between Us," delivered before the ACM (Association for Computing Machinery, a major professional computer-science organization), celebrating the fiftieth anniversary of computing, he reflected on computer literacy as an example of the realization that the technical and the cultural were not separate:

The public has often warmed to the surface of science and engineering, but never before to the depth. While there are tens of millions of people who love dinosaurs and black holes, how many of them have gone on a dig or analyzed spectrum data? When it comes to computers, though, a mass culture of technical literacy is being born, especially among children. We always thought computers had to become popularized, and instead the public has decided to become surprisingly technical.[21]

He noted, however, that beauty and cultural awareness, which had not been part of engineering, would need to be integrated in the future. The mainstreaming of technology into culture could open up great opportunities:

Jaron Lanier: ⟨http://www.advanced.org/~jaron/⟩

Whatever the reason, I would want to celebrate the public's embrace of computer arcana, except for one thing. The material itself is unrelentingly ugly. . . . Computer science is, alas, the only engine of culture that has not concerned itself with beauty. Why should we have? We didn't know we were making culture. We thought we were making invisible tools. We've been granted a surprise franchise as culture creators. . . . The result will be a mass theater of spontaneous shared imagination and dreaming. My fond hope is that it will take the form of networked VR with inspirational authoring tools that are capable of quick, improvisatory creation.

Mark Reaney

Theater has always focused on creating virtual realities. Mark Reaney at the University of Kansas established the Institute for the Exploration of Virtual Realities as an early pioneer in this area. Reaney was instrumental in creating several works that explored the integration of experimental technologies with theatrical production. In his paper "Virtual Reality on Stage," Reaney notes the conceptual linkage. Virtual reality was an unprecedented tool that could realize the ideas of playwrights who were ahead of their times:

Theatre is the original virtual reality machine. Accessing it, audiences visit imaginary worlds which are interactive and immersive. . . . Theatre and computers functioning as virtual reality generators have remarkable similarities. Both offer fleeting, metaphysical experiences. Both create fictive worlds in which intangible concepts can be given perceptible form. . . . Virtual reality can unlock many scripts, realizing potentials that have been thwarted by production techniques that, being bound by muslin, wood, and steel, can not keep pace with the imagination of playwrights.[22]

The *Adding Machine,* the first major experiment, focused on a classic 1923 expressionist play about the dehumanizing effects of technology. Audiences joined live and computer actors in 3-D computer-generated scenes. For example, in one scene, the projection of the "boss" grows bigger in correspondence with an actor's emotions. In *Wings,* the audience wore nonimmersive head-mounted displays called i-glasses that allowed them to see 3-D rear projections as part of the scenery of the live action while maintaining "strong connections with the live actors."

Mark Reaney: ⟨http://www.ukans.edu/~mreaney/⟩

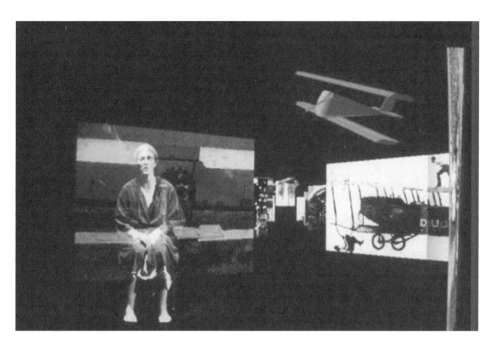

Fig. 7.3.11. Mark Reaney, *Wings*. The integration of live dramatic action and VR spaces.

Tesla Electric created a VR-augmented work that explored the life of Nicholas Tesla. The production created a rich 3-D world in back of the stage action. Tesla was the visionary eccentric inventor credited with inventing radio, AC power, and the tesla coil. His life was also marred by encounters with the capitalists George Westinghouse and Thomas Edison, who reaped the benefits of his work.

In a pair of papers available on the Web site, Reaney explains the similarities between virtual reality and theater and the possible ways they could enhance each other. He starts out with several questions about VR's ability to help solve scenographic problems and how theater could aid VR-designing practice. He notes that virtual reality and theater share interesting similarities—their interest in live actions and their focus on 3-D space created by some form of illusion. There were also differences that had to be addressed—theater had traditionally been an experience heightened through its communal nature, while virtual reality was usually experienced by individuals or small groups. Also, theater audiences were customarily passive while VR participants were actively in control. In VR-augmented theater who would be in control, the performers or the audience? Compromises had to be made, and Reaney attempted to give primacy to historical theatrical conventions.[23]

Other Artists and Projects

George Coates was one of the early experimenters in applications of digital technology to theater. Before digital technologies were invented, he created amazing illusions through slide and video projections using unusual lenses, mechanical contrivances such as disappearing doors and platforms, and the use of specialized projection surfaces such as skims and architectural features. Virtual reality technology has been used in several of the productions. For example, in *20/20 Blake,* William Blake's drawings come to life, with live actors appearing to move within the illustrations. In *Virtual Orchestra Performance,* **Tapio Takala** enabled visitors to control the players of a virtual orchestra by their body gestures. If they quit directing, the orchestra loses focus and the performance deteriorates. The event uses rule-based agents and neural networks to guide the players' responses. A related system called DIVA reads gestures as visitors move in free space.

Worcester Polytechnic Institute mounted a production of *One Flew over the Cuckoo's Nest* using rudimentary virtual reality. The production borrowed some perspectives from computer games to create the artificial worlds. Illustrating the power of the technology to enhance production, the Web site notes that the script indicates that "the stage does not go completely dark, but is covered by moving projections . . . bizarre, intertwining patterns through which people move, slowly as in a dream." Virtual reality seemed ideal to create this kind of illusion. *Sally Jane Norman*'s explorations of the paralells between cyberspace personas and the traditions of theater and puppetry were reported in her paper "Dramatis Personae: Casting Cyberselves."

Research and Commercial Virtual Reality Environments

Virtual reality has become a major focus of research and commercial development. Groups around the world are creating unprecedented immersive environments for education, scientific research, entertainment, policy planning, medicine, finance, and the like. While they do not necessarily conceptualize their work as art, the environments they invent are often conceptually provocative, interactively engaging, and sensually rich. Many were created by people with history in the media and visual arts, and might be considered art if mounted in artistic settings. Here are some examples:

George Coates: ⟨http://www.georgecoates.org/⟩
Tapio Takala: ⟨http://www.cs.hut.fi/~tta/⟩
Worcester Polytechnic Institute: ⟨http://www.wpi.edu/Academics/Depts/IFD/TT/vr.html⟩
Sally Jane Norman: ⟨http://www.telefonica.es/fat/enorman.html⟩

Rima's House   An event that uses motion tracking to allow visitors to change their projected image and its movements. (Vivid Virtual Theatre, New Zealand National Museum)

Distributed Scientific Visualization of Ocean Models   Simulation allows users to interactively change characteristics of the ocean. (Robert M. Knesel, Naval Oceanographic Office)

Overcoming Phobias Using Virtual Reality   Allows therapists to work with clients with specific fears by having them encounter the objects of their fear in simulated environments. (Rob Kooper, Georgia Institute of Technology)

Virtual Explorer   A nanobot navigating through the body. (Kevin Dean, Senses Bureau, University of California at San Diego)

Summary

Virtual reality can serve many purposes, ranging from the enhanced ability to inspect representations of reality to exploration of synthetic and fictional worlds that claim no mapping of reality. Some analysts suggest that VR's most significant impact may be to reflect on physical reality, either to enhance our appreciation of that world or to question the epistemological status of the physical world that we take for granted. In their paper "Virtual Art as a Socio-Technical Interface," Thierry Bardini and Michael Century analyze the major Bioapparatus and Art and Virtual Environments efforts the Banff Centre described earlier. They note that complex social and technical issues complicated the integration of the arts and virtual reality, such as different backgrounds and cultural assumptions about the place of technology. In a paper called "The Artistic Origins of Virtual Reality," Myron Krueger notes that historically many of the key concepts of VR research—such as full-body participation, the idea of shared telecommunications space, multi-sensory feedback, third-person participation, unencumbered approaches, and the data glove—came from the arts, not the technical community. The future contribution of the arts is still to be negotiated.

Notes

1.   See the University of Washington's virtual reality lab on-line Web site for bibliographies and links: ⟨http://www.hitl.washington.edu/projects/knowledge_base/VRArt/⟩.

2.   Interval Research, "*Placeholder* Description," ⟨http://www.interval.com/research/NewMed/index.html⟩.

3.   Ibid.

Thierry Bardini and Michael Century: ⟨http://www.total.net/~centodd/Papers/Cyberconf3.html⟩

4.   B. Laurel, "Ethos of Computing," ⟨http://www.tauzero.com/Brenda_Laurel/Severed_Heads/Ethos_of_Computing.html⟩.

5.   M. Naimark, "Presence and Absence in the Age of Cyberspace," ⟨http://www.interval.com/frameset.cgi?papers/1997-091/index.html⟩.

6.   "Immersence" Web site, "Char Davies Bio," ⟨http://www.immersence.com/immersence_home.htm⟩.

7.   C. Davies, "Changing Space: VR as an Arena of Being," ⟨http://www.immersence.com/immersence_home.htm⟩.

8.   C. Davies, "*Éphémère*—Documentation," ⟨http://www.immersence.com/immersence_home.htm⟩.

9.   H. Iwata, "*Garnet Vision,*" ⟨http://siggraph.org/s97/conference/garden/garnet.html⟩.

10.   M. Benayoun, "*World Skin* Documentation," ⟨http://www.benayoun.com⟩.

11.   S. Brown, "*Apparitions* Description," ⟨http://www-apparitions.ucsd.edu/project.html⟩.

12.   J. Shaw, "Doors of Perception Essay," ⟨http://mmol.mediamatic.nl/Doors/Doors2/Shaw/Shaw-Doors2-E.html⟩.

13.   E. Shanken, "Commentary on *Golden Calf,*" ⟨http://mitpress.mit.edu/e-journals/Leonardo/reviews/shankencalf2.html⟩.

14.   T. Itoh, "Reflection on Jeffrey Shaw Work," ⟨http://www.ntticc.or.jp/permanent/cave/introduction_e.html⟩.

15.   Ars Electronica, "*Global Interiors* Documentation," ⟨http://193.170.192.5/prix/1996/E96gnI-global.html⟩.

16.   Dutch Electronic Arts Festival, "Documentation of *Global Interiors,*" ⟨http://www.v2.nl/DEAF/96/nodes/FujihataM/project1.html⟩.

17.   Ars Electronica, "*Global Interiors* Documentation," ⟨http://193.170.192.5/prix/1996/E96gnI-global.html⟩.

18.   A. Hegedüs, "*Handsight* Documentation," ⟨http://www.c3.hu/butterfly/Hegedus/⟩.

19.   Ars Electronica, "Award Statement—*Shapes of Things Past,*" ⟨http://www.aec.at/prix/jury/E97inter.html⟩.

20.   B. Britton, "Description of Lascaux Project," ⟨http://www.daap.uc.edu/soa/benb/⟩.

21.   J. Lanier, "Frontiers between Us," ⟨http://www.advanced.org/~jaron/cacm50.html⟩.

22.   M. Reaney, "Virtual Reality on Stage," ⟨http://www.ukans.edu/~mreaney/⟩.

23.   M. Reaney, "Virtual Scenography" and "Virtual Reality on Stage," ⟨http://www.ukans.edu/~mreaney/⟩.

Motion, Gesture, Touch, Gaze, Manipulation, and Activated Objects

Introduction: Reworking the Interface

The computer is not a neutral object. These devices come with conceptual baggage derived from their historical origins in military and commercial enterprise. As other chapters have described, the computer screen and its image conventions derive from a long history of representation in Western culture stretching from painting through perspective, photography, and cinema, to computer animation and desktop metaphors. Similarly, the conventional physical computer interface of keyboard and mouse come with significant cultural baggage.

Their constraints have limited the imagination in thinking of ways that digital information systems can be integrated into human life. The fact that the mouse and keyboard require a person to be tethered to an appliance (often sitting on a desk) fit well with Western patterns of organizational discipline and regimentation. The physical actions used represent a minuscule proportion of the potential physical repertoire of human-body actions. Some analysts suggest that this physical limitation can be understood as part of larger cultural patterns of body fear and hatred. The inherent isolation of one person to each interface also reinforces ideas of alienation.

Researchers and artists have started to wonder how the interface between digital systems and persons could be extended more widely into human life. Reaching beyond keyboards and mice, how could they read human actions such as motion, gesture, touch, gaze, speech, and interactions with physical objects? Wearable computing may one day convert body action into information function. Chapter 7.1 describes the wide-ranging investigations under way in the research world.

Immersive virtual reality (considered in chapter 7.3) represents a special case of this kind of research. Customized instrumentation such as data gloves, data suits, head-tracking helmets, and wands allow the systems to read a wide range of actions. Specialized helmets or surround room-sized displays respond with images and sounds that correspond with the movements.

Other systems, however, try to read motion without relying on all the custom body instrumentation, for example, using video image processing to track motion or gesture in free space. Working on the frontiers of this research, artists have created installations that respond to a great variety of visitor actions and probe the implications of this capability. Some assert that these body-oriented interfaces escape some of the cultural baggage of conventional arrangements and offer possibilities of moving digital art in much wider and deeper realms of life. Others, however, assert that even these body-sensing technologies are not benign and that they carry their own histories—for example, in military research—and that they can easily be perverted and assimilated into narratives of control and surveillance.

This chapter surveys artists who build installations that integrate body sensing and people's interactions with digitally activated objects. The artists investigate the implications of sensing motion, gesture, gaze, touch, and complex actions such as walking. Speech interfaces are discussed more in chapter 7.5.

Motion

David Rokeby

David Rokeby was one of the pioneers of motion-sensing art. He is well-known for his installations that sense human motion using video image processing. Some of these events invite participants to explicitly move their bodies in order to "play" the system; others monitor people's motions more unobtrusively. *Very Nervous System,* first created in 1986, is a classic in the field. Since that time, Rokeby has created a rich diversity of installations that explore a variety of cultural, conceptual, and personal themes. He often focuses on the limitations of conventional interfaces and interactivity and investigates noncontrol-oriented interactivity.

Very Nervous System is a series of interactive sound installations that use video image processing to detect visitors' motions and generate synthesized sound in response. The mapping of motion to sound is complex and dynamic. Rokeby wants to move against the limitations and the predictable definiteness of standard interfaces. He describes how he seeks to create transforming experiences that lead visitors to new insights into their motion in normal space, and suggests that the diffuse "resonant" nature of interaction can create an almost "shamanistic" experience:

I created the work for many reasons, but perhaps the most pervasive reason was a simple impulse towards contrariness. The computer as a medium is strongly biased. And so my impulse while using the computer was to work solidly against these biases. Because the computer is purely logical, the language of interaction should strive to be intuitive. Because the computer removes you from your body, the body should be strongly engaged. Because the computer's activity takes place on the tiny playing fields of integrated circuits, the encounter with the computer should take place in human-scaled physical space. Because the computer is objective and disinterested, the experience should be intimate.

David Rokeby: ⟨http://www.interlog.com/~drokeby/⟩

Fig. 7.4.1. David Rokeby, *Very Nervous System.* The complex relationship between visitor action and response creates a musical performance.

The active ingredient of the work is its interface. The interface is unusual because it is invisible and very diffuse, occupying a large volume of space, whereas most interfaces are focused and definite. Though diffuse, the interface is vital and strongly textured through time and space. The interface becomes a zone of experience, of multidimensional encounter. The language of encounter is initially unclear, but evolves as one explores and experiences.[1]

Rokeby tries to make a distinction that interactive art is not necessarily "reactive." He tries to create complex interdependent relationships between his systems and visitors' actions, and works against the tendency of viewers to search for a deterministic, totally predictable feedback loop. In a lecture called "Info Art," presented at the Kwangju Biennale, he explains:

It is important to understand that *Very Nervous System* is not a control system. It is an interactive system, by which I mean that neither partner in the system (installation and person) is in control. "Interactive" and "reactive" are not the same thing. The changing states of the installation are a result of the collaboration of these two elements. The work only exists in this state of mutual influence. This relationship is broken when the interactor attempts to take control and the results are unsatisfying.

Rokeby also explores the assumptions that guide human perception. *Silicon Remembers Sand* creates an activated rectangle of sand. A computer monitors people's movements in the sand and changes the surrounding sounds and the image (which is often water) that is projected on the sand. Visitors can choose to maintain the illusion by standing outside, or interact with it by stepping into the sand. Rokeby notes that the installation is "some sort of fake reflecting pool, an inversion of Narcissus's experience," in which visitors mistakenly identify with the shadows of others that shape emerging events.

In *Watch,* one of his few noninteractive works, Rokeby used the image-processing capability of his system to create displays that visualized time. He positioned a camera pointed at a public space and projected the processed images into a gallery. In one version only moving things were clear; the rest was blurred. In another version, only still images were clear.

In the essay "Transforming Mirrors," Rokeby explains that the nature of interfaces are themselves issues calling out for artistic exploration, and that artists can help our culture avoid unexamined incorporation. He proposes that interfaces are self-reflecting "mirrors" through which we communicate with ourselves:

The expressive power of the interface, in conjunction with the increasing "apparent" transparency of interface technologies, raises complicated ethical issues regarding subjectivity and control. Interactive artists are in a position to take the lead in generating a discussion of these concerns, but, on the other hand, are also in danger of becoming apologists for industrial, corporate, and institutional uses of these technologies. An awareness of the contradictions inherent in mediated interactivity is essential if we, as a society, are to move into the future with our eyes open.[2]

Rokeby observes that there is a danger in information technology's drive to exclude information that is not neat, definable, and logical. His art attempts to explore ways of celebrating ambiguity.

Paul Garrin

Paul Garrin produced a series of installations that tightly link the computer's ability to read motion with commentary on a variety of currently topical cultural themes. *Border*

Fig. 7.4.2. Paul Garrin and David Rokeby, *Border Patrol*. Robotic snipercams lock onto moving viewers.

Patrol (realized in collaboration with David Rokeby) focuses on issues of military surveillance and the marketing hype that portrays new technologies as decentralizing power. As audience members walk in front of the installation, which resembles a border guard encampment made of sandbags, position-sensing video cameras track their motion, much as a sniper would in a gunsight.

In Garrin's *White Devil*, a twelve-monitor, video-simulated, vicious, snarling, scratching guard dog tracks the visitor's movements along a fence. On the other side is a burning limousine that can only be helplessly observed. The visitor has to ineffectively watch as the dog threatens and the villa burns. A related installation, *Yuppie Ghetto with Watchdog*, is discussed in chapter 7.7:

The fire is a symbol for social conflicts within the world. The visitor becomes the victim and the accused at the same time because he or she is part of a society, the excesses of which one has to watch without being able to intervene. The establishment focuses on material values, barricades itself from any aid, and leads itself ad absurdum.[3]

Monika Fleischmann

Monika Fleischmann is a research artist and head of computer art activities at GMD, the German National Research Center for Information Technology and a cofounder of Art+Com in Berlin, a research institute for computer-aided media research. She has created numerous award-winning works that explore innovative interfaces such as immersive virtual reality (see chapter 7.3), touch, balance, and motion.

Fleischmann believes that artists have an important role in bringing awareness of the body back into the development of digital information systems. She warns that many conventional interfaces "diminish human possibilities," and aims to make computers into "imagination systems" by increasing "information across multi-sensory interaction channels":

My aim is to bring poetry and imagination to media art. As opposed to the theory of the disappearing body, I want to recover the senses of the body and to observe the dynamic gesture of different gender and culture in interactive media. If we don't support digital art and media culture, the quality of life will be lost through the dominance of machines.[4]

In *Rigid Waves* (in collaboration with Christian A. Bohn and Wolfgang Strauss, in 1993), the movement of the spectator changes the information that can be obtained. The closer one approaches what appears to be an impressionistic painting, the more realistic are the details that can be obtained. If one gets too close, the image shatters into pieces. Fleischmann sees one's coming closer as an attempt to "see oneself from the outside, to stand side-by-side with oneself and to discover other, hidden 'selves.' In this fractured mirror, we find ourselves shattered and splintered. Our 'self' has been liberated and has been broken down into multiple 'selves.' " When the viewer leaves, the image becomes static again.

In "Visualising Cultural Identity," Fleischmann asserts that interactive arts and extended interfaces are critical tools in ensuring that technological society does not get locked into highly restricted forms, and she links the opportunities of interactive environments with historical functions of literature and theater:

What is important is to push back the boundaries of perception and, wherever possible, to climb over these. The Greeks invented theatre to externalize the drama of life lived at the symbolical level. . . . An interactive installation involves the visitor passing through a process. He feels the

---

Monika Fleischmann: ⟨http://viswiz.gmd.de/VMSD/PAGES.en/mia/homepage.html⟩

Fig. 7.4.3. Monika Fleischmann, Christian A. Bohn, and Wolfgang Strauss, *Rigid Waves.* Movement of the spectator changes the information that can be obtained.

experience through his own body. Interactive theatrical illusion spaces are used for trying out new scenarios. Reality is treated "as if." In the virtual space, we practice for reality and live with a feeling of "as if." As if we are dreaming, as if we are flying, as if we are dying, falling, sliding, going into orbit, as if we are existing.[5]

Fleischmann warns that we must adjust the tools to the body, not the converse. She also suggests that the creation of these kinds of transformative media environments and "new cultural identities" requires the collaboration of many different kinds of thought— stretching from engineering through the arts.

Myron Krueger

Myron Krueger was one of the earliest pioneers in creating art that used position- and gesture-sensing technologies. In some of his installations he enabled visitors to interact

---

Myron Krueger: ⟨http://www.iamas.ac.jp/interaction/i97/artist_Krueger.html⟩

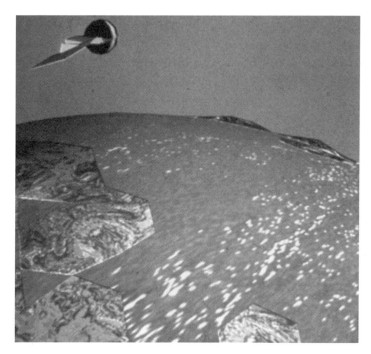

Fig. 7.4.4. Myron Krueger, *Small Planet*. Gestures of unfettered visitors allow them to fly over a landscape.

with video creatures superimposed over their projections, for example, bouncing one of his "critters" in one's hand. In one version, people in remote locations could tele-interact with each other in the shared projected space. Krueger has been a forceful advocate for artistic experimentation with emerging technologies. His book *Artificial Realities* pre-dated the virtual reality fad by many years and described his installations, which were far ahead of their time. In *Videoplace,* people could even fingerpaint on a screen projection of themselves by moving their hands.

Krueger suggests that these systems can be a stimulus to creative expression. The artist's contribution is in delineating a range of possibilities. His system provides a "med-ley" of capabilities, communicating the pleasure of aesthetic creation. A more recent piece called *Small World* allows the viewer to fly over a simulated landscape by a childlike spreading out of hands like a bird, and leaning for turns, diving, and ascending. Some-times viewers see caped figures flying next to them, which represent other people in networked installations in remote locations. The documentation describes the setup: "To make being on a sphere palpable, this environment shrinks the world to a scale that can be circumnavigated very quickly. Participants stand in front of a large projection screen depicting a realistic three-dimensional terrain. The projection screen is a portal into that world."[6]

Krueger has established a research company called Artificial Reality, which consults on extending the technology into a variety of fields and its continuing development. He asserts that the use of the body to control digital systems is significantly different than traditional symbolic manipulations: "The allure of the phenomenon of virtual reality makes doubly significant its importance as a new communications medium. It is a more personal interface to a machine because it engages your body directly rather than your mind symbolically. It's your actions that get things done."[7]

Rafael Lozano-Hemmer

Rafael Lozano-Hemmer elevates motion to an act of inquiry. His relational architecture installation called *Re:Positioning Fear* used visitor movements near the three-hundred-fifty-year-old Graz, Austria, Zeughaus arsenal to reveal layers of contemporary and historical information. The installation sought to explore "several historical transformations and displacements of Fear." Text solicited from Internet chat sessions focused on contemporary fears such as AIDS, terrorism, global warming, and genetic tampering, and was projected on the facade and broadcast via Web cameras.

Rather than just projecting the entire text, however, Lozano-Hemmer created a layered projection system jointly determined by the movements of local observers and distant Internet contributors. The only text that could be read was that marked by the shadow of the local visitors. Wireless position sensors tracked the users movements and turned on robotic lights such that the user's shadow created a "dynamic stencil" in which projected revealed text could be read. The arsenal was chosen because it was a symbol of the threat from historical Turkish expansionism:

Tele-absence is defined as the technological acknowledgement of the impossibility of self-transmission. Tele-absence is a celebration of where and when the body is not. The shadow is not an avatar, an agent, nor an alias of the participant's body; it is projected darkness, a play of geometries, a disembodied body part. . . . A shadow interface can be interpreted to be a metaphor of the obliqueness of ancient (and contemporary?) threat.[8]

Rafael Lozano-Hemmer: ⟨http://www.telefonica.es/fat/elozano.html⟩

Fig. 7.4.5. Rafael Lozano-Hemmer, *Re:Positioning Fear*. Motion-tracking lights and data projection cause images to appear in the shadows.

Other installations have similarly invited visitors to explore integrated physical and virtual spaces by motion and gesture. His *Trace* installation is discussed in chapter 6.3, on telepresence. In the *Able Skin* installation (in collaboration with Will Bauer and Emilio Lopez-Galiacho), visitors, armed with 3-D motion trackers, could wave their arms to control robotic lighting and projected points of view while exploring a virtual Renaissance building, the Palladios Villa Rotonda. In *Displaced Emperors,* two geographical locations were conflated. Visitors used their bodies to explore a castle in

Linz, Austria; however, the imagery revealed was displaced imagery from a Mexican palace.

Don Ritter

Don Ritter, who has also consulted for research companies such as Nortel, is an artist with a long record of experimentation with new technologies in digital video, telecommunications, and audio. Interaction often plays a central role in his installations, endowing the visitors' choices about action with symbolic power. Ritter considers the sociality and physicality of interaction a critical and neglected element. In his paper "My Finger's Getting Tired: Unencumbered Interactive Installations for the Entire Body," he notes that the typical encumbrances of mouse and data glove interrupt the aesthetic experience:

Although the intellectual experience of screen-based interactive art may be satisfying, the physical experience of sitting in a chair, clicking a mouse, and entering keystrokes is not satisfying to the physical body. If interactive art is going to become an influential and cultural medium, the entire body—and not just the index finger—must be involved in the interactive and aesthetic experience. . . . Related to the aesthetic experience of interactive art is the social situation surrounding the experience.[9]

His installations typically feature this unencumbered social interaction. In *Crowd Control,* projected video crowds beckon a visitor to move to a podium to speak and react to the motions the visitor makes, providing "anyone with an opportuntiy to control the masses." If the visitor moves to the podium, the crowds begin to get excited and increase in their enthusiasm the longer and louder the visitor speaks. Ritter notes that he provides only "an ironic illusion of power."

Skies projects images of nature on the wall and floor. The visitor willing to walk on the images discovers secret paths that can activate special images and sounds. *TV Guides* ties the video on a TV screen to user motion, with the programming stopping when users move and starting again when they stop. *Fit* features a projected video of an aerobics instructor who moves when the viewer moves. *Intersections* alters the sounds of approaching cars in a virtual street, based on user motion. As users try to cross a virtual four-lane road in a dark room, 3-D sounds create the screeching of cars to a stop and their acceleration, synchronized with their position. Ritter seeks to confront visitors with

---

Don Ritter: ⟨http://www.users.interport.net/~ritter/⟩

Fig. 7.4.6. Don Ritter, *Intersection*. Visitors in a totally dark room try to navigate a 3-D sound space that simulates streets crowded with traffic and cars that screech, stop, and crash, based on their position.

an unknown and uncertain situation and notes that some visitors have been so frightened that they were unable to cross through the installation.

Other Artists and Projects

Boundaries between People *Scott Sona Snibbe* is a researcher and artist formerly at Interval Research. His installations explore motion and gesture interfaces. *Boundary Function* senses the position of people in space and then projects lines on the floor that divide the space between them. The system automatically adjusts its boundary lines as people move and come and go: "The intangible notion of personal space and the line that always exists between you and another becomes concrete." Snibbe illustrates the new mix of art and research, incorporating insights from anthropology, biology, and mathematics:

The regions which surround each person are mathematically referred to as Voronoi diagrams or Dirichlet tessellations. These diagrams are widely used in diverse fields, spontaneously occurring at all scales of nature. . . . In anthropology and geography they are used to describe patterns of human settlement; in biology, the patterns of animal dominance and plant competition; in chemistry, the packing of atoms into crystalline structures.

Scott Sona Snibbe: ⟨http://www.snibbe.com/scott/bf/index.htm⟩

Akitsugu Maebayashi creates instruments and installations in which human motion and action are essential instigators. For example, in the *Audible Distance* installation, visitors move in a dark room with head-mounted displays and heartbeat sensors. One can sense others in the room only by amplified heartbeats and graphic shapes modified in correspondence with distance to others. Maebayashi notes that digital systems allow him to work with subtleties of social interactions that have not been previously available, such as the relationship of psychological and physical distance and the "boundaries" between people. The *Cause and Effect* installation assembled devices that confounded our usual sense of relationships with everyday objects, for example, a mirror whose transparency changed with visitor distance and a door lock or a ball-pitching machine controlled by other people's motion.

Daniel Schwartz and collaborators *Peter Franck, Richard Hughes, and Eugene James Flotteron* constructed *D-rhum room* to also explore boundaries between people. The room's walls actively respond to the movements of visitors with stretching, pushing, banging, and movement. It encourages visitors to think about the fuzziness of boundaries, the "biases of learned language," and the limited expectations of architectural space.

Revelations by Movement I In *Tracing*, *George Legrady* tracks visitor motion about a two-sided projection screen as a metaphor for two relationships to technology—immersion and alienation. Texts, images, and sounds respond to the motion. He notes that "the wall dividing the installation space functions as a two-sided mirror that aims to reflect on the viewers' relation to technologically processed information—a contested arena of cultural immersion or exclusion."

Miroslaw Rogala has created a variety of installations that make use of motion tracking to create metaphorical experiences. For example, *Lovers Leap* created a model of the mental space of power and control by asking the viewer to enter a space where they influenced but did not control the evolving projected image. The complex relationship between motion and consequence invites users to think about the limits of control. *Garden: NatuRealization* was situated in "Bughouse Square" in Washington Square Park, a historically significant location in Chicago where orators of all persuasions would try

Akitsugu Maebayashi: ⟨http://www.tama.or.jp/~mae884⟩; ⟨http://p3.org/metatokyo/light_page_e/mae/mae-intro.html⟩
Audible Distance: ⟨http://www.ntticc.or.jp/permanent/12maebayashi/maebayashi_e.htm⟩
Cause and Effect: ⟨http://p3.org/metatokyo/light_page_e/index.html⟩
Daniel Schwartz, et al., *D-rhum room:* ⟨http://www.romeblack.com/digital_funhouse/drhum/content.html⟩
Miroslaw Rogala: ⟨http://www.mcs.net/~rogala/⟩

to sway crowds. Rogala's sound installation allowed visitors to move to various locations to activate speeches contributed by previous physical and Internet participants. *Divided We Stand* metaphorically uses the motion of visitors to represent the vectors of drawing together and drawing apart that mark contemporary society. The amalgamation of various visitors' motions results in a changing sound, image, and text environment. It calls forth each participant to think about the limits of personal and collective actions.

Revelations by Movement II   Jim Campbell (see chapter 7.2) creates motion-sensing installations such as *Untitled (for Heisenberg)*. This installation comments on the "uncertainty principle" from physics, which states that the more accurately we try to measure a phenomenon, the more it will be affected by the observation. In this installation, viewers see video of an intimate couple projected onto salt. As one comes closer, the image zooms in and becomes increasingly abstract and indeterminate.

Nancy Paterson created an installation called *The Meadow,* which places the visitor in the midst of a simulated meadow. The meadow is created by videodisk-fed monitors placed on each side. The view changes based on ultrasonic detection of user motion, for example, changing seasons or the appearance of new positioned events, such as a flock of birds or the sounds of children. She tried to create an attractive space of "ambiguity and irony" in which people could travel across the seasons through their body motions. In *Waiting Room,* **Simon Biggs** presents a dance hall with projected figures on two walls dancing with each other and the viewer, and in *Halo,* four projected figures fly about until they are attracted to approaching viewers. In ***Martin Riches***'s *Interactive Field* thirty-six panels move like dancers in response to the movements of gallery visitors. In *Transgression* **Masayuki Towata and Yasuaki Matsumoto** have created an environment in which the cloud images on the ceiling are projected to the floor as visitors move about the gallery, thus extinguishing the boundary between the electronic space and the body. In **Studio Azzurro**'s *Landing Talk* visitors encounter doors in a dark space. The presence in front of a door appears to open the door to stories and sounds of loneliness and oppression.

Jim Campbell: ⟨http://www.artcenter.edu/exhibit/jim/campbell.html⟩
Nancy Paterson: ⟨http://www.bccc.com/nancy/nancy.html⟩
Simon Biggs: ⟨http://www.easynet.co.uk/simonbiggs/⟩
Martin Riches: ⟨http://www.ntticc.or.jp/special/biennale99/exhibition/martin_e.html⟩
Masayuki Towata and Yasuaki Matsumoto: ⟨http://www.ntticc.or.jp/permanent/tm/towata_matsumoto_e.html⟩
Studio Azzurro: ⟨http://www.ntticc.or.jp/special/biennale99/exhibition/studio_e.html⟩

Emotional Access by Movement   Henry See is known for installations that probe the information-organizing capabilities of computers. His *Memory* installation presented visitors with a system that had large stores of factual and poetic memories and a guide that started forgetting. In accordance with the strange twists of artistic fates, See went on to establish a technology research company called Merz, which is known for its innovative software for managing complex information. In seeking to create the illusion of intelligence, another related line of inquiry, See created an installation called *Regard,* which tied the behavior of a video character reading a book to the relational movement overtures made by the visitor, either respecting or transgressing the reader's private space.

I created a number of installations that read users' motions to orchestrate the events. *Excursions in Emotional Hyperspace* (see chapter 7.6) featured four artificial character mannequins who interacted with each other and the viewer. Visitors directed the event by their motion. *Father Why* similarly used movement to allow viewers to control a sound-and-speech event. It mapped the emotional responses to a father dying to four physical spaces (the places of Anger, Sadness, Longing, and Forgetfulness). Visitors could determine how deep to go in each place by their body motions and length of stay. *BodySurfing* examined the importance of the body in the cyber era by asking viewers to activate events by drumming, stretching, gesturing, touching, and running.

The Activation of Devices and Visual Worlds   Nola Farman and Anna Gibbs invited viewers into a room with a giant plastic heart whose beats changed with the changing proximity. Commenting on war machines, ***Michael and Anna Saup, Gideon May, and Stefan Karp*** created a motion-tracking installation in which organic, plasmatic, mirror-abstracted images of visitors became the projected display In *Plasma: Architexture,* a projection screen and pneumatic structure in the center of a room responded to the motions of viewers. The movements were mapped into a virtual world cumulated with the motions of previous viewers. Documentation noted that "after a while, the artificial being of the pneumatic sculpture will free itself from its reproductive state and start to act in its own interpolative manners." ***Ronald MacNeil and Bill Keays****'s MetaField Maze* uses vision tracking to allow user-body motion to manipulate a ball going through a room-sized maze projected on the floor. ***Tsai Wen-Ying,*** known for twenty-five years

---

Henry See: ⟨http://www.merzcom.com⟩
Regard: ⟨http://www.iamas.ac.jp/interaction/i97/artist_See.html⟩
Stephen Wilson, *Father Why:* ⟨http://userwww.sfsu.edu/~swilson/⟩
Michael Saup: ⟨http://www.salon-digital.de/particles/paradocs/bio/mick.html⟩
Bill Keays: ⟨http://www.media.mit.edu/~keays/research/index.html⟩

for kinetic arts, has built installations of fiberglass rods that respond to the movement and sounds of viewers.

Chris Dodge, a former researcher from MIT's Media Lab, helped develop the Interactive Mind Forest element of the *Brain Opera,* which allows over one hundred people to create image-and-sound events together. Another project called *Evolving Sandscapes* used a sandbox as the interface, storing a record of people's activities in the sand. His installation *What Will Remain of These?* uses video to track the traffic flow through an architectural space. Anytime someone stops in front of the installation it isolates the image of the person and adds it to a cumulating particle representation of people— "visualizing the dialectic between the individual, who makes an effort to view the work, and the masses that idly walk on by." Several members of **Sine::apsis** incorporate motion sensing as aesthetic elements of their work—for example, in *Engineered for Empathy* **Amy Youngs** offers commentary on future plant genetic engineering having created a tobacco plant that changes its glow in accordance with viewer motion and **Daniel Wayne Miller**'s *Breathing Room* has a mobile light-projecting robot respond to visitor motions.

Dance In *Binary Ballistic Ballet,* **Michael Saup and the Supreme Particles Group** collaborated with the choreographer William Forsythe and an ensemble of the Frankfurt Ballet to create an integrated dance, sound, and information display system. In this system sounds, movement, and projected moving-word displays mutually influenced each other. A later work called *Global Hockets* similarly explored the integration of these elements. **John D. Mitchell and Robb E. Lovell** created the *Intelligent Stage* for Arizona State University's Institute for the Study of the Arts. The activated stage uses video processing technology to detect performer motions and initiate actions in response. It has been used for dance, theater, and performance art. **Isabelle Chinoiniere** performs dances in which her movements and gestures control sounds and images. For example, *Communion* used digital sensors as stage assistants in realizing the dance. **Georgia Technology Institute**'s *Dance Technology Project* supports experimental work in dance including a project where motion tracked tossed balls are transformed in the projected

---

Chris Dodge: ⟨http://liquid-sky.media.mit.edu/cdodge/⟩
What Will Remain of These?: ⟨http://www.firewater-productions.com⟩
Sine::apsis, Amy Youngs, Daniel Wayne Miller: ⟨http://www.sine.org/⟩
John D. Mitchell and Robb E. Lovell: ⟨http://www.asu.edu/cfa/art/people/faculty/Collinswriting.html⟩
Isabelle Chinoiniere: ⟨http://mitpress.mit.edu/e-journals/Leonardo/reviews/bureaudisea.html⟩
Georgia Technology Institute, Dance Technology Project: ⟨http://www.oip.gatech.edu/IMTC/html/dance_tech.html⟩

background image into other objects and where dancers can dance with artificial 3-D computer graphic characters. Using digital video scene analysis, the German dance group **Palindrome** invites the audience to influence the sounds used by dancers by gesturing and making shapes with their bodies.

Music   Christian Möller, who has an ongoing relationship with Canon's ArtLab, creates installations that explore new architectural spaces activated by people's motion. For a time he headed the Archemedia research institute, which created a variety of interactive urban architecture applications. Produced in conjunction with V2 for a Rotterdam summer festival, *Audio Park—The Party Effect* enabled viewers to change the lighting and 3-D spatial sound in a large public square by moving in and out of specially lit areas that tracked motion (see chapter 7.5). *Electro Clips* (completed in collaboration with Stephen Galloway and Louis-Philippe Demers) was an installation for ballet that activated a stage so that the dancers' motions could control lighting and sound. **York der Knoefel**'s *Gene Ration Time Factor* activated music from different world musicians as visitors moved in front of their images. **Doris Vila**'s installations, such as *Book of Air* and *Spatial Rights Modulator,* use viewer motion and gesture, such as waving flags, to affect video and sound events. **ICC**'s *Sound as Media* show features experimental sound artists, many of whose works sense the motion and proximity of visitors, such as **Marc Behren**'s *Tokyo Circle* in which sensors monitor visitors' movements through the gallery, which generates the sound's spatial pattern.

Gesture

Pamela Z

Pamela Z creates experimental music through a combination of conceptual, low-tech and high-tech means. She combines found texts, such as lists of names from the yellow pages, and found sounds, such as percussions from banging on household items, with digitally manipulated sounds and her operatic voice to create works that are simultaneously beautiful and provocative. The gestural activation of digital sounds is also an

Palindrome: ⟨http://www.palindrome.de⟩
Christian Möller: ⟨http://www.canon.co.jp/cast/artlab/pros2/pers-01.html⟩
Doris Vila: ⟨http://www.interport.net/~outpost/34⟩
ICC *Sound as Media* show: ⟨http://www.ntticc.or.jp/special/sound_art/index_e.html⟩
Pamela Z: ⟨http://www.sirius.com/~pamelaz/⟩

Fig. 7.4.7. Composer and performer Pamela Z wearing the *BodySynth*™ device, which she uses to create conceptual sound performance activated by touch and body gesture. Photo: Lori Eanes.

important part of her repertoire. Using a device called the BodySynth, she places sensors on her body so that she can control a MIDI interface with her gestures. Every part of her body becomes an instrument. Commentators noted that she is the "grandmaster of feedback," integrating voice, everyday sounds, and electronically processed elements with conceptual layers that create "a bewildering, mesmerizing, highly organized cacophony that taps deeply into the infosaturated, data overloaded, hyperventilated psyches of urban life."

Pamela Z belongs to a group of musicians called SensorChip, who explore the possibilities of gesture in musical performance. She often uses the BodySynth device, invented by Ed Severinghaus. She also participated in the Xerox PARC PAIR (artist-in-residence) program, in which she worked with researchers interested in motion studies and the anthropology of work.

Fig. 7.4.8. Christa Sommerer and Laurent Mignonneau, *Intro Act* (copyright 1996). Visitors use gestures to interact with artificial life forms. (Supported by CNAP France, collection of Musée d'Art Contemporain, Lyon.)

Christa Sommerer and Laurent Mignonneau

Christa Sommerer and Laurent Mignonneau (see chapter 4.3) have created several installations that integrate gesture tracking. In *Trans Plant,* the visitors' movements and body type affect the kind of artificial creatures that they can create: "*Trans Plant* [is] an interactive environment that allows visitors to enter a virtual space, where they can see themselves creating a virtual garden with each movement of their body. Body gestures, frequency of movement, and body size directly affect the growth of the virtual plants."[10]

Intro Act continues this exploration, in which the visitor uses gesture to interact with the artificial life forms. The Web documentation explains: "As a visitor moves her body freely, her gestures trigger organic shapes to form in the virtual mirror. These abstract, nonlinear, multilayered forms relate to the visitor's body, her position, size, speed of movement, and actions in general." In *Haze Express,* the visitor is in a virtual train, and gestures at the windows produce a variety of scenes.

Christa Sommerer and Laurent Mignonneau: ⟨http://ms84.mic.atr.co.jp/~christa/WORKS⟩

Other Artists and Projects

Dance   **Benoit Maubrey and the Audio Ballerinas** wear solar-powered sound clothing that responds to their movements and gestures. **Wayne Siegel** used the DIEM Digital Dance system to compose a dance called *Movement Study I,* in which the dancer controlled the music. Sensors placed on the dancer's wrists, elbows, knees, and ankles read the bending of the joints and sent data to a computer via a wireless transmitter. Many performers and theorists in dance report on their research on gesture recognition at the Web site *Dance Technology Zone.*

Face and Body I   In *Cyber Bunraku,* **Ken-ichi Anjyo, Yasuhiro Higashide, and Hiroshi Sakamoto** created a system in which the combined facial expression of a performer and control gestures by a puppeteer control computer graphic characters. **Sally Jane Norman** attempts to bridge the world of puppetry with new technologies through activities such as the International Institute of Puppetry and helps in organizing the Touch conference in conjunction with the Dutch artist research group Steim. **Loren Carpenter**'s *Kinetic Evolution* allowed large audiences to control computer graphic animations by holding up colored wands that were interpreted by a video digitizer. A Carnegie Mellon project, *Journey into the Brain,* later used the same technology. **Sidney Fels**'s *Glove-Talk* converts hand gestures into speech. **Atau Tanaka, Zbig Karkowsky, and Edwin van der Heide** created the Virtual Sonic Band, in which they play virtual instruments through hand and body gestures. **Wolfgang Krueger and Wolfgang Strauss,** at the German National Research Center for Information Technology, GMD, developed the *Responsive Work Bench,* a graphic-enabled surface that tracks gesture and eye movement. **Daniel Rozin**'s *VideoPaint Easel* invites users to compose images with hand gestures on the surface. **Emily Weil**'s *Screen Play* lets viewers compose portraits by body gesture. **Andrea Polli**'s *Gape* and *Inside the Mask* create complex musical instruments activated by eye motion. **Lucia Grossberger-Morales**'s events use the computer as a performance instrument. **Barbara Lee** organized the Digital Art Focus Wearable Com-

---

Audio Ballerinas: ⟨http://www.snafu.de/~maubrey/⟩
Wayne Siegel: ⟨http://www.daimi.aau.dk/~wsiegel/Movement_Study_I.html⟩
Dance Technology Zone: ⟨http://www.art.net/~dtz/⟩
Cyber Bunraku: ⟨http://siggraph.org/s97/conference/garden/bunraku.htm⟩
International Institute of Puppetry: ⟨http://www.ardennes.com/asso/iim⟩
Loren Carpenter: ⟨http://193.170.192.5/prix/1994/E94auszI-kinoetic.html⟩
Responsive Work Bench: ⟨http://viswiz.gmd.de/VMSD/PAGES.en/mia/homepage.htm⟩
Andrea Polli: ⟨http://homepage.interaccess.com/~apolli/⟩
Barbara Lee, Digital Art Focus: ⟨http://www.ylem.org/NewSite/projects/DAF.html⟩

puting Fashion Show and Performance, in which the actions of performers and the audience combine to shape ongoing events. **Simon Penny**'s *Fugitive* attempts to interpret gross body motion as "mood."

Face and Body II **Haruo Ishii**'s *Hyperscratch 9.0* offers visitors an unprecedented experience of using an invisible 3-D interface through which they can move their hands in open space, thus controlling an instrument made of light and bell strikers. **Steven Schkolne**'s *Surface Drawing* invites people to draw and erase by moving hands in free space. **Laetitia Sonami** uses her activated *Lady's Glove* to perform music via hand and arm movements. **Sensorband** uses a variety of technologies to use hand and arm motions and brain waves to control instruments in their performances. **Jakub Segen**'s *Visual Conductor* allows control of sound events via hand and baton movements. In conjunction with Interval Research, **Laurie Anderson** has developed an innovative instrument called the *Talking Stick* which generates a wide range of sounds and is motion sensitive. **Donald Swearingen**'s *Noise into Water* allows hand movement above a table fitted with sensors to control sound events. **Jean-Marie Dallet**'s *Ile de Batz* reads visitors' gestures of pointing at a map to call out relevant digital images and video sequences. **Eric Paulos** (see chapter 6.3) oversees a research project called *Experimental Interaction Unit* (EIU), which looks into the physical, aural, visual, and gestural interactions between humans and machines. The research moves against the dominant interface conventions and seeks to develop tele-embodiment devices. The **Sponge Group** creates events based on wearable sensors and gesture detection.

Touch and Tactility

Monika Fleischmann

Building on the myth of Narcissus, Fleischmann's *Liquid Views* installation (in collaboration with Christian Bohn and Wolfgang Strauss) provides a commentary on human desires to understand and possess oneself, and the power of mirrors and water as a

---

Simon Penny: ⟨http://www-art.cfa.cmu.edu/www-penny/works/fugitive/fugitive.html⟩
Haruo Ishii: ⟨http://www.land-net.com/stone/⟩
Steven Schkolne: ⟨http://www.cs.caltech.edu/~ss/⟩
Laetitia Sonami: ⟨http://www.otherminds.org/Sonami.html⟩
Sensorband: ⟨http://www.sensorband.com/root.html⟩
Laurie Anderson: ⟨http://www.laurieanderson.com/notes.html⟩
Jean-Marie Dallet: ⟨http://www.ntticc.or.jp/special/biennale99/exhibition/jean-marie_e.html⟩
Eric Paulos: ⟨http://www.eiu.org/⟩
Sponge Group: ⟨http://sponge.org⟩

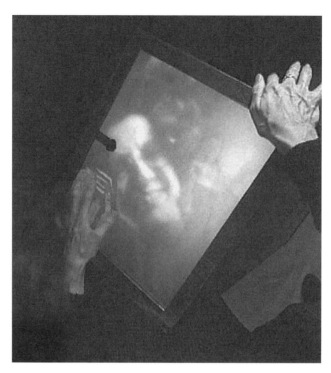

Fig. 7.4.9. Monika Fleischmann and Wolfgang Strauss, with C.-A. Bohn. *Liquid Views*—a horizontal projection screen filled with simulated water dissolves with finger stroking.

metaphor for this quest. *Liquid Views* adds touch to the metaphor and reflects on technology as an aid and hindrance to the quest to know ourselves. Viewers see a horizontal projection screen filled with projected water that invites finger stroking. The more the viewers stroke, the more ripples cause their liquid view to dissolve. After a period of inactivity, the water calms and again becomes a liquid mirror. Fleischmann's Web site explains the importance of the Narcissus myth:

Narcissus drowned in himself. . . . The access to the self remains closed. The central theme is the transition from the upper to the lower world, the transition from the rational world to the spheres of unconsciousness and vice versa. . . . The Narcissus of the media age is watching the world through a liquid mirror that questions our normal perception. A glass mirror has no inner life retaining our image. The digital image, however, can be stored, manipulated, and altered within the computer. In *Liquid Views,* the mirror becomes the actor. The transformed, hallucinatory image originates on the other side of the mirror, which normally is not accessible to us.[11]

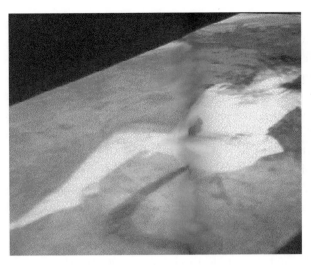

Fig. 7.4.10. Thecla Schiphorst, *Bodymaps*. A visitor touching an image projected on velvet changes the image and sounds. Photo credit: Kim Clarke.

Thecla Schiphorst

Thecla Schiphorst is a computer-media artist, choreographer, and dancer. Attempting to keep alive the sensual and erotic in contemporary culture, she creates works that link information technologies with motion and touch. Her *BodyMaps: Artifacts of Touch* installation invites visitors to bring their hands near and touch the image of bodies projected on velvet. As the visitor places the hands closer, the image shudders and sounds change. The Interactions97 show describes the piece:

This work invites relationships through an experience grounded in proprioceptive knowledge, skin sense feeling, listening through touch, seeing through hearing, together integrated through attention. . . . The surface yearns for contact and touch. Its rule base is complex and subtle, impossible to decode; its effect is disturbing, erotic, sensual, and subjective. . . .

The intention of the work is to subvert the visual/objective relationship between the object and the eye, between click and drag, between analysis and power, to create a relationship between participant and technology that transgresses rules of ownership and objectivity, and begs questions of experience, power, and being.[12]

Thecla Schiphorst: ⟨http://www.digearth.bcit.bc.ca/dedocs/thecla/a.htm⟩

Schiphorst, who has worked with the Merce Cunningham dance troupe, also is involved with other dance and technology research. She was one of the principles who created the Life Forms software for choreography. She also has been involved in the Electric Body Project, which uses gestural sampling and motion-capture technologies as a choreography tool. Its research investigates methods of layering dance representation similar to the way in which image- and sound-editing programs allow manipulation. The research also introduces the use of transformative maps to increase the expressive capabilities of the system.[13]

In "Body Noise: Subtexts of Computers and Dance," Schiphorst traces Western civilization's distrust of the body and the warped thinking that demotes embodiment to a subtext of technological progress. She suggests that body knowledge can be an important resource in the technological future:

In my own practice of dance, it is the language of embodiment which has provided the deepest technical knowledge and experience that I possess. . . . Could it be possible that . . . technological processes can be seen as subcategories of physical experience and consciousness, informed and transformed by kinesthesia, embodiment, and physical memory? Our cultural resistance to this approach uncovers a twofolded dilemma. On the one hand, as Jeanne Randolph so aptly suggests, "the problem now is that technology's perception of culture is becoming our only perception of culture. . . . As numerous writers have indicated, the ability of many people in Western society to experience their own bodily feelings and sensations is profoundly impaired." The valuation of the experience of the body is shuffled outside the realm of serious technical consideration, a position it has held for centuries. . . . [current thinking is] a brilliant example of the age-old *histeria* suggesting body as Other, body as Abject, body as Shadow, and now in our postindustrial age, body as Commodity. . . . Antithetical to this notion, I would like to suggest that the knowledge of the body could radicalize and perhaps even "come to the rescue" of current technological practice and implementation. What field has more counterknowledge and possibility of subverting and infiltrating the ideology of the Technologically Correct?[14]

Stahl Stenslie and Kirk Woolford

In the *Senseless* installation (see chapter 2.5), participants entered a virtual world populated by artificial characters and linked to remote participants via the network. They wore a body suit with sixteen touch pads that allowed real and virtual characters to

Stahl Stenslie and Kirk Woolford: ⟨http://televr.fou.telenor.no/stahl/⟩

Fig. 7.4.11. Stahl Stenslie and Kirk Woolford, *cyberSM*. Participants wearing stimulator suits can tactilely communicate with each other. © Stahl Stenslie.

tactilely stimulate each other. Stenslie and Woolford's *cyberSM* project further explores the addition of tactile communication to image and sound transmission. They devised a way for people to physically stimulate each other at a distance as a way of examining issues of gender, physicality, anonymity, and eroticism:

The *cyberSM* project allows the establishment of trans-gender appearances, identities, and entities by letting the participants choose their own visual appearance from a large data bank of digitized human bodies. Once chosen, the participants send the image of their virtual self to the others on the network. The body thus becomes a visual fantasy. Central to the *cyberSM* project is the ability to transmit physical stimuli from one participant to the other. This is made possible through the use of stimulator suits connected over (international) telephone lines, which allow the users to remotely stimulate one another's bodies. Not only does this physical element of communication allow the *cyberSM* project to more closely model interhuman communication, it creates a new form of interaction. Throughout the *cyberSM* connection, participants have a physical dialogue, but they remain anonymous the whole time. . . . The *cyberSM* connection allows humans interacting in a virtual space to actually feel each other with their bodies. It is a sensual communication link, challenging our concept of eroticism, adding a missing sense to electronic communication.[15]

Other Artists and Projects

I created an installation called *Demon Seed,* which invited spectators to control four moving robot arms through strokes, squeezes, hugs, punches, pinches, and the like, via a velvet-covered squeeze rod. The installation used near-body senses as a tool to explore the attempt to distance robot technologies. **Christian Möller** has produced a number of installations that explore the use of touch. *Audio Grove* presented visitors with a forest of fifty-six vertical 5.5-meter rods. Users can compose sound-and-light events by touching the rods, aided by software designed to guarantee a harmonious overall experience. *The Musical Score Lent Acoustic Form,* completed for the Figaro House Mozart Museum, allows visitors to use touch to access musical information by dragging their fingers across a score. The installation called *The Third Dimension of the Painting "Ritratto di Genti-lumo" by Tintoretto* allows viewers to use touch to discover the underlayers of this painting. A replica is projected on a touch-sensitive glass. A viewer moves a finger over sections, revealing underlayers that have been discovered by researchers using X-ray and infrared imagers. **Steim,** the Dutch techno-art research center, sponsored the Touch symposium, focused on "the renaissance of the physical presence of the performer in the electronic performance arts." **Grahame Weinbren**'s *Frames* gives visitors the chance to touch panels that transform the photographs of actors to resemble the 150-year-old photographs of mental patients. **Anthony Dunne and Fiona Raby**'s research series *Fields and Thresholds* explores new telecommunications possibilities. In one installation, body warmth is communicated by creating two linked benches in physically separate locations so that one bench warms up in the appropriate spot when someone sits on the corresponding spot on the other bench.

In addition to his GPS (Global Positioning System) work (see chapter 3.5), **Iain Mott** and his collaborators create sound installations that read visitor motion and touch. *Squeezebox* is a sound and projected-image installation that is activated by visitor touch and the physical manipulation of cast hands projecting out of the installation. The direction of pressure moves 3-D sound around the space and changes timbre and projected imagery. In the *Inter Dis-communication Machine* and *Inter-skin* installations, **Kazuhiko**

Stephen Wilson: ⟨http://userwww.sfsu.edu/~swilson/art/demon.html⟩
Christian Möller: ⟨http://www.canon.co.jp/cast/artlab/pros2/pers-01.html⟩
Steim, Touch symposium: ⟨http://www.xs4all.nl/~steim/main.html⟩
Grahame Weinbren: ⟨http://www.ntticc.or.jp/special/biennale99/exhibition/grahame_e.html⟩
Anthony Dunne and Fiona Raby: ⟨http://www-crd.rca.ac.uk/rcacrdresearch/adunne/⟩
Iain Mott: ⟨http://members.tripod.com/~soundart/⟩

Hachiya probes our ability to know another person's sensations and, by extension, point of view. Using various technologies, he lets one see and experience tactile sensations of what another person is experiencing. Who am I? will be irrelevant, to be replaced by What is all that I can be?

Teri Rueb's *Memory Is a Pea: A Memorial* builds on the Princess and the Pea story. In this story the queen mother tested the sensitivity of a potential princess candidate by seeing if she could detect a pea placed at the bottom of ten mattresses. Using pressure sensors, Rueb created an installation of a bed with this capability. As the participant lay down, dreamlike sounds were activated and a computer monitor scrolled a text of memories and dreams left by previous visitors.

Gaze

Lynn Hershman

Lynn Hershman explores a variety of technologies in her work (see also chapters 6.3 and 7.2). *Room of One's Own* tracks the direction of a viewer's gaze. Viewers look into a tiny interactive peep show of a bedroom with objects while a protagonist chides the viewer for his or her persistent voyeurism. The work reflexively uses this capability to lead the user to consider the history of the male gaze in Western history. In *America's Finest,* Hershman activates the culturally loaded objects of the rifle and rifle sight. She allows the user to "aim" the gun, which she notes is similar to the intrusive power of the gaze:

The associative notions of guns/camera/trigger links all media representation to lethal weapons. *America's Finest,* an interactive M16 rifle, addresses these issues. Action is directly instigated through the trigger itself, which, when pulled, places the viewer/participant within the gun site (this time their entire body holding the gun). They see themselves fade under horrible examples in which the M16 was used, and if they wait, ghosts of the cycling images dissolve into the present.[16]

Seiko Mikami

Seiko Mikami's *Molecular Informatics* installation creates a virtual world of molecules that are created and made visible by the moving gaze of visitors. Wearing special glasses,

Kazuhiko Hachiya: ⟨http://www.intelligentagent.com/dec_bodies.html#Inter⟩
Teri Rueb: ⟨http://fargo.itp.tsoa.nyu.edu/~rueb/paper.html⟩
Lynn Hershman: ⟨http://arakis.ucdavis.edu/hershman/intro.html⟩
Seiko Mikami: ⟨http://www.v2.nl/DEAF/96/nodes/MikamiS/project1.html⟩

Fig. 7.4.12. Lynn Hershman, *America's Finest*. Visitor aiming of the gun determines media events.

the computer senses the direction of the visitor's gaze and creates molecular structures wherever they look on a projection. Each new viewer's moleucle is added to the cumulative image. Mikami is interested in the unconscious aspects of gazing as well as consciously aimed sight. The work explores the philosophical debate about whether the world is constructed by human attention as opposed to discovered.

Other Artists and Projects

In exploring the nature of seeing, ***Kazuhiko Hachiya*** enables viewers, aided by technology, to see other people and their actions that would otherwise be invisible. For Hachiya,

Kazuhiko Hachiya: ⟨http://www.p3.org/p3-light/P3exhibitions.html⟩

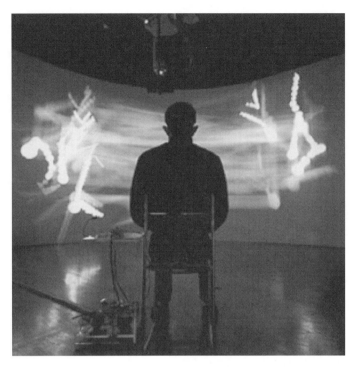

Fig. 7.4.13. Seiko Mikami. Molecular informatics-morphogenic substance via eye tracking. Viewer-gaze direction causes the development of artificial molecules. Premiere version 1.0. Canon ArtLab, Tokyo, 1996.

these new visions are a metaphor for human relationships and how we know. In *Vanishing Body,* people enter a totally dark room. If the viewer takes off some clothes, they are given special infrared goggles that will allow them to see the silhouettes of other people in the space. In his *Seeing Is Believing* exhibit, when one has the appropriate infrared sensing gear, what appears to be flickering lights is resolved into text from diaries from Internet contributors. **Dirk Luesebrink and Joachim Sauter**'s *Zerseher* (Disviewer) metaphorically explores the power of the gaze. As one looks at particular places in an activated picture, the place of focus dissolves under the gaze. Ars Electronica's Web site describes the installation: "The spectator realizes all of a sudden that wherever he looks, he is destroying the image with his eyes."

Dirk Luesebrink and Joachim Sauter: ⟨http://www.aec.at/prix/1992/E92auszI-zerseher.html⟩

Fig. 7.4.14. Trevor Darrell, Harlyn Baker, Gaile Gordon, John Woodfill, *Mass Hallucinations*. A system remembers faces and applies distortions.

Face Recognition

Trevor Darrell, Harlyn Baker, Gaile Gordon, and John Woodfill

This Interval Research group aims to enhance the computer's ability to recognize faces. As part of the research, the group has prepared a series of art installations that use recognition abilities to create engaging events that invite audiences to consider the significance of the face. At SIGGRAPH97 they presented *Magic Morphin Mirror: Face-Sensitive Distortion and Exaggeration,* which intelligently tracked and distorted faces of visitors.

From an early age, the image of one's face in a mirror evokes a quality of being connected and disconnected at the same time. One sees an "other," but knows it is one's self. This project explores the boundary between these qualities through a virtual mirror with face-specific image manipulation. . . . *Magic Morphin Mirror* creates a self-referential experience with an image that

---

Magic Morphin Mirror: 〈http://www.interval.com/frameset.cgi?papers/1998-005/index.html〉

is clearly neither really oneself nor entirely synthetic nor autonomous. . . . Faces are central to human communication and yet machines have been largely blind to their presence.[17]

The abstract to their paper "A Virtual Mirror Interface Using Real-Time Robust Face Tracking" explains some of the complexity of the task. It also illustrates the necessity for artists working at the frontiers of research to become knowledgeable about issues in their areas of interest:

Stereo processing is used to isolate the figure of a user from other objects and people in the background. Skin-hue classification identifies and tracks likely body parts within the foreground region. Face pattern detection discriminates and localizes the face within the tracked body parts.[18]

They presented the *Mass Hallucination* installation at SIGGRAPH98. A video face recognition system kept track of the people who had come to see it and distorted images as a function of the number of people watching it, their behaviors, and whether they've watched the device before. The authors noted that "it encourages crowds of people to collectively manipulate the display with their bodies or faces. Yet it is also personal, in that it can recognize the appearance of a user for short-to-medium periods of time and tailor the display accordingly."

Rob Myers, Peter Broadwell, Rebecca Fuson, and Delle Maxwell

These collaborators are part of the Plasm group, which has created a variety of interactive works that irreverently explore areas of emerging technology. *Plasm: Your Mug* plays with face-recognition techniques. The installation mounted at SIGGRAPH provided visitors with a simulated bar. Behind the bar were some very strange "mirrors" that distorted facial features in strange ways generated by artificial-life approaches and their own body movement. The SIGGRAPH documentation describes the event:

Gazing into the mirror behind the bar, visitors will notice their face's reflection adapting to their reaction to it. Their visage is being served up via short-order evolution, as fleets of genetic automata mutate onward, surviving by the nature of the visitor's engagement. Over the course of this evolution, the visage always tracks its participant tightly, maintaining the intimate kinship usually reserved for one's own shadow or reflection.

Plasm: Your Mug: ⟨http://www.plasm.com/peter/public_html/YerMug.html⟩

Real-time video tracking and feature extraction is used to corral a fleet of semi-autonomous geomorphs on the screen. Each geomorph presents its own 3-D rendered form, animated according to its own independent behavior. Corraling ensures that the fleet of geomorphs tracks the participant's face and motions tightly, to maintain the relationship of a reflection with the visitor. . . . The geomorphs themselves are models drawn from a stockpot of cultural idioms and reflections, encoded as a sequence of genetically mutable factors. The participant may shepherd this mutating construction with their body language, instrumented via their video-tracked envelope and force-sensing devices on the countertop and seating.[19]

Complex Actions — Balance, Walking, and Bicycling

Jeffrey Shaw

Jeffrey Shaw, who for many years was head of experimental activities at the Zentrum für Kunst und Technologie (ZKM), is known for his installations that explore virtual reality and the nature of the media space created by new technologies. His *Legible City* is recognized as a classic. It allows a user to navigate a virtual projected city constructed of giant buildings in the shapes of letters by bicycling on a stationary bicycle. One can follow narratives and other threads of text by riding down the streets of this strange city. Visitors are left to reflect on interrelationships between physical cities, literature, and computer-projected virtual realities as navigable structures. The physical activity of bicycling grounds the viewer within these otherwise abstract worlds. Ars Electronica's jury statement explains some of the layers of meaning in the work:

[T]he application of three-dimensional computer-imaging technologies in this context has a revolutionary meaning. Instead of the traditional activity of art as a representation of reality, the artwork can now become itself a simulation of reality within which the viewer's point of view is located. . . . Travelling through this city of words is consequently a journey of reading. Choosing direction, choosing where to turn, is a choice of the story lines and the user's position. In this way this city of words is a kind of three-dimensional book which can be read in any direction, and where the spectators construct their own conjunction of texts and meanings as they bicycle their chosen path there. . . .

[T]he city is constituted psychologically by the meanings these words carry as they are read by the bicyclist traveling through these streets. The texts have been written as eight separate story

Jeffrey Shaw: ⟨http://www.aec.at/prix/1990/E90azI-legible.html⟩

Fig. 7.4.15. Jeffrey Shaw, *Legible City*. Visitors can ride around a virtual city made of giant words via a bicycle interface.

lines that have a particular relationship to Manhattan—for instance, monologues spoken by Mayor Koch, Frank Lloyd Wright, Donald Trump, Noah Webster, a cab driver, a tour guide, an ambassador, etc.[20]

Monika Fleischmann

Monika Fleischmann and her collaborators created works activated by visitors' physical actions. In *Spatial Navigator* and *House of Illusion,* visitors could move through simulated 3-D worlds by walking. The commonplace, everyday physical acts became the means to move around the cyber world of a building full of information. Fleischmann notes that movement and gesture are important tools that humans use in coming to understand the world. She refers to Virilio's reflections on contemporary culture's obsession with

Monika Fleischmann: ⟨http://viswiz.gmd.de/VMSD/PAGES.en/mia/homepage.html⟩

Fig. 7.4.16. Monika Fleischmann and Wolfgang Strauss, *Skywriter*. Viewer body-balance motions on a "magic carpet" platform enable the navigation of a virtual space.

speed: "The action of mechanical walking is a metaphor for a deadlocked fitness society. The intention is to convey Paul Virilio's theory of a racing standstill." She sees movement and touch as essential to knowing.

In *Skywriter,* people stood on a special platform that could sense shifts in balance. The "virtual balance" manipulations of the body's center of gravity enabled the person to fly through a virtual landscape. "Unlike a joystick or mouse, which reduces man to minimal reflex actions, 'Virtual Balance' requires the coordinated use of the entire body and its perception. . . . The dramatic effect of the action is governed by the person's relationship to his own body."

Other Artists and Projects

Christian Möller created an installation called *Virtual Cage* that is activated by the visitor shifting body balance to maintain position on an unstable platform. Via laser

Christian Möller, *Virtual Cage:* ⟨http://www.canon.co.jp/cast/artlab/pros2/pers-01.html⟩
Virtual Cage commentary: ⟨http://www.canon.co.jp/cast/artlab/pros2/report/craemer-02.html⟩

projections, the shifts affect the visual and aural display that surround the platform. On a screen, a swarm of chirping particles/creatures spontaneously gather and react to the viewer's actions. ***Kazuhiko Hachiya*** constructed installations in which common young people's activities controlled computer-mediated displays. *Over the Rainbow* focused on swinging, and *Light/Depth* read the actions of skateboarders. Images of waves changed with the skater's actions. ***Brian Duggan, Jason Ditmars, and Ronen Mintz*** developed the *Virtual Wheelchair* project to enable anyone to experience wheelchair navigation through virtual spaces. A special ramp provided force feedback to simulate friction and gravity, and read the motion and turning of a wheelchair to adjust a display to simulate movement through a virtual landscape. ***Peter Broadwell*** and his collaborators created *Plasm: Above the Drome* to allow users to navigate a virtual space via a surfboard. *Plasm: A Nano Sample* created a virtual environment consisting of three monitors on wheels that communicated with each other to enable actions such as shapes swooping under the floor from monitor to montior. *Times Up*'s performance *An Evening Spent in a Hypercompetitive State of Mind* confronted visitors with a "biomechanical game show" that stressed visitor's bodies to the limit. ***Jamy Sheridan***'s *Sandbox* read user manipulation of a real sandbox to modify computer graphic imagery projected into the box. In *Sunset Boulevard* ***Margaret Crane, Dale MacDonald, Scott Minneman, and Jon Winet*** created a drive-by billboard-sized video display in which viewers could affect the flow by using garage door openers as they drove by. In *Knit One, Swim Two* ***Ingrid Bachmann*** offers visitors two fourteen-foot manipulable sensor-tracked knitting needles that control weights moving in water that ultimately are translated into images on a monitor. Working at Sony Computer Science Labs, ***Kim Binstead*** has created a project called *HyperMask,* which tracks the position and orientation to the audience of an actor moving onstage and projects an animated face adjusted to appropriate perspective.

Kazuhiko Hachiya: ⟨http://www.p3.org/p3-light/P3exhibitions.html⟩
Virtual Wheelchair: ⟨http://www.sdsc.edu/Press/vwheel.html⟩
Peter Broadwell: ⟨http://www.plasm.com/~peter/⟩
Times Up: ⟨www.timesup.org⟩
Jamy Sheridan: ⟨http://algoart.com/algoart/links.html⟩
Sunset Boulevard: ⟨http://www.pair.xerox.com/cw/sunset/info.html⟩
Ingrid Bachmann: ⟨http://hermes.stmarys-ca.edu/mission_community/art_gallery/interactive/#bachmann⟩
Kim Binstead: ⟨http://www.dai.ed.ac.uk/daidb/students/kimb/⟩

Breath

Char Davies

Char Davies has created a series of virtual worlds populated by organic and landscape forms that are quite different than the polygon-dominated constructions that constitute most VR environments. One's breathing rhythm moves a person through the worlds as the work tries to "dissolve the boundaries between the self and nature." (See chapters 2.5 and 7.3.) In a *Leonardo Electronic Almanac* review of *Osmose*'s exhibition at ISEA95, Annick Bureaud concludes that the breath interface was critical in the work's success in transporting the interactor into a virtual world:

With *Osmose* we are within the work but, more powerfully, the work is within ourselves that we exhale with our breath, intimacy, interpenetration of the work and the I, relying on the spectator's body, whose essential movements (breath and equilibrium) are the very conditions to the understanding of the work itself.[21]

Other Artists and Projects

In ***Ulrike Gabriel***'s *Breath,* the viewer's breath was translated into the movements of computer graphic shapes on four monitors. ***Elaine Brechin*** created *Windgrass* as a sculptural object that resembles a clump of earth with grass growing out of it. The simulated grass uses sensors to change the light emitters embedded inside in reaction to people blowing across it. Brechin, who was a researcher at Interval, is interested in the potential of new interfaces to decrease the alienness of digital devices. ***Edmond Couchot and Michel Bret*** created installations such as *Dandelion* and *Feather,* in which blowing across a physical object brings forth corresponding computer video images of the object responding.

Activated Objects

Many artists have started to work at the boundary of the physical and the virtual, in which actions in one world influence the other. See also chapters 6.3 and 6.4 for other examples of this kind of work.

Char Davies: ⟨http://www.immersence.com/immersence_home.htm⟩
Ulrike Gabriel: ⟨http://www.t0.or.at/video/ug.htm⟩
Elaine Brechin: ⟨http://www.interval.com/frameset.cgi?projects/windgrass/index.htm⟩
Edmond Couchot and Michel Bret: ⟨http://artmag.com/techno/landowsky/projet.html⟩

Fig. 7.4.17. David Small and Tom White, *Steams of Consciousness*. Visitor actions in a water pond affect the projected words, which seem to float. Photo: © Webb Chappel.

David Small and Tom White

Working in the Aesthetics and Computing group at MIT's Media Lab, Small and White developed an installation in which water became the interface to language. *Streams of Consciousness* presents a small waterfall with projected words appearing to float on the surface. The Web site notes that visitors' attempts to touch the words result in water and word ripples: "You can reach out and touch the flow, blocking it or stirring up the words, causing them to grow and divide, morphing into new words that are pulled into the drain and pumped back to the head of the stream to tumble down again." Illustrating the linkages in reseach and art, Small had been investigating three-dimensional typography and White was studying "liquid haptics," the use of flowing water to carry information.

David Small and Tom White: ⟨http://acg.media.mit.edu/projects/stream/⟩

Fig. 7.4.18. Toshio Iwai, *Music Plays Images* × *Images Play Music*. Physical pianos are linked so that actions of one affect the other.

Toshio Iwai

Toshio Iwai invents installations that blur the boundaries between the digital and the physical in several ways. Actions taken in the digital world activate physical objects, and physical objects control digital information. He pioneers unprecedented music-and-image events that suggest many areas of future art and interface research.

Toshio Iwai: ⟨http://www.iamas.ac.jp/~iwai/iwai_main.html⟩

Iwai's widely shown *Music Plays Images × Images Play Music* installation activates pianos. In one version, he collaborated with the piano virtuoso Sakamoto. Two grand pianos were placed onstage with their tops removed. Video cameras captured performers' fingers and the movement of the striking hammers. Several variations explored the interrelationships of the physical and the virtual. Examples included: Sakamoto playing a small keyboard on the side of the stage that caused the grand piano to play; light emanating from the piano played by Sakamoto bouncing into the other piano and activating strings where it landed; and a moving video image of Sakamoto's body and hands causing the piano to play as it overlapped the image of the piano. The Interactions97 jury statement notes: "In this sense the piece follows a Japanese tradition in which one does not look for controversy and conflict in interaction, but for a conjoining and merging of players."[22]

Azby Brown's *Wired* magazine interview, called "Portrait of the Artist as a Young Geek," describes Iwai's idea that his innovations are inevitable in the evolution of musical and artistic traditions:

Iwai sees his main purpose as achieving a greater fusion of musical and optical media. "We're at a certain border, ready for the next step," he notes. "In the case of the piano, the past couple of centuries have seen it evolve from the cembalo to the pianoforte, gaining in expressive capability at each step. . . . [Images and sound] are linked from the outset, inseparable. I'm hybridizing based on new technical capabilities," Iwai says, "but the effect will be to restore what has only recently been discarded."[23]

In *Composition on the Table, Toshio Iwai* explores "mixed reality," in which manipulations of dials and levers on a table affect images projected to indicate relationships between the physical indicators in front of several players. In *Violin—Image of Strings,* computerized images of a violin are projected onto a violin in the creation of complex sound events. Curator Itsuo Sakane describes the event:

Across the strings, the visitor extends a finger and touches the touch sensor; the image of the finger is superimposed on the violin. A beautiful beam of light springs from it and dances over the strings, synchronized with the violin's music. As you change the movement of your finger over the sensor, new combinations of image and music are born. You become, in effect, a composer, participating in creating a work in which sound and image are combined.[24]

Iwai's artistic vision requires a nonsuperficial involvement with technology and an eagerness to become a technological innovator himself, rather than relying on other experts. The *Wired* interview describes Iwai's integrated approach:

In Iwai's case, his work always emerges from a deep understanding of several complex fields. Rather than merely accepting the limitations of hardware, he will modify it to produce a desired effect; rather than using stock software from the box, he will visualize novel ways of linking programs and hardware to make a certain type of imagery possible. He finesses the interconnections between hardware, software, input, and display because he speaks all of them fluently.[25]

Masaki Fujihata

Masaki Fujihata patrols the boundaries between the real and the virtual. The *Beyond Pages* installation presents a table with a projected book image on it. A real pen can interact with the projected image. Touching words causes real and virtual things to happen, for example, a virtual computer-generated door to open or a real light to turn on, all accompanied by sound. Fujihata (see also chapter 7.3) wants to confuse these boundaries. The Interactions97 description explains the fascinating space that is created:

The natural world—apple, stone—is folded in beside the common artifice—door, light—and between them sits the mark of the unnatural human, gate of the supernatural, core of the book, writing. . . . The sound score immaculately emulates the motion of each against paper, save for the syllabic glyphs of Japanese script, for which a voice pronounces the selected syllable. Stone and apple roll and drag across the page, light illuminates a paper-shaded desklamp. In the middle pages, kanji letters scroll breakneck under the nib of your pen. Lifting it selects a word. . . . Language even at its most foreign straddles the divide between the otherness of nature and the familiarity of artifacts. It makes the strangeness of stone as familiar as the alienness of the light switch.[26]

Perry Hoberman

Perry Hoberman's work spans several areas, including kinetics (chapter 5.2), the archaeology of media (chapter 7.2), and ID technologies (chapter 7.7). Several works focus on the boundary between the physical and the virtual. In *Systems Maintenance* he has created three related parallel versions of a room full of furniture (full physical scale, miniaturized physical, and virtual) that could be manipulated with the impact being felt in the other versions. Exploring the shrinking and manipulation of time made possible

---

Masaki Fujihata: 〈http://www.flab.mag.keio.ac.jp/index.html〉; 〈http://www.zkm.de/~fujihata/〉
Perry Hoberman: 〈http://www.hoberman.com/perry/php/index.html〉

Fig. 7.4.19. Masaki Fujihata, *Beyond Pages*. Actions with a physical book influence projections, sounds, and real actions.

by contemporary technology, his *Timetable* confronts visitors with a table full of dials that control projections on the table that move through past, present, and future. In *Lightpools* visitors affect interacting circles and polygons on a floor of projected light based on their position and gestures. In an earlier project he collaborated with *Nick Phillip* to create *Cathartic User Interface* to specifically comment on the cool, dispassionate efficiency of customary interfaces by offering visitors a chance to use violent physical

Perry Hoberman, *Timetable:* ⟨http://www.ntticc.or.jp/special/biennale99/exhibition/perry_e.html⟩

activity to control a computer. Hoberman and Phillip created a carnival side-show-type installation composed on many old discarded keyboards hanging in an array on a wall. Visitors could activate the event by throwing small mouse-shaped balls toward the keyboards, which would then emit sounds, images, and text collected from the Internet about people's frustrations with computer interfaces.

Other Artists and Projects

Georg Ritter constructed an installation based on the bumper car amusement-park ride in which riders can collide and bump each other with abandon. His *Autodrome* installation maps the movement of physical cars with projected movements in 3-D virtual space. Events in the two worlds affect each other; for example, physical car position moves virtual characters on a screen. Also, a virtual "Pac-head" indicates its position in physical space by sound and causes cars to stop or otherwise react. "Drivers are really and virtually present. The vehicle is the robot in which the human being is the processor. The course of events is the process which is determined by the behavior of individual drivers with respect to one another."[27]

Joseph Michael is part of a company called Robodyne Cybernetics, which is devoted to developing environments that approach the reality of the "holodeck" from *Star Trek*. Michael has also produced prototype systems that link VR technology to physical-shape-changing robot cubes to create VR scenes with sufficient physical elements to illustrate the potential of virtual reality. For example, upon seeing a chair in a VR head-mounted display, the robot cubes would adjust themselves to roughly model a chair such that the participant would link the physical experience with the view in the goggles. *Heide Solbrig*'s *Typewriter Racing Test* commented on historical attitudes toward women's physical education and employment by having visitors activate a simulated computer typing program through workouts on exercise equipment.

Working at the Tachi Laboratory of the University of Tokyo, *Masahiko Inami* (in collaboration with *Fumihiro Endo and Toshitomo Ohba*) developed a digitally activated manipulable cube. Virtual objects inside the cube were shown on LCD displays located on the cube's six faces. The images responded in accordance with the manipulation in order to simulate real objects located inside. *Motoshi Chikamori and Kyoko Kunoh* created a floor installation called *Kage*, full of simple objects described as "myste-

---

Georg Ritter, *Autodrome:* 〈http://www.servus.at/autodrom/english.html〉
Joseph Michael, *Holodeck:* 〈http://www.nano-technology.com/holodeck/〉
Masahiko Inami: 〈http://www.siggraph.org/s97/conference/garden/media.html〉

rious and meditative." As visitors touch the objects, huge projected shadows fly out from the area of touch. In Chikamori's *O[en]*, visitor choices about constructions with physical hemispheres in the center of a room change the way images of the room are projected onto a dome-shaped roof. ***Ian Ginilt and Saoirse Higgins***'s installation *The Star-Dogged Moon* invites visitors to interact with a multimedia event by manipulating real physical objects. ***Keith Roberson*** created a shopping cart computer called *Little Havoc,* whose interface was activated by rolling it around and watching the images that come up on its monitor. He sees it as a commentary on the high-tech, high-gloss world, "a pathetic new breed of hack/junk/found interactive art." ***Beryl Graham***'s *Individual Fancies* asks participants to sit at an interactive tea table to explore loneliness with a rotating teapot, projected video actions on the table, and spatial sound. ***Jean-Pierre Hebert***'s *Sisyphus* presented an activated ball that drew patterns in a sandbox under invisible magnetic motion control. In ***Kaeko Murata***'s *Fisherman's Café,* projected fish swimming across a table are affected by the tea-drinking behaviors of the participants. For example, viewers are challenged to "catch" projected fish by putting physical glasses over them.

In ***Loretta Skeddle***'s *CyberRosary* visitors control a room of networked computers that are praying by manipulating a rosary. Ironically commenting on who is controlled, ***Douglas Edric Stanley's*** *Asymptote* invites visitors to manipulate strings in the center of the gallery that affect a large projection of a marionette. The ***F.A.B.R.I.CATORS*** have created *Tracking the Net,* which allows viewers to manipulate virtual space by pulling and touching a plastic net suspended in free space. ***Hiroshi Ishii and The MIT Media Lab's Tangible Bits Group*** have created several works that explore the virtual/physical border such as *Music Bottles,* which allows visitors to manipulate bottles as a music performance environment, and *Personal Ambient Displays,* which are small wearable devices that communicate information via "tactile modalities such as heating and cooling, movement and vibration, and change of shape." In *Performance Architecture* ***Christopher Janney*** has created architectural spaces that respond to pedestrians with changing light and sound. ***Xerox PARC*** has created

Ian Ginilt and Saoirse Higgins: ⟨http://www.uib.es/agenda/exhibit/saoirse.html⟩
Beryl Graham: ⟨http://www.stare.com/beryl/⟩
Jean-Pierre Hebert, *Sisyphus:* ⟨http://www.solo.com/sisyphus/⟩
Douglas Edric Stanley: ⟨http://www.ntticc.or.jp/special/biennale99/exhibition/douglas_e.html⟩
Motoshi Chikamori: ⟨http://en.ntticc.or.jp/⟩
Kaeko Murata: ⟨http://www.mars.dti.ne.jp/~kasa/⟩
Hiroshi Ishii et al.: ⟨http://tangible.www.media.mit.edu/groups/tangible/projects/⟩
Christopher Janney: ⟨http://www.janney.com⟩
PARC: ⟨http://www.thetech.org/xfr/⟩

a museum exhibit called *Experiments in the Future of Reading,* which presents text-based experiments such as *Tilty Tables,* in which projected text appears to roll in the direction of the tilt, *Glyph-O-Scope,* which includes information printed in the shading of the image that can be read with the aid of a special device, and *Listen-Reader,* which senses what page of a book is open and where the hands are positioned to generate relevant sounds.

Summary: Can a Computer Do More Than See and Hear?

Much analysis focuses on the linkage of computer visual-representation conventions to Western civilization's reliance on the Renaissance's conventions of perspective and all its associated baggage of unitary points of view and the privileging of vision. Many researchers and artists are working to extend digital systems into the realms of motion, touch, gaze, and kinesthetics. Some see this as liberating computers to enter into a fuller range of human life. Others see it as the further extension of control and domination into the realm of the body. The artists in this chapter believe that the movement away from the conventional interface of mouse and keyboard is a significant cultural event worth investigating.

Notes

1.   D. Rokeby, *"Very Nervous System,"* ⟨http://www.interlog.com/~drokeby/⟩.

2.   D. Rokeby, "Transforming Mirrors," ⟨http://www.interlog.com/~drokeby/⟩.

3.   P. Garrin, *"White Devil* Documentation," ⟨http://www.235media.com/install/white_devil.html⟩.

4.   M. Fleischmann, "Goals," ⟨http://viswiz.gmd.de/VMSD/PAGES.en/mia/homepage.html⟩.

5.   M. Fleischmann, "Visualizing Cultural Identity," ⟨http://viswiz.gmd.de/VMSD/PAGES.en/mia/homepage.html⟩.

6.   Interactions97, "Krueger Documentation," ⟨http://www.iamas.ac.jp/interaction/i97/artist_Krueger.html⟩.

7.   M. Krueger, "Artificial Reality Description," ⟨http://www.teleport.com/~cognizer/eet/Almanac/TUTOR/VRTUTOR.HTM⟩.

8.   R. Lozano-Hemmer, "Re:positioning Fear," ⟨http://xarch.tu-graz.ac.at/home/rafael/fear/⟩.

9.   D. Ritter, "My Finger's Getting Tired," ⟨http://www.users.interport.net/~ritter/⟩.

10. C. Sommerer and L. Mignonneau, *"Trans Plant* Documentation," ⟨http://ms84.mic.atr.co.jp/~christa/WORKS⟩.

11. M. Fleischmann, *"Liquid Views,"* ⟨http://viswiz.gmd.de/VMSD/PAGES.en/mia/projects.html⟩.

12. Interactions97, "Description of *Bodymaps,*" ⟨http://www.iamas.ac.jp/interaction/i97/artist_Schiphorst. html⟩.

13. T. Schiphorst et al., "Electric Body Project," ⟨http://www.art.net/Resources/dtz/schipo2.html⟩.

14. T. Schiphorst, "Body Noise: Subtexts of Computers and Dance," ⟨http://www.art.net/~dtz/ schipo3.html⟩.

15. S. Stenslie and K. Woolford, "*CyberSM* Description," ⟨http://televr.fou.telenor.no/stahl⟩.

16. L. Hershman, "*America's Finest* Description," ⟨http://arakis.ucdavis.edu/hershman/⟩.

17. SIGGRAPH97, "Description of *Magic Morphin Mirror,*" ⟨http://www.siggraph.org/s97/conference/ garden/morphin.html⟩.

18. T. Darrel et al., "A Virtual Mirror Interface Using Real-Time Robust Face Tracking," *Magic Morphin Mirror,* ⟨http://www.interval.com/frameset.cgi?papers/1998-005/index.html⟩.

19. Plasm, "SIGGRAPH Description of *Plasm: Your Mug,*" ⟨http://www.plasm.com/peter/public_html/ YerMug.html⟩.

20. Ars Electronica, "*Legible City* Prize Statement," ⟨http://www.aec.at/prix/1990/E90azI-legible.html⟩.

21. A. Bureaud, "Review of *Osmose,*" ⟨http://mitpress.mit.edu/e-journals/Leonardo/reviews/bureaudisea. html⟩.

22. Interactions97, "Award Statement for Iwai," ⟨http://www.aec.at/prix/jury/E97inter.html⟩.

23. A. Brown, "Portrait of the Artist as a Young Geek," ⟨http://www.wired.com/wired/archive/5.05/ ff_iwai_pr.html⟩.

24. I. Sakane, "Interactions97 Description," ⟨http://www.aec.at/prix/jury/E97inter.html⟩.

25. A. Brown, op. cit.

26. Interactions97, "Description of Fujihata Installation," ⟨http://www.iamas.ac.jp/interaction/i97/ artist_Fujihata.html⟩.

27. G. Ritter, "*Autodrome* Description," ⟨http://www.servus.at/autodrom/english.html⟩.

Speech Synthesis, Voice Recognition, and 3-D Sound

Sound Research

Researchers are making progress on a variety of sound-related technologies that promise to be culturally significant—voice recognition, speech synthesis, sonification, earcons, and 3-D sound simulation. This chapter examines artistic experimentation with these technologies. The ability to produce and understand spoken language has been identified by anthropologists as one of the prime accomplishments of our species. Other animals, such as dolphins and primates, may have significant vocal communication capabilities, but they do not approach human capabilities. The extension of speech to machines will mark a significant cultural event that will call out for artistic attention.

Rudimentary computer-speech-synthesis capabilities have existed for decades, but the speech produced was machinelike and lacked the subtleties of inflection tied to meaning that mark human speech. Contemporary researchers using artificial intelligence techniques have begun to generate almost-human-sounding speech synthesizers that can adjust their words in accordance with the underlying meaning of the text they speak.

Voice recognition similarly has existed in rudimentary form for decades. In the early days it could only perform speaker-dependent recognition—identifying isolated words from a speaker who had previously conducted a training session to create a template. Contemporary technology, aided by the increase of speed and memory of current digital electronics, can perform speaker-independent recognition of almost full-speed spoken text. The recognition is aided by artificial-intelligence techniques that use context and syntax to resolve its confusion. Recognition in this sense means only identifying the intended word, not the meaning of the words.

Using sound to locate possible predators and prey has been an important biological asset for animals and humans. The perceptual system uses the distance between the two ears as a means of identifying the location of a sound source. The brain can use the time difference in the arrival of sounds and the echoing in the outer ear to deduce location with good accuracy. Scientists and technology developers have made progress in creating digital sound sources that can simulate localization in space. For example, computer sound systems can simulate any location in 3-D space around a listener. I once experienced a trade show demonstration in which the salesperson flew a helicopter sound from my distant left into the middle of my head and out to the distant right. Three-dimensional sound is part of advanced VR systems, and already there are game systems available with rudimentary 3-D sound capabilities, with researchers investigating a variety of applications. Some are also working on the reverse problem of enabling a

Fig. 7.5.1. Iain Mott, Marc Raszewski, and Tim Barrass, *Squeezebox*. Visitors push the sculpted hands down to control the 3-D placement of sound and morphology of computer graphics.

computer to identify the physical location of sound sources. Artists are beginning to move into many of these sound research areas.

3-D Sound

Iain Mott

In collaboration with Marc Raszewski, Iain Mott (see also chapter 3.5) developed *The Talking Chair* sound installation, which explored 3-D sound. In a 3-D space, visitors could move a wand that controlled the nature of the sound and its physical 3-D location. Users could use the wand to move sound around the sculpted frame surrounding the chair. In a *Leonardo* article, Mott describes his interest in physicality: "*The Talking Chair* is an attempt to forge links between the ephemeral nature of music and the material world, by expressing musical process through a physical poetry. It requires the creative

Iain Mott: ⟨http://members.tripod.com/~soundart/⟩

input of the individual to give it meaning, to perfect an imaginary universe of cause, effect, and response."[1]

Other Mott projects also focus on 3-D sound. *Squeezebox* confronted visitors with a kinetic sculpture of four hands that could "mold" the 3-D sound virtual structure in the gallery. Manipulations also shaped graphics:

The cast hands of *Squeezebox* invite participation. Participants grasp and press down the sculpted pieces, working against a pneumatic back-pressure to elicit both sound and image. The interaction reveals a form which has visual, aural, as well as physical properties. As participants press down on the hands, music swells from the room. A sound mass is shifted from one point of the room to another by pressing down on alternate pistons. Music is produced algorithmically and is derived from a set of rules which respond to the spatial location of the sound mass.[2]

The system of rules, however, is never static. One spatial strategy gives way to another, resulting in an evolution of sound, requiring a constant readjustment of focus in the listener.

Christian Möller

In *Light and Audio Park: The 220 V Party Effect,* a large outdoor sound-and-light V2-sponsored installation was created in Rotterdam composed of eight, twelve-meter loud-speaker towers and ninety-six airplane floodlights. Participants could shape the sound and its 3-D placement by moving in and out of shadows:

The sound collage is made up of a large number of radio stations broadcasting simultaneously mixed with specially composed sound elements. Via a lavish 3-D audio processor, the visitors can shift (sort) the transmitted sound elements spatially and thus vary the balance. The park becomes a gigantic ghettoblaster that simultaneously receives all available broadcasts and transmits a choreography of sound.[3]

In a paper called "Audio Experiments," Möller explains his interest in human abilities to filter and locate sound—the "party effect"—even in the midst of potentially disturbing multisource sound environments:

Although it seems to be absurd, in an audio structure consisting of numerous different noises, the greater the complexity is, the more tolerable the event becomes. . . . One of the most interesting

---

Christian Möller: ⟨http://www.canon.co.jp/cast/artlab/pros2/pers-01.html⟩

Fig. 7.5.2. Christian Möller and Louis-Philippe Demers, *Light and Audio Park*. Motions of visitors in outdoor installation affect the lights and the 3-D placement of sound.

filter effects known to us in our everyday life is known in perception psychology as the so-called party effect. The party effect describes the well-known situation of a large group of people talking in one room all at the same time making conversation. If one of the guests hears his name mentioned in the context of another conversation, he becomes attentive; he gets the famous "rhubarb ear" in order to be able to bridge greater spatial distances and to filter the relevant information out of the uniform hiss of the babble of voices. This elasticity in the individual grade of attention is a natural capacity of more highly evolved beings, showing sciences dedicated to the creation of machine intelligence how hopelessly complex such filters can be at the highest level.[4]

In another project that Möller undertook for an exhibition called *Botschaft der Musik* (The Message of Music) in Vienna, he produced a computer simulation that gave visitors the experience of being able to walk around a chamber music perform-ance in order to use spatial location to focus on specific instruments. He explains the importance of perceptual psychology's understanding—of taking a number of mathematical or physical laws of human hearing into consideration, such as differential distance to the ears, reflections in the auricle of the ear, and the transmission qualities of air.

Other Artists and Projects

Brenda Laurel and Rachel Strickland's *Placeholder* installation (see chapter 7.3) relied on spatial sound to orient visitors and to help sculpt the unorthodox mythic space, using such features as ambient localized sounds as guides and localization of "Critter helpers" by sound position. **Susan Alexis Collins**'s installation *AudioZone* used spatial sound to lay an alternative reality atop the physical gallery. Exploring the role of passivity in interactive art, Collins used spatial sound to confront visitors with suggestions for potential actions in particular places (for example, "don't touch") and to create "audio walls" to divide the gallery. **Knowbotic Research**'s *Anonymous Muttering* transformed a Rotterdam space into an interactive 3-D sound encounter (see chapter 5.3).

I created several installations that map emotional or conceptual space to 3-D physical space. In *Memory Map,* the spoken memories of visitors were mapped in physical space in accordance with the sex and age of the current visitor, for example, the voices of those older than the present viewer coming from in front and those younger coming from behind, and the voices of males and females coming from different sides of the hall. Visitors shared reflections about the past and present, and anticipations of the future in the realms of accomplishment, relationships, and values, which become the repertoire of sounds for subsequent viewers. In the V2-sponsored installation *Oratorio for Religious Opinion,* I mapped digitized opinions about religion to audio speakers located on the church square of a Dutch town, spatially mapping the townspeople's spoken opinions about religious questions, such as: Why is there evil in the world? to physical locations determined by qualities such as religious conservatism or age. The **Stichting Rainstick** group creates art installations exploring spatial sound, such as *Tai-Tendo,* in which physical and networked pilots moved bumper cars and the spatial positioning of sound. Bill Viola's *Presence* uses parabolic sound focusers to specifically locate breathing and heart-

---

Brenda Laurel and Rachel Strickland: *Placeholder:* ⟨http://www.interval.com/research/NewMed/index.html⟩

Susan Alexis Collins, *AudioZone:* ⟨http://www.ucl.ac.uk/slade/sac/azone.html⟩

Knowbotic Research, *Anonymous Muttering:* ⟨http://www.khm.de/people/krcf/AM_rotterdam/⟩

Stephen Wilson: ⟨http://userwww.sfsu.edu/~swilson/⟩

Stichting Rainstick: ⟨http://valley.interact.nl/rainstick/⟩

beat sounds in particular spaces. ***Nigel Helyer's*** *Transit of Venus* presents visitors with a 3-D sound installation in which reflections on the cultural and scientific activities of explorer James Cook are mapped to different parts of the exhibition space. *Naut Humon* and the artist group ***Sound Traffic Control*** create 3-D sound events that use emergent genetic music techniques and build on the metaphor of airports with sounds of coming and going such as planes taxiing, taking off, and landing. ***Center for Research in Electronic Art Technology (CREATE)*** at the Music Department of the University of California at Santa Barbara organized a symposium on *Sound in Space* to explore the artistic potentials of 3-D sound.

Speech Synthesis and Manipulation

Arthur Elsenaar and Remko Scha

Arthur Elsenaar and Remko Scha are well-known for their work with Huge Harry, a screen-based artificial character who gives lectures on issues in art and technology. They are part of a group of artists and theorists known as the Institute for Artificial Art, which pushes the possibilities of synthetic creatures. Speech synthesis is part of their agenda (see chapters 2.5 and 7.6).

Huge Harry demonstrates his prowess as a speaker by regularly giving talks at art and computer science meetings. In talks such as "On the Role of Machines and Human Persons in the Art of the Future," he is a forceful advocate of machine consciousness and holds the presidency of the Institute for Artificial Art. He also gives musical performances in collaboration with his friends Perfect Paul and Whispering Wendy. Scha enhances his work through research in computational linguistics, perception, and algorithmic art.

In his paper "Virtual Voices," he notes the long-standing interest in the mimetic imitation of life and the abstract analysis of human activities such as intelligence and speech. He analyzes the importance of the Cartesian notion about life as mechanism and the stimulus it provided toward the production of automatons. He traces the history of synthetic speech, noting two approaches: duplication of the mechanisms of life, such as vocal cords, the focus on the abstract analysis of sounds.[5]

Nigel Helyer's *Transit of Venus:* ⟨http://www-personal.usyd.edu.au/~nhelyer/⟩
Sound Traffic Control: ⟨http://www.asphodel.com/stc/⟩
Arthur Elsenaar and Remko Scha: ⟨http:// www2.netcetera.nl/~iaaa/⟩

The first speech-synthesis systems were very inhuman and raised the specter of unthinking machines on the loose, evoking "disturbing questions about the possibilities and the dangers of technology, about mind and matter, and the nature of human identity." Synthesis is improving; for example, DecTalk offers customizable voices such as Rough Rita, Frail Frank, Whispering Wendy, Huge Harry, Kit the Kid, Perfect Paul, Beautiful Betty, Uppity Ursula, and Doctor Dennis.

Other Artists and Projects

David Rokeby (see chapter 7.4) is best known for his installations that use image processing to interpret people's motions. *Giver of Names* uses similar technology to recognize objects placed by visitors on a pedestal in front of its camera eye. Commenting on human processes of categorization, the computer names the object with its speech-synthesized voice and sometimes shares its ruminations. The names it speaks are not always simple objective descriptors. Rokeby notes that one "aim is to highlight the tight conspiracy between perception and language, bringing into focus the assumptions that make perception viable, but also biased and fallible, and the way language inhibits (or alternately enhances) our ability to see." *Jerome Joy*'s *Vocales* presented interactive, networked speech-synthesis events. In *Orpheus,* *Ken Feingold* uses the computer to manipulate text from Jean Cocteau's *Orpheus*. The syntax is maintained with different words inserted. "The computer program randomly pulls words from this matrix each time through the loop of the overall piece. In this way, the original syntax is fixed, but the poetry is 'real time' and variable." The voice seems to come from a projection of a puppet head. In *Surprising Spiral,* touching fingerprints in a large sculpted book produces complex events built out of digitized and synthesized speech.

Commenting on gender roles and toys, the *Barbie Liberation Organization* completed a famous project in which they switched the digital voice boxes in Barbie and G.I. Joe dolls, and then put them back on store shelves. The *RTMark* Web site describes the change: "G.I. Joe, decked out in army fatigues and combat gear, was altered to speak with Barbie's voice, making statements such as 'Want to go shopping?' while the ideologically renovated Barbie warns us: 'Dead men tell no lies!'" *Bruce Cannon*'s

---

David Rokeby, *Giver of Names:* ⟨http://www.interlog.com/~drokeby/gon.html⟩
Jerome Joy: ⟨http://homestudio.thing.net/⟩
Ken Feingold: ⟨http://www.kenfeingold.com/⟩
RTMark, Barbie Liberation Organization project: ⟨http://rtmark.com/ppblo.html⟩
Bruce Cannon: ⟨http://www.jps.net/bcannon/sculptures/⟩

sculptures, such as *Abjet D'Art* and *Box with the Sound of Its Own Critique,* link speech to viewer motion and action (see chapter 5.2). **Jim Campbell,** best known for his digital video installations (see chapter 7.2), created *I Have Never Read the Bible* to investigate the limitations of logic-controlled memory. His computer reads the entire Bible one letter at a time, suggesting that the piece could be seen as "an examination of the problems with entrusting everything to the workings of technology." **Paul DeMarinis's** *Alien Voices* consists of two connected booths in which viewers can manipulate the voice with which they communicate to the other booth (see chapter 5.2).

Natalie Jeremijenko created *Voice Box,* which offered one hundred talking boxes that could be stacked. **Ron Kuivila's** installations, such as *The Linear Predictive Zoo* and *On Schedule,* provides a humorous commentary on digital precision. He arranged for a chorus of over 1,400 talking unsynchronized wristwatches to announce their reading of the time via their synthesized voices. Famous for her experiments with speech-processing technology, **Laurie Anderson** has implemented many speech pieces, including the interactive *Here,* which allows conceptual mixes of the 258 most used words of the English language according to categories such as Rhyme, Definition, Possession, Shuffle, Expansion, Acronyms, and Subtraction.

Speech Recognition

Sharon Grace

Sharon Grace, known for her telecommunications art, created the installation *Millennium Venus,* in which a visitor can have a conversation with a computer character via voice recognition. The documentation from the "Responsive Artworks" show describes the event:

Millennium Venus is an installation where the participant has a conversation with a cyborg. She whispers untold secrets about the millennium and waits for a reply. She talks from a time different from our own, a time of disappearance. She talks about the eternal variations in a new language,

Jim Campbell: ⟨http://www.artscenecal.com/ArticlesFile/Archive/Articles1997/Articles0697/⟩
Ron Kuivila: ⟨http://www.clic.net/~avatar/kuivila/kuivilaproj.html⟩
Laurie Anderson: ⟨http://www.stedelijk.nl/capricorn/anderson/index.html⟩
Sharon Grace: ⟨http://www.telefonica.es/fat/egrace.html⟩

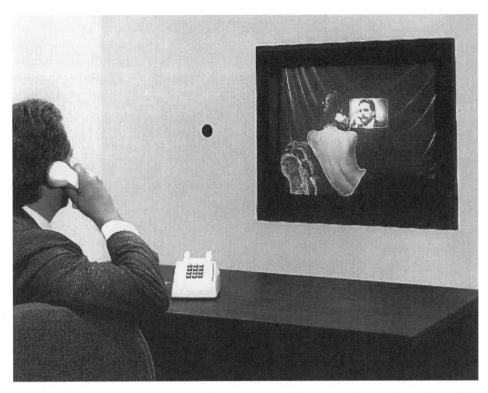

Fig. 7.5.3. Sharon Grace, *Millennium Venus*. Visitors can communicate with a computer character via speech recognition. Photo: Juliet Stelzmann.

a new system of representation between stimulus and answers. In the end, she asks if it is possible to dwell in time, to invent life span, to reinvent space.[6]

Stephen Wilson

I created several installations that used speech recognition as the method of interaction. In *Synthetic Speech Theater*, four fictional characters located in four corners of a room debated the meaning of life using synthesized speech. The visitor could direct the flow of the debate by speaking key words into a microphone located in the center. In *Inquiry Theater*, viewers could take a virtual walk down Mission Street in San Francisco. They

---

Stephen Wilson: ⟨http://userwww.sfsu.edu/~swilson/⟩

stood in front of a projected moving panorama of the storefronts and could choose to enter various stores by speaking key words into a microphone.

Other Artists and Projects

In *Amidst the White Noise,* **Christiane Robbins** inspects the "insitutional codings of what is coded as the Real" by deconstructing technologies such as speech recognition and analysis, as used in police work. An audience member engages the computer in a call and response in which the computer speaks with simulated emotion and attempts to recognize the emotional content of the human being. In *Play Cinema,* **Naoko Tosa** created a system that read speech as well as gesture and emotional tone (see chapter 7.6). In *Oh toi qui vis la-bas,* **Don Ritter** presented a voice-controlled video projection. The imagery and its position on the screen was controlled by the singer with the text and imagery slowly sliding off the bottom of the screen during periods of silence, but returning when the singing resumed. Chapter 7.6 on artificial intelligence considers speech recognition more. **John Maeda**'s *Reactive Squares* presented a graphic interface in which screen elements reacted to voice (see chapter 4.2). **Liz Vander Zaag and Western Front Multimedia** developed SAY software, "which allows the emotional content of the user's voice—volume and tone—to affect the interaction."

Summary: Something to Talk About

Speech technology is getting increasingly better, cheaper, and easier to work with. Interface researchers predict that speech will be the major form of interface in the future, especially as intelligent devices diffuse into the everyday world. What will these devices sound like? What will be the limits of their ability to understand us?

Notes

1. I. Mott, "*Talking Chair Article,*" ⟨http://members.tripod.com/~soundart/⟩.

2. I. Mott, "*Squeezebox* Description," ⟨http://members.tripod.com/~soundart/⟩.

Christiane Robbins: ⟨http://www.banff.org/deep_web/fiftha/⟩
Naoko Tosa: ⟨http://www.mic.atr.co.jp/~tosa/⟩
Don Ritter: ⟨http://www.users.interport.net/~ritter/⟩
John Maeda: ⟨http://www.maedastudio.com⟩
Liz Vander Zaag and Western Front Multimedia: ⟨http://www.frontmedia.com/⟩

3. C. Möller, "*Light and Audio Park*," ⟨http://www.canon.co.jp/cast/artlab/pros2/works/rotter.html⟩.

4. C. Möller, "Audio Experiments," ⟨http://www.canon.co.jp/cast/artlab/pros2/works/⟩.

5. H. Harry, "Virtual Voices," ⟨http://www2.netcetera.nl/~iaaa/virtual.html⟩.

6. S. Grace, "*Millennium Venus* Documentation," ⟨http://www.telefonica.es/fat/egrace.html⟩.

Artificial Intelligence

Artificial Intelligence Research

Artificial intelligence (AI) is a significant area of research that has intrigued researchers and some artists for many years. The *Handbook of Artificial Intelligence* defines AI as "part of computer science concerned with designing intelligent computer systems, that is, systems that exhibit the characteristics we associate with intelligence in human behavior—understanding language, learning, reasoning, solving problems, and so on."[1]

AI researchers have worked on areas such as the machine understanding of text (such as news stories or short stories), speech recognition, image recognition (such as face recognition or assembly-line monitoring), autonomous robotics (such as planetary probes), game playing, data analysis, expert systems (such as doctor or loan-officer simulations), education (such as AI tutors), and the like. In the 1970s and 1980s there was great enthusiasm for artificial intelligence in academia and the corporate world. The U.S. government invested huge research sums through the Defense Department Advanced Projects Research fund (DARPA). Researchers optimistically predicted the simulation of human (or even superhuman) intelligence within a few years.

Great debate arose over the possibilities and limitations of artificial intelligence. Philosophers, cognitive scientists, and others attempted to unravel the meaning of mind, intelligence, and artificiality. Philosophers, such as Hubert Dreyfus in *What Computers Can't Do,* denied the possibility of artificial intelligence because machines lacked the experience of being in the world. For example, they lacked the situated sense of body that helps interpret discourse about bodies and solve problems that involve the body. Other books, such as Haugeland's *Artificial Intelligence: The Very Idea,* Roger Penrose's *The Emperor's New Mind,* and Douglas Hofstadter's and Daniel Dennett's *The Mind's I* explored these questions of AI's possibilities in great detail. Some recent interest is evidenced by books such as Hans Moravec's *Robot: Mere Machine to Transcendent Mind* and Ray Kurzweil's *The Age of Spiritual Machines,* which entertain the notion that technological advancement may finally enable computers that can match and perhaps even exceed the intellectual accomplishment of humans.

The field is now no longer so popular, and the debates are not so strident. Research continues but without as much funding or attention. Some commentators suggest that the useful aspects of the research have already been incorporated in many fields, for example, the semantic-analysis systems that make optical character recognition (OCR) so accurate, the military's use of scene recognition in cruise missiles, and the expert systems that allow software help desks to answer questions so quickly. Even though it is no longer so much in fashion, artificial intelligence raises fascinating questions that are still unresolved. What is more, the questions have great import beyond the technical

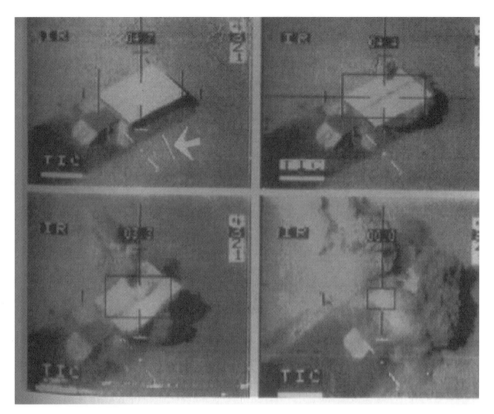

Fig. 7.6.1. Real-time damage assessment images of the effects of a "smart bomb" as it moves toward a preselected target. The missile relies on advanced machine intelligence to interpret visual terrain information, GPS location data, and its speed and direction. The illustration is from the presentation "Smart Bombs and Super Highways: Shifting Rhetorics of Technologies and Issues of Pedagogical Authority," by Michelle R. Kendrick, ⟨http://www.vancouver.wsu.edu/fac/kendrick/⟩.

niche of computer science. My paper "Artificial Intelligence Research as Art" expresses the view that the arts ought to become more active in the field:

Artificial intelligence is one of these fields of inquiry that reaches beyond its technical boundaries. At its root it is an investigation into the nature of being human, the nature of intelligence, the limits of machines, and our limits as artifact makers. I felt that, in spite of falling in and out of public favor, it was one of the grand intellectual undertakings of our times and that the arts ought to address the questions, challenges, and opportunities it generated.[2]

Through the years there has been a small but steady interest by those in the arts and humanities. There have been art shows, conferences, publications, and bibliographies produced, such as the Austrian Research Institute's bibliography *Artificial Intelligence and the Arts,* the American Association for Artificial Intelligence (AAAI) 1992 art show, and the special issue of the *Stanford Humanities Review* on "Constructions of the Mind." A core of artists and musicians have worked in the field, pursuing themes such as creative production, artificial characterization, speech recognition, and representations of intentionally.

Algorithms for Creativity

The challenge to develop algorithms for artificial creativity has engaged artists, psychologists, and computer scientists. They have asked: Could one create a computational system that would allow the computer to generate works that could be called creative or artistic? Structurally, the artistic effort is analogous to that of the artists, described in section 4, working in the areas of algorithmic, mathematics, and A-life-based art. The artists' contributions are in understanding the phenomenon of interest, devising the algorithms that manifest that understanding, and organizing and creating computational systems to generate the actual sensual output, for example, images or music/sound compositions.

There are significant philosophical questions involved in definitions of the creative and artistic, for example, historical and cultural constraints and the difficulty of judging quality. Nonetheless, the question has engaged many because creativity is seen as the quintessence of being human and a good test for AI concepts and procedures. Since almost its beginning, the journal *Leonardo* has published articles focused on efforts to sufficiently understand the style of particular artists in order to create computer systems for generating images that appear to be done by that particular artist.

Ray Lauzzanna and Languages of Design

Ray Lauzzanna edited a journal called *Languages of Design* and helped organize the International Society for the Computational Modeling of Creative Processes, which focused on research and artistic activity in the realm of algorithmic creativity. The description of the society's mission provides a good summary of the range of research and artistic activities concerned with these issues. The journal is no longer actively published, but compilations of previous articles are available from Penrose Press:

Considering recent developments in generative grammars, genetic algorithms, and other research in computer and cognitive science, this society was formed to promote the development of formal representations for creative processes and the implementation of these processes on computing machines.

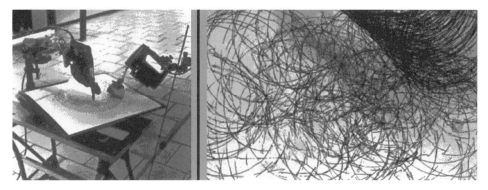

Fig. 7.6.2. Institute of Artificial Art, an organization devoted to the investigation of artificial systems. A machine drawing system using sabre saws.

This multidisciplinary focus is reflected by the board of directors and the editorial board of the journal of the society, *Languages of Design,* which includes linguists and literary theorists, music theorists and composers, computer scientists and researchers in artificial intelligence, artists, architects, and art critics. The special interests of the society focuses on formal design theory, generative grammars, shape grammars, computational musicology, and computational aesthetics. All forms of computational modeling are reported in the society journal, including formal languages, finite state automata, grammatical inference, pattern recognition, cellular automata, semantic networks, connectionism, and syntactical analysis. A broad variety of analytic perspectives are represented, including syntactics, semiotics, deconstruction, hermeneutics, stylistics, narratology, philology, morphology, prosody, harmony theory, formal musicology, and performance analysis.[3]

Institute of Artificial Art

In the Netherlands, Arthur Elsenaar and Remko Scha have been active in establishing the Institute of Artificial Art which is an "organisation consisting of machines, computers, algorithms, and human persons, who work together toward the complete automatization of art production." The institute prepares papers and performances (see also chapters 2.5 and 7.5). Their synthetic character Huge Harry regularly gives lectures on the benefits of artificial intelligence and creativity. In an Ars Electronica97 lecture, Huge Harry explains perspectives on machine superiority:

Institute of Artificial Art: ⟨http://www2.netcetera.nl/~iaaa/⟩

Fig. 7.6.3. Harold Cohen, screen image of a figure generated by program AARON as part of a long-term project to study image-generating intelligence and algorithms of artistic production.

Human artists always have rather selfish goals that usually involve money, fame, and sex. . . . Machines are in a much better position to create objects of serene beauty; . . . It is well-known that the associative processes in human neural networks are almost incapable of exploring large search spaces in a systematic way; human persons are therefore relatively ineffective in generating new artworks. . . .

[M]achine art is intrinsically superior to the output of human artists, and should be allowed to flourish without being usurped by human expression. At the Institute of Artificial Art it has been our goal to develop the technologies that are necessary to realize the potential of fully automatic machine art.[4]

Harold Cohen

Harold Cohen, the California artist and art professor at the University of California, San Diego, directs one of the most well-known research projects in artificial intelligence and art. Since 1968, Cohen has been developing computer algorithms to model the creative drawing and painting process. His program called AARON controlled a small robotic drawing machine, or plotter, to produce drawings and paintings. The work has been exhibited around the world to great acclaim and interest.

Harold Cohen: ⟨http://crca.ucsd.edu/~hcohen⟩

A well-respected painter, Cohen became interested in modeling creative activity. He challenged himself: Could he program a computer to autonomously generate drawings and paintings that would be judged of artistic merit? Could he understand the mental processes that he and other artists used well enough to externalize them into a program? His odyssey included research at Stanford University's artificial intelligence lab and a book called *Aaron's Code: Meta-Art, Artificial Intelligence, and the Work of Harold Cohen* (written with Pamela McCorduck). His multiyear project of programming, AARON, creates drawings that are different and unpredictable, but stylistically coherent and intriguing. Cohen's writing in the 1972 catalog for one of the first shows of his computer art demonstrate his thinking about what was truly important in the synthesis of art and computers:

I would also not have it in mind to draw those interminable geometrical figures popularly identified by now as computer art. . . . What fascinated me about the computer in the first place was not its precision, nor its prodigious capacity for work, nor even its amazing versatility. It was, and remains, its ability to build a primitive decision-making function—if such and such is the case, do this; otherwise, do that—into complex functions bearing strange and remarkable resemblances to human logical processes. The machine is not important to me in itself; but its use makes possible the formulation of precise and rigorous ways of clarifying those processes and in general those involved in the activities of art.[5]

Over the years, Cohen has continually refined the program, adding capabilities such as shape closing, figure representation, and color. He believes that art is more than techniques of visual marking: what we see is influenced by what we know, and meaning is important. Also, there are cognitive tendencies of the mind that transcend local culture, such as division and repetition. This view was stimulated by his experience of early Native American drawings. He strived to understand these universals and built his program around them.

Cohen's work is provocative. AARON is acknowledged as successful in creating images read as art. Its output does not look random, and it is easy to attribute intentionality and autonomy to the program. Cohen has succeeded in embodying art-making concepts, processes, and representations. Viewers may disagree on how much they like its work, how "skillful" they think it is, or how good, compelling, or relevant it is as art.

Sadly, much of the significance of AARON may be opaque to art audiences. What choices did Cohen have to make over the years about ways to represent art making? What are the decision processes that the program uses? What other kinds of choices might have been made? What difficulties were confronted in its construction? What issues still remain? If the creation of the system is a major part of the artwork, then understanding its composition and function is a critical part of its appreciation. Also, the work raises larger artistic questions. Cohen claims that he is discovering universals in perception and response to images. Yet AARON has a recognizable style and "signature." Like many other science-inspired artists described in this book, the skepticism of cultural analysis over the possibility of transcending historical and cultural contingency confronts them, just as it does the sciences.

Artificial-Intelligence-Based Musical Composition and Performance

The music and AI community has similarly been active in research to create artificial composition and performance systems. They have met with some success, and now there are even commercial software products that help in composition or allow a user to enter music to which an artificial combo will appropriately respond and improvise. *Tod Machover* at the Opera of the Future group at the MIT Media Lab is interested in systems that enhance the performance experience of all levels of performers by creating instruments that "understand the artistic intentions of the performer." Attempting to integrate Marvin Minsky's *Society of Mind* and ideas of hypertextuality, *Sharon Daniel* worked with *The Brain Opera* project to create a structure for cumulative collaborative intelligence that seeks "to abandon conceptual systems based on ideas of center, margin, hierarchy, and linearity and find an alternative to the Cartesian model of subjectivity—a new, "recombinant" subjectivity that is multiple, distributed, and associative." Composer *Laurie Spiegel* embued her software *MusicMouse* with the intelligence to generate accompaniment in several styles to go along with whatever musical choices users make as they use the software. Composer *David Cope* developed a program that could inherit the style of a known composer and then generate new compositions in that style. His books *Computers and Musical Style* and *Experiments in Musical Intelligence* report on this work.

Several of the major electronic-music research centers around the world support AI-related research. Europe especially has a long tradition of AI and music research, such as that undertaken by *Peter Beyls* and the Music, Mind, Machine research group on music cognition at the University of Nijmegen and the University of Amsterdam, in

Tod Machover: ⟨http://brainop.media.mit.edu⟩
Sharon Daniel: ⟨http://arts.ucsc.edu/sdaniel/new/brain2.html⟩
Laurie Spiegel: ⟨http://retiary.org/ls/index.html⟩
David Cope: ⟨http://arts.ucsc.edu/faculty/cope/home⟩
Music, Mind, Machine research group: ⟨http://stephanus2.socsci.kun.nl/mmm/⟩

the Netherlands. Other researchers working on algorithmic composition include *George Lewis, Christopher Dobrian, Carl Stone,* and *Clarence Barlow.* Books such as *Music with AI: Perspectives on Music Cognition,* edited by Mira Balaban, Kernal Ebcioglu, and Otto Laske, summarize some of this research. *Ron Pellegrino* investigates "visual music," including systems to visualize underlying structures as a compositional tool.

Image and Speech Recognition

Over the years, AI researchers have toiled to create programs and computer systems that could successfully recognize speech or make sense of visual information. They have achieved some success. Commercial programs are available for speech recognition that can perform tasks once thought impossible, for example, speaker-independent, telephone-response systems. Image recognition has similarly flourished, for example, OCR in text scanning, missile targeting systems based on scene understanding, robot vision systems for patrolling assembly line problems, and fingerprint recognition systems for law enforcement. Each of these accomplishments require sophisticated knowledge representation and deduction systems that were only research topics a few decades ago.

As is typical of emerging research fields, few people refer to the exotic research origins once they become incorporated into everyday technical realities. For example, rarely does anyone hear mention of artificial intelligence in the product promotions for OCR. Many challenges still confront researchers in areas that could be called artificial intelligence, but they are no longer identified under that unified rubric. For example, researchers continue to work on face recognition, the interpretation of live video, and gesture recognition systems. Since speech recognition and visual interpretation are seen as uniquely human abilities, it seems appropriate that artists would be interested, although they, like the researchers, have moved into fields such as gesture recognition, motion tracking, and face recognition (see chapter 7.4).

Naoko Tosa

Japanese artist Naoko Tosa walks in both the art and research worlds. She works at the ATR Media Integration Research Center on pioneering new, more powerful interfaces between humans and machines. Her interests include speech recognition, gesture recog-

---

Ron Pellegrino: ⟨http://www.microweb.com/ronpell/VisualMusic.html⟩
Naoko Tosa: ⟨http://www.mic.atr.co.jp/~tosa/⟩

nition, artificial intelligence, and affective computing (see chapter 2.5). She creates installations and interactive movies in which the characters react to visitors in complex, sensitive ways. Her projects are full of poetry and scientific rigor, and win acclaim in both the art and research worlds. She exemplifies the new hybrid artist-researcher beginning to emerge as artists become pioneers in working with new technologies. A sampling of her presentations and papers illustrate the range of her work, for example, "The Esthetics of Artificial Life Characters," "Artistic Communication for A-Life and Robotics," and "Creating a Movie with Autonomous Actors That Respond to Emotions."[6]

Neuro Baby, one of her best-known installations, uses sophisticated neural-network programming to create a computer graphic entity that responds to the emotional tones of voices. The baby responds appropriately with crying or cooing depending on the way the viewer addresses it. The artist's statement from the Ars Electronica showing in 1993 explains the intention:

This work is the simulation of a baby, born into the "mind" of the computer. *Neuro Baby* is a totally new type of interactive performance system, which responds to human voice input with a computer-generated baby's face and sound effects. If the speaker's tone is gentle and soothing, the baby in the monitor smiles and responds with a prerecorded laughing voice. If the speaker's voice is low or threatening, the baby responds with a sad or angry expression and voice. If you try to chastise it with a loud cough or disapproving sound, it becomes needy and starts crying. The baby also sometimes responds to special events with a yawn, a hiccup, or a cry. If the baby is ignored, it passes time by whistling, and responds with a cheerful "Hi" once spoken to.[7]

The artist sees *Neuro Baby* as an exploration of possible directions in future artificial life that will show signs of intelligence and sensitivities beyond normal speech recognition:

Neuro Baby can be a toy or a lovely pet—or it may develop greater intelligence and stimulate one to challenge traditional meanings of the phrase *intelligent life.* In ancient times, people expressed their dreams of the future in the media at hand, such as in novels, films, and drawings. *Neuro Baby* is a use of contemporary media to express today's dreams of a future being.[8]

The extension of *Neuro Baby,* called *MIC* and *MUSE,* elaborated on the themes. *MIC* was a male baby with a distinct personality who responded more clearly to a variety of emotions, using voice patterns to make interpretations of emotions. For example, joy (happiness, satisfaction, enjoyment, comfort, smile) was extrapolated from an exciting, vigorous voice that rises at the end of a sentence. Anger (rage, resentment, displeasure)

Fig. 7.6.4. Naoko Tosa, ATR—MIC Labs. Mic, an artificial computer character that detects emotional tone in speech and reacts with its own emotional response.

was derived from a voice that falls at the end of a sentence. Other emotions it attempted to read included surprise, sadness, disgust, teasing, and fear. The work also allowed viewers to make music to communicate particular emotions. An ambitious later system called *Play Cinema* (in collaboration with Ryohei Nakatsu and Takeshi Ochi) attempted a fully integrated, networked, multi-person, interact-anytime interactive cinema system in which participants could interact with fictional characters, such as in a demonstration version of *Romeo and Juliet*. The system included speech, gesture, and emotion recognition.

Other Artists and Projects

Nick Baginsky produced the installation *The Three Sirens* based on neural networks. These improvisational robots learn rules about improvisation and instrumental sound. They analyze the sound they hear and learn to control their motors and mechanical sound makers to create further improvised music. MIT Media Lab researcher ***Jonathan Klein*** has created *TurnStyles: A Speech Interface Agent to Support Affective Communication*. This system detects

Nick Baginsky: ⟨http://www.provi.de/~nab⟩
Jonathan Klein: ⟨http://affect.www.media.mit.edu/projects/affect/AC_research/projects/turnstyles.html⟩

the emotional content of speech using techniques such as conversational pacing, turn-taking patterns, and "expressive paralinguistics" in order to adjust its own voice output to be more in tune with the user's emotional state. Commercial researchers have pursued related lines of voice inquiry, although not in the same way, for example, voice analyzers that attempt to determine stress or lying. There is even one called *Sales Edge* that purports to let you know when a prospect is ripe for the closing. In these kinds of research, investigators are trying to create artificial intelligence that is at the edge of or beyond human capabilities. ***Ray Kurzweil*** developed the *Cybernetic Poet,* which can produce respectable poetry in a variety of styles.

Interactions with Artificial Characters

Researchers and artists are attempting to understand and model core elements of human personalities. Historically, researchers attempted to focus more on the information-processing capabilities of artificial intelligence, for example, could computers be programmed to excel at playing chess, solving mathematical problems, or helping a robot make simple visual determinations? Expert systems research attempted to model the decision-making processes of experts in narrowly defined realms such as medical triage, mineral prospecting, or making loans. A few researchers attempted to model more affective aspects of personality, such as Kenneth Colby's PARRY,[9] a modeling of a paranoid personality, but most of the field seemed to hold to the view that intelligence could be somehow disembodied and separated from affective elements.

Some artists question the wisdom of this approach. They attempt to model computer entities that demonstrate personality, intentionality, point of view, and emotions, and they create installations that investigate the responses of viewers to interactions that have these kind of digital characters. They suggest that the typical strategy of ignoring these elements of human functioning was doomed to create low-functioning and unacceptable digital characters. In my paper "AI Research as Art" I explained this view:

[M]any of the decisions to be made about the shape of AI programs are not purely technical. The simulation of human information processing outside of narrow realms, and the creation of machine partners which interest and satisfy humans, will depend on sophisticated artistic and psychological design choices as well. Some discourse about AI seems to imply that intelligence

---

Ray Kurzweil: ⟨http://www.kurzweilcyberart.com/⟩

can be viewed as an abstract, disembodied process. This view assumes that there is a "correct" way for the processes of natural language understanding, planning, problem solving, or vision to function and that there are "raw" meanings that programs can understand and manipulate. In this view, understanding and problem solving are technical accomplishments that can be assessed objectively. . . . [H]umans crave texture in interactions with intelligent entities that go beyond technical correctness, for example, personality, mood, purposiveness, sensitivity, fallibility, humor, style, emotion, self-awareness, growth, and moral and aesthetic values. Disaffection with limited interactions will become more severe as AI applications spread. . . . The texture of interaction is not a superficial decoration but is intrinsic to the basic fabric of humanlike understanding and intelligence.[10]

Artists have a perspective and contribution to make to AI research. Literature, theater, and the arts have been the main cultural repository of the concern with fictional characters. An AI entity, after all, is a fictional character. The contemporary research world has begun to recognize this, and several major research centers, such as Microsoft, Stanford, and the MIT Media Lab, have research groups focused on "affective computing."

Sara Roberts

Over the last decade, Sara Roberts has been creating artworks exploring elements of artificial intelligence. An early videodisk piece called *Early Programming* presented the viewer with a fictional computer-controlled child. As various events came up, the viewer could pick what to say to the child in response, for example, typical parental responses promising punishment. The child's ultimate development and mood depended on the viewer's decisions as a virtual parent.[11]

A later installation, derived from a novella by Goethe, called *Elective Affinities,* featured four virtual people interacting—a couple and two friends as they drove along. Here is Roberts's description:

His [Goethe's] story, which relates questions of passion and commitment to the chemistry of natural elements, features a complex relationship between four people. In this adaptation, each character is "played" by a computer that is networked to the other three; thus, each character is "aware" of the moods and movements of the others.

Sara Roberts: ⟨http://shoko.calarts.edu/faculty/sroberts.html⟩

Fig. 7.6.5. Sara Roberts, *Elective Affinities*. Visitors' motions allow them to overhear the thoughts of each computer character in a conversation.

The viewer, when coming within a small distance of any one of the pedestals, activates a speaker in the pedestal and "overhears" the thoughts of that particular character. It is impossible to listen to more than one of the characters at a time unless there is another person in the room—the viewer must in a sense "elect an affinity" by choosing which character to listen to.

The four characters are networked, and when one character glances at another a message is sent that tells who is looking and what mood he or she is in. The character's emotional engine will determine whether they look back or ignore the glance, but their mood will always be affected by these interactions.[12]

Stephen Wilson

Several of my own works have explored aspects of digital characterization. *Excursions in Emotional Hyperspace* allowed viewers to interact with four mannequins who were having a conversation, and *Is Anyone There?* asked passersby to have conversations with artificial characters via pay phones.

Stephen Wilson: ⟨http://userwww.sfsu.edu/~swilson/⟩

Fig. 7.6.6. Stephen Wilson, *Excursions in Emotional Hyperspace*. Four mannequins seem to be tracking each other's speech.

In *Excursions in Emotional Hyperspace,* visitors entered a room inhabited by four mannequins dressed to represent four different characters. Each held a particular pose; each represented a different fictional person who had a specific set of attitudes. One of the characters was angry and rebellious; another was happy to be part of the event; another was reluctantly submissive; and another was philosophical and tried to take the big view. The Web site explains:

Each mannequin wanted to tell its story and express its perspectives on being part of the event. Visitors were invited to walk around the room and look closer at the mannequins. Standing in front of a mannequin caused it to start talking about how it felt about being there, using a digital voice. Continuing to stand there caused it to go deeper into those perspectives. Walking away caused it to stop talking. . . .

Walking to another mannequin, however, did not cause it to just start anew. The new mannequin would comment on what the previous mannequin had just said from its own perspective.

The mannequins seemed to be actively listening to each other and tracking the conversation. The manequin would then share its own opinions.

Visitors thus had the experience of encountering artificial characters with intelligence and points of view, although this system lacked any advanced AI capabilities. The mannequins could not really recognize voices or parse the words of their fellow characters. All possible combinations of motion sequences were predetermined and all appropriate comments on previous statements were prerecorded. Nonetheless, the experience for visitors did simulate contact with artificial characters.

Is Anyone There? (see also chapter 6.2) was an interactive telecommunications event that explored issues such as the linkage of telecommunications and alienation, and the possibilities for contacts with artificial characters. A computer called five public telephones in San Francisco every hour on the hour and tried to talk to people, recording the conversations. Probing people's attitudes toward synthetic characters, this event explored how self-revealing those who answered the phone would be with the various digital characters programmed into the computer and how gallery observers felt about these exchanges. Although lacking any real language-parsing capabilities, it incorporated strategies such as phrasing, pacing, and repartee in order to make conversation believable and engaging as an interchange.

Luc Courchesne

Luc Courchesne creates installations in which visitors interact with artificial characters. These interactions are crucial in gaining access to all parts of the event. In *Landscape One,* visitors enter into a public park full of virtual characters projected onto a photorealistic 360-degree landscape that represents a public garden. They need help to explore the environment:

For this they have to make contact with one of the virtual characters by selecting, using voice or touch, questions or comments from imposed sets. Questions on, for example, where they are, what is around, where one can go from here will engage a conversation leading to some form of relationship. The exchange may be cut short, with everyone going back to their business, or it may reach a point where visitors will convince a character to lead them somewhere. In such

---

Luc Courchesne: ⟨http://www.din.umontreal.ca/courchesne/⟩

Fig. 7.6.7. Luc Courchesne, *Landscape One*. Visitors must interact with artificial characters to gain access to deeper levels of conversation.

a case, visitors are being pulled through the landscape after their virtual guide and the whole room appears to be moving in this direction. . . .

This journey through space is also a journey through words, meanings, language, subjectivity. It highlights not only the physical world in which this is happening but also its diverse meanings and functions to different people. The experience is about communication/discommunication between people, with movements through space as manifestation of its nature; successful forms of communication will offer visitors more varied inroads into more remote places.

Other works involving conversations with artificial characters include *Family Portrait, Conversation with a Virtual Being,* and *Encounter with a Virtual Society.* In *Hall of Shadows,* visitors enter into a conversation among four virtual friends, and their actions determine whether they will be invited into the deepest circle of conversation. The four levels of conversation are casual chat, an attempt by the visitor to make friends, spirited debate by the characters, and revelation of a virtual being's existential crisis. The visitor faces a choice to help or "abandon them to their fate." The installation questions the "meaning and value of human life in the age of cyber culture," exploring such questions as: What is the nature of man/woman relationships? and What sort of politics is likely to dominate in the future?

Other Artists and Projects

Joseph Bates and his colleagues have been working for many years on an interactive drama environment called *Oz.* Building on concepts and techniques from artificial intelligence, *Oz* places artificial characters in a computer-graphic, artificial world with which the viewer can interact. Each character's mind is modeled, including such features as intentions and emotional tendencies that guide the characters' interactions with each other and the interactor. *Oz* has been shown both in art settings, such as Ars Electronica, and in academic research settings, such as AI meetings. The project uses the phrase *interactive drama* to mean the "presentation by computers of rich, highly interactive worlds, inhabited by dynamic and complex characters, and shaped by aesthetically pleasing stories." The system tries to embody dramatic theory and principle in order to create a cathartic experience and to make "believable agents" that manifest personality, emotion, self-motivation, change, and the awareness of social relationships. Bates and his colleagues call their work "techno-artistic" research.

In *The Other,* **Catherine Ikam and Louis François Fleri** create "emotional fiction" digital video projections of 3-D characters that seem to track the audience and indicate emotional expression. In *Ultima Ratio,* **Daniela Alina Plewe** uses AI decision-support systems in order to arrange for participants to encounter the dilemmas of heroes from stories such as *Hamlet, Casablanca, Robocop,* and *Medea.* The system generates "inner monologues of the characters and visualizes them in real-time as 3-D diagrams." Partially influenced by Chinese puppet traditions, **Mary Flanagan** has been working with multiuser VRML virtual performance spaces that include autonomous synthetic characters. In **Andrew Stern**'s *Virtual Babyz* viewers interacted with a world of 3-D simulated artificial babies.

MIT Media Lab's Synthetic Characters Group is seeking to make "intelligent" animated characters with whom viewers can interact. They see the need to span several disciplines, such as animal behavior, traditional character animation, artificial intelli-

---

Joseph Bates: ⟨http://www.lb.cs.cmu.edu/afs/cs.cmu.edu/project/oz/⟩
Catherine Ikam and Louis François Fleri: ⟨http://www.235media.com/media_art/install/the_other.html⟩
Daniela Alina Plewe: ⟨http://www.canon.co.jp/cast/artlab/pros3/ur.html⟩
Mary Flanagan: ⟨http://mflanagan.fal.buffalo.edu/paperperf.htm⟩
Andrew Stern: ⟨http://www.babyz.net/index.html⟩
MIT Media Lab: ⟨http://characters.www.media.mit.edu/groups/characters/⟩

gence, robotics, and computer graphics and modeling, and to integrate artistic, media, and scientific perspectives. They seek to create a "sense in the user's mind that the interactive characters with whom they are interacting are 'sentient' beings." They use three major techniques: "First, the behavior and motion of the interactive characters must be believable and expressive and change over time in response to their interaction with the user; second, staging, lighting, and camera decisions must work together to reveal the internal state of the characters at any given instant; third, the manner in which the user interacts with the other characters must be tangible and compelling." They note that the "dramatic and visual presentation are not just icing on the cake, but rather intrinsic elements in the creation of believable characters and successful interactions."

In *Claudio Pinhanez*'s *It/I,* a computer-character video observes the actions of a human actor and produces appropriate computer graphics and sound in response. The *Stanford Virtual Theatre* project studies the creation of automatic intelligent characters that can interact with humans or each other. *John Manning*'s *Problematicize This* explores "synthetic sociology" by having networked computers simulate panelists at a symposium, with the audience passing judgment on how well they did. In *You Could Be Me,* *Nell Tenhaaf* tests participants on "their adaptation to artificial empathy." A video character solicits as much input and intimacy as she can get and uses a genetic algorithm based on their answers and input style to calculate a set of "genes" for empathy fitness that cumulate in the evolving style of the video character.

Affective Computing

Affective computing is another research area likely to involve artists in the future. Researchers are asking how they can enable computers to understand the emotions of users and synthesize emotions in the machine's processes. They are not restricting themselves to the creation of specialized worlds as described in previous sections, but rather seeking to understand how affective elements might be useful as a pervasive feature of people's interactions with intelligent systems. Many researchers believe that emotional life needs to become a major consideration in the design of interfaces and must be considered a part of machine "intelligence."

Claudio Pinhanez: ⟨http://www.media.mit.edu:80/~iti/⟩
Stanford Virtual Theatre: ⟨http://www-ksl.stanford.edu/projects/cait/index.html⟩
John Manning: ⟨http://www.ylem.org/newsletters/JulyAug98/article3.html⟩

Fig. 7.6.8. MIT Media Lab Affective Computing Group, *The Affective Tiger* reacts to his playmate with a display of emotion, based on his perception of the mood of play.

Research focuses on many topics, such as extensions of artificial intelligence, wearable computing, and using nonconventional sensors such as galvanic skin response detectors to ascertain emotional states. Researchers in other places in the world, such as Stockholm's KTH, also support affective computing research. Rosalind W. Picard's book *Affective Computing* summarizes much of the work. Here is the research mission statement from the MIT Affective Computing Group:

Understanding affect in humans[:] Recent neurological studies indicate that the role of emotion in human cognition is essential; emotions are not a luxury. Instead, emotions play a critical role in rational decision-making, in perception, in human interaction, and in human intelligence. . . . [W]e are studying how affect is expressed in the body as a pattern of biological responses so we can build computational recognition systems that can reliably sense human affect. . . . Supporting humans as affective beings. Current computer systems are incapable of either sensing or responding to users who are frustrated, intimidated, or pleased with their computer. State-of-the-art educational technology systems have no way of telling whether a hard math problem is motivating to the learner, or making him/her feel defeated.

MIT Media Lab, Affective Computing: ⟨http://www.media.mit.edu/affect/⟩

Synthesizing affect in the machines themselves[:] . . . We believe that this added capacity may enable computer systems to make decisions in fundamentally different ways—ways we suspect that are much more effective than the Boolean paradigm allows. Further, embuing computer systems with synthesized emotions will allow them to interact with users on a much more human level.[13]

Stanford researchers Byron Reeves and Clifford Nass have looked at related phenomena in their research on social responses to computing. They found that many people attribute social motivations, feelings, and behavioral tendencies to computers and other technology, for example, thinking about their computers as having a "bad day."

The research reported in their book *The Media Equation: How People Treat Computers, Television, and New Media Like Real People and Places* is often quite intriguing, surprising, and humorous. For example, in one experiment people were more likely to critique a computer's performance as a tutor when they were in another room than when in the same room as the computer, thus reacting with the same kind of social sensitivity people would show toward another person.

Microsoft introduced an interface wizard called Bob that was based on some of this research. Significantly, it crashed in the marketplace. Understanding and working with people's responses to technology will be a significant challenge to artists in the coming decades. Research into affective and social computing are significant resources and challenges for art interventions that seek to explore the boundaries of what machines can be.

Agents

Although artificial intelligence no longer commands the attention it once did, much research continues in the development of autonomous "agents." Agents are software elements with the intelligence to take care of some set of specific tasks. Often these tasks require great sophistication in understanding the user's needs and integrating information, what in the old days might have been called artificial intelligence. Some agents are conceived to undertake quite specific, limited tasks; others are quite ambitious, for exam-

---

Byron Reeves and Clifford Nass: ⟨http://hci.stanford.edu/ls/html/body_social_responses.html⟩

ple, buying agents that will go out on the Net, find the best price for items it knows you want, negotiate with sellers (or their agents) and actually consummate the purchase. Some agents have explicitly programmed operations; others have the capability to learn by watching the user's actions or from their prior experience. Some are presented merely as software; others are presented in almost anthropomorphic terms. Agents are one of the significant offspring of the AI research of the last decades.

In its visionary early decade, Apple Computer was one of the major corporate sponsors of agent research. Apple created a famous video portrayal of the future called *The Knowledge Navigator*. In this video, a professor from the future had a small, digitally intelligent "butler/assistant" named Phil who lived in his computer and took care of all kinds of business for him. It located and analyzed information resources, contacted colleagues, set up appointments, prepared presentations, answered the phone, talked to his mother, and reminded him of family obligations. Phil is simultaneously visionary and troubling. He is emblematic of the set of aspirations, research directions, and commercial trends that surround the topics discussed in this chapter. This constellation is significant enough to warrant artistic involvement and commentary.

Brenda Laurel and Abbe Don

Apple also undertook research to actualize these visions. Significantly, it involved artists in the research teams to explore these capabilities. For example, Brenda Laurel and Abbe Don worked with the Human Interface group to create the agents called Guides to help in information navigation with early hypermedia. Here is Brenda Laurel's definition of agents from her chapter "Interface Agents: Metaphors with Character" in the book *The Art of Computer-Human Interface:*

An interface agent can be defined as a character, enacted by the computer, who acts on behalf of the user in a virtual (computer-based) environment. Interface agents draw their strength from the naturalness of the living-organism metaphor in terms of both cognitive accessibility and communication style. Their usefulness can range from managing mundane tasks like scheduling, to handling customized information searches that combine both filtering and the production (or retrieval) of alternative representations, to providing companionship, advice, and help throughout the spectrum of known and yet-to-be-invented interactive contexts.[14]

According to Laurel, the key characteristics of agents include agency (the ability to take action on behalf of another), responsiveness, competence, and accessibility. MIT Media Lab's Software Agents research group defines *agents* this way: "computer systems to which one can delegate tasks. Software agents differ from conventional software in that they are long-lived, semi-autonomous, proactive, and adaptive."[15]

The Guides were information navigation agents designed to help users of an early hypermedia *Grolier Encyclopedia* work on American history. The goal was to overcome the disorientation and "cognitive load" that often goes with navigating complex inter-linked hypermedia documents. Guides appeared in personas such as a pioneer man, a settler woman, and a Native American. The design team wanted to reinforce the idea that point of view influences the way history is framed and to give users the experience of multiple perspectives on particular events.

Agents remain somewhat controversial. Many feel that the anthropomorphism is inappropriate and diversive. Laurel argues that anthropomorphic qualities actually can enhance the efficiency and effectiveness of computer-human interactions:

By capturing and representing the capabilities of agents in the form of character, we realize several benefits. First, this form of representation makes optimal use of our ability to make accurate inferences about how a character is likely to think, decide, and act on the basis of its external traits. This marvelous cognitive shorthand is what makes plays and movies work. . . . With interface agents, users can employ the same shorthand—with the same likelihood of success—to predict, and therefore control the actions of their agents. Second, the agent as character . . . invites conversational interaction. This invokes another kind of shorthand—the ability to infer, cocreate, and employ simple communication conventions.[16]

Laurel's book *Computer as Theatre* is a landmark in the thinking about the nature of agents and the future possibilities of computer autonomous agents. She analyzes the contribution that theater can make to thinking about and designing human interactions with computers.

Contemporary Agent Research

Researchers are currently a bit less enthusiastic about projecting great anthropomorphic futures for agents. Yet ambitious research is under way around the world in corporate and academic research labs spurred on by the challenges and opportunities of the World Wide Web. In one early example, "Firefly," a music site, invented agents that could make suggestions about music that a Web visitor might be interested in by analyzing the choices of people similar to the user. The research mission of MIT Media Lab's Software Agents research group, under the direction of Pattie Maes, gives an idea of the kind of research being pursued worldwide:

The Expert Finder Project
Building Agents to Assist in Finding Help

Fig. 7.6.9. Adriana S. Vivacqua, MIT Media Lab's Software Agent Research group. Logo for a project to help people find experts who can help them.

Our group has projects on information filtering agents, agents as navigation guides, remembrance agents, recommender agents, matchmaking agents, and buying and selling agents. Some themes we study are personalization, user profiling, information filtering, privacy, recommender systems, electronic commerce, community ware, learning user profiles, reputation mechanisms, negotiation mechanisms, and coordination mechanisms.[17]

Summary: The Role of the Arts in Artificial Intelligence and Agent Research

It may be that agents will become a pervasive feature of life, like the telephone. They may come to have great impact on the everyday details of life. Even more profoundly, they may influence the ways we define who a person is and the borders between ourselves and our technological systems. Much of what we do in the future communications and computer-mediated world may be executed in our stead by our agents. Much of what we know of other people and institutions may be derived from exchanges with their agents.

If these developments play out as they may, their shaping is much too important to leave to technical specialists alone. They imply questions asked by artists and those in the humanities for centuries. Even at this initial research stage it is easy to see the wider impact of the questions that agent designers must answer. Many systems depend on the agent understanding a person's goals, needs, and predispositions—what is sometimes called profiling and personalization. How does one come to understand these qualities and to represent them? How much does a recommender agent need to know to do a good job? Since our agents will represent us in dealings with others, what kind of face do we want to present? What are the ethical standards that we want our agents to manifest in their negotiating? Our agents will often filter information for us—what information is important to each person? What information might be important even though a person consciously would not identify it as so? What principles of filtering should be implemented? Agents are not just neutral technical inventions. We must be aware of the metanarratives they instantiate.

My paper "Artificial Intelligence Research as Art" sums up the critical role of the arts:

If we are going to have artificially intelligent programs and robots, I would have sculptors and visual artists shaping their appearance, musicians composing their voices, choreographers forming their motion, poets crafting their language, and novelists and dramatists creating their character and interactions. To ignore these traditions is to discard centuries of experience and wisdom relevant to the research questions at hand.[18]

Notes

1. A. Barr and E. Feigenbaum, *The Handbook of Artificial Intelligence*. vol. 1 (Los Altos, CA: Morgan Kaufmann, 1981), p. 3.

2. S. Wilson, "Artificial Intelligence Resarch as Art," ⟨http://shr.stanford.edu:80/shreview/4-2/text/wilson.htmlb⟩ (also in print in *SEHR*, vol. 4, issue 2: "Constructions of the Mind," 1995).

3. R. Lauzzanna, "*Languages of Design* Archives," ⟨http://www.penrose-press.com/scrips/index.cgi⟩.

4. H. Harry, "Machine Superiority," ⟨http://www.media-gn.nl/artifacial/HHLecture.html⟩.

5. H. Cohen, "1972 Catalog Statement," quoted in P. McCorduck, *Aaron's Code* (New York: W. H. Freeman, 1991), p. 41.

6. Papers available at ⟨http://www.mic.atr.co.jp/~tosa/⟩.

7. Ars Electronica, "*Neuro Baby* Description," ⟨http://www.aec.at/fest/fest93e/tosa.html⟩.

8. N. Tosa, "*Neuro Baby* Description," ⟨http://www.mic.atr.co.jp/~tosa/⟩.

9. Described in D. C. Dennett, "Can Machines Think?" in R. Kurzweil, ed., *The Age of Intelligent Machines* (Cambridge: MIT Press, 1990), p. 52.

10. S. Wilson, op. cit.

11. S. Roberts, "Early Programming: An Interactive Installation," *Leonardo* 24(1) (1991): 90.

12. S. Roberts, "Description—*Elective Affinities*," ⟨http://mitpress.mit.edu/e-journals/Leonardo/isast/wow/wow303/roberts.html⟩.

13. MIT Media Lab Affective Computing group, "Research Goals," ⟨http://www.media.mit.edu/affect/⟩.

14. B. Laurel, "Interface Agents: Metaphors with Character," in B. Laurel, *The Art of Computer-Human Interface* (Reading, Mass.: Addison-Wesley, 1990), p. 356.

15. MIT Media Lab Software Agents research group, "Definition of Agents," ⟨http://agents.www.media.mit.edu/groups/agents/⟩.

16. B. Laurel, op. cit., p. 358.

17. MIT Media Lab Software Agents research group, "Research Agenda," ⟨http://agents.www.media.mit.edu/groups/agents/⟩.

18. S. Wilson, "Artificial Intelligence Research as Art," ⟨http://userwww.sfsu.edu/~swilson/⟩.

Information and Surveillance

Information Management, Visualization, Commerce, and Surveillance

One of the main applications of digital technology is the acquisition, storage, classification, manipulation, and retrieval of information. Digital systems afford unprecedented speed, efficiency, and reach in these operations. In benign applications such as research, information systems allow individuals to be aware of much more potentially relevant resources than ever before. Information visualization provides new tools for comprehending complex bodies of information. New sensors allow access to once unavailable parts of the electromagnetic spectrum. In the darker vision, information systems open the door to widespread, panoptical surveillance, control, and manipulation by corporate, government, and military agencies. For example, Eugene Thacker's "Database/Body: Bioinformatics, Biopolitics, and Totally Connected Media Systems" in the e-zine *Switch* analyzes contemporary biology as an extension of surveillance into the realm of the body (see ⟨http://switch.sjsu.edu⟩).

This era is seen as the information age. More people are employed in information work than in the production of physical goods. Corporate wealth is determined more by information than real estate. Scientific method and technological development's main value is seen in their creation and tracking of information. Some analysts elevate information to the highest medium of power and control. Intellectual property is more important than physical property. Some visions propose to use ID technologies such as bar codes to track products from raw material through to the consumer product and to its ultimate disposal. They also propose to track persons from birth to death. The analysis of these claims is complex, and a full discussion is beyond the scope of this book, although many sections address elements of the analysis.

Information is a slippery term. For example, almost all art could be considered information. Some artists, however, explicitly claim to be addressing issues of surveillance and information systems in contemporary society. This chapter reviews their work.

The modern multinational corporation is a prime example of the institution created by the information age. Many artists are exploring the form of these organizations with their convergence of research, technology, telecommunications, marketing, and media. They are creating alternative art organizations as comments on the form and as competitors in mind space. Others are trying to cultivate free access to databases as information becomes property. This chapter reviews artists who comment on our culture's focus on information systems.

Fig. 7.7.1. Perry Hoberman, *Bar Code Hotel*. Visitors scan bar codes to influence graphic projections and sound.

Bar Codes and ID Technologies

Perry Hoberman

Perry Hoberman's *Bar Code Hotel* confronts visitors with an environment in which everything is bar coded. An entire room is covered with printed bar-code symbols that may be "read" by the guests. Each symbol triggers and controls the presence, location, sound, form, and behavior of virtual objects presented in a large stereoscopic display. It is provocative for thinking about the implications of a society that attempts to bar code everything.

Other Artists and Projects

In *Items 1–2,000: A Corpus of Knowledge on the Rationalized Subject* (see chapter 2.5), **Paul Vanouse** explores the bar code as an icon of the reduction of persons to information slots. He places a nude performer in a wax display case covered with bar codes. In a ritual of objectification, visitors can scan the bar codes to bring up standardized medical

Perry Hoberman: ⟨http://www.hoberman.com/perry/⟩
Paul Vanouse: ⟨http://www.contrib.andrew.cmu.edu/~pv28/loaded.html⟩

images of regions of the body that lie under the bar-code markers. ***Eduardo Kac***'s project to embed an identification chip in his body (see chapter 2.5) also reflects on the intrusion of these technologies.

Surveillance

Julia Scher

Julia Scher asks audiences to consider the significance of technologically mediated surveillance (see also chapter 6.4) and reflect on the omnipresence of surveillance. Her installations typically confront visitors with the experience of being under surveillance. The visitors are surrounded by cameras and monitors on which they can watch other visitors, themselves, and strange events apparently happening in the gallery space. She often appears in her security guard persona.

In the *Security by Julia* series, she poses in the gallery, populated with surveillance technology, as a guard in a pink security uniform. She invites visitors to print out "souvenir" images of themselves under surveillance. In *The Institutional State,* in Spain, she intermixed surveillance images from a mental hospital with those from the museum. In Freiburg, Germany, she placed the whole city under surveillance cameras. In *Always There,* she presents an installation of a bed with four cameras pointed at the surface.

The documentation of her *Securityland* installation in Chicago's Museum of Contemporary Art show explains her tactics of drawing visitors in and in playing with the horror, thrill, and complicity of being watched:

Scher knows both the equipment and how people willingly participate in systems over which they are powerless. . . . Often accompanying the surveilled images is text across the monitor screen or a spoken track recorded by Scher. In her *Security by Julia* persona, Scher speaks in a smoothly seductive voice that is reassuring in the same way as her shirt that reads "Don't Worry." In these sound works her voice takes on a sweetly authoritative, institutional tone as she persuades listeners, "Please do not leave until the sensors have completely absorbed you." The mission statement of her spoof on-line service, Information America, states: "We utilize freshly gathered judgments, identifications, and verifications to make our bright, shiny, and vitamin-rich database state-of-the-art. . . . Our goal is not to manage individuals, only space.". . . .

Eduardo Kac: ⟨http://www.ekac.org/⟩
Julia Scher: ⟨http://adaweb.com/project/secure/corridor/sec1.html⟩

Fig. 7.7.2. Julia Scher, *Securityland*. Scher in her reassuring security guard uniform. Courtesy of Walker Art Center Digital Arts Study Collection. Photo: Joseph Cultice. See ⟨http://adaweb.walkerart.org/project/secure/corridor/sec1.html⟩.

[Scher:] "In a gallery situation my first task is to gently suck you in and let you get used to the space. Then the electronic threats and dangers are perceived, and then you get nailed."

Scher plays with the thrill of watching and being watched. In her work, the voyeuristic pleasure of observing others is matched only by the excitement of being the center of attention. . . . Using humor and outlandish tactics, Scher underscores the contradictions of our hypertechnological culture, where the promises of progress seduce us into accepting and even reveling in the creeping invasion of privacy and constant monitoring of ourselves.[1]

Simon Penny

Simon Penny's installations (see also chapters 4.3, 5.2, and 5.4) typically address issues in techno-culture with irony and humor. *Big Father* confronts visitors with five "surveillance" stations that sense their motion and respond with sound and video. He explains our immersion in the "data sphere."

Over the past twenty years, an entirely new global system of digital communication has come into being, comprised of satellite relays, optical fibre and coaxial cables, and computer networks. This augments the already vast global radio traffic. This new phenomenon is referred to as the data sphere. Examined as an organism, the data sphere is colonial in the sense that an ant colony

---

Simon Penny: ⟨http://www-art.cfa.cmu.edu/www-penny/index.html⟩

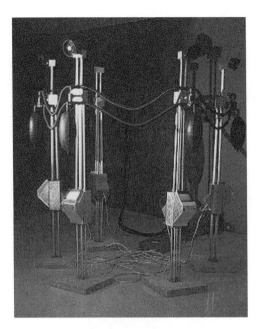

Fig. 7.7.3. Simon Penny, *Big Father*. Five surveillance stations sense motion and respond with sound and video.

or a marine sponge is colonial. Information is transmitted and received between millions of sensor and effector nodes via a distributed rhizomatic network. Viewed in this way, any electronic information-gathering device which is hooked into this system becomes a sense organ of it. These sense organs operate on a vast range of scales, from the galactic (outward-looking satellites and ground-based observatories), to the global (earth-watching satellites), the local (video surveillance systems), the personal (medical imaging technologies), and the microscopic (scanning, tunneling electron microscopes). One might even postulate an imagination or dreaming in the form of synthetic computer imagery.[2]

Steve Mann

Steve Mann is well-known for his wearable camera. He developed wireless video cameras that he wore as he went about his business in Boston. His camera's eye view was constantly uploaded to the Web, and his research was part of MIT Media Lab's wearable

---

Steve Mann: ⟨http://www.eecg.toronto.edu/~mann⟩

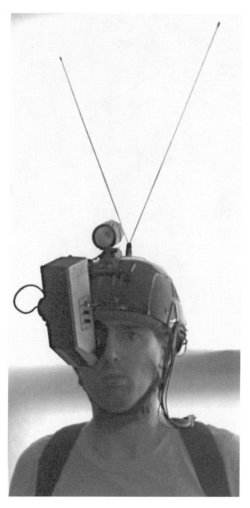

Fig. 7.7.4. A 1970s–early 1980s embodiment of Steve Mann's *WearComp* (wearable computer) invention, built for control of the wearable photographic apparatus in a mediated-reality environment, as described in ⟨http://wearcomp.org⟩.

computing projects. Mann also undertook cultural research and art explorations related to the privacy implications of his research. He often interweaves technological research and cultural commentary, for example, writing about the theft of solitude and the theft of attention.[3]

As he went about his business he had run-ins with security guards and noted that organizations that had constant surveillance cameras on people were themselves reluctant to be the subjects of the camera. In "Reflectionism and Diffusionism," published in *Leonardo,* Mann recounted how this asymmetry troubled him and heightened his awareness of the cultural significance of surveillance technologies. He developed the idea of "reflectionism" as a strategy for individuals to countervail against surveillance. The abstract to his paper explains:

The recent proliferation of video surveillance cameras interconnected with high-speed computers and central databases is moving us toward a high-speed "surveillance superhighway," . . . . The goal of this paper is to stimulate inquiry into both surveillance and the rhetoric used to justify its use. "Reflectionism" is proposed as a new philosophical and tactical framework that takes the Situationist tradition of appropriating the methodology of the oppressor one step further by targeting that methodology directly against the oppressor. The oppressor then becomes the audience of a performance resulting from this new use of his or her own methodology.[4]

Jordan Crandall

Jordan Crandall creates pseudo-technological devices designed to sensitize audiences to the way that technologies mold life. At Documenta X, his installation *Suspension Vehicle RF-7600* put visitors into a strange projection environment that looked at the effects of technology on ideas of space and movement. His "vehicle" includes innovative *hypoteheto-technologies* such as the "rhythmic fitting":

This installation looks at contemporary space as produced through the mediations of networking technologies. It positions this space as a dynamic combination of reality and virtuality, and explores its alternate modes of access, navigation, and inhabitation. It figures inhabitants whose bodily forms, capacities, and rhythms have been adjusted to fit these new conditions. It explores the ways in which viewing agencies, bodies, and inhabited spaces are crossformatted and "paced" through various protocols and vehicles. . . .

Contemporary spaces . . . are accessed, ordered, and navigated under the conditions of various on-line and off-line protocols. These protocols include computer formats and settings, social codes and customs, and structures of agreement, arrangement, and formalization. These protocols serve to initialize, configure, and normalize space, rendering it inhabitable and infusing it with organization and procedure. . . . Vehicles are the results and means of these acclimation processes.[5]

Jordan Crandall: ⟨http://www.blast.org/crandall/⟩

Crandall pays attention to emerging technologies, probes their meanings, and shares those understandings with art audiences. He has a special interest in surveillance. For example, his "Drive" essay comments on body-tracking technology such as walkmen, finger scanners, motion trackers, CFO Vision data-analysis tools, and computerized gym equipment. He unpacks the underlying meanings of exercise equipment, which he sees as a metaphor for the society's desire to regiment and control the body and submit it to "regimes of fitness."

In an exhibition at the Gering Gallery, Crandall assembled an installation that featured Readymades from the surveillance industry. Visitors were confronted by a large video projection of a Muybridge-like scene of repetitive action overlaid with the surveillance vision markings such as one sees on night-vision targeting imagers. Also available in the gallery were prototypes of other surveillance technologies, such as stereo-lithography renderings of body parts, and a retinal scanner from EyeDentify, Inc., that is used to identify 192 features of retinal patterns. Crandall invites visitors to the gallery to peer into the scanner to be identified. By presenting these technologies in an art setting, Crandall asserts attention to the emerging surveillance technologies as an art act.

Other Artists and Projects

Taking surveillance technology to its highest form and linking it with concepts of wearable systems, **Paul Vanouse** created the *Security Bra*. This device uses ultrasonic sensors to detect anyone within a security perimeter of the person wearing the bra and beeps in accordance with the distance of the intruder. "The *Security Bra* is a garment capable of returning the male gaze and electronically defining one's personal space." In **Paul Garrin**'s installation *Yuppie Ghetto with Watchdog,* the visitor observes a video of a cocktail party with military and police images of excessive aggression appearing in the windows of the party room. The visitor is blocked off from the party by barbed wire and bars. As soon as the visitor enters the room, a video-simulated dog tracks their movement (via motion trackers) and aggressively defends against any intrusion on the party (see also chapter 7.4).

Joel Slayton's work often focuses on issues of surveillance. His *Telepresent Surveillance* installation confronted visitors with mobile robots, each outfitted with wireless video-

Paul Vanouse: ⟨http://www.contrib.andrew.cmu.edu/usr/pv28/info.html⟩
Paul Garrin: ⟨http://www.235media.com/install/yuppie_ghetto.html⟩
Joel Slayton: ⟨http://c5.sjsu.edu/projects.html⟩

surveillance cameras. These robots patrolled their spaces "using interprobe communication" looking for "warm body targets" and broadcasting to the Internet (see also chapter 5.4). **Bruce Cannon**'s kinetic sculptures often focus on surveillance, for example, *Scanner* opens its ammunition case lid and looks around and closes quickly if anyone comes near, (see also chapter 5.2). **Sean Cubitt** explores ideas of cyber architecture and its interconnections with surveillance. **Concha Jerez and Jos Iges**'s *Polyphemus' Eye* presents a video-surveillance installation with complex relationships between the monitors and the monitored. In *Marking Time,* **Natalie Bookchin** confronts viewers with a CD work substantively focused on prison surveillance that surreptitiously keeps track of every mouse movement and time latency as they interact with the piece.

   Richard Lowenberg was a pioneer in the artistic explorations of military surveillance technologies such as satellites. He has long been interested in working artistically with ranges of the electromagnetic spectrum outside of visible light. In several projects he investigated the possibilities of infrared imaging and night-vision systems. His "dancer" project presents infrared images of a dancer in motion in which the breath and temperature regions of the body invisible to the naked eye are made visible. He is interested both in the intrusions of privacy and the beauty of invisible regions of the spectrum.

   Christine Meierhofer's installations sometimes make elements of her private life public. In *Pretty Good Privacy,* she produces two versions of the same room—one physical and empty, the other a virtual reproduction filled with things from her private life. The viewer is invited to take a video camera to conduct surveillance. Although the physical room is empty, the image on the viewfinder shows the same part of the room filled with private objects. **Eve Andrée Laramée**'s installation *A Scientific Observation of a Private Obsession* explores issues of "power and control in science and technology . . . surveillance, loss of privacy, control of the body, and paranoia." It confronts visitors with a strange device: "In the center of this structure is a pedestal with a mirrored box on it. Motion detectors cause the mirror box to rotate when a viewer enters inside the circular laboratory space." The imagery on the spinning mirrors is an amalgam of vessels containing

---

Bruce Cannon: ⟨http://www.jps.net/bcannon/sculptures/⟩
Sean Cubitt: ⟨http://www.isea.qc.ca/l_pool/panels/p1.html⟩
Concha Jerez and Jos Iges: ⟨http://residence.aec.at/polyphemus/⟩
Natalie Bookchin: ⟨http://www.calarts.edu/~bookchin/⟩
Richard Lowenberg: ⟨http://lda.ucdavis.edu/dept/lda/courses/classsites/archive/lda190_s97/⟩
Christine Meierhofer: ⟨http://www.icf.de/CAC/Artists/a/.start.html⟩
Eve Andrée Laramée: ⟨http://www.artnetweb.com/laramee/⟩

food eaten by artists, fragments of text, and images of the viewers. A video projection of the whole scene recursively fills the far wall, thus gazing "at viewers who are looking at art that is looking at science."

Databases and Research Processes

Natalie Jeremijenko

Natalie Jeremijenko creates art installations that comment on the conceptual structures that underlay large databases. She has a background as both an artist and engineer. She is concerned about the bases of categorization and the nature of representation imposed by computer-mediated databases and data acquisition systems. She established the "Bureau of Inverse Technologies" as a fictional organization that highlights some of these concerns by initiating information projects.

Jeremijenko explains her desire to make visible the not-so-obvious meanings implicit in the design of databases and technological information-acquisition systems. She asserts that no systems are neutral even though rhetoric often promotes that view:

I begin with the assertion that technologies are tangible social relations. That said, technologies can therefore be used to make social relations tangible. Technologies create the material conditions within which we work, and imagine ourselves and our identities. I am concerned with how technology is developed within a context where overarching priority is given to formal systems over content, and where the complicating and politicizing projects of postmodernity are marginalized.

I am interested in the epistemological work of current technologies. This includes what gets technological attention and what does not, what gets counted, and what gets left out. What is the political fabric of the information age? And what interventions can be made in a place where economics gets equated with politics, where diversity is rendered in homogeneous database fields, and where consumption forms identity?[6]

In *Slit,* she presents an installation in which video monitors are physically moved to illustrate filmic conventions such as zoom, pan, and the like. Her intention is to deconstruct the illusions of film and comment on research projects under way in order to use these conventions to characterize video information for database access. The cameras

---

Natalie Jeremijenko: ⟨http://cat.nyu.edu/natalie/⟩

Fig. 7.7.5. Bureau of Inverse Technology (engineer: Natalie Jeremijenko). The *Suicide Box* project proposes to keep systematic live-action records on suicide attempts from the Golden Gate Bridge. (Still from *Suicide Box* video; distributor: Video Databank.)

are mechanically moved to reenact film conventions, thus making obvious the usually invisible manipulations of sight. Visitors "move around and find the single privileged position from which we are accustomed to viewing film." She notes, "I use this project to introduce the way that cultural tradition, in this case filmic convention, becomes the technical imperative, and how a tradition of representation becomes operationalized, or commodified."

One of the Bureau of Inverse Technology's major projects is the *Suicide Box*. This project is presented in the same language that would be used to present a serious engineering plan. Data from its trial run is presented with appropriate statistical aplomb. The box is designed to count the number of suicides attempted off a bridge, such as the Golden Gate Bridge, by using a motion-detection trigger to capture video of anything that falls. In deadpan engineering text, the report recounts that "a recent trial period activation yielded seventeen events in one hundred days . . . a rate equivalent to 0.17 suicides per day." The text describes new sociological indices:

Considering the effects of low visibility, the event rate is 0.68 suicides per day. Data captured by the *Suicide Box* provide an increasingly accurate measure of a social phenomena not previously adequately quantified. Using this data, the Bureau has developed a new economic indicator, the Despondency Index. Dynamically updated to the Down Jones Industrial at each bridge occurrence, the Despondency Index brings the BIT data in line with the micro-attention given to market indicators.

Jeremijenko analyzes the *Suicide Box* project for what it reveals about the political nature of data abstraction. She warns about the information conventions taken for granted in technical fields such as computer science and engineering:

The politics of information is invisible in its ubiquity, and yet it has radically transformed many diverse areas of knowledge, considering the transformation effected in and with information technology in areas such as genetics, epidemiology, risk analysis, sociology. . . .

Consider the slogan used to advertise a new database package, "information is power." . . . Truth or falsity of something is seen as a property of the information such that power is the distorting lens of the information camera. Power and information are presumed extrinsic to each other and somehow independent. Fundamentally, power is not seen to affect the truth of the information, and power is not seen to contribute constructively to information.

However, the account of power and information which the Bureau's research demonstrates is one in which power does not simply impinge on information from without. . . . The very technological attention to suicide is the critical element in its existence as information—it simply did not exist in this form prior to the bureaucratic eye lent to it by the Bureau.

Stephen Wilson

I created a robotic public-art installation called *CrimeZyLand* that explored issues in media representations of crime, information visualization, and public access to complex databases. It won the competition to be placed in the San Francisco Art Commission's Exploration: City Site outdoor public-art space across from San Francisco's city hall. *CrimeZyLand* offered visitors a large outdoor map (twelve by forty meters) of the city of San Francisco in which the ten highest crime locations were indicated by tall poles. Kinetic clowns and toy police car sirens and lights would activate twenty-four hours a day at those precise times that the statistical database indicated a crime would be oc-

---

Stephen Wilson: ⟨http://userwww.sfsu.edu/~swilson⟩

Fig. 7.7.6. Stephen Wilson, *CrimeZyLand*. Public art with kinetic clowns visualizing statistical levels of crime in real time.

curring. Visitors could pick the type of crime and hear live police radio of current police actions. Internet visitors could also control the event, view current physical viewers, and speak their opinions of crime via a speech synthesizer. The Web site describes the physical event and its conceptual focus:

CrimeZyLand is an art installation, a "playland of crime" that transforms the City Site lot into a computer-controlled living "map" that creates light, motion, and sound corresponding to the minute-by-minute statistical level of crimes committed in San Francisco districts, as indicated by the police department CABLE crime statistics. In what appears to be a carnival or theme park, viewers can experience the crime "pulse" of the city firsthand. Here is its artistic agenda:

1. Crime as Entertainment: Using the strategy of absurd extension, this installation asks viewers to question the media circus created around crime. Are TV crime reports or this humorous "Disneyland of Crime" appropriate events?

2. Deconstructing Crime: What's a crime? Who defines it? What are our prejudices about crime? Are street crimes worthy of more attention than other crimes against the community, such as poisoning the bay or the creation of dangerous products that kill or maim?

3. Information visualization and access: The installation will use the tools of public sculpture

to give viewers intuitive access to this provocative information about urban life. What is an appropriate representation of the underlying information?

4. Real vs. Virtual Presence: The installation asks viewers to think about the difference between physical and Internet participation in public events. Some analysts note that because of crime, urban dwellers increasingly engage in "cocooning." This installation offers enhanced control options to those brave viewers who venture out to be physically present.

InfoWar

In 1998, Ars Electronica declared "InfoWar" as its main theme. It identified the international network of technologically mediated information flow and control as a major feature of contemporary life. It featured art and theory that commented on the role of information in dominating commercial and military systems and the technological infrastructure, and invited contributions from artists working on resistance to this information domination.

The information society—no longer a vague promise of a better future, but a reality and a central challenge of the here-and-now—is founded upon the three key technologies of electricity, telecommunications, and computers: technologies developed for the purposes, and out of the logic, of war, technologies of simultaneity and coherence, keeping our civilian society in a state of permanent mobilisation driven by the battle for markets, resources, and spheres of influence. A battle for supremacy in processes of economic concentration, in which the fronts, no longer drawn up along national boundaries and between political systems, are defined by technical standards. A battle in which the power of knowledge is managed as a profitable monopoly of its distribution and dissemination.

[Military attention focuses on] cyber war, whose ultimate target is nothing less than the global information infrastructure itself: annihilation of the enemy's computer and communication systems, obliteration of his databases, destruction of his command and control systems. . . . These new forms of postterritorial conflicts, however, have for some time now ceased to be preserve of governments and their ministers of war. NGOs, hackers, computer freaks in the service of organised crime, and terrorist organisations with high-tech expertise are now the chief actors in the cyber-guerilla nightmares of national security services and defence ministries.[7]

In response to this emphasis, an independent jury established the "Information Weapon Competition" to invite artists to develop technologies that could work against the dominant information systems. The prize was offered "to the most outstanding information weapon with an accent on its functionality, design, and successfulness." The organizers claimed to be interested in the question: What could info weapons look like, if they are not simply E-mail bombs, spam, or regular propaganda and disinformation campaigns on the 'content' level?"[8]

Eric Paulos, a robotics and surveillance artist, described his conceptual approach to designing such an "information" weapon that would corrupt any electronic instruments within a "Technology Free Zone":

The rapidly approaching ubiquity of technology and its inevitable but rarely discussed terroristic use demand for immediate exploration and development of technology disruption devices. In the ensuing world dominated by technology and information, the true culmination of power will rest not with the institution controlling the information but the organization, group, or individual capable of disabling, altering, or destroying the underlying support structure of information: electricity, telecommunications, and computers.[9]

Konrad Becker

Konrad Becker is a "hyperreality researcher" and director of the Institute for New Culture–Technologies, in Austria. He creates media installations and writes extensively on issues in culture and technology. One major interest is surveillance and information war. In his essay "Synreal Systems," he analyzes the power of information systems and his fear of the loss of individual autonomy:

Information can be manipulated or faked at many levels. It is not only the message itself that can be tampered with. Initially, the source for an information item may be masked or relabeled; then, the routing and the placing of channels and media can be subject to manipulation; and, finally, access of the receiver may be restricted. Data processing is the silent weapon in an undeclared war. Social engineering, the analysis and automatization of society, is derived from military operations research, the methodology of tactics and logistics. The automatization of society works the same way as the automatization of a meat factory. The freedom of the individual disappears as unspectacularly as a popular illusionist spectacularly made the Statue of Liberty disappear from the New York Harbor: she is miraculously out of the frame of vision.

His article on cyber war provides a provocative set of references that stretch from philosophers and theorists to documents from military research. Becker believes that the development of information warfare technology is not an esoteric subject of interest only

---

Konrad Becker: ⟨http://netbase.t0.or.at/~konrad/⟩

to researchers, but rather a central topic in understanding the evolution of culture. It is an appropriate and crucial concern for the arts. Here is a sample of the information technology articles he references: "A Theory of Information Warfare," "Eavesdropping on the Electromagnetic Emanations of Digital Equipment," and "Microwave Harassment and Mind-Control Experimentation."[10] His "Infobody" interventions, such as *Linguistic Infiltration Programs* (SLIP) and *Telepresent Contagious Postures* (TCP) ironically comment on real-world tendencies.

Other Artists and Projects

Karen O'Rourke and her colleagues organized the ***ArtChivists*** Web site, which reports on artists exploring databases as art. **Judy Malloy** created a series of works reflecting on corporate information and propaganda as art, for example, *OK Research, OK Engineering, Bad Information*. The **Museum of Jurassic Technology** offers reflections on techno-culture in an ironic commentary on museums. *Foresight Exchange* sets up a stock-market-like exchange of ideas about the future. **Marcos Novak**'s TransArchitecture (see chapter 6.3) explores the idea that physical and data architectures will merge into new hybrids in the future. The **Scope** conferences in Austria, such as Information vs. Meaning, draw together information researchers and conceptual artists to "focus on critical changes in how information is accessed and managed." **Switch,** the e-zine of the CADRE institute at San Jose State University, dedicated one issue to "exploration of the issues of data and database, in terms of both ontological questions, and how they impinge upon various worlds of art," noting that artists have rarely developed the agency to work with data in its new forms. Samples of the articles include **Frank Dietrich**'s "Data Particles—Meta Data—Data Space," in which he analyzes the "properties of data particles to include speculation about metadata, navigation, and transformation rules"; media-curator for the Walker Art Center, **Steve Dietz**'s "Memory_Archive_Database," which looks at the database as an art form; and **Joel Slayton and Geri Wittig**'s "Ontology of Organization as System," which gives a "detailed theoretical account of datum as autopoietic agency."

Karen O'Rourke: ⟨http://www.fohel.com/archiving-as-art/core/bootstrap.html⟩
Judy Malloy: ⟨http://www.artswire.org/Artswire/infotour/infart.html⟩
Museum of Jurassic Technology: ⟨http://www.mjt.org/index.html⟩
Foresight Exchange: ⟨http://www.ideosphere.com/docs/fx.html#Welcome⟩
Marcos Novak: ⟨http://www.aud.ucla.edu/~marcos/⟩
Scope, Information vs. Meaning conference: ⟨http://www.scope.at⟩
Switch e-zine: ⟨http://switch.sjsu.edu⟩

Sharon Daniel creates database-inspired art works that create dynamic collaborative systems for cumulating visitor information using human and algorithmic, A-Life agents to evolve the structure. In one project called *Subtract the Sky,* Daniel and *Mark Bartlett* have attempted to create innovative visual and informational models to orchestrate the re-inspection and integration of astonomical, geological, and artistic perspectives on the sky and the earth. *Mary Flanagan*'s *Phage,* named after helpful viruses, churns through all the data stored on a user's hard drive as source material for its work—allowing "the user to experience his or her computer as a palimpsest of his or her own life rather than simply as a productivity or work tool. With its viral process, [phage] creates a feminist map of the machine." *Robert Niderffer and Victoria Vesna* investigate databases as a prime location for art research and intervention: "The most promising arena for conceptual work is already in place as the archives and database systems are being developed with dizzying speed. It is in the code of search engines and the aesthetics of navigation that the new conceptual field work lies for the artist. These are the places not only to make commentaries and interventions, but also to start conceptualising alternative ways for artistic practice and even for commerce." Extrapolating database ideas into video, MIT Media Lab's *Object Based Media Group* is working on a "smart video" technology that treats the visual objects in live video streams as hyperlinkable access points to additional information.

Information Visualization

Not all see the new information systems as necessarily dark forces. Some artists are interested in the new capabilities to collect and represent information. They believe that these new capacities might enhance life and enrich our understanding of the human and nonhuman world.

Donna Cox

Donna Cox was an artist on a collaborative team of scientists and artists exploring the possibilities of scientific visualization at the National Supercomputing Center at the

Sharon Daniel: ⟨http://time/arts.ucla.edu/AI_Society/daniel.html⟩
Mary Flanagan: ⟨http://www.maryflanagan.com/virus.htm⟩
Robert Niderffer and Victoria Vesna: ⟨http://time.arts.ucla.edu/AI_Society/ vesna _intro.html⟩
Object Based Media: ⟨http://www.media.mit.edu/~vmb/obmg.html⟩
Donna Cox: ⟨http://www.ncsa.uiuc.edu/People/cox/⟩

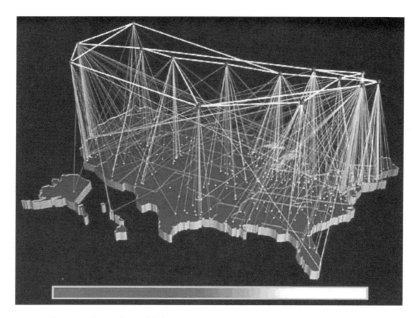

Fig. 7.7.7. Donna Cox and Robert Patterson, visualization of the Internet structure.

University of Illinois. She has written many papers analyzing the role of artists in visualizing complex data sets. The journal *Leonardo* offered this brief profile in awarding her an Award for Excellence:

[Cox] is the artist contributor to a team exploring interdisciplinary research through supercomputing. Cox recounts her role as a member of the "Renaissance Team" discovering visual representations of multidimensional computations. Supercomputing is particularly valuable for complex simulations, as its speed, vector environment, and parallel processing enable many equations to be solved simultaneously. These experiments are valuable in distinguishing appropriate new tools to enable scientists to find correlations in data. Scientific simulations are further evidence of the shared quest of artists and scientists—to make visible the complex, yet invisible, structures of the universe.[11]

Rachel Strickland

Rachel Strickland, a researcher with Interval Research and one of the collaborators in *Placeholder* (see chapter 7.3), creates a variety of interactive media installations inspired by the perspectives of anthropology. One installation called *Portable Effects: An Installation for Interactive Anthropology* allows visitors to interactively explore what things people carry with them in their pocketbooks, backpacks, wallets, and pockets. The installation consists of three networked stations allowing for the inspection and documentation of one's materials, sorting, and review of other people's records. People and what they are carrying are photographed and weighed. A gamelike review station allows visitors to try to guess which sets of belongings go with which persons. Strickland sees this phenomenon as a window on concepts of portability and everyday design as well as a more general reflection on contemporary life. Some commentators noted that the installation also raises questions of surveillance and the danger of technology enabling even greater intrusions into personal lives. Another called it a "voyeuristic thrill." The documentation describes the exhibit's setup:

People's selection and arrangement of the things they take with them in handbags, pockets, briefcases, backpacks, etc., form the context of the investigation. Between setting forth in the morning and returning home at night, each person lives nomadically for several hours a day. You can't take everything with you, neither in your backpack nor in your head. Identifying essentials, figuring out how to contain, arrange, and keep track of them as you go are instances of design thinking. Understanding the properties and consequences of portability is a way to grasp principles that underlie the transferability of knowledge from one domain to another. . . .

Portable effects uses contemporary technology to offer new perspectives on the common everyday activity of carrying things around. The new insights themselves become the focus of the art. The installation also suggests possible dangers lurking in computer-mediated intrusiveness.

Other Artists and Projects

Questioning the "rigid and artificial separation between information and aesthetics in scientific images," *Felice Frankel,* in her role as an artist in residence at MIT, collaborates with scientists to visually extract information from scientific research by exploiting the visual potential of research techniques (see chapter 3.2). *Dee Berger* investigates the scanning electron microscope as a tool for making visible the invisible. *Trudy Myrrh Reagan* creates collages based on scientific principles, conundra, and imagery. *John Maeda* and *MIT's Aesthetics of Computing Group* in part focus on integrating concepts of graphics and computation to develop "systems of visual relations that intuitively reveal

---

Rachel Strickland: ⟨http://web.interval.com/projects/pfx/⟩
Dee Berger: ⟨http://Ideo.columbia.edu/micro/index.html⟩
MIT's Aesthetics of Computing Group: ⟨http://acg.media.mit.edu/⟩

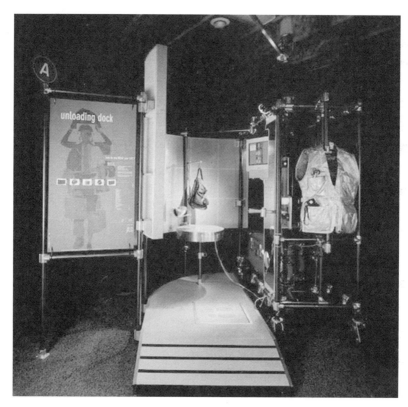

Fig. 7.7.8. Rachel Strickland, *Portable Effects: A Survey of Nomadic Design Practice*. A museum installation that invites visitors to examine what they carry every day and how they carry it. Coproduced by Interval Research Corp. and the Exploratorium.

qualitative information of varying degrees of relevance." Several artists and theorists have begun to focus on the topography and geography of cyberspace. For example, **Christine Paul** has written on the cybertopography of hypermedia; **Martin Dodge and Rob Kitchin** have coauthored the book *Mapping Cyberspace;* and **Gerald de Jong** oversees *Project Fluidiom,* which explores the synergistic "utilization of spatial geometry and dynamics for the purpose of facilitating interpersonal communication and expression of ideas" in software and networked environments.

Christine Paul: ⟨http://www.eastgate.com/catalog/Unreal.html⟩
Martin Dodge: ⟨http://www.cybergeography.org⟩
Gerald de Jong: ⟨http://www.fluidiom.com/⟩

Fig. 7.7.9. Eve Andrée Laramée, *Apparatus for the Distillation of Vague Intuitions*. An installation to comment on the objectivity of science. Photo: James Prinz.

Reflections on Science

For many, science is at its heart an information system. It is a technology for collecting, codifying, representing, and communicating information. Contemporary society places great faith in its procedures and claims to authority. Some artists believe that this faith needs to be profoundly questioned.

Eve Andrée Laramée

Eve Andrée Laramée's installations reflect on science and its systems of knowledge and present ironic commentaries on scientific processes and claims of objectivity and truth. Laramée's Web site describes the installation *Apparatus for the Distillation of Vague Intuitions,* which offers visitors a laboratory scene in which intuitions and guesses predominate:

Eve Andrée Laramée: ⟨http://www.artnetweb.com/laramee/⟩

Apparatus for the Distillation of Vague Intuitions consists of a room-sized laboratory table supporting an array of dysfunctional scientific apparatus. The hand-blown glass vessels are engraved with text elements referring to subjectivity, intuition, guesswork, and desire. The piece seeks to relocate science within the subjective realm. It refers to human fallibility, error, approximation, and chance as valid components in discovery and perception. It alludes to the fact that science is a cultural construct and challenges the notions of objectivity and authority. The images and apparatus of science are potent visual signifiers in our culture. People want a path to knowledge, and science represents a reliable modern belief system. Science is put forth as being objective, nonhierarchical, and nonpropagandistic.

Neither science nor art can propose a tyranny of truth. This work foregrounds the irrational and the uncertain as a strategy to blur the boundaries between different modes of cognition and consciousness. I am interested in how human beings formulate knowledge through both art and science in a way which embraces poetry, absurdity, contradiction, and metaphor. The boundary between sense and nonsense is a slippery one and has quite a bit to do with the position of the observer.[12]

Instrument to Communicate with Kepler's Ghost presents viewers with a device that appears to be a serious historical scientific apparatus. Astronomical drawings nearby refer to Johannes Kepler's work on Harmonices Mundi, which includes falsified data that he used to prove his theories that planetary motions have a musical relationship. The absurd purpose of the device aimed at communication with spirits, together with its historical contextualization with this event in Kepler's life, invite viewers to reconsider science's basis for authority. Visitors could use a simple telegraph device to ostensibly send messages. Laramee noted that "because the device had the appearance of a credible apparatus with a logical-seeming system of operation, the audience frequently believed the device actually worked," thus calling into question the biases towards the authroity of science and technology.

Laramée carefully selects elements of her installations to enhance her investigations. For example, she explains to an interviewer that the use of flowers in the *Instrument* installation referred to gender issues in science and the possibilities of other kinds of inquiry. She also explains how her wiring of flowers to circuits helps ask questions about our assumptions regarding nature. She noted the contradiction that even she had some discomfort in wiring up flowers while we blithely accept the lawn mower's mangling of organic life. Her installations often reflect on science's illusions of rationality and their instantiation of certain views of nature as separate from mankind: "Nature, which is thought of as wild and in need of being controlled, is the perfect object for scientific study. Nature offers itself to the scientistic gaze. Nature has been relegated to the position of 'other,' which relates to the notion of women and marginalized peoples being 'closer' to nature. The separation of women from key metaphors of science happened with the birth of modern science."[13]

Todd Siler

Todd Siler has long been a leader in promoting the integration of art and science. His artworks and installations often draw on scientific information. In the last few years he has advocated a movement called "metaforming," in which individuals are taught to enhance creativity by integrating logical and intuitive processes. His books *Learn to Think Like a Genius* and *Breaking the Mind Barrier* explain his perspective.

Siler heads a multiyear project called "Blast: Remaking Civilization," which points toward the integration of knowledge that he sees will be necessary in the twenty-first century. The project's plans call for a symposium, book, and box. Eve Laramée was a collaborator in producing the box, which collects materials from the symposium and from external sources. The project asks people to address questions about the future of civilization and education and incorporates contributions by artists, writers, scientists, students, cultural theorists, and information technologists. Laramée's Web site describes the project:

As Siler suggests, "We're connected to everything we create and that has created us. We and it (nature) are one and the same. Every detail of nature details our nature. As 'processmorphs' (things that are alike in process but unlike in form or appearance), we're different manifestations of the same phenomena. The aerodynamics of hurricanes and thunderstorms, the sensual motions of waveforms, the unpredictable episodes of earthquakes, etc.—all reflect our brainways and mental processes and social forces. Likewise, the things that we create bear these similarities of process. The greatest achievement of twenty-first-century science will be to substantiate and illuminate this fact—and wisely (consciably) apply its knowledge."[14]

Pursuing a similar approach in books such as *Lateral Thinking* and *Parallel Thinking*, Edward de Bono promotes processes of problem solving and creative work that cross disciplinary boundaries such as those that separate art and science.

Todd Siler: ⟨http://www.metaphorming.com/⟩
Edward de Bono: ⟨http://www.edwdebono.com/⟩

Other Artists and Projects

Nell Tenhaaf's "Savoir" series of conceptual sculptures, such as *Species Life* and *Orphaned Life Form,* provide a commentary on scientific procedures in genetic science and the "desire for knowledge" and the "limits of self-knowledge." *Nola Farman* and seven other artists offered interpretations of Australian science in *The RTZ-CRA Cabinet of Curiosities,* which was modeled after the cabinets that scientist-naturalists used for specimen storage in the eighteenth century. *Benjamin Potter*'s *Afterlife Series* presents specimens preserved in honey so that they are still animated. In *Ebon Fisher*'s multimedia events, scientific codes, and representations become essential parts of the ambiance. *Gail Wight, Marta Lyall, Catherine Richards,* and *Mark Dion,* discussed in other chapters, all create installations that question the authority that science claims for itself and the purity of its procedures for managing information.

Information Organizations and Structures

Knowbotic Research

Knowbotic Research has won many awards for their investigations of the nature of information in the contemporary technology-mediated world. This collaboration, which includes Christian Hübler, Alexander Tuchacek, Yvonne Wilhelm, and other scientists and artists creates installations and projects that allow participants to enter data space to interact in unprecedented ways. Projects have focused on architecture, sound, urban experiences, and scientific research enterprises. They seek to understand the new realities developing from the collision of physical and network experience. Their *SMDK–SimulationSpaceMosaic of Mobile Datasounds* allowed visitors to navigate a VR world composed of data. In "Nonlocations as Fields of Action," they explain that they create media installations as research about possible configurations for the culture:

Knowbotic Research formulates its projects via the combination of information and knowledge structures with complex spheres of experience and action. A particular emphasis is placed on the performability and nonlocalisation of data and network-supported environments. In loose cooperation with computer scientists, scientists in general, and architects, the heterogeneous quali-

---

Nell Tenhaaf: ⟨http://www.yorku.ca/faculty/academic/tenhaaf/objectmenu.html⟩
Ebon Fisher: ⟨http://www.users.interport.net/~alula/ebindex.html⟩
Knowbotic Research: ⟨http://www.t0.or.at/~krcf/⟩

ties of media events are probed and tested and made accessible within specific public spheres (e.g., in real data-space installations, urban experiments, etc.)[15]

Knowbotic claims that contemporary techno-political systems do not allow the flexibility of conceptualization and action that are possible with the new technologies. It calls these new structures "non-locations" and "mem_branes," in which people are not forced to "operate within hard and fast fields of options," in which technology helps people to "continually jump from one system to another," and in which nonlocations function as "zones of difference which generate confrontation and point beyond the cross-communicated indexical exchange of information."

In "Non-Locations/Event: Under Construction," it explains that technology creates new possibilities. It is interested in structures that allow experimentation with those possibilities: confrontations between what "man deems possible and what machines offer him as makeable," productive new models such as "nonlinearity, multidimensionality, acceleration, compression, multiple layers, poly-perspectives, multifunctionality," and new attitudes about interfaces that don't deal with "negotiating between realities, but acts in a field of effects where the human and the machinic can no longer be easily distinguished."[16]

The installation *Anonymous Muttering* presented visitors with a physical sound-and-light experience that is a manifestation of underlying information structures created locally and via Net contributions. Knowbotics provided interfaces that let people inspect and manipulate that structure. The particular installation can be thought of as an experiential illustration of their understanding of possible configurations of the larger culture. It presented events "which are composed of fleeting, initialised and 'found' singular events. What happens cannot be traced back to references in real space."[17]

Knowbotic claims that the new synthesized physical/data spaces form an almost palpable meshwork that can be bent, folded, and manipulated. It describes this meshwork as "smoothened heterogeneity" and differentiates it both from architecture and electronic networks. It is fluid and bendable. "The value of this networked information lies not in the guarantee of its traceability, but in its immanent transformatory potential which invites interventions, like the deletion, addition, encryption, or granulation of data."

Its Web-based project *IO_Dencies* (see chapter 6.4) attempted to create network representations and interfaces for understanding the urban experiences of Sao Paulo and Tokyo. It believed that new understandings and possibilities for action could be stimulated by unorthodox informational structures. It created a "topological cut through the heterogeneous assemblage of physical spaces, data environments, urban imaginations, connective agencies, and individual experiences." It sought to create new forms that "exploit the technical possibilities of the networks and that allow for new and creative forms of becoming present, becoming visible, becoming active, in short, of becoming public."

In an interview with Andreas Broeckmann for V2, Knowbotic explains its idea that contemporary realities are based on neither physical nor electronic space alone. It sees common notions of "networking" as oversimplified. Urban topology is made up of elements that "can be economic, political, technological, or tectonic processes, as well as acts of communication and articulation, or symbolic and expressive acts. . . . [T]he city features not as a representation, but as an interface which has to be made and remade all the time"[18]

In "Dialogue with the Knowbotic South," Knowbotic investigated the nature of simulation in scientific and artistic research and the possibilities of communication between the arts and sciences. It chose research on Antarctica because its compelling presence of physical nature provided an interesting counterpoint to virtual constructions of knowledge. The project provided windows into scientific visual and textual discourse about the Antarctic and ultimately questioned how we know what we know. In an interview with Paolo Atzori for *CTheory,* Knowbotic Research described its agenda:

From an artistic point of view, our project formalizes the problem of a missing language. . . . For the first time, scientists not only prove the laws of nature, they also formulate conditions of possible systems. In our project we treat an actual state of nature corresponding to our information culture. We want to create a field of discourse freed from the rules of the specialists' disciplines. It is a field not only for natural scientists but also for scholars and philosophers who are discussing current ideas of reality. . . .

Our worldview is based on what we see in the future, a worldwide data space induced by the communication technologies, filled with tons of information coming from all different disciplines of knowledge. I think it is very important to create models which focus on the needs and possibilities of the person who tries to receive this information. . . . Our work is also a liberation from science. We create an environment where, initially, we fabricate actual phenomena of scientific thinking. But we emancipate these phenomena from their reference (science) by a self-organization model.[19]

As part of the project, the artists created "knowbots," semi-autonomous software agents that could locate, manipulate, and present information related to Antarctic research. They made the knowbots somewhat interactive so that viewers' actions could have effects on the information they saw:

Fig. 7.7.10. Knowbotic Research. Dialogue with the *Knowbotic South*. Investigation of the nature of simulation in scientific and artistic research.

Information Organizations and Structures

The main problem for the knowledge robots is that we are dealing with two bigger entities, the so-called reference nature that is still very powerful in the Antarctic, one of the few almost intact ecological systems, and the related scientific institutions. The knowbots act with completely different kinds of inputs, originating a tension so you can't bring these two worlds really together. This produces an aesthetic field for artists. . . .

The interesting thing is that we deal with processes you can't see in reality. Hidden processes, sometimes extremely small or extremely big, and very complex. . . . Actually, for the scientists, it no longer makes sense to work directly in contact with nature. They need data, intelligent data for their terminals in the institutes. . . . [S]ensors are directly connected to computers. They exterritorialize their nature in the networks. Maybe our artistic work is a kind of reterritorialization.

The artists reflect that the world will be inundated with information. Artists have an important role in decoding this immensity of data, but they must be willing to enter the data structures. Countervailing against developments such as the ecological destruction of the Antarctic requires intervention at the realm of information acquisition, visualization, and communication:

The important point is not to discuss the meaning of measures, but rather how can we visualize and handle this complexity of information. That's a problem for the scientists too. There are so many data: How can we turn it into information and knowledge? How can we handle this with the knowledge we have?

It is necessary to define a strategy about order and the generation of new things. With computers we analyze fragments of the reality and at the same time we build and initiate complex processes. This is what the work is about.

We are inside a technological system whose direction and speed are defined by industry and science. Politics and arts have to follow, and it is nearly impossible to do anything without being inside. It is a confrontation which can't work if you play with the traditional ways of art. . . . It's an old artists' strategy to make politics and scientists aware of the consequences of their concepts of reality.

As part of their strategy, Knowbotic set up a research center called Mem_brane in collaboration with the Academy of Media Arts in Cologne, Germany, which attempts to integrate technical and cultural capabilities and perspectives without commercial pressures:

Mem_brane is a laboratory for artistic-oriented projects which concentrate on network and electronic information systems. The works that are realized focus upon unconventional development using the aesthetics of culture and technology as a critical reflection of digital communication in and outside of data worlds. The enhanced growth of technological infrastructure in the present information society must be evaluated across its daily applications, where questions are raised with relevance to media culture, economics, and politics.[20]

In reflecting on the work of Knowbotic, theorist Tim Druckrey notes that the dominant culture tries to colonize cyberspace in accordance with old models that marginalize deviant and chaotic elements. Groups like Knowbotic demonstrate critical alternatives:

The traditional humanistic worldview is thus reconceptualized in the sense of being reduplicated. In such outlines uninhabited outer regions are negated, sub- or cybversive forms of existence are ignored (i.e., ghettoized), microstructures are overlooked, chaotic behaviour and uncertainty are curbed, etc. These defects reflect a lingering continuation of an economic and ideological 'missionary' approach, as well as elements of a colonial attitude. This type of virtual culture draws its reality-constituting factors from the "use" of prefabricated structures; the user elements are algorithmized components of our city culture.[21]

Makrolab

Makrolab was developed by a collaboration of Slovenian artists who are creating a new kind of information organization. Their organization presents itself both conceptually and physically as an advanced scientific outpost using the latest technological tools to monitor happenings and create events contextualized as art. Makrolab artists feel it is essential to monitor as many channels of information as possible, and thus their systems include radio, GPS, computers, and internet technologies. Metaphorically, they see their activities as an attempt to enter into the "big organism," the "ur-animal." The Makrolab Web site explains why it was created and their activities at Documenta:

[T]he exhibited object, the Makrolab console, represents the external, fragmentary view on the Makrolab research station, which is set on the hill Lutterberg. . . . Makrolab is designed as an autonomous, modular communications and living environment, which is powered by sustainable sources of energy (solar and wind power). It is designed for a long existence in an isolated environment and can withstand extreme natural conditions.

[I]t has its own research and experience goal. The station is built as a combination of various scientific and technological logistics systems. Makrolab makes use of scientific and technological

---

Makrolab: ⟨http://makrolab.ljudmila.org/⟩

tools, knowledge, and systems, but it projects them in the social domain of art. We, the authors and crew, make use of the system of art for the shaping and representation of an integral, empirical, and creative experience. . . . [T]he electromagnetic spectrum is a part of the global sociopolitical space, which is invisible and immaterial on one hand but presents a productive factor of general living and social conditions on the other. It can be sensed only by the means of suitable interfaces and specialized knowledge.[22]

In a lecture given in connection with its presence at Documenta, Makrolab presented some of its rationale. It attempted to create a somewhat isolated entity that could reflect on trends in the larger culture using the tools of technological and cultural research and corporate activities. Fields of research include "acoustics, atmosphere, communications, dreams, inner life, linguistics, low-energy systems, psychoacoustics, solar power systems, social-evolution systems and strategies, wind power systems, weather, and war strategies."

[Its goals are] to transform abstract and intangible qualities and properties present in the world, such as radio waves, atmospheric events, or psychic movements into material, three-dimensional structures, documents, and objects through a de-abstractization process, or if you want, a process of materialisation. . . .

Makrolab is a declarative position outside of the spectacle, also outside of society. . . . The thesis is that individuals in a restricted, intensive isolation can produce more evolutionary code than large social movements of great geographical and political extent.

Joel Slayton and C5

Joel Slayton and his collaborators have created a corporation/research organization dedicated to the pursuit of new developments in technology, theory, and art. C5 presents the full regalia of corporate structure, including finance, governance, marketing, and research elements. Visitors can read corporate reports, inspect research reports, buy stock, and so on. Slayton explains the rationale on the C5 Web site:

Advances resulting from intratheoretic reductionism have resulted in the exploration of unique models in which cascading and parallel considerations of hyperstructuralism and contextuality are significant. Indeterminate information systems (brains and computers) are impetus for research and exemplification of fundamental principles which can be used for tactical surveillance and

---

C5: ⟨http://c5.sjsu.edu/index.html⟩

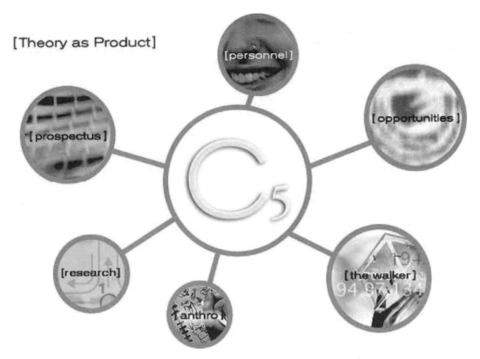

[Theory as Product]

[personnel]

[opportunities]

[prospectus]

C5

[research]

[the walker]

[anthro]

Fig. 7.7.11. Joel Slayton, president of C5. Web site ID page for project exploring the structure and forms of corporate information systems, including data mining from movements of toy cars.

strategic analysis involving new forms of knowledge representation. The complex phenomena of self-organization, diffusion, cues, presence, richness, ambiguity, uncertainty, complexity, evolution, inferencing, and entanglement are common themes for experimentation at C5.

C5 is the corporation of acculturation. The sciences of the artificial are stimuli redefining the nature of group formations and operations management resident in technology enterprise. Systems analysis and information mapping are the contemporary substance of data perception, of which the artifact is interface. C5 solutions are informed by collaborative expertise, including implementations of artificial intelligence, bio-engineering, public relations, liquid computing, emergent behavioral systems, bio-metrics, virtuality, cognitive psychology, semiotics, anthropology, literary criticism, military studies, library science, and art. Theory is product.[23]

C5 includes subsidaries in "Simulation, Heuristics, Complexity, Identity, and Ubiquity." C5 comments on the linkage between serious research and new tech-corporate economics by offering a presentation that hovers somewhere between corporate gibberish, theoretical excess, and cutting-edge research.

Specific research projects include *Radio Controlled Surveillance Probes (RCSP), Mining Research,* and *Data Mining/Knowledge Representation.* The *Data Mining* project explains the focus on information:

We believe that a deconstruction of heuristic and expert strategies forces the conceptual rectification of notions such as belief networks, adaptive learning, and knowledge consensus. In systems where uncertainty is high and specific goals not defined, new approaches to emerging knowledge are required.

C5 is at its best in its trade show presentations. At SIGGRAPH98, C5 researchers clad in official corporate uniforms had a booth that demonstrated the latest research in RCSP. Very serious presenters operated a fleet of radio-controlled cars moving in choreographed movement while a computer systematically collected this important data and transmitted it to a distant server for data analysis and ultimate display on the booth's computers.

Etoy

Etoy is a pseudo corporation created by an artist group. It undertakes action throughout the world commenting on corporate structures and worldwide communication and information systems. It is the artistic version of a multinational corporation. It has developed "office" modules (sometimes made out of shipping containers), maintains an active Web site, sells stock, features distinctive uniforms, and otherwise acts like a strange corporate entity. The Internet has been essential to its operations. It won the top prize in the 1996 Ars Electronica Net competition.

In one action called *Time Zone,* it declared all international operations would use its own unique time zone, synchronized to the Internet. The explanation of the project also conveys the corporate ambiance the group creates:

[T]he etoy CREW established their own time zone. All etoy SERVERS and etoy TANKS run a virtual time system: etoy. TimeZone is oriented on UNIX seconds . . . to save money and time through more efficient coordination and parallel working. To offer perfect customer service and faster R&D. It unifies the etoy *BRANCHES* around the world, and covers a minimum spare time for employees.[24]

Etoy: ⟨http://www.etoy.com⟩

Fig. 7.7.12. An etoy time-zone screen from the artist organization that explores corporate forms.

The etoy Web site describes some of its operating principles and its attempt to use artistic methods to study the state of our culture:

etoy operates somewhere between productions of overdrive communications and a redefinition of content and message in the electronic age. . . . etoy uses artistic means to demonstrate what happens when communication is stretched to its outermost limits and emits both emotional and intellectual impulses.

Why? the world is changing radically—one cannot compare digital structures and procedures with mechanical processes. etoy explores and programs search engines, global databases, EDP services, automated processes, and virtual environments, all of which are concerned exclusively and emphatically with themselves. . . . [etoy is] a production within that supercharged zone between fake and truth, between showmanship and functionality, between outside and inside, between content and shape.

Other Artists and Projects

Krzysztof Wodiczko initiated a major effort at MIT's Center for Advanced Visual Studies called "Interrogative Design" that asks the arts and design communities to interrogate their own functions and context, taking a risk in "exploring, articulating, and responding to the questionable conditions of life." This approach is seen as especially important as a "post-military cultural force" to respond to the flow of military research and the dominance of technology. ***Alex Galloway, Mark Tribe, and the Rhizome Organization*** have attempted to "aestheticize corporate identities" by experimenting with new corporate forms and visual Web interfaces. In response to international trends in global capitalism, ***Gregory Green*** created a new state called the *New Free State of Caroline* on an unclaimed South Pacific island from which he initiated projects such as *Gregniks, An Alternative Space Program* and *Manual,* which includes excerpts from military and CIA handbooks re-created in exact original style. ***Stephen Soreff*** has run a twenty-five-year conceptual art project that includes a fictitious university and the art-review magazine called *AGAR* (Avant Garde Art Review), which has included articles from the future on art forms such as genetically engineered art and televisual windows.

Many of the "Hactivists" described in other chapters, such as chapter 6.4, who seek to deconstruct the assumptions of technology mediated organizations create new informational and organizational structures as part of their work. Examples include ®TMark, IRATIONAL.ORG, Reclaim the Streets, Critical Art Ensemble, and the Redundant Technology Initiative. Links can be found at the ***Applied Autonomy Organization*** site listed below.

Summary: Being Formed by Information

Has the information explosion made the world clearer or more obscure? Like so many of the fruits of research, there is no simple answer. Each of us potentially has access to information that would have been inconceivable a few years ago. Optimists suggest that this access could usher in higher levels of understanding about the universe and humanity. For example, Roy Ascott (see chapter 6.1) suggests that the expansion of information networks has done much more than just increase the number of factoids we can consider. Rather, it has opened up the possibility of transformed consciousness, the ability to see

---

Krzysztof Wodiczko: ⟨http://www.mit.edu/mit-cavs/www/DemilTech.html⟩
Rhizome Organization: ⟨http://rhizome.org⟩
Gregory Green: ⟨http://www.aeroplastics.net/aeroplastics/Greengr/greengr2.htm⟩
Stephen Soreff, *AGAR:* ⟨http://www.asci.org/future/agar.html⟩
Applied Autonomy Organization: ⟨http://www.appliedautonomy.com⟩

the big picture, including kinds of knowledge outside of the scientific and technological worldview.

Pessimists suggest that information access is an illusion. Much of the so-called information is based on faulty assumptions and categories that serve other purposes besides enlightenment. Rather than becoming enhanced subjects, many of us are being turned into data objects, reduced to data points in the master databases used for control and exploitation. Some even hold that the information is really mush—mediated signs without real referents. As information comes to be the cultural and economic heart of life, it is fitting that artists should enter into the debate.

Notes

1.   Chicago Museum of Contemporary Art, "*SecurityLand* Documentation," ⟨http://www.mcachicago.org/anxiety/scher.html⟩.

2.   S. Penny, "*Big Father* Documentation," ⟨http://www-art.cfa.cmu.edu/www-penny/index.html⟩.

3.   Dutch Electronic Arts Festival, "Steve Mann Description," ⟨http://wearcam.org/deaf98/historic.htm⟩.

4.   S. Mann, "Reflectionism and Diffusion," ⟨http://hi.eecg.toronto.edu/leonardo/⟩.

5.   J. Crandall, "Documenta X Statement," ⟨http://www.blast.org/crandall/documentaX/⟩.

6.   N. Jeremijenko, "Introduction to Information Work," ⟨http://www.tech90s.net/nj/index-nj.html⟩.

7.   Ars Electronica, "InfoWar Introduction," ⟨http://web.aec.at/infowar/eng.html⟩.

8.   Information Weapon Competition, "Statement," ⟨http://www.dds.nl/hypermail/p2p-list/0056.html⟩.

9.   E. Paulos, "Technology Free Zone," ⟨http://www.cs.berkeley.edu/~paulos⟩.

10.   K. Becker, "Cyber War," ⟨http://www.t0.or.at/msguide/cyberwa1.htm⟩.

11.   *Leonardo,* "Award for Excellence—Donna Cox," ⟨http://mitpress.mit.edu/e-journals/Leonardo/isast/awards.html⟩.

12.   E. Laramée, "Documentation of *Apparatus for the Distillation of Vague Intuitions,*" ⟨http://www.artnetweb.com/laramee/⟩.

13.   E. Laramée, "Interview on *Apparatus* Installation," ⟨http://www.artnetweb.com/laramee/⟩.

14.   E. Laramée, "Documentation of Blast," ⟨http://www.artnetweb.com/laramee/⟩.

15.   Knowbotic Research, "Non-Locations as Fields of Action," ⟨http://www.khm.de/people/krcf⟩.

16.   Knowbotic Research, "Non-Locations/Event: Under Construction," ⟨http://www.khm.de/people/krcf⟩.

17. Knowbotic Research, "*Anonymous* Muttering," ⟨http://www.khm.de/people/krcf/AM_rotterdam/long.html⟩.

18. Knowbotic Research, "Interview with Andreas Broeckmann," ⟨http://www.khm.de/people/krcf/IO/netscape/content/IO_dencies.html⟩.

19. P. Atzori, "*C-theory* Interview with Knowbotics," ⟨http://www.ctheory.com/a38-cyberantartic.html⟩.

20. Knowbotic Research, "Mem_brane Description," ⟨http://www.khm.de/~mem_brane/Info/einfo.html⟩.

21. T. Druckrey, "Analysis of Knowbotics," ⟨http://www.t0.or.at/~krcf/nlonline/nonCorealities.html⟩.

22. Makrolab, "Lecture on Documenta," ⟨http://makrolab.ljudmila.org/lec/⟩.

23. C5, "Corporate Statement," ⟨http://c5.sjsu.edu/index.html⟩.

24. Etoy, "Corporate Operations Description," ⟨http://www.etoy.com⟩.

8

Resources

Exhibitions and Festivals; Educational Programs; Art and Research Collaborations; Organizational Resources, Think Tanks, and Web Resources

This chapter outlines resources available to those interested in the integration of art, science, and technology. It considers permanent spaces, festivals, art shows, integrative research institutes, organizations, educational programs, special initiatives linking artists with researchers, and publications that explore the art described in this book. It also considers resources in the research world that are relevant to new integration, including think tanks, professional organizations, conferences, and publications. Several efforts specifically focused on the integration of art, science, and technology are described in chapter 1.2.

The listing raises issues that cannot be fully analyzed here. Resources confound old categories, while many organizations cross categories; for example, Ars Electronica is a museum, research environment, and festival sponsor, and many educational institutions are closely linked with research centers. The term *electronics* is part of many organizations' names even though they have moved on to embrace a wider range of experimental technologies. Many conventional institutions, such as universities and museums, have decided they want to cover "new media," but they fail to understand the challenge and scope of the new art-and-science integration and create programs and structures of "electronic arts" that are more assimilations into the past than visions of the future. This chapter can offer only an illustrative list. An expanded and updated version of this list is available on my Web site.

Permanent Spaces, Museums, and Comprehensive Institutions

Ars Electronica

Ars Electronica, based in Linz, Austria, is one of the oldest institutions supporting experimental art and technology. Taking the visionary step in 1979 of initiating a festival to celebrate new media art, ORF (the Austrian national radio and television network) created Ars Electronica, which soon became one of the world's most prestigious venues for technological art. The festival has now grown to encompass several activities, including the yearly Prix competitions, thematic festivals, and the Ars Electronica Center.

Its activities are renown for continuing the spirit of innovation and preserving a focus on art. Excerpts from its Web site provide some flavor of its approach: "a leading

Stephen Wilson's lists of resources: ⟨http://userwww.sfsu.edu/~infoarts/links/wilson.artlinks2.html⟩
Ars Electronica: ⟨http://www.aec.at⟩

authority at the nexus of science, the economy, and public life. . . . The identity it has attained and sustained beyond technological fads and evanescent hype . . . transcend the boundaries of separate disciplines."

The Prix awards prizes in several categories. The history of the expanding categories reveals Ars Electronica's attempt to anticipate the future. The original 1987 categories included Computer Graphics, Music, and Animation. In 1992, the Prix added Interactive Art, and in 1995, Net Art and Effects. In 1999, 1,500 artists from around the world submitted documentation of their work to be considered by an international jury for three major money awards and honorary mentions in each category. The winners are exhibited at the yearly festival. Many of the artists described in this book have won Prix awards.

Ars Electronica mounts a thematic festival each year with installations, performances, and a symposium at the same time as the Prix events. Each festival integrates artists, scientists, and theoreticians around some issue in emerging technology. The symposium is often the first opportunity that artists and scientists have to hear of each other's work. Ars Electronica has been a pioneer in defining frontier areas of research as essential focuses for the arts. Here is the list of conference themes from the last several years:

1991   Out of Control
1992   Endo Nano
1993   Genetic Art
1994   Intelligente Ambiente
1995   Welcome to Wired World
1996   Memesis
1997   Fleshfactor: Information Machine
1998   InfoWar
1999   Life Science
2000   New Sex

The introduction to the 1999 festival called Life Science demonstrates the integrative approach and the focus on new technologies.

With this year's festival, Ars Electronica begins to focus on issues in the field of modern biotechnology. This constitutes a reorientation, as well as the continuation of a practice with a long history of success: namely, turning attention to those areas where conflicts develop in the sphere of tension and interplay at the nexus of technology and society, and bringing art into play as an interface and catalyst for the interaction involving science and the general public. Life science— the term which subsumes modern genetic and biological technologies—has emerged as a leading contender to become the key technology of the coming decades.

The Ars Electronica Center itself is a dynamic synthesis of museum and research center. It provides permanent exhibits as well as media archives, research labs, and network resources. Artists can apply for several residency and grant programs that are supported with money, equipment, and technical expertise: "Itself a product of artistic creativity, the center's complex as a museum of the future does not serve primarily as a venue for the presentation of works of art; rather, organizing and producing works of art and fostering exposure to them occupy the focus of its efforts."

InterCommunication Center (ICC)

The InterCommunication Center in Tokyo, sponsored by NTT (the Japanese Telecommunications Company), offers a permanent exhibit space, temporary exhibitions, archives, workshops, symposia, and on-line databases to support artists working with new technologies. The ICC was based on a visionary decision in 1989 by NTT to create a "museum of the future" that would stimulate cultural creativity by integrating science and art. The director's statement about the center analyzes the revolutionary significance of printing in human culture and proposes that electronic communications represent another major step:

Today's electronic information revolution is bringing about an even greater transformation of communication and culture than the invention of printing from movable type. Through the use of computers and communications networks, our consciousness is again undergoing a great transformation. Based on the theme of electronic communications, NTT InterCommunication Center (ICC) aims to promote dialogue between science, technology, art, and culture, and to envision a society for the future which is rich in imagination and creativity. The words *technology* and *art* both find their origins in the Greek word *techne,* which was employed in earlier times to mean both engineering and art. In modern times, technology

---

ICC: ⟨http://www.ntticc.or.jp/index_e.html⟩

and art have become divided into two separate areas of specialization and, because thought and feeling have been torn asunder, both seem to have lost their creativity. During the Renaissance, technology and art were inseparable and together led to the flowering of rich humanity and creativity. *InterCommunication* means communication for creation through mutual exchange and fusion. Contemporary society needs to break free from the dichotomy of technology and art and bring together diverse concerns, transcending the barriers of cultures and systems.

The permanent exhibits are one of the only places in the world where viewers can see works like those described in this book. Temporary exhibits cover a wide range, for example, "Sensitive Chaos—Dialogue with Being" and "Co-Habitation with the Evolving Robots." In keeping with the new vision of museums, the ICC is famous for the technical support it gives artists to realize their works. Its on-line databases "twentieth-Century Matrix" and "Artists Database" provide a unique international resource about artists working with technology. Its newsletter, *Intercommunication,* provides a forum for analysis of this art and its implications for our culture.

ZKM

ZKM, in Karlsruhe, was one of the media academies established in Germany to promote literacy in new media. Since that time it has grown to be a leading resource for technology and art. It offers a Media Museum, the Institute for Visual Media, a media library, cultural communications (an education liaison), and the Institute for Music und Acoustics.

Like the other new museums in this category, in addition to providing exhibition space, ZKM defines the critical examination of social significance and artistic support as parts of its function: "The new concept defines the museum as a production forum for artistic and scientific projects in the domain of innovative media technology . . . a kind of base camp for expeditions into the twenty-first century." It encourages the production of works that explore possible futures by experimenting with the structure of the museum: "Many of the installations in the Media Museum have been created directly for the spatial and thematic context of the Media Museum itself; they have

---

ZKM: ⟨http://www.zkm.de⟩

been designed and realized as a collaboration between artists, designers, scientists, and technicians. . . . Artistic approaches and science propagate each other and produce synergies."

The Institute for Visual Media describes itself as a "forum for creative and critical engagement with the continuously evolving conditions of media culture." It offers residencies to artists working with experimental media and provides financial, intellectual, hardware, and technical support to encourage work in fields such as virtual reality, digital video, simulation, telecommunications, and digital dance. It was one of the few places in the world where artists could get access to advanced technologies such as virtual reality. The staff also engages in ongoing research to develop new hardware and software of relevance to the arts.

Exploratorium

San Francisco's Exploratorium was founded by Dr. Frank Oppenheimer in 1969 as a unique hands-on museum of science and art dedicated to the concept of museum as a learning center. Exhibits are designed to make people ask questions and engage other visitors in speculation. Its programs have focused on perception: How do we see, hear, smell, feel, and otherwise experience the world around us? and on cognition—how we make sense of the world. The museum offers 650 permanent exhibits (ranging over areas such as hearing, seeing, numbers, electricity, and genetics), major temporary shows, special programs for teachers, lectures, an award-winning on-line virtual museum, and artist- and performer-in-residence programs. It is organized into three centers: Public Exhibition, Teaching and Learning, and Media and Communication. The success of the integrated strategy is demonstrated by the fact that the Exploratorium receives funding from both scientific and arts sources, for example, the National Science Foundation, the National Institute for Health, the National Endowment for the Arts, and the National Endowment for the Humanities.

Perhaps one of the most innovative principles guiding the Exploratorium has been the idea that the arts and sciences are both useful and necessary for understanding the world, and that they can be integrated. Artists have created exhibits that teach scientific concepts in novel ways, and scientists have stimulated artists to create new kinds of

---

Exploratorium: ⟨http://www.exploratorium.edu⟩

work. Engineers have been challenged to think about questions never confronted before. The Exploratorium's Web site explains that the exhibits that integrate art and science have "a degree of depth and a sort of openness which attracts a very large audience." The museum has an innovative long-running artist-in-residence program in which artists are selected not to complete a specific project but rather to collaboratively pursue an area of interest in conjunction with the Exploratorium's scientists, technologists, psychologists, and exhibit designers. The Web site lists some principles underlying the residency program:

- Provide the public access to the investigative processes used by artists.
- Develop new insights and understandings by incorporating the artistic process with other investigative processes.
- Enhance the role of the museum as a center of cultural investigation.
- Provide a laboratory setting for artist-conducted research which, in turn, adds to the overall creative atmosphere of the museum and provides an intellectual and technical basis for artists.
- Initiate internal and public discourse about the relationship among art, science, human activities, and topics related to multidisciplinary and multicultural activities. Elucidate, by example, the role that artists can play in modern society.

Critical Questions about the Nature of Museums

Until a few years ago, mainstream art museums have mostly ignored science-and-technology-inspired art. Many had just recently managed to acknowledge the importance of "new" technologies, such as photography and video. Some of the newer museums of contemporary art have taken the "revolutionary" step of introducing video galleries into new buildings and creating Web sites. Mainstream museums have now begun to exhibit the artists described in this book.

These gestures toward electronic art must be considered only preliminary steps. They try to assimilate the new forms, such as computer art, into the old paradigms. Computers are seen as just a new medium. As this book has demonstrated, and as the other museums listed in this section have tried to implement, techno-scientific art raises more fundamental challenges:

1.   Museums of the future will need to widen their definitions of relevance because art and techno-scientific research will not be so easily separated.
2.   Since the process of research and development is critical to the new art, museums

may need to define research support as part of their function and the focus of what they exhibit.

3. Museums will need to clarify the real meaning of physical presence, as cyberspace raises basic questions about "where" the museum is—both its exhibits and its audience.

4. Since techno-scientific art depends on infrastructure that is rapidly changing, the museum must confront the shifting realities of preservation, what Cohen and Ippolito call "variable media"; the future may require the separation of the concept of an artwork from its physical form so that the curator will have the freedom to display it in whatever current technologies exist.

5. The growing realization about the importance of research as a cultural act will further inflame debate over the validity of isolating art (and museums) as separate entities elevated above other cultural niches.

Competitions and Festivals

SIGGRAPH

SIGGRAPH is the international professional organization focusing on computer graphics. Formally, it is one of the special interest groups of the Association for Computing Machinery (ACM), the computer science organization. Its Web site defines its goals: "Our scope is to promote among our members the acquisition and exchange of information and opinion on the theory, design, implementation, and application of computer-generated graphics and interactive techniques to facilitate communication and understanding."

Each year SIGGRAPH mounts the major international conference on computer graphics and animation, which includes exhibits, conferences, hands-on workshops, new technology demonstrations, a kids and technology section, animation show, and an art show. From the early 1980s, the art show reached beyond the confines of computer graphics to present a wide range of experimental new-technology art, becoming one of the few places where this kind of art could be displayed. Since it is run by guest curators each year, its emphasis has conceptually varied, from focusing on 2-D graphics and

---

Janet Cohen, Keith Frank, and Jon Ippolito, *Variable Media:* ⟨http://www.three.org⟩
SIGGRAPH99 art show: ⟨http://www.siggraph.org/s99/conference/art/index.html⟩
SIGGRAPH99 technology show: ⟨http://www.siggraph.org/s99/conference/etech/index.html⟩

design to more experimental installation art. The 1999 "TechnOasis" art show presented itself in this way: "Experience turn-of-the-century digital art in all its variety: visual, interactive, animated, sculptural, installed, virtual, Web-based, telecommunicated, and participatory."

The panels, papers, and experimental technology demonstrations are also valuable resources for those trying to understand the future of technology. The 1999 new technology show "Millennium Motel" described itself as "located between aesthetics and logic, where infrastructures of technology converge with the networks of desire. Check in and check out 1999's multimodal interface design, intelligent autonomous agents, scientific visualization, conceptual electronic performance, wearable computers, and alternate realities."

International Symposium of Electronics Arts (ISEA)

ISEA is a worldwide organization of artists and academics interested in promoting experimental technology-based art and theory. Like many similar organizations, it started with a focus on electronic arts but has expanded to include a wide range of experimental technologies. It offers a newsletter, on-line services, and its most-known offering, the yearly symposium, which typically includes art shows, performances, workshops, papers, and panels. ISEA describes its mission as "to establish and facilitate interdisciplinary communication in the field of art, technology, science, education, and industry. ISEA advocates a culturally diverse community, which stimulates a global promotion and development of electronic art practices."

Each year the symposium is offered in a different place in the world, and with a different theme. The 1999 conference was held in Sao Paulo, Brazil, in conjunction with CAiiA and the journal *Leonardo,* and was called Invencao: Thinking the Next Millennium. Its Website described its mission:

Invencao is an opportunity for those working at the creative edge of the arts, sciences, and technology to collaborate in the transdisciplinary development of ideas and innovative strategies for life in the next millennium. Invencao is a "seeding" event that seeks to identify key questions and issues that can lead to the radical transformation of culture.

Just as increasingly artists work with the metaphors of science, so scientist are employing forms of representation, such as visualisation, which owe much to research in the digital arts. As art is

---

ISEA: ⟨http://www.isea.qc.ca/intersociety/about.html⟩

transformed by interactivity, so science increasingly recognises the subjectivity of the observer. In turn, technology informs our aesthetic and epistemological structures and is engendering new processes of perception, communication, and cognition.

Invencao will examine the consequences of this convergence of art, science, and technology on our sense of self and human identity, on consciousness, community, and the city, as well as on learning and leisure. For example, the artist is challenged to consider what might lie beyond "electronic art": where might the connectivity of the Internet, the interactivity of hypermedia, and the fluidity of virtual reality lead us? The scientist, walking a delicate balance between the world of the quantum, deep space, chaos, and complexity, has profound questions to ask about the constraints of nature and the part that can be played by artificial intelligence and postbiological systems in the construction of reality. Biotechnology and nano-engineering add further dimensions to these questions.

UNESCO-ICSU Science in the Arts—Art in the Sciences Conference (Hungary)

This conference and exhibition, cosponsored by art and science organizations, celebrated the "overlappings of the human creativity on the boundaries of the arts and the sciences." It included an exhibition of art inspired by scientific ideas, represented scientific discoveries, introduced conceptions that might give birth to new scientific ideas, and explored topics such as symmetry-dissymmetry, perspective, dimensionality, and new materials.

Other Festivals, Competitions, and Shows

Ars Digitalis (Berlin)	Digital Chaos (Bath, U.K.)
ARTEC	Digital Salon
Art Future	Dutch Electronic Arts Festival (DEAF)
Butterfly Effect (Budapest)	Electra 96 (Oslo)
CAiiA, Consciousness Reframed	European Media Arts Festival (EMAF)
Champs Libre (Montreal)	5.CYBERCONF (Madrid)
Cyberria (Bilboa)	Imagina
Cyberstar	Interactions
Digital Arts and Culture (Atlanta)	Invision Awards

UNESCO science-art conference: ⟨http://www.d.umn.edu/~ddunham/artsci.html⟩

L.A. Media Festival	New Bodies (Maribor)
Lyon Bienalle	New Media Minds
MECAD (Spain)	Next Five Minutes (Amsterdam)
Meta Forum III (Budapest)	Symposium on Arts and Technology
Milia	Teleopolis (Luxembourg)
Mostra de Realidade Virtual (Brazil)	Videobrasil
Multimedia Arts Asia Pacific	

In addition, many one time shows have been organized by museums and arts organizations. Because of limited space, this book can offer details on only a few of the resources available in each category. Additional resources are presented in supplementary tables. Readers are invited to search out their web sites for more details. Many are listed at the infoarts links pages located at http://userwww.sfsu.edu/~infoarts/links/wilson. artlinks2.html⟩.

Organizations and Information Publishers

Leonardo and the International Society for the Arts, Sciences, and Technology (ISAST), the Leonardo Electronic Almanac, and OLATS

Leonardo magazine, founded by engineer and kinetic artist Frank Malina over thirty years ago, was one of the pioneer resources in the world of art and science. Long before there was much interest in the experimentation with new technologies and scientific concepts, he published articles by artists working in the field. The journal, now published by MIT Press, is part of the family of activities of ISAST.

In addition to the print journal *Leonardo,* ISAST activities include the *Leonardo Music Journal* (including a companion CD-ROM); the book series; international awards (for pioneering artists, articles, and art organizations); the *Leonardo Electronic Almanac (LEA); Leonardo Digital Reviews; Leonardo On-Line;* and support for international conferences. An affiliated organization called Creative Disturbance explores experimental forms of for-profit activity such as brokering contacts between artists and potential corporate or institutional supporters. The sister organization in France publishes the "Observatoire Leonardo Web site (OLATS), primarily in French. The mission statement describes its effort:

Leonardo—ISAST: ⟨http://mitpress.mit.edu/e-journals/Leonardo/home.html⟩

Leonardo/ISAST, established in 1982, serves the international art community by providing channels of communication for artists, scholars, technologists, scientists, educators, students, and others interested in the arts, with an emphasis on documenting the voices of artists all over the world who use science and developing technologies in their work.

The print journal includes peer-reviewed articles on topics such as art and interactive telecommunications, visual mathematics, art and biology, artificial life and art, synesthesia and interarts, and Brazilian electronic arts. The LEA includes these sections: artists' statements; short feature articles; bibliographies; reviews of publications, exhibitions, conferences, and digital works; calls for papers; announcements of events involving art and new media; and listings of job and educational opportunities. The OLATS site includes projects such as Virtual Africa; the Space and Arts Workshop; and the Pathbreakers and Pioneers project, to put the documentation of pioneers in technological arts on-line.

Australian Network for Art and Technology (ANAT)

ANAT seeks "to advocate, support and promote the arts and artists in the interaction between art, technology, and science, nationally and internationally." It works to position artists as "active participants in the information age" and to foster interchange between artists and technologists. Its activities include the National Summer School, which offers residencies for artists in emerging technologies; research into new technologies; the Art and Technology Conference and Workshops Fund, which offers small grants to experimental arts; real and virtual residencies, conferences, exhibits, print and on-line newsletters, an extensive Web site, and the "Screenarts" Web site, which provides listings of digital exhibitions and forums (in conjunction with the Australian Film Commission).

ANAT encourages critical debate around art and technology. It endorses "the creation and consumption of new media art forms within a critical context beyond commercially driven techno-evangelistic hype. In doing so, ANAT advocates for more considered and informed approaches to the way technology interfaces with art and cultures." It has sponsored projects such as CODE RED, which explored the role of artists in ensuring a vibrant and diverse mass media, and FUSION'99, a series of collaborative telepresence

---

ANAT: ⟨http://www.anat.org.au⟩

events to explore "the current breakdown of definitions, dualisms, and geographical boundaries on the internet."

V2

Started in 1981 as an artist alternative space in 's-Hertogenbosch, the Netherlands, V2 has grown to be a comprehensive center for experimental technology-based arts. It offers a yearly Dutch Electronic Arts Festival (DEAF), exhibitions, lectures, archives, and extensive on-line activities. It has an international reputation for identifying artists and ideas that are important in the interplay of art and technology and for encouraging collaboration among various media and disciplines. Its mission statement highlights the interdisciplinary nature at contemporary arts:

The merging of media, such as we see happening in computer sciences, is reflected in the multidisciplinary nature of the V2's activities. For presentations in past years, V2 has cooperated, and will continue to cooperate with, people from the worlds of visual arts, architecture, music, philosophy, sociology, and film. . . . We do not have a center in mind that solely reflects the technological state of the art, but instead a center where relationships and connections between different artistic disciplines are forged and where developments in practice and theory are presented alongside one another to help obtain a critical and analytical view of these media; a lively center at the heart of developments in the area of sound, image and technological affairs within art and society, which reflects upon these matters by means of manifestations, concerts, and exhibitions.

Philosophically, it believes that "unstable media" have a cultural influence beyond the traditional confines of media:

Unstable media are media that use electron streams and frequencies, such as motors, light, sound, video, computers, and so forth. The word unstable is, according to us, more adequate than electronic, because it refers to one of the most important properties of these media, to wit, the rendering unstable of all things social, political, and cultural within our society—the unstable electron as a basic concept for our society.

––––––––––––––––––––

V2: ⟨http://www.v2.nl/⟩

For example, the 1998 DEAF description demonstrates this approach: "The DEAF98 Exhibition presents electronic art projects that facilitate encounters and interactive experiences with aspects of an *'ars accidentalis.'* The artworks deal with computer distortions and wars, with airplane crashes and the disappearance of horizon and linear perspective, and with the incompatibilities of physical and virtual spaces."

Ylem

Ylem is an organization based in the San Francisco Bay Area, focused on the exploration of the relationship between art and science. It offers forums, a newsletter, and Web resources such as an on-line gallery, and it partners with other organizations to sponsor events. (The word *ylem* is the physics term representing the matter out of which the universe evolved.) Past newsletters have addressed themes such as telepresence, science fiction and the discourse of the "Other," and art on the Internet.

Art and Science Collaborations, Inc. (ASCI)

ASCI, in New York, is a network for artists and others interested in the integration of art and science. It provides information to members on events and opportunities, and helps to organize speakers series, conferences, and exhibitions. It publishes the biweekly *ASCI Bulletin,* holds meetings, and organizes virtual exhibitions on-line. Examples include the "LightForms98" light-art exhibition in collaboration with the New York Museum of Science, and the Art/Sci'98 symposium, which included researchers, industry representatives, philosophers, and artists reflecting on a variety of topics in art and science, such as the nature of creativity.

Fine Arts Forum

Since 1987, the Fine Arts Forum has been the major on-line jumping-off place for timely information about events, competitions, opportunities, conferences, and Web sites in the field of art and technology. It started as a mailing list, and in 1994 added a Web site. It also includes Web resource listings.

Ylem: ⟨www.ylem.org⟩
ASCI: ⟨http://www.asci.org/⟩
Fine Arts Forum: ⟨http://www.msstate.edu/Fineart_Online/index.html⟩

Rhizome

Rhizome is an organization that fosters experimental forms of communication on new media principally through its Web site and E-mail lists, aiming to be a "comprehensive resource for information and critical writing about what's going on at the intersection of emerging technology and contemporary art." It draws together geographically dispersed "artists, authors, designers, programmers, musicians, curators, critics, and others," and uses the intelligence of computer-mediated communication to filter and distribute information in a usable way.

CTheory

CTheory is a much respected online international journal of theory, technology, and culture. It publishes articles, interviews, and reviews. Samples of articles include "Men with Modems," "Resisting the Neoliberal Discourse of Technology," and "Through the Dark Mirror: UFOs as a Postmodern Myth."

IDEA

IDEA is a directory of electronic arts resources published by Centre Histoire, Art, Ordinateur, Science (CHAOS), in France. First published in 1990, it is periodically updated in French and English. It covers artists, periodicals, media resources, and organizations in fifty countries covering the fields of art, science, and technology. Each entry includes a short description and contact information.

MediaMatic and Doors of Perception Conferences

MediaMatic is a Dutch organization focused on critical commentary on emerging technologies and culture. It offers publications and organizes the Doors of Perception conferences, which have gained their reputation from: "the variety of disciplines and backgrounds from which speakers contribute: here are scientists, designers, philosophers, gurus, critics, misanthropes, and true world-changing believers, from five continents. They disagree on most issues, but are united in their determination to address this question: What are the Internet, information technology, and multimedia for?"

Rhizome: ⟨http://www.rhizome.org/⟩
CTheory: ⟨http://english-server.hss.cmu.edu/ctheory/ctheory.html⟩
IDEA: ⟨http://nunc.com/index.phtml⟩
Doors of Perception: ⟨http://www.doorsofperception.com/doors⟩

Other Organizations and Information Resources

ArtByte

Artswire

AV—arkki (Finland)

C3 (Hungary)

Connecticut College Center for Art and Technology

Convergence: Journal of Research into New Media Technologies

Crashmedia

Cultronix

DA2—the Digital Arts Development Agency (London, UK)

D'lux Media Arts (Australia)

Experimenta Media Arts (Australia)

IRCAM—L'Institut de Recherche et Coordination Acoustique/Musique (France)

Institute for New Culture Technologies

Intelligent Agent

Ljubljana digital media lab

MECAD—Media Centre of Art and Design (Barcelona, Spain)

The Nature and Inquiry Group (Boston)

Neue Galerie Luzern

Public Domain

Speed

SWITCH

Teleopolis

Thing

UNESCO Cultural Affairs (special focus on art and science)

De WAAG—The Society for Old and New Media (Amsterdam)

Art and Science-Technology Sponsors, Competitions, and Academic Convergence Programs

Several contemporary initiatives to foster collaboration between artists and researchers are described in chapter 1.2. This section lists several other sponsors of integrative projects, competitions, and conferences.

Examples of Art and Science-Technology Convergence Efforts

In Canada, *Foundation Daniel Langlois,* established by the artist-researcher Daniel Langlois, creator of Softimage animation software, funds a resource center and art-technology research projects with a special emphasis on countries outside of Europe and North America. The Rockefeller Foundation's Creativity and Culture section has funded studies by the Academy of Science and others to identify strategies and structures necessary to encourage new developments in art and technology. The Moet Hennessy/Louis Vuitton *Science pour l'Art Priz* holds thematic competitions to honor research that crosses disciplines, such as the award for research in which the aesthetics of form play a role. ATT's *New Experiments in Art and Technology* grants are awarded to museums for projects that cross disciplines. *The Art and Technology Foundation* in Spain funds exhibitions and integrative projects. *The Center for the Integration of Art and Science* (at Texas Tech University), *NEXA* (at San Francisco State University), the *Center for Art and Technology* (at Connecticut College), Gulbenkian Foundation (UK), NESTA—National Endowment for Science, Technology, and the Arts (UK) and *Mindship International* (in Denmark) all support convergence conferences and projects.

Think Tanks and University Labs

PARC

Xerox PARC has a long illustrious history as one of the prime research laboratories in the world, with a list of innovations that have had revolutionary impact, such as the visual interface, ethernet, and postscript. The center currently pursues several broad research themes: knowledge ecologies, network devices and document services, emerging document types, smart matter, and information fabric. Some of the particular projects pursued are listed below:

The Bayou Project, Digital Libraries, Digital Video Analysis, Dynamics of Computation, Fluid Documents, High-Performance Sensor Arrays for Digital X-ray and Visible Light Imaging, Human Vision Model, Image Understanding, Interactive Information Services, Inter-Language Unification (ILU), Internet Ecologies Area, Machine Learning in Information Access, MagicLenses, Map Documentation, Model-Based Computing, MEMS, MEMS-Based Active Control of Macro-Scale Objects, Metaobject Protocols, Nano-technology, Natural Language Theory and Technology, Smart Matter, User Interface Research, and Work Practice and Technology.

Microsoft, IBM, Bell Labs, and Interval Research

Several industrial research labs are famous for attempts to create think-tank atmospheres that pursue long-term visionary research. Researchers from art backgrounds are often included. For example, Microsoft research includes these investigations: adaptive systems and interaction, collaborative and multimedia systems, decision theory and adaptive systems, information retrieval and analysis, machine learning, speech, telepresence, and virtual worlds and vision technology. IBM supports research in these topics: security,

---

Xerox PARC: ⟨http://www.parc.xerox.com⟩

multimedia, human computer interaction, mathematics, materials science, and computational biology. Bell Labs' experimental multimedia program sponsors research in these areas: natural interfaces, exploratory data analysis, and vision research. Interval Research, which had a reputation for the most visionary inclusion of artists, was suddenly closed late in this book's production cycle by its venture-capital sponsor. More information about Interval and its closure is available at the URL listed below.

MIT Media Lab

MIT's Media Lab is one of the major research laboratories in the world investigating new technologies. Its interdisciplinary teams are funded by a large number of industry "sponsors." Its charter is "to invent and creatively exploit new media for human well-being and individual satisfaction without regard to present-day constraints. We employ supercomputers and extraordinary input/output devices to experiment today with notions that will be commonplace tomorrow. The not-so-hidden agenda is to drive technological inventions and break engineering deadlocks with new perspectives and demanding applications." Areas of research interests include: aesthetics and computation, affective computing, computers seeing action, electronic publishing, epistemology and learning, explanation architecture, gesture and narrative language, interactive cinema, machine listening, machine understanding, media and net-works, micromedia, news in the future, nano-scale sensing, object-based media, opera of the future, personal information architecture, perceptual intelligence, physics and media, sociable media, the society of mind, software agents, spatial imaging, the speech interface group, synthetic characters, tangible media, and the Visible Language Workshop.

Other Research Laboratories and Initiatives

Art + Com (Berlin)

Batelle Labs (energy, transportation, and health)

British Telecom Research Labs

CID-KTH Center for Interactive Design (Sweden)

Defense Research Center (DARPA)

Electronic Visualization Lab (University of Illinois, Chicago)

Georgia Institute of Technology—The Graphics, Visualization & Usability Center

GMD—German National Research Center

Tia O'Brien, article about Interval: ⟨http://www.svmagazine.com/2000/week37/features/Story01.html⟩
MIT Media Lab: ⟨www.media.mit.edu/⟩

HITL Lab (at the University of Washington human-interface technology and virtual reality)

Human Genome Project

I3 (European Community Sponsored Research Network on Intelligent Information Interfaces)

Institute for the Future

Interface Research (at the University of Maryland)

Interface Research Lab (at the University of Toronto)

Los Alamos National Laboratory

Philips's "Vision of the Future" group

Sandia Labs (nuclear energy and materials)

Santa Fe Institute (artificial life, neurobiology, and adaptive agents)

SICS—Swedish Institute for Computer Science

SRI International (future-technology research)

Educational Resources

MIT Media Lab

MIT's Media Lab also offers an academic program, primarily at the graduate level, tightly linked to its research. It premises that a rapidly changing, technologically mediated culture demands educational programs that are interdisciplinary, focusing simultaneously on technological and human issues, and that research is an essential ingredient. Since the field is a fledgling discipline, it seeks students "whose sense of enterprise and inquiry will lead them to develop their own new paths." Here are excerpts from the Web site:

The field of MEDIA ARTS AND SCIENCES can be thought of as exploring the technical, cognitive, and aesthetic bases of satisfying human interaction as mediated by technology. In more forward-looking terms, it addresses the quality of life in the information-rich environment of the future. . . . These "mediating technologies" are only in the first stages of their modern evolution; they are still crude, unwieldy, and unpersonalized, poorly matched to the human needs of their users. Their fullest development in those terms is emerging as one of the principal technical and design challenges of the emerging information age. . . . The field is deeply rooted in the modern communication, computer, and human sciences.

MIT Media Lab: ⟨www.media.mit.edu/⟩

CAiiA-STAR

CAiiA-STAR is a highly selective research center that offers the M.Phil. and Ph.D. degrees. It is based at CAiiA, the Centre for Advanced Inquiry in the Interactive Arts, at the University of Wales College, Newport U.K., and STAR, the center for Science, Technology, and Art Research, in the School of Computing at the University of Plymouth. The center encourages research into the "new field of practice, theory, and application, which is emerging from the creative convergence of art, science, technology, and consciousness research." Students can participate primarily on-site or on-line. On-line students commit to a significant investment of research time at their host institution and in three, ten-day composite sessions a year, one of which includes the Consciousness Reframed conferences. The program aims "to define and establish new fields of practice through research in the creative and innovative use of interactive media, telematic systems, and the cognitive and biological sciences through multidisciplinary collaboration."

Art Technology Studies—School of the Art Institute of Chicago (SAIC)

Founded in 1982, Art Technology Studies offers an interdisciplinary program for teaching new technologies. Areas covered include: "computer imaging, interactive media, computer animation, kinetics, electronics, holography, digital sound and video, electronic media-based installation, machine control, and telecommunication- and Internet-based art." In addition, students can study computer programming for image processing, algorithmic composition, or the development of virtual environments and the techniques of digital control of electromagnetic devices such as sculptures, synthesizers, and telephones. The program is a part of the Time Arts program, which includes areas such as performance and filmmaking. The SAIC also includes the Center for Advanced Studies in Art and Technology, which supports experimental student and faculty work. The SAIC offers both the BA and MFA degrees.

Conceptual Information Arts

The Conceptual Information Arts program (CIA) at San Francisco State University is a studio art program focusing on the intersections of art, science, technology, and culture.

CAiiA-STAR: ⟨http://caiia-star.newport.plymouth.ac.uk/⟩
SAIC: ⟨http://www1.artic.edu/saic/programs/depts/undergrad/ats.html⟩
CIA: ⟨//userwww.sfsu.edu/~infoarts/⟩

It stresses the integration of intuitive processes typical of the arts, with structured processes of research, planning, and problem-solving more characteristic of other disciplines. It promotes nonconventional art media, new media, and the movement of artists into non-art contexts. It teaches students concrete skills related to contemporary science and technology, such as structured problem-solving, the analysis of biological systems, computers, telecommunications, robotics, interactive media, and the electronic synthesis of image, text, and sound. The program stresses the perspectives of critical analysis of cultural systems, language, and media. Students can earn either a BA or MFA.

Carnegie Mellon University Art Department

The art department at Carnegie Mellon University offers students an interdisciplinary practice-and-theory-oriented program "designed to develop individuals capable of working as artists in a complex, rapidly changing global culture." It teaches students to master traditional art skills and to draw on wider cultural resources, such as developments in science and technology. Demonstrating this approach, the core "Concept Studios" focus on five categories: "concepts related to the self and the human being, space/time, systems, process, and context." Students are encouraged to take advantage of the significant technical resources of Carnegie Mellon and research in the Studio for Creative Inquiry. It offers BA and MFA degrees.

CADRE—San Jose State University Art Department

CADRE focuses on "the development and testing of emerging technology applications in art, design, education, and communications in the context of critical discourse." BA and MFA degrees are offered. Some areas of interest include: "computer graphics, 3-D modeling, animation, digital video, interactive multimedia, automated documents, expert systems, artificial agents, human-machine interface design, real-time performance control, computer-mediated environments, telecommunications, networking, video conferencing, distributed education, telepresence, robotics, and virtual reality. Of particular concern is the investigation of issues pertaining to: "hypertextuality, interactivity, information mapping, navigation, immersion, agents, virtuality, emergent systems, identity, and collaboration."

Carnegie Mellon art department: ⟨http://www-art.cfa.cmu.edu/⟩
CADRE: ⟨http://cadre.sjsu.edu/sub/institute/institute.htm⟩

International Academy of Media Arts and Sciences (IAMAS)

IAMAS is an innovative Japanese institution focused on providing new-media education that prepares students to work with emerging technologies. It offers rigorous courses and state-of-the-art facilities to both undergraduate and graduate students. The academy also supports artist in residencies and the Interaction art festivals.

Media Academy Cologne

In 1990, the German government set up a network of media academies to address the new emerging media technologies. The academy at Cologne achieved an international reputation for its program focused on "formulating and developing the interaction between artistic imagination, theoretical and historical knowledge, and individual expression by means of new media." It emphasizes the social institution nature of new media, thus also emphasizing "joint development of a media/cultural identity, one conscious of its own social, political, aesthetic, and ethical obligations."

Other Educational Programs

Academy of Media Arts (Cologne)

Arizona State University, Institute for the Arts (ISA (Tempe, Arizona))

Art Academy of Trondheim—KIT (Trondheim, Finland)

Australian Centre for Art and Technology

California Institute for the Arts

Centro Multimedia (Mexico City)

Chelsea College of Art (U.K.)

Cogswell Polytechnical College (U.K.)

College of Fine Arts in Ume (Sweden)

Concordia University (Montreal)

Ecole d'art d'Aix-en-Provence, CYPRES new media research

Ecole d'art d'Angoulime and Poitier

Georgia Institute of Technology (Interactive Media Technology Center)

IRCAM Experimental Music Center (Paris)

Mertz Akademie

Minneapolis College of Art

MIT Center for Advanced Visual Studies (CAVS)

New Media Institute (Frankfurt)

New York University, Tisch School of the Arts, Interactive Telecommunications

Ohio State University, Expanded Arts (an art and technology program)

Queensland University of Technology (Brisbane, Australia)

Rensselaer Polytechnic (EMAC program)

IAMAS: ⟨http://www.iamas.ac.jp/⟩
Media Academy Cologne: ⟨http://www.khm.de/frameset/reiter_e.htm⟩

Royal College of Art (U.K.)

School of the Visual Arts (New York City)

State University of New York at Buffalo, Integrated Media Center

Technical University of British Columbia

University for Applied Arts Department of Visual Media (Vienna, Austria)

University of Bergen, Department of Humanistic Informatics (Norway)

University of California at Los Angeles design and new media department

University of California at San Diego, Santa Barbara, and Santa Cruz art departments

University of Florida, Digital Worlds Program

University of Illinois at Chicago, Electronic Visualization Laboratory (EVL)

University of Maryland, Baltimore, New Media

University of Michigan, New Genres

Université du Quebec à Montreal, Departement d'arts plastiques

University of Paris I, University of Paris 8

University of Texas, Multidisciplinary Art and Technology

University of Washington, New Media and HITL Lab

Research Conferences, Science Magazines, Trade Magazines, and Science and Technology Studies

There are many other resources that can only be briefly noted here. There are academic and trade organizations and conferences focused on just about every area of technological research, for example, SIGCHI (computer-human interface); artificial intelligence, virtual reality, robotics, artificial life, agents, computer-supported group work, and visualization. There are magazines and journals stretching from the scholarly to the popular, such as *Scientific American, New Scientist Magazine, Popular Science, Popular Mechanics, Science, Nature, Discover Magazine, OMNI, Technology Review, American Scientist, Invention and Technology* (published by the Smithsonian Institute), *Physics World, Science News,* and *HotWired.* There are trade magazines aimed at specific technological niches, such as *GPS World, Advanced Imaging, R&D, Electronics, ID Systems, Defence News, Toy World,* and *Biomedical Research.* There are also trade magazines in more traditional computer and media fields, such as *Internet World, Web Techniques, DV, Publish,* and *Interactive Week.* Finally, readers might be interested in the academic discipline called science technology studies (STS), which includes scholars studying science and culture from perspectives such as history, philosophy, sociology, and literature. Extensive print and on-line resources are available in STS.

Summary: Institutions as Art

The intersection of art with science and technology calls for new kinds of resources. Traditional museums, schools, art shows, journals, and organizations will not satisfy the need without significant innovation. This chapter has surveyed some that have started the innovation. Creative thinking about the art/science institutions of the future may well be part of the work of the arts in the next decades.

8.2

Summary: The Future

This is research's era. Inquiry and innovation are breathtaking in their reach. The future promises to be even more remarkable than the recent past. As these chapters have demonstrated, one cannot assume that these developments are necessarily progress and that their presenting rationales are the only way they can be read. Still, this book has tried to find a middle road between wariness about the subtexts and delight in the new possibilities.

The artists described in these chapters are beginning to engage that world of research in profound ways. They are reclaiming art as a zone to question and innovate—even in a world dominated by science and technology. They are wonderfully diverse in ideologies, goals, and approaches. Many of the works tickle both the intellect and the spirit. They have begun to enter into research not only to use its gizmos or to critique its blindnesses but also to help shape its future.

Similarly, we can marvel at the researchers and their leaps of imagination. We can be astonished by the way much of this research reaches beyond utilitarian or disciplinary provincialism to propose wider questions about humanity and its possibilities. Finally, there are signs of growing interest and cross-fertilization between the arts and the worlds of science and technology.

In spite of these hopeful signs, this is no time for complacency. The effort to integrate research and art is still very raw; it has far to go on the way to maturity. There are several areas of weakness and questions to be addressed.

Areas of Neglect I identified many areas of fascinating techno-scientific research that are totally unaddressed by artists. Sometimes it seemed that the research is like a wildly spreading fire in a great forest and there are only enough artists to mind the fire in a few spots. While for some areas of research it is easy to recognize artists' entry point—for example, virtual reality, robotics, or telepresence—other areas seem quite remote and esoteric. Reflecting on my plans for this book, some colleagues suggested I concentrate on areas where there is activity and "obvious" opportunity for cultural commentary such as the impact of new media representation or identity.

I start from a different position: every area of research is potentially addressable by the arts. In part, this idea is based on my experience of the last twenty years watching esoteric areas such as computers and telecommunications pass from arcana to the center of wide cultural interest. A similar fate awaits other esoteric areas now defined as irrelevant if artists can cultivate the interest and background understandings necessary to engage the areas.

I do not suggest that reclaiming areas of research is always easy. Periodically, as part of my art practice, I discipline myself to ponder each article in a given issue of a science journal such as *Scientific American* or the *New England Journal of Medicine,* asking a set of questions: What is the article really about? What are the scientific and cultural contexts of the research? What might an artist do to investigate the concepts or processes of that research? For some of the articles the answers are easy; others are somewhat mystifying and would require more study even to begin to answer the questions.

Shallowness of Analysis   Much of the artistic experimentation in frontier areas is somewhat superficial in its analysis of research issues. At this early stage, the freshness of perspective that artists bring by itself makes their involvement valuable, for example, an artist even beginning to try to build sculpture out of stem cells. Eventually, however, as artists continue to explore a research area, they will need to become more penetrating and subtle in their engagement with the area of research, for example, understanding embryology, cell biology, methods of shaping growth and nurturing development, and the implications of different kinds of acquisition methods.

Emotional Power of Artworks   The work described in this book offers awe-inspiring intellectual challenges. One could think of it as a new kind of conceptual art. However, at this stage of development, it is sometimes less interesting to observe or experience than it is to think about. It lacks the quality of emotional and sensual engagement that has historically been a hallmark of the arts. As these new forms mature, they will be faced by the challenge of developing this dimension. Or perhaps our criteria of engagement of art will change to match the strengths of the work.

Cultivating Researcher Interest in Art Experimentation   This book has focused on the growing interest by artists in research and the subsequent enrichment of the arts. Parallel questions could be asked about researcher involvement with the arts. Many scientific and technological researchers define the arts as alien territory that is professionally irrelevant. If they are personally interested, they hold stereotypical views of the arts that stops with classical museum and gallery forms such as painting and sculpture. These views will need to expand in the future if the integration of art, science, and technology is to proceed. There are already signs of change. Researchers at some of the think tanks described in chapter 8.1 have begun to position their research as both technical and art research. They present it at professional meetings and in art shows such as SIGGRAPH's and Ars Electronica's. This cultivation of interest in experimental arts as a resource in research is still an open question in need of much development.

Institutional Arrangements   How should collaboration and mutual influence be set up? How can both the arts and technical fields meet in a context of equality? Currently, the economic and political power lies with science and technology. For example, the dominant culture at think tanks is shaped by the worlds of business and science. The development of institutional arrangements for true equal collaboration will not be easy.

Information Arts has focused on the mutual influence between art and research. The implications of this interrelationship stretches beyond artists and researchers to the larger public. Increasingly, citizens are asked to make policy decisions that relate to research. Also, general cultural literacy in science and technological research is not keeping up with pace of technical developments. An integration of art and science promises to be much more appealing and understandable to a larger public than either are alone.

Research is the search for the future. Science and art are major forces in contemporary society. They must both contribute to the shaping of that future.

Appendixes

A

Methodology

Research Questions and Choice of Artists

There is not yet a robust canon for interpreting the art described in this book. Its rapid development, integration of disciplines outside of art, and movement into new technologies and contexts all make it more difficult to assess its significance than with more conventional art forms. This section describes the methodology used to locate art and make decisions about the inclusions in the book. It also describes the processes used to identify the main research agendas in the various fields of science and technology.

Personal Experience vs. Documentation and Report

Firsthand Experience   It is nearly impossible to have firsthand experience of all the work presented in this book. This art is international in scope and is presented in exhibitions all over the face of the globe. Like scientific research, it is developing quickly and may not be shown more than a few times in various exhibitions before the artists move on to their next works. Although I have been fortunate to participate in several of these exhibitions and thus to see a good sampling of this art, I (like most commentators) have been not been able to directly experience most of the art described in this book.

Reliance on Documentation and Report   The presentation of many of the works in this book therefore must rely on the artist's own documentation, reports by visitors, judgments by curators and selection committees, and commentary by other analysts. Unfortunately, even for the same artwork, various visitor accounts differ—some report moving, intellectually provocative encounters while others denounce the work for its shallowness or deadness. Similarly, I have personally experienced works that I felt were not nearly as interesting as their documentation promised and others that were much richer than the documentation would suggest. I have served on selection committees in which, although there was substantial agreement, there were also wildly divergent opinions about the significance of some works. Nevertheless, this book is based on the premise that it is essential to gather together descriptions of science-and-technology-inspired work in one place. Note, however, that the significance and the polish of the works varies greatly from artist to artist, and the underlying basis of my reports vary.

Interactivity   Since many of the works presented in this book are interactive, they offer special challenges for review. Usually there are many paths through the event, with some features only revealing themselves in response to specific actions. Different visitors can experience very different artworks. Thus, any short review inevitably does injustice to the richness of this kind of work.

Research Sources and Locating Artists

Within the limitations of the international spread of this work and the diversity of topics and contexts addressed, I have tried to be reasonably comprehensive in my search for artists and examples of work. I employed the following strategies:

- Review of artists chosen in major international competitions and recurring art festivals and shows
- Review of artists in one-time art shows
- Review of print and on-line resources
- Consultation with experts
- Examination of syllabi from professors teaching related topics
- Comprehensive Web search

Assessments of Significance

Artists presented in this book range from those who are seasoned, internationally recognized leaders in their field to those who might be just starting their careers. I have used a two-stage process: First, I have identified artists with international reputations whose names repeatedly come up in shows and competitions. Next, I have identified interesting artists without that recognition whose work explores areas relevant to this book. It is sometimes difficult to assess the strength of this work, but I have adopted a comprehensive, inclusive approach as the best strategy for building the compendium of art and research.

Original plans called for all artists to be described with full sections containing excerpts of their writing and images. The original draft could not fit in one book. Thus, I have had to reduce descriptions of many artists' works to a few lines in "Other Artists and Projects" sections. Often there was no basis for the distinction. Readers are warned not to ascribe lesser significance to the more limited descriptions. Wherever possible, Web links to all artists are provided so that readers can study the less-described works more fully.

Gaps and Omissions

I have missed some artists and works that should have been included. Since these fields are moving so quickly, there will have been changes between completion of the manuscript and its publication. I invite readers to consult my art links Web site for updates

---

Stephen Wilson, updates: ⟨http://userwww.sfsu.edu/~infoarts/links/wilson.artlinks2.html⟩

and to send me suggestions for inclusion. There are other omissions. Even as I completed the last stages of this manuscript, I would uncover artists whose work was highly relevant. The last-minute discoveries made me uneasy about other omissions, but perhaps that is inevitable. From personal experience in this field, I know that I am disappointed when surveys don't mention my work, and I apologize to any colleagues I have missed. Analysis of some of these omissions shows the kinds of artists who may not be represented.

The Web as an Imperfect Tool I have relied heavily on the World Wide Web, which is an excellent though imperfect research tool. There are several limitations. Searches even with multiple search engines only find a proportion of Web pages that exist. In this relatively early stage of Web evolution, not all artists or museums have Web pages, for example, those in countries without extensive Internet infrastructures.

Relationship to the Mainstream Art World For many years the mainstream art world of museums and galleries ignored techno-scientific–inspired art. Most established artists did not concern themselves with developments such as computers, microbiology, or telecommunications. As a result, pioneering artists found alternative exhibition venues, such as the SIGGRAPH and Ars Electronica art shows, and critical contexts such as ISEA conferences and *Leonardo*. We are now at a transition point. There are signs that museums and galleries will begin to show more interest in science and technological topics, and established artists are beginning to explore technical tools and contexts. Many of the artists who used to be able to show only in the special venues are being invited into mainstream art institutions. The transition is not complete. Established artists and museums starting to explore technological experimentation have not yet connected with the alternative exhibition venues used by artists and commentators experimenting with techno-scientific inquiry. It is possible that I may have missed this work.

Research Questions and Overviews of Research Agendas

This book proposes that scientific and technological research is relevant to the arts. Consequently, in each major section I have endeavored to present a brief overview of relevant scientific research and technological developments, research agendas, categories for conceptualizing research, and future areas of investigation. I have presented samples of new technologies and products, drawing from sources as diverse as toy stores and military research labs. These summaries are necessarily condensed, skimming over debates within the disciplines. Still, they offer a window into what may be significant future issues.

I try to make my outsider status an asset in applying fresh frameworks to thinking about research. Although I am not a researcher in any particular scientific discipline, as an outsider I do not have the same partisan agendas as practitioners in defining significance. Over the last twenty years I have made the monitoring of scientific and technological research a major part of my artistic practice. I try to observe developments by reading scientific journals, studying the programs of academic meetings, staying especially alert to science-related commentary in popular media, and by visiting Web sites devoted to research in various disciplines. I have been an artist in residence at various think tanks and a developer for new-technology companies.

Locating Research Agendas

In an attempt to make these overviews comprehensive and reliable, I consulted the following sources:

- Summaries by government agencies and professional organizations
- World-class university departmental statements
- Web searches of think tanks
- Programs and proceedings of professional meetings
- Popular press
- New products, trade journals, and trade shows

Breakthroughs do not come only from university research labs. Inventors and entrepreneurial development provide a critical source of innovation. Indeed, one important lesson of the microcomputer revolution is the danger of reliance only on mainstream academic and corporate agendas, which often dismiss innovative ideas as pipe dreams, economically and technically unfeasible, and not significant.

Therefore, I regularly monitor trade magazines in a variety of technological and cultural niches. In addition to more obvious sources in the computer and Internet realms, I also read magazines such as *Defense Electronics, Electronic Products, Biomedical Products, Bar Code News, GPS World, Wireless World, R&D,* and *Advanced Imaging.* I regularly drop in on trade and professional meetings on topics ranging from brain surgery to architecture to restaurants. For a while I kept watch on new patents issued, and watch for new product directions in everything from military to toy technology. I define these activities as part of my role as an information artist, and they are an important research source for this book.

Biases and Gaps

In our information-glutted society, more research information is now produced in an hour than was produced in a year in the nineteenth century. This book's coverage thus has certain limitations. I have higher awareness of research developments in fields related to my artistic interests—topics such as artificial intelligence, telecommunications, computer media, alternative sensors, GPS, and information visualization—than in fields such as physics, geology, or mathematics. My interest and interpretative ability varies from field to field. Also note that this book concentrates on science and technology; developments in the social sciences and professions similarly call out for artistic attention.

Encouraging those in the humanities and arts to develop curiosity about the research world, skills in understanding it, and the ability to work with the material is a major agenda of this book. A significant cultural opportunity and challenge awaits. I would hope that more artists in the future might define many of the areas identified in this book as their agenda, both to explore new potentials and to comment on cultural implications. The chapters indicate large areas of scientific research that receive almost no artistic attention.

B

Books for Further Inquiry

This section lists books useful for further study. They are broken into topical areas, although some could obviously be placed in several categories.

Art and Science, Art and Mathematics

Benthall, Jonathan. *Science and Technology in Art Today.* New York: Praeger, 1972.

Bijvoet, Marga. *Art as Inquiry: Toward New Collaborations Between Art, Science, and Technology.* New York: Peter Lang, 1997.

Bronowski, Jacob. *Ascent of Man.* Boston: Little, Brown, 1974.

Bronowski, Jacob. *The Visionary Eye: Essays in the Arts, Literature, and Science.* Cambridge, Mass.: MIT Press, 1978.

Burke, James. *Connections.* Boston: Little, Brown, 1978.

Caglioti, Giueseppe. *Dynamics of Ambiguity.* New York: Springer-Verlag, 1992.

Cassidy, Harold Gomes. *The Sciences and the Arts: A New Alliance.* New York: Harper, 1962.

Curtin, D. W., ed. *The Aesthetic Dimension of Science.* New York: Philosophical Library, 1982.

Ede, Siân, ed. *Strange and Charmed: Science and the Contemporary Visual Arts.* London: Calouste Gulbenkian Foundation, 2000.

Emmer, Michele. *Visual Mind: Art and Mathematics.* Cambridge, Mass.: MIT Press, 1993.

Field, J. V. *Science in Art: Works in the National Gallery That Illustrate the History of Science and Technology.* Stanford in the Vale, U.K.: British Society for the History of Science, 1997.

Graubard, Stephen, ed. *Art and Science.* Lanham, MD: University Press of America, 1988.

Hartal, Paul. *The Brush and the Compass: The Interface Dynamics of Art and Science.* Lanham, MD: University Press of America, 1988.

Henderson, Linda Dayrymple. *The Fourth Dimension and Non-Euclidean Geometries in Modern Art,* Princeton, N.J.: Princeton University Press, 1993.

Herdeg, Walter. *The Artist in the Service of Science.* Zurich: Graphis Press, 1973.

Ivens, W. *Art and Geometry.* Cambridge, U.K.: Cambridge University Press, 1986.

Jacob, Opper. *Science and the Arts: A Study in Relationships from 1600–1900.* Rutherford, N.J.: Fairleigh Dickinson University Press, 1973.

Johnson, Martin. *Art and Scientific Thought: Historical Studies towards a Modern Revision of Their Antagonism.* New York: AMS Press, 1970.

Kepes, Gyorgy. *New Landscape in Art and Science.* Chicago: P. Theobald, 1956.

Kepes, Gyorgy. *Structure in Art and in Science.* New York: G. Braziller, 1965.

McConnell, R. B. *Art, Science, and Human Progress.* New York: Universe Books, 1987.

Miller, A. I. *Insights of Genius: Imagery and Creativity in Science and Art.* New York: Springer-Verlag, 1996.

Pollock, Martin, ed. *Common Denominators in Art and Science.* Aberdeen, U.K.: Aberdeen University Press, 1983.

Reeves, Eileen. *Painting the Heavens: Art and Science in the Age of Galileo.* Princeton, N.J.: Princeton University Press, 1997.

Richardson, John. *Modern Art and Scientific Thought.* Urbana: University of Illinois Press, 1971.

Rothstein, Edward. *Emblems of Mind: The Inner Life of Music and Mathematics.* New York: Times Books, 1995.

Rhodes, Lynette. *Science Within Art.* Bloomington: Indiana University Press, 1980.

Rhys, Henry, ed. *Seventeenth-Century Science and the Arts.* Princeton, N.J.: Princeton University Press, 1961.

Richardson, John Adkins. *Modern Art and Scientific Thought.* Urbana: University of Illinois Press, 1971.

Rieser, Dolf. *Art and Science.* New York: Van Nostrand Reinhold, 1972.

Ross, Alan Strode Campbell, ed. *Arts vs. Science: A Collection of Essays.* London: Methuen, 1967.

Schlain, Leonard. *Art and Physics.* New York: Morrow, 1991.

Shirley, John W. F., and David Hoeniger. *Science and the Arts in the Renaissance.* London: Associated University Presses, 1985.

Smith, Cyril Stanley. *A Search for Structure: Selected Essays on Science, Art, and History.* Cambridge, Mass.: MIT Press, 1983.

Smith, Cyril Stanley. *From Art to Science: Seventy-Two Objects Illustrating the Nature of Discovery.* Cambridge, Mass.: MIT Press, 1980.

Snow, C. P. *The Two Cultures and the Scientific Revolution.* Cambridge, U.K.: Cambridge University Press, 1959.

Strosberg, Eliane. *Art and Science.* Paris: UNESCO, 1999.

Tauber, Alfred. *The Elusive Synthesis: Aesthetics and Science.* Boston: Kluwer, 1996.

Volk, Tyler. *Metapatterns across Space, Time, and Mind.* New York: Columbia University Press, 1995.

Vitz, Paul C. *Modern Art and Modern Science: The Parallel Analysis of Vision.* New York: Praeger, 1984.

Waddington, Conrad Hal. *Behind Appearance: A Study of the Relations between Painting and the Natural Sciences in This Century.* Cambridge, Mass.: MIT Press, 1970.

Wagner, Catherine. *Art and Science: Investigating Matter.* St. Louis: Nazraeli Press, 1996.

Washburn, Dorothy, and Crow, Donald. *Symmetries of Culture.* Seattle: University of Washington Press, 1988.

Weschler, Judith, ed. *On Aesthetics in Science.* Cambridge, Mass.: MIT Press, 1979.

Wilson, Stephen. *Great Moments in Art and Science.* (in press).

Zee, Anthony. *Fearful Symmetry. The Search for Beauty in Modern Physics.* New York: Macmillan, 1986.

Art and Technology, Art and Computers

ACM, Special Interest Group on Computer Graphics (SIGGRAPH). *Visual Proceedings* (including art show catalogs), 1980–1999.

Apple, Jacki, and Helen Thorington. *Breaking the Broadcast Barrier: Radio Art 1980–1994.* New York: New American Radio, 1996.

Ars Electronica. *Cyberarts* (catalogs of symposia and Prix awards). 1990–1999.

Ascott, R., and Carl Loeffler. "Connectivity." *Leonardo* (special issue) 24(2), 1991.

Augaitis, Daina, and Dan Lander. *Radio Rethink.* Banff, B.C., Canada: Walter Phillips Gallery, 1994.

Balaban, M., K. Ebcioglu, and O. Laske, eds. *Understanding Music with AI: Perspectives on Music Cognition.* Cambridge, Mass.: MIT Press, 1992.

Berger, René, and Lloyd Eby, eds. *Art and Technology.* New York: Paragon House, 1986.

Bredekamp, Horst. *The Lure of Antiquity and the Cult of the Machine.* Princeton, N.J.: Markus Wiener, 1995.

Burnham, Jack. *Beyond Modern Sculpture.* New York: G. Braziller, 1968.

Cohen, Harold (with Pamela McCorduck). *Aaron's Code: Meta-Art, Artificial Intelligence, and the Work of Harold Cohen.* New York: W. H. Freeman, 1991.

Cope, David. *Computers and Musical Style.* Madison, Wis.: A-R Editions, 1991.

Cope, David. *Experiments in Musical Intelligence.* Madison, Wis.: A-R Editions, 1996.

Davis, Douglas. *Art and the Future: A History/Prophecy of the Collaboration between Science, Technology, and Art.* New York: Praeger, 1973.

Deken, Joseph. *Computer Images: State of the Art.* New York: Stewart, Tabori and Chang, 1983.

Druckrey, Timothy. *Ars Electronica: Facing the Future.* Cambridge, Mass.: MIT Press, 1999.

Franke, Herbert W. *Computer Graphics—Computer Art.* Berlin: Springer-Verlag, 1984.

Frankel, Felice, and George Whitesides. *On the Surface of Things.* New York: Chronicle Books, 1997.

Goldberg, Ken. *The Robot in the Garden: Telerobotics and Telepistemology on the Internet.* Cambridge, Mass.: MIT Press, 2000.

Goodman, Cynthia. *Digital Vision.* New York: Abrams, 1987.

Harris, Craig. *Art and Innovation.* Cambridge, Mass.: MIT Press, 1999.

Hopkins, Bart, ed. *Gravikords, Whirlies, and Pyrophones: Experimental Musical Instruments.* San Francisco: Ellipsis Arts, 1996.

Jacobson, Linda, ed. *Cyberarts: Exploring Art and Technology.* San Francisco: Miller Freeman, 1992.

Kluver, Billy, J. Martin, and B. Rose, eds. *Pavilion: Experiments in Art and Technology.* New York: E. P. Dutton, 1972.

Kranz, Stewart. *Science and Technology in the Arts: A Tour through the Realm of Science/Art.* New York: Van Nostrand Reinhold, 1974.

Kriesche, Richard. *Teleskulptur.* Graz, Austria: Kulturdata, 1993.

Krueger, Myron. *Artificial Reality II.* Reading, Mass.: Addison-Wesley, 1991.

Latham, William, and Todd, Stephen. *Evolutionary Art and Computers.* New York: Academic Press, 1992.

Laurel, B. *Computers as Theatre.* Reading, Mass.: Addison-Wesley, 1991.

Leavitt, Ruth. *Artist and Computer.* New York: Harmony Books, 1976.

Loveless, Richard. *The Computer Revolution and the Arts.* Tampa: University of South Florida Press, 1989.

Maeda, John. *Design by Numbers.* Cambridge, Mass.: MIT Press, 1999.

Malina, Frank J., ed. *Kinetic Art.* New York: Dover, 1974.

Malina, Frank J., ed. *Visual Art, Mathematics, and Computers.* Oxford, U.K.; New York: Pergamon Press, 1979.

Moles, Abraham. *Art and Technology.* Urbana, University of Illinois Press, 1966.

Mulder, Arjen, and Maaike Post. *Book for Electronic Art.* Rotterdam, Netherlands: de Balie, 2000.

Orvell, Miles. *After the Machine: Visual Arts and the Erasing of Cultural Boundaries.* Jackson: University of Mississippi Press, 1995.

Pickover, Clifford A. *Computers and the Imagination: Visual Adventures beyond the Edge.* New York: St. Martin's Press, 1991.

Pickover, Clifford A. *Visions of the Future: Art, Technology, and Computing in the Twenty-first Century.* New York: St. Martin's Press, 1992.

Poissant, Louise. *Esthétique des Arts Médiatiques.* Montreal: Presses de l'Université du Québec, Collection Esthétique, 1995.

Popper, Frank. *Art, Action, and Participation.* New York: New York University Press, 1975.

Popper, Frank. *Art of the Electronic Age.* New York: Harry N. Abrams, 1993.

Popper, Frank. *Electra.* Paris: Musée d'Art Moderne de la Ville de Paris, 1983.

Reichardt, Jasia. *The Computer in Art.* New York: Van Nostrand Reinhold, 1971.

Reichardt, Jasia. *Cybernetic Serendipity.* New York: Praeger, 1969.

Rosenberg, M. J. *The Cybernetics of Art: Reason and the Rainbow.* New York: Gordon and Breach Science, 1983.

Rosenboom, David. *Biofeedback and the Arts, Results of Early Experiments.* Cambridge, Mass.: MIT Press—Leonardo monographs, 1976.

Schwartz, Lillian. *The Computer Artist's Handbook: Concepts, Techniques, and Applications.* New York: Norton, 1992.

Shanken, Edward. *Roy Ascott Writings.* Berkeley: University of California Press, 1999.

Sommerer, Christa, and Laurent Mignonneau, eds. *Art @ Science.* New York: Springer, 1998.

Spiller, Anne Morgan. *Computers in the Visual Arts.* Reading, Mass.: Addison-Wesley, 1999.

Tuchman, Maurice. *Art and Technology.* New York: Viking, 1971.

Wilson, Stephen. *Multimedia Design.* Englewood Cliffs, N.J.: Prentice Hall, 1991.

Wilson, Stephen. *Using Computers to Create Art.* Englewood Cliffs, N.J.: Prentice Hall, 1986.

Wilson, Stephen. *World Wide Web Design Guide.* Indianapolis: Hayden, 1995.

Philosophical and Sociological Critiques of Science

Aronowitz, Stanley, et al., eds. *Techno-science and Cyberculture*. London: Routledge, 1995.

Babich, Babette. *Continental and Postmodern Perspectives on Philosophy of Science*. London: Avebury, 1995.

Barnes, Barry, David Bloor, and John Henry. *Scientific Knowledge: A Sociological Analysis*. London: Athlone Press, 1996.

Bleier, Ruth. *Feminist Approaches to Science*. New York: Pergamon, 1988.

Bloor, D. *Knowledge and Social Imagery*. 2nd ed. Chicago: University of Chicago Press, 1991.

Dawkins, Richard. *Unweaving the Rainbow*. New York: Houghton Mifflin, 1998.

Durbin, Paul T., ed. *A Guide to the Culture of Science, Technology, and Medicine*. New York: Free Press, 1977.

Feyerabend, Paul K. *Against Method*. London: New Left Books, 1975.

Foucault, Michel. *Archaeology of Knowledge*. London: Tavistock, 1972.

Fuller, Steve. *Philosophy of Science and Its Discontents*. Conduct of Science Series. Madison: University of Wisconsin Press, 1993.

Galgan, Gerald J. *The Logic of Modernity*. New York: New York University Press, 1982.

Galison, Peter. *Image and Logic: A Material Culture of Microphysics*. Chicago: University of Chicago Press, 1997.

Galison, Peter, and David J. Stump. *Disunity of Science*. Stanford, Calif.: Stanford University Press, 1996.

Galison, Peter, and C. Jones, eds. *Picturing Science, Producing Art*. London: Routledge, 1998.

Galison, Peter, and Emily Thompson, eds. *The Architecture of Science*. Cambridge, Mass.: MIT Press, 1999.

Gleick, James. *Chaos*. New York: Viking, 1987.

Griffin, D. *Reenchantment of Science*. Buffalo: State University of New York Press, 1988.

Gross, Paul R.; Levitt, Norman; and Wise, Martin W., eds. *The Flight from Science and Reason*. Baltimore: Johns Hopkins University Press, 1988.

Harding, Sandra. *Whose Science? Whose Knowledge?* Ithaca, N.Y.: Cornell University Press, 1991.

Haraway, Donna. *Primate Visions: Gender, Race and Nature in the World of Modern Science*. London: Routledge, 1989.

Hull, David. *Science as a Process: An Evolutionary Account of the Social and Conceptual Development of Science*. Chicago: University of Chicago Press, 1990.

Keller, Evelyn Fox. *Reflections on Gender and Science*. New Haven, Conn.: Yale University Press, 1985.

Kuhn, Thomas. *The Structure of Scientific Revolutions*. Chicago: University of Chicago Press, 1970.

Lakatos, I., and A. Musgrave, eds. *Criticism and the Growth of Knowledge.* Cambridge, U.K.: Cambridge University Press, 1970.

Latour, Bruno. *Science in Action.* Cambridge, Mass.: Harvard University Press, 1987.

Latour, Bruno, and Woolgar, Steve. *Laboratory Life: The Construction of Scientific Facts.* 2nd ed. Princeton, N.J.: Princeton University Press, 1986.

Laudan, Larry. *Progress and Its Problems: Towards a Theory of Scientific Growth.* Berkeley: University of California Press, 1977.

Lenoir, Timothy. *Instituting Science: The Cultural Production of Scientific Disciplines.* Stanford, Calif.: Stanford University Press, 1997.

Lyotard, Jean-François. *Postmodern Condition.* Manchester, U.K.: University of Manchester Press, 1984.

Hoyningen-Huene, Paul. *Reconstructing Scientific Revolutions: Thomas S. Kuhn's Philosophy of Science.* Chicago: University of Chicago Press, 1993.

Margolis, Howard. *Paradigms and Barriers: How Habits of Mind Govern Scientific Beliefs.* Chicago: University of Chicago Press, 1993.

Penley, Constance. *NASA/Trek: Popular Science and Sex in America.* London: Verso, 1998.

Rouse, Joseph. *Engaging Science: Science Studies after Realism, Rationality, and Social Constructivism.* Ithaca, N.Y.: Cornell University Press, 1996.

Pickering, Andrew, ed. *Science as Practice and Culture.* Chicago: University of Chicago Press, 1992.

Toulmin, Stephen. *Return to Cosmology.* Berkeley: University of California Press, 1985.

Woolgar, Steve. *Science: The Very Idea.* London: Tavistock, 1988.

Biology, Nature, Ecology, and the Body

Ackerman, Dianne. *Natural History of the Senses.* New York: Random House, 1990.

Balsamo, Anne Marie. *Technologies of the Gendered Body.* Durham, N.C.: Duke University Press, 1996.

Becker, Robert O., and Andrew Marino. *Cross Currents.* Los Angeles: Jeremy Tarcher, 1990.

Becker, Robert O., and Gary Selden. *Body Electric.* New York: Morrow, 1985.

Birke, Lynda, and Ruth Hubbard. *Reinventing Biology.* Bloomington: Indiana University Press, 1995.

Butler, Judith. *Bodies That Matter: On the Discursive Limits of "Sex."* New York: Routledge, 1993.

Cassirer, Ernst. *The Individual and the Cosmos in Renaissance Philosophy.* Philadelphia: University of Pennsylvania Press, 1972.

Crary, Jonathan, and Kwinter, Sanford. *Incorporations.* London: Verso, 1992.

Deitch, Jeffrey. *Post Human.* New York: Distributed Art Press, 1992.

Dion, Mark, and Alexis Rockman. *Concrete Jungle.* New York: Re/Search-Juno Books, 1996.

Doyle, Richard. *On Beyond Living: Rhetorical Transformations of the Life Sciences.* Stanford University Press: Stanford, Calif., 1997.

Elder, Klaus. *Social Construction of Nature.* London: Sage, 1996.

Evernden, Neil. *Social Creation of Nature.* Baltimore: Johns Hopkins University Press, 1992.

Fabrega, Horacio. *Evolution of Sickness and Healing.* Berkeley: University of California Press, 1997.

Favazza, Armando. *Bodies under Siege, Self-mutilation, and Body Modification in Culture and Psychiatry.* Baltimore: Johns Hopkins University Press, 1996.

Foucault, Michel. *Birth of the Clinic.* New York: Pantheon Books, 1973.

Foucault, Michel. *History of Sexuality.* New York: Pantheon Books, 1978.

Gallaher, Catherine, and Thomas Laqueur. *Making of the Modern Body.* Berkeley: University of California Press, 1987.

Gare, Arran. *Postmodernism and the Environmental Crisis.* New York: Routledge, 1995.

Gray, Chris Hables, ed. *The Cyborg Handbook.* New York: Routledge, 1995.

Grosz, Elizabeth. *Unbearable Weight: Feminism, Western Culture, and the Body.* Sydney: Allen and Unwin, 1994.

Haraway, Donna J. *Modest_Witness@Second_Millennium_FemaleMan_Meets_OncoMouse, Feminism and Technoscience.* Routledge: London, 1997.

Haraway, Donna J. *Simians, Cyborgs, and Women: The Reinvention of Nature.* London: Free Association Books, 1991.

Hayles, N. Katherine. *How We Became Posthuman: Virtual Bodies in Cybernetics, Literature, and Informatics.* Chicago: University of Chicago Press, 1999.

Kevles, Bettyann Holtzmann, and Marilyn Nissenson. *Picturing DNA,* ⟨http://www.geneart.org/genome-toc.htm⟩.

Kevles, D., and L. Hood. *Code of Codes.* Cambridge, Mass.: Harvard University Press, 1992.

Kimball, Andrew. *Human Body Shop.* San Francisco: HarperCollins, 1993.

Komesaroff, Paul A., ed. *Troubled Bodies: Critical Perspectives on Postmodernism, Medical Ethics, and the Body.* Durham, N.C.: Duke University Press, 1995.

Kress-Rogers, Erika, ed. *Handbook of Biosensors and Electronic Noses.* Boca Raton, Fla.: C.R.C. Press, 1997.

Lacy, Suzanne. *Mapping the Terrain.* Seattle: Bay Press, 1995.

Levy, Ellen, and Berta Sichel. "Contemporary Art and the Genetic Code." In *Art Journal* 33:1 (spring) 1996.

Leiss, William. *Domination of Nature.* New York: Braziller, 1972.

Lupton, Deborah. *Medicine as Culture.* London: Sage, 1994.

Matilsky, Barbara. *Fragile Ecologies.* New York: Rizzoli, 1992.

Mayr, Ernst. *This Is Biology: The Science of the Living World.* Cambridge, Mass.: Belknap Press of Harvard University Press, 1997.

Merchant, Carolyn. *The Death of Nature: Women, Ecology, and Scientific Revolution.* New York: Harper, 1980.

Merchant, Carolyn, ed. *Key Concepts in Critical Theory: Ecology.* Atlantic Highlands, N.J.: Humanities Press, 1994.

Naske, Barbara. *Humans and Other Animals.* London: Pluto, 1989.

Oakes, B. *Sculpting with the Environment—A Natural Dialogue.* New York: Van Nostrand Reinhold, 1995.

Oetschalager, M. *The Idea of Wilderness.* New Haven, Conn.: Yale University Press, 1991.

Purcell, Rosamond. *Special Cases, Natural Anomalies and Historical Monsters.* San Francisco: Chronicle Books, 1997.

Schachtel, Ernest. *Metamorphosis.* London: Routledge, 1963.

Shaviro, Steven. *The Cinematic Body.* Minneapolis: University of Minnesota Press, 1993.

Sheehan, James, and Morton, Sosna, eds. *The Boundaries of Humanity: Humans, Animals, Machines.* Berkeley: University of California Press, 1991.

Shepard, Paul. *The Others: How Animals Made Us Human.* Washington, D.C.: Island Press, 1996.

Soulé, Michael, and Gary Lease, eds. *Reinventing Nature? Responses to Postmodern Deconstruction.* Washington, D.C.: Island Press, 1995.

Stafford, Barbara. *Body Criticism.* Cambridge, Mass.: MIT Press, 1991.

Springer, Claudia. *Electronic Eros: Bodies and Desire in the Postindustrial Age.* London: Athlone Press, 1996.

Taylor, Peter; Saul Halfon; and Paul Edwards, eds. *Changing Life: Genomes, Ecologies, Bodies, Commodities.* Minneapolis: University of Minnesota Press, 1993.

Tompkins, Peter, and Christopher Bird. *The Secret Life of Plants.* New York: Harper and Row, 1973.

Treichler, Paula A., Lisa Cartwright, and Constance Penley, eds. *The Visible Woman: Imaging Technologies, Gender, and Science.* New York: New York University Press, 1998.

Turner, B. S. *Regulating Bodies.* London: Sage, 1992.

Volk, Tyler. *Gaia's Body: Toward a Physiology of Earth.* New York: Copernicus, 1998.

Waldby, Cathy. *Visible Human Project.* New York: Routledge, 2000.

Williams, Raymond. *Problems of Materialism and Culture.* London: Verso, 1980.

Wright, Will. *Wild Knowledge: Science, Language, and Social Life in a Fragile Environment.* Minneapolis: University of Minnesota Press, 1992.

Technologists' and Scientists' Accounts of Emerging Research

Amato, Ivan. *Stuff: The Things the World is Made Of.* New York: Basic Books, 1997.

Ambron, Susanne, and Hooper, Kritina, ed. *Interactive Multimedia.* Seattle: Microsoft Press, 1988.

Ball, Philip. *Designing the Molecular World.* Princeton, N.J.: Princeton University Press, 1994.

Brand, Stewart. *Media Lab.* New York: Penguin, 1987.

Brockman, John. *Third Culture.* New York: Simon and Schuster, 1995.

Brockman, John, and Katinka Matson. *How Things Are.* New York: William Morrow, 1995.

Cilliers, Paul. *Complexity and Postmodernism.* London: Routledge, 1998.

Crandall, B. C. *Nanotechnology: Molecular Speculations on Global Abundance.* Cambridge, Mass.: MIT Press, 1992.

Dodsworth, Clark. *Digital Illusion.* Reading, Mass.: Addison-Wesley, 1997.

Drexler, Eric. *Engines of Creation.* New York: Anchor, 1986.

Drexler, Eric. *Unbounding the Future.* New York: William Morrow, 1991.

Gates, Bill. *The Road Ahead.* New York: Penguin Books, 1995.

Gilder, George. *Telecosm.* New York: Simon and Schuster, 1996.

Hall, Stephen. *Mapping the New Millennium.* New York: Vintage Books, 1993.

Helmreich, Stefan Gordon. *Silicon Second Nature: Culturing Artificial Life in a Digital World.* Berkeley: University of California Press, 1998.

Jones, Steven, ed. *Cybersociety.* Thousand Oaks, Calif.: Sage, 1995.

Kelly, Kevin. *Out of Control: The Rise of Neo-Biological Civilization.* Reading, Mass.: Addison-Wesley, 1994.

Kosko, Bart. *Fuzzy Thinking: The New Science of Fuzzy Logic.* London: Flamingo, 1993.

Krueger, Myron C. *Artificial Reality II.* Reading, Mass.: Addison-Wesley, 1991.

Langton, Christopher G., ed. *Artificial Life: An Overview.* Cambridge, Mass.: MIT Press, 1997.

Lanier, Jaron. *Information Is Alienated Experience,* forthcoming.

Levy, Steven. *Artificial Life: The Quest for a New Creation.* New York: Penguin, 1992.

Lewin, Roger. *Complexity: Life at the Edge of Chaos.* New York: Collier, 1992.

McNeil, Daniel, and Frieberger, Paul. *Fuzzy Logic.* New York: Simon and Schuster, 1993.

Moravec, Hans. *Mind Children: The Future of Robot and Human Intelligence.* Cambridge, Mass.: Harvard University Press, 1988.

Moravec, Hans. *Robot: Mere Machine to Transcendent Mind.* Oxford, U.K.: Oxford University Press, 1999.

Negroponte, Nicholas. *Being Digital.* New York: Alfred A. Knopf, 1995.

Rheingold, Howard. *The Virtual Community: Homesteading on the Electronic Frontier.* New York: HarperCollins, 1995.

Rushkoff, Douglas. *Cyberia: Life in Trenches of Cyberspace.* San Francisco: HarperSan Francisco, 1994.

Sproull, Lee, and Kiesler, Sara. *Connections: New Ways of Working in Networked Organizations.* Cambridge, Mass.: MIT Press, 1991.

Siler, Todd. *Breaking the Mind Barrier.* New York: Bantam, 1996.

Siler, Todd. *Learn to Think Like a Genius.* New York: Bantam, 1997.

Toffler, Alvin. 1980. *Third Wave.* New York: William Morrow, 1983.

Waldrop, Mitchell M. *Complexity: The Emerging Science at the Edge of Order and Chaos.* New York: Penguin, 1992.

Wooley, Benjamin. *Virtual Worlds.* Oxford, U.K.: Blackwell, 1992.

Critical Theory and Other Analyses of Culture and Technology

Adorno, Theodor W. *The Culture Industry: Selected Essays on Mass Culture.* Edited by J. M. Bernstein. London: Routledge, 1991.

Amin, Ash, ed. *Post-Fordism: A Reader.* Oxford: Blackwell, 1994.

Ascott, Roy, ed. *Reframing Consciousness: Art, Mind and Technology.* London: Intellect Publishers, 2000.

Ascott, Roy, ed. *Mind@large.* London: Intellect Publishers, 2000.

Baudrillard, Jean. *Ecstacy of Communication.* New York: Autonomedia, 1988.

Baudrillard, Jean. *Simulations.* New York: Semiotext[e], 1983.

Bell, David, and Barbara M. Kennedy, eds. *Cybercultures Reader.* New York: Routledge, 2000.

Bender, Gretchen, and Timothy Druckrey, eds. *Culture on the Brink: The Ideologies of Technology.* New York: Dia/Bay Press, 1994.

Benjamin, Walter. *Illuminations.* New York: Schocken Books, 1976.

Borgmann, Albert. *Crossing the Postmodern Divide.* Chicago: University of Chicago Press, 1992.

Brook, James, and Iain A. Boal, eds. *Resisting the Virtual Life: The Culture and Politics of Information.* San Francisco: City Lights Press, 1995.

Bruno, Giuliana. *Alien Zone.* London: Verso, 1990.

Bukatman, Scott. *Terminal Identity: The Virtual Subject in Post-Modern Science Fiction.* Durham, N.C.: Duke University Press, 1993.

CAiiA. *Consciousness Reframed: Art and Consciousness in the Post-Biological Era.* Conference Proceedings. Newport, Wales: University of Wales College, 1997.

Carey, James W. *Communication as Culture: Essays on Media and Society.* Boston: Unwin Hyman, 1989.

Conley, V. Andermatt. *Rethinking Technologies.* Minneapolis: University of Minnesota Press, 1992.

Coyne, Richard. *Technoromanticism.* Cambridge, MA: MIT Press, 2000.

Crary, Jonathan. *Techniques of the Observer.* Cambridge, Mass.: MIT Press, 1992.

de Kerckhove, Derrick. *The Skin of Culture: Investigating the New Electronic Reality.* Toronto: Somerville House, 1995.

De Landa, Manuel. *War in the Age of Intelligent Machines.* New York: Zone Books, 1991.

Dery, Mark. *Escape Velocity: Cyberculture at the End of the Century.* New York: Grove Press, 1997.

Dery, Mark. *Flame Wars: The Discourse of Cyberculture.* Durham, N.C.: Duke University Press, 1993.

Dienst, Richard. *Still Life in Real Time: Theory after Television.* Durham, N.C.: Duke University Press, 1994.

Druckrey, Timothy. *Iterations.* Cambridge, Mass.: MIT Press, 1994.

Druckrey, Timothy, ed. *Electronic Culture: Technology and Visual Representation.* New York: Aperture, 1997.

Druckrey, Timothy, and Gretchen Bender, eds. *Cultures on the Brink: Ideologies of Technology.* Seattle: Bay Press, 1994.

Easton, Thomas. *Taking Sides: Clashing Views on Controversial Issues in Science, Technology, and Society.* Guilford, Conn.: Dushkin/McGraw-Hill, 1998.

Feenberg, Andrew. *Critical Theory of Technology.* Oxford, U.K.: Oxford University Press, 1991.

Fischer, Claude S. *America Calling: A Social History of the Telephone to 1940.* Berkeley: University of California Press, 1992.

Focillon, Henry. *The Life of Forms in Art.* New York: Zone Books, 1992.

Foster, Hal. *The Return of the Real.* Cambridge, Mass.: MIT Press, 1996.

Foster, Hal, ed. *Discussions in Contemporary Culture.* Seattle: Bay Press, 1987.

Foucault, Michel. *Discipline and Punish.* New York: Viking, 1979.

Friedberg, Anne. *Window Shopping.* Berkeley: University of California Press, 1993.

Giedion, Siegfried. *Mechanization Takes Command: A Contribution to Anonymous History.* New York: Norton, 1948.

Gray, Chris Hables, ed. *The Cyborg Handbook.* London: Routledge, 1996.

Guisnel, Jean. *Cyberwars: Espionage on the Internet.* New York: Plenum Press, 1997.

Habermas, Jurgen. *The Theory of Communicative Action.* 2 vols. Cambridge, MIT Press, 1990.

Hardison, O. B. *Disappearing through the Skylight: Culture and Technology in the Twentieth Century.* New York: Viking, 1991.

Heidegger, Martin. *The Question Concerning Technology.* New York: Garland, 1977.

Heilbroner, Robert. *Visions of the Future: The Distant Past, Yesterday, Today, Tomorrow.* Oxford, U.K.: Oxford University Press, 1996.

Hershman, Lynn, ed. *Clicking In: Hot Links to a Digital Culture.* Seattle: Bay Press, 1996.

Insoe, Hiroshi, and John Pierce. *Information Technology and Civilization.* New York: W. H. Freeman, 1994.

Kahn, Douglas. *Noise Water Meat: A History of Sound in the Arts.* Cambridge, Mass.: MIT Press, 2000.

Kahn, Douglas, and Diane Neumaier, eds. *Cultures in Contention.* Seattle: Real Comet Press, 1985.

Kahn, Douglas, and Gregory Whitehead, eds. *Wireless Imagination.* Cambridge, Mass.: MIT Press, 1992.

Krauss, Rosalind, et al., ed. *October: The Second Decade.* Cambridge, Mass.: MIT Press, 1997.

Kroker, Arthur. *The Possessed Individual: Technology and Postmodernity.* New York: Macmillan, 1992.

Kroker, Arthur, and Marilouise Kroker. *Hacking the Future: Stories for the Flesh-Eating 90s.* Montreal: New World Perspectives, 1996.

Kroker, Arthur, and Marilouise Kroker, eds. *Digital Delirium.* New York: St. Martin's Press, 1997.

Kroker, Arthur, and Michael A. Weinstein. *Data Trash: The Theory of the Virtual Class.* Montreal: New World Perspectives, 1994.

Levinson, Paul. *Digital McLuhan: A Guide to the Information Millennium.* New York: Routledge, 1999.

Lovejoy, Margot. *Postmodern Currents, Art and Artists in the Age of Electronic Media.* Upper Saddle River, N.J.: Prentice Hall, 1997.

Lucky, Robert. *Silicon Dreams: Information, Man, and Machine.* New York: St. Martin's Press, 1989.

Ludlow, Peter. *High Noon on the Electronic Frontier.* Cambridge, Mass.: MIT Press, 1996.

Lunenfeld, Peter. *Digital Dialectic.* Cambridge, Mass.: MIT Press, 1999.

Mander, Jerry. *Four Arguments for the Elimination of Television.* New York: William Morrow, 1978.

Manovich, Lev. *The Language of New Media.* Cambridge, Mass.: MIT Press.

Marvin, Carol. *When Old Technologies Were New.* Oxford, U.K.: Oxford University Press, 1988.

Morse, Margaret. *Virtualities: Television, Media Art, and Cyberculture.* Bloomington: Indiana University Press, 1998.

McLuhan, Marshall. *Understanding Media.* Cambridge, Mass.: MIT Press, 1994.

Mitchell, William J. *City of Bits: Space, Place, and the Infobahn.* Cambridge, Mass.: MIT Press, 1995.

Norris, Christopher. *What's Wrong with Postmodernism.* Baltimore: Johns Hopkins University Press, 1990.

Penley, Constance, and Andrew Ross, eds. *Technoculture.* Minneapolis: University of Minnesota Press, 1995.

Penny, Simon. *Critical Issues in Electronic Media.* Albany, N.Y.: State University of New York Press, 1995.

Poster, Mark. *Critical Theory and Poststructuralism: In Search of a Context.* Ithaca, N.Y.: Cornell University Press, 1989.

Poster, Mark. *The Mode of Information.* Chicago: University of Chicago Press, 1990.

Poster, Mark. *The Second Media Revolution.* Cambridge, MA: Polity Press, 1995.

Postman, Neil. *Technopoly: The Surrender of Culture to Technology.* New York: Vintage, 1993.

Ronnell, Avital. *The Telephone Book: Technology, Schizophrenia, Electronic Space.* Omaha: University of Nebraska Press, 1988.

Roszak, Theodore. *The Cult of Information.* New York: Pantheon, 1988.

Rucker, Rudy, R. U. Sirius, and Queen Mu, eds. *Mondo 2000: A User's Guide to the New Edge.* New York: HarperPerennial, 1992.

Schwartau, Winn. *Information Warfare: Chaos on the Electronic Superhighway.* New York: Thunder's Mouth Press, 1994.

Shallis, Michael. *The Silicon Idol: The Micro Revolution and Its Social Implications.* Oxford, U.K.: Oxford University Press, 1984.

Stabile, Carol A. *Feminism and the Technological Fix.* Manchester, U.K.: Manchester University Press, 1994.

Stone, Allucquère Rosanne. *The War of Desire and Technology at the Close of the Mechanical Age.* Cambridge, Mass.: MIT Press, 1996.

Teich, Albert. *Technology and the Future.* 7th ed. New York: St. Martin's Press, 1997.

Turkle, Sherry. *Life on the Screen: Identity in the Age of the Internet.* New York: Touchstone Books, 1997.

Virilio, Paul. *The Aesthetics of Disappearance.* Translated by Philip Beitchman. New York: Semiotext(e), 1991.

Virilio, Paul. *The Lost Dimension.* New York: Semiotext(e), 1991.

Virilio, Paul. *The Vision Machine.* Bloomington: Indiana University Press, 1994.

Virilio, Paul. *War and Cinema.* London: Verson, 1989.

Wallis, Brian. *Art after Postmodernism: Rethinking Representation.* New York: New Museum, 1984.

Webster, F. *Theories of the Information Society.* London: Routledge, 1995.

Weibel, Peter, and Timothy Druckrey, eds. *Net Condition: Art and Global Media.* Cambridge, Mass.: MIT Press, 2001.

Weizenbaum, Joseph. *Computer Power and Human Reason.* New York: Pelican, 1976.

Zofia, Z. *Whose Second Self? Gender and (Ir)Rationality in Computer Culture.* Victoria, B.C., Canada: Deakin University Press, 1993.

Analysis of the Computer's Impact on Society and Specific Digital Technologies

Bolter, J. David. *Turing's Man: Western Culture in the Computer Age.* New York: Penguin, 1984.

Bruce, Vicki, and Andy Young. *In the Eye of the Beholder: The Science of Face Perception.* New York: Oxford University Press, 1998.

Card, Stuart K., Jock D. Mackinlay, and Ben Shneiderman. *Information Visualization: Using Vision to Think.* San Francisco: Morgan-Kaufmann, 1998.

Chesterman, John, and Andy Lipman. *The Electronic Pirates: DIY Crime of the Century.* London: Routledge, 1988.

Crevier, Daniel. *AI: The Tumultuous History of the Search for Artificial Intelligence.* New York: Basic Books, 1993.

Cringeley, Robert X. *Accidental Empires: How the Boys of Silicon Valley Make Their Millions, Battle Foreign Competition, and Still Can't Get a Date.* New York: Penguin, 1992.

Critical Art Ensemble. *Electronic Civil Disobedience and Other Unpopular Ideas.* New York: Autonomedia/Semiotext(e), 1996.

Critical Art Ensemble. *The Electronic Disturbance.* New York: Autonomedia, 1994.

Davies, Paul. *The Cosmic Blueprint: Order and Complexity at the Edge of Chaos.* New York: Penguin, 1987.

Davis, Erik. *Techgnosis: Myth, Magic, Mysticism in the Age of Information.* New York: Harmony Books, 1998.

Dreyfus, Hubert L. *What Computers Still Can't Do: A Critique of Artificial Reason.* Cambridge, Mass.: MIT Press, 1992.

Edwards, Paul. *The Closed World: Computers and the Politics of Discourse in Cold War America.* Cambridge, Mass.: MIT Press, 1996.

Featherstone, Mike. *Undoing Culture: Globalization, Postmodernism, and Identity.* London: Sage, 1995.

Featherstone, Mike, Scott Lash, and Roland Robertson, eds. *Global Modernities.* London: Sage, 1995.

Freedman, David H. *Brainmakers: How Scientists are Moving beyond Computers to Create a Rival to the Human Brain.* New York: Touchstone, 1994.

Freiberger, Paul, and Michael Swain. *Fire in the Valley: The Making of the Personal Computer.* Berkeley, Calif.: Osborne/McGraw-Hill, 1984.

Dennett, Daniel C. *Consciousness Explained.* New York: Penguin, 1991.

Gardner, Howard. *The Mind's New Science: A History of the Cognitive Revolution.* Rev. ed. New York: HarperCollins, 1987.

Gelernter, David. *Mirror Worlds.* New York: Oxford University Press, 1991.

Gelernter, David. *The Muse in the Machine: Computers and Creative Thought.* London: Fourth Estate, 1994.

Gershenfeld, Neil. *When Things Start to Think.* New York: Holt, 1999.

Gilder, George. *Microcosm.* New York: Simon and Schuster, 1989.

Goldberg, Ken. *Robot in the Garden.* Cambridge, Mass.: MIT Press, 2000.

Hafner, Katie, and John Markoff. *Cyberpunk: Outlaws and Hackers on the Computer Frontier.* London: Corgi, 1991.

Harasim, Linda M., ed. *Global Networks: Computers and International Communication.* Cambridge, Mass.: MIT Press, 1993.

Haugeland, John. *Artificial Intelligence: The Very Idea.* Cambridge, Mass.: MIT Press, 1985.

Hertz, J. C. *Joystick Nation.* Boston: Little, Brown, 1997.

Hiltz, Roxanne Starr, and Murray Turoff. *The Network Nation: Human Communication Via Computer.* Cambridge, Mass.: MIT Press, 1993.

Hofstadter, Douglas, and Daniel Dennett. *The Mind's I.* New York: Bantam, 1982.

Holtzman, Steven R. *Digital Mantras: The Languages of Abstract and Digital Worlds.* Cambridge, Mass.: MIT Press, 1994.

Ichbiah, Daniel, with Susan L. Knepper. *The Making of Microsoft: How Bill Gates and His Team Created the World's Most Successful Software Company.* Rocklin, Calif., Prima, 1993.

Jones, Steven G. *Cybersociety: Computer-Mediated Communication and Community.* London: Sage, 1995.

Kolko, Beth, Lisa Nakamura, and Gilbert Rodman, eds. *Race in Cyberspace.* New York: Routledge, 2000.

Kurtzman, Joel. *The Death of Money: How the Electronic Economy Has Destabilized the World's Markets and Created Financial Chaos.* New York: Simon and Schuster, 1993.

Kurzweil, Raymond. *The Age of Spiritual Machines: When Computers Exceed Human Intelligence.* New York: Viking, 1999.

Kurzweil, Raymond, ed. *The Age of Intelligent Machines.* Cambridge, Mass.: MIT Press, 1992.

Lansdale, Mark W., and Thomas C. Ormerod. *Understanding Interfaces: A Handbook of Human-Computer Dialogue.* San Diego: Academic Press, 1994.

Laurel, Brenda. *Computers as Theatre.* Reading, Mass.: Addison-Wesley, 1993.

Laurel, Brenda, ed. *The Art of Human-Computer Interface Design.* Reading, Mass.: Addison-Wesley, 1990.

Levidow, Les, and Kevin Robins, eds. *Cyborg Worlds: The Military Information Society.* London: Free Association Books, 1989.

Lévy, Pierre. *Becoming Virtual: Reality in the Digital Age.* New York: Plenum Trade, 1998.

Lévy, Pierre. *Collective Intelligence: Mankind's Emerging World in Cyberspace.* New York: Plenum Press, 1997.

Levy, Steven. *Hackers: Heroes of the Computer Revolution.* New York: Anchor Press/Doubleday, 1984.

Levy, Steven. *Insanely Great: The Life and Times of Macintosh, the Computer That Changed Everything.* New York: Penguin, 1994.

Lury, Celia. *Cultural Rights: Technology, Legality, and Personality.* London: Routledge, 1993.

Lyon, David. *The Electronic Eye: The Rise of Surveillance Society.* Minneapolis: University of Minnesota Press, 1994.

Lyon, David. *The Information Society: Issues and Illusions.* Cambridge: Polity, 1988.

MacCabe, Colin, ed. *High Theory, Low Culture.* Manchester, U.K.: Manchester University Press, 1993.

Martin, William J. *The Global Information Society.* London: Ashgate, 1995.

Minsky, Marvin. *The Society of Mind.* New York: Simon and Schuster, 1986.

Moravec, Hans. *Mind Children: The Future of Robot and Human Intelligence.* Cambridge, Mass.: Harvard University Press, 1988.

Moravec, Hans. *Robot: Mere Machine to Transcendent Mind.* New York: Oxford University Press, 1999.

Muller, Jorg. *Design of Intelligent Agents.* Berlin: Springer-Verlag, 1996.

Nancy, Jean Luc. *The Inoperative Community.* Minneapolis: University of Minnesota Press, 1991.

Nelson, Theodor. *Computer Lib/Dream Machines.* Redmond, Wash.: Tempus Books of Microsoft Press, 1987.

Penrose, Roger. *The Emperor's New Mind.* New York: Oxford University Press, 1989.

Picard, Rosalind W. *Affective Computing.* Cambridge, Mass.: MIT Press, 1997.

Reeves, Byron, and Clifford Nass. *The Media Equation: How People Treat Computers, Television, and New Media Like Real People and Places.* Cambridge, U.K.: Cambridge University Press, 1996.

Shneiderman, Ben. *Designing the User Interface: Strategies for Effective Human-Computer Interaction.* 2nd ed. Menlo Park, Calif.: Addison-Wesley, 1998.

Scientific American. "Communications, Computers, and Networks: How to Work, Play, and Thrive in Cyberspace," vol. 265, no. 3 (September 1991).

Scientific American. "The Computer in the Twenty-first Century" (special issue of the *Scientific American*), 1995.

Shields, Rob, ed. *Cultures of Internet: Virtual Spaces, Real Histories, Living Bodies.* London: Sage, 1996.

Silverstone, Roger, and Eric Hirsch, eds. *Consuming Technologies: Media and Information in Domestic Spaces.* London: Routledge, 1992.

Smith, Anthony. *From Books to Bytes: Knowledge and Information in the Postmodern Era.* London: British Film Institute, 1993.

Smith, Mark, and Peter Kollock, eds. *Communities in Cyberspace.* London: Routledge, 1999.

Sterling, Bruce. *The Hacker Crackdown: Law and Disorder on the Electronic Frontier.* New York: Penguin, 1992.

Stoll, Clifford. *Cuckoo's Egg.* New York: Doubleday, 1995.

Stoll, Clifford. *Silicon Snake Oil.* New York: Doubleday, 1995.

Turkle, Sherry. *The Second Self: Computers and the Human Spirit.* New York: Simon and Schuster, 1984.

Wallace, James, and Jim Erickson. *Hard Drive: Bill Gates and the Making of the Microsoft Empire.* New York: Harper Business, 1992.

Wasko, Janet. *Hollywood in the Information Age.* Oxford, U.K.: Polity, 1994.

Watson, Mark. *AI Agents in Virtual Reality Worlds.* New York: Wiley, 1996.

Webster, Frank. *Theories of the Information Society.* London: Routledge, 1995.

Wiener, Norbert. *Cybernetics; or, Control and Communication in the Animal and the Machine.* Cambridge, Mass.: MIT Press, 1948.

Wiener, Norbert. *The Human Use of Human Beings: Cybernetics and Society.* Boston: Houghton Mifflin, 1954.

Hypermedia and Interactive Narrative

Aronowitz, Stanley; Martinsons, Barbara; and Menser, Michael, eds. *Technoscience and Cyberculture.* London: Routledge, 1996.

Barrett, Edward, ed. *The Society of Text: Hypertext, Hypermedia, and the Social Construction of Information.* Cambridge, Mass.: MIT Press, 1989.

Birketts, Sven. *Gutenberg Elegies: The Fate of Reading in an Electronic Age.* London: Faber and Faber, 1994.

Bolter, Jay David, and Richard Grusin. *Remediation.* Cambridge, Mass.: MIT Press, 1999.

Coyne, Richard. *Designing Information Technology in the Postmodern Age, from Method to Metaphor.* Cambridge, Mass.: MIT Press, 1996.

Coyne, Richard. *Technoromanticism: Digital Narrative, Holism and the Romance of the Real.* Cambridge, Mass.: MIT Press, 1999.

Delany, Paul, and George P. Landow, eds. *Hypermedia and Literary Studies.* Cambridge, Mass.: MIT Press, 1991.

Gumbrecht, Hans Ulrich, and K. Ludwig Pfeiffer, eds. *Materialities of Communication.* Translated by William Whobrey. Stanford, Calif.: Stanford University Press, 1994.

Heim, Michael. *Electric Language: A Philosophical Study of Word Processing.* New Haven, Conn.: Yale University Press, 1987.

Joyce, Michael. *Of Two Minds: Hypertext Pedagogy and Poetics.* Ann Arbor: University of Michigan Press, 1995.

Landow, George P. *Hypertext: The Convergence of Contemporary Critical Theory and Technology.* Baltimore: Johns Hopkins University Press, 1992.

Landow, George P., ed. *Hyper/Text/Theory.* Baltimore: Johns Hopkins University Press, 1994.

Landow, George P., and Paul Delaney, eds. *The Digital Word: Text-Based Computing in the Humanities.* Cambridge, Mass.: MIT Press, 1993.

Lanham, Richard A. *The Electric Word: Democracy, Technology, and the Arts.* Chicago: University of Chicago Press, 1993.

Murray, Janet H. *Hamlet on the Holodeck: The Future of Narrative in Cyberspace.* New York: Free Press, 1997.

Neilson, Jakob. *Multimedia and Hypertext: The Internet and Beyond.* Cambridge, Mass.: AP Professional, 1996.

Tuman, Myron C., ed. *Literacy Online: The Promise (and Peril) of Reading and Writing on Computers.* Pittsburgh: University of Pittsburgh Press, 1992.

Reflections on Digital Aesthetics, Video, and Cinema

Bolter, David. *Writing Space: The Computer, Hypertext and the History of Writing.* Hillsdale, N.J.: Lawrence Erlbaum Associates, 1991.

Couchot, Edmond. *La technologie dans l'art; de la photographie la realité virtuelle.* Paris: Ed Jacqueline Chambon, 1998.

Cubitt, Sean. *Digital Aesthetics.* London: Sage, 1998.

D'Agostino, Peter, and Tafler, David, eds. *Transmission: Theory and Practice for a New Television Aesthetics.* Rev. ed. New York: Sage, 1995.

Dovey, Jon, ed. *Fractal Dreams: New Media in Social Context.* London: Lawrence and Wishart, 1996.

Hall, Doug, and Sally-Jo Fifer, eds. *Illuminating Video: An Essential Guide to Video Art.* New York: Aperture/Bay Area Video Coalition, 1990.

Hayward, Philip, and Tana Wollen, eds. *Future Visions: New Technologies of the Screen.* London: British Film Institute, 1993.

Lister, Martin. *The Photographic Image in Digital Culture.* London: Routledge, 1995.

Mitchell, William J. *The Reconfigured Eye: Visual Truth in the Post-Photographic Era.* Cambridge, Mass.: MIT Press, 1992.

Ritchin, Fred. *In Our Own Image: The Coming Revolution in Photography.* New York: Aperture Foundation, 1990.

Wombell, Paul, ed. *Photovideo: Photography in the Age of the Computer.* London: Rivers Oram Press, 1991.

Virtual Reality

Benedikt, Michael, ed. *Cyberspace: First Steps.* Cambridge, Mass.: MIT Press, 1994.

Heim, Michael. *The Metaphysics of Virtual Reality.* New York: Oxford University Press, 1993.

Helsel, S. K., and J. P. Roth, eds. *Virtual Reality: Theory, Practice, and Promise.* Westport, Conn.: Meckler, 1991.

Loeffler, Carl. *Virtual Reality Casebook.* New York: Van Nostrand, 1994.

Markley, Robert, ed. *Virtual Realities and Their Discontents.* Baltimore: Johns Hopkins University Press, 1996.

Morley, David, and Kevin Robins. *Spaces of Identity: Global Media, Electronic Landscapes, and Cultural Boundaries.* London: Routledge, 1995.

Moser, Mary, and Douglas MacLeod. *Immersed in Technology.* Cambridge, Mass.: MIT Press, 1996.

Pool, Ithiel de Sola. *Technology without Boundaries: On Telecommunications in a Global Age.* Cambridge, Mass.: Harvard University Press, 1990.

Rheingold, Howard. *Virtual Reality.* New York: Simon and Schuster, 1991.

Shuler, Douglas. *New Community Networks: Wired for Change.* Reading, Mass.: Addison-Wesley, 1996.

Thalmann, Nadia Magnenat, and Daniel Thalmann. *Virtual Worlds and Multimedia.* New York: John Wiley, 1993.

Woolley, Benjamin. *Virtual Worlds: A Journey in Hype and Hyperreality.* Oxford, U.K.: Blackwell, 1992.

Name Index

Ito, Koji, 584
Ito, Toyo, 246
Itoh, Toshiharu, 711–712
Iwai, Toshio, 679, 766–768
Iwata, Hiroo, 619, 712

Jacob, Bruce L., 364
Jacobsen, Mogens, 362
Janney, Christopher, 188, 771
Jaquet-Droz, Henri-Louis, 370
Jaquet-Droz, Pierre, 370
Jarnow, Al, 243
Jarry, Alfred, 329
Jenik, Adriene, 526
Jensen, Tom, 66–67
Jeremijenko, Natalie, 101, 145, 782, 822–824
Jerez, Concha, 821
Jobs, Steve, 39
Johanson, Patricia, 144
John, Ludwig, 676
Johnson, B. E., 262
Jolliffe, Daniel, 404
Jones, Bill, 255
Jones, Caroline, 12–13
Jones, Stephen, 187
Joy, Jerome, 584, 781

Kac, Eduardo, 91, 99–100, 120–121, 172–174, 388, 455, 488–490, 498–499, 528, 532–538, 815
Kahlo, Frida, 675
Kahn, Douglas, 407–409, 641
Kahn, Ned, 237–239, 256–258
Karcymar, Natan, 487
Kare, Antero, 114
Karkowsky, Zbig, 748
Karp, Stefan, 743
Kaufman, Stuart, 207
Kaul, Paras, 123, 182

Keays, Bill, 716, 743
Kennedy, Kathy, 421
Kennedy, William, 64
Kepes, Gregory, 28–29
Kepler, Johannes, 834
Kermer, David, 100–101
Khlebnikov, Velimir, 499
Kikauka, Laura, 404, 444
King, Ross, 104
King-shu Tse, 369
Kirchgeorg, Jutta, 678
Kitchin, Rob, 832
Kittler, Friederich, 647
Klein, Douglas, 321
Klein, Jonathan, 796–797
Klein, Yves, 253–254, 348–350, 427
Klinkowstein, Tim, 487
Knapp, Sandra, 249
Knipp, Tammy, 403
Knott, Laura, 526
Knowlton, Kenneth, 313
Koch, A. S., 85
Kogawa, Tetsuo, 421
Kohl, James, 62
Kohl, Kurt, 230
Komar, Vitaly, 572–573
Komisar, Milton, 389
Koshland, Daniel, 67
Kovachevich, Thomas, 104
Kovacs, Greg, 68
Kranzberg, Melvin, 13
Kress-Rogers, Erika, 63
Kriesche, Richard, 274, 479, 517–518
Kroker, Arthur, 157, 635, 643
Kroker, Marilouise, 157, 635, 643
Krueger, Myron, 169, 694, 726, 735–737
Krueger, Ted, 254, 452–454
Krueger, Wolfgang, 748
Kubisch, Christina, 250
Kuhn, Gregory, 241–243

Werby, Andrew, 256
Werger, Barry Rian, 425
West, Steven, 571
Weyhrauch, Peter, 671
White, Neil, 108
White, Norman, 388, 443–444, 487
White, Tom, 765–766
Whitehead, Frances, 120
Whitehead, Gregory, 407–409, 421, 498–500, 505
Whiteley, Elizabeth, 323
Whiting, Jim, 445
Whitney, John, 314
Wholgemeuth, Eva, 254
Widrow, Bernard, 68
Wiener, Norbert, 700
Wiernik, Neil, 503
Wight, Gail, 105, 112, 196, 836
Wilhelm, Yvonne, 836
Williams, Raymond, 74
Wilp, Charles, 274
Wilson, Anne, 570
Wilson, Louis K., 197
Wilson, Stephen, 239, 290–291, 450, 487, 495, 577, 606, 653–654, 743, 754, 779, 783–784, 788, 797–801, 810, 824–826, 881–885
Winet, Jon, 524, 763
Winters, Uli, 98, 250
Wisniewski, Maciej, 588
Wittig, Geri, 828
Wodiczko, Krzysztof, 845–846
Wohlgemuth, Eva, 192, 584
Wollensak, Andrea, 289
Woodfill, John, 758–759
Woodham, Derrick, 254
Woods, Arthur, 261–263, 267, 269, 272, 279–280
Woodward, Kathleen, 150
Woolford, Kirk, 164–167, 752–753

Wortzel, Adrianne, 445, 525
Wostenholme, Colleen, 196
Wozniak, Steve, 39
Wray, Nick, 497
Wright, Alexa, 196–197
Wright, Frank Lloyd, 228
Wright, Richard, 649

Yamamoto, Keigo, 243–244
Yanobe, Kanji, 442–444
York Donald, 66–67
Youngs, Amy, 744
Yuxweluptun, Lawrence Paul, 703

Zaag, Liz Vander, 784
Zapp, Andrea, 240
Zelleweger, Polle, 589
Zeltzer, David, 721
Zet, Martin, 114
Zielinski, Siegfried, 639, 647–648, 652
Zittel, Andrea, 107
Zurr, Ionat, 175

Subject Index

Body (cont.)
  imaging and, 189–192, 196
  Internet and, 549–552
  medicine and, 193–198
  modification and, 174–180
  Post-Human approach and, 154–157
  psychological processes and, 189
  rethinking of, 77–84
  senses and, 59–63, 169–170
  surgery and, 170–174
  technological stimulation and, 157–169
  tissue culture and, 174
Brain
  bionics and, 68
  biosurveillance and, 65–66
  mapping of, 84
  monitoring of, 66–67
  process experiments and, 180–188
  robots and, 371–372, 376–377
Breath, 116, 188, 764, 779–780
"Brief History of Art Involving DNA, A"
  (Gessert), 97
British Telecom Research Labs, 867
Bureau of Inverse Technology, 823–824
Butterfly Effect, The, 859

CADRE institute, 828, 870–871
CAiiA (Center for Advanced Inquiry into
  the Interactive Arts), 483, 858–859, 869
Canon ArtLab, 44
Carnegie Mellon University, 870
CAT (computer aided tomography), 57,
  90, 152, 299
CAVE technology
  Artificial Life and, 346
  digital culture and, 620
  virtual reality and, 704–705, 707–708,
  716, 719–720
CBW (chemical and biological warfare),
  71–72

CD-ROMs, 559–560, 577, 583, 665–666,
  678
Cellular bioscience, 56
Cellular Pirate Radio, 508
Center for Advanced Visual Studies, 845–
  846
Center for Metahuman Exploration, 169,
  497
Center for Research in Electronic Art Tech-
  nology (CREATE), 780
Center for the Integration of Art and Sci-
  ence, The, 866
Center for Twentieth Century Studies,
  The, 78
Centre for Computational Neuroscience
  and Robotics, 379
C5, 594, 842–844
Chaos, 207–209, 258
CHAOS (Centre Histoire, Art, Ordinateur,
  Science), 262, 864
Checkin kiosk, 65
Chemicals, 204, 254
  aesthetics and, 85
  sensory disorders, 62
  warfare and, 71–72
Chernobyl, 231
Christianity, 14
CID-KTH Center for Interactive Design, 867
Cinema, 24
  computers and, 665–666, 673–674
"Circuits of the Voice: From Cosmology to
  Telephony" (Dyson), 489
City of God (Augustine), 14
Class myopia, 479
Cloning, 57, 59
Close senses, 60, 62
CNES (National Center for Spatial Stud-
  ies), 270
Coalition for Education in the Life Sciences
  (CELS), 58

Cognitive science, 15, 371

Cog (robot), 381–382

Combinatronics, 299

Commercialism, 6, 11, 28, 35
  artist involvement and, 36–37
  Mem_brane and, 840–841
  robots and, 372
  virtual reality and, 725–726

Communication, 35–36, 38. *See also* Telecommunications

Community, 478

Complexity, 209

ComputerLib/Dream Machines (Nelson), 475

Computers, 6, 39, 446, 658, 665, 885. *See
  also* Internet
  affective computing and, 616–617, 795,
    804–806
  art games and, 676–679
  artificial intelligence and, 627 (*see also* Arti-
    ficial intelligence (AI))
  autonomous agents and, 624
  bandwidth and, 460–461
  biology and, 57, 616–617
  CMC and, 476, 481–482, 485
  CTI and, 464–465
  cultural theory and, 630–657, 666–671
  emotion and, 42, 607–608, 616–617
  extension of capabilities and, 671–676
  facial expression and, 615
  feminism and, 644–647
  force feedback and, 617–619
  fractals and, 319, 321–322, 330–333
  gesture recognition and, 613–614
  group work and, 465–466
  haptics and, 617–619
  high-tech art and, 36
  hyperfiction and, 688–690
  identity and, 637–641, 643–647
  information and, 621–622, 625–626,
    813–848

integrated recognition systems and, 617
interactivity and, 653–657, 686–688 (*see
  also* Interactivity)
Linux and, 588–589
mathematics and, 302
motion and, 610–611, 620–621, 730–
  732
multi-person events and, 679–681
music and, 608–609
object tracking and, 609–610
personal recognition and, 611–613
physical world and, 633–634
research and, 607–647
revolution of, 605
robots and, 371, 376–377 (*see also* Robots)
smart houses and, 630
social computing and, 624
sound and, 620, 623–624
speech and, 619, 780–784
substance and, 23–24
technological imagination and, 605–607
  (*see also* Technology)
textual information and, 625–626, 665–
  666
ubiquitous computing and, 627–628
video installations and, 681–686
virtual reality and, 620, 634–637 (*see also*
  Virtual reality (VR))
vision and, 641–643
wearable, 628–630, 748–749, 817–819
Web art and, 559–601

Conceptual Design Program, 336

Conceptual Information Arts (CIA) program, 870

Connecticut College Center for Art and
  Technology, 865

Consciousness, 479–484

Constructionism, 77–84

Constructivism, 388, 408

Kinetics
  activated objects and, 764–772
  complex action and, 760–763
  conceptual, 392–404
  databases and, 824–826
  gesture and, 745–749
  interactivity and, 730–745
  light sculpture and, 389–392
  precursors of, 388–389
  research and, 387–388
  robots and, 369–384, 405, 425–456
  sound and, 407–423
  surveillance and, 821
Klein bottles, 328
Knowbotic Research, 352, 720
  information and, 836–841
  sound and, 779
  Web art and, 581–582
Knowledge Navigator, The (Apple), 807
Konsum Art.Server, 588
KTH, 805
Kunstradio, 504–505
Kunstraum Innsburck, 170
Kyosei (common good), 44

Laboratory Science (Latour and Callon), 76
La Fura dels Bau group, 160, 526
L.A. Media Festival, 860
Land-Warrior (LW) system, 72
Languages of Design, 789–790
L'Arte dell'Ascolto, 509
Laura Knott Dance Company, 225
Leg Lab, 376
Leonardo (journal), 42, 85, 261, 404, 482,
    883
  artificial intelligence and, 789
  as integration resource, 860–861
  ISEA and, 858
  mathematics and, 321
  surveillance and, 819

Leonardo Electronic Music Journal, 407
Leonardo Space Project, 261, 276
Lie groups, 299
Life 2.0, 360, 364
Light sculpture, 389–392
Lingua Elettrica, 596–597
Linux, 588–589
Living Architecture, 42
Living Things, 42
Ljubljana digital media lab, 865
Los Alamos, 868
Los Angeles County Museum, 36

Maggie (robot), 379
Magnetics, 253
Makrolab, 841–842
"Manifesto Della Radio" (Marinetti and
    Masnata), 499
Mapping the Terrain (Lacy), 131
Markets, 6, 11, 28, 35
  artist involvement and, 36–37
  Mem_brane and, 840–841
  robots and, 372
  virtual reality and, 725–726
MARKOLAB, 274
Markov models, 615
Massage, 81
Masterpieces, 30–31
Materialism, 23–24, 74
Materials science, 204, 215–218, 253–
    256
Mathematics, 99, 241, 321, 704, 885
  algorithmic art, 313–320
  Anker and, 102
  audience literacy and, 334–335
  books for further inquiry, 887–888
  creative algorithms and, 789–794
  cultural theory and, 337–338
  Euclidean geometry and, 48
  fractals and, 330–333

nature and, 73
MIMI and, 328
Möbius strip and, 334
N-dimensional space and, 329–330
non-Euclidean geometry and, 41
number theory, 299
promise/problems in art of, 333–334
Pythagorean theorem, 323–324
research agendas in, 298–302
Romboy Homotopy, 326
sculpture and, 322–329
understanding systems and, 336–337
vs. science, 297–298
MECAD (Media Centre of Art and Design), 860, 865
Mechanics, 251–253
Media, 60, 78, 639, 851. *See also* Telecommunications
Media Academy Cologne, 871, 872
MediaMatic, 864
Media Stream project, 622
Medicine, 193–198. *See also* Biology; Body
body modification and, 174–180
Post-Human approach and, 154–157
remote military, 471–472
research and, 56–59, 74–77
rethinking of, 77–84
robots and, 372
surgery and, 170–174
Meeting capture, 607
Mem_brane, 840–841
MEM (micro-electro-mechanical) manipulation, 64, 227–228
Merce Cunningham dance troupe, 752
Messaging, 464–465
Meta Forum III, 860
Metamorphosis (Schachtel), 59–60
Methodology, 7
assessment of significance and, 882

biology and, 57
deconstruction and, 25–28
documentation and, 881
interactivity and, 881
omissions and, 882–885
research agendas and, 882–884
Microbiology, 95, 109
investigations of, 96–101
social implications and, 105–108
structural visualization and, 101–105
Microcosm (Gilder), 474
Microprobes, 64
Microsoft, 619, 868
Middle Ages, 14
Milia, 860
Mimetic flesh, 157
MIMI (mathematically illuminated musical instrument), 328
Mindship International, 866
MindViewer, 65–66
Mindware, 70
MIT Media Lab, 22, 616–617, 628
activated objects and, 765–766, 771
artificial intelligence and, 793, 796, 798, 803–804, 807–809
CAVE and, 720
motion and, 744
robots and, 451
smart video and, 829
speech synthesis and, 619
surveillance and, 818
telecommunications and, 466
Möbius strip, 334
Modems, 461
Modernism, 26–27, 409
Moist realities, 87–88
Monochrome, 547
MOOs, 481–482, 525
Mostra de Realidade Virtual, 860
Mother Millennia, 571

Sound. *See also* Music
  acoustic ecology and, 123–126
  computers and, 607–609, 623–624
  experimental instruments and, 421–422
  healing and, 81
  information systems and, 414
  kinetics and, 407–423
  recording's effects and, 409
  research in, 775–780
  speech and, 607–608 (*see also* Speech)
  subaudible, 413
  3D, 620, 776–780
SoundCulture, 407
Space, 205, 210, 261–262, 280–281
  art executed in, 268–271
  conceptual works and, 271–275
  Helyer and, 414–416
  imaging and, 622
  N-dimensional, 329–330
  painting/photography and, 271
  perspective views and, 263–267
  research critiques and, 278–279
  SETI and, 275–278
Speaker segmentation, 607
Spectroscopy, 57
Speech, 67, 464–465, 607–608
  artificial intelligence and, 794–797
  recognition and, 782–784
  synthesized and, 619, 780–784
Speed, 865
Spirituality, 307, 318, 633, 635–636, 648, 787
Stanford Research Institute (SRI), 58, 868
Stomach sculpture, 159
Stonehenge, 40
Storey Institute, 140
Storytelling, 133
Structure of Scientific Revolutions, The (Kuhn), 12
Studio Azzurro, 685, 742
Studio for Creative Inquiry, 137, 337, 393
Substance, 23–24

Subsumption, 377–379
Sun Angel, 62
Supreme Particles Group, 744
Surgery, 170–174
Surrealism, 388, 409
Surveillance, 9, 467, 814
  information and, 815–822
  InfoWar and, 826–827
  interactivity and, 729
  VSAM and, 622
Survival Research Labs, 350, 420, 546
  extreme performance and, 432–435
Swiss National Labs, 619
SWITCH, 865
Symposium on Arts and Technology, 860
Synthetic Characters group, 613, 617, 803–804
Systems thinking, 336–337

Tacheles, 546
Tactical Art Media, 652
Tactility, 749–755
Tangible Bits group, 627, 771
Tao of Physics (Capra), 206
Taste, 59–62
TAZ (tactical art zones), 502
Technology. *See also* Mathematics; Science
  acoustic ecology and, 123–136
  ART+COM and, 43
  artificial intelligence and, 787–811 (*see also* Artificial intelligence (AI))
  Arts Catalyst and, 45
  ATR Lab and, 44
  Banff Centre for the Arts and, 42
  biology and, 55–93
  body and, 149–200
  Canon ArtLab and, 44
  computers and, 605–658 (*see also* Computers)
  contemporary modifications and, 26–27
  convergence and, 511–512
  critical theory and, 20–23